Birds

Princeton Field Guides

Rooted in field experience and scientific study, Princeton's guides to animals and plants are the authority for professional scientists and amateur naturalists alike. **Princeton Field Guides** present this information in a compact format carefully designed for easy use in the field. The guides illustrate every species in color and provide detailed information on identification, distribution, and biology.

Recent Titles

Reptiles and Amphibians of Europe, by E. Nicholas Arnold
Marine Mammals of the North Atlantic, by Carl Christian Kinze
Birds of Chile, by Alvaro Jaramillo
Birds of the West Indies, by Herbert Raffaele, James Wiley, Orlando Garrido, Allan Keith, and Janis Raffaele
Birds of Northern India, by Richard Grimmett and Tim Inskipp
Birds of Africa South of the Sahara, by Ian Sinclair and Peter Ryan
Reptiles of Australia, by Steve Wilson and Gerry Swan
Birds of Australia, Seventh Edition, by Ken Simpson and Nicolas Day
Birds of the Middle East, by R. F. Porter, S. Christensen, and P. Schiermacker-Hansen

Birds *of the Middle East*

R. F. Porter, S. Christensen, and
P. Schiermacker-Hansen

Princeton University Press
Princeton and Oxford

Published in the United States, Canada and the Philippine Islands
by Princeton University Press, 41 William Street, Princeton,
New Jersey 08540

In the United Kingdom and European Union, published by
Christopher Helm, an imprint of A&C Black Publishers Ltd.,
37 Soho Square, London, W1D 3QZ

First Princeton Field Guides Edition, 2004

Library of Congress Control Number 2004106816

ISBN-13: 978-0-691-12104-8
ISBN-10: 0-691-12104-4

nathist.princeton.edu

Printed and bound in Hong Kong

10 9 8 7 6 5 4 3

Dedicated to

P A D Hollom

a pioneer of modern bird identification

and a friend of the Middle East

Contents

INTRODUCTION

This field guide was a natural evolution from *Birds of the Middle East and North Africa* which two of us, Steen Christensen and Richard Porter, compiled with Phil Hollom and Ian Willis. That work was not comprehensive—omitting species already covered by the European guides. We hope that this book, with its full coverage of all species, will overcome that problem and in so doing greatly assist newcomers to birding in the region.

The region and species covered

A map of the region covered is shown on the inside covers and comprises the following countries:

> Bahrain, Cyprus, Iran, Iraq, Israel, Jordan, Kuwait, Lebanon, Oman, Qatar, Saudi Arabia, Syria, Turkey, West Bank and Gaza, the United Arab Emirates and the Republic of Yemen (including the island of Socotra).

We have attempted to include all species recorded in the region up to the beginning of 1996. For regularly occurring birds and those authenticated by published accounts the selection has been straightforward. However, for vagrants which have only first been recorded in the very recent past, it has been more difficult; for these we have often relied on personal communications and thus the record may not have been scrutinized by any formal assessment process such as that established by a country records committee. In the case of some species, for example White-bellied Storm-petrel, Madeiran Storm-petrel, Asiatic Dowitcher, Güldenstädt's Redstart and Great Rosefinch their claimed occurrence in the region is now considered doubtful; however we have felt it helpful to include them as they are certainly contenders for the Middle East list. Where such species have been included, an appropriate note appears under 'status'.

It should, therefore, be remembered that this book is purely a guide. Anyone wishing to establish the known status of a species in a particular country should consult a country avifauna; those readily accessible are listed under 'References and Further Reading' on pages 429–432.

Escapes and introduced birds

Throughout the world more and more birds are being trapped and transported to foreign lands where they are kept in captivity and are liable to escape, or even be deliberately released. Many will not survive in an alien environment but others will live for many years and may even establish a feral breeding population. The Middle East is no exception and in the past few decades several escaped birds have established breeding populations in the wild, notably parakeets and mynahs. We have included all such species known to us at the time of preparing the plates, but inevitably since then escaped species newly established in the wild have been drawn to our attention, but too late for inclusion. For example, we now understand that Grey Crowned Crane *Balearica regulorum*, Indian Peacock *Pavo cristatus*, Barbary Dove *Streptopelia risoria*, Budgerigar *Melopsittacus undulatus*, Sulphur-crested Cockatoo *Cacatua galerita*, Brahminy Mynah *Sturnus pagodarum*, Baya Weaver *Ploceus philippinus*, Streaked Weaver *P. manyar* and Lesser Masked Weaver *P. intermedius* are breeding in parts of Saudi Arabia and the United Arab Emirates. Most occur in very small numbers, often just one or two pairs. The potential for them spreading is, however, considerable.

Sequence and taxonomy

This has become a very confusing, even contentious, area of ornithology and newcomers to birding will have every reason to be mystified by the various treatments given by different books. There are those taxonomists who prefer to 'lump' species and those who prefer to 'split' them (so that individual subspecies become species in their own right). We have largely followed the sequence, and the treatment of subspecies, used in *Birds of the Middle East and North Africa* and *Birds of the Western Palearctic*, but on a few occasions for clarity of presentation we have treated subspecies as if they were full species. Where this has been done, it is clearly indicated. In the case of species not covered by these books, notably the endemics on the island of Socotra, we have in nearly all cases followed the taxonomic treatment used by C.G. Sibley and B.L. Monroe in *Distribution and Taxonomy of Birds of the World* (1990).

English names

These have presented problems. A number of lists of English names of birds have been published recently which have attempted to find names acceptable to a wide majority. The lists differ, though there are many names common to each. Thus, choosing names for species has not been easy.

In this book we have consulted the relevant lists as well as standard works that encompass the region and have chosen the names that we feel are most familiar or helpful in a Middle East context. We have, however, also given alternative names, where relevant, but only those that we feel are fairly often used or are a help in preventing confusion. The treatment given to alternative names is not comprehensive but does include those used in America. For example, in the case of *Caprimulgus europaeus* we have used the name European Nightjar (with the alternative name Nightjar) to prevent confusion with the other six species that occur in the region. If a problem arises, then the scientific name should prevent ambiguity.

Illustrations

All species are illustrated in colour showing, where relevant, the range of plumage variations created by sex and age. Subspecies are also illustrated where it is helpful and are labelled accordingly. The plates are in largely systematic order, thus grouping species of similar type. In general this has enabled birds depicted on any one plate to be drawn to scale. The last two plates—111 and 112—cover a mixture of species, mostly vagrants recorded in the region after the rest of the plates were completed.

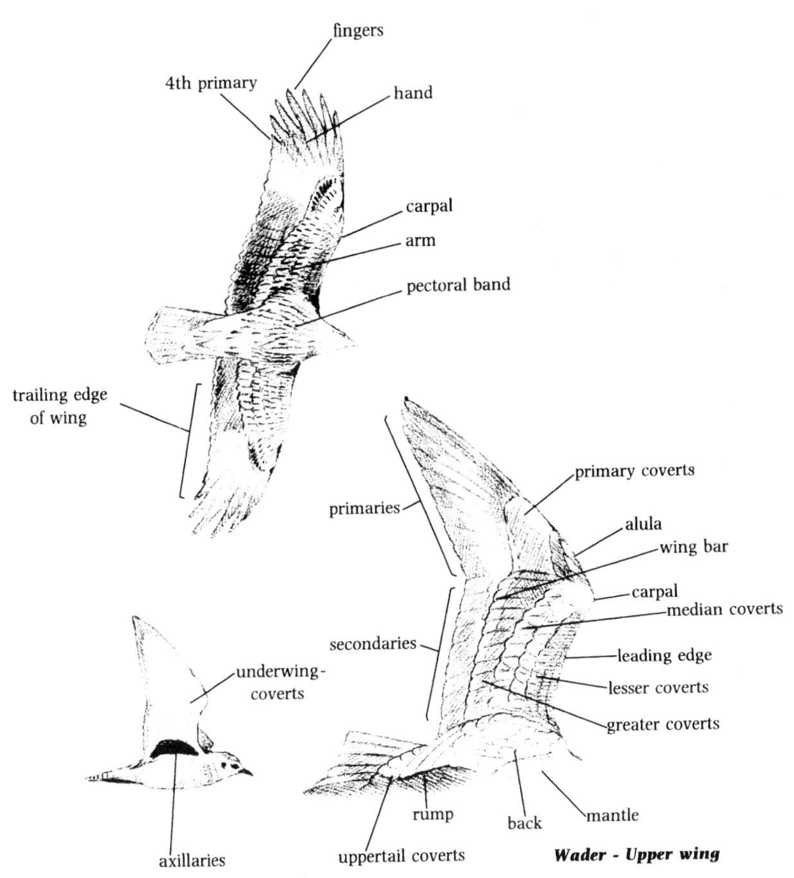

Wader - Upper wing

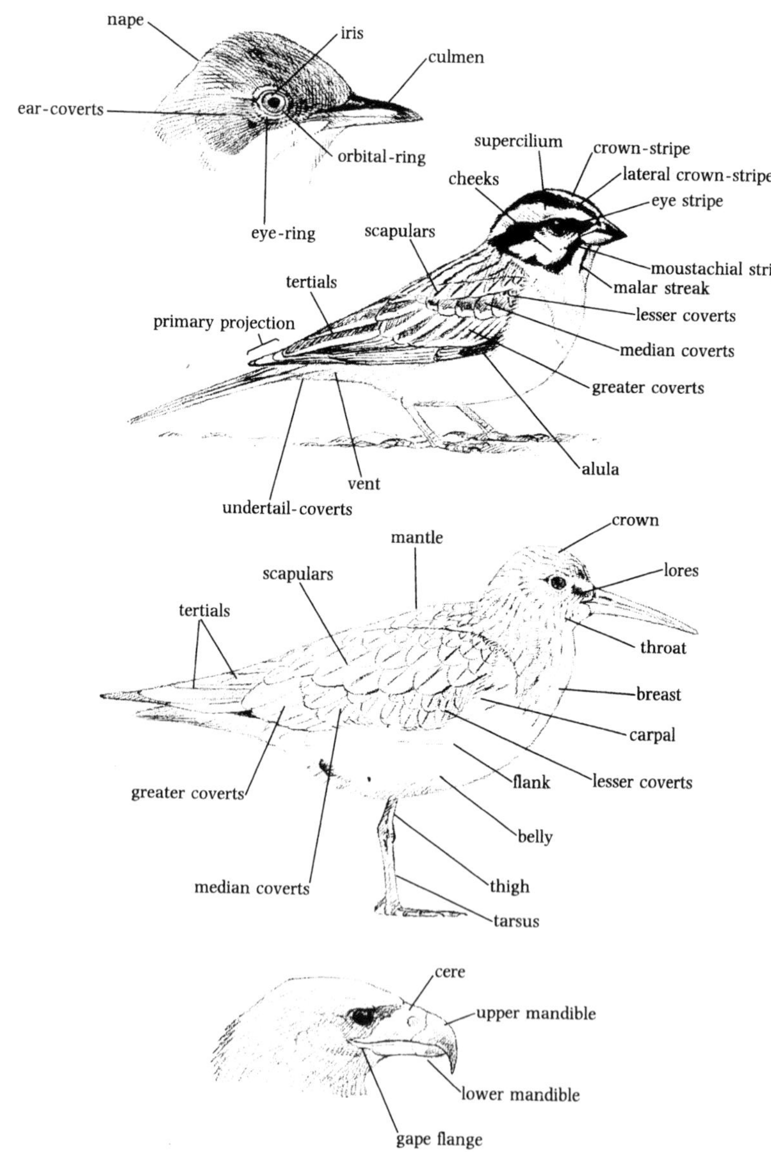
nape
iris
culmen
ear-coverts
orbital-ring
eye-ring
supercilium
crown-stripe
lateral crown-stripe
cheeks
eye stripe
scapulars
moustachial stripe
malar streak
tertials
lesser coverts
primary projection
median coverts
greater coverts
alula
vent
undertail-coverts
crown
mantle
lores
scapulars
tertials
throat
breast
carpal
flank
lesser coverts
greater coverts
belly
thigh
median coverts
tarsus
cere
upper mandible
lower mandible
gape flange

The identification text

Our aim has been to make these accounts of help to beginner and expert alike. We have concentrated on those features which are important for identification and these are highlighted in italics. For each species, length from bill tip to tail (L) is given in centimetres; for larger birds wing-span (W) is also given. Where the identification of a species does not present a problem, the texts are short.

The illustrations opposite and on p. xi show the various features of a bird—its topography—used in the identification texts. Knowledge of these, especially the feather tracts, is vital for describing a bird and its plumage.

Voice

Generally voice is only given where a species occurs regularly in the region or where it might be helpful for identification.

Maps

These show breeding distribution only. Details of occurrence on passage and in winter are given under 'Status'.

We have taken a cautious approach to mapping distribution, relying on proven breeding records in the case of rare or less common species. For commoner species, however, mapping has also relied heavily on sightings during the breeding season, even though breeding may not have been proven. We have made extensive use of the excellent maps prepared for the *Atlas of Arabian Breeding Birds* to which many observers have contributed, and on the country avifaunas listed in 'References and Further Reading' (pages 430–432). Others have given freely of their knowledge and are listed under 'Acknowledgements'.

In general, the breeding, passage and winter distributions of birds are relatively well-known for the Middle East with the exception of Iraq, parts of the deserts of Arabia (notably the Empty Quarter) and to some extent eastern Turkey, Iran, Syria and Lebanon. Nonetheless, there is much awaiting discovery.

Status and habitat

The status text complements the map, or, in cases where a species does not breed, gives a brief account of its wintering, passage or vagrancy status in the region. In this we have sometimes used broad expressions to denote areas:

Near East	the broad eastern Mediterranean area of Syria, Lebanon, Jordan, Israel, West Bank and Gaza
SW Arabia	The Republic of Yemen and adjacent areas of Saudi Arabia
E Arabia	Oman, UAE and adjacent areas of Saudi Arabia

We make no apologies for using information on status collected during the preparation of *Birds of the Middle East and North Africa*, though this has been extensively revised. Only the habitats used by the species within the Middle East are given.

ACKNOWLEDGEMENTS

In preparing this book we have received a great deal of help from many people, all of whom have given freely of their advice and encouragement. First, our thanks are due to Phil Hollom and Ian Willis whose co-authorship with two of us of *Birds of the Middle East and North Africa* provided the inspiration to tackle this project.

For discussion of identification issues, we would particularly like to thank Jens Eriksen, Rob Hume, Thomas Johansen, Tony Marr, Rod Martins, Neil Morris, Dick Newell, Klaus Nielsen and Peter Symens.

The preparation of the maps and status sections was greatly assisted by advice from Ian Andrews, Simon Aspinall and Charles Pilcher. Guy Kirwan and Mike Jennings were especially helpful in providing much unpublished data for Turkey and the *Arabian Breeding Bird Atlas* respectively. Extensive use was made of the publications listed under 'References and Further Reading' on pages 429–432 and we thank the authors for producing them. Unfortunately, *The Birds of Israel* did not appear until after the text had been finalized and therefore we have not been able to make use of information cited in that important work.

Many others helped in numerous ways from commenting on the text, providing general information and support, through to assisting the artists with advice and reference material. We would like to thank Omar Al-Saghier, Colin Bath, David Bolton, Duncan Brooks, Margaret Burton, John Cox, Mike Curzon, Ian Dawson, Mike Evans, Mike Everett, Cliff Foley, Fay Gale, Neil Gale, Paul Gale, Michael Gallagher (and the Oman Bird Records Committee), Robert Gillmor, Chris Harbard, Derek Harvey, Adrian Hickman, Phil Hollom, Chris Jones, Martyn Jones, Alan Kitson, Bryan Meloy, Tom McJannet, Colin Richardson, Gary Robins, Mustafa Sari, Jo Thomas, Effie Warr and Peter Wilkinson. Our thanks also go to the British Museum of Natural History, the Royal Albert Memorial Museum in Exeter (UK), the Zoological Museum of Copenhagen and Bird World in Rhode (UK) for allowing access to their collections.

We thank Ben Hoare and Sylvia Sullivan for their editorial comments and patient proof reading, Andy Richford, Carol Parr and the team at T&AD Poyser for their enthusiastic and professional support and Shell International Petroleum Company Limited for their generous sponsorship.

Colour Plates

1 Red-throated Diver *Gavia stellata*

Status: rare winter visitor Turkey and S Caspian. Vagrant Mediterranean and Gulf of Aqaba. **Habitat:** mainly coastal waters.

2 Black-throated Diver *Gavia arctica*

Status: winter visitor. Widespread Turkey (scarcer on S coast). Scarce S Caspian. Vagrant E Turkey, Israel and Gulf of Aqaba. **Habitat:** mainly coastal waters.

3 Great Northern Diver *Gavia immer*

Status: vagrant Mediterranean, and Black Sea coast. **Habitat:** coastal waters.

4 Little Grebe *Tachybaptus ruficollis*

Status: partial migrant. Winters in most breeding areas, also Black Sea, Mediterranean coast and Arabia. **Habitat:** lakes and pools, sometimes very small or densely vegetated. In winter, also estuaries. Nests in water on heap of vegetation.

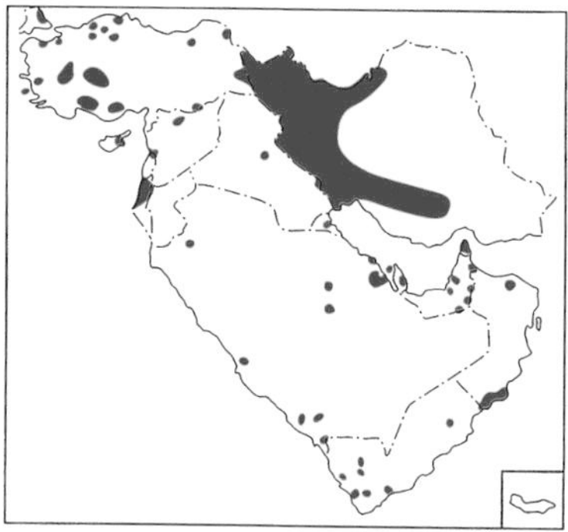

5 Great Crested Grebe *Podiceps cristatus*

Status: has bred Cyprus; partial migrant. Winter range includes Turkey (especially Black Sea coast), south to central Red Sea, east to Iran and Arabian Gulf, but scarce UAE and rare Oman and Jordan. **Habitat:** open areas of fresh water throughout year, also coastal waters in winter. Nest is anchored to vegetation.

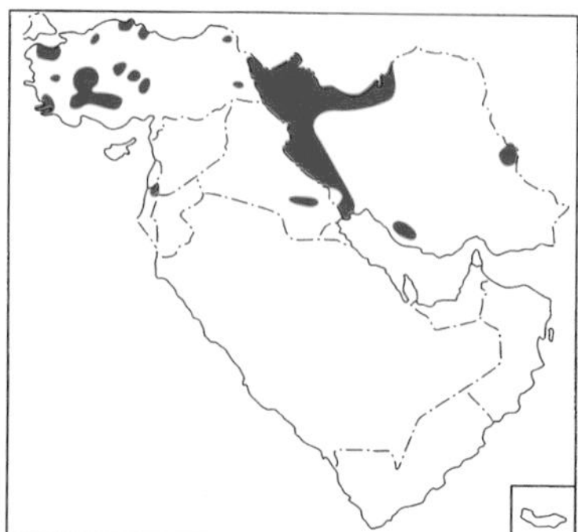

6 Red-necked Grebe *Podiceps grisegena*

Status: partial migrant. Winters uncommonly Turkey and S Caspian. Vagrant Cyprus, Syria and Arabian Gulf. **Habitat:** like Great Crested (also nest), but breeding pools usually smaller and more overgrown.

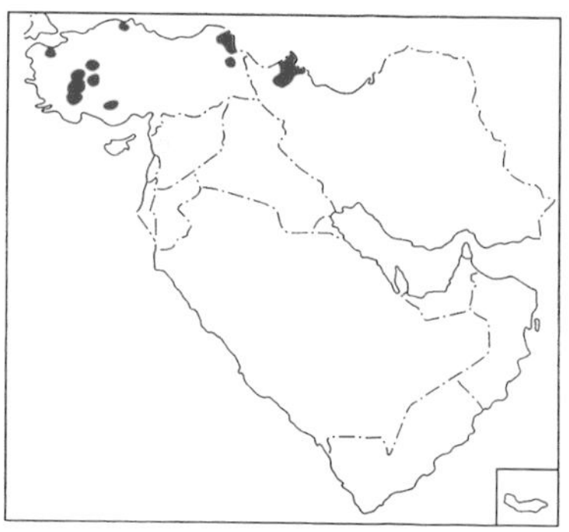

7 Slavonian Grebe *Podiceps auritus*

Status: winter visitor to S Caspian, extremely rare Turkey. Vagrant to Cyprus, Israel, S Iran and Kuwait. **Habitat:** fresh and coastal waters.

8 Black-necked Grebe *Podiceps nigricollis*

Status: has bred Cyprus; partial migrant. Winters south to coasts of Mediterranean and Arabian Gulf; also on some inland waters (mainly in or near breeding range) and sparingly in Arabia. **Habitat:** fresh water throughout the year, also coastal in winter. Nests at water level in well vegetated shallow pools.

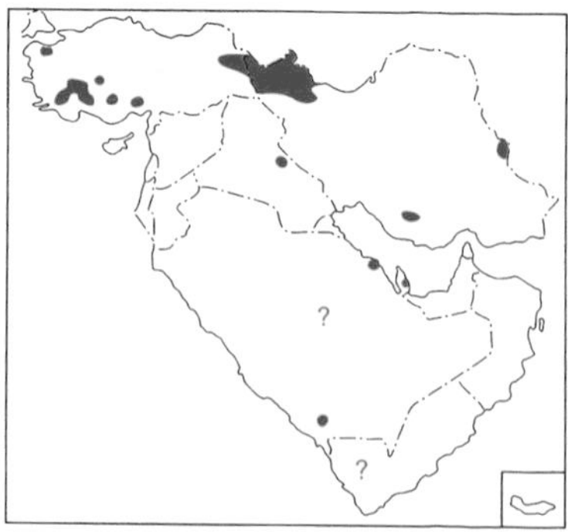

1 Red-throated Diver
ad. winter
juv.
2 Black-throated Diver
ad. winter
juv.
3 Great Northern Diver
ad. winter
juv.
4 Little Grebe
winter
summer
5 Great Crested Grebe
winter
summer
8 Black-necked Grebe
winter
summer
6 Red-necked Grebe
winter
summer
7 Slavonian Grebe
winter
ML

9 **Shy Albatross** *Diomedea cauta*

Status: vagrant Gulf of Aqaba from southern oceans. **Habitat:** maritime.

---

11 **Atlantic Petrel** *Pterodroma incerta*

Status: vagrant Gulf of Aqaba; from S Atlantic. **Habitat:** maritime.

---

12 **Soft-plumaged Petrel** *Pterodroma mollis*

Status: vagrant Israel, probably from Indian Ocean. **Habitat:** maritime.

---

13 **Jouanin's Petrel** *Bulweria fallax*

Status: in summer concentrations around Kuria Muria islands, Oman. Otherwise widely dispersed in Arabian Sea off Yemen, Socotra and E Arabia where often the commonest seabird. **Habitat:** maritime.

---

20 **Mediterranean Shearwater** *Puffinus yelkouan*

Status: throughout the year, often in flocks, in E Mediterranean, Black Sea and Bosphorus. (Note Manx Shearwater has not been recorded in the Middle East.) **Habitat:** maritime. Colonial nesting probably in burrows and under rocks, though breeding grounds undiscovered.

21 **Little Shearwater** *Puffinus assimilis*

Status: vagrant to Mediterranean coast of Israel (from N Atlantic) and to Gulf of Aqaba (from southern oceans). **Habitat:** maritime.

---

22 **Persian Shearwater** *Puffinus (lherminieri) persicus*

Status: outside breeding sites occurs in varying concentrations in W Arabian Gulf, Arabian Sea and S Red Sea throughout most of the year. **Habitat:** maritime. Colonial nesting in holes and rock crevices.

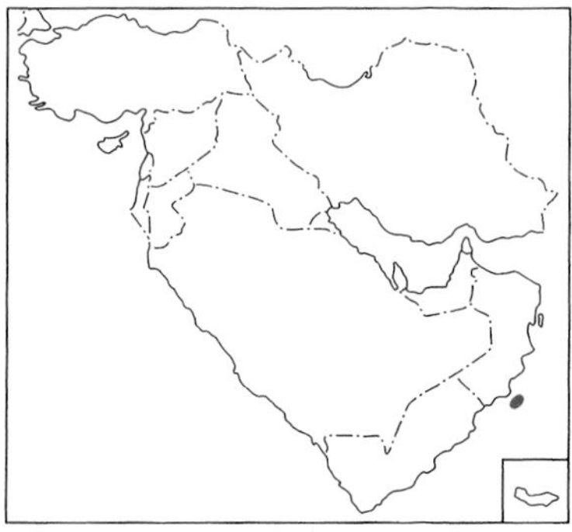

9 Shy Albatross
(not to scale)
11 Atlantic
Petrel
13 Jouanin's
Petrel
12 Soft-plumaged
Petrel
20 Mediterranean
Shearwater
yelkouan race
21 Little
Shearwater
22 Persian
Shearwater
JG

10 **Cape Petrel** *Daption capensis*

Status: vagrant Arabian Sea from southern oceans. **Habitat:** maritime.

---

14 **Cory's Shearwater** *Calonectris diomedea*

Status: mainly summer visitor to the seas off W Turkey, but some remain in E Mediterranean throughout the year. Regular Gulf of Aqaba, rare Turkish Black Sea coast, occasional Arabian Sea. Vagrant Arabian Gulf. **Habitat:** maritime.

---

15 **Streaked Shearwater** *Calonectris leucomelas*

Status: vagrant from W Pacific to Gulf of Aqaba and Arabian Sea. **Habitat:** maritime.

---

16 **Pale-footed Shearwater** *Puffinus carneipes*

Status: non-breeding visitor from S Indian Ocean to the Arabian Sea, common off Oman, occasional Gulf of Oman, rare UAE. Vagrant Gulf of Aqaba and Yemen. **Habitat:** maritime.

17 **Great Shearwater** *Puffinus gravis*

Status: vagrant Lebanon and Israel (from Atlantic Ocean). **Habitat:** maritime.

---

18 **Wedge-tailed Shearwater** *Puffinus pacificus*

Status: uncommon gulf of Oman (but rare UAE), Arabian Sea (Gulf) and S Red Sea. **Habitat:** maritime.

---

19 **Sooty Shearwater** *Puffinus griseus*

Status: regular, but scarce in N Red Sea and Gulf of Aqaba. Vagrant Oman, UAE and E Mediterranean. **Habitat:** maritime.

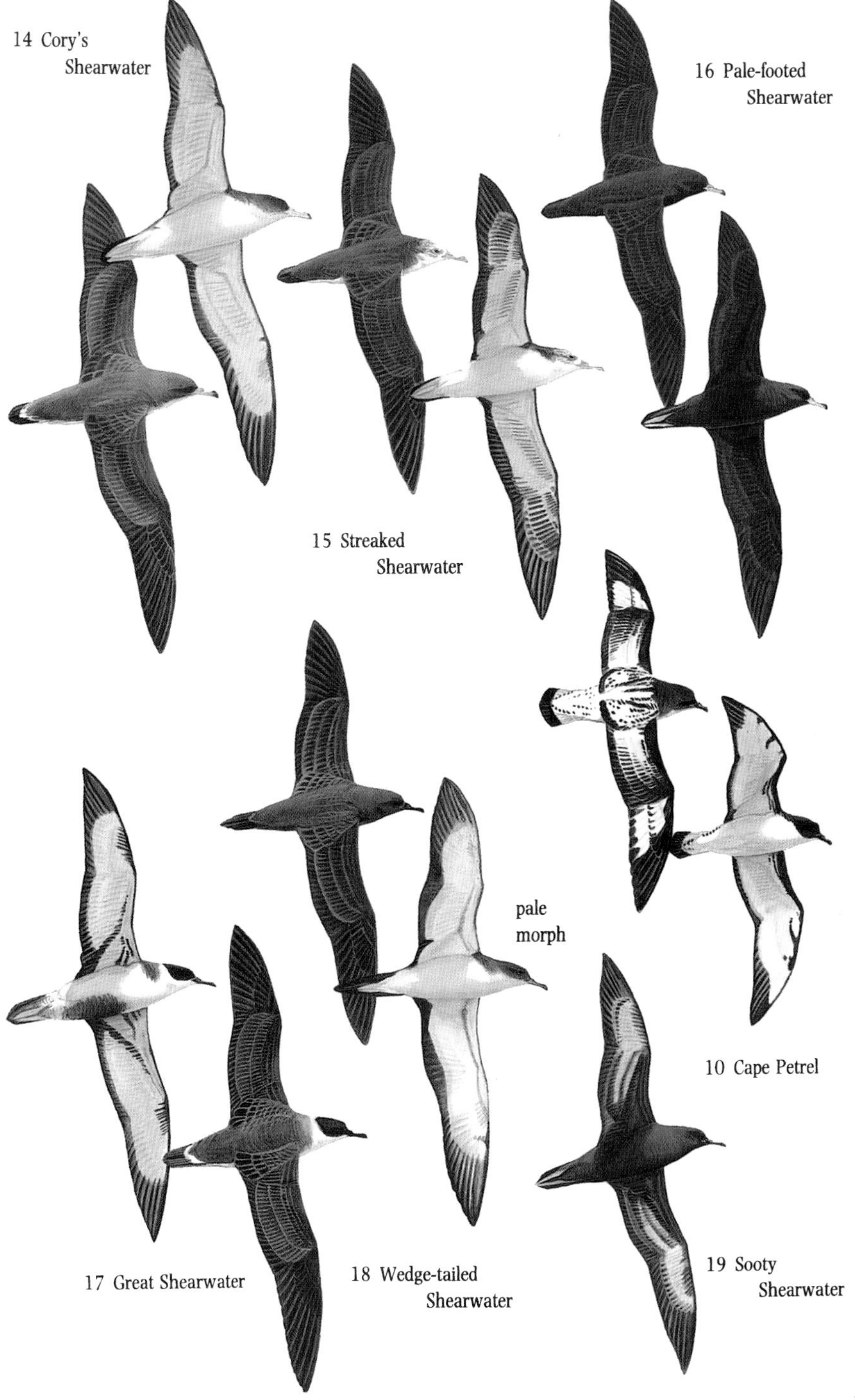
14 Cory's Shearwater
15 Streaked Shearwater
16 Pale-footed Shearwater
pale morph
10 Cape Petrel
17 Great Shearwater
18 Wedge-tailed Shearwater
19 Sooty Shearwater
JG

23 **Wilson's Storm-petrel** *Oceanites oceanicus*

Status: an Antarctic species occurring in summer in the N Hemisphere. Recorded Red Sea (occasionally north to Gulf of Aqaba) and off coasts of E Arabia (common Oman), and Iran, with few into Arabian Gulf. **Habitat:** maritime.

24 **Black-bellied Storm-petrel** *Fregetta tropica*

Status: regular in summer from southern oceans to central Arabian Sea, but only occasionally reaches coastal waters (e.g. off Oman). **Habitat:** maritime.

25 **White-bellied Storm-petrel** *Fregetta grallaria*

Status: unconfirmed records from Oman, from southern oceans. **Habitat:** maritime.

26 **White-faced Storm-petrel** *Pelagodroma marina*

Status: regular of SE coast of Arabia in summer, from southern oceans. **Habitat:** maritime.

27 **European Storm-petrel** *Hydrobates pelagicus*

Status: dispersing birds from colonies in Atlantic and Mediterranean reach E Mediterranean coasts and even Black Sea. Unconfirmed observation from Red Sea. Vagrant Turkey. **Habitat:** maritime.

28 **Leach's Storm-petrel** *Oceanodroma leucorhoa*

Status: vagrant to Mediterranean Israel, Gulf of Aqaba and UAE. **Habitat:** maritime.

29 **Swinhoe's Storm-petrel** *Oceanodroma monorhis*

Status: vagrant to Gulf of Aqaba and Arabian Sea in summer, from N Indian Ocean to which it migrates from NW Pacific. **Habitat:** maritime.

30 **Madeiran Storm-petrel** *Oceanodroma castro*

Status: claimed as a vagrant to Gulf of Aqaba. **Habitat:** maritime.

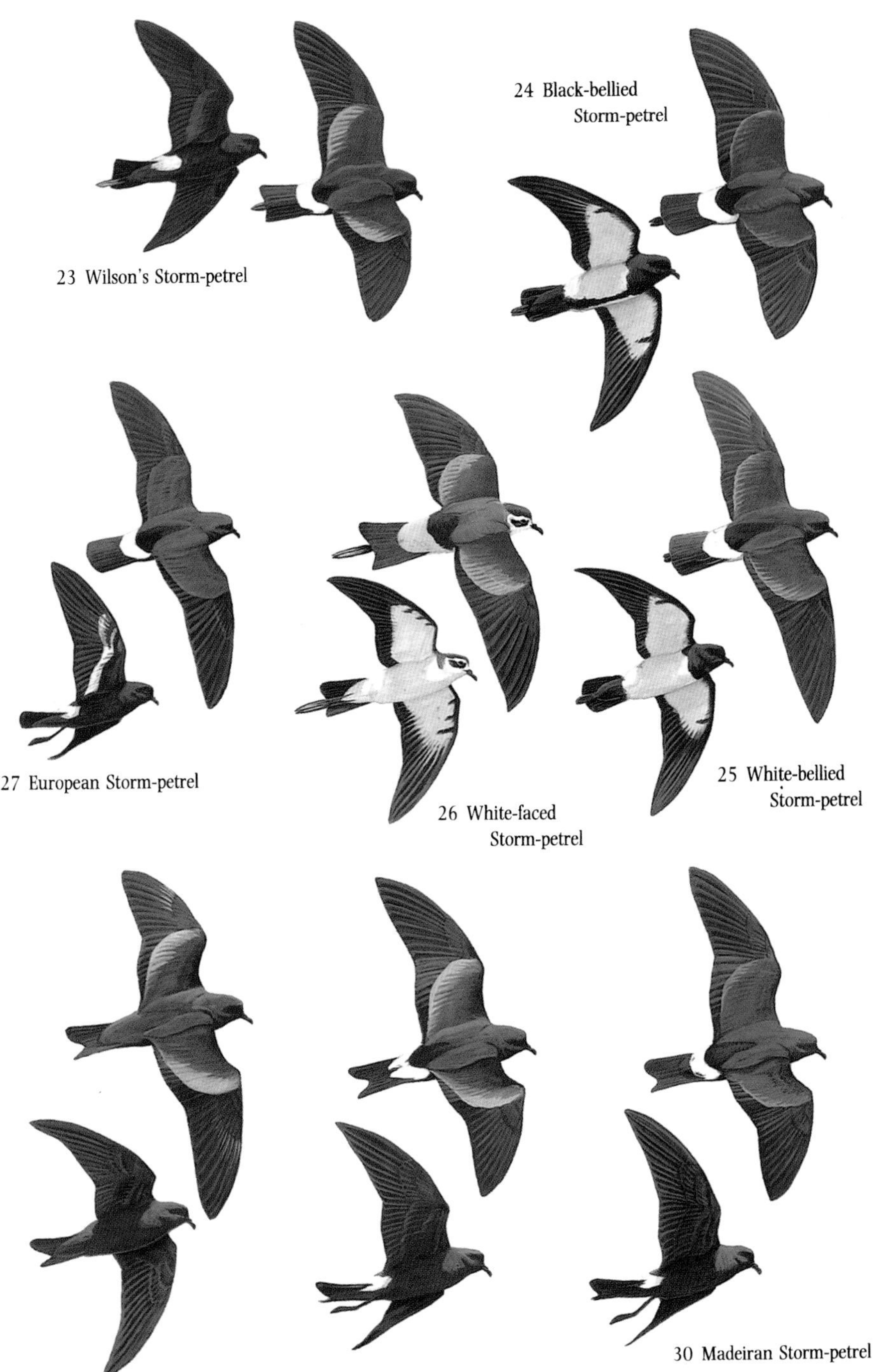
23 Wilson's Storm-petrel
24 Black-bellied Storm-petrel
27 European Storm-petrel
26 White-faced Storm-petrel
25 White-bellied Storm-petrel
29 Swinhoe's Storm-petrel
28 Leach's Storm-petrel
30 Madeiran Storm-petrel
JG

31 **Red-billed Tropicbird** *Phaethon aethereus*

Status: resident with dispersal over Arabian Sea and throughout Red Sea to Gulf of Aqaba (where rare). **Habitat:** maritime. Breeds mostly on rocky cliff faces, even inland.

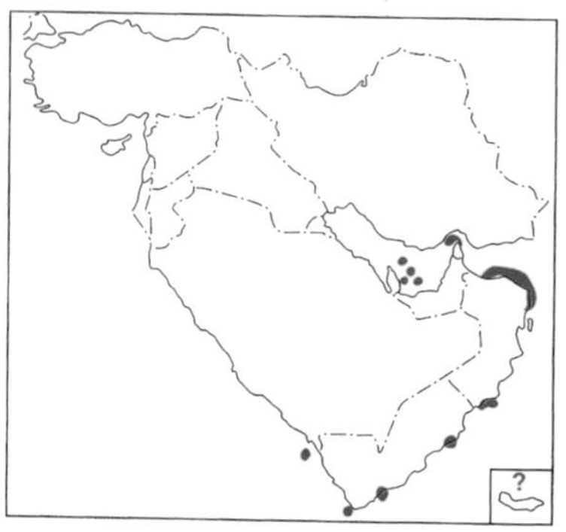

32 **Red-footed Booby** *Sula sula*

Status: vagrant, from Indian Ocean to Arabian Sea and Red Sea. **Habitat:** maritime.

33 **Masked Booby** *Sula dactylatra*

Status: mainly resident; occurs elsewhere off coasts of S and E Arabia and S Red Sea but not commonly seen. Vagrant UAE. **Habitat:** maritime. Nests colonially on sandy or rocky islands.

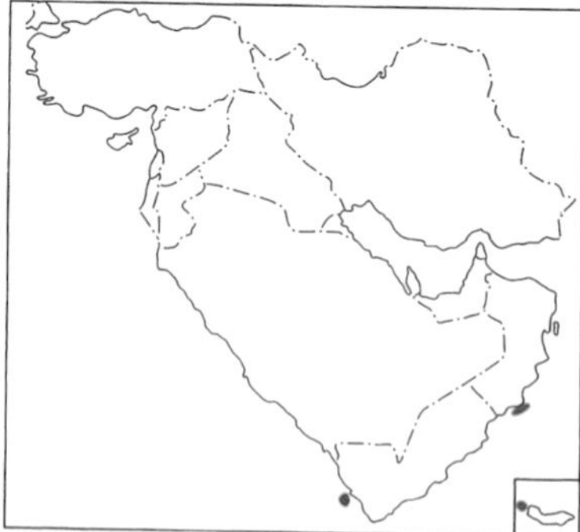

34 **Brown Booby** *Sula leucogaster*

Status: resident and dispersive, occurring widely in Red Sea and off S Arabia. Regular S Oman but vagrant N Oman and Arabian Gulf. **Habitat:** maritime. Nests in colonies on rocky islands.

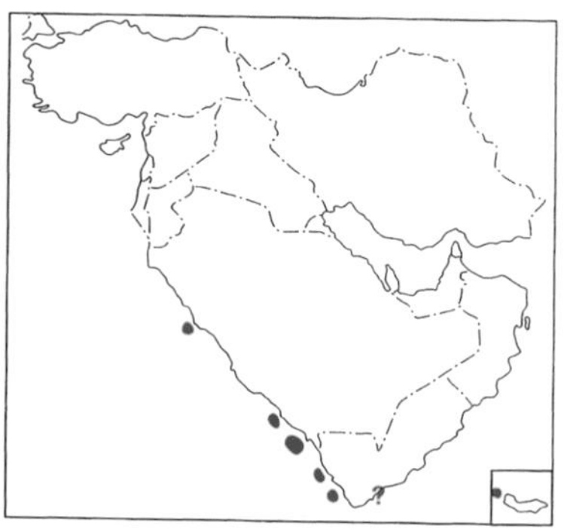

35 **Gannet** *Sula bassana*

Status: rather scarce visitor to E Mediterranean coasts from Atlantic breeding areas. **Habitat:** maritime.

45 **Great Frigatebird** *Fregata minor*

Status: unrecorded in region but unidentified frigatebirds, possibly this species, have been seen Oman and Gulf of Aqaba. **Habitat:** maritime.

46 **Lesser Frigatebird** *Fregata ariel*

Status: vagrant Oman and possible UAE. **Habitat:** maritime.

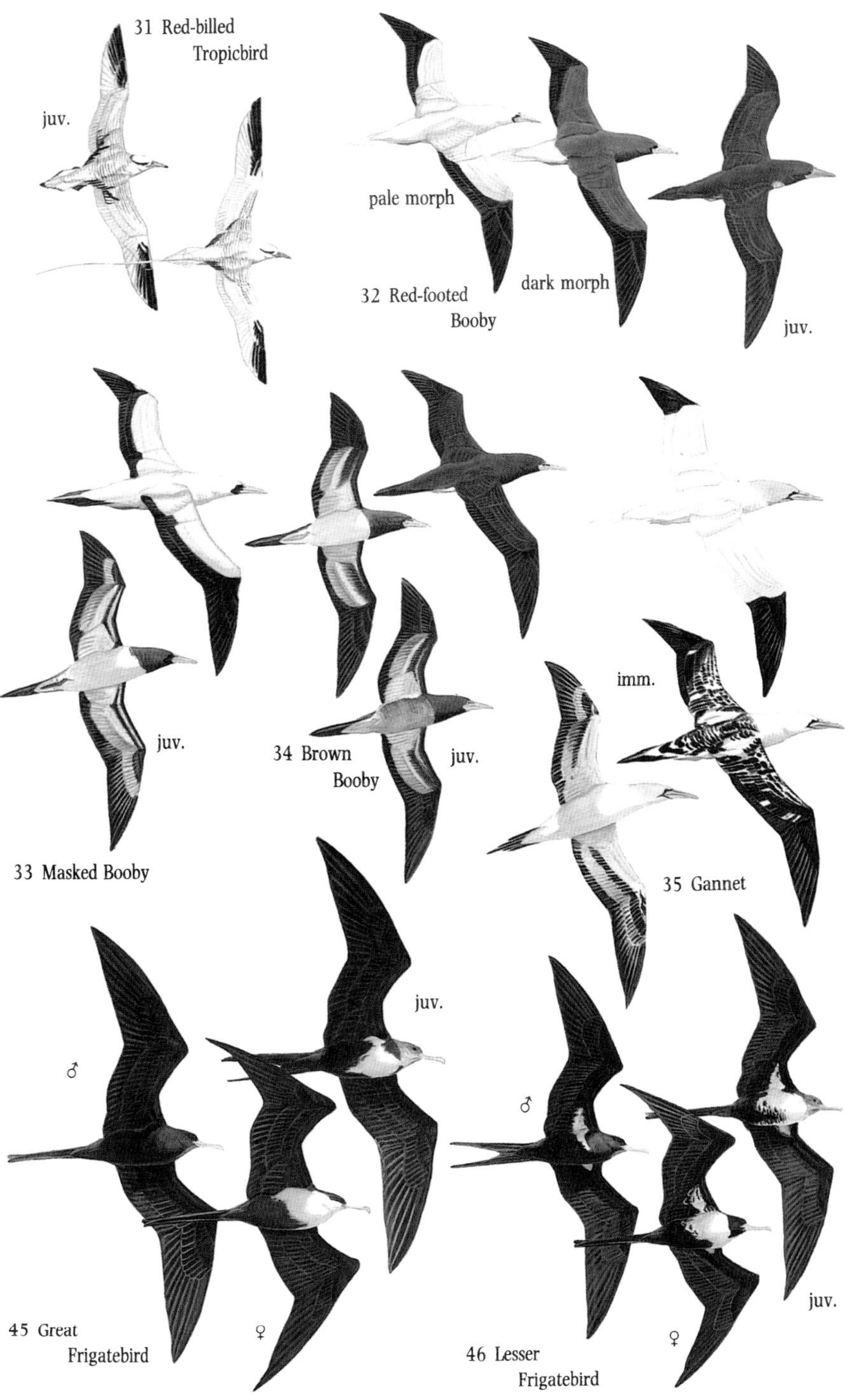
31 Red-billed Tropicbird
juv.
pale morph
dark morph
32 Red-footed Booby
juv.
juv.
34 Brown Booby
juv.
imm.
33 Masked Booby
35 Gannet
juv.
♂
♂
45 Great Frigatebird
♀
46 Lesser Frigatebird
♀
juv.
JG

36 **Cormorant** *Phalacrocorax carbo*

Status: winters on coasts of Mediterranean, N Red Sea, Arabian Gulf and Oman; also occasionally inland in parts of N Arabia. Scarce to Jordan and S Red Sea. **Habitat:** mainly coastal but also inland on passage. Nests in tree colonies in Turkey.

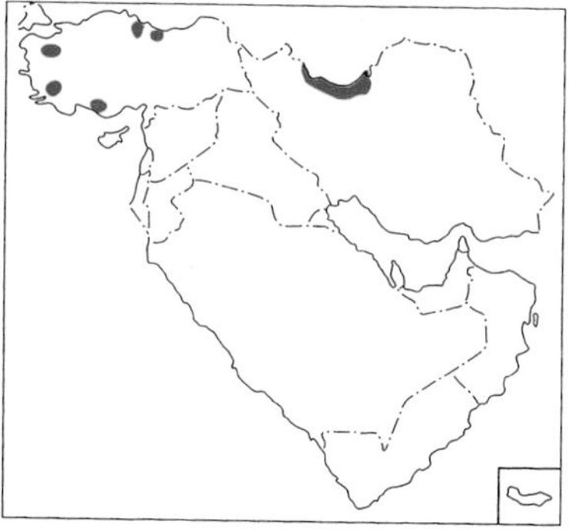

37 **Shag** *Phalacrocorax aristotelis*

Status: resident with some dispersal. Vagrant Iraq and SE Mediterranean. **Habitat:** maritime. Breeds colonially on rocky coasts and steep cliffs. Occasionally inland in winter.

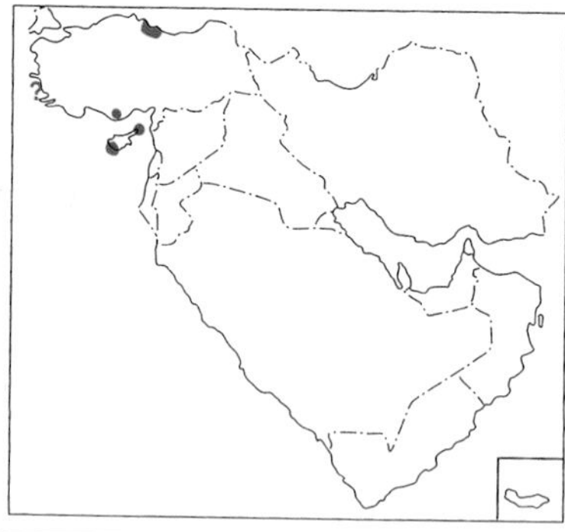

38 **Socotra Cormorant**
Phalacrocorax nigrogularis

Status: resident and dispersive in Arabian Gulf and off SE coasts of Arabia to S Red Sea. Vagrant N Red Sea and Israel. **Habitat:** maritime, largely coastal. Nests in packed colonies among boulders on rocky islands, or on sand or gravel.

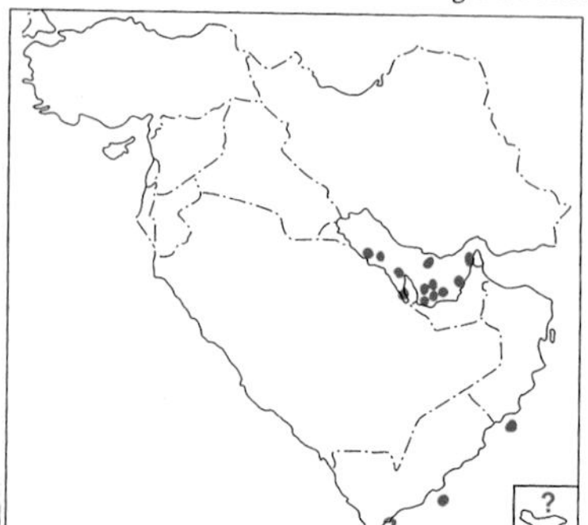

39 **Pygmy Cormorant** *Phalacrocorax pygmeus*

Status: partial migrant, dispersing locally outside breeding season. Occasionally reaches Lebanon, is becoming regular winter visitor Cyprus and Israel (where now breeding). **Habitat:** freshwater lakes and rivers with extensive reeds or trees to provide nesting areas.

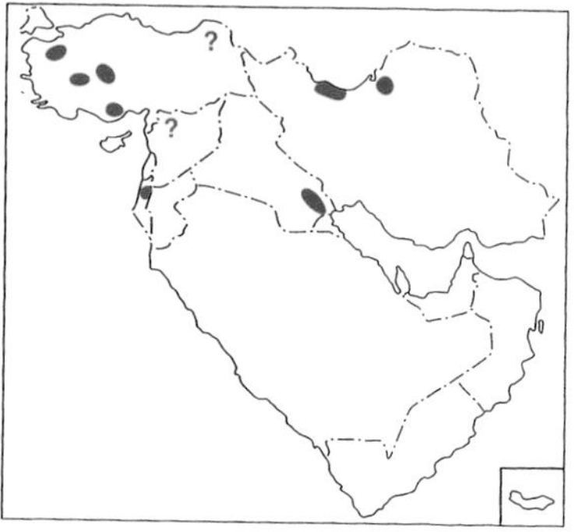

40 **Long-tailed Cormorant**
Phalacrocorax africanus

Status: vagrant to Yemen, from Africa. **Habitat:** rivers and lakes, often far inland; also sea coasts.

41 **Darter** *Anhinga rufa*

Status: mainly resident, but has occurred in SW Iran in winter. **Habitat:** large lakes and swamps with much vegetation and scattered trees. Nests colonially often among cormorants and herons.

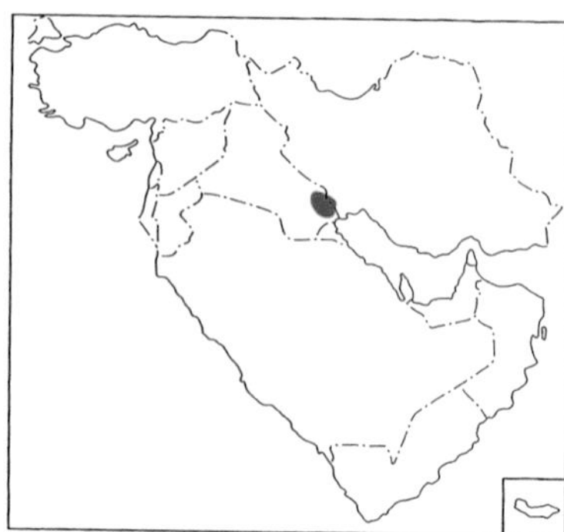

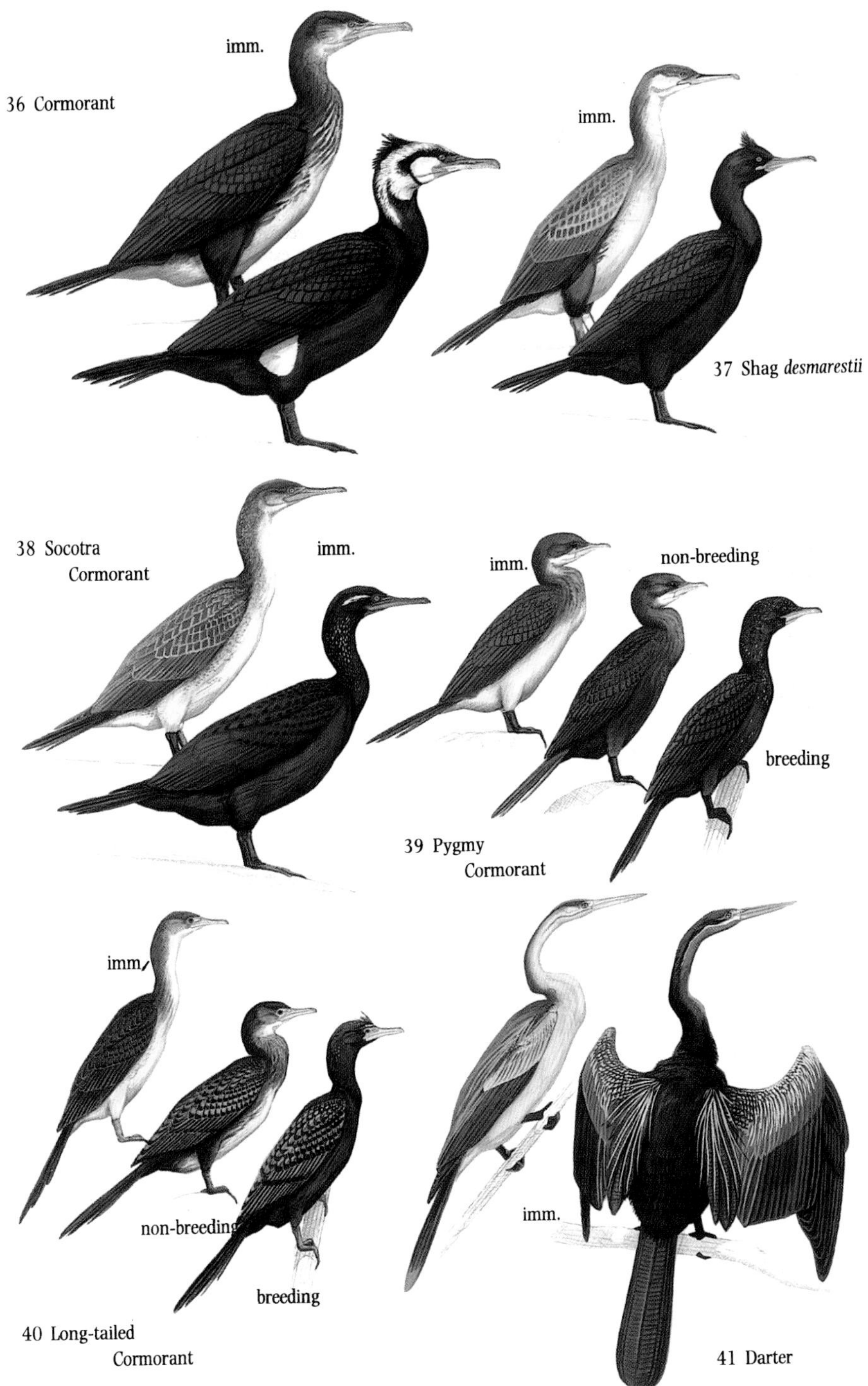
36 Cormorant
imm.
imm.
37 Shag desmarestii
38 Socotra Cormorant
imm.
imm.
non-breeding
breeding
39 Pygmy Cormorant
imm.
non-breeding
breeding
40 Long-tailed Cormorant
imm.
41 Darter

42 **White Pelican** *Pelecanus onocrotalus*

Status: partial migrant. Large passage Turkey and Israel. Some winter S Turkey east to Iran and south to Israel; scarce Cyprus and all Arabia, particularly in east. **Habitat:** large inland waters, marshes and shallow coastal lagoons. Nests colonially among reeds.

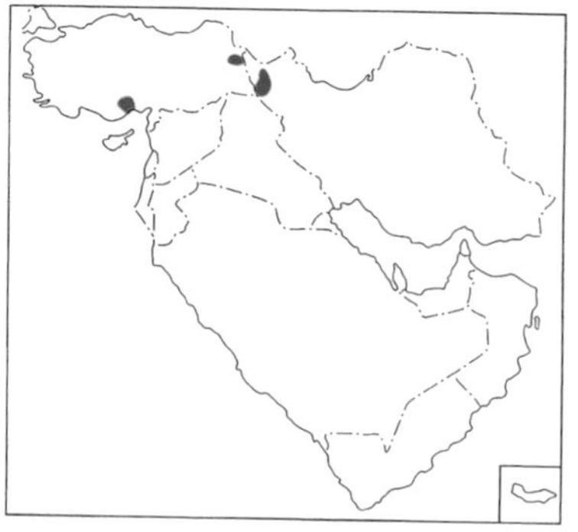

43 **Dalmatian Pelican** *Pelecanus crispus*

Status: declining. Partial migrant, some winter W and S Turkey, Syria, Iraq, S Caspian; rare Israel, vagrant Cyprus and UAE. **Habitat:** as White Pelican. Nests like White but sometimes uses trees.

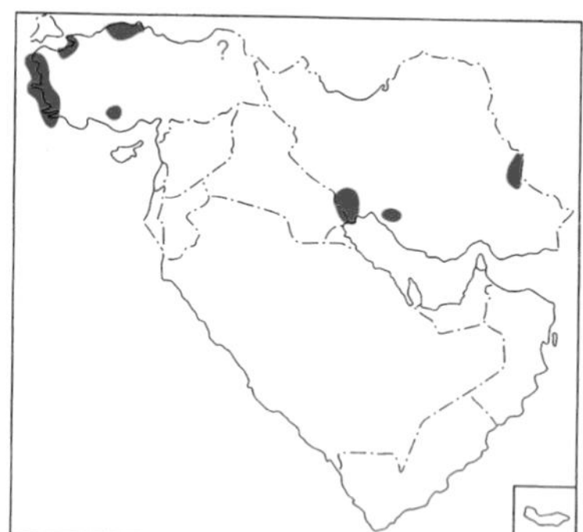

44 **Pink-backed Pelican** *Pelecanus rufescens*

Status: resident; occurs regularly in area enclosed by dotted-line. Rare Israel. **Habitat:** coasts, lagoons, sandbanks and inland lakes. Nests in mangroves and on sandy islands.

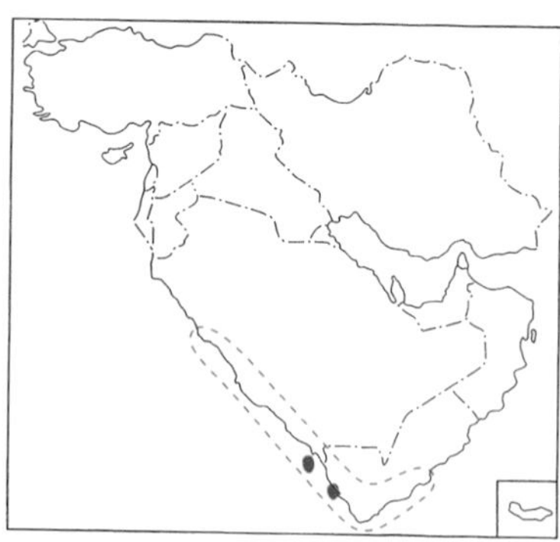

42 White Pelican
non-breeding
juv.
43 Dalmatian Pelican
non-breeding
juv.
breeding
44 Pink-backed Pelican
non-breeding
juv.
juv.
JG

47 **Bittern** *Botaurus stellaris*

Status: partial migrant dispersing throughout Turkey and Iran. In winter occurs south to Yemen, but generally uncommon in Arabian peninsula. **Habitat:** reedbeds (usually large) in breeding season; also wetlands with marginal vegetation.

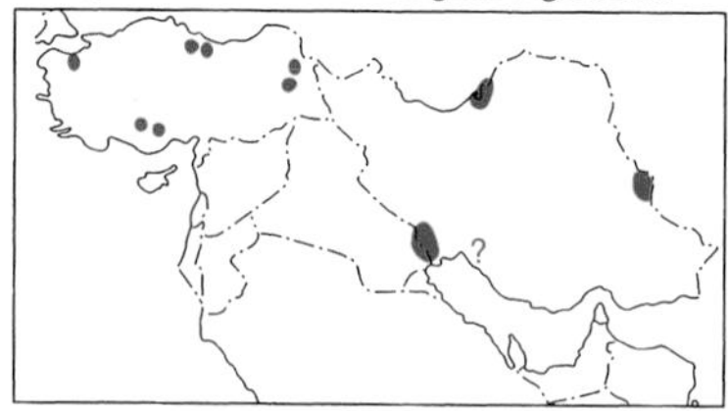

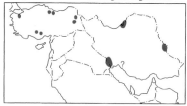

48 **Little Bittern** *Ixobrychus minutus*

Status: Summer visitor, though occasional in Oman in winter. Passage throughout the region. **Habitat:** well vegetated rivers, ponds and lakes. Nests near water often in loose colonies.

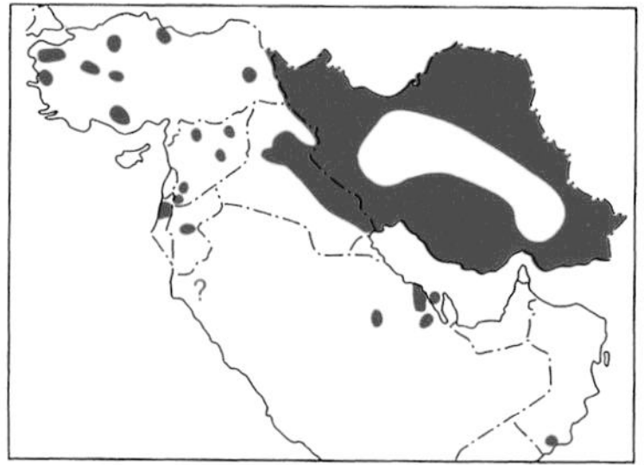

49 **Night Heron** *Nycticorax nycticorax*

Status: mainly a summer visitor. Widespread on passage (but less common S Arabia). Some winter in Oman. **Habitat:** rivers, lakes and marshes with areas of dense vegetation. Nests colonially in trees, bushes and reedbeds.

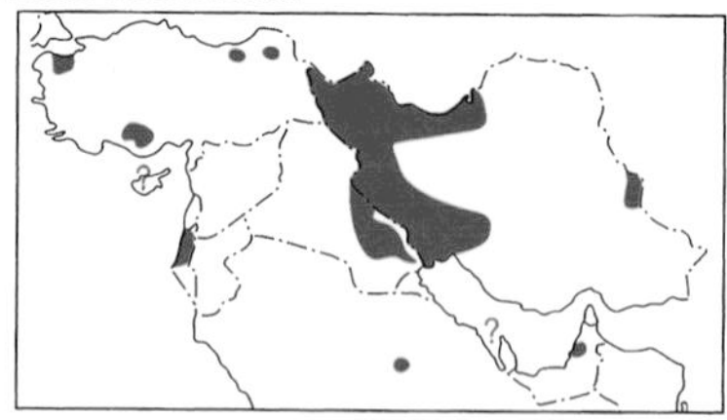

50 **Striated Heron** *Butorides striatus*

Status: mainly resident but apparently some passage in Oman. Rare Israel. Vagrant Bahrain. **Habitat:** coastal mudflats and reefs, especially with mangroves, in which it usually nests.

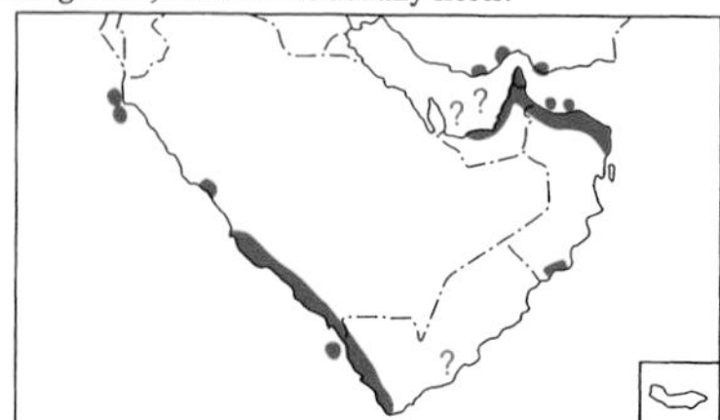

51 **Squacco Heron** *Ardeola ralloides*

Status: mainly summer visitor. Some winter on Arabian Gulf to Oman. Widespread on passage. **Habitat:** wetlands, ditches and rivers with good vegetation. Nests colonially (often with other herons) in trees and reeds.

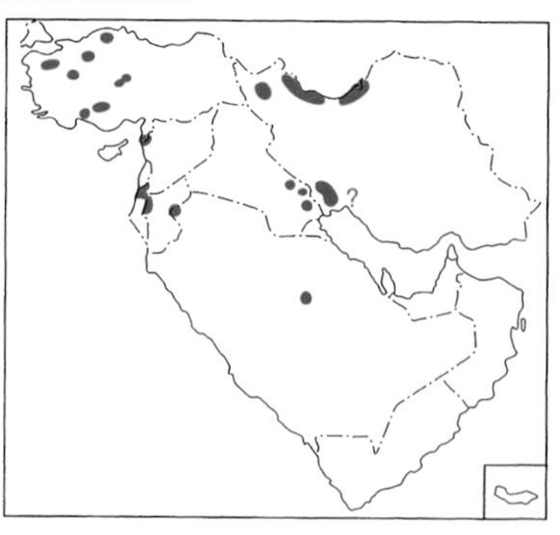

52 **Indian Pond Heron** *Ardeola grayii*

Status: mainly resident; also regular visitor in very small numbers to Oman, but uncommon UAE. **Habitat:** fresh and saltwater marshes, especially with dense cover; rivers, ponds and mangroves. Nests in trees.

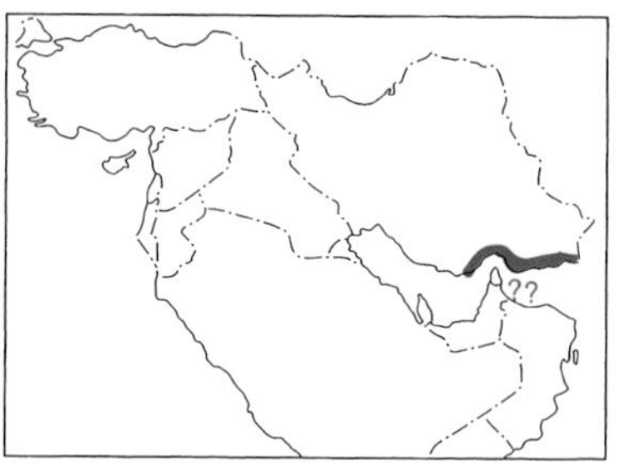

53 **Cattle Egret** *Bubulcus ibis*

Status: resident or dispersive; but summer visitor to S Caspian. Passage (with some wintering) Arabia and non-breeders may summer (e.g. in S Oman). **Habitat:** fields, marshes, generally in drier habitats than other herons, and infrequently on the coast. Nests colonially in reedbeds, bushes or trees.

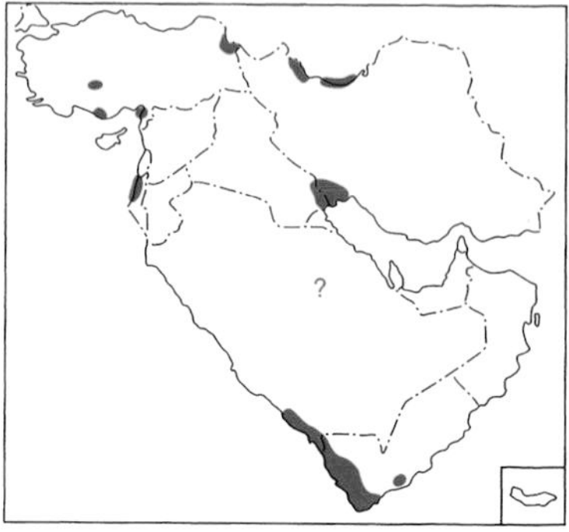

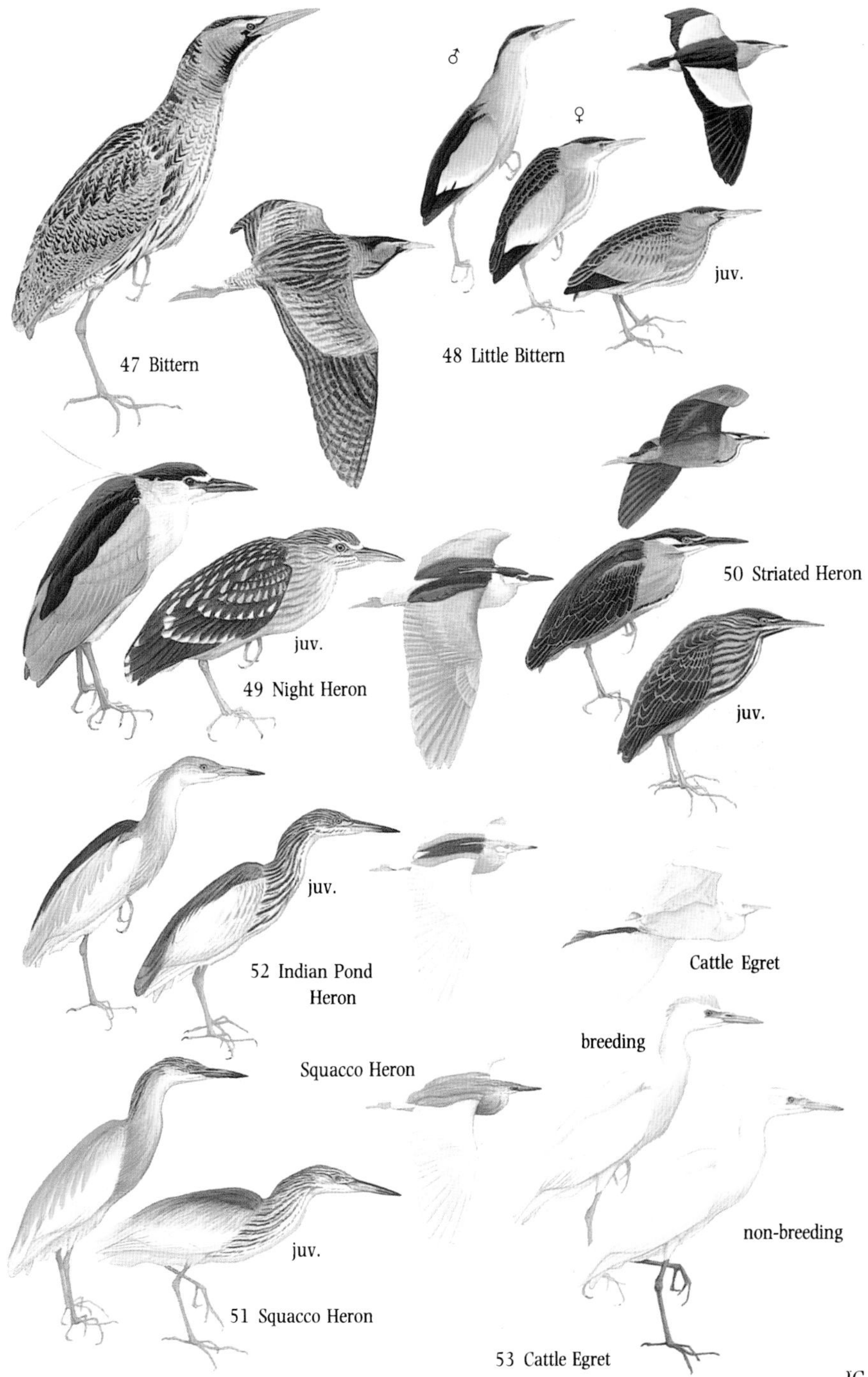
♂
♀
juv.
47 Bittern
48 Little Bittern
50 Striated Heron
juv.
49 Night Heron
juv.
juv.
52 Indian Pond
Heron
Cattle Egret
breeding
Squacco Heron
non-breeding
juv.
51 Squacco Heron
53 Cattle Egret
JG

54 **Western Reef Heron** *Egretta gularis*

Status: mostly resident. Occurs in non-breeding areas throughout year and map does not distinguish between breeding and non-breeding range. **Habitat:** almost entirely coastal, especially where mangroves and tidal flats. Can straggle inland. Nests colonially in mangroves and on offshore islands.

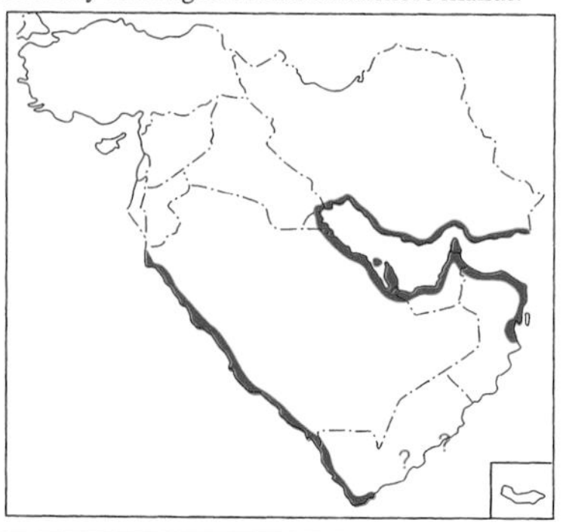

55 **Little Egret** *Egretta garzetta*

Status: summer visitor to Turkey and N Iran, resident or partial migrant elsewhere. Passage throughout region, in smaller numbers Arabia. **Habitat:** wetlands. Colonial nester in marshes, and trees.

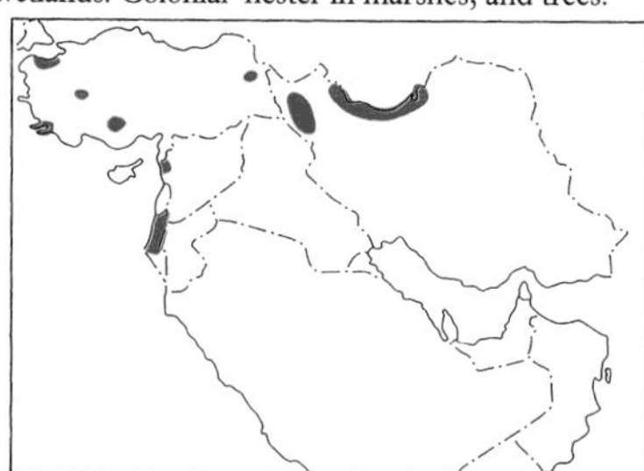

56 **Intermediate Egret** *Egretta intermedia*

Status: rare Oman, vagrant Jordan and UAE (from Africa and Asia). **Habitat:** wetlands, coastal and inland.

57 **Great White Egret** *Egretta alba*

Status: largely migratory. In winter, W and S Turkey south to Israel and east to Iran,. E Arabia and Oman. Passage throughout region. **Habitat:** wetlands and rivers. Usually nests colonially in reedbeds or trees.

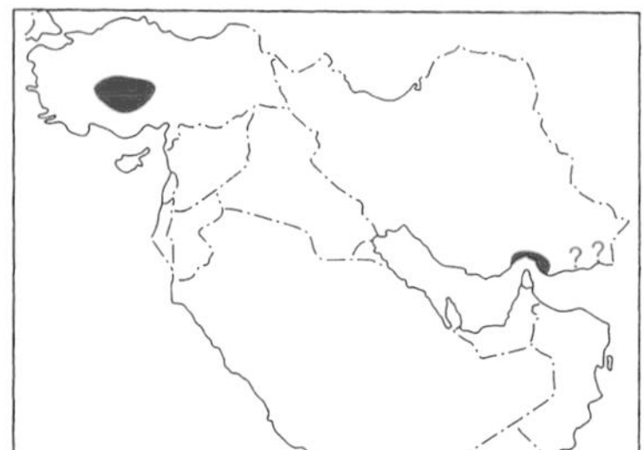

58 **Black-headed Heron** *Ardea melanocephala*

Status: vagrant Yemen, Oman and Israel (from Africa). **Habitat:** inland or coastal wetlands.

59 **Grey Heron** *Ardea cinerea*

Status: partial migrant. Occurs widely on passage and winter on most coasts and some inland waters. **Habitat:** can be found in any wetland area, including coastal. Nests colonially in trees and reedbeds.

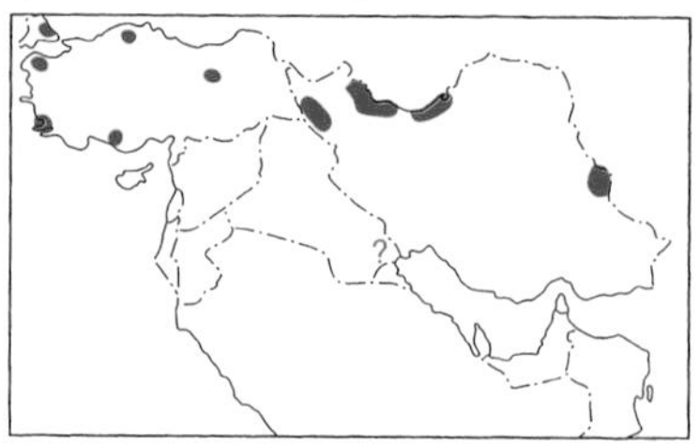

60 **Purple Heron** *Ardea purpurea*

Status: mainly summer visitor, some winter Iraq and S and E Arabia. Occurs widely on passage. **Habitat:** marshes, reedbeds, overgrown ditches. Nests colonially in reedbeds, sometimes bushes.

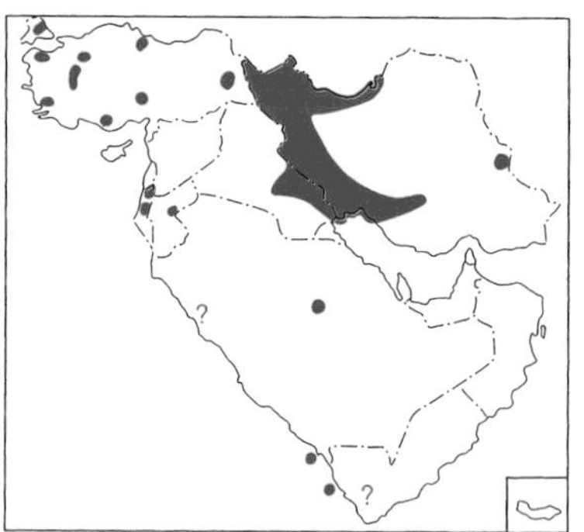

61 **Goliath Heron** *Ardea goliath*

Status: resident and dispersive. Vagrant Israel and Jordan (Gulf of Aqaba), Syria and Oman. **Habitat:** reedbeds in wetlands and rivers; also reefs, mudflats and coastal islands. Nests in mangroves and reeds.

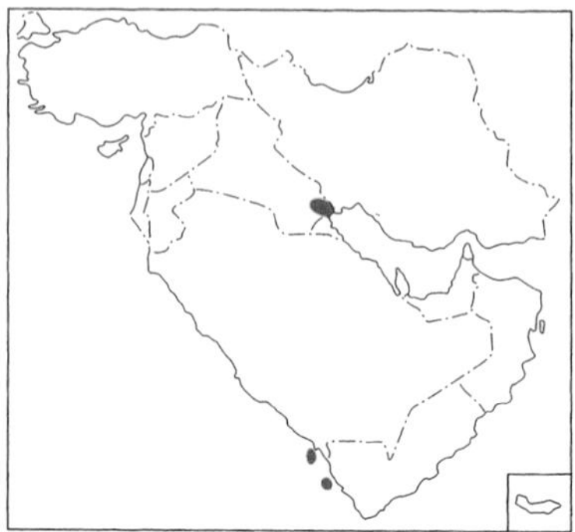

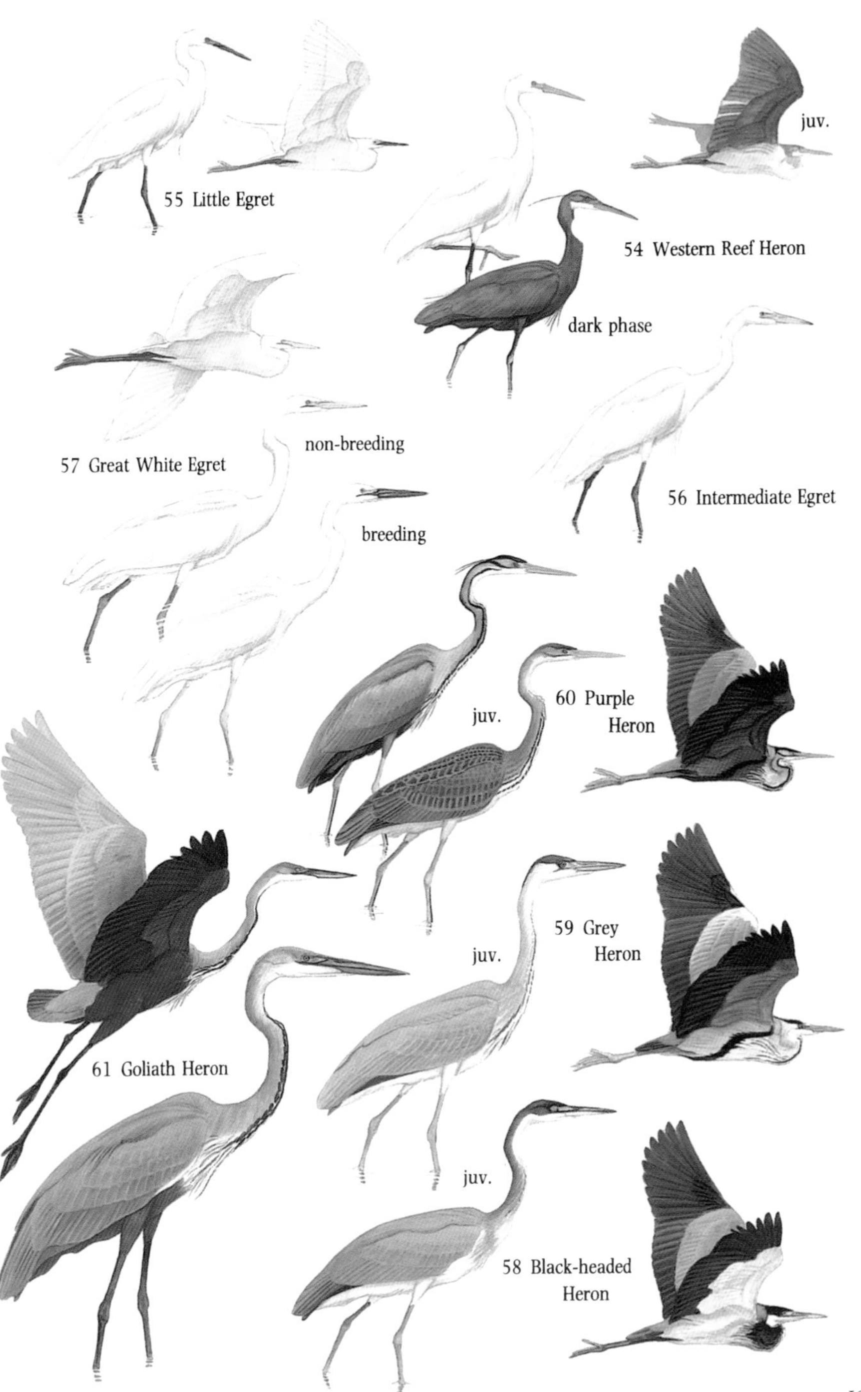
55 Little Egret
juv.
54 Western Reef Heron
dark phase
57 Great White Egret
non-breeding
breeding
56 Intermediate Egret
juv.
60 Purple
Heron
61 Goliath Heron
juv.
59 Grey
Heron
juv.
58 Black-headed
Heron

JG

62 **Hamerkop** *Scopus umbretta*

Status: resident; some dispersal outside breeding season. **Habitat:** well vegetated rocky wadis with running water from *c.* 300–3000 m. Bulky domed nest in tree or on cliff.

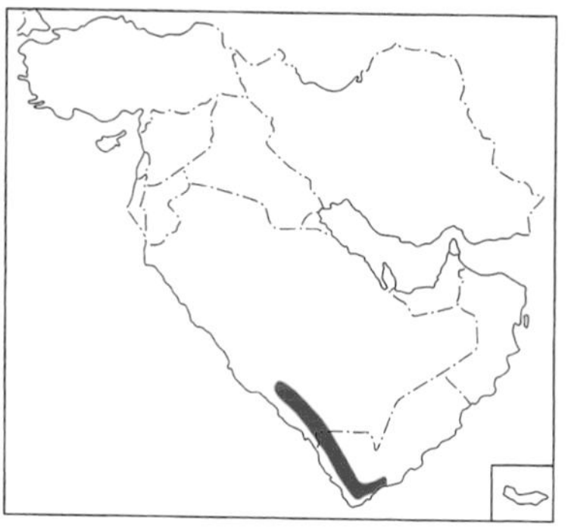

63 **Yellow-billed Stork** *Mycteria ibis*

Status: rare Israel, vagrant S Turkey (from Africa). **Habitat:** inland and coastal wetlands.

64 **Black Stork** *Ciconia nigra*

Status: summer visitor with main passage through Red Sea and E Mediterranean coast. Winters in small numbers SW Arabia, occasionally Israel and C Arabia but vagrant Oman. **Habitat:** damp forests, marshes and rocky river margins. Nests in woodlands or on crag.

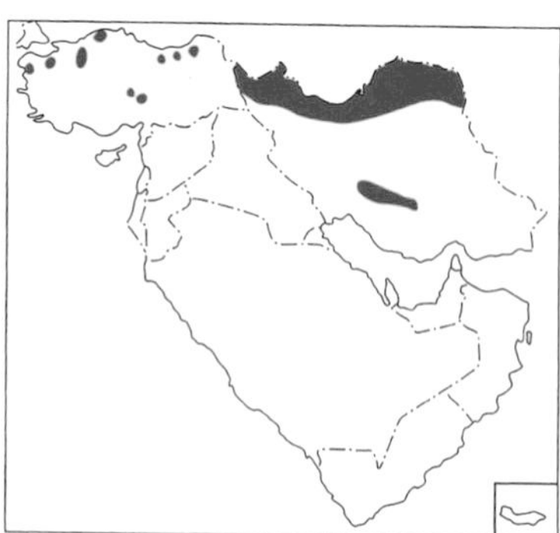

65 **Abdim's Stork** *Ciconia abdimii*

Status: summer visitor. Irregular, usually rare, Oman. **Habitat:** dry plains, semi-deserts and foothills. Nests singly or in colonies in trees, on roof tops or pylons.

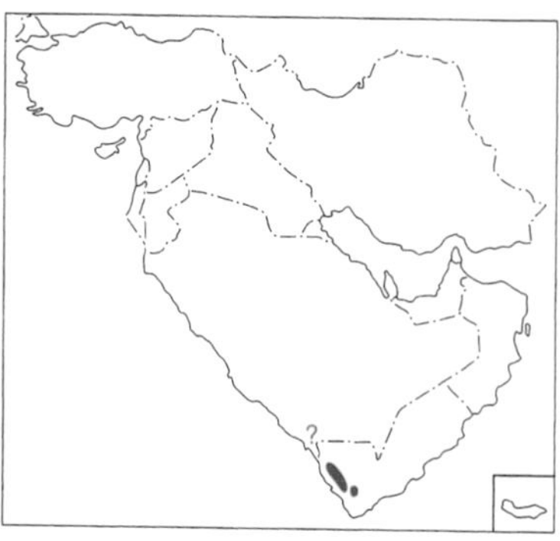

66 **Woolly-necked Stork** *Ciconia episcopus*

Status: vagrant Iran. Range includes Africa and India. **Habitat:** wetlands, river margins, coastal mudflats and coral reefs.

67 **White Stork** *Ciconia ciconia*

Status: summer visitor to breeding areas. Large passage through Turkey and Near East, but few through Arabia. Winters S Iraq, SW Iran, SW Arabia but rarer Near East, UAE and S Turkey. **Habitat:** wetlands, plains and arable. Nests on buildings and trees.

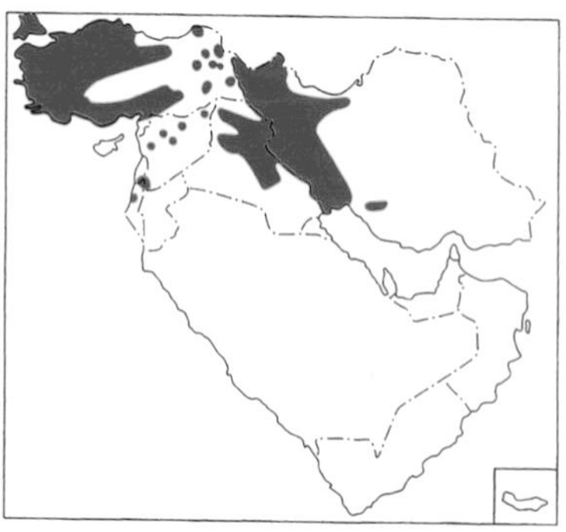

68 **Marabou Stork** *Leptoptilos crumeniferus*

Status: vagrant Israel. Range Africa. **Habitat:** savannas and wetlands.

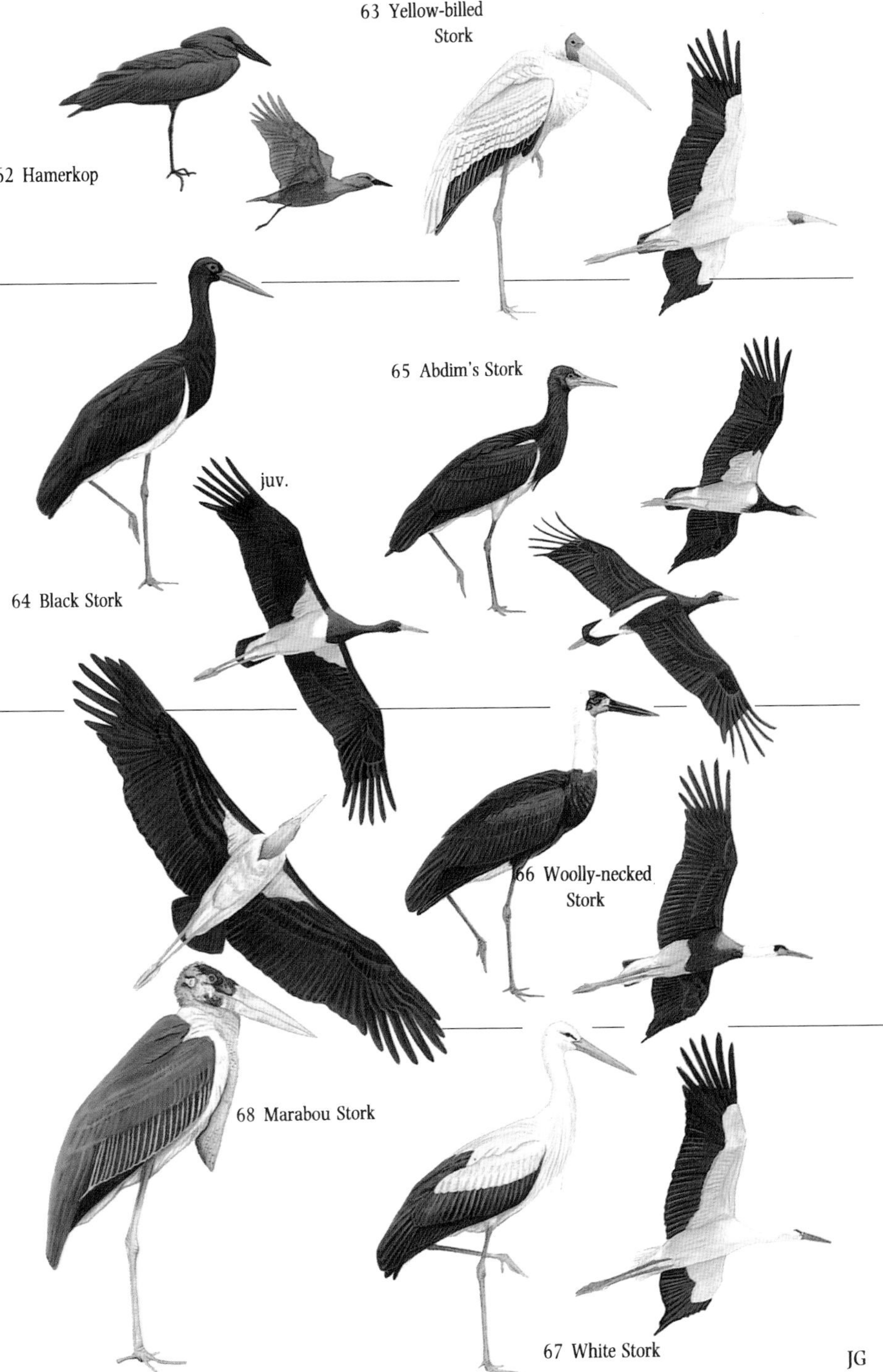

62 Hamerkop
63 Yellow-billed Stork
64 Black Stork
juv.
65 Abdim's Stork
66 Woolly-necked Stork
67 White Stork
68 Marabou Stork
JG

69 **Glossy Ibis** *Plegadis falcinellus*

Status: largely summer visitor, widespread on passage but scarce E Arabia. Winters in small numbers Iraq, Israel and Arabia. **Habitat:** wetlands, mudflats. Colonial nester in reedbeds, occasionally in trees.

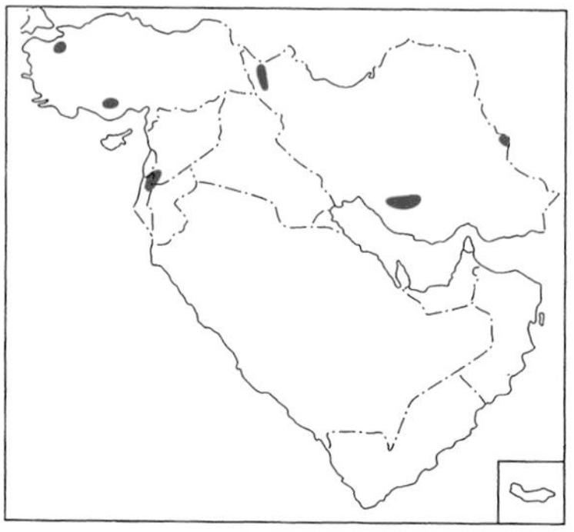

70 **Bald Ibis** *Geronticus eremita*

Status: used to breed until recently in SE Turkey. Status uncertain in SW Arabia where regular occurrence of a small flock and individuals suggests possible nesting. Vagrant Israel. **Habitat:** rocky, arid regions, close to wet grassland, marshes or rivers for feeding. Nests colonially on rocky ledges.

71 **Sacred Ibis** *Threskiornis aethiopicus*

Status: mainly resident in Iraq, also regular in SW Iran and Yemen where may breed. Vagrant Oman, Bahrain, Kuwait, UAE (where also free-flying feral population) and Israel. **Habitat:** wetlands, cultivated areas, coastal marshes. Nests colonially in trees or even on ground.

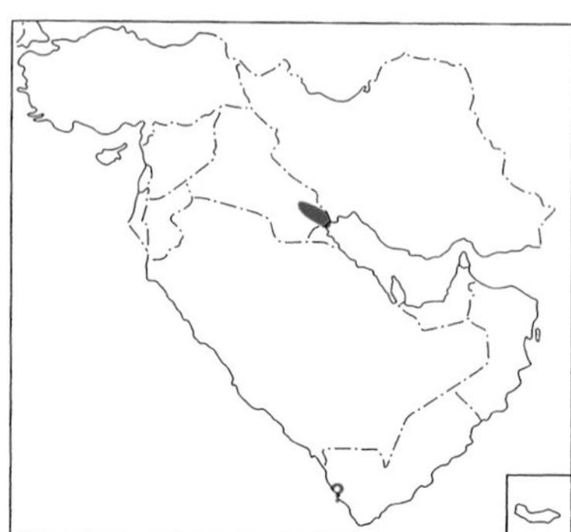

72 **Spoonbill** *Platalea leucorodia*

Status: partial migrant occurring on passage through most of region; scarce E Arabia but regular UAE. Winters from S Turkey and S Iran southwards. Rare Jordan. **Habitat:** shallow open water and mudflats. Breeds colonially in reedbeds, islands and trees.

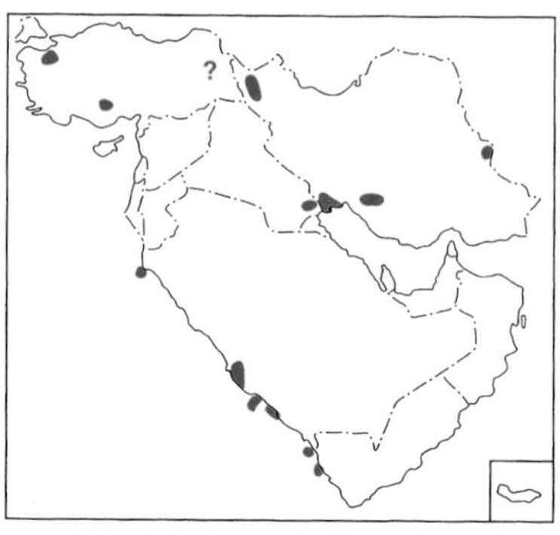

73 **African Spoonbill** *Platalea alba*

Status: vagrant S Oman and Yemen. **Habitat:** lakes, marshes, coastal lagoons and mudflats.

74 **Greater Flamingo** *Phoenicopterus ruber*

Status: partial migrant. Winters Turkey, Near East, Cyprus, also coasts of Arabian Gulf, Red Sea and Arabian Sea. Rare Jordan and vagrant Lebanon. **Habitat:** saline coastal lagoons, salt-lakes, mudflats. Breeds colonially on mud banks or in shallow water of salt-lakes building mud-heap nest, a few centimetres above water.

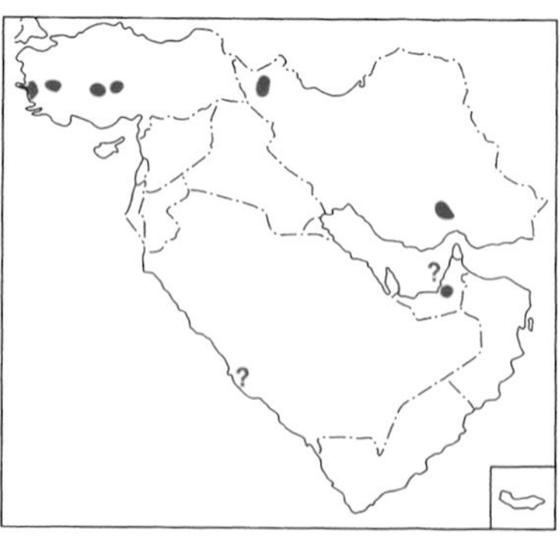

75 **Lesser Flamingo** *Phoenicopterus minor*

Status: regularly recorded in Yemen and nest building observed in south. Vagrant Oman and possibly S Iran. **Habitat:** coastal lagoons and salt-lakes.

juv.
70 Bald Ibis
juv.
69 Glossy Ibis
juv.
71 Sacred Ibis
juv.
juv.
73 African Spoonbill
72 Spoonbill
Greater
Lesser
juv.
juv.
74 Greater Flamingo
75 Lesser Flamingo
JG

76 **Mute Swan** *Cygnus olor*

Status: partial migrant. Winters Turkey, N Iran and irregularly in Cyprus. Vagrant Syria, Iraq and Arabian Gulf to Oman. **Habitat:** large lakes, marshes, deltas; in winter also on sheltered sea-coasts. Nests on ground near water, or in reeds and other vegetation in water.

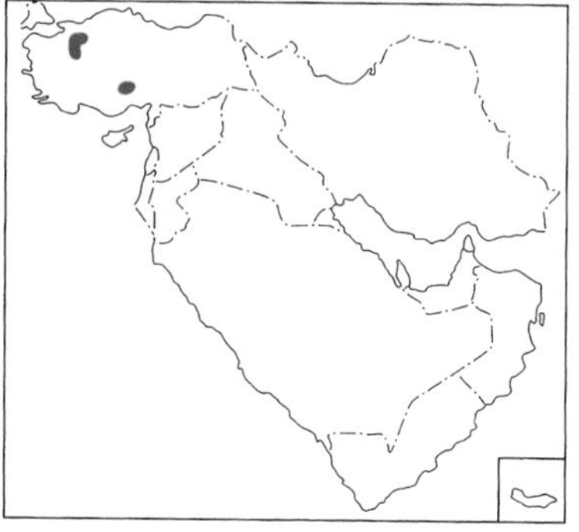

77 **Bewick's Swan** *Cygnus columbianus*

Status: winter visitor N Iran and very rarely Turkey. Vagrant Israel, Iraq, NW Saudi Arabia, UAE and Oman. **Habitat:** lowland and coastal wetlands.

78 **Whooper Swan** *Cygnus cygnus*

Status: winter visitor N Iran and Turkey (mainly W and central). Vagrant Cyprus. **Habitat:** sea-coasts, tidal waters, lakes, large rivers.

79 **Bean Goose** *Anser fabalis*

Status: occasional in winter Iran. Vagrant Turkey, Cyprus, Israel. **Habitat:** grasslands near fresh water.

80 **White-fronted Goose** *Anser albifrons*

Status: winter visitor Turkey, N Syria, Iraq, N and W Iran, sometimes Israel, Cyprus, scarcer Gulf States and Oman. Vagrant Saudi Arabia, Jordan. Occasionally summers W Turkey. **Habitat:** grasslands, marshes, estuaries.

81 **Lesser White-fronted Goose** *Anser erythropus*

Status: winter visitor Iraq marshes, now rare N and W Iran. Vagrant Turkey, Syria, Israel, Kuwait, Oman. **Habitat:** as White-fronted Goose.

76 Mute Swan
imm.
78 Whooper
Swan
juv.
77 Bewick's
Swan
juv.
79 Bean Goose
juv.
80 White-fronted
Goose
81 Lesser
White-fronted Goose
juv.
ML

82 **Greylag Goose** *Anser anser*

Status: resident and winter visitor, wintering Turkey, Iran, Iraq, south to Jordan, rarely reaching NW and E Arabia, Israel. **Habitat:** grasslands, arable fields near coasts, marshes, estuaries. Breeds sociably in marshes, reedbeds, boggy thickets, islets.

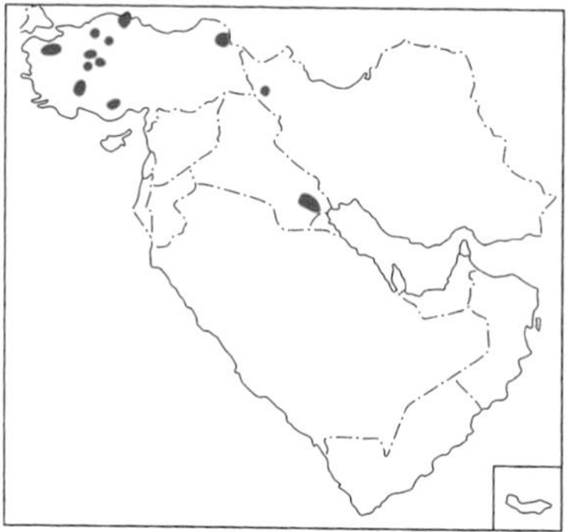

83 **Brent Goose** *Branta bernicla*

Status: vagrant Cyprus. **Habitat:** maritime outside breeding season, frequenting coasts and estuaries.

84 **Red-breasted Goose** *Branta ruficollis*

Status: winter visitor to Iraq marshes, rare Turkey. Occasional Iran. Vagrant Cyprus, Israel. **Habitat:** normally winters on grassy steppes, roosting along sea coasts.

85 **Egyptian Goose** *Alopochen aegyptiacus*

Status: occasional in winter Cyprus. Vagrant NW Saudi Arabia. Breeds ferally in UAE. **Habitat:** freshwater marshes, lakes and rivers with nearby grazing land.

86 **Ruddy Shelduck** *Tadorna ferruginea*

Status: partial migrant. Winter concentrations in Turkey, Iran, Iraq and E Saudi Arabia, scarce or irregular Jordan; rare Arabian Gulf, Oman and Yemen. **Habitat:** much more terrestrial than Shelduck. In winter frequents sandy lake shores, river banks, fields and even arid steppes. Breeds in holes in dunes, cliffs, old trees and walls.

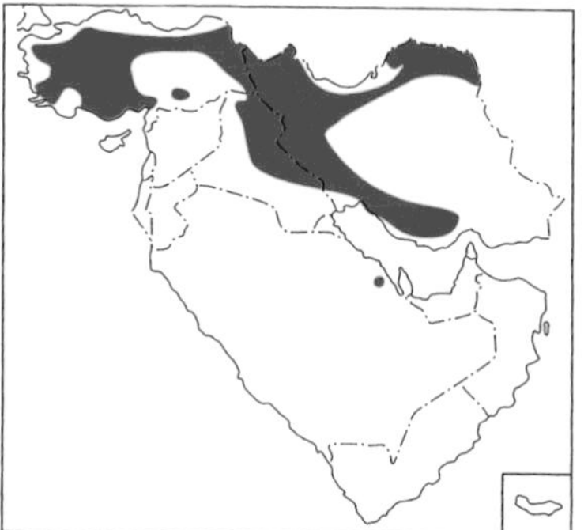

87 **Shelduck** *Tadorna tadorna*

Status: has bred Jordan; breeding birds mainly resident. Migrants from farther north winter widely in most countries in our Asian area, but in Arabia scarce or irregular south to Yemen. **Habitat:** sandy and muddy coasts, and salt-lakes and marshes in steppes and semi-desert. Nests in hole in ground or tree or under thick cover.

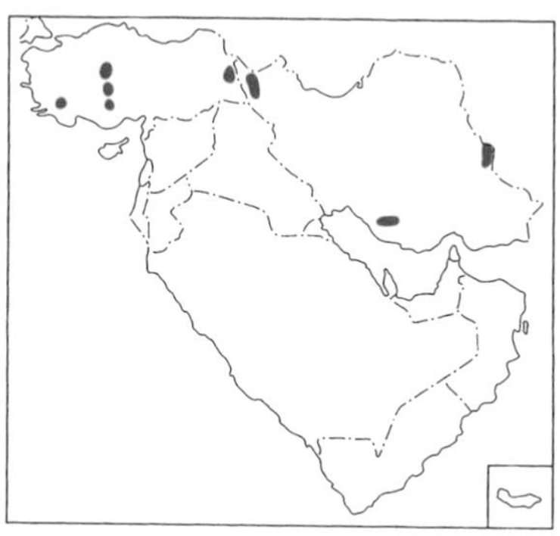

89 **Cotton Teal** *Nettapus coromandelianus*

Status: vagrant Iran, Bahrain and UAE, scarce but regular in winter Oman. Range includes India. **Habitat:** well vegetated ponds and lakes.

90 **Wigeon** *Anas penelope*

Status: winter visitor to most of the Asian area, even to parts of S and SE Arabia. Some passage evident in for example, E Mediterranean, slight in E Arabia. **Habitat:** coastal mudflats and salt marsh, and inland marshes, lakes.

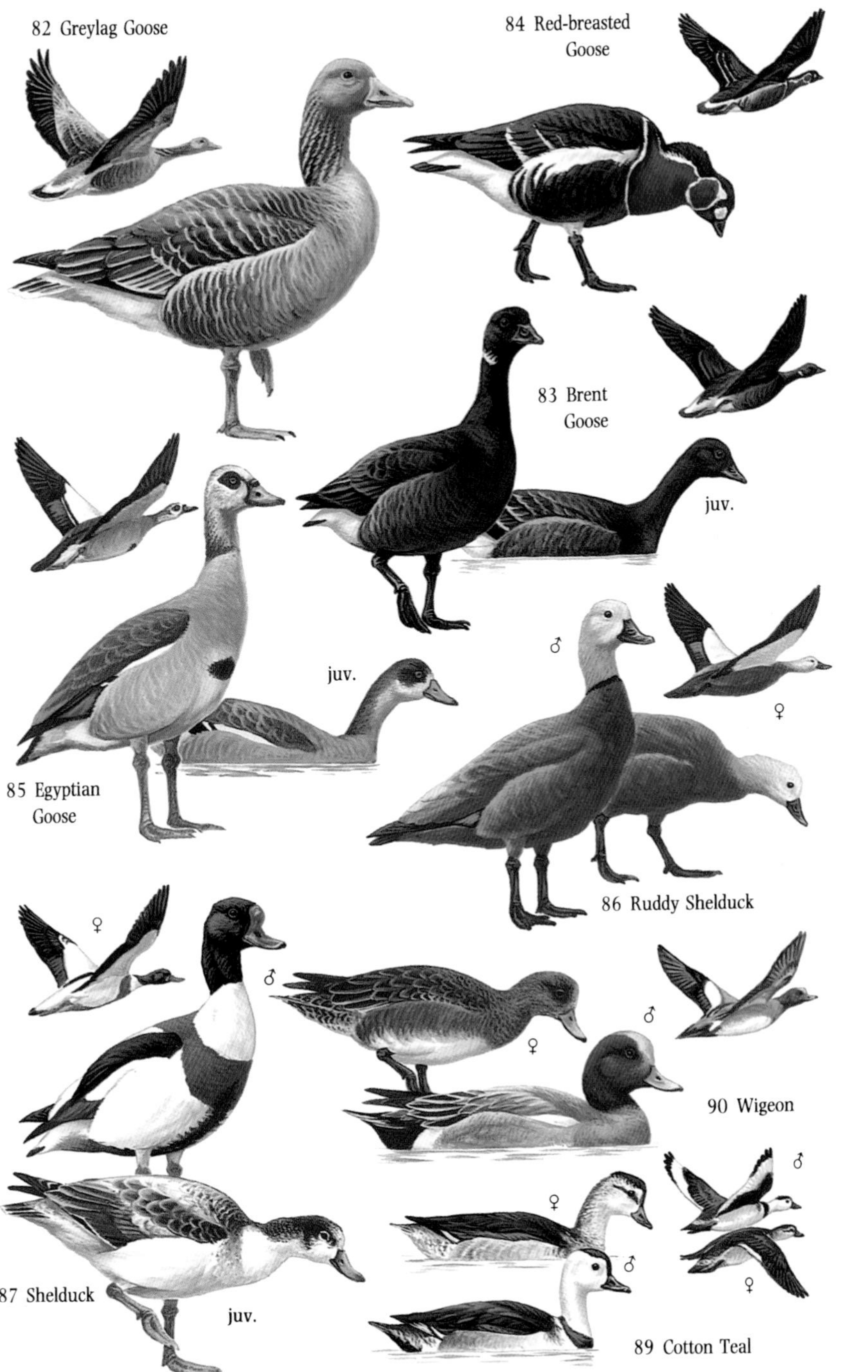
82 Greylag Goose
84 Red-breasted Goose
83 Brent Goose
juv.
♂
♀
juv.
85 Egyptian Goose
86 Ruddy Shelduck
♀
♂
♂
♀
90 Wigeon
♂
♀
♂
♀
87 Shelduck
juv.
89 Cotton Teal
ML

91 **Falcated Duck** *Anas falcata*

Status: vagrant Iraq, Iran, Jordan, Turkey. Range E Asia, India. **Habitat:** large rivers, lakes, barrages, floodlands, also coastal.

92 **Gadwall** *Anas strepera*

Status: partial migrant. Winter visitor from Turkey and Iran southwards, but scarce in Arabia. **Habitat:** like Mallard, but seldom occurs on sea coasts. Nests in thick vegetation by water.

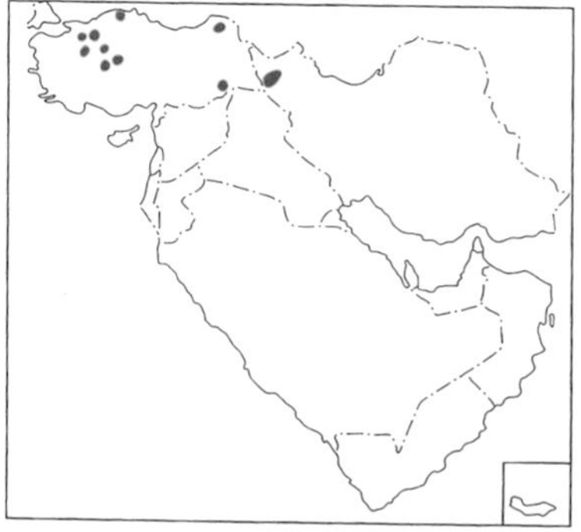

93 **Teal** *Anas crecca*

Status: breeding in Turkey probably more extensive; almost entirely a winter, or passage, visitor to coasts and inland waters. Widely in Iraq, Iran, W Turkey, parts of Near East, and Arabian Gulf; fairly common Arabia. **Habitat:** wide range of wetlands from saltmarshes to lakes and ditches. Nests in thick ground cover.

94 **Mallard** *Anas platyrhynchos*

Status: possible feral origin for birds breeding in Arabia; partial migrant. Winters throughout region, but less common in SW Arabia. **Habitat:** almost any wetland, including sea-coasts and estuaries in winter. Nests on ground in thick cover or hollow in tree.

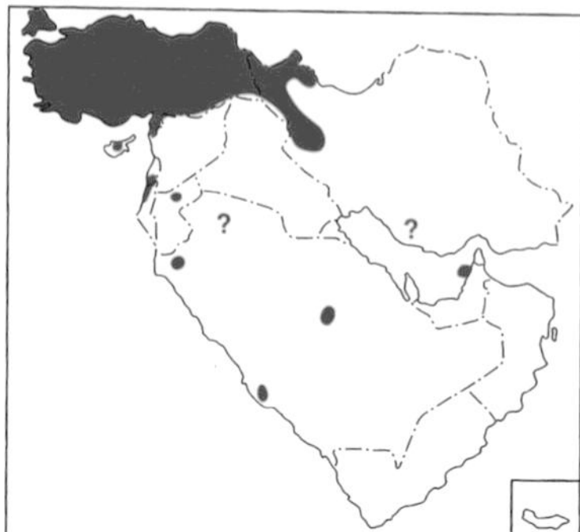

95 **Pintail** *Anas acuta*

Status: mainly winter visitor and passage migrant, Mediterranean, W Iran, Iraq, also fairly common in Arabia. **Habitat:** chiefly sheltered, shallow coasts and estuaries in winter, also open, shallow inland waters. Nests on ground.

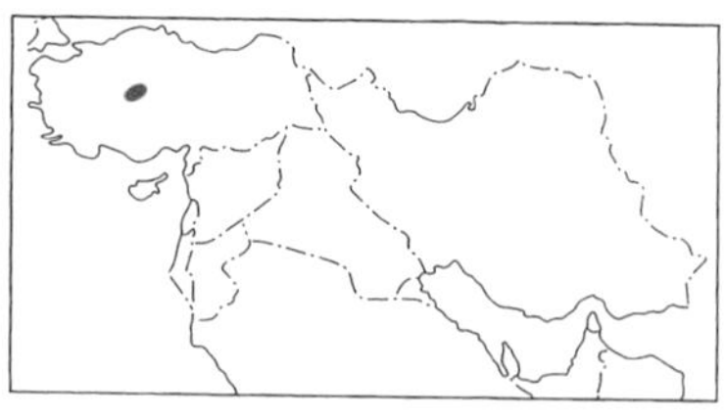

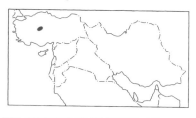

96 **Garganey** *Anas querquedula*

Status: summer visitor Turkey and NW Iran; otherwise passage mainly evident in Arabia and to some extent Gulf of Aqaba and Near East. Some winter Arabia. **Habitat:** as Teal, but seldom on salt water. Nests in long grass or rank vegetation near water.

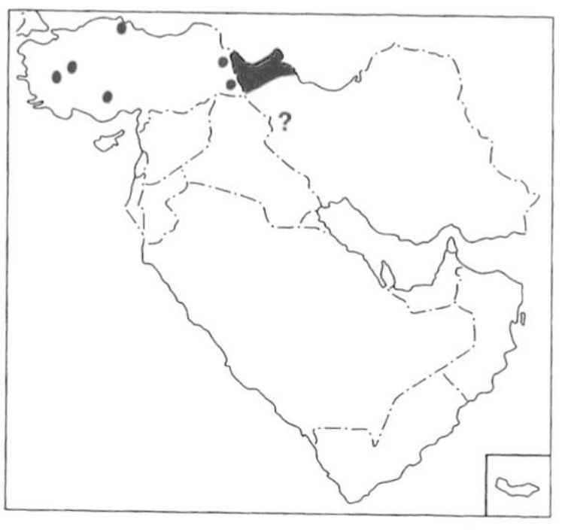

97 **Shoveler** *Anas clypeata*

Status: almost entirely winter visitor and passage migrant; Mediterranean countries, Iran, Iraq, coasts and inland Arabia; occasionally over-summers. **Habitat:** less maritime than other surface-feeders. Usually in marshes, floodlands and overgrown ponds. Nests on ground.

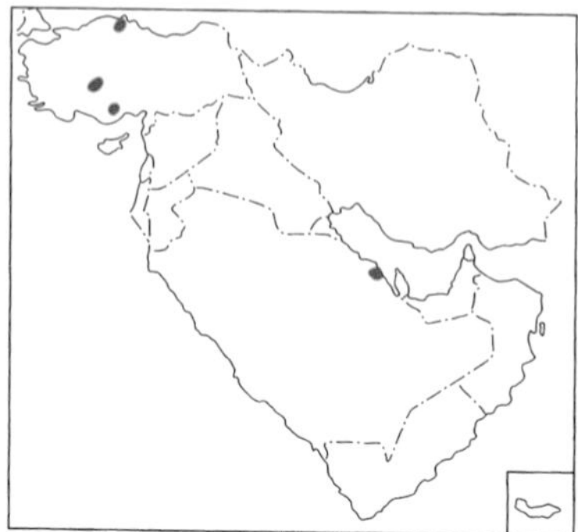

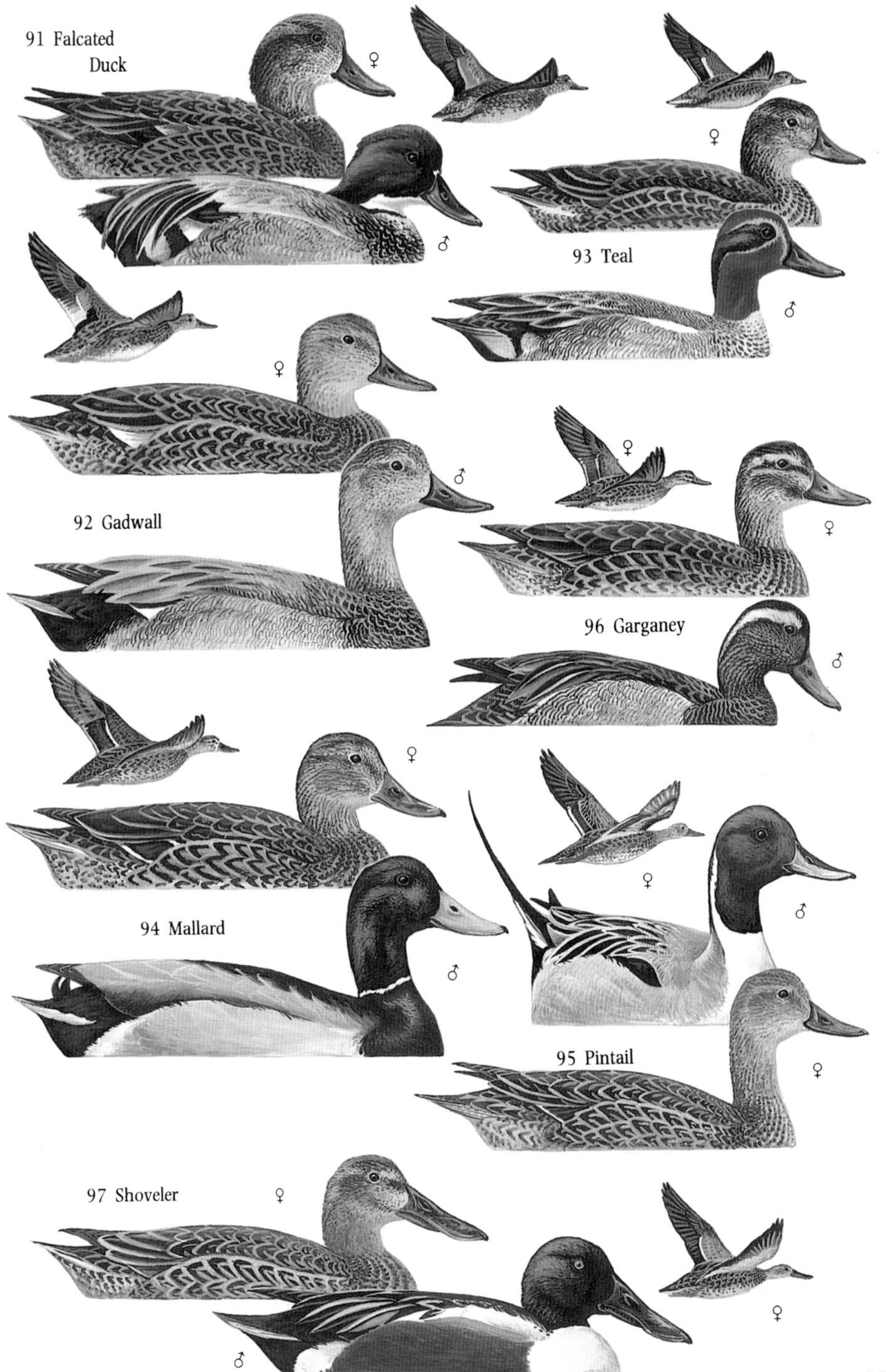
91 Falcated Duck
♀
♂
♀
93 Teal
♂
♀
♂
92 Gadwall
♀
♀
96 Garganey
♂
♀
♀
♂
94 Mallard
♂
95 Pintail
♀
97 Shoveler
♀
♀
♂

98 **Marbled Teal** *Marmaronetta angustirostris*

Status: has bred Jordan. Partial migrant; winter numbers and movements irregular. Winters S Turkey, SE Iraq, many SW Iran, few Israel; rare E and central Arabia; vagrant Cyprus, Kuwait, Bahrain, UAE and Oman. **Habitat:** lakes and reservoirs rich in vegetation; nests near water. In winter also on salt-lakes.

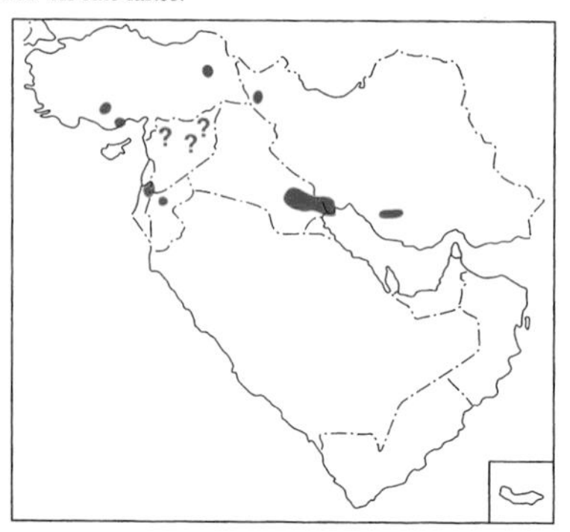

99 **Red-crested Pochard** *Netta rufina*

Status: partial migrant; winters commonly Turkey and Iran, few in Near East and Iraq; irregular E, central and NW Arabia. Rare UAE. Vagrant Bahrain. **Habitat:** large reedy freshwater lakes, brackish lagoons but seldom on the sea. Breeds among vegetation on islands in lagoons.

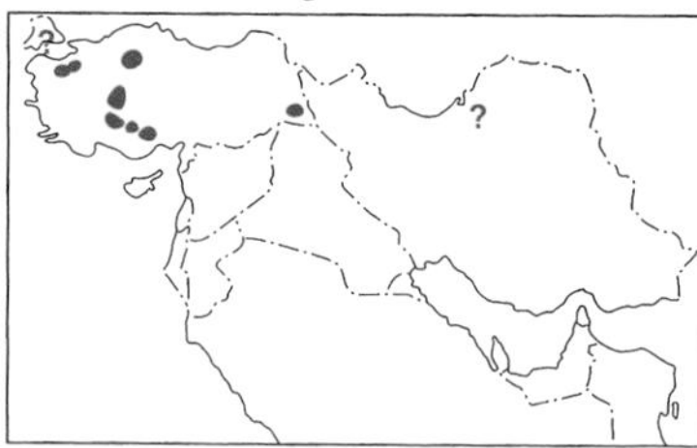

100 **Pochard** *Aythya ferina*

Status: partial migrant; on passage and in winter in Near East, Turkey, Cyprus, Iran, Iraq; scarce and irregular in Arabia, both coastal and inland. **Habitat:** seldom at sea. Large and small vegetated lakes, and other enclosed waters; breeds in dense reeds.

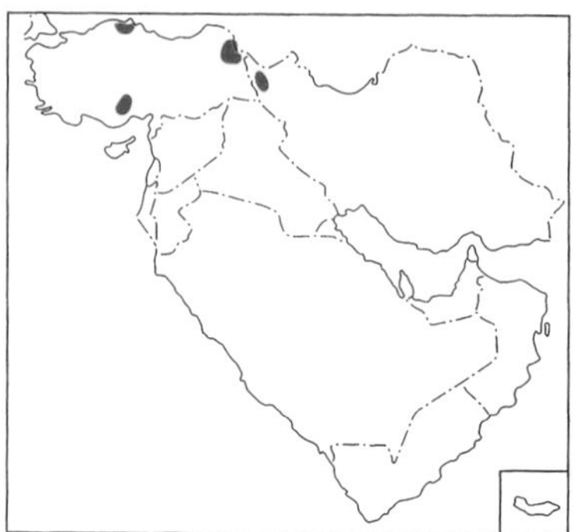

101 **Ferruginous Duck** *Aythya nyroca*

Status: partial migrant. Nesting E Arabia follows recent creation of ponds. In winter scattered over most of the region, including S Arabia; rather scarce UAE; passage through Oman. Vagrant Bahrain. **Habitat:** shallow, well vegetated freshwater pools or lakes, sometimes saline; in winter also open lagoons or on coast. Nests in dense reeds, and other tall vegetation.

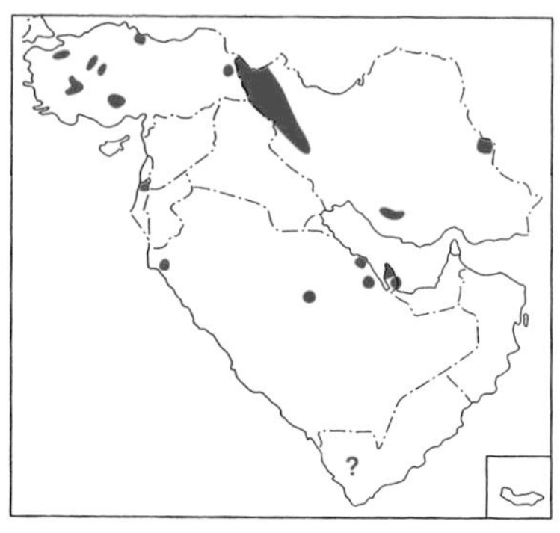

102 **Tufted Duck** *Aythya fuligula*

Status: migrant and winter visitor from the north in numbers to Turkey, Iran, Iraq and Near East; less common and in smaller numbers in most Arabia, scarcer in south. **Habitat:** large and small lakes, also coastal. Nests on ground in vegetation near water.

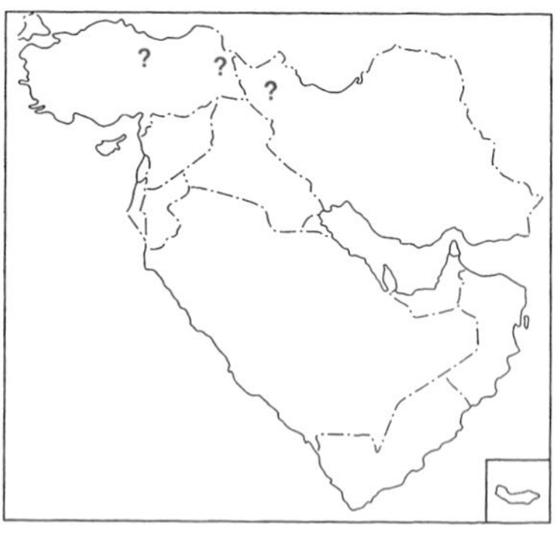

103 **Scaup** *Aythya marila*

Status: winter visitor in small numbers to N and W Turkey, scarce in N Iran; vagrant Cyprus, Israel and Iraq. **Habitat:** maritime, usually in bays and estuaries.

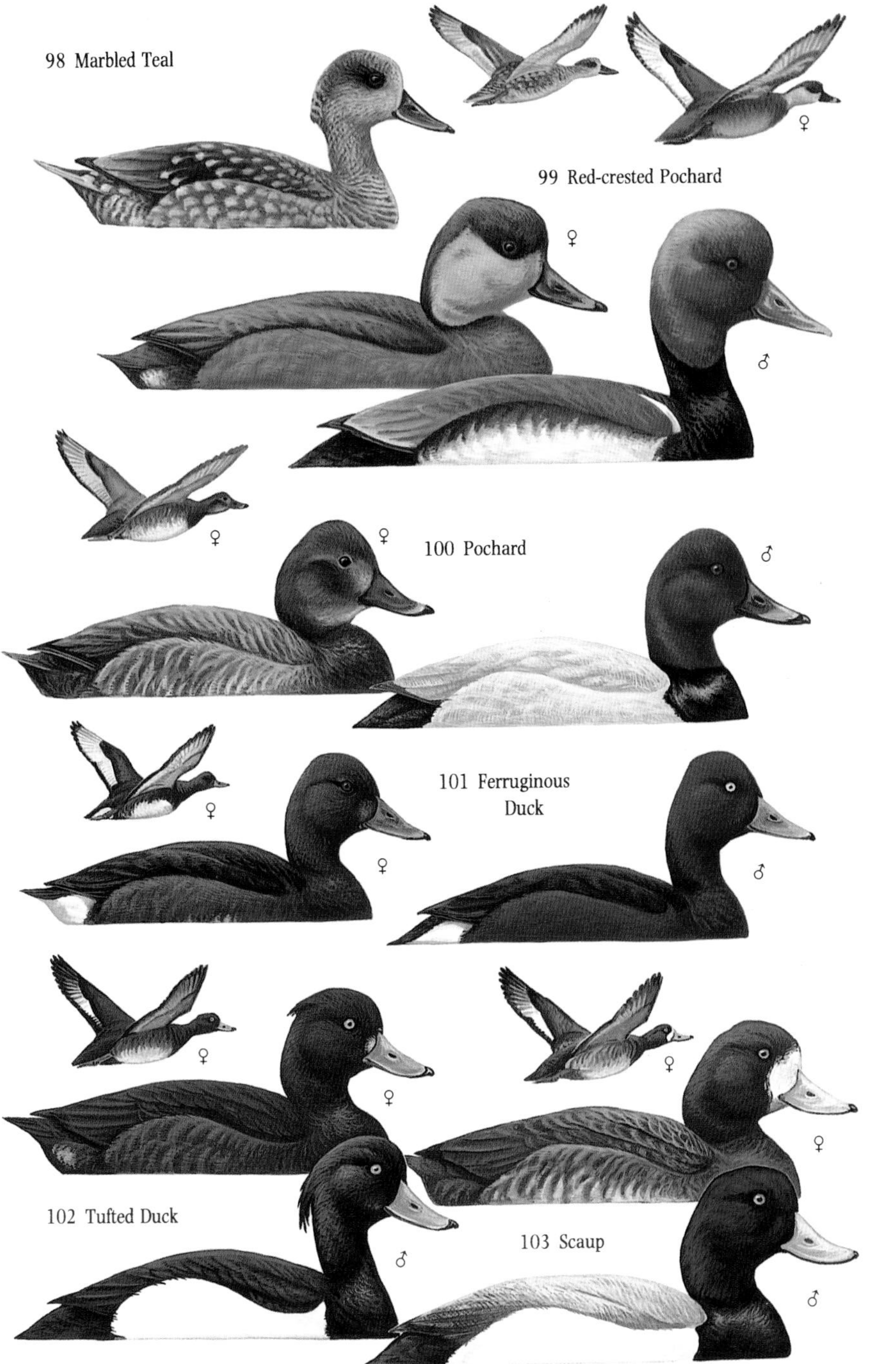

98 Marbled Teal
♀
99 Red-crested Pochard
♀
♂
♀
♀
100 Pochard
♂
♀
101 Ferruginous
Duck
♀
♂
♀
♀
♀
♀
102 Tufted Duck
♂
103 Scaup
♂
ML

105 **Long-tailed Duck** *Clangula hyemalis*

Status: vagrant N Iran, Turkey, Gulf of Aqaba and Kuwait. **Habitat:** mainly maritime.

106 **Common Scoter** *Melanitta nigra*

Status: vagrant Turkey, Iran and Cyprus. **Habitat:** mainly maritime.

107 **Velvet Scoter** *Melanitta fusca*

Status: summer visitor to Turkish breeding waters. Winters N and S coasts Turkey. Vagrant Israel and Kuwait. **Habitat:** mainly maritime in winter. Nests on ground in thick vegetation near freshwater lakeside.

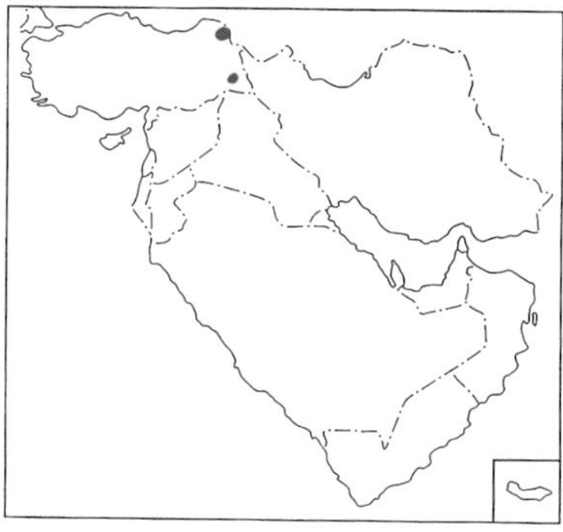

108 **Goldeneye** *Bucephala clangula*

Status: winter visitor to Iran (especially Caspian), and Turkey. Vagrant Iraq, Near East and Cyprus. **Habitat:** coastal waters, often also on inland lakes.

109 **Smew** *Mergus albellus*

Status: winter visitor Iran (especially Caspian), Iraq, Turkey. Vagrant most other countries in E Mediterranean area. **Habitat:** lakes, reservoirs and rivers, occasionally on sheltered coasts.

110 **Red-breasted Merganser** *Mergus serrator*

Status: winter visitor in small numbers to Black and Caspian Sea, a few regularly reaching S Turkey and S Iran. Straggles southwards to Near East area. In Arabia recorded as vagrant in several places on Arabian Gulf, also Red Sea and Gulf of Aqaba. **Habitat:** chiefly maritime.

111 **Goosander** *Mergus merganser*

Status: winter visitor in small numbers to Turkey, Iran and Iraq. Vagrant Israel and Cyprus. **Habitat:** large rivers, lakes, reservoirs.

112 **White-headed Duck** *Oxyura leucocephala*

Status: partial migrant, generally scarce. Winters Turkey, fewer Iran, Iraq, Syria and Israel. Rare Cyprus, vagrant Saudi Arabia. **Habitat:** reedy, generally shallow, inland waters and brackish lagoons. Nests among reeds and aquatic vegetation in water.

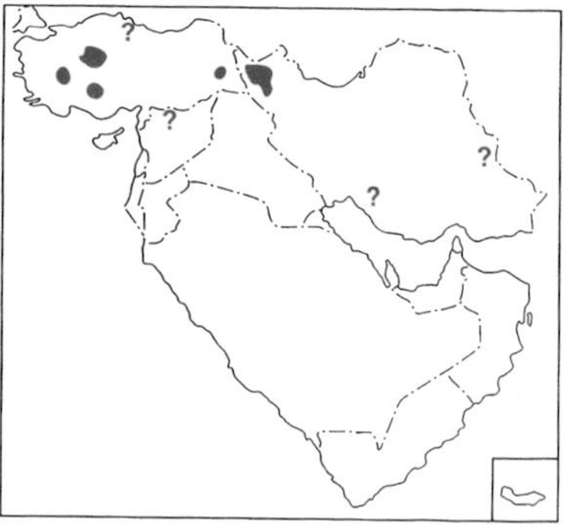

♀ winter
♂ winter
105 Long-tailed Duck
♀
♂
106 Common Scoter
108 Goldeneye
♀
♂
♀
♀
107 Velvet Scoter
♂
♀
109 Smew
♂
♀
111 Goosander
♂
♀
♀
112 White-headed Duck
♂
110 Red-breasted Merganser
♂
ML

117 **Black Kite** *Milvus migrans*

Status: mainly resident SW Arabia, migratory elsewhere; noticeable passage at the Bosphorus (autumn), large passage Israel (spring). Winters over most of region, commonly Arabia, scarcer in north. **Habitat:** woodland (N & W) or semi-desert, often near wet areas or rubbish dumps. Nests in trees or cliff, often sociably.

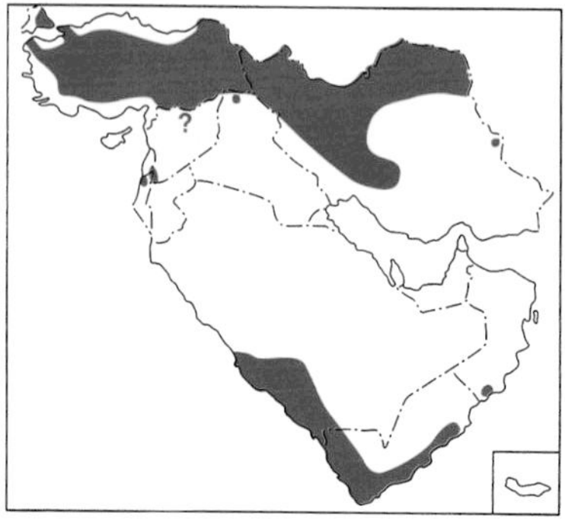

118 **Red Kite** *Milvus milvus*

Status: uncommon or scarce on passage and as a winter visitor to Turkey and N Iran, more rarely in E Mediterranean countries to Iraq. **Habitat:** open country with scattered woods.

120 **Pallas's Fish Eagle** *Haliaeetus leucoryphus*

Status: rare visitor from Asia to Iran, E and SW Arabia. Vagrant Oman, UAE and Israel. **Habitat:** wetlands and rivers; including coastal.

121 **White-tailed Eagle** *Haliaeetus albicilla*

Status: adult resident; young disperse in winter, rarely reaching Cyprus, Israel, Iraq and S Iran. **Habitat:** wetlands, rivers, lakes and coasts. Nests in tall tree or cliff-ledge.

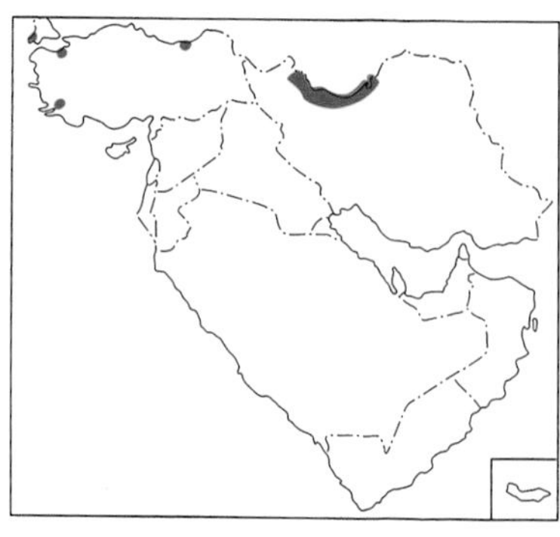

122 **Lammergeier** *Gypaetus barbatus*

Status: resident. Vagrant Cyprus. **Habitat:** desolate mountain ranges, large rocky wadis.

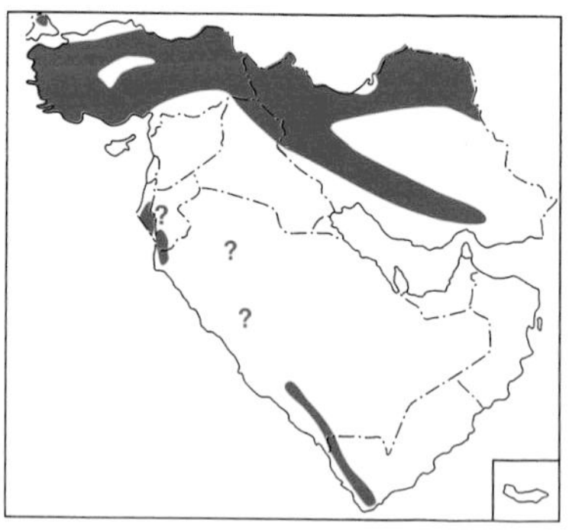

juv.
aegyptius
117 Black Kite
118 Red Kite
juv.
120 Pallas's Fish Eagle
juv.
juv.
juv.
121 White-tailed Eagle
122 Lammergeier

JG

123 **Egyptian Vulture** *Neophron percnopterus*

Status: migratory to breeding areas in north of range, but sedentary Arabia and S Iran where migrants also winter. Passage evident through countries bordering E Mediterranean, Oman and Yemen. Vagrant Cyprus. **Habitat:** mostly mountains with gorges and wadis; frequents village refuse dumps. Nests in recess of cliff.

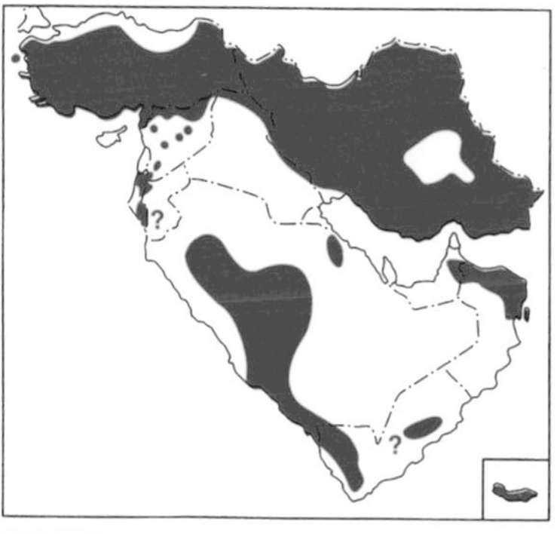

124 **Indian White-backed Vulture** *Gyps bengalensis*

Status: resident. **Habitat:** open regions, foothills, farmland. Nesting colonies in trees.

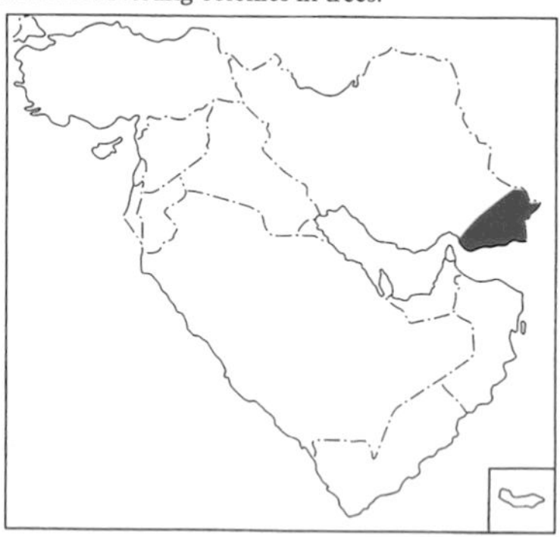

125 **Griffon Vulture** *Gyps fulvus*

Status: mainly resident, but partial migrant Turkey. Some passage along E Mediterranean countries. Irregular E Arabia. **Habitat:** mountains; occurs over all types of country in search for food. Breeds socially in caves or on cliff-ledges.

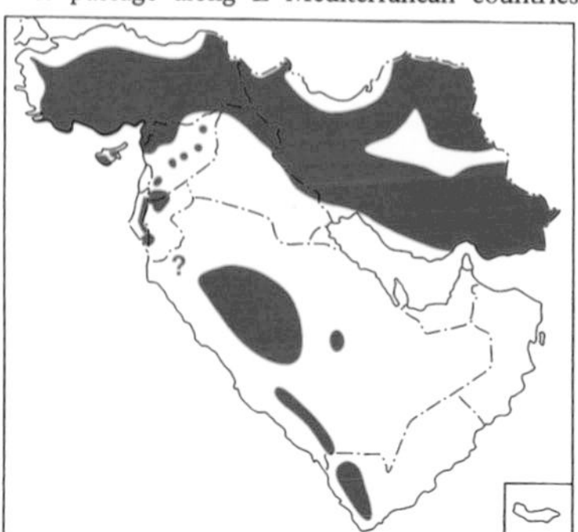

126 **Rüppell's Vulture** *Gyps rueppellii*

Status: doubtfully recorded in SW Arabia (from Africa). **Habitat:** arid mountains and surrounding open savanna.

127 **Lappet-faced Vulture** *Torgos tracheliotos*

Status: mainly resident. Vagrant Syria. **Habitat:** semi-desert steppe, savannas, rocky wadis with acacias in top of which huge nest is built.

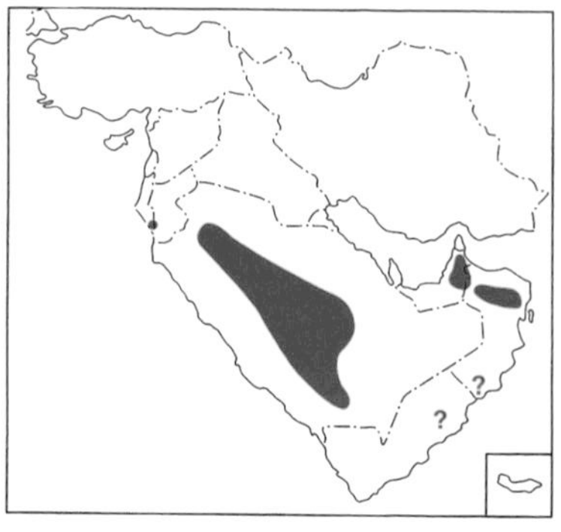

128 **Black Vulture** *Aegypius monachus*

Status: formerly bred Cyprus. Resident, immatures disperse in winter south to many Middle East countries, but always very scarce. Vagrant Oman. **Habitat:** desolate mountains (often extensively wooded), foothills and plains but also semi-deserts. Nests in trees, sometimes on cliff.

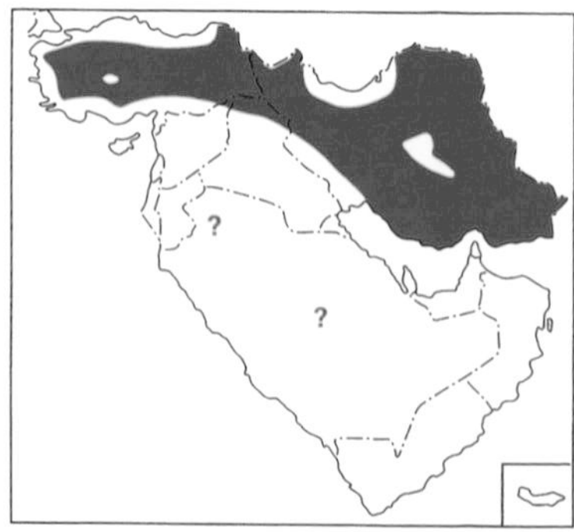

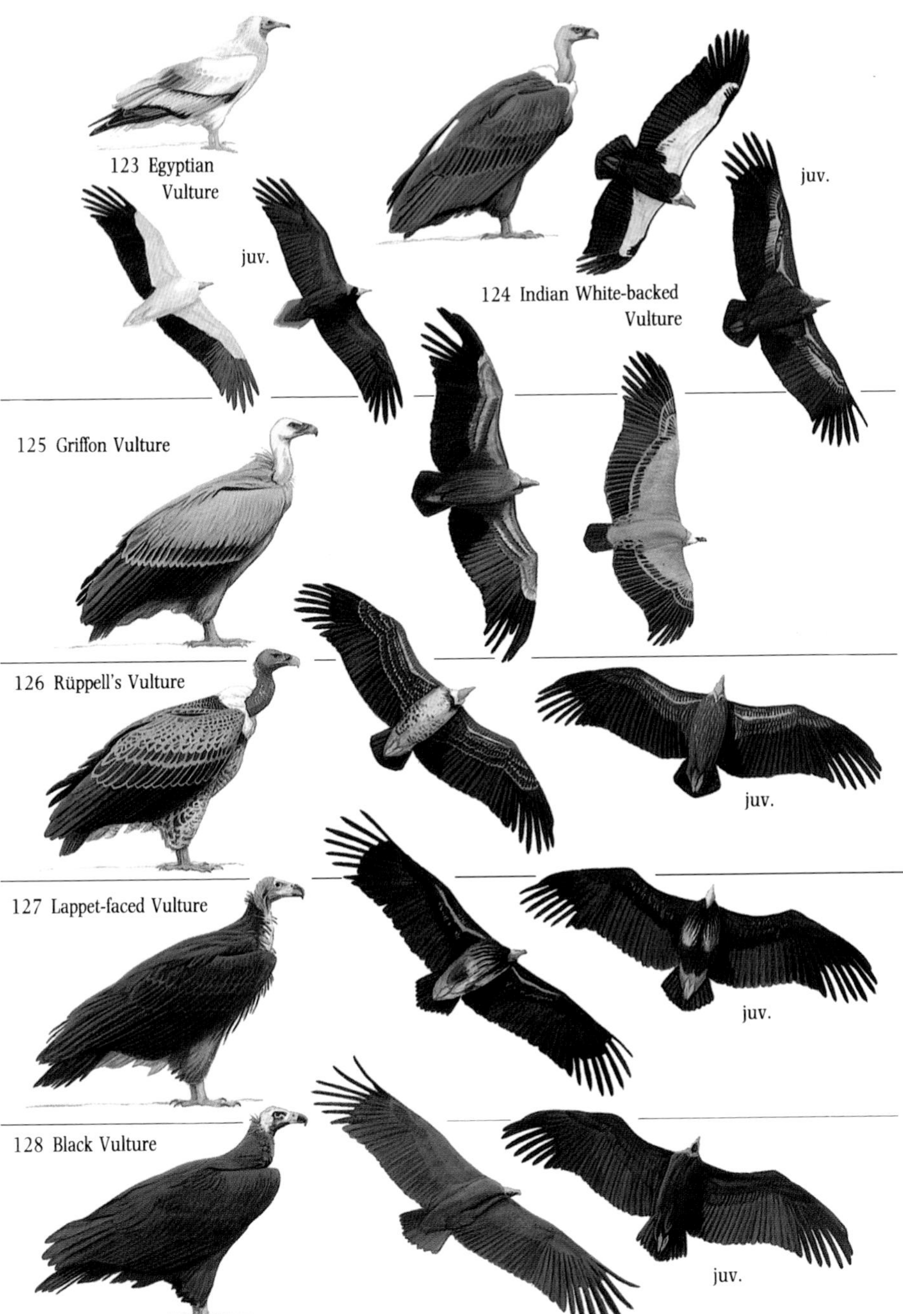
123 Egyptian Vulture
juv.
124 Indian White-backed Vulture
juv.
125 Griffon Vulture
126 Rüppell's Vulture
juv.
127 Lappet-faced Vulture
juv.
128 Black Vulture
juv.

JG

130 **Bateleur** *Terathopius ecaudatus*

Status: resident. Rare Israel, vagrant Iraq and central Saudi Arabia. **Habitat:** open savanna, foothills, thornbush country and woodland. Nests in large tree.

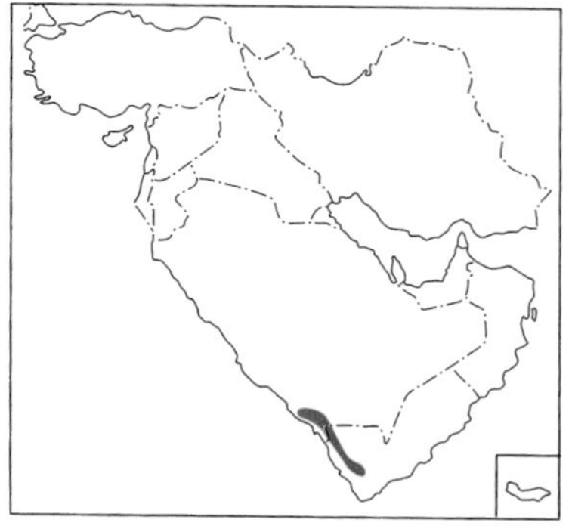

131 **Marsh Harrier** *Circus aeruginosus*

Status: has bred Lebanon. Partial migrant; occurs widely on passage and in winter over most region. **Habitat:** swamps and marshes with reeds, neighbouring farm- and wastelands. Nests in reedbeds.

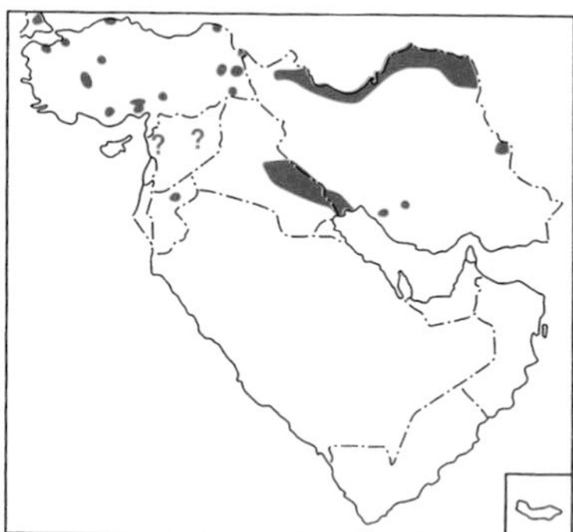

132 **Hen Harrier** *Circus cyaneus*

Status: regular winter visitor to northern regions south to Israel; rarer farther east (Gulf States) and most Saudi Arabia; vagrant Bahrain, UAE and Yemen. **Habitat:** marshes, meadows and open farmland.

133 **Pallid Harrier** *Circus macrourus*

Status: may breed on N Iran steppes. Widely on passage; some winter near East, Iran and UAE; fewer S Turkey, Cyprus and south side of Arabian Gulf. **Habitat:** dry steppes, open grassland, agricultural fields.

134 **Montagu's Harrier** *Circus pygargus*

Status: summer visitor. Widely on passage, but scarcer Arabia, including Yemen and Oman where a few also winter. **Habitat:** marshes and agricultural land; in winter over any open country. Nests in wet vegetation or low scrub.

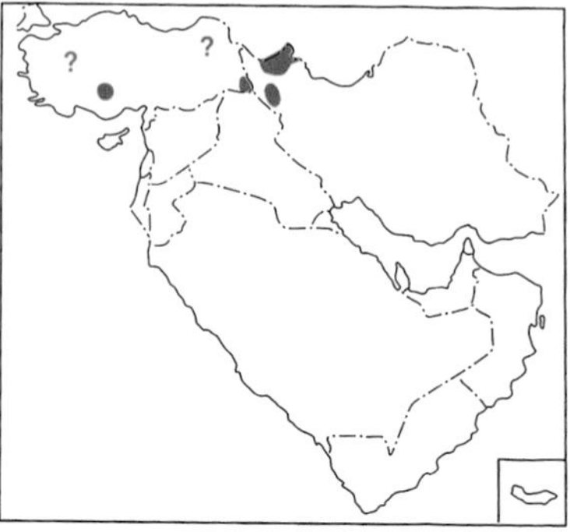

130 Bateleur
juv.
♀
♂
131 Marsh Harrier
♂
♀
132 Hen Harrier
♂
♀
juv.
133 Pallid Harrier
♂
♀
juv.
134 Montagu's Harrier
♀
♂
juv.
JG

135 **Dark Chanting Goshawk** *Melierax metabates*

Status: resident with some dispersal. Vagrant Israel and UAE. **Habitat:** semi-desert with bushes, savanna or other open areas with trees and scrub. Nests in tree.

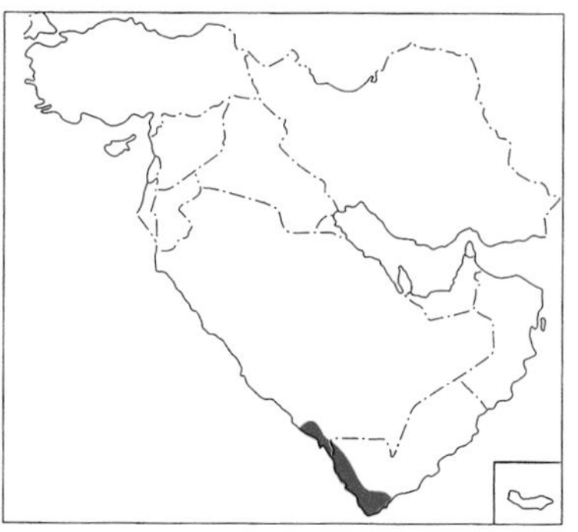

136 **Gabar Goshawk** *Micronisus gabar*

Status: probably resident in SW Saudi Arabia and Yemen. **Habitat:** savanna and bush country with acacias. Nests in tree.

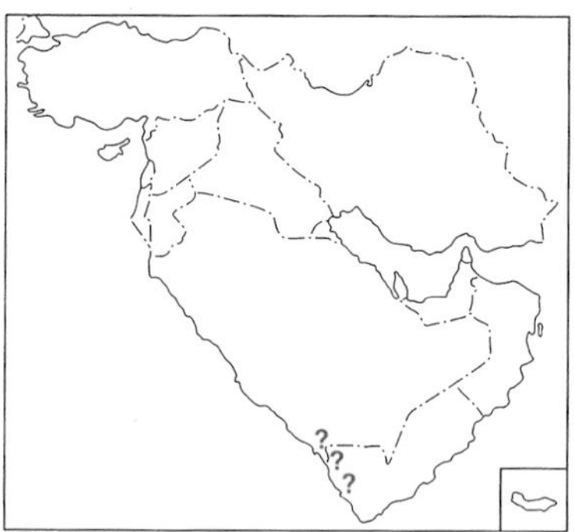

137 **Goshawk** *Accipiter gentilis*

Status: mainly resident with immatures dispersing. Winter visitor N Iran; irregular Cyprus and Near East, rare E Arabia and apparently SW Arabia. **Habitat:** woods, particularly coniferous, often near open country. Nests in large trees.

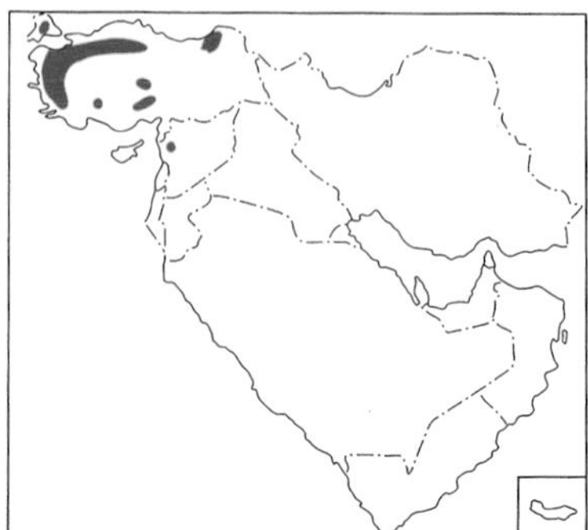

138 **Sparrowhawk** *Accipiter nisus*

Status: partial migrant, widely on passage, particularly Bosphorus, Levant but also SW Arabia. Winters over most of region. **Habitat:** dense woods or open country with trees. Nests in tree.

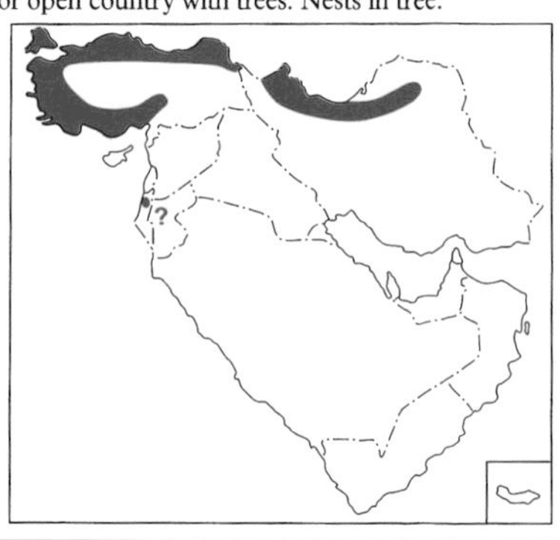

139 **Shikra** *Accipiter badius*

Status: summer visitor N Iran; some winter SE Iran. Resident Yemen. Vagrant Israel, Saudi Arabia, Kuwait and Oman. **Habitat:** lightly wooded regions, wooded slopes and wadis. Nests in tree.

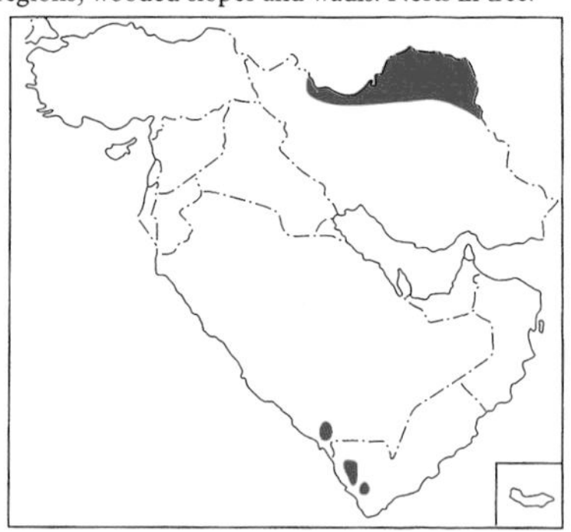

140 **Levant Sparrowhawk** *Accipiter brevipes*

Status: probably breeds more extensively in Turkey. Summer visitor. Large passage Bosphorus and Levant, thinner passage Iran. Rare Cyprus, Saudi Arabia, Kuwait and Yemen. Vagrant UAE. **Habitat:** open country, often with deciduous woods, more varied on passage. Nests in tree.

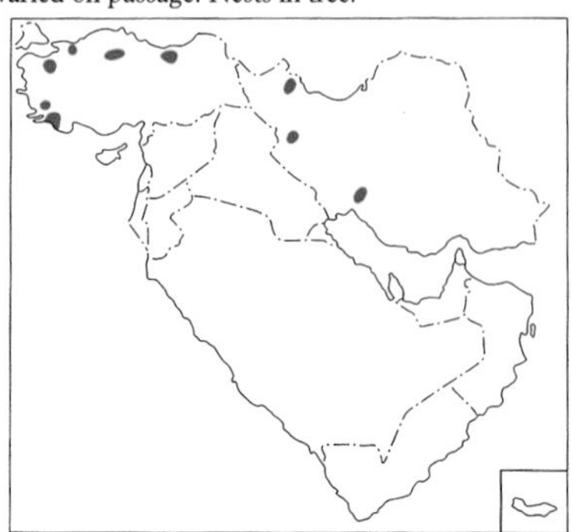

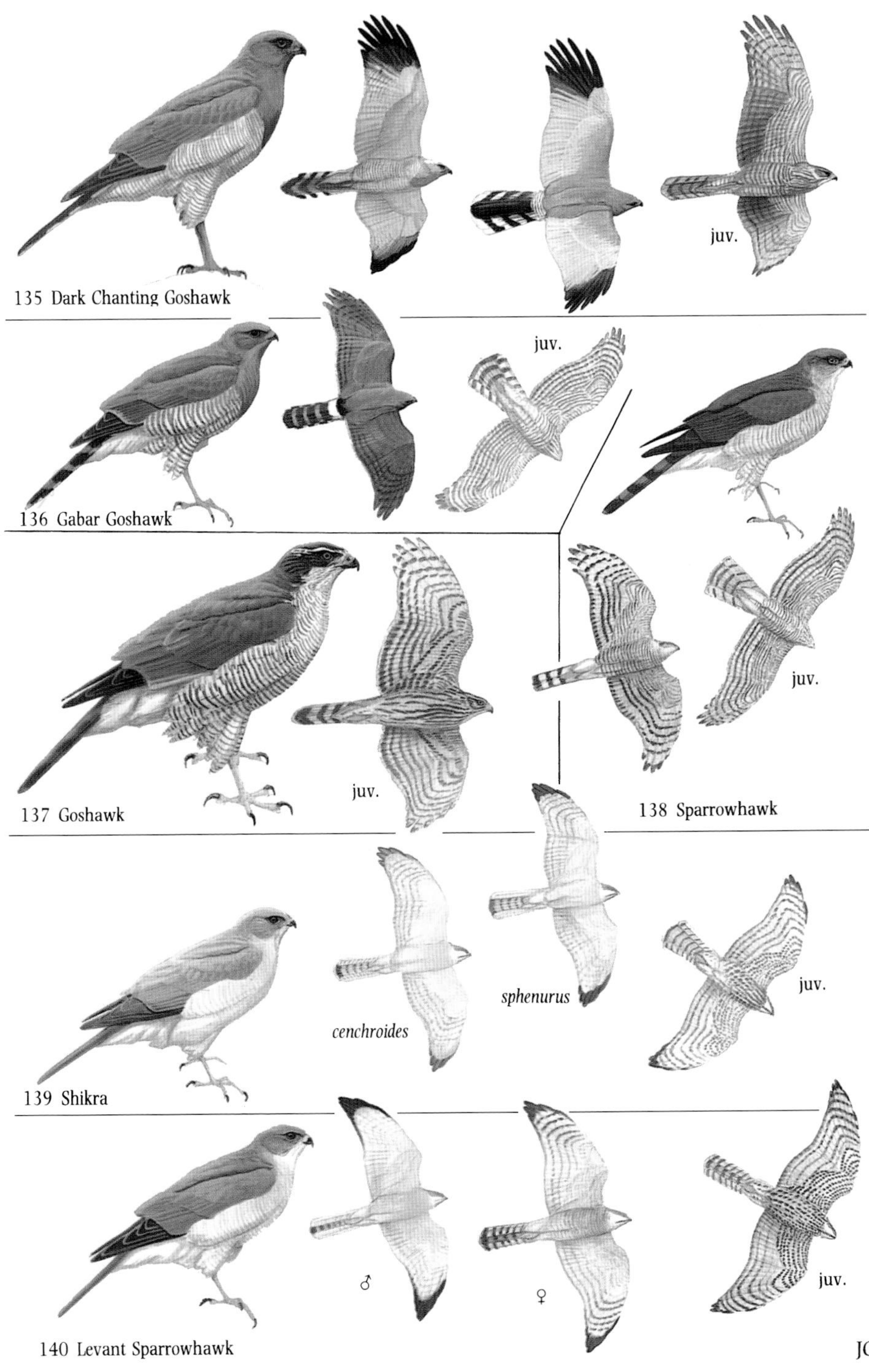
135 Dark Chanting Goshawk
juv.
136 Gabar Goshawk
juv.
137 Goshawk
juv.
juv.
138 Sparrowhawk
139 Shikra
cenchroides
sphenurus
juv.
140 Levant Sparrowhawk
♂
♀
juv.
JG

114 **Crested Honey Buzzard** *Pernis ptilorhynchus*

Status: vagrant UAE, Israel and Saudi Arabia. **Habitat:** on passage over any country.

141 **White-eyed Buzzard** *Butastur teesa*

Status: resident. **Habitat:** dry low open country with scrub and few trees, lightly wooded foothill slopes. Nests in tree.

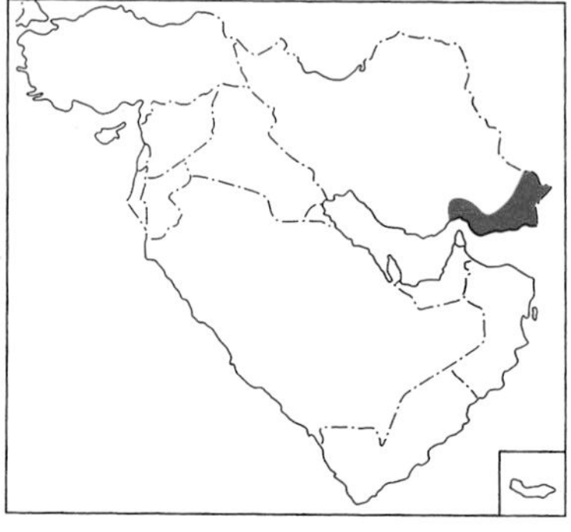

142 **Common Buzzard (includes Steppe Buzzard)** *Buteo buteo*

Status: occurs on passage and in winter over most of the region. Large migrations NE Turkey, Bosphorus, Israel and SW Arabia. **Habitat:** forested and wooded regions, plains and mountain slopes, often near wetlands. Nests in tree or on cliff.

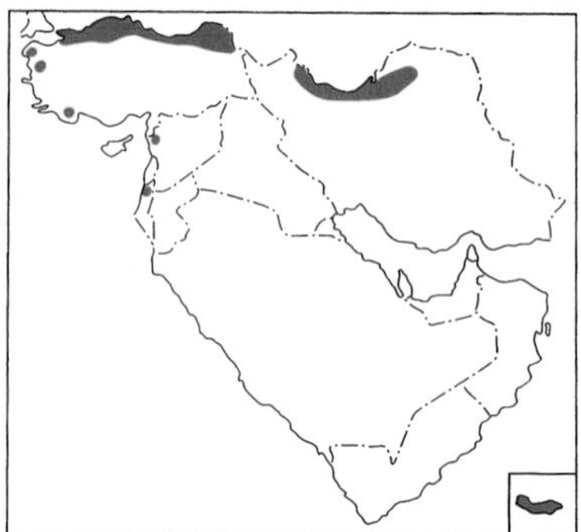

143 **Long-legged Buzzard** *Buteo rufinus*

Status: recent breeding records Cyprus and UAE. Resident and partial migrant, more widespread in winter. Small, but important passage Israel and SW Arabia. **Habitat:** open plains and steppes, deciduous woodland; also mountains. More widespread in winter. Nests on rock ledge or tree.

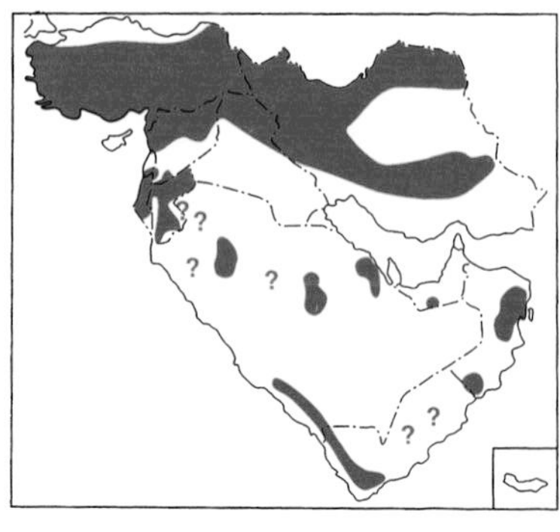

144 **Rough-legged Buzzard** *Buteo lagopus*

Status: uncommon winter visitor NW and central Turkey and N Iran. Vagrant Cyprus and Near East. **Habitat:** open wasteland, bushy plains with trees, marshes.

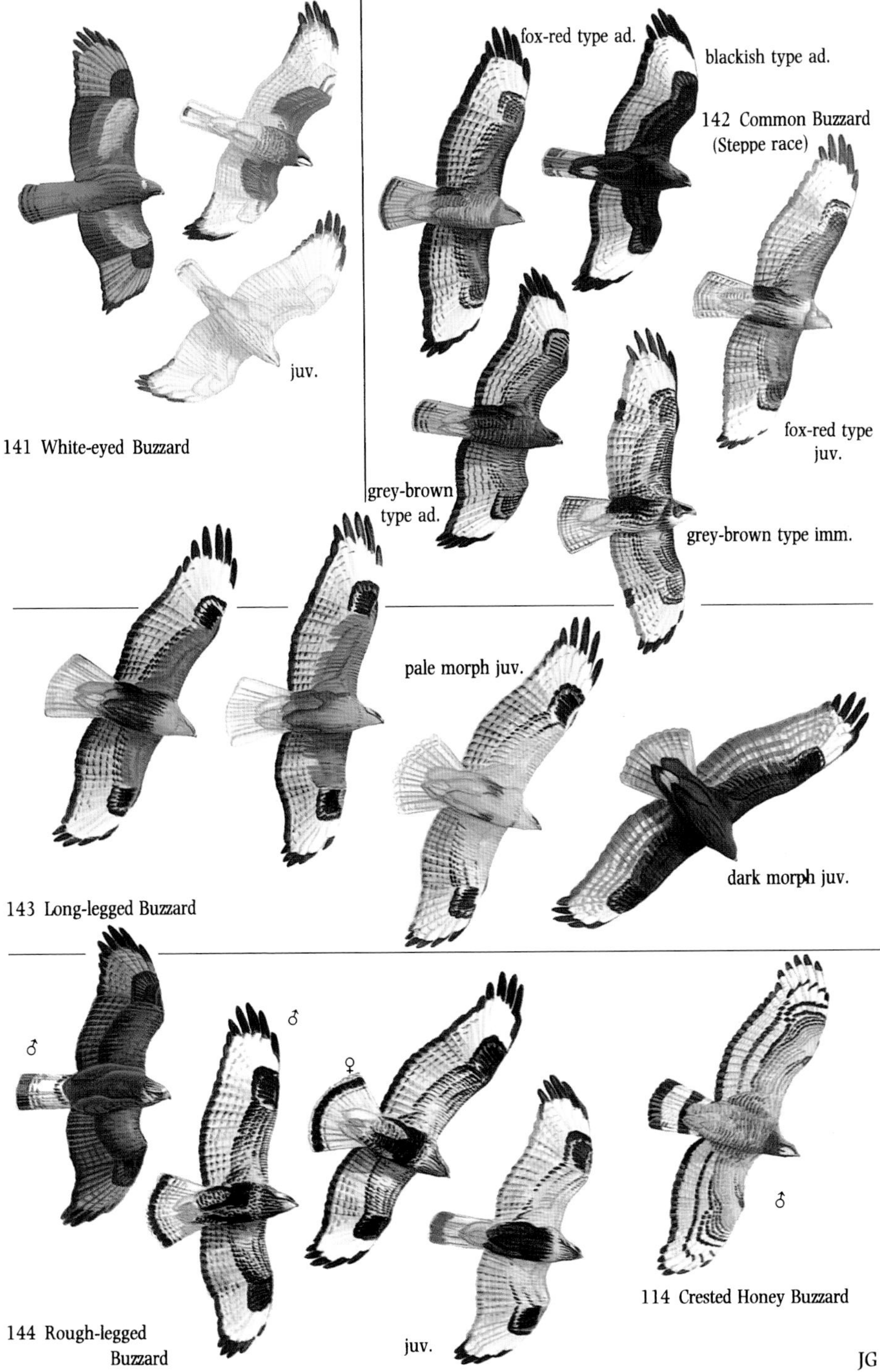

juv.
141 White-eyed Buzzard
fox-red type ad.
blackish type ad.
142 Common Buzzard (Steppe race)
fox-red type juv.
grey-brown type ad.
grey-brown type imm.
pale morph juv.
dark morph juv.
143 Long-legged Buzzard
♂
♂
♀
juv.
144 Rough-legged Buzzard
♂
114 Crested Honey Buzzard
JG

145 **Lesser Spotted Eagle** *Aquila pomarina*

Status: migratory, wintering in Africa. Heavy passage Israel, noticeable passage Bosphorus and NE Turkey; rare Cyprus and Oman, vagrant UAE, Kuwait and possibly Yemen. **Habitat:** breeds in forests in lowlands and mountains; often near water. Nests in tree.

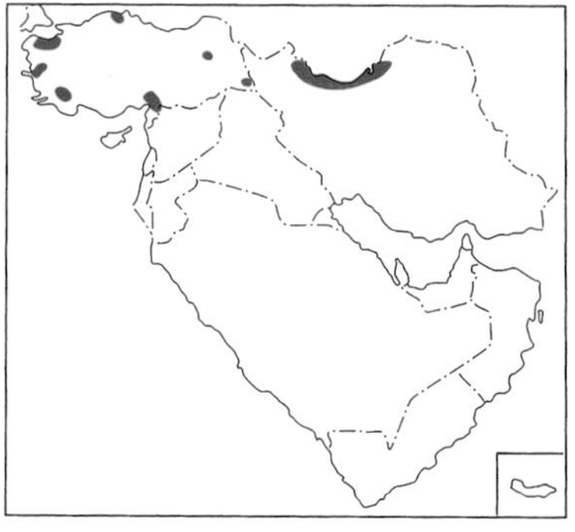

146 **Greater Spotted Eagle** *Aquila clanga*

Status: scarce migrant Near East and SW Arabia, scarcer elsewhere. Winters in small numbers over most region at suitable habitats, but scarcer Turkey; vagrant Cyprus. **Habitat:** usually near water, in marshes with trees, lakes, river deltas and swamps.

147 **Tawny Eagle** *Aquila rapax*

Status: mainly resident. Vagrant Oman and Israel. **Habitat:** dry regions in mountains or plains with scattered trees, savanna, often at rubbish dumps. Nests on mound, ruin or tree.

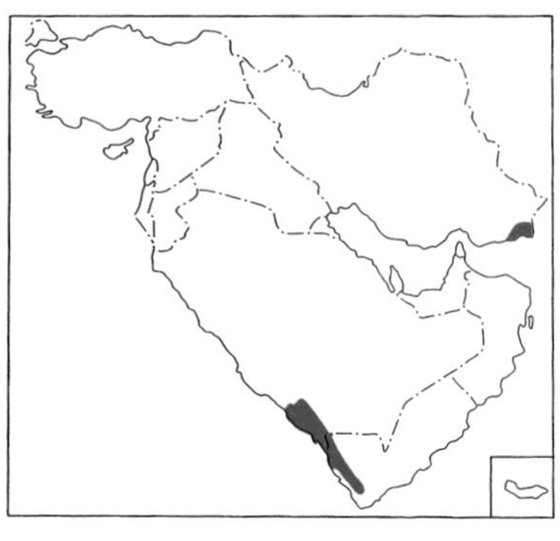

148 **Steppe Eagle** *Aquila nipalensis*

Status: passage migrant; heavy passage Israel, Jordan and SW Arabia, moderate NE Turkey, smaller passage E Arabia. Winters widely over Arabia, scarcer farther north, rarely Turkey. **Habitat:** open steppe and semi-desert, wadis and foothills, near marshes; also rubbish dumps in desert towns.

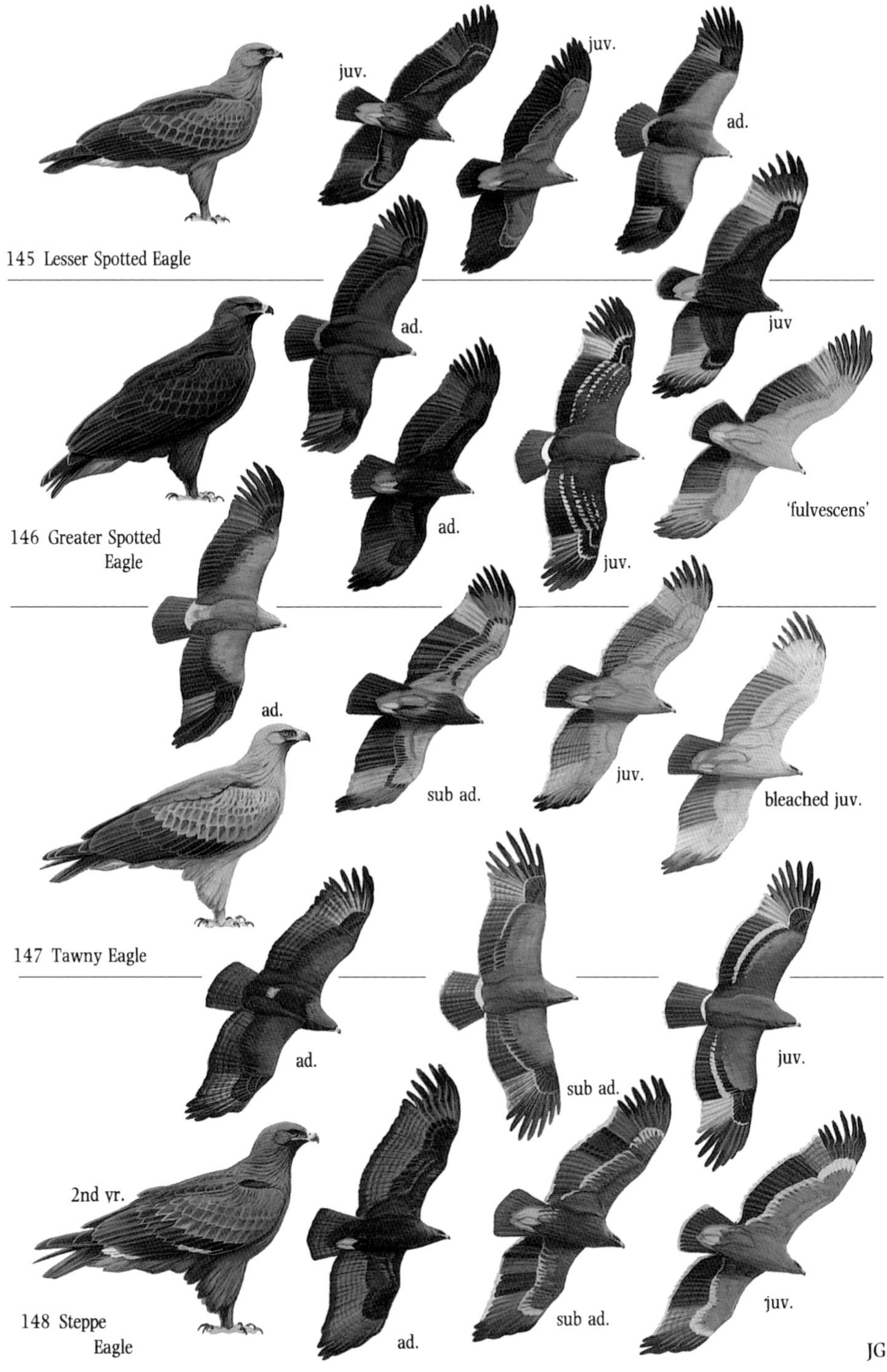
juv.
juv.
ad.
145 Lesser Spotted Eagle
ad.
juv
146 Greater Spotted Eagle
ad.
juv.
'fulvescens'
ad.
sub ad.
juv.
bleached juv.
147 Tawny Eagle
ad.
sub ad.
juv.
2nd yr.
148 Steppe Eagle
ad.
sub ad.
juv.
JG

149 **Imperial Eagle** *Aquila heliaca*

Status: partial migrant; small passage Israel and SW Arabia, scarcer elsewhere. Winters over most region where suitable habitat. **Habitat:** open plains with woods, forested foothills, in winter also steppes, marshes and other open regions. Nests in large tree.

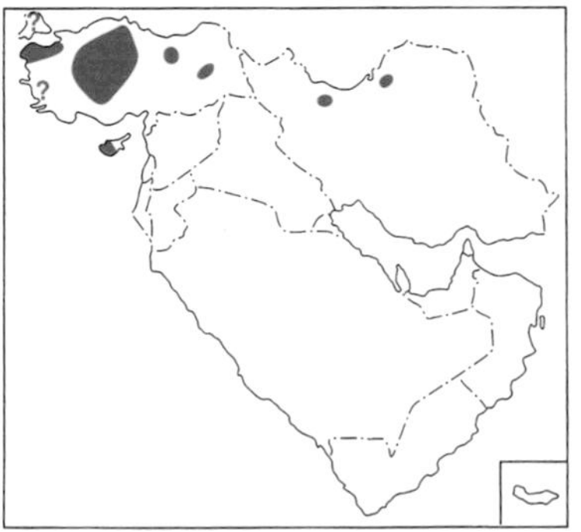

150 **Golden Eagle** *Aquila chrysaetos*

Status: has bred Jordan. Resident, but immatures disperse. Vagrant Cyprus. **Habitat:** barren or wooded mountains, wooded lakes, plains and semi-deserts with trees. Nests on rock ledge, sometimes in tree.

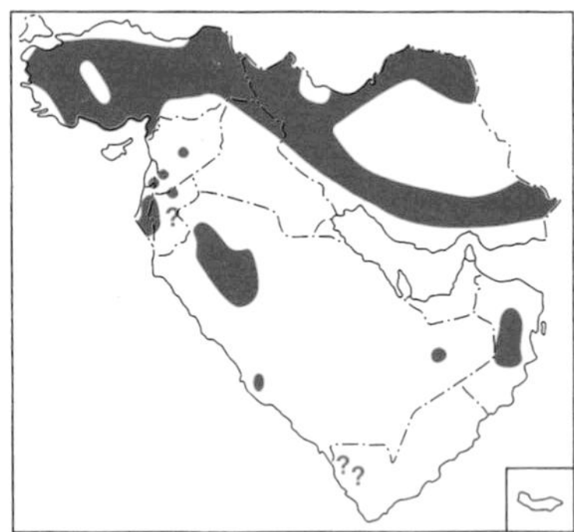

151 **Verreaux's Eagle** *Aquila verreauxii*

Status: resident. May breed Israel. **Habitat:** wild mountains. Nests on cliff, rarely in tree.

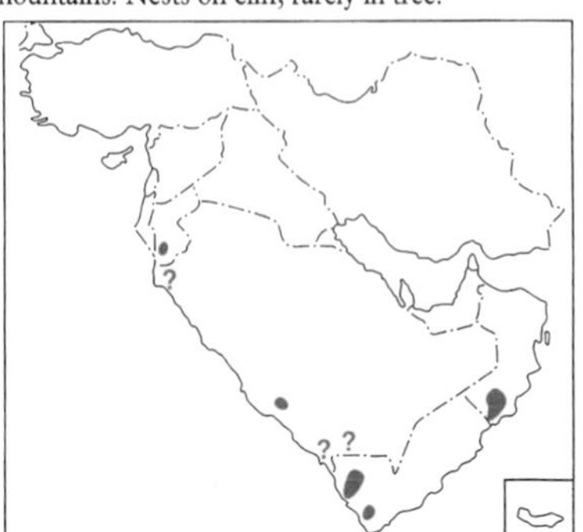

152 **Booted Eagle** *Hieraaetus pennatus*

Status: summer visitor. Passage E Mediterranean countries, fewer in Kuwait, UAE, Oman and Yemen, scarce elsewhere in Arabia and Iraq; rare Cyprus. Few winter Yemen, occasionally E Mediterranean and Oman. **Habitat:** deciduous and pine forest. Outside breeding season also cultivated areas with trees and marshes. Nests in tree, but also on cliff.

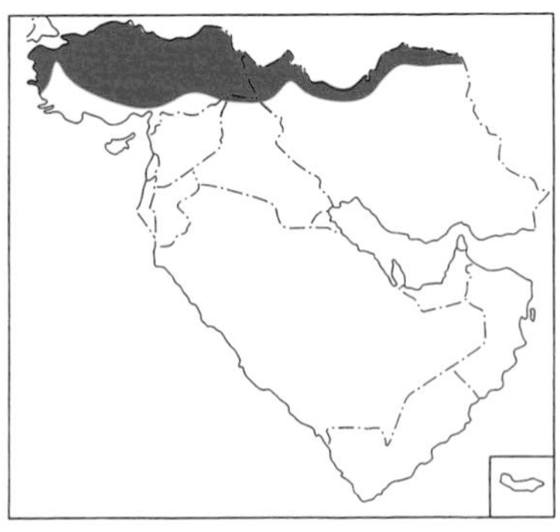

153 **Bonelli's Eagle** *Hieraaetus fasciatus*

Status: resident, but immatures disperse. **Habitat:** rocky mountains, forested steep foothills. In winter on plains and semi-deserts. Nests on rock-ledge, occasionally in tree.

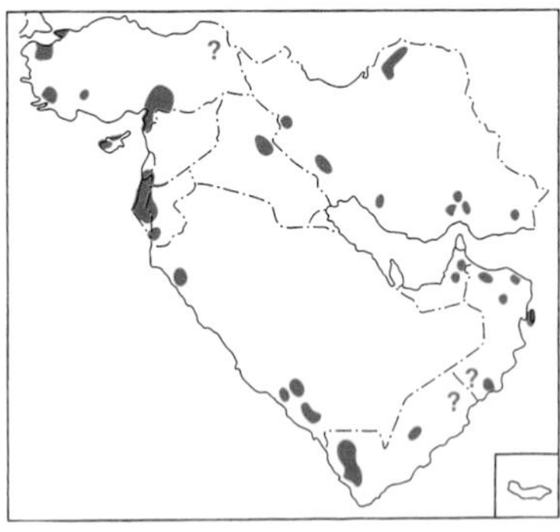

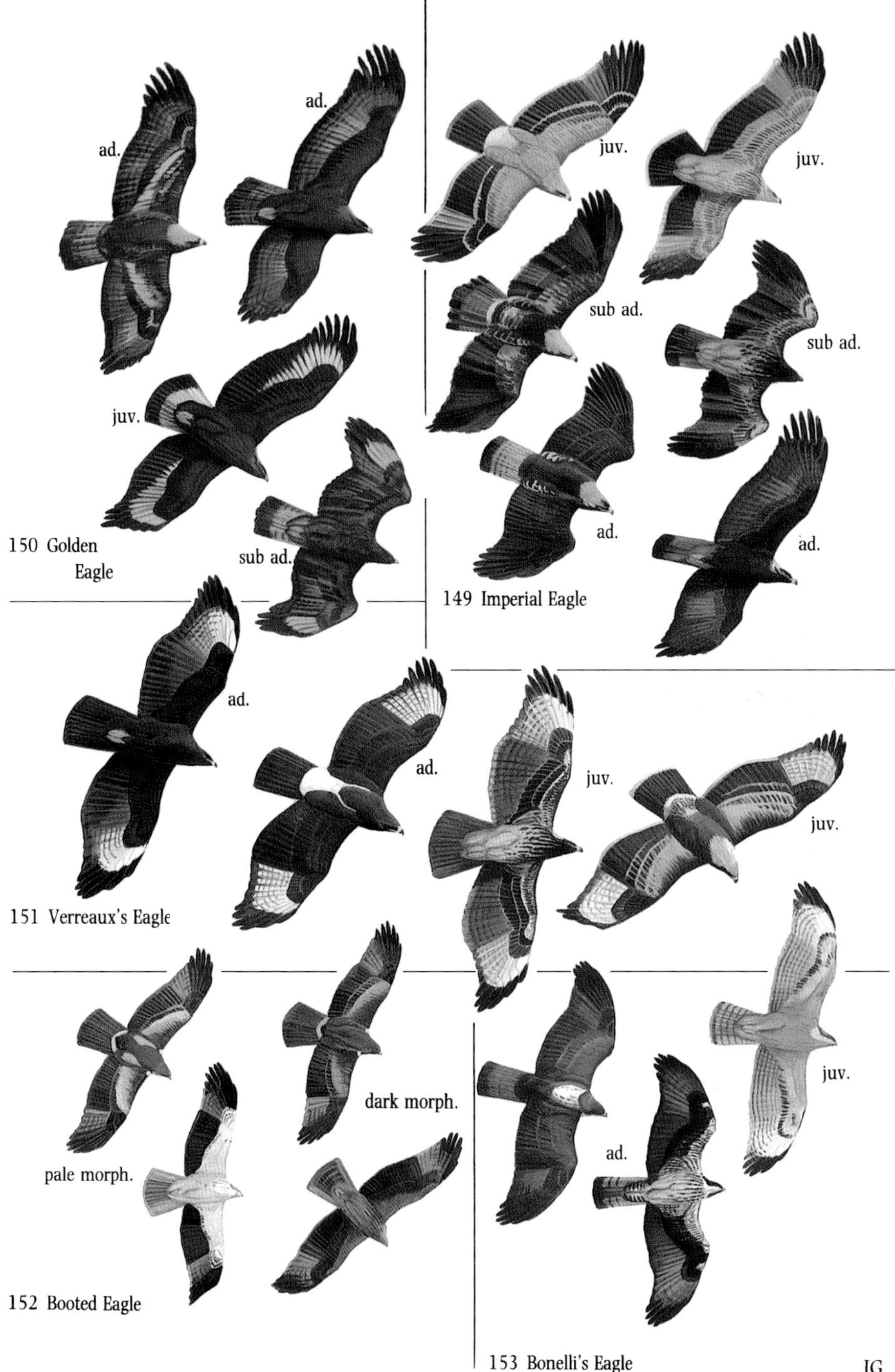
ad.
ad.
juv.
juv.
juv.
sub ad.
150 Golden Eagle
sub ad.
sub ad.
ad.
ad.
149 Imperial Eagle
ad.
ad.
juv.
juv.
151 Verreaux's Eagle
dark morph.
pale morph.
juv.
ad.
152 Booted Eagle
153 Bonelli's Eagle
JG

113 **Honey Buzzard** *Pernis apivorus*

Status: summer visitor. Perhaps breeds Syria. Heaviest passage NE Turkey and Near East, large numbers Bosphorus; scarce but regular Arabia. Few winter records E Arabia. **Habitat:** woodland; more varied on passage. Nests in tree.

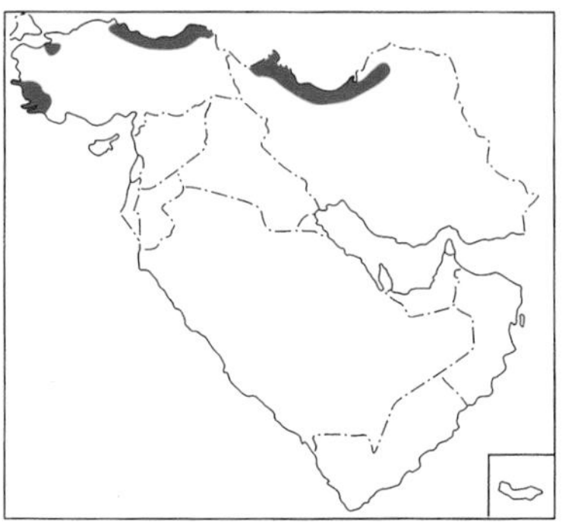

129 **Short-toed Eagle** *Circaetus gallicus*

Status: summer visitor, few winter Arabia. Large passage in Near East, scarcer farther east. **Habitat:** open wooded plains, stony foothills, semi-desert. Nests in tree.

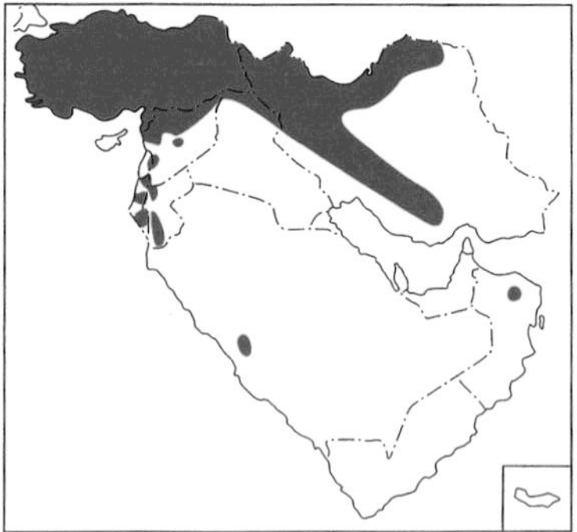

154 **Osprey** *Pandion haliaetus*

Status: occasionally breeds N Turkey. Resident and migrant. Widespread on passage. Winters coasts of Near East, Arabia, S Iran and marshes of Iraq. **Habitat:** always near water—inland or coastal. Nests in trees, sea-cliffs, remote islands (sometimes on ground), ruins, old wrecks, sometimes in scattered groups.

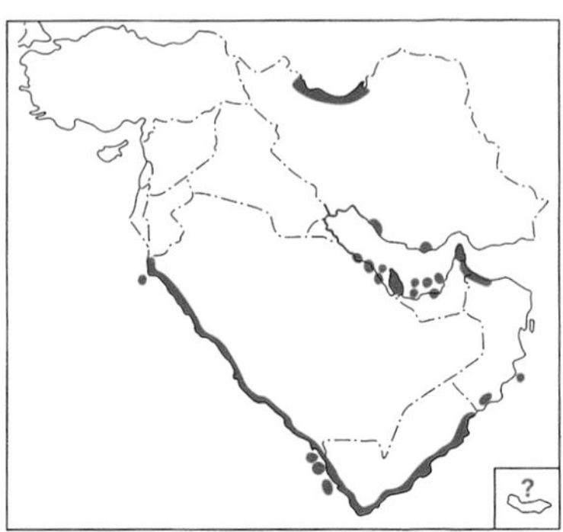

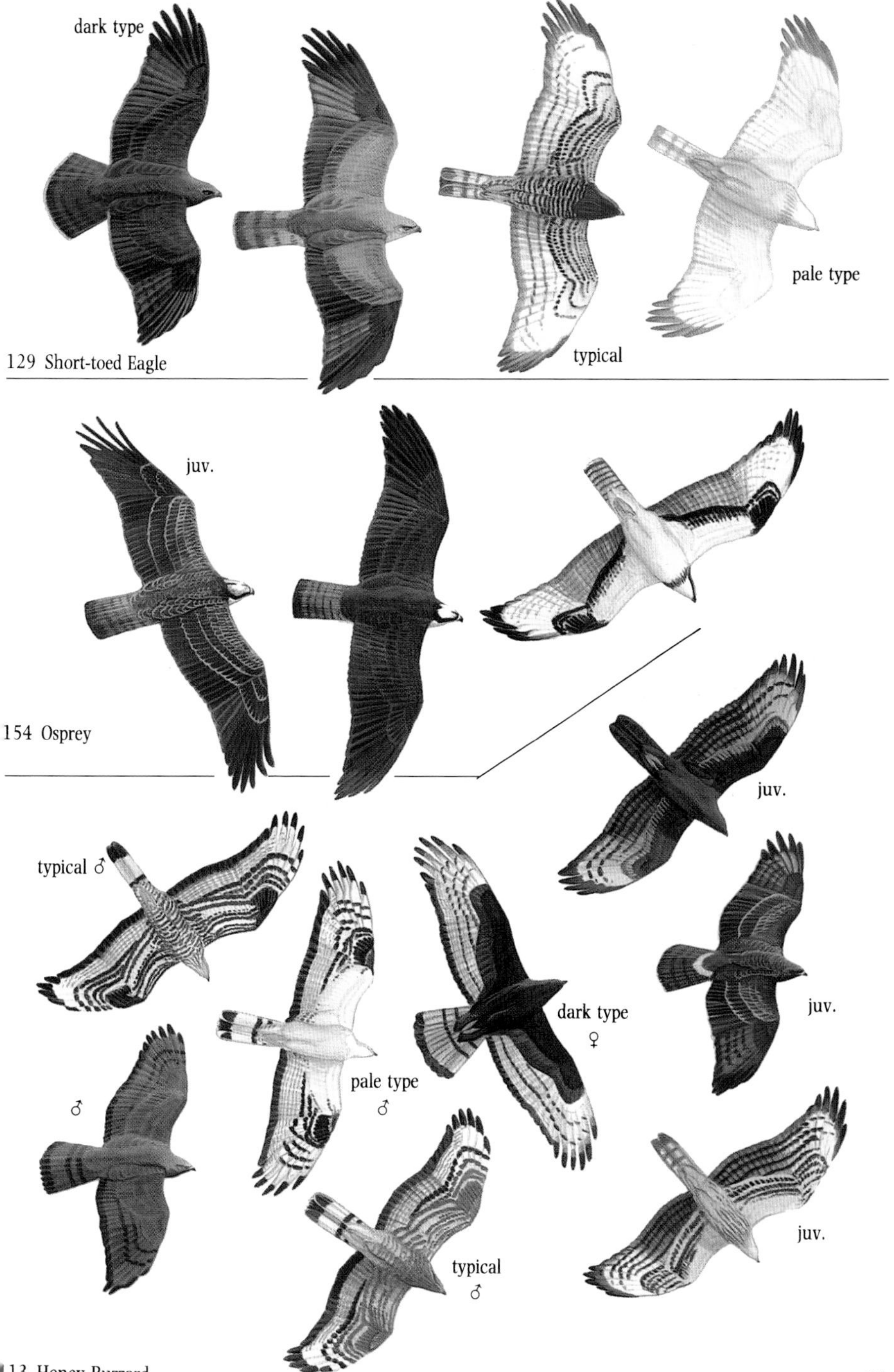
dark type
pale type
129 Short-toed Eagle
typical
juv.
154 Osprey
juv.
typical ♂
dark type
♀
juv.
♂
pale type
♂
typical
♂
juv.
13 Honey Buzzard
JG

155 **Lesser Kestrel** *Falco naumanni*

Status: summer visitor, but few winter S Turkey and Oman. Widely on passage over most of region. **Habitat:** old buildings, rocky gorges, but hunts over open country. Colonial, nesting in holes in roofs, walls or rock crevices.

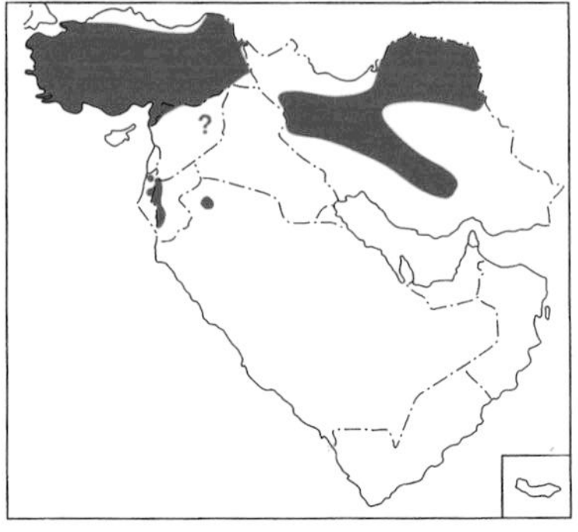

156 **Kestrel** *Falco tinnunculus*

Status: resident and partial migrant; in winter widespread over most of region. **Habitat:** open country with trees, wetlands, mountains and semi-deserts. Nests in hole or ledge on cliff or building; old nests of other species in trees.

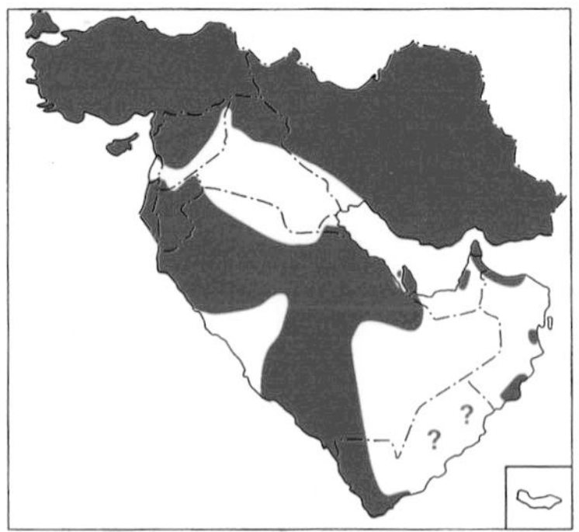

157 **Red-headed Merlin** *Falco chicquera*

Status: resident. **Habitat:** bare open plains with some trees in which it nests.

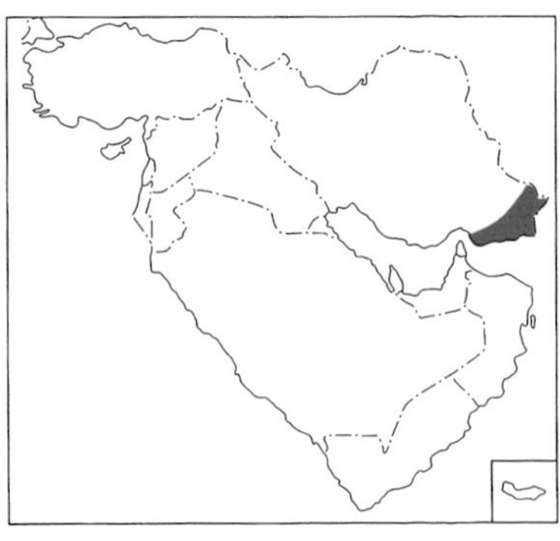

158 **Red-footed Falcon** *Falco vespertinus*

Status: passage migrant, mainly Turkey, Cyprus and countries bordering E Mediterranean, but uncommon Iraq; vagrant Iran. **Habitat:** plains with trees, cultivation and bushy wastelands.

159 **Amur Falcon** *Falco amurensis*

Status: rare migrant Oman where occasional in winter; vagrant SW Arabia and UAE. Breeds E Asia, winters SE Africa. **Habitat:** open regions, cultivated fields with trees, lightly wooded areas.

160 **Merlin** *Falco columbarius*

Status: winter visitor in north of region, scarce S Iraq, Near East, Cyprus and NW Arabia; rare Saudi Arabia, including Yemen. Vagrant Bahrain, UAE and Oman. **Habitat:** open country; steppes and semi-deserts, marshes, farmland and plains.

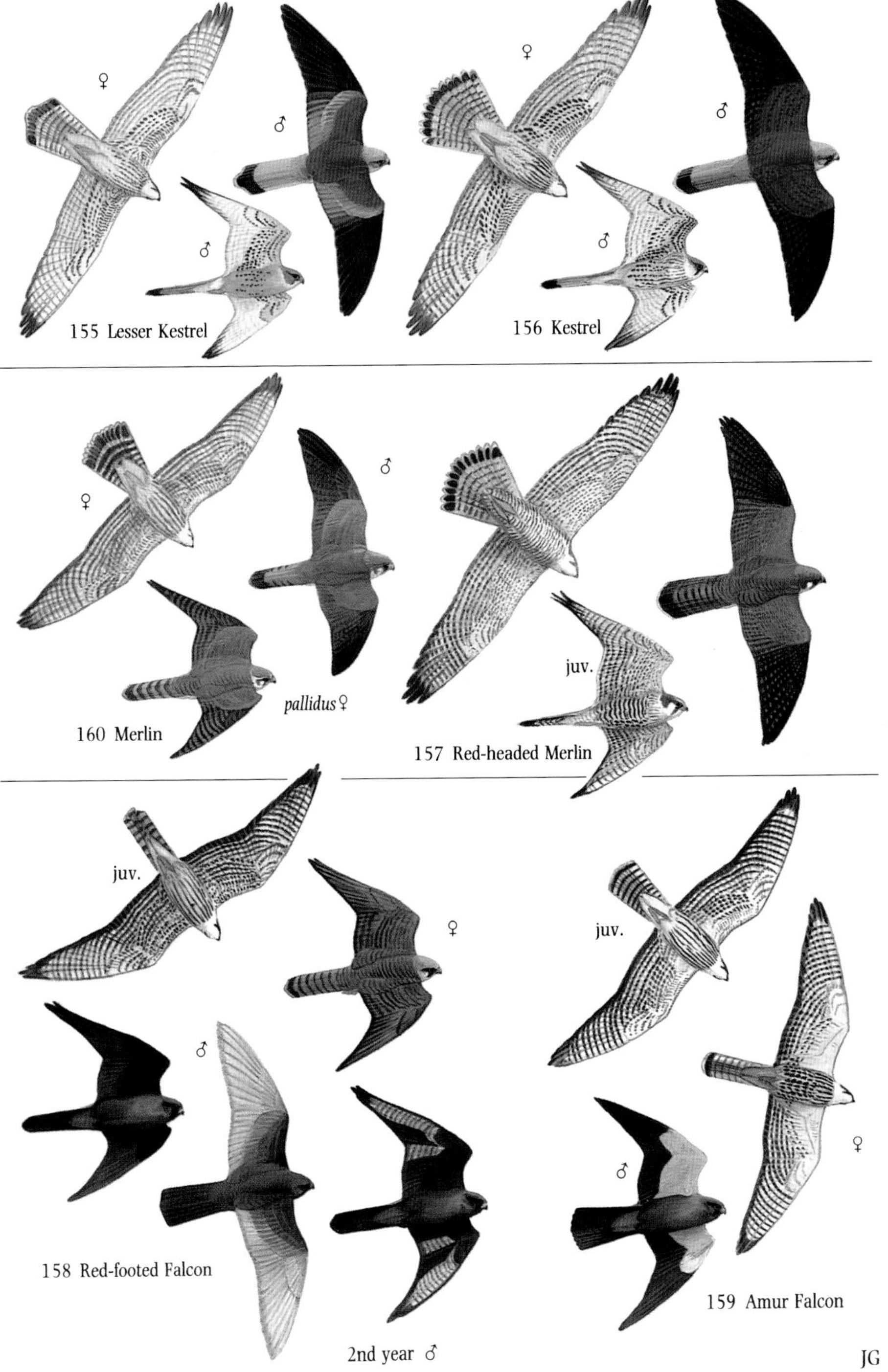
♀
♂
♂
155 Lesser Kestrel
♀
♂
♂
156 Kestrel
♀
♂
pallidus ♀
160 Merlin
juv.
157 Red-headed Merlin
juv.
♀
♂
158 Red-footed Falcon
2nd year ♂
juv.
♀
♂
159 Amur Falcon
JG

115 **Black-winged Kite** *Elanus caeruleus*

Status: resident, but some dispersal. Vagrant Turkey, Israel, W Saudi Arabia, UAE and Oman. **Habitat:** cultivated areas with some trees, edges of woodland. Nests in trees.

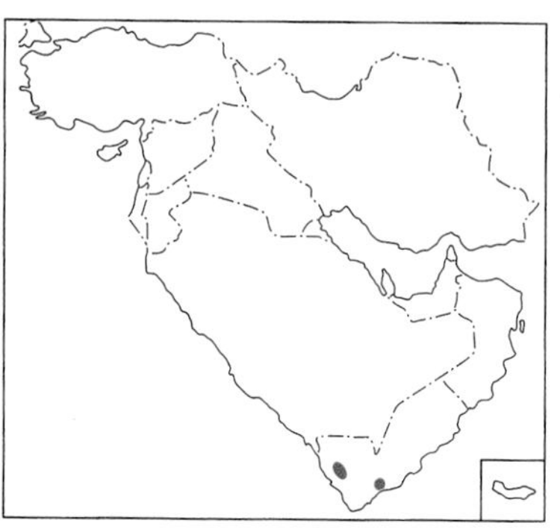

116 **African Swallow-tailed Kite** *Chelictinia riocourii*

Status: vagrant Yemen from tropical Africa. **Habitat:** semi-desert and savanna grassland.

161 **Hobby** *Falco subbuteo*

Status: summer visitor, wintering tropical Africa. Has bred Cyprus. Widespread on passage all over region. **Habitat:** open, scattered woodland, clumps of trees in cultivated areas. Breeds in trees in old nest, usually of Crow.

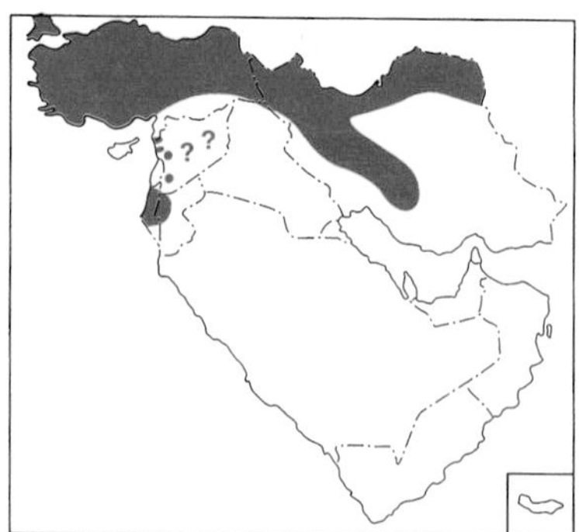

162 **Eleonora's Falcon** *Falco eleonorae*

Status: summer visitor, wintering Madagascar. Seen regularly throughout W Turkey (scarcer in SE) well away from breeding areas. Passage Near East. **Habitat:** colonial, nesting in holes on rocky islands and sea-cliffs. Often hunts over neighbouring wetlands.

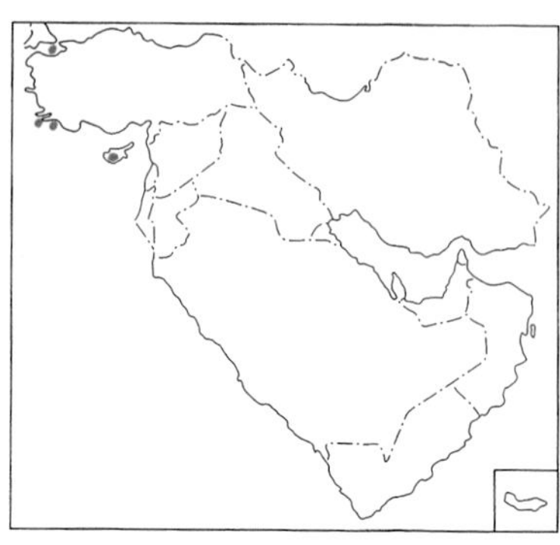

163 **Sooty Falcon** *Falco concolor*

Status: summer visitor, wintering Madagascar. Scattered records in Arabia away from breeding areas. Vagrant Cyprus, Syria, Lebanon, Iran and Turkey. **Habitat:** coral islands, inland cliffs in desert. Nests in holes in coral or rock, rarely under bush.

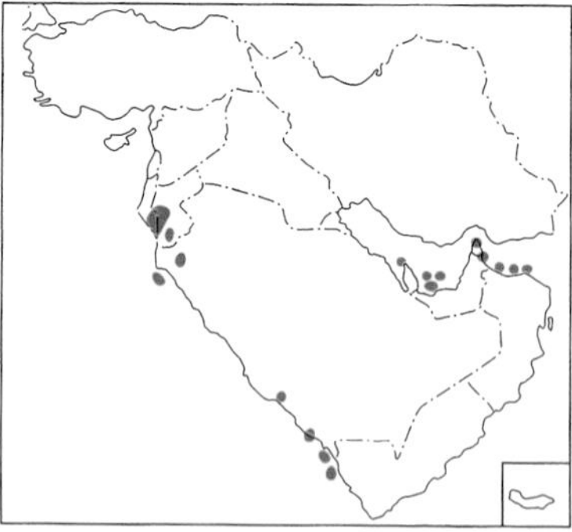

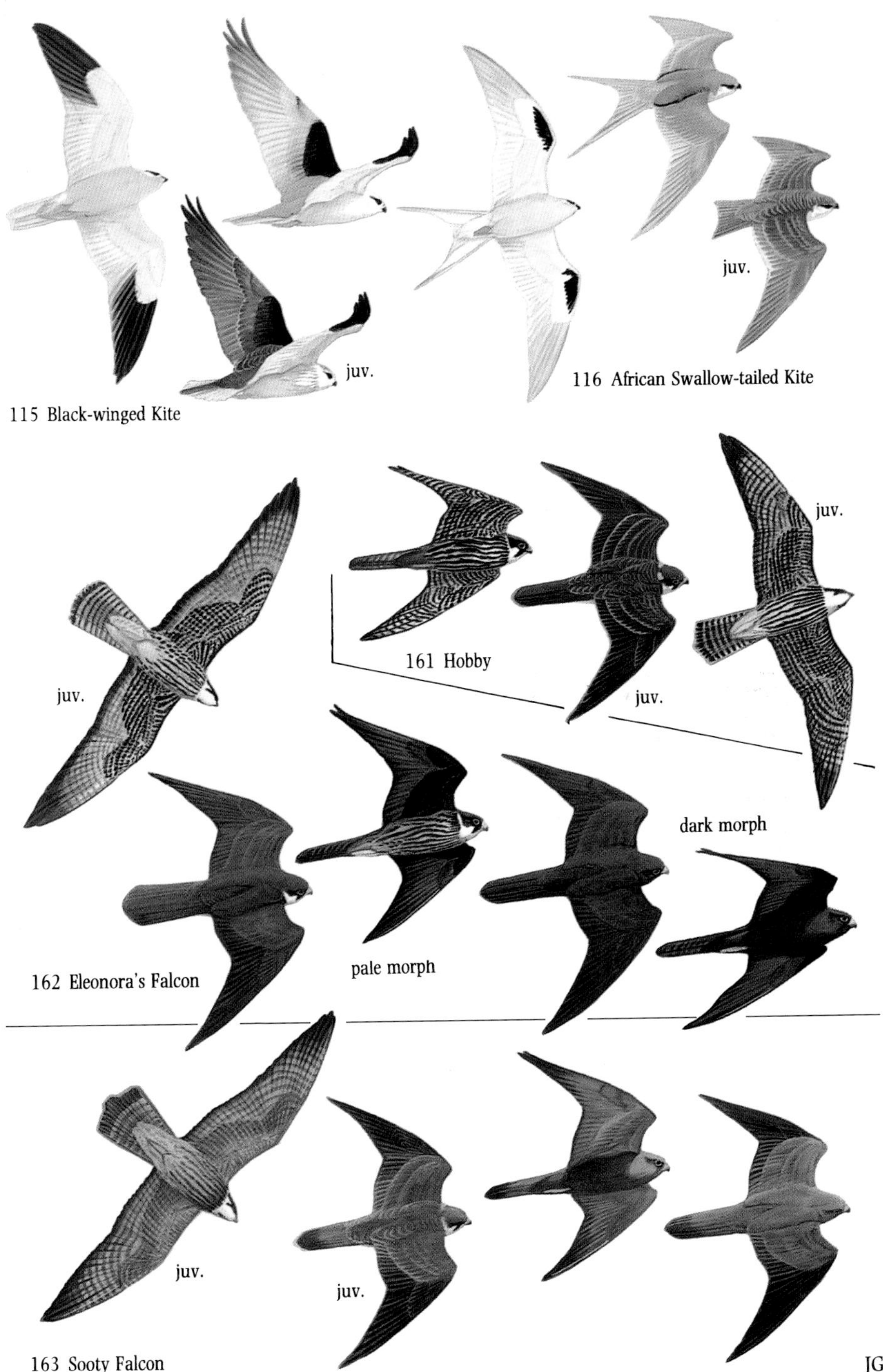
juv.
juv.
115 Black-winged Kite
116 African Swallow-tailed Kite
juv.
juv.
juv.
161 Hobby
dark morph
pale morph
162 Eleonora's Falcon
juv.
juv.
163 Sooty Falcon
JG

164 **Lanner Falcon** *Falco biarmicus*

Status: status unclear but recent population decrease; resident and dispersive. Regular Syria, Gulf States, Oman. Vagrant Cyprus, Iraq. **Habitat:** rocky mountains, cliffs and ruins, stony plains and semi-deserts. Nests on cliff or ruin, but sometimes in nest of other species.

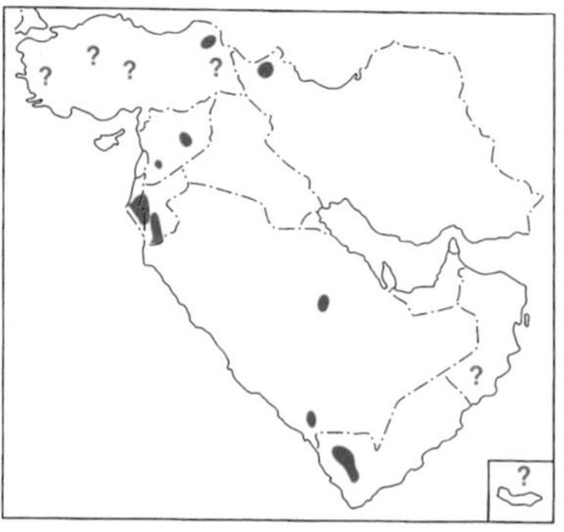

165 **Saker Falcon** *Falco cherrug*

Status: partial migrant. Widely spread, but scarce, on passage and in winter over most of the region. **Habitat:** lightly wooded steppe, foothills; also mountains, semi-deserts. Nests on cliff-ledge or tree, often in old nest.

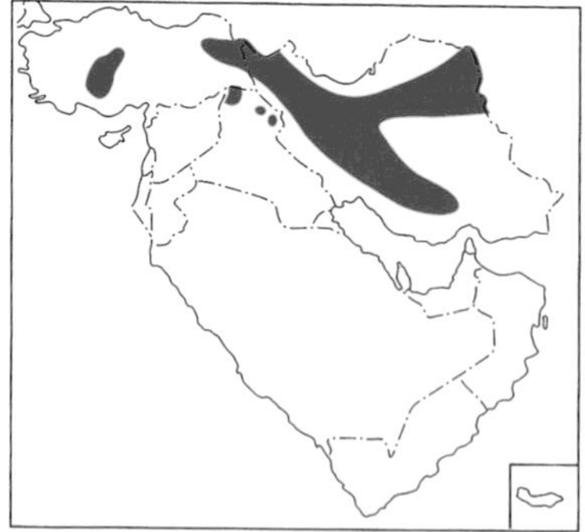

166 **Peregrine Falcon** *Falco peregrinus*

Status: partial migrant, widely on passage and as winter visitor over most region. **Habitat:** mountains and forested regions, cliffs; outside breeding season also marshes, wastelands. Breeds in cliff-ledge, building or old nest in tree.

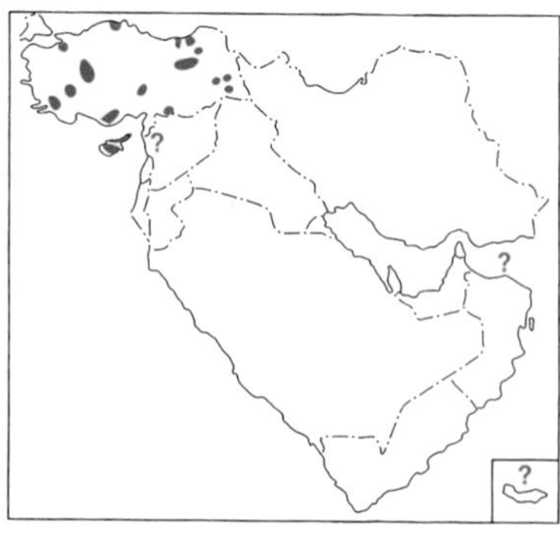

167 **Barbary Falcon** *Falco pelegrinoides*

Status: resident, dispersive or partial migrant; spreads to Iraq and Turkey in winter. Status uncertain in large parts of UAE and Oman. **Habitat:** arid mountains, semi-deserts, cliffs and wadis. Nests on cliff-ledge.

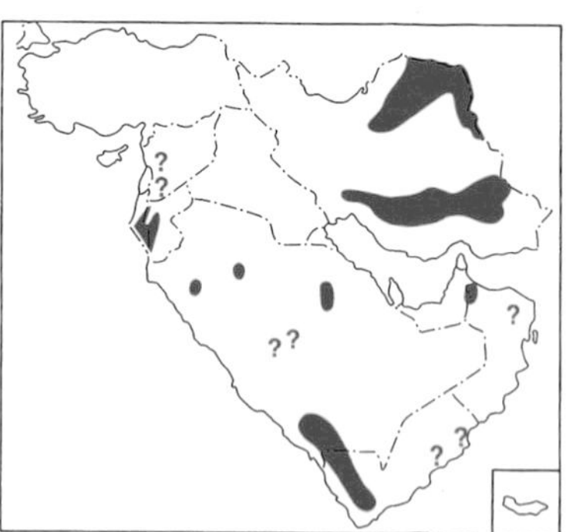

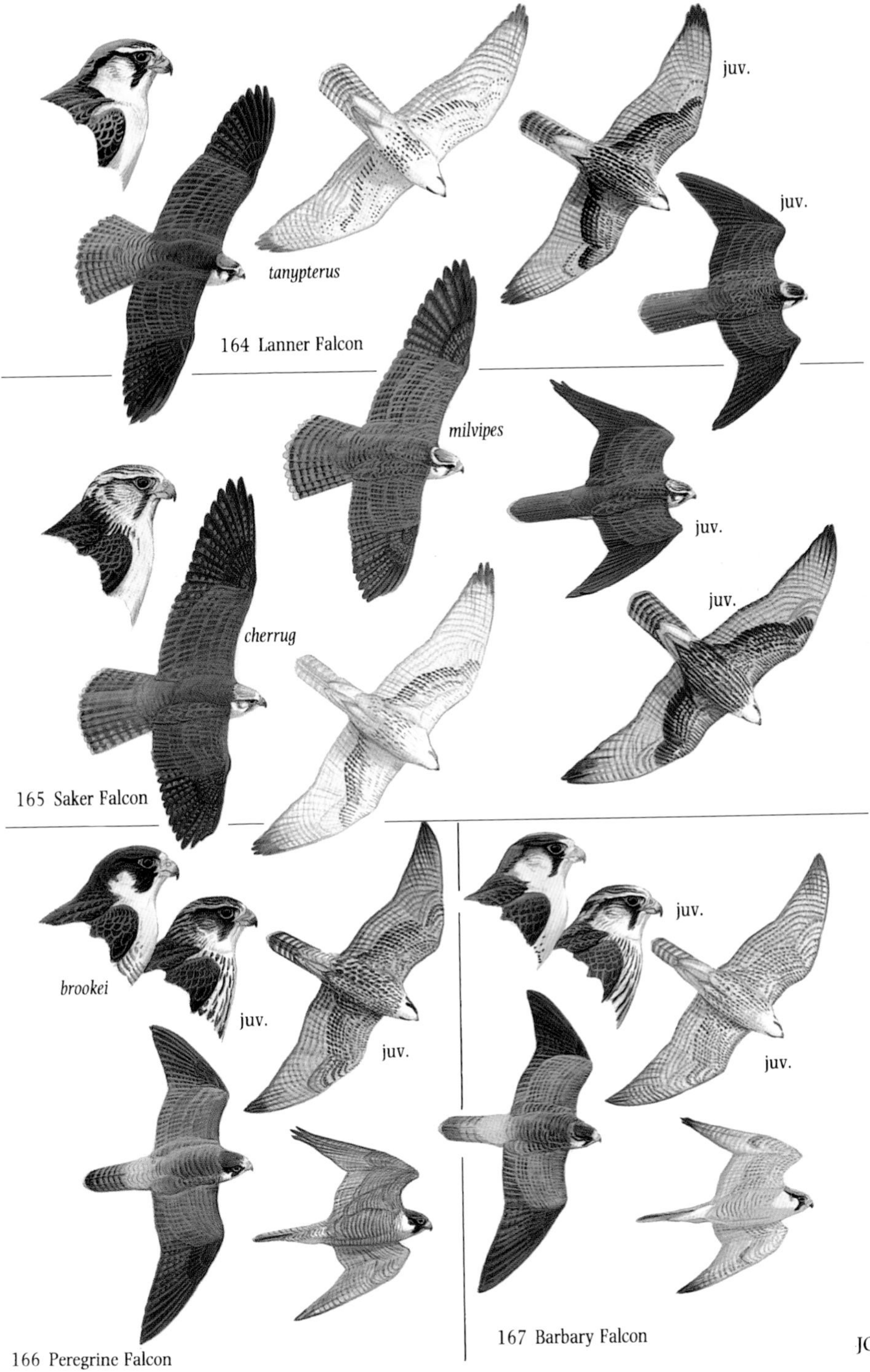
juv.
juv.
tanypterus
164 Lanner Falcon
milvipes
juv.
juv.
cherrug
165 Saker Falcon
juv.
brookei
juv.
juv.
juv.
166 Peregrine Falcon
167 Barbary Falcon
JG

168 **Caucasian Black Grouse** *Tetrao mlokosiewiczi*

Status: resident. **Habitat:** alpine and sub-alpine meadows with rhododendron and birch thickets, between 1500–3000 m, but descending to fir forest down to 700 m in winter. Nests on ground in dense cover.

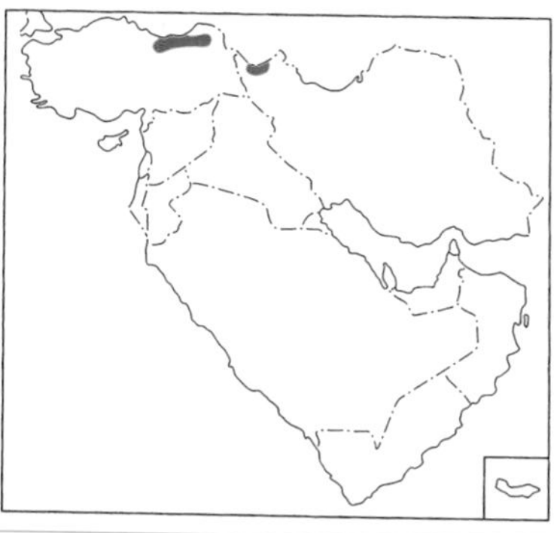

169 **Caspian Snowcock** *Tetraogallus caspius*

Status: resident. **Habitat:** steep rocky slopes with meadows and bushes in alpine and sub-alpine zones above 1800 m. Rarely descends to tree limit. Nests on ground.

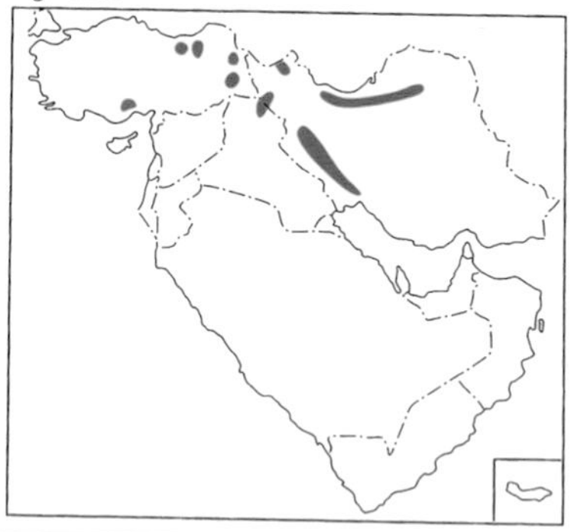

170 **Chukar** *Alectoris chukar*

Status: resident. **Habitat:** rocky slopes with some vegetation, also semi-deserts and wasteland with agriculture. Nests on ground under cover.

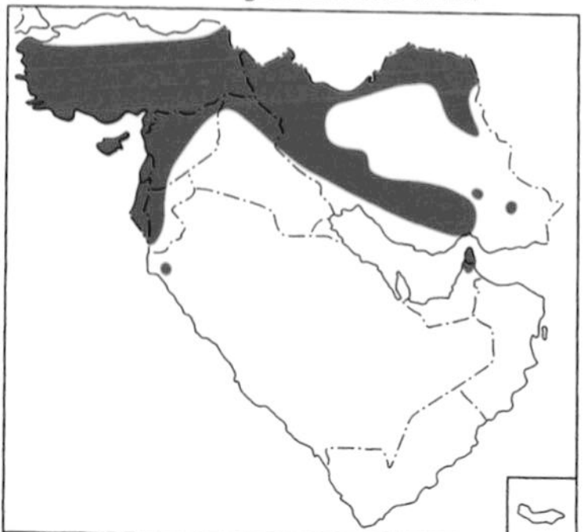

171 **Philby's Partridge** *Alectoris philbyi*

Status: resident. **Habitat:** barren rocky hillsides, sometimes with bushes; above 1500 m but usually higher than 2500 m. Nests under bush or cover of rock.

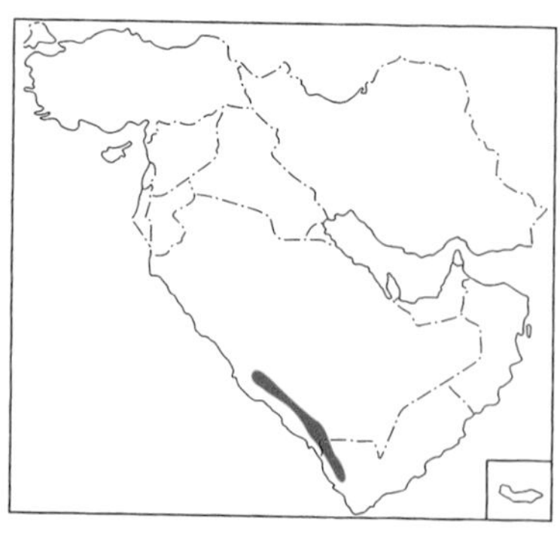

172 **Arabian Partridge** *Alectoris melanocephala*

Status: resident. **Habitat:** rocky slopes with bushes between 250–2800 m (lower altitudes than Philby's). Nests on ground in cover.

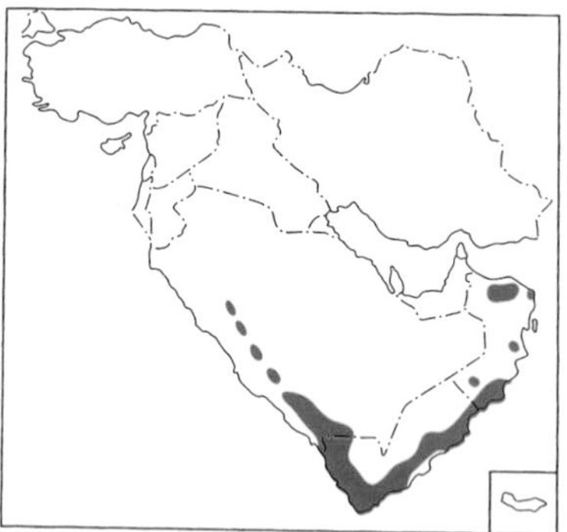

♀
168 Caucasian
Black Grouse
♂
display
169 Caspian
Snowcock
♀
♂
170 Chukar
171 Philby's Partridge
172 Arabian Partridge

ML

173 **See-see Partridge**
Ammoperdix griseogularis

Status: resident. **Habitat:** barren stony or gravelly hill-sides, open desolate tracts or broken ground, steep rocky banks. Avoids tall or thick vegetation, but needs access to drinking water. Up to 2000 m. Nests on ground.

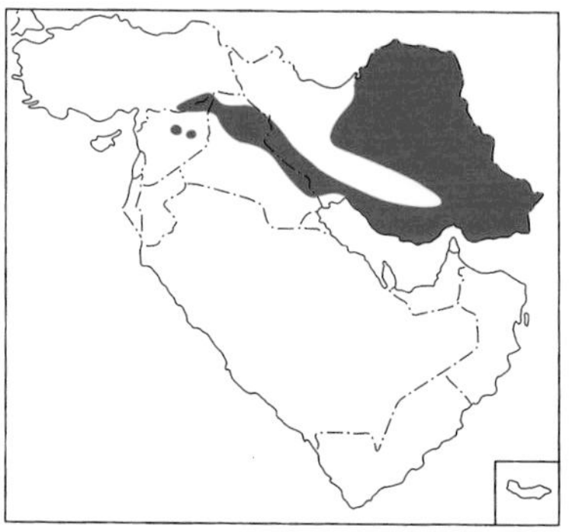

174 **Sand Partridge** *Ammoperdix heyi*

Status: resident. **Habitat:** desolate, rocky and stony slopes and wadis, even cliffs, not far from water. Breeds on ground, under overhanging rock or bush.

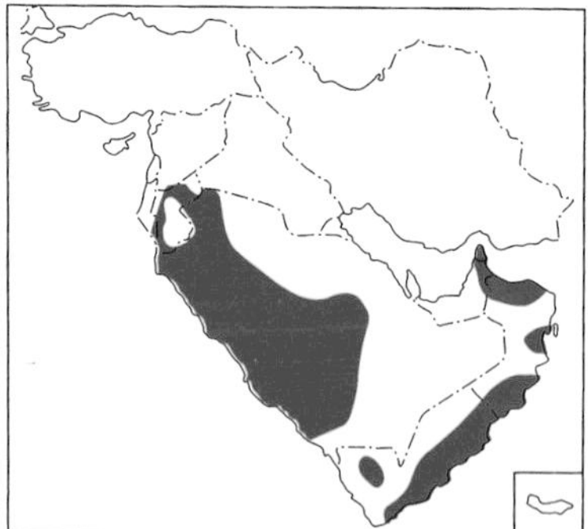

175 **Black Francolin** *Francolinus francolinus*

Status: resident. A few records Hofuf (E Saudi Arabia) and UAE suggest possibility of feral breeding. **Habitat:** densely scrub-covered lowlands and wadis, often not far from water, dunes with scattered vegetation, also reedy flood-plains. Nests on ground.

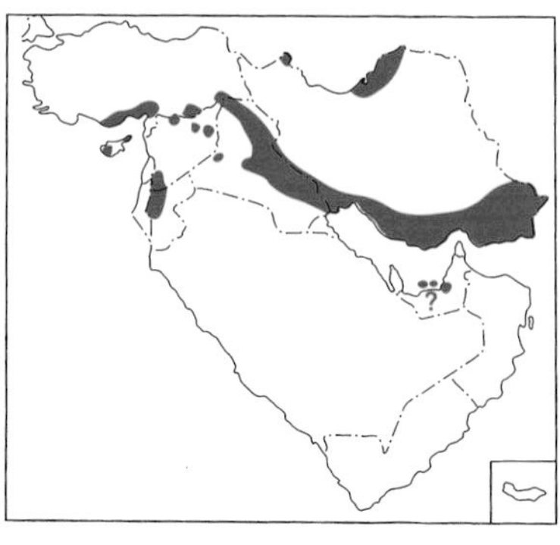

176 **Grey Francolin** *Francolinus pondicerianus*

Status: resident; introduced to Bahrain and several UAE islands. Vagrant Saudi Arabia. **Habitat:** thorn scrub, edges of cultivation and semi-desert. Nests on ground under tuft or bush.

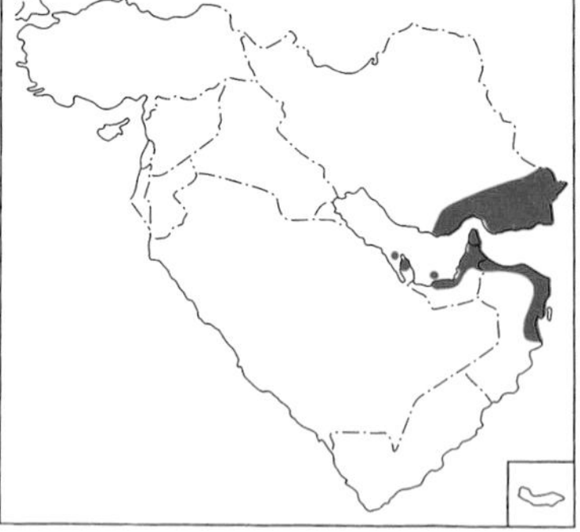

intermedia
♂
intermedia
♀
heyi
♀
174 Sand Partridge
heyi
♂
♀
♂
173 See-see
Partridge
♀
175 Black
Francolin
176 Grey
Francolin
♂
♀
ML

177 **Grey Partridge** *Perdix perdix*

Status: resident. **Habitat:** farmland, pastures, wastelands, sand dunes. Nests on ground under bush or in long grass.

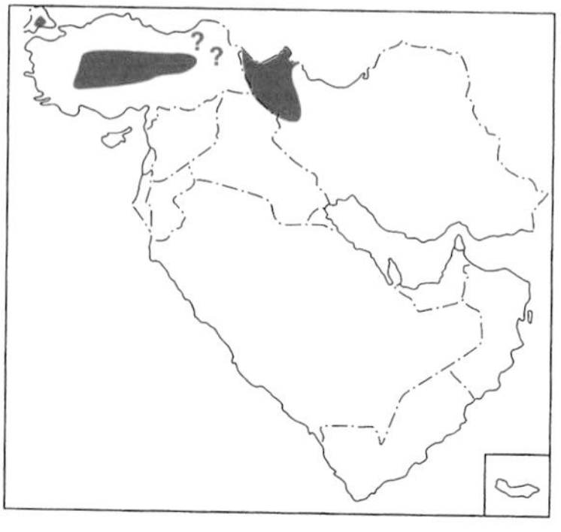

178 **Quail** *Coturnix coturnix*

Status: mainly summer visitor. Passage widespread. Some winter Turkey, Iraq, Iran, but apparently not Near East or Arabia (except Yemen and, scarcely, Bahrain and Oman). **Habitat:** farmcrops, rough grassland, avoiding trees and bare ground. Nests on ground.

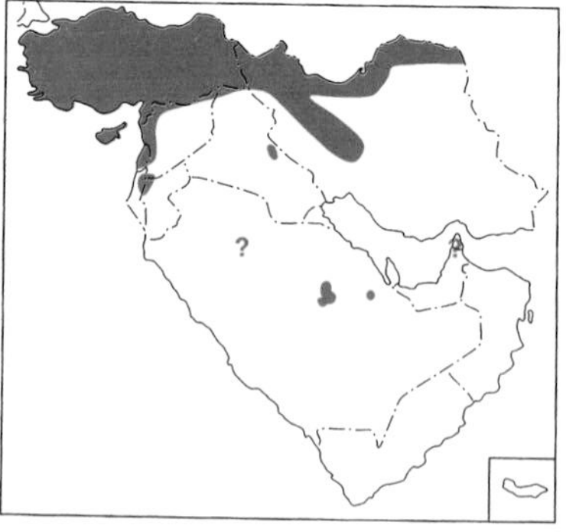

179 **Harlequin Quail** *Coturnix delegorguei*

Status: uncertain; apparently an irregular visitor but may breed SW Saudi Arabia and Yemen where recorded in numbers a century ago; vagrant S Oman. Range Africa. **Habitat:** open grassland and cultivated fields. Nests on ground.

180 **Pheasant** *Phasianus colchicus*

Status: resident. **Habitat:** woodland borders and thickets, interspersed with farmland in lowland or rolling country, also reedbeds. Nests among thick ground cover.

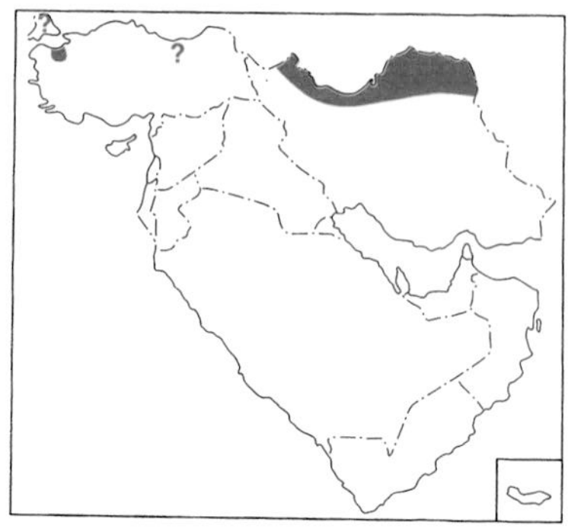

181 **Helmeted Guineafowl** *Numida meleagris*

Status: resident. **Habitat:** savanna, grassy plains, hills, wadis, often near cover. Nests on ground in thick cover.

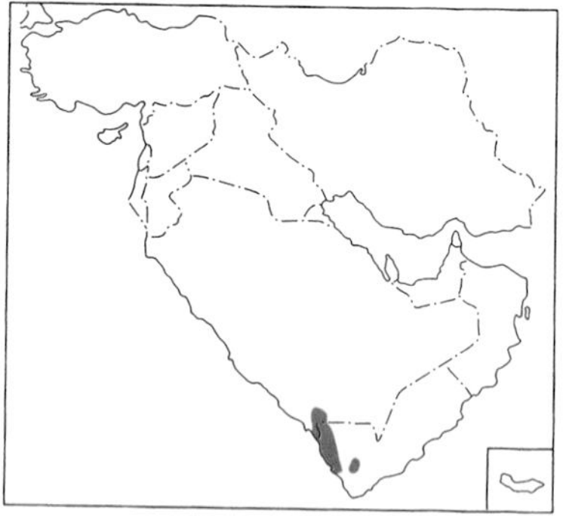

182 **Little Button Quail** *Turnix sylvatica*

Status: has bred Yemen; apparently rare and irregular visitor to SW Arabia. Vagrant Iran and Oman. Range mainly Africa and India. **Habitat:** sandy plains with palmetto scrub, brush-covered wastes, extensive low thickets, stubble and neglected cultivation. Nests in dense vegetation.

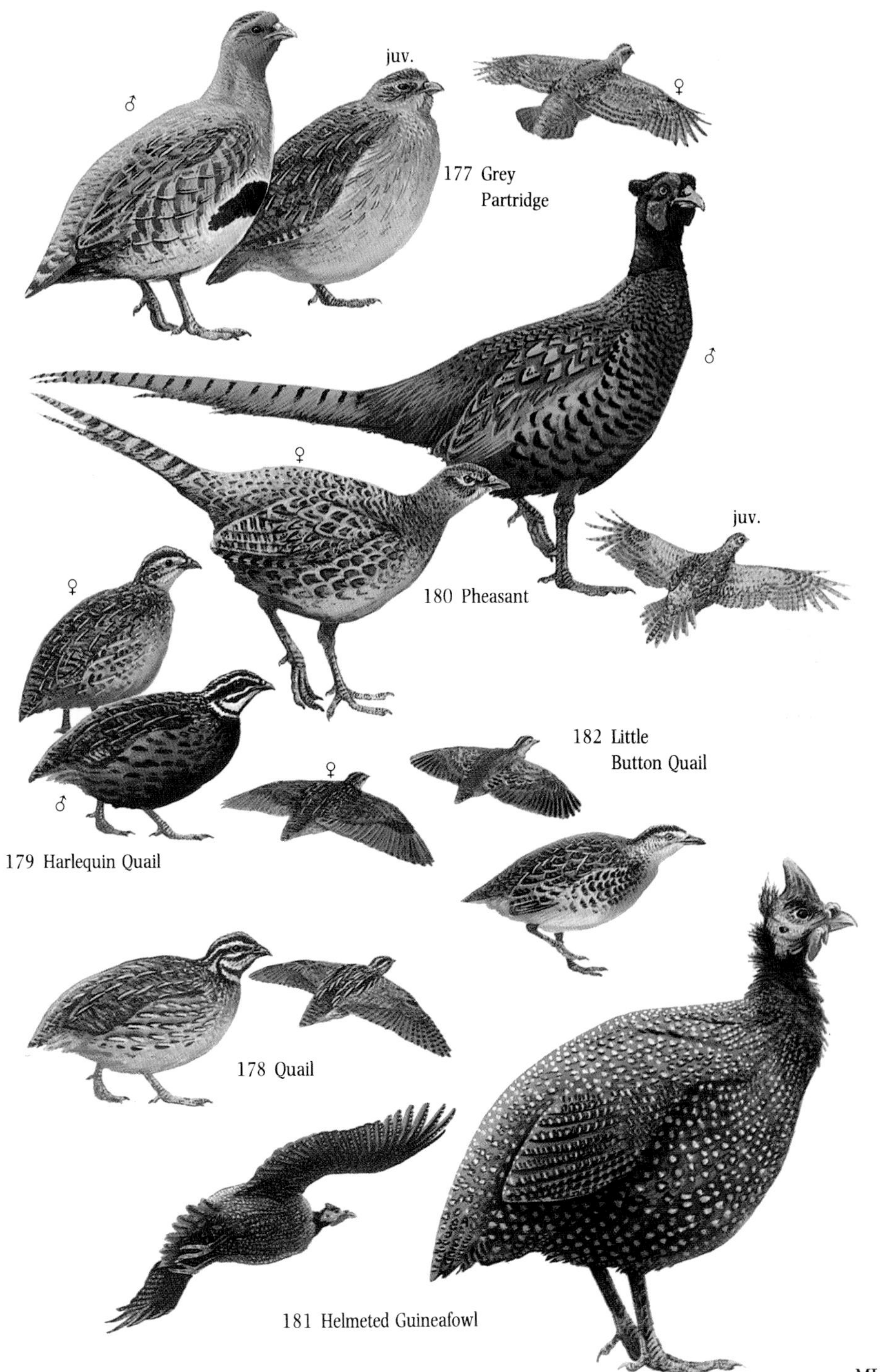
♂
juv.
♀
177 Grey
Partridge
♂
♀
juv.
180 Pheasant
♀
182 Little
Button Quail
♀
♂
179 Harlequin Quail
178 Quail
181 Helmeted Guineafowl

ML

183 **Water Rail** *Rallus aquaticus*

Status: has bred Israel. Partial migrant, but breeders mainly resident in Middle East. Winter visitor Mediterranean countries, Iraq, Iran, Arabian Gulf and, scarcely, south to Oman, rarer Yemen. **Habitat:** dense aquatic vegetation, reed- and osier-beds, overgrown ponds, ditches, river banks. Nests among reeds or sedges above shallow water.

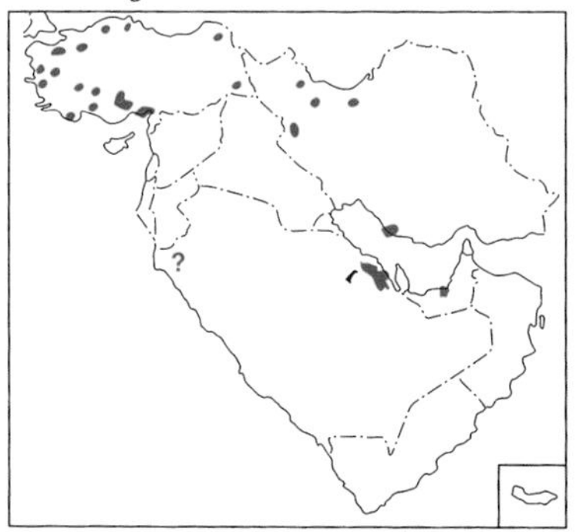

184 **Spotted Crake** *Porzana porzana*

Status: mainly migrant. Some winter Near East, Oman, perhaps Saudi Arabia. On passage may occur almost anywhere in the region though vagrant Yemen and Syria. **Habitat:** swamps and fens, overgrown ditches, outer margins of ponds, rivers, etc. Nests in thick vegetation by or in very shallow water.

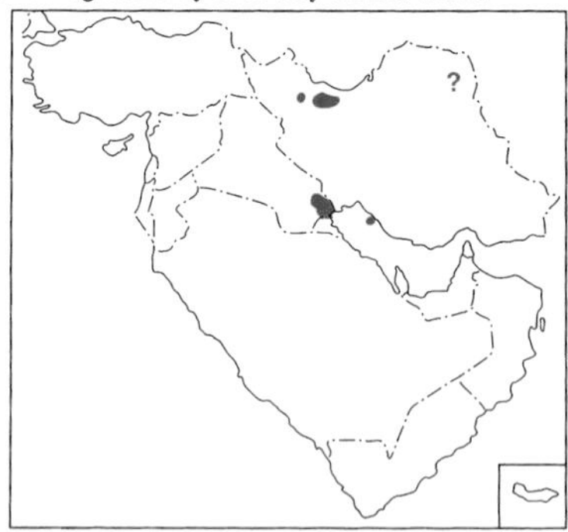

185 **Little Crake** *Porzana parva*

Status: almost certainly breeds widely in Turkey, but not proven. Inconspicuous passage, throughout whole region, wintering Near East and Oman. Vagrant Kuwait and Syria. **Habitat:** as Spotted, but with fondness for high reeds in deeper water and lagoons with floating vegetation. Nests in thick vegetation close to or in water.

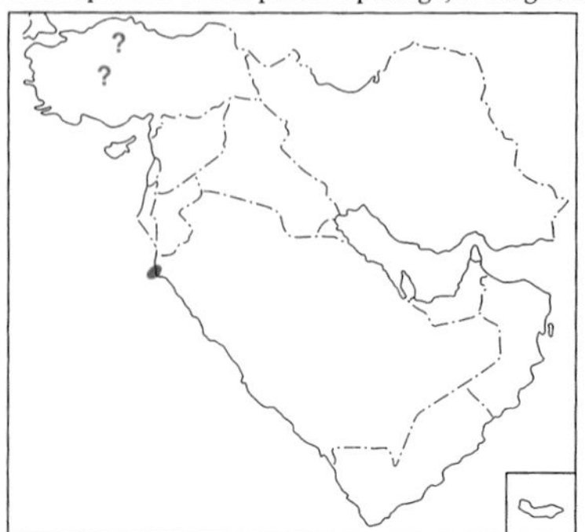

186 **Baillon's Crake** *Porzana pusilla*

Status: inconspicuous passage through whole region, though vagrant Syria and Yemen. Winters Near East, Iraq, Oman. **Habitat:** usually prefers lower, denser vegetation (sedges, rushes) and smaller pools than Little, in floodlands, fens, water margins with abundant cover. Nests on ground or tussock in water.

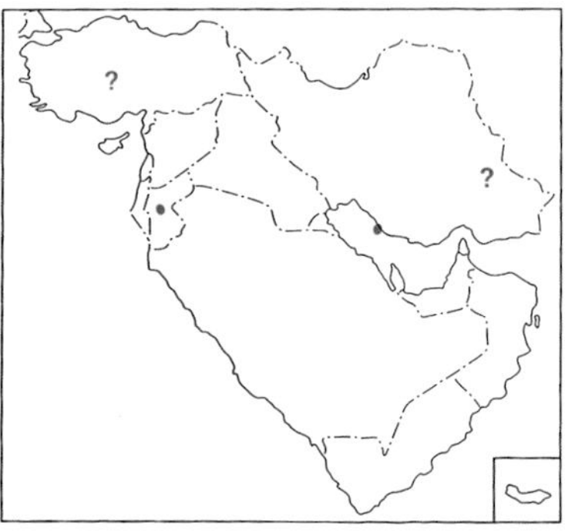

187 **Corncrake** *Crex crex*

Status: has perhaps bred Turkey. Passage migrant, occurring chiefly in Mediterranean countries; scarce Iraq, Arabia though concentrations recorded in recent agricultural developments in E Saudi Arabia. May winter occasionally in Oman. Vagrant Syria. **Habitat:** meadows, lush vegetation, and crops. Avoids standing water.

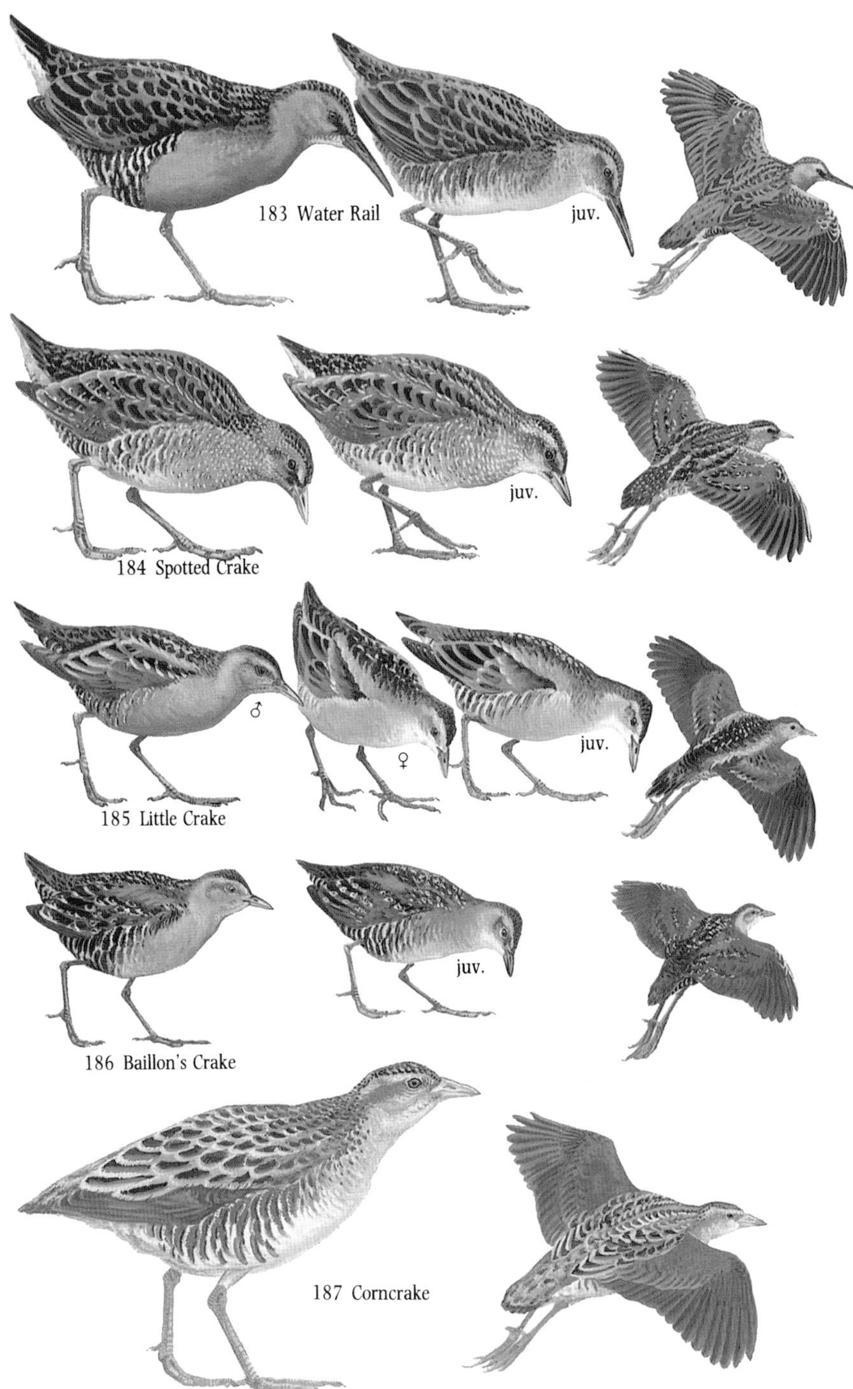
183 Water Rail
juv.
184 Spotted Crake
juv.
185 Little Crake
♂
♀
juv.
186 Baillon's Crake
juv.
187 Corncrake
ML

188 **White-breasted Waterhen** *Amaurornis phoenicurus*

Status: scarce passage migrant and winter visitor Oman. Vagrant Yemen, UAE. Range includes Indian subcontinent. **Habitat:** swamps, water holes with dense vegetation, mangrove edges.

189 **Moorhen** *Gallinula chloropus*

Status: resident and winter visitor. Range expanding in Arabia to occupy newly created waters. **Habitat:** freshwater wetland and river margins with cover. Nests in aquatic vegetation or bushes near water.

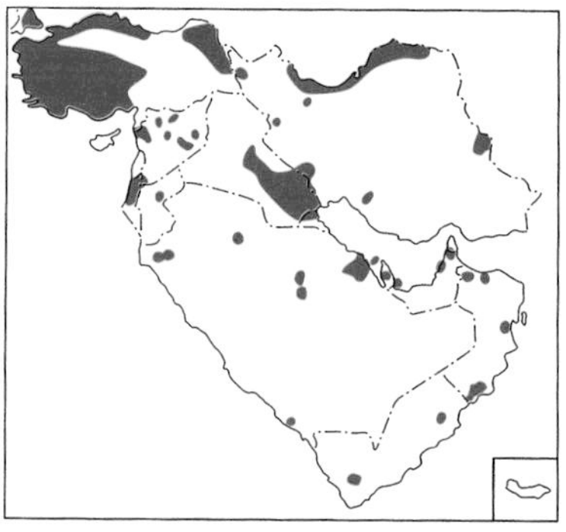

191 **Allen's Gallinule** *Porphyrula alleni*

Status: vagrant Cyprus, Oman. Range Africa. **Habitat:** dense swamps and nearby rank grass or thick bushes.

192 **Purple Gallinule** *Porphyrio porphyrio*

Status: may breed Kuwait. Mainly resident, but some wandering. Vagrant Cyprus, Near East (e.g. Syria), Arabian Gulf (e.g. UAE) and Oman. **Habitat:** swamps with extensive reedbeds, borders of lakes and rivers fringed with dense tall cover, in which it nests in shallow water.

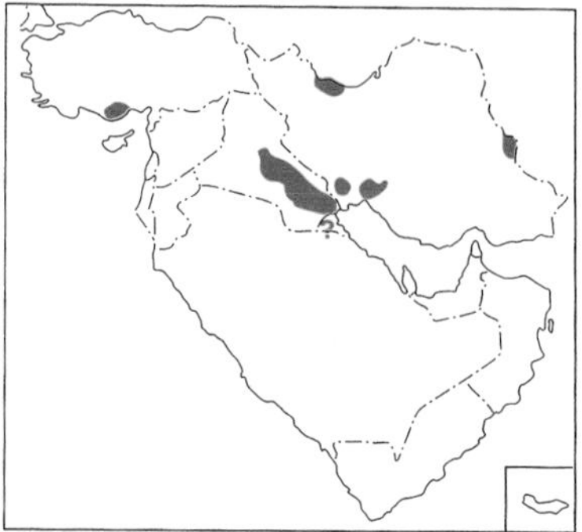

194 **Coot** *Fulica atra*

Status: sometimes breeds Cyprus; range expanding in Arabia to occupy newly created waters. Partial migrant. In winter in most breeding areas, also reaches Cyprus, Arabia south to Oman and Yemen. **Habitat:** prefers larger areas of open water than Moorhen. Packs occur on lakes, reservoirs and salt water in winter. Nests in aquatic vegetation in shallow water.

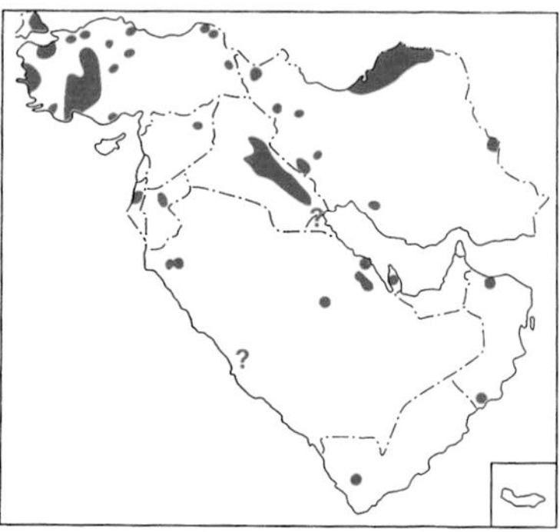

189 Moorhen
juv.
juv.
194 Coot
188 White-breasted
Waterhen
191 Allen's Gallinule
juv.
juv.
192 Purple
Gallinule

ML

195 **Common Crane** *Grus grus*

Status: partial migrant. Passage mainly Near East; a few W Arabia and vagrant E and S Arabia including UAE and Yemen. Winters Turkey, Iran, Iraq, Syria, Israel and Jordan. **Habitat:** bogs, lightly wooded swamps, reedbeds. In winter avoids wooded regions, occurring on river banks, lagoons, fields and steppes. Nests on ground, sometimes in shallow water.

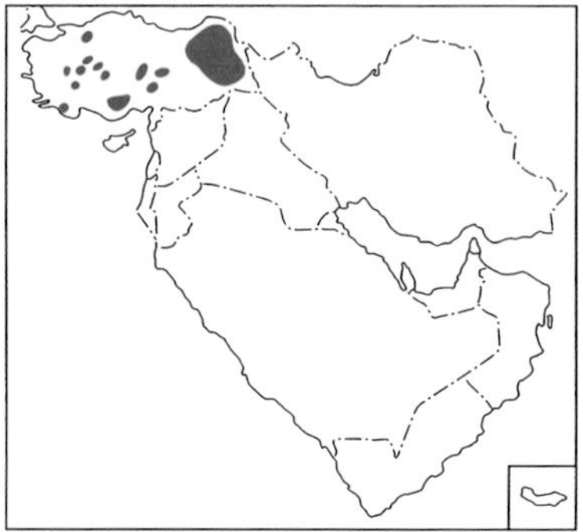

196 **Siberian White Crane** *Grus leucogeranus*

Status: rare winter visitor N Iran. Range Asia. **Habitat:** marshes, lagoons, and flooded meadows.

197 **Demoiselle Crane** *Anthropoides virgo*

Status: summer visitor. Breeding tenuous in Turkey. Passage over Cyprus and numerous W Saudi Arabia, seldom Iran. Rare E and S Arabia and Near East. **Habitat:** open or shrubby plains with access to water; will feed on arable land. In winter, fields, lake margins, sandbanks. Nests on ground.

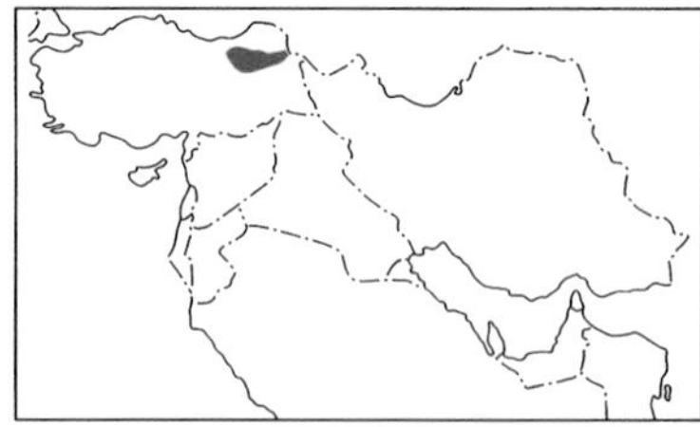

198 **Little Bustard** *Tetrax tetrax*

Status: partial migrant. Resident Iran with winter immigration. May also winter Turkey. Vagrant Cyprus, Near East, Iraq, Oman. **Habitat:** rough, grassy plains, large fields of corn or fodder crops. Nests on ground.

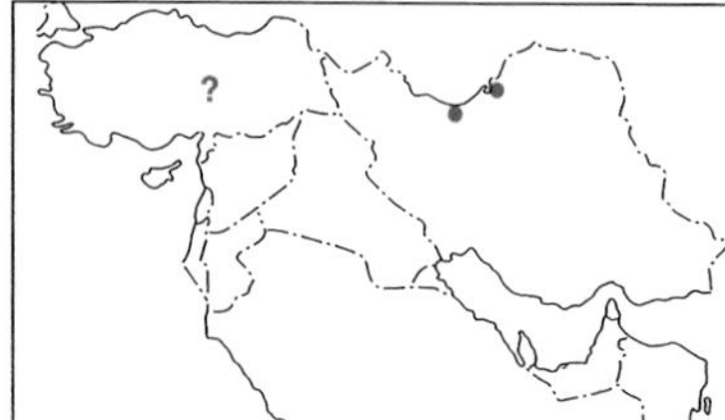

199 **Houbara Bustard** *Chlamydotis undulata*

Status: formerly widespread resident in E Turkey. Summer visitor to Iran breeding grounds; elsewhere mainly resident in declining numbers. Winter visitor in declining numbers to coastal Iran and Iraq. Rare or scarce winter visitor to Arabia away from breeding areas. Vagrant Cyprus. **Habitat:** stony or sandy steppes or semi-desert, open or with scattered shrubs and grass. Also occurs in marginal corn and other crops. Nests on ground.

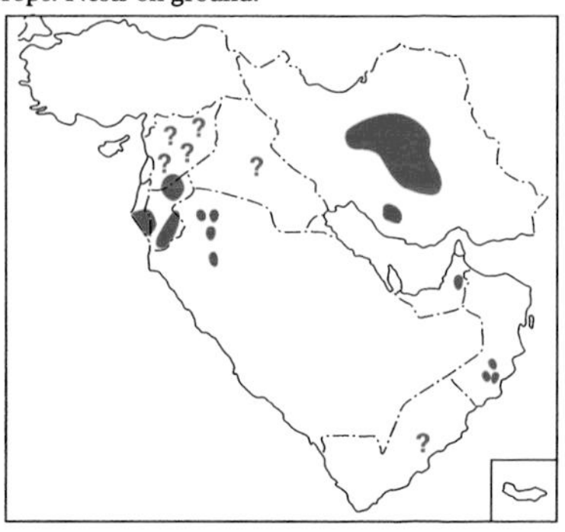

200 **Arabian Bustard** *Ardeotis arabs*

Status: resident. **Habitat:** open savanna; in Yemen largely confined to cereal fields in sandy plains. Nests on ground.

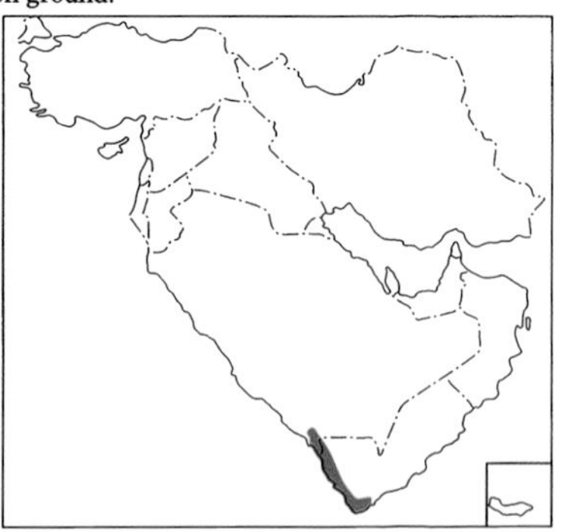

201 **Great Bustard** *Otis tarda*

Status: in winter occurs outside breeding areas in N Iran, N Iraq, N Syria, Turkey. Vagrant Cyprus, Near East. **Habitat:** open plains, grassy steppes, extensive cereal fields. Nests on ground.

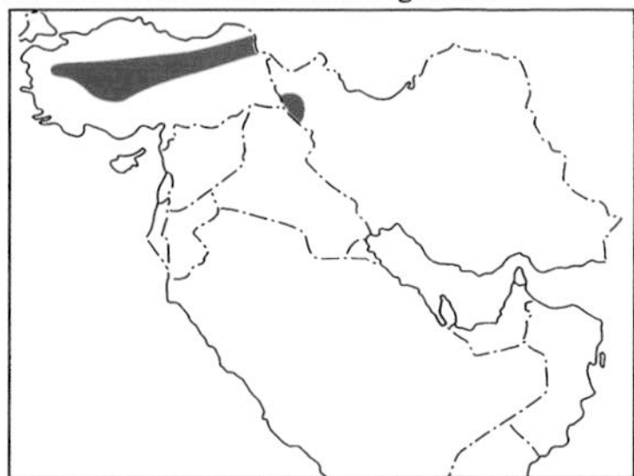

juv.
195 Common
Crane
196 Siberian
White Crane
juv.
juv.
197 Demoiselle
Crane
198 Little
Bustard
♂
♀
199 Houbara
Bustard
♂
winter
♀
200 Arabian
Bustard
♂
201 Great
Bustard
ML

202 **Pheasant-tailed Jacana**
Hydrophasianus chirurgus

Status: passage migrant and winter visitor (from Indian sub-continent) Oman, where occasionally over-summers (one breeding record). Vagrant Yemen. **Habitat:** swamps, ponds, creeks and marshes with patches of open water and floating vegetation.

203 **Painted Snipe** *Rostratula benghalensis*

Status: has bred Israel where also occasional in winter. Vagrant Oman and Yemen. **Habitat:** well vegetated swamps with muddy patches, reedbeds, and will feed on open fields.

204 **Oystercatcher** *Haematopus ostralegus*

Status: mainly summer visitor to breeding areas. Winters coasts of Arabia, Iran, Iraq and a few S Turkey. Slight passage several parts E Mediterranean. Some non-breeders spend summer in wintering areas. Vagrant Jordan and Syria. **Habitat:** margins of lakes and rivers; in winter mainly coastal. Nests on ground in open.

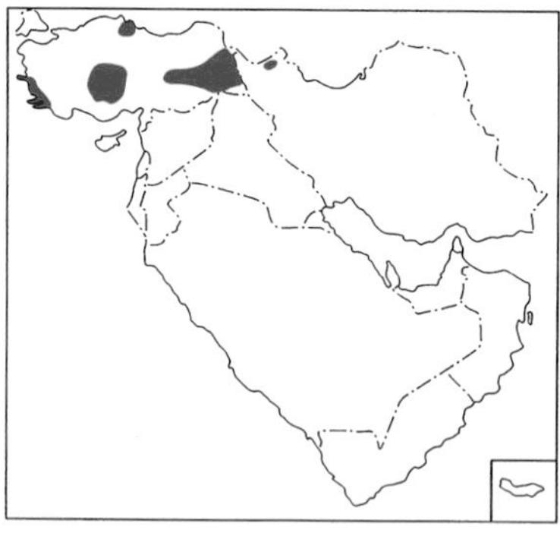

205 **Black-winged Stilt**
Himantopus himantopus

Status: mainly summer visitor Turkey, N Iran, otherwise resident or partial migrant in breeding areas. Irregular breeder Cyprus. Widespread on passage, both coastal and inland. Winters Near East, S Iraq and Arabia. **Habitat:** shallow freshwater or brackish marshes, lagoons, estuaries. Nests on ground, usually in open.

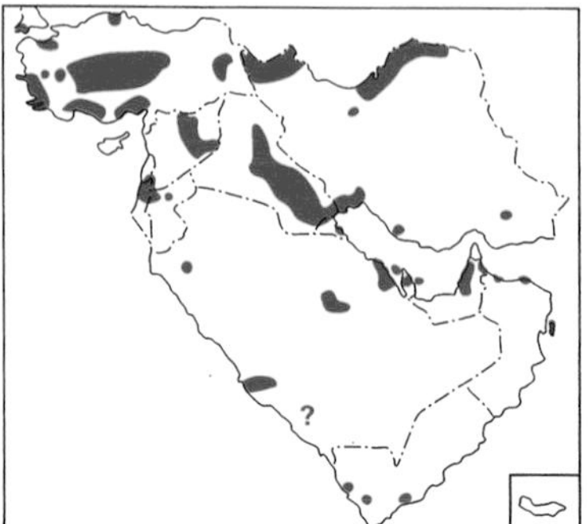

206 **Avocet** *Recurvirostra avosetta*

Status: partial migrant. Summer visitor central and E Turkey. Winters most of E Mediterranean countries, Iraq marshes, Iran; scarce on all coasts of Arabia with little inland passage. Some over-summer in S Arabia. **Habitat:** exposed, often saline, mudflats, estuaries and sandbanks. Breeds colonially near shallow water.

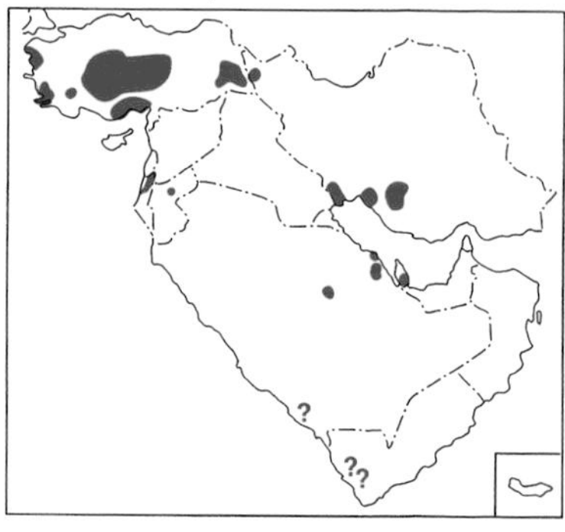

208 **Crab Plover** *Dromas ardeola*

Status: partial migrant, or dispersive. Present all year in parts of Arabian Gulf (including Kuwait where may still breed on off-shire islands), coasts of Gulf of Oman, Arabian Sea and S Red Sea. Vagrant Turkey. **Habitat:** sea-coasts, mudflats, coral reefs; never inland. Nests in tunnel excavated in sandy ground; colonial.

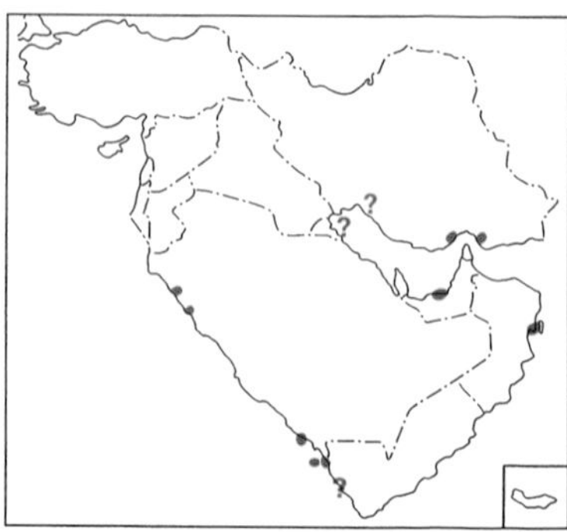

winter
202 Pheasant-tailed Jacana
juv.
summer
juv.
♂
203 Painted Snipe
♀
♀
204 Oystercatcher
♂
206 Avocet
♀
205 Black-winged Stilt
♂
juv.
208 Crab Plover

ML

209 **Stone Curlew** *Burhinus oedicnemus*

Status: partial migrant; mainly summer visitor in Turkey, Iran, N Iraq. In Arabia uncommon or irregular passage and winter in all areas. **Habitat:** stony, sandy, or dry mud, open plains or slopes with scant vegetation, including steppe, semi-desert and, locally, extensive arable; occasionally among scattered trees and light thorn scrub in hotter climates. Nests on ground.

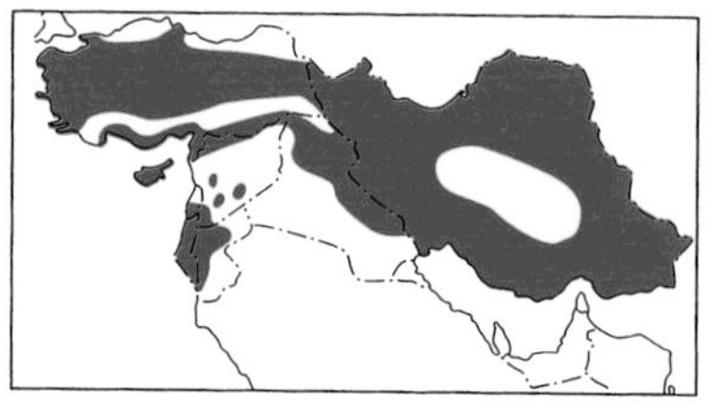

210 **Senegal Thick-knee** *Burhinus senegalensis*

Status: vagrant Saudi Arabia. **Habitat:** rarely far from water; savannas, wadi-beds, occasionally on coast.

211 **Spotted Thick-knee** *Burhinus capensis*

Status: resident. **Habitat:** savannas and scrub, rocky river beds, broken ground, more bushy than Stone Curlew frequents. Nests on ground, usually near cover of bushes.

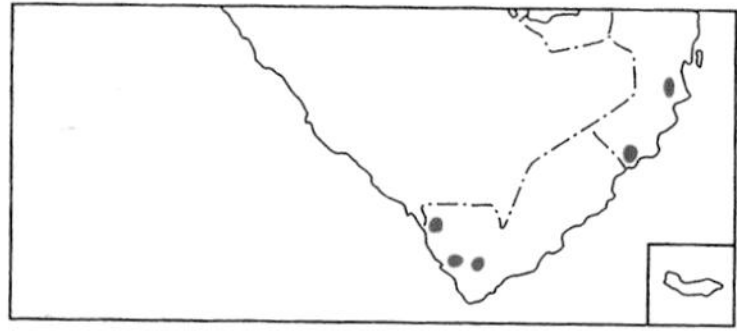

212 **Great Stone Plover** *Esacus recurvirostris*

Status: resident. Vagrant Oman. **Habitat:** rocky river beds and bars and their barren environs, also coastal reefs and rocky beaches, estuaries and salt-pans. Nests on ground, unsheltered.

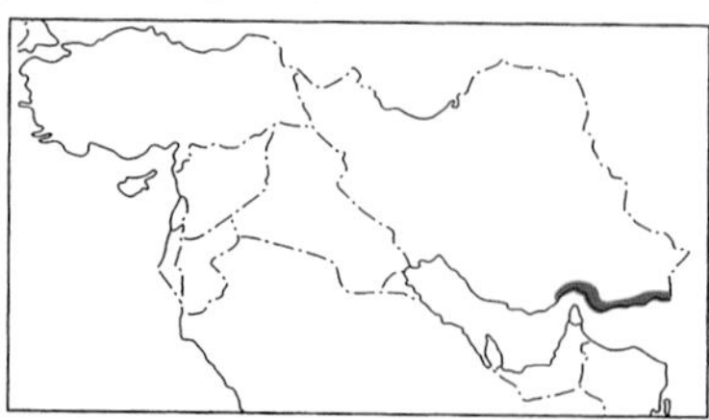

213 **Cream-coloured Courser**
Cursorius cursor

Status: may be more widespread as a breeding bird in Arabia. Partial migrant. In winter, breeding areas are vacated in Iran, N Iraq, Syria and Jordan and it becomes more widespread in Arabia, a few passage or wintering birds reaching Oman, Yemen. Passage E and SE Turkey. Rare Cyprus. **Habitat:** sandy or stony semi-desert with thin, scanty vegetation, marginal cultivation, arid flat rolling, but not broken, country. Nests on ground in open.

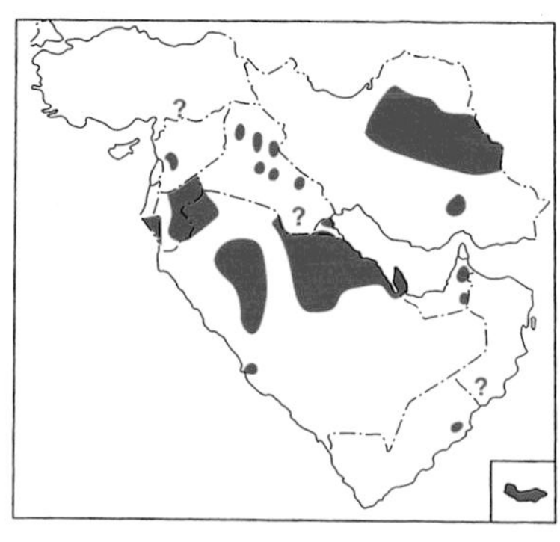

214 **Collared Pratincole** *Glareola pratincola*

Status: probably breeds more extensively in Turkey. Summer visitor. Has bred Cyprus. Passage outside breeding areas includes Cyprus, Syria, all Arabia. **Habitat:** sun-baked mudflats and flat, firm plains with low vegetation, often near water. Breeds colonially on ground.

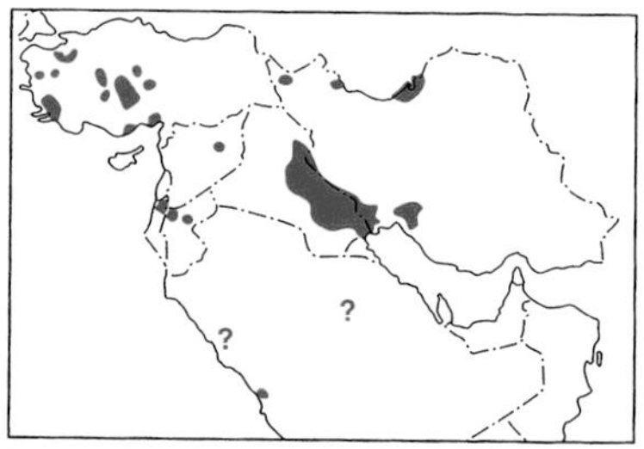

215 **Oriental Pratincole** *Glareola maldivarum*

Status: vagrant Cyprus and UAE. Range S and E Asia. **Habitat:** as Collared.

216 **Black-winged Pratincole**
Glareola nordmanni

Status: scarce passage migrant Turkey, Iran, Cyprus, Near East, but scarcer or vagrant Syria, Jordan, Bahrain, UAE, Yemen. **Habitat:** much as Collared, but favours steppes.

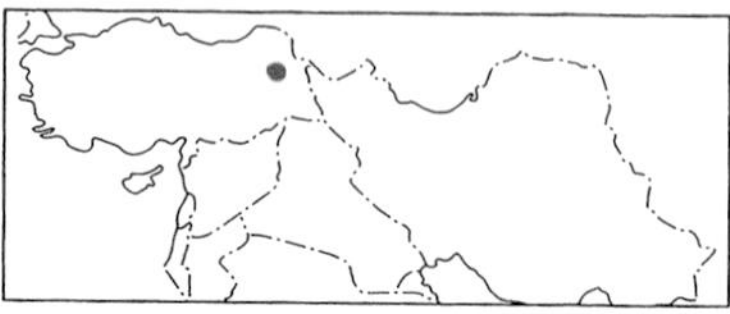

217 **Little Pratincole** *Glareola lactea*

Status: scarce winter visitor Oman; vagrant Bahrain, UAE and Yemen. Range includes Indian subcontinent. **Habitat:** wide sandbanks of rivers, or mudflats by lakes, hawking over wetlands.

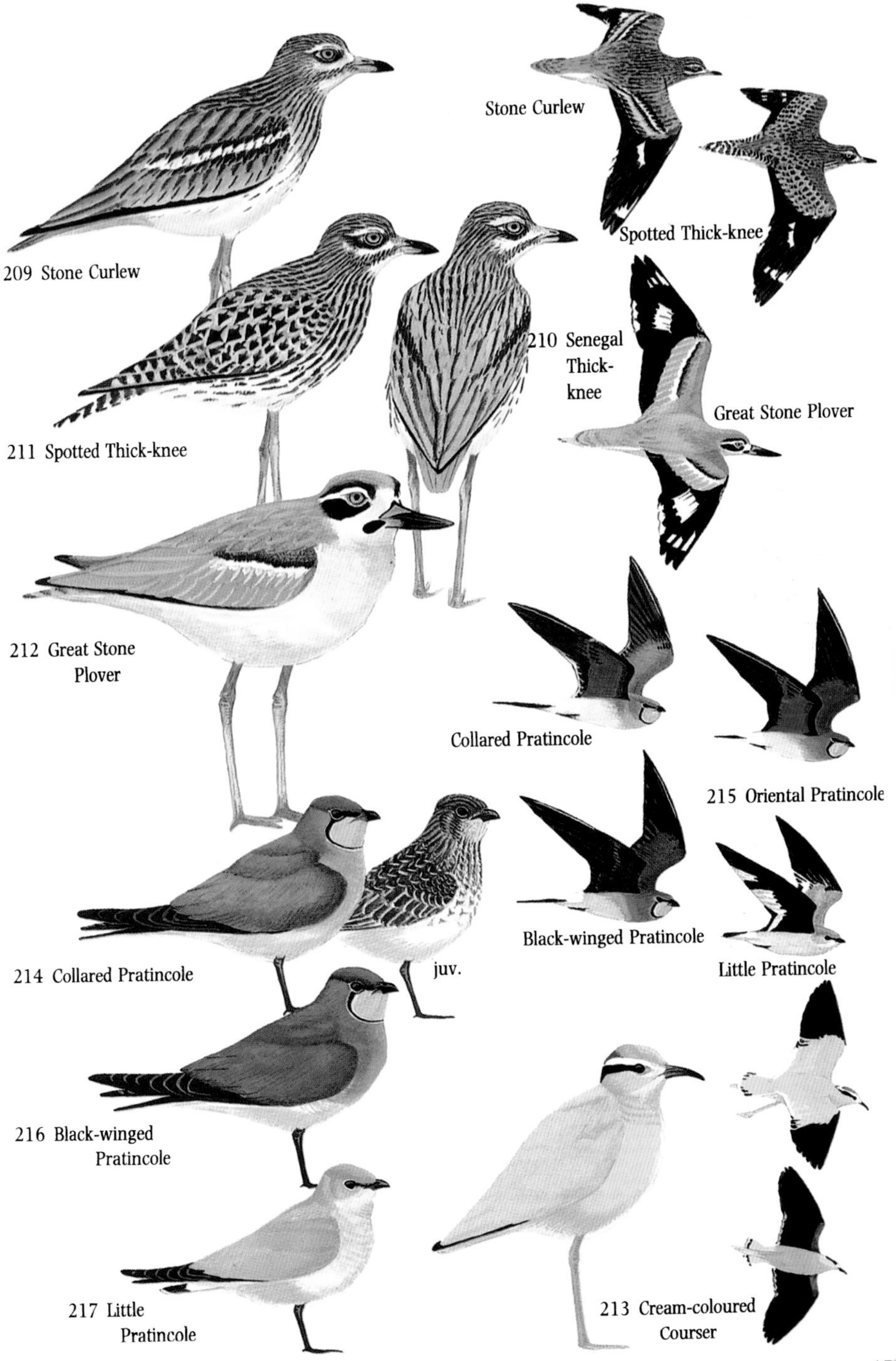
Stone Curlew
Spotted Thick-knee
209 Stone Curlew
210 Senegal Thick-knee
Great Stone Plover
211 Spotted Thick-knee
212 Great Stone Plover
Collared Pratincole
215 Oriental Pratincole
Black-winged Pratincole
Little Pratincole
214 Collared Pratincole
juv.
216 Black-winged Pratincole
217 Little Pratincole
213 Cream-coloured Courser
AB

218 **Little Ringed Plover** *Charadrius dubius*

Status: has bred Cyprus. Widely scattered throughout region on passage. Winters in small numbers southwards from Turkey (rare) and Arabian Gulf. **Habitat:** mainly freshwater localities, in particular gravelly river islands and sandy borders of lakes; also some brackish areas. In winter coastal as well as inland. Nests on ground.

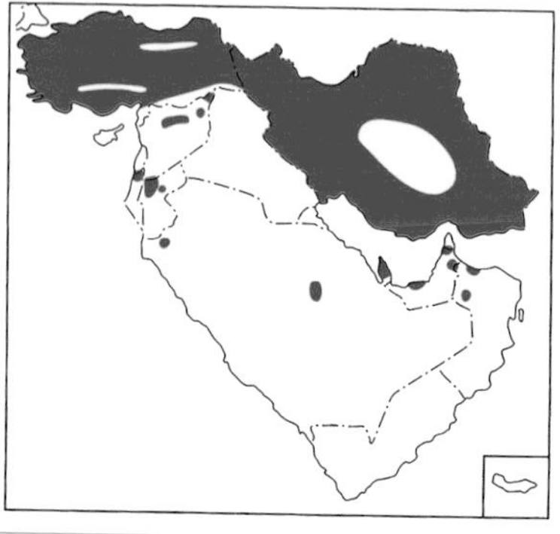

219 **Ringed Plover** *Charadrius hiaticula*

Status: widespread passage migrant, relatively uncommon inland. Winters on nearly all coasts in the region, more widely on some inland sites. A few remain in summer, for example Oman and UAE. **Habitat:** sandy, muddy and stony shores, both coastal and inland.

220 **Kittlitz's Plover** *Charadrius pecuarius*

Status: regular winter visitor in small numbers to Israel. Vagrant Cyprus, Bahrain and UAE. **Habitat:** muddy margins of lakes, rivers and other wetland areas.

221 **Kentish Plover** *Charadrius alexandrinus*

Status: partial migrant or dispersive, seen along all coasts. In winter, a large degree of overlap with breeding range, but it extends inland in SW Arabia, while vacating central Turkey, and inland N Iran. **Habitat:** shingle, sandy and muddy beaches and dry mudflats, mainly coastal but also inland, often by saline lagoons. Nests on ground.

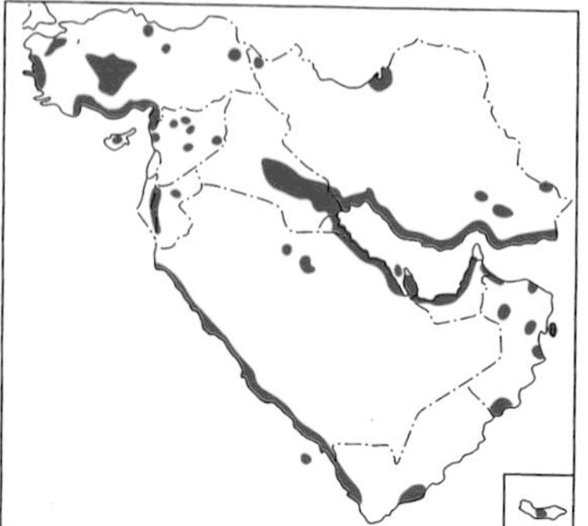

222 **Lesser Sand Plover** *Charadrius mongolus*

Status: winter and passage visitor to coasts of S Iran and Arabia. Passage central Saudi Arabia. Some over-summer E Arabia. Vagrant Near East, Turkey. **Habitat:** tidal mudflats and sandy coasts.

223 **Greater Sand Plover**
Charadrius leschenaultii

Status: summer visitor to breeding areas. Winter and passage on coasts of S Iran, Arabia, and Near East north to Cyprus; some remaining in summer. Also passage central Saudi Arabia. **Habitat:** in the region breeds on inland sand- and mudflats, mainly near water, nesting on ground. Otherwise essentially coastal, on intertidal flats and dry foreshore.

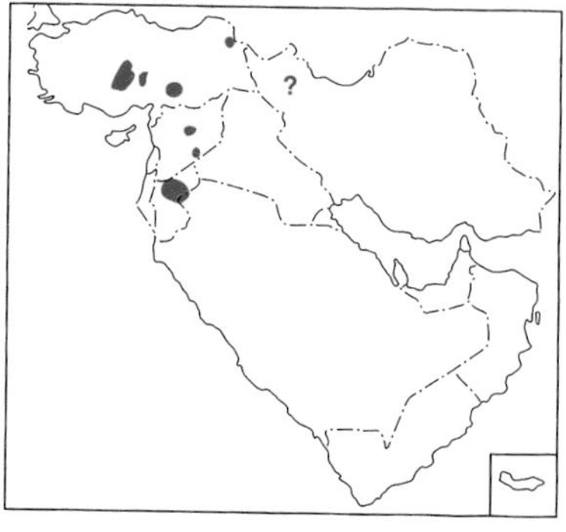

224 **Caspian Plover** *Charadrius asiaticus*

Status: passage migrant Iran, Iraq, Arabia (scarce in S), fewer Israel and Jordan, rare UAE, E Turkey. Has occurred in winter Iraq, Gulf of Aqaba and Yemen. Vagrant Cyprus, Syria. **Habitat:** fields, grassy plains, semi-desert; also coastal areas.

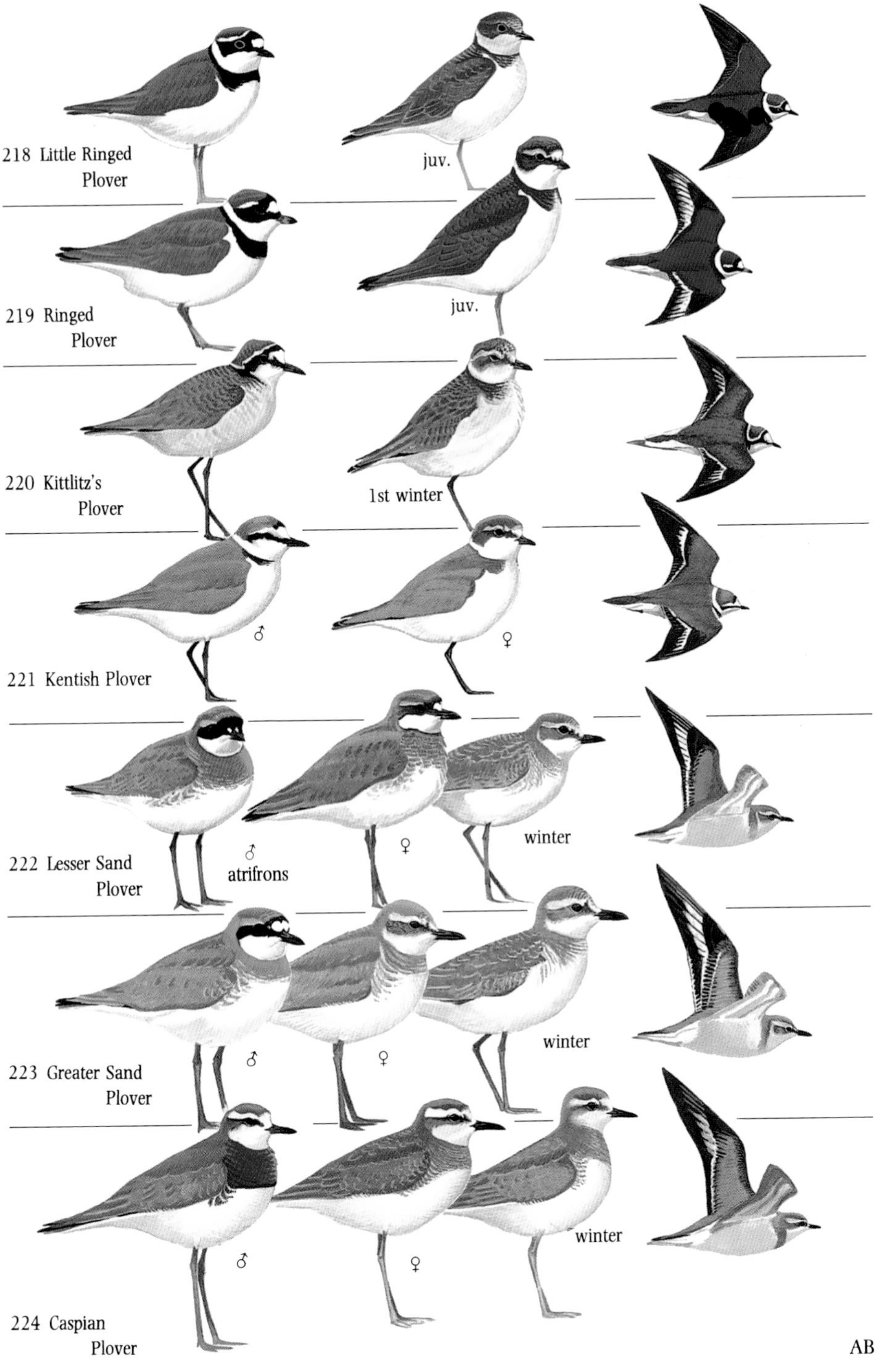
218 Little Ringed Plover
juv.
219 Ringed Plover
juv.
220 Kittlitz's Plover
1st winter
221 Kentish Plover
♂
♀
222 Lesser Sand Plover
♂ atrifrons
♀
winter
223 Greater Sand Plover
♂
♀
winter
224 Caspian Plover
♂
♀
winter
AB

225 **Dotterel** *Charadrius morinellus*

Status: winter visitor Near East, Iraq, E and NE Saudi Arabia and W Iran. Rare Gulf States and Cyprus. Vagrant E Arabia. Also passage Turkey. **Habitat:** stony steppe, semi-desert and poor arable land often far from water.

---

226 **American Golden Plover** *Pluvialis dominica*

Status: vagrant Turkey. **Habitat:** in winter on open short-grass fields, cultivated farmland, wetlands and mudflats, also sandy beaches and savanna ponds.

---

227 **Pacific Golden Plover** *Pluvialis fulva*

Status: winter or passage Arabian Gulf, S and E Arabia including Yemen and Oman. Vagrant Iran, Iraq, Israel, Jordan, Turkey and Cyprus. **Habitat:** more often mudflats and coasts than European Golden; also open grassland, cultivated fields.

---

228 **European Golden Plover** *Pluvialis apricaria*

Status: winter visitor to most E Mediterranean countries especially Turkey, scarcer N Iran and Syria; rare or vagrant Lebanon, Iraq, Arabia, Gulf States and Yemen but perhaps some confusion with Pacific Golden Plover. **Habitat:** dry and wet grassland, plough and stubble, and sometimes shores.

---

229 **Grey Plover** *Pluvialis squatarola*

Status: winter and passage to coasts of most E Mediterranean countries and coasts of Arabia and Iran. Some non-breeders remain through summer. **Habitat:** coastal, on intertidal flats and saltings; occasionally inland around shallow lakes.

---

230 **Spur-winged Plover** *Hoplopterus spinosus*

Status: summer visitor Turkey and Syria. Farther south mainly resident or dispersive, with winter and breeding ranges very similar. Passage Cyprus (has bred), less commonly Saudi Arabia. Vagrant Oman, Iran. **Habitat:** edges of fresh and saline marshes and irrigation, with short vegetation, nesting on dried-out mud or nearby sandy areas.

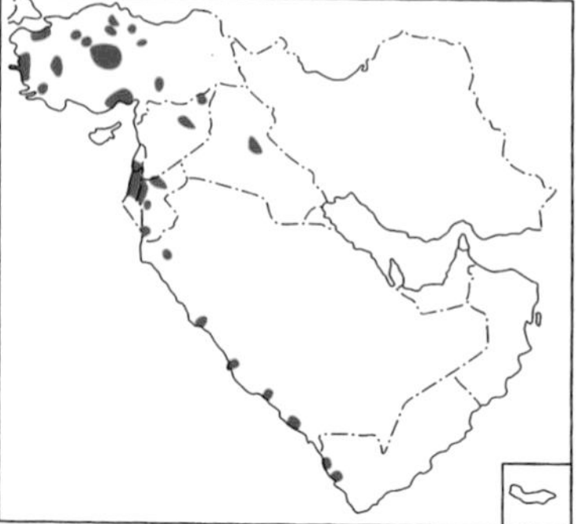

---

231 **Black-headed Plover** *Hoplopterus tectus*

Status: old record from Jordan; vagrant Israel. Range Africa. **Habitat:** dry thorn-bush country with open spaces.

---

232 **Red-wattled Plover** *Hoplopterus indicus*

Status: resident and dispersive with scattered records in Gulf States away from breeding areas. Vagrant central Saudi Arabia, Qatar, Syria and Israel. **Habitat:** open country, usually near fresh water, and showing preference for agricultural land. Nests on ground.

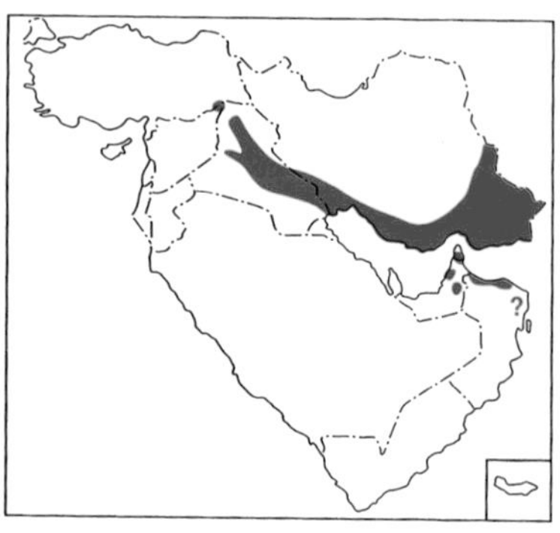

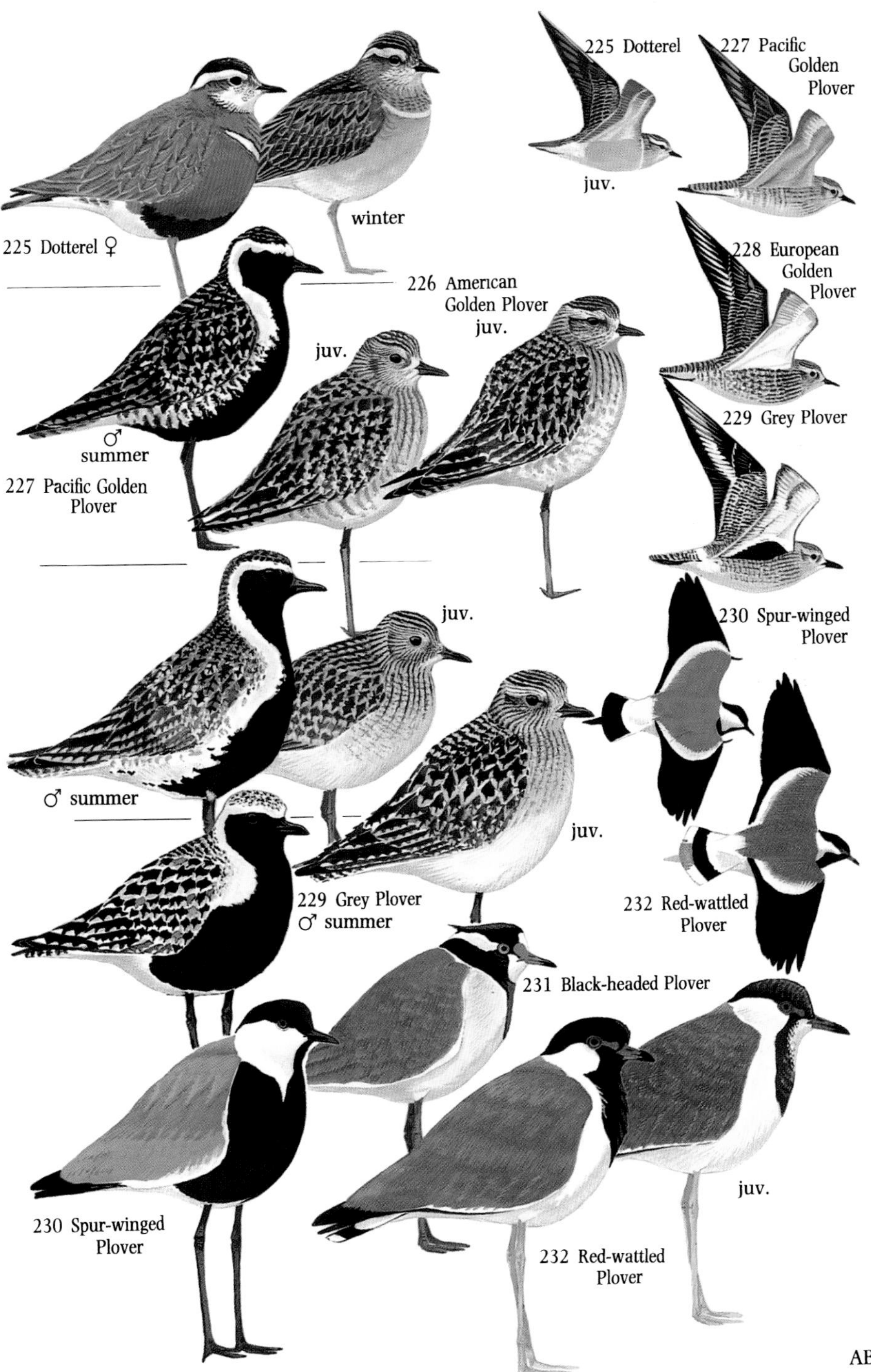
225 Dotterel
juv.
227 Pacific Golden Plover
225 Dotterel ♀
winter
228 European Golden Plover
226 American Golden Plover juv.
juv.
229 Grey Plover
♂ summer
227 Pacific Golden Plover
juv.
230 Spur-winged Plover
♂ summer
juv.
229 Grey Plover ♂ summer
232 Red-wattled Plover
231 Black-headed Plover
juv.
230 Spur-winged Plover
232 Red-wattled Plover
AB

233 **Sociable Plover** *Chettusia gregaria*

Status: regular winter visitor to Iraq and Israel; scarce Saudi Arabia, Oman, UAE, Bahrain and Qatar. Vagrant Yemen. Passage Turkey (rare), Iran, Near East. **Habitat:** semi-deserts, steppes and bare or cultivated fields (e.g. with winter cereals); rare on coast.

234 **White-tailed Plover** *Chettusia leucura*

Status: partial migrant; summer visitor Turkey, N Iran, central Iraq, summers on Arabian Gulf; a few winter E Arabia and Yemen. Scattered passage throughout region. Rare Cyprus. **Habitat:** fresh or saline pools, marshes and wet plains. Nests on ground, sometimes colonially.

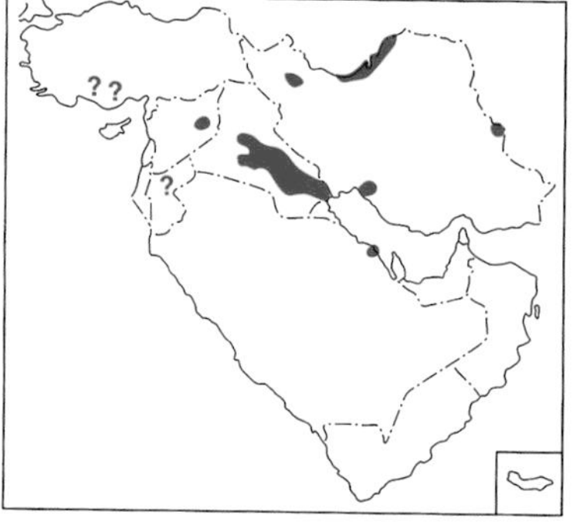

235 **Lapwing** *Vanellus vanellus*

Status: partial migrant. Winters Turkey, Iran, rarer south to central Arabia and Oman. **Habitat:** open fields, marshes, shallow pools and coastal flats. Nests on ground.

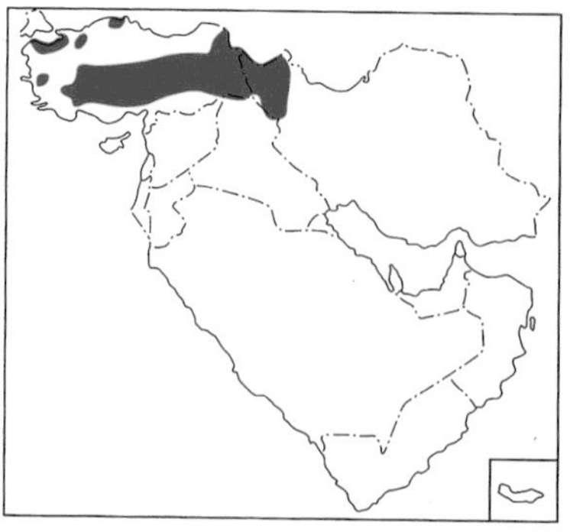

236 **Great Knot** *Calidris tenuirostris*

Status: scarce migrant and winter visitor to E Arabian Gulf south to central Oman. Vagrant Israel and Bahrain. (Migrant from Asia to Australia). **Habitat:** coastal mudflats.

237 **Knot** *Calidris canutus*

Status: scarce passage migrant Turkey. Rare visitor to Near East, Iran. Vagrant Jordan, E Arabia and Yemen. **Habitat:** coastal mudflats.

238 **Sanderling** *Calidris alba*

Status: passage and winter visitor to most coasts; occasionally inland. Vagrant Syria, Lebanon, Jordan. **Habitat:** sandy beaches and wetlands.

244 **Pectoral Sandpiper** *Calidris melanotos*

Status: vagrant Israel. **Habitat:** marshes, pools with grassy edges.

245 **Sharp-tailed Sandpiper** *Calidris acuminata*

Status: vagrant Yemen. Range Asia/Australia. **Habitat:** as Pectoral Sandpiper.

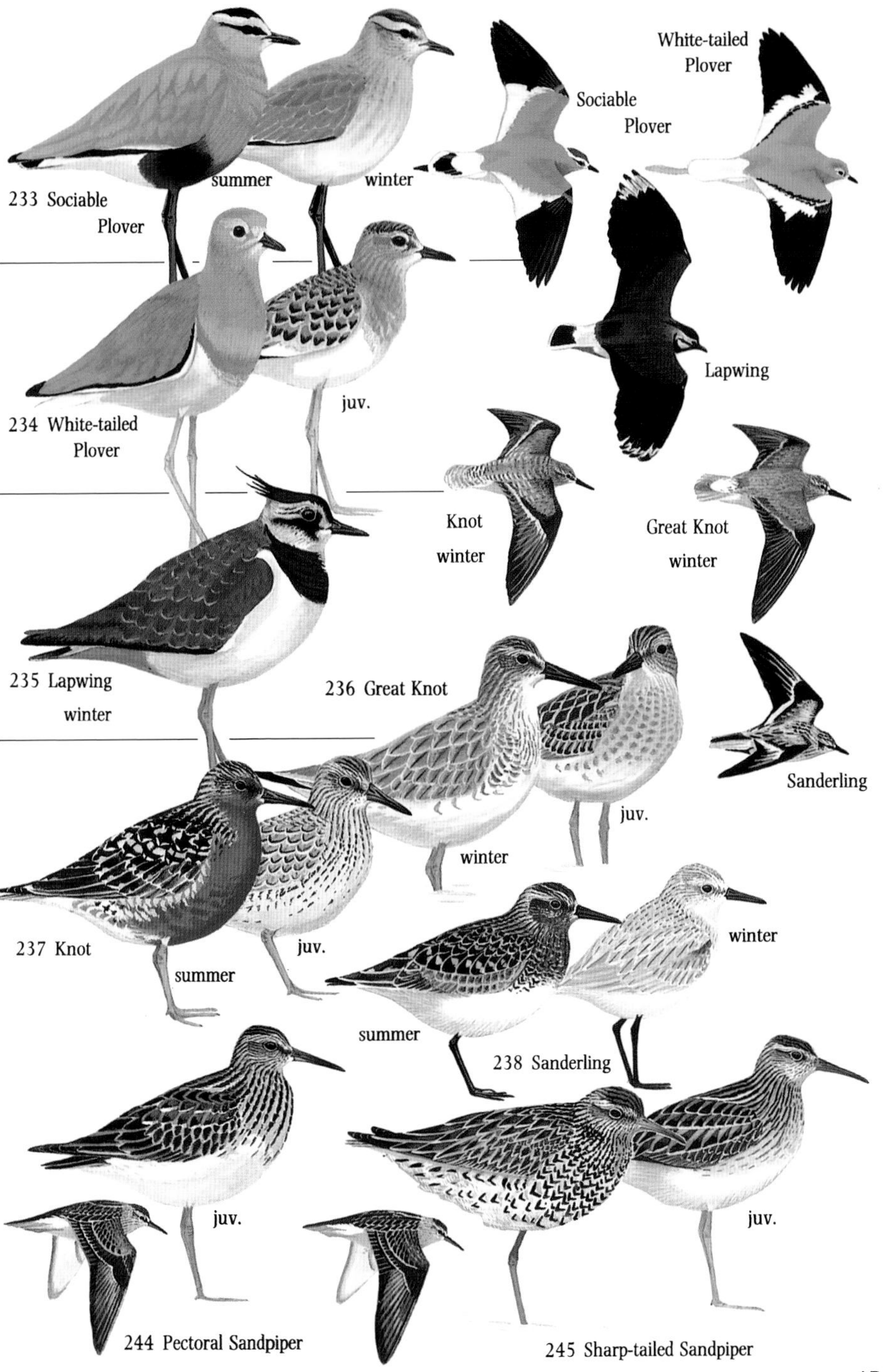
233 Sociable Plover
summer
winter
Sociable Plover
White-tailed Plover
234 White-tailed Plover
juv.
Lapwing
235 Lapwing winter
Knot winter
Great Knot winter
236 Great Knot
winter
juv.
Sanderling
237 Knot
summer
juv.
238 Sanderling
summer
winter
juv.
juv.
244 Pectoral Sandpiper
245 Sharp-tailed Sandpiper

AB

240 **Little Stint** *Calidris minuta*

Status: passage through whole region; winters mainly in Arabia. **Habitat:** coastal flats and inland wetlands.

241 **Temmick's Stint** *Calidris temminckii*

Status: passage through most of region. Winters Near East Iraq, SW Iran and Arabia. **Habitat:** in more muddy coastal areas than Little Stint and also more often on inland pools and marshes.

242 **Long-toed Stint** *Calidris subminuta*

Status: rare passage and winter UAE, and Oman. Vagrant Bahrain, Saudi Arabia, Iran, Yemen and Israel. From Asia/Australia. **Habitat:** freshwater margins; coastal pools and flats.

246 **Curlew Sandpiper** *Calidris ferruginea*

Status: passage through most of region; winters most coasts of Arabia but few Israel; occasional in summer. **Habitat:** similar to Dunlin.

247 **Dunlin** *Calidris alpina*

Status: passage and winter to all coasts of the region; occasional in summer. **Habitat:** coastal mudflats and estuaries; also inland wetlands.

248 **Broad-billed Sandpiper** *Limicola falcinellus*

Status: passage throughout region, but often uncommon. Winters S Iran, Bahrain, UAE, Oman, Yemen. Vagrant Cyprus, Jordan. **Habitat:** as Dunlin.

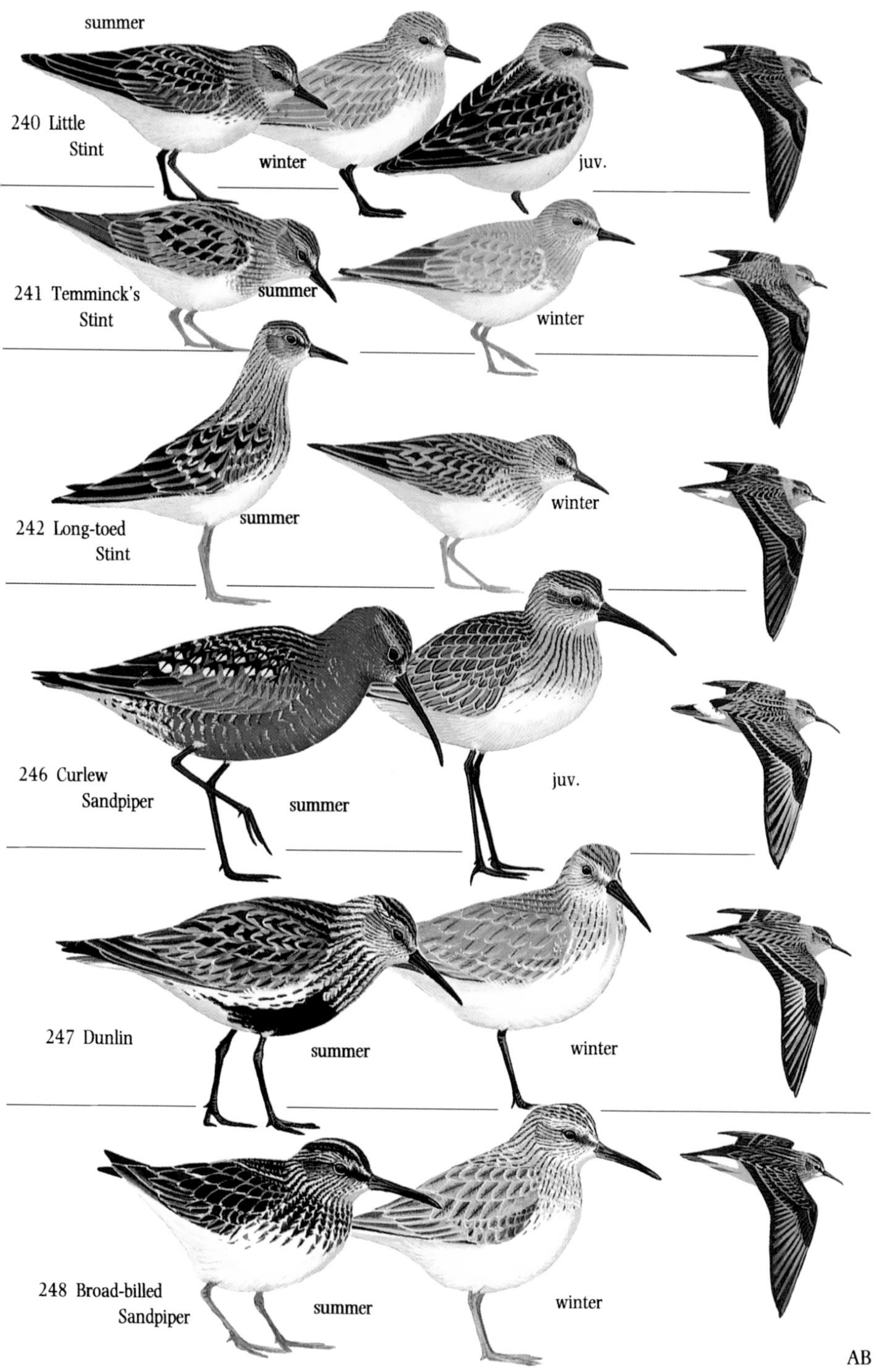
summer
240 Little
Stint
winter
juv.
241 Temminck's
Stint
summer
winter
242 Long-toed
Stint
summer
winter
246 Curlew
Sandpiper
summer
juv.
247 Dunlin
summer
winter
248 Broad-billed
Sandpiper
summer
winter

AB

251 **Jack Snipe** *Lymnocryptes minimus*

Status: passage and winter visitor in very small numbers to most countries in the region, including Oman. Vagrant Syria, Yemen. **Habitat:** much as Common Snipe.

252 **Common Snipe** *Gallinago gallinago*

Status: passage and winter visitor to most parts of the region including S Arabia. **Habitat:** wet grasslands, ricefields, marshy water-margins.

253 **Great Snipe** *Gallinago media*

Status: generally scarce passage migrant Turkey/Iran south to Israel; rare in Arabia but locally scarce Oman. Occasional winter records include Cyprus, Yemen. **Habitat:** drier localities than Common Snipe such as stubble fields, rough grassland; also marshes.

254 **Pintail Snipe** *Gallinago stenura*

Status: vagrant or rare passage/winter visitor recorded Israel, E Iran, Arabian Gulf, Oman (regular), Yemen. Range Asia. **Habitat:** wet open fields, damp boggy areas, water-margins but more often on dryer ground than Common Snipe.

255 **Solitary Snipe** *Gallinago solitaria*

Status: may nest in NE Iran, and reported as migrant in small numbers E Iran. Vagrant central Saudi Arabia. **Habitat:** alpine and subalpine zones above 2000 m; upper limit of moist forest and mountain streams. Descends in winter to lower levels, frequenting scrubby upland bogs and ricefields.

256 **Long-billed Dowitcher** *Limnodromus scolopaceus*

Status: vagrant Oman. **Habitat:** muddy freshwater pools with marginal vegetation, much as Common Snipe.

257 **Asiatic Dowitcher** *Limnodromus semipalmatus*

Status: reported Yemen. Range Asia/Australia. **Habitat:** mudflats, sandbanks and estuaries.

258 **Woodcock** *Scolopax rusticola*

Status: winter visitor to Turkey, Iran and Near East. Vagrant Arabian Gulf and N Arabia. **Habitat:** moist woodland, scrub, thick cover.

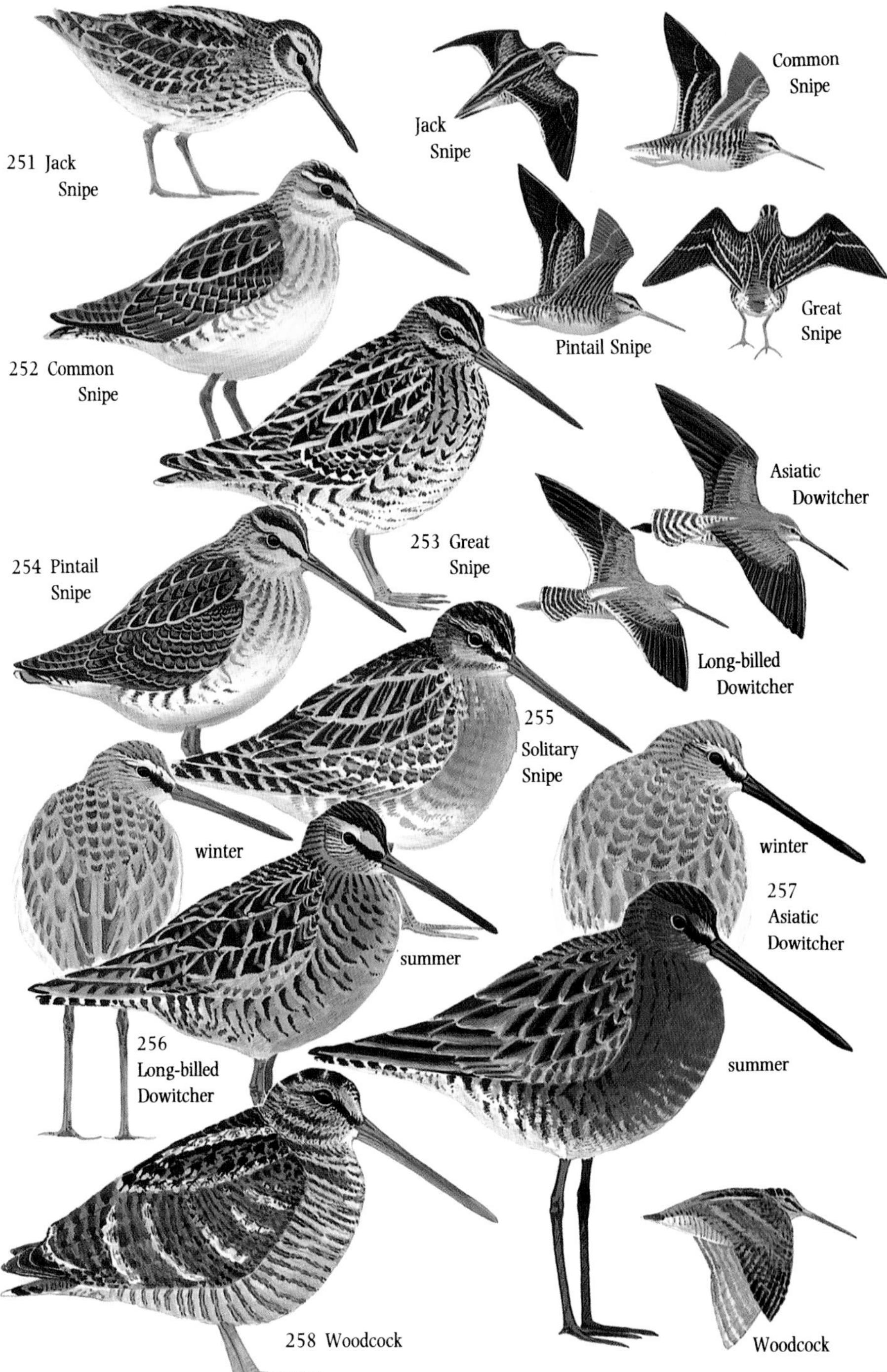
251 Jack Snipe
Jack Snipe
Common Snipe
252 Common Snipe
Pintail Snipe
Great Snipe
253 Great Snipe
Asiatic Dowitcher
254 Pintail Snipe
Long-billed Dowitcher
255 Solitary Snipe
winter
winter
257 Asiatic Dowitcher
summer
256 Long-billed Dowitcher
summer
258 Woodcock
Woodcock
AB

259 **Black-tailed Godwit** *Limosa limosa*

Status: passage/winter visitor to most of region but less common E Arabia. Some non-breeders over-summer. **Habitat:** muddy freshwater margins, floodlands, marshes, also estuaries and tidal creeks.

---

260 **Bar-tailed Godwit** *Limosa lapponica*

Status: occurs on passage and in winter throughout region but scarcer in Turkey and Near East. Vagrant Syria. **Habitat:** usually coastal, on mudflats, sandy beaches or estuaries.

---

261 **Whimbrel** *Numenius phaeopus*

Status: winters Arabian Gulf to E and SW Arabia, rarer elsewhere. More numerous and widespread on passage, but less common Turkey. Few remain summer. Vagrant Jordan, Lebanon. Unrecorded Syria. **Habitat:** muddy and sandy beaches, creeks, rocky shores, coral reefs.

262 **Slender-billed Curlew** *Numenius tenuirostris*

Status: very rare or vagrant Turkey, Jordan, Iraq, Kuwait, Oman, Yemen and Israel. **Habitat:** mudflats, shores and margins of freshwater pools.

---

263 **Curlew** *Numenius arquata*

Status: winter and passage visitor to most coasts in the region, also some inland wetlands; some non-breeders over-summer. Vagrant Syria. **Habitat:** mudflats, tidal sands and rocks, coral reefs and mangrove swamps; also inland on muddy or grassy margins of rivers and other waters.

---

264 **Far Eastern Curlew** *Numenius madagascariensis*

Status: may occasionally occur in E Iran, but no recent records. Range Asia/Australia. **Habitat:** marshes, sea-coasts and mudflats.

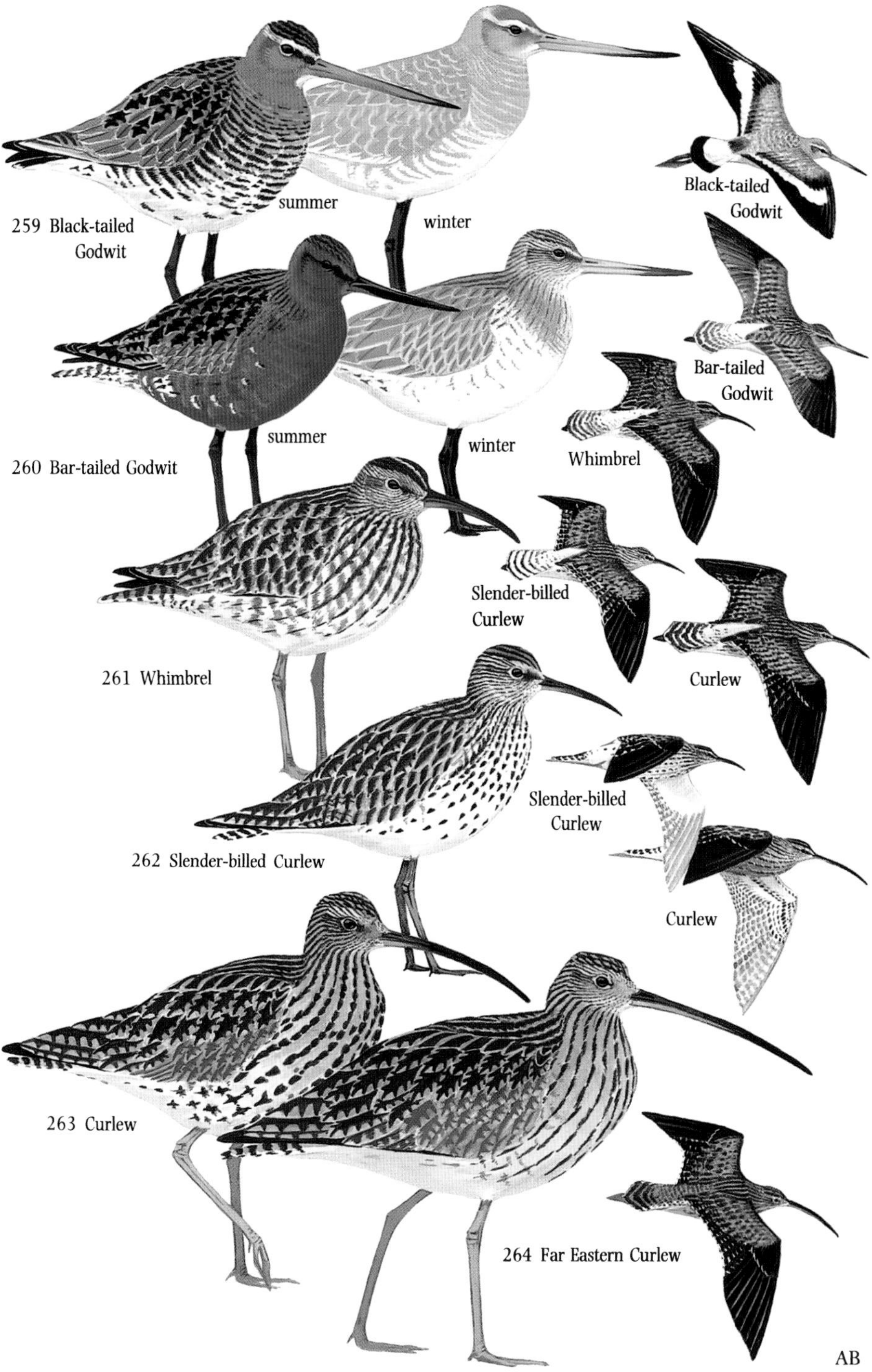

summer
winter
259 Black-tailed Godwit
Black-tailed Godwit
summer
winter
260 Bar-tailed Godwit
Bar-tailed Godwit
Whimbrel
261 Whimbrel
Slender-billed Curlew
Curlew
262 Slender-billed Curlew
Slender-billed Curlew
Curlew
263 Curlew
264 Far Eastern Curlew
AB

250 **Ruff** *Philomachus pugnax*

Status: passage and winter visitor to nearly all countries in the region, but not wintering Cyprus, N Turkey, N Iran. Some non-breeders over-summer. **Habitat:** inland marshes, lake shores, occasionally estuaries.

265 **Spotted Redshank** *Tringa erythropus*

Status: on passage widespread on most coasts, also inland marshes. Uncommon or scarce Arabia and a few coastal Iran. Scarcer in winter but range still wide. Occasional over-summering. **Habitat:** as Redshank, but less coastal.

266 **Redshank** *Tringa totanus*

Status: widespread passage migrant and winter visitor throughout region, occurring most coasts and some inland marshes; some over-summering. **Habitat:** breeds inland in damp grassland or short sedges near open fresh or saline waters. Nests on ground. Winters mainly on coast on sheltered shores, mudflats, creeks; also inland on muddy shallow watersides without much vegetation.

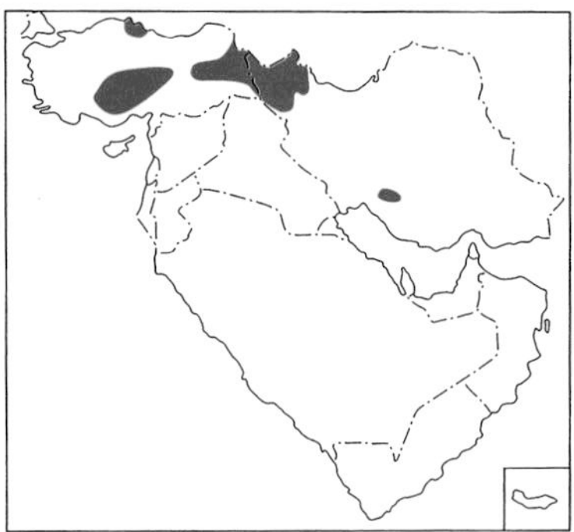

267 **Marsh Sandpiper** *Tringa stagnatilis*

Status: widely scattered on passage, mainly in small numbers. Fewer in winter generally, when rare north of Arabian Gulf and Iraq marshes. **Habitat:** swampy fresh waters and marshes; also, to lesser extent, tidal creeks and flats.

268 **Greenshank** *Tringa nebularia*

Status: occurs throughout the region. Mainly passage in Mediterranean countries, also Iraq, Iran and Arabia; less common but still widespread in winter. Few non-breeders over-summer. **Habitat:** much as Redshank.

269 **Green Sandpiper** *Tringa ochropus*

Status: broad-fronted passage, and winter visitor to fresh waters throughout the region. Seldom concentrated, but sometimes flocks Arabia in autumn. **Habitat:** muddy banks of streams and channels, edges of floodlands, in marshes may use mud patches smaller than most waders tolerate, often with sheltering banks; sometimes brackish creeks.

270 **Wood Sandpiper** *Tringa glareola*

Status: widespread passage across the whole region. Small numbers winter, but only rarely in NE Mediterranean countries. **Habitat:** freshwater marshy areas with muddy margins, less confined than sometimes chosen by Green Sandpiper; also at times saltmarsh.

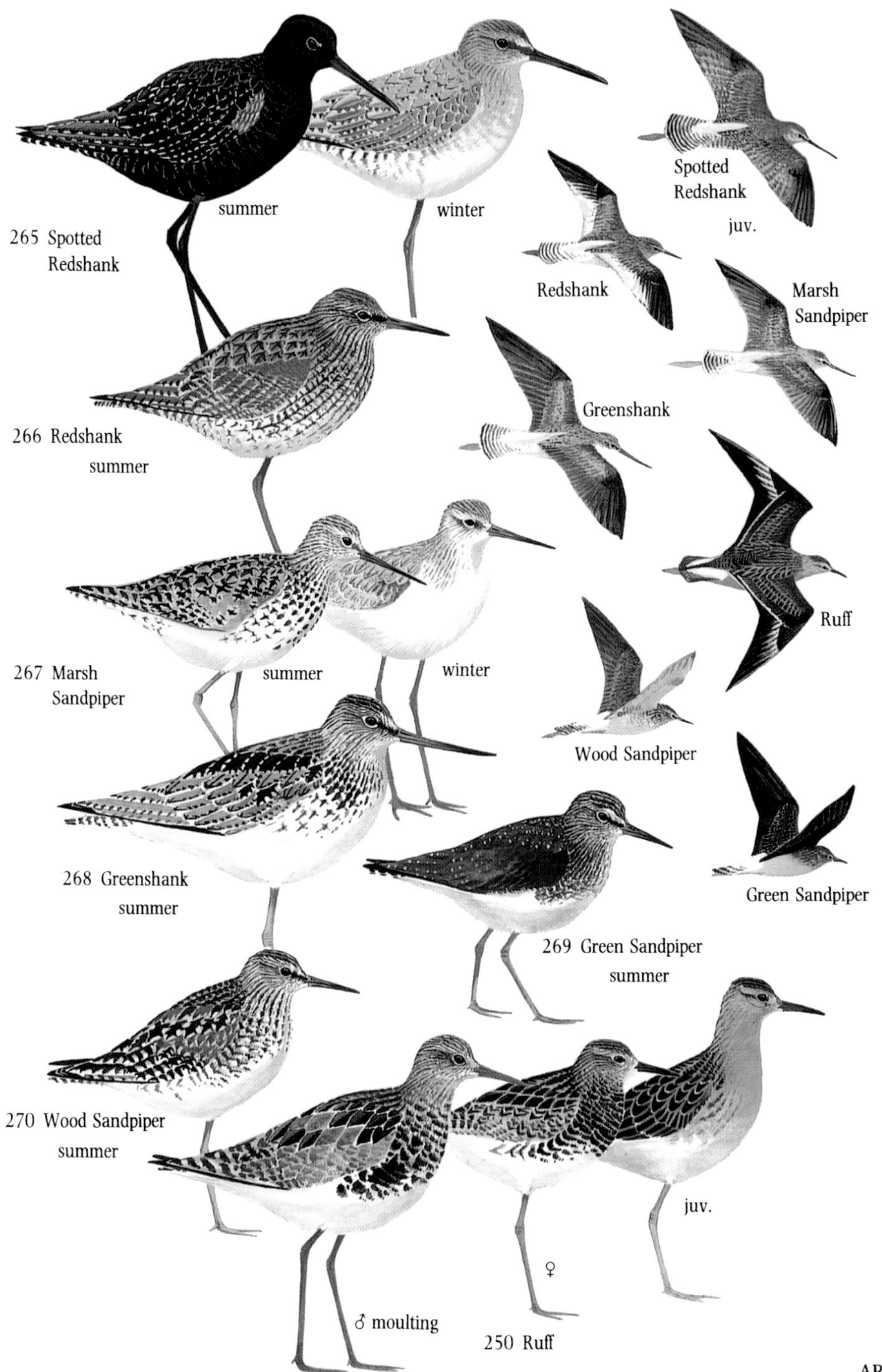
summer
winter
Spotted
Redshank
juv.
265 Spotted
Redshank
Redshank
Marsh
Sandpiper
266 Redshank
summer
Greenshank
Ruff
267 Marsh
Sandpiper
summer
winter
Wood Sandpiper
268 Greenshank
summer
Green Sandpiper
269 Green Sandpiper
summer
270 Wood Sandpiper
summer
juv.
♀
♂ moulting
250 Ruff
AB

249 **Buff-breasted Sandpiper** *Tryngites subruficollis*

Status: status uncertain, possibly vagrant Turkey. Range America. **Habitat:** dry fields with very short grass, in preference to shores.

---

271 **Terek Sandpiper** *Xenus cinereus*

Status: winters on coasts of S Iran and Arabia; common on passage there, occurring inland also. Passage declines westwards, becoming scarce at north end of Red Sea, but still regular Iraq, scarcer E Turkey and Israel. Vagrant Cyprus, Jordan and Syria. Some non-breeders oversummer. **Habitat:** mainly coastal on tidal mudflats, saltmarsh and mangrove creeks, coral reefs, estuaries. Scarce inland on passage.

---

272 **Common Sandpiper** *Actitis hypoleucos*

Status: passage through whole of the region. In Near East, widespread in small numbers, common in Arabia to S Iran but rare Turkey. Some over-summer in winter quarters. **Habitat:** breeds along clear-running rivers, hill-streams and lakes, nesting on ground. Outside breeding season, water-margins of most kinds, running, standing or coastal, but seldom open shore or flats.

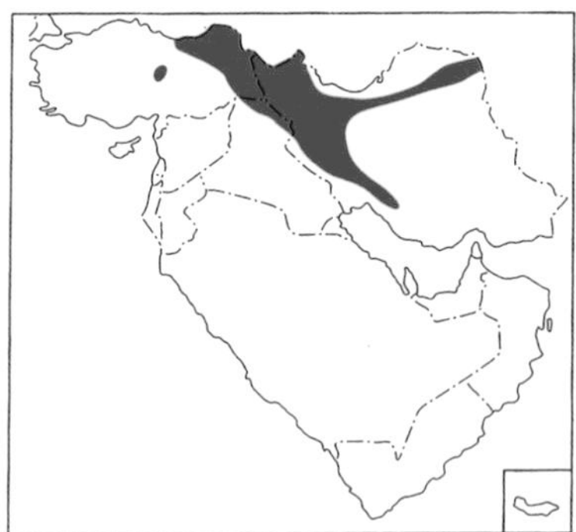

273 **Spotted Sandpiper** *Actitis macularia*

Status: vagrant Turkey. Range America. **Habitat:** similar to Common.

---

274 **Turnstone** *Arenaria interpres*

Status: passage includes coasts of whole region. In winter, Arabia and Iran. Occasional inland. Some over-summer. **Habitat:** rocky or sandy coasts. Rarely inland lakes on passage.

---

275 **Wilson's Phalarope** *Phalaropus tricolor*

Status: vagrant Turkey, Oman. Range America. **Habitat:** margins of fresh and brackish waters.

---

276 **Red-necked Phalarope** *Phalaropus lobatus*

Status: winters in flocks at sea off coasts of Oman and Yemen. On passage common in E Arabian Gulf and Oman, and heavy concentrations can occur on inland lakes in Iran, Iraq and to lesser extent E Turkey; scarce SW Arabia and Near East. Vagrant in rest of Mediterranean area. **Habitat:** maritime; inland lakes on passage.

---

277 **Grey Phalarope** *Phalaropus fulicarius*

Status: recorded occasionally in Iran. Vagrant Saudi Arabia, Iraq, Israel, Turkey and E Arabia. Winters at sea off W Africa. **Habitat:** essentially maritime.

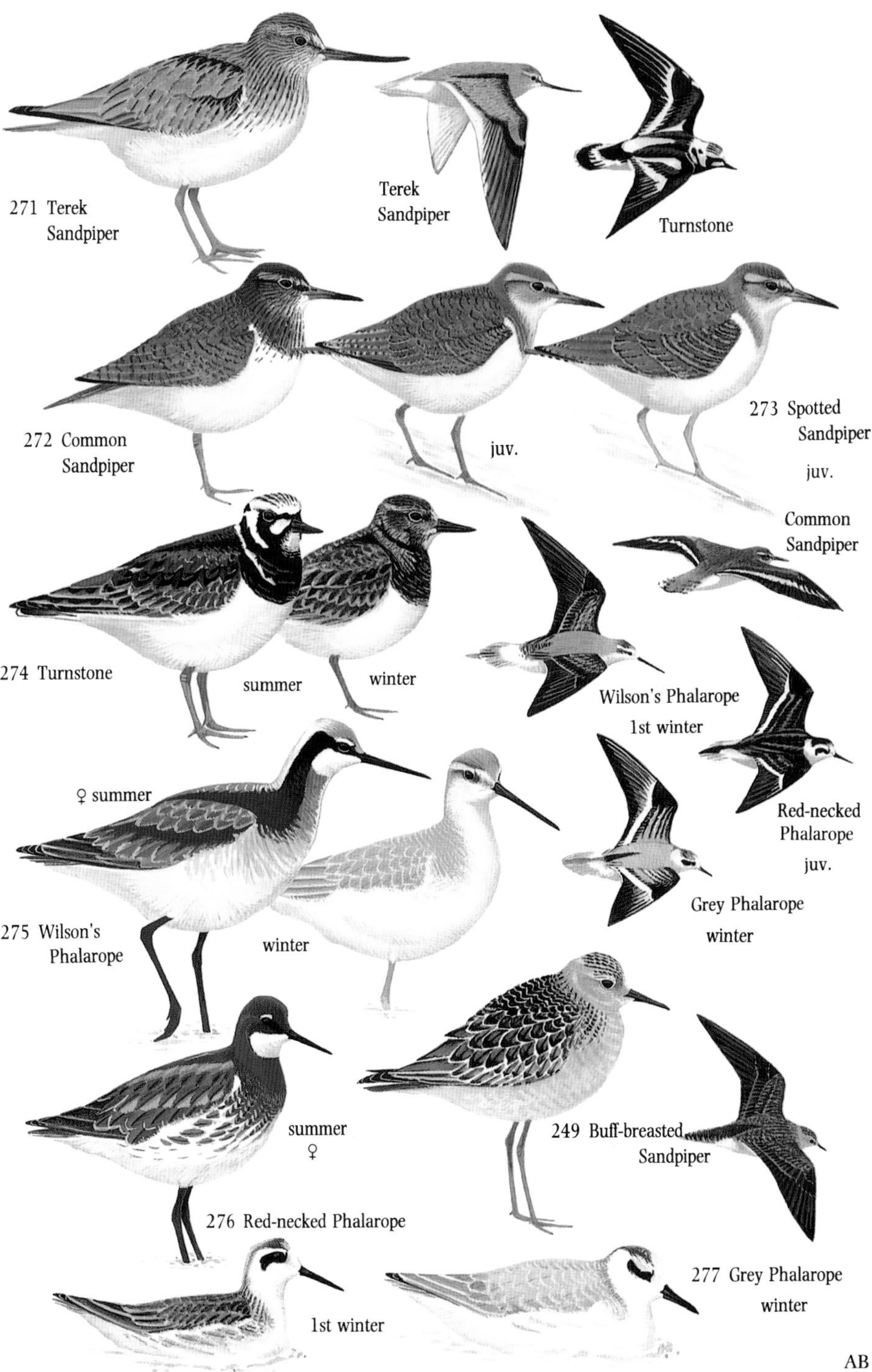

271 Terek Sandpiper
Terek Sandpiper
Turnstone
272 Common Sandpiper
juv.
273 Spotted Sandpiper
juv.
Common Sandpiper
274 Turnstone
summer
winter
Wilson's Phalarope
1st winter
♀ summer
Red-necked Phalarope
juv.
Grey Phalarope
winter
275 Wilson's Phalarope
winter
summer
♀
249 Buff-breasted Sandpiper
276 Red-necked Phalarope
277 Grey Phalarope
winter
1st winter
AB

278 **Pomarine Skua** *Stercorarius pomarinus*

Status: may winter Arabia Sea where regular on passage. Scarce migrant E Mediterranean, Arabian Gulf, S Caspian; sometimes many in N Red Sea. Very rare Turkey. **Habitat:** maritime, but will pass overland on migration.

279 **Arctic Skua** *Stercorarius parasiticus*

Status: some winter off S and SE Arabia. Scarce migrant coasts of N and W Turkey (where some summer), Arabia, more common N Red Sea and Gulf of Aqaba; scarcer E Mediterranean and S Caspian. **Habitat:** maritime, but will pass overland on migration.

280 **Long-tailed Skua** *Stercorarius longicaudus*

Status: scarce migrant Gulf of Aqaba. Vagrant Oman, Iran, Turkey and Saudi Arabia. Vagrant Kuwait. **Habitat:** maritime, but will pass over land on migration.

281 **Great Skua** *Catharacta skua*

Status: vagrant Turkey; very occasional records around coasts of Arabia may refer to South Polar Skua, which see. **Habitat:** maritime, but may pass overland on migration.

282 **South Polar Skua** *Catharacta maccormicki*

Status: vagrant Gulf of Aqaba; uncommon summer and occasional winter records off Oman, and vagrant 'Great Skuas' from Arabian seas may refer to South Polar Skua. **Habitat:** maritime.

319 **African Skimmer** *Rhynchops flavirostris*

Status: vagrant Yemen and Israel. **Habitat:** river sandbanks, marshes with open water, rarely coastal.

320 **Indian Skimmer** *Rhynchops albicollis*

Status: vagrant Oman; an old record from Iran. **Habitat:** rivers with sandbanks and marshes with open water; rarely coastal.

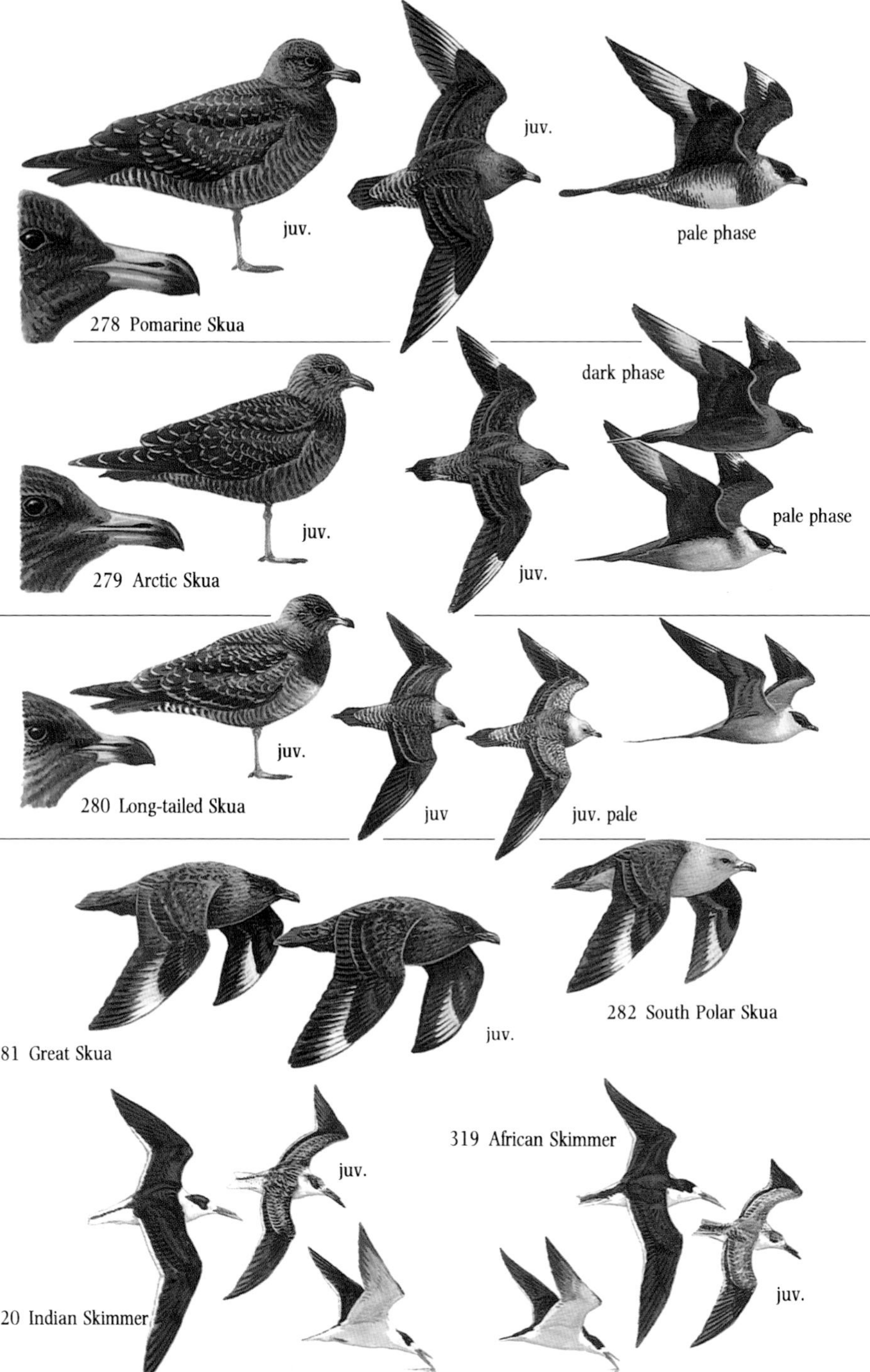
juv.
juv.
pale phase
278 Pomarine Skua
dark phase
juv.
pale phase
juv.
279 Arctic Skua
juv.
juv
juv. pale
280 Long-tailed Skua
juv.
282 South Polar Skua
281 Great Skua
319 African Skimmer
juv.
juv.
320 Indian Skimmer
ML

283 **Sooty Gull** *Larus hemprichii*

Status: present all year in breeding areas, but spreads over much of Red and Arabian Seas; including Iranian coast. Rare inner Arabian Gulf and Gulf of Aqaba. **Habitat:** coastal; often near ports and settlements. Nests mainly on islands on ground or cliffs, solitary or colonial.

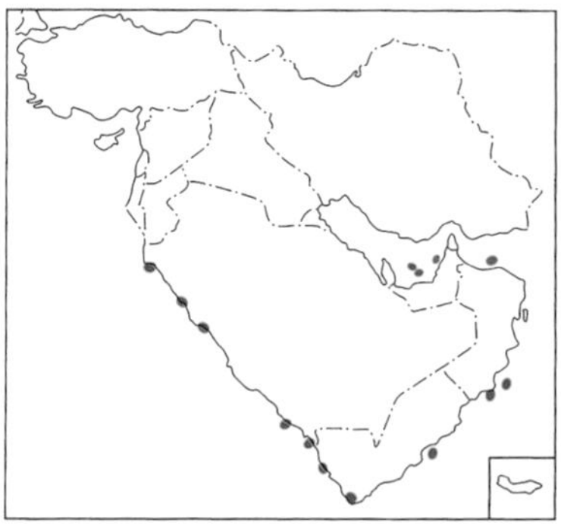

284 **White-eyed Gull** *Larus leucophthalmus*

Status: outside breeding season extending into Gulf of Aqaba; vagrant UAE, Oman, Iran and Turkey, Mediterranean coast of Israel. **Habitat:** as Sooty Gull. Nests on ground near shore of low-lying islands; colonial.

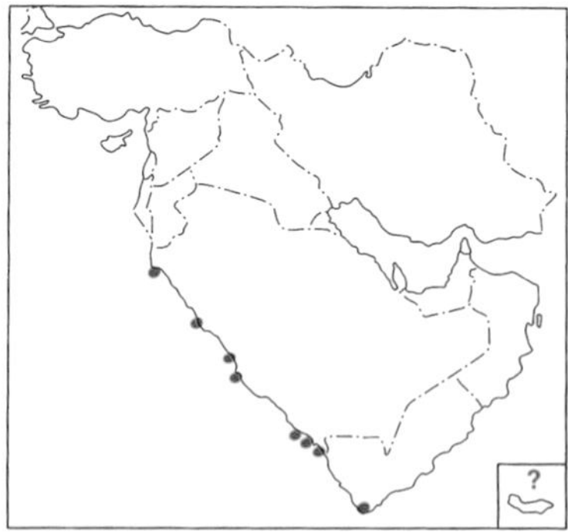

286 **Mediterranean Gull** *Larus melanocephalus*

Status: outside breeding season scarce E Mediterranean and W Gulf of Aqaba. Vagrant, Iraq and Arabian Gulf. **Habitat:** coastal; in breeding season on inland lakes. Nests on islands in lagoons.

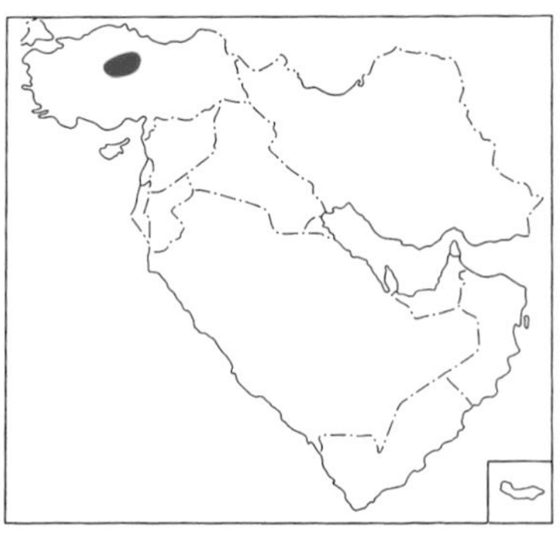

287 **Little Gull** *Larus minutus*

Status: winters and passage E Mediterranean, Black Sea and S Caspian. Rare or vagrant Arabian Gulf, N Red Sea, Gulf of Aqaba and Syria. **Habitat:** maritime in winter, but visits coastal and sometimes inland waters.

288 **Sabine's Gull** *Larus sabini*

Status: vagrant Israel and UAE (summer). Range Arctic. **Habitat:** maritime; sometimes blown onto coasts.

300 **Kittiwake** *Rissa tridactyla*

Status: very scarce or rare E Mediterranean, Gulf of Aqaba. Vagrant Turkey, Iran, Syria and Oman. **Habitat:** coastal and off-shore waters.

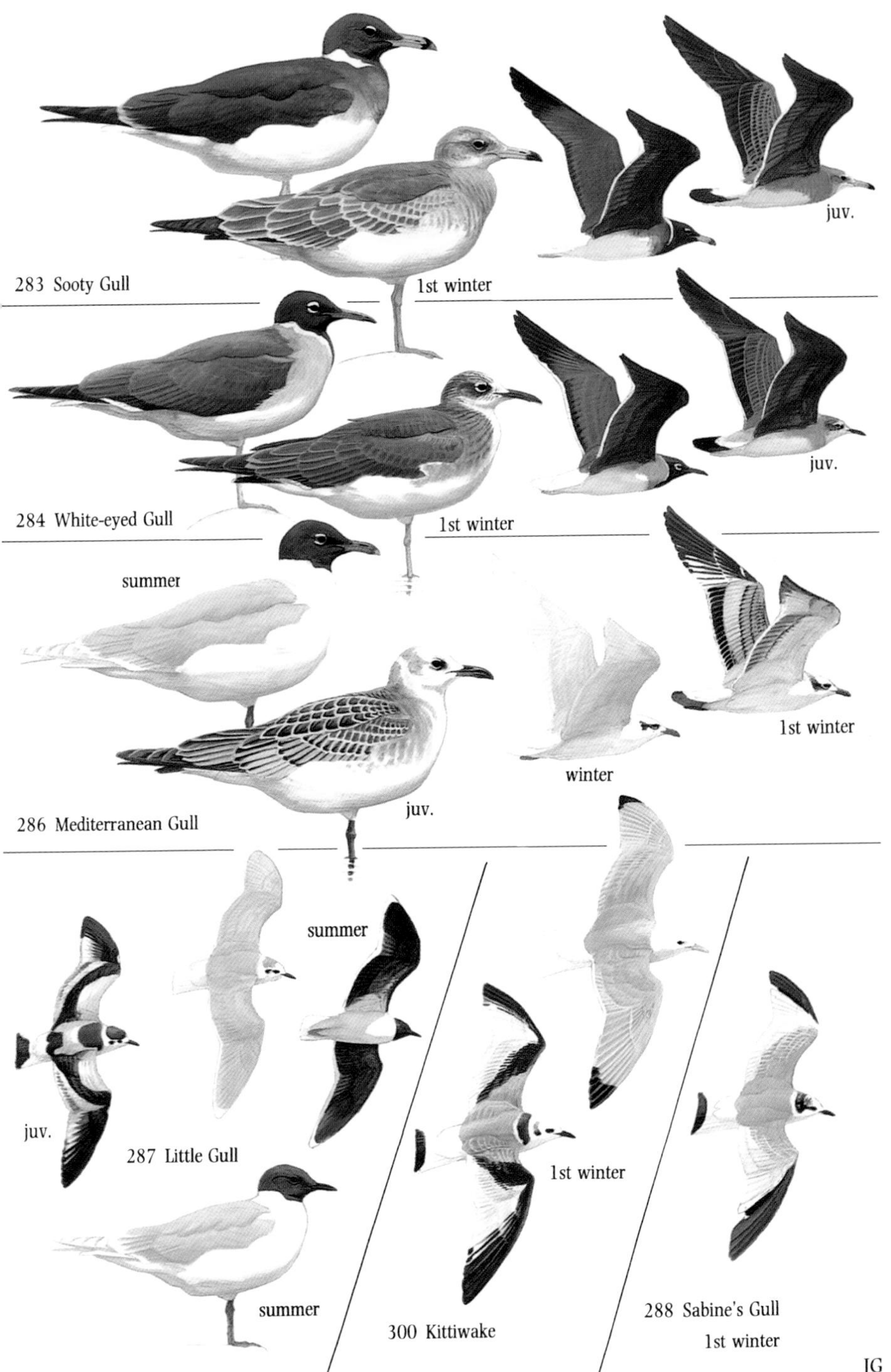
283 Sooty Gull
1st winter
juv.
284 White-eyed Gull
1st winter
juv.
summer
winter
1st winter
286 Mediterranean Gull
juv.
summer
juv.
287 Little Gull
summer
1st winter
300 Kittiwake
288 Sabine's Gull
1st winter
JG

289 **Black-headed Gull** *Larus ridibundus*

Status: widespread on passage and in winter on all coasts and most inland waters. Immatures often summer outside breeding range. **Habitat:** coastal and inland waters. Nests in colonies in reedbeds, vegetated banks or lake islands.

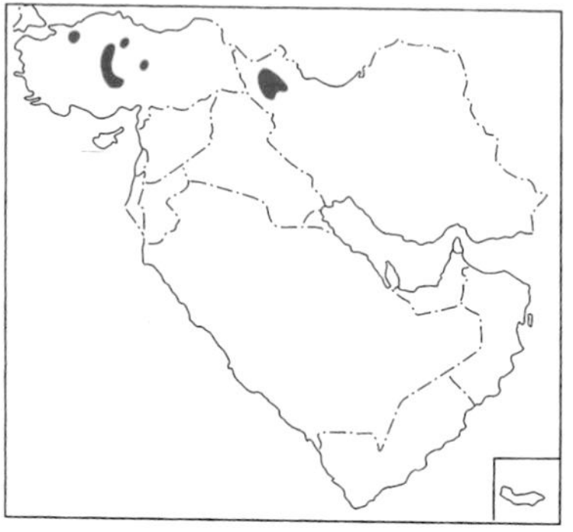

290 **Brown-headed Gull** *Larus brunnicephalus*

Status: vagrant Iran, Oman, and perhaps Arabian Gulf. Record from Gulf of Aqaba may be erroneous. Range Asia. **Habitat:** coastal waters, large inland lakes.

291 **Grey-headed Gull** *Larus cirrocephalus*

Status: vagrant Gulf of Aqaba (both Israel and Jordan). Record from E Saudi Arabia may be erroneous. Nearest range Africa. **Habitat:** coastal and inland waters.

292 **Slender-billed Gull** *Larus genei*

Status: mainly summer visitor to breeding area. Winters E Mediterranean coasts, sometimes inland, and all coasts of Arabia (where it sometimes over-summers). **Habitat:** coastal and inland waters. Nests on edge of wetland, marshes and lake islands in small colonies.

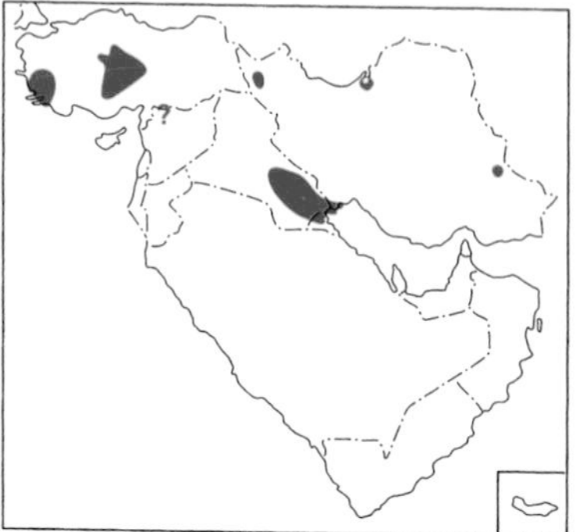

293 **Audouin's Gull** *Larus audouinii*

Status: regular S coast of Turkey. Vagrant Gulf of Aqaba and Mediterranean Israel and Lebanon. **Habitat:** coastal; more maritime in winter. Nests colonially on rocky cliffs.

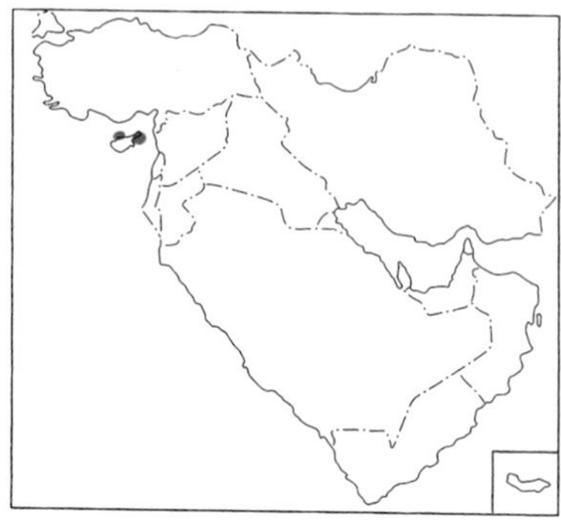

294 **Common Gull** *Larus canus*

Status: small numbers winter Black Sea coast of Turkey and S Caspian; scarce or rare E Mediterranean, Jordan, Iraq, S Iran. Vagrant on southern side of Arabian Gulf, NW Saudi Arabia, Syria and Oman. **Habitat:** coastal and inland waters; sometimes in ploughed fields.

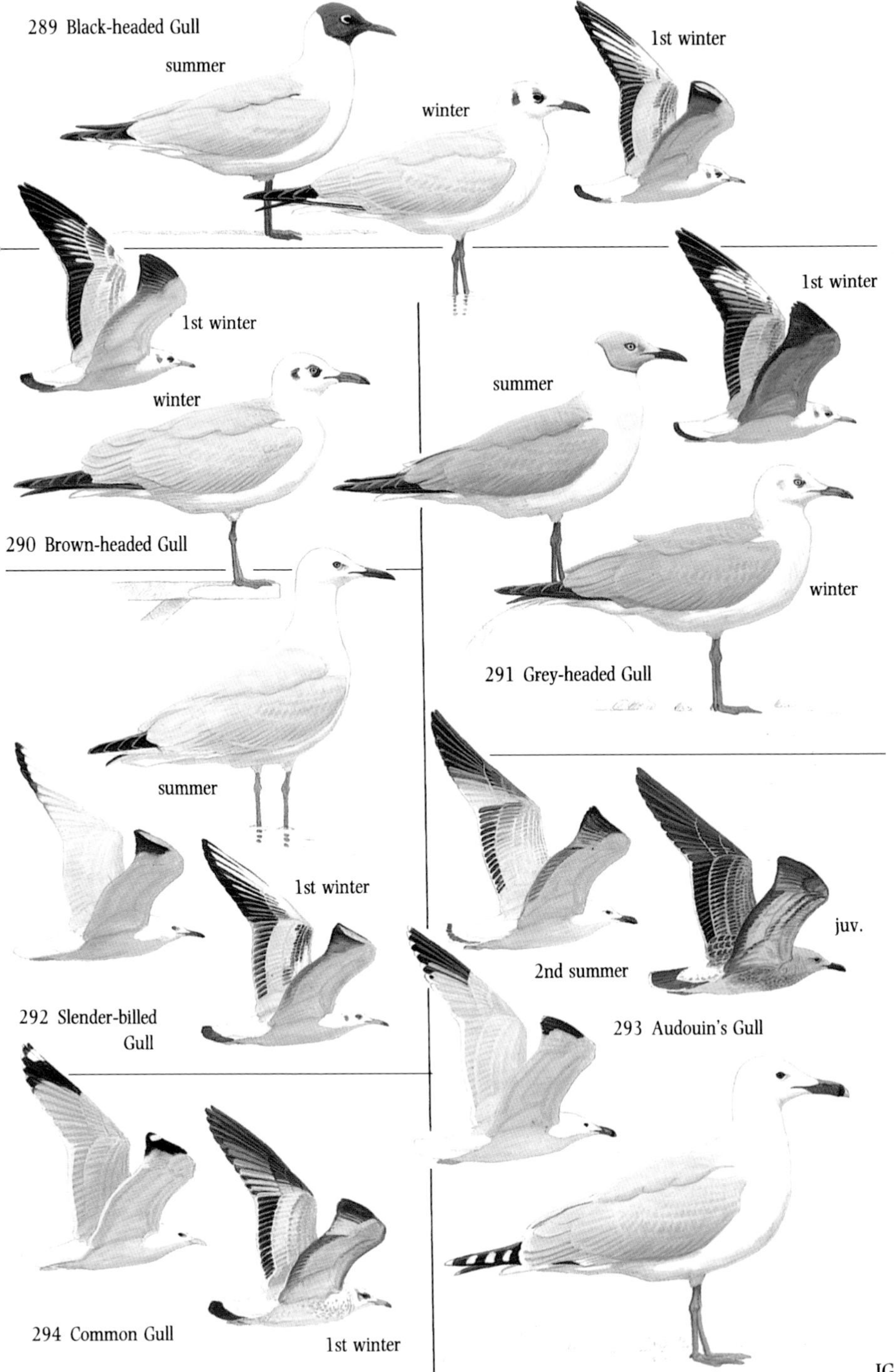
289 Black-headed Gull
summer
winter
1st winter
1st winter
winter
290 Brown-headed Gull
1st winter
summer
winter
291 Grey-headed Gull
summer
1st winter
292 Slender-billed Gull
2nd summer
juv.
293 Audouin's Gull
294 Common Gull
1st winter
JG

285 **Great Black-headed Gull**
Larus ichthyaetus

Status: winter visitor to Turkey, Iran, Syria, all coasts of Arabia and Mediterranean Israel; scarce Gulf of Aqaba; vagrant Cyprus. **Habitat:** mainly coastal, rare on inland lakes.

295 **Lesser Black-backed Gull** *Larus fuscus*

Status: mainly passage and winter visitor but immatures present in some areas throughout year. Winter range includes all sea coasts within the region. Race *fuscus* widespread; *heuglini* and *taimyrensis* poorly understood but mainly E and SW Arabia, rarer Near East. **Habitat:** coastal and less frequently inland waters.

296 **Yellow-legged Gull** *Larus cachinnans*

Status: resident or dispersive. Passage and wintering Mediterranean, Black and Caspian seas and southwards to all coasts of Arabia. Some oversummer south to Oman. **Habitat:** coastal, scarce inland waters. Nests usually colonially on rocky coasts or on islands in inland lakes.

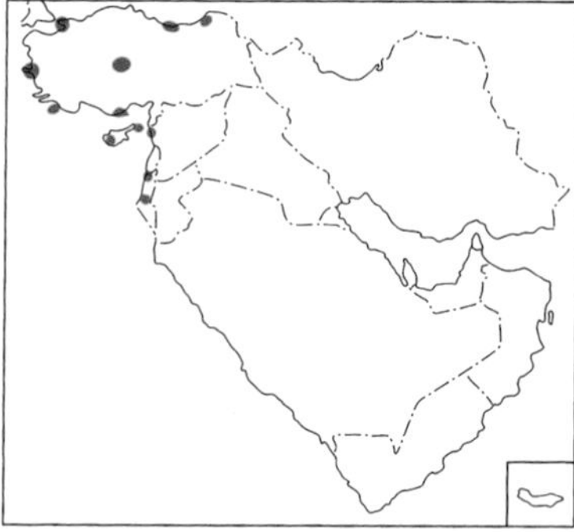

297 **Armenian Gull** *Larus armenicus*

Status: dispersive and migratory; winters south to E Mediterranean, central Syria and Arabian Gulf (e.g. UAE); possibly Yemen. **Habitat:** coastal and inland waters. Nests colonially on islands in inland lakes.

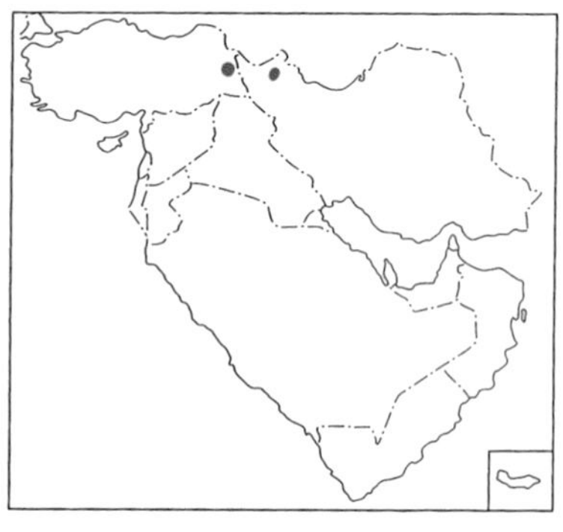

298 **Glaucous Gull** *Larus hyperboreus*

Status: vagrant Turkey; also old record from Gulf of Aqaba. Range Arctic. **Habitat:** coastal waters.

299 **Great Black-backed Gull** *Larus marinus*

Status: vagrant Turkey, Iran, Israel; doubtfully elsewhere in E Mediterranean. **Habitat:** coastal waters.

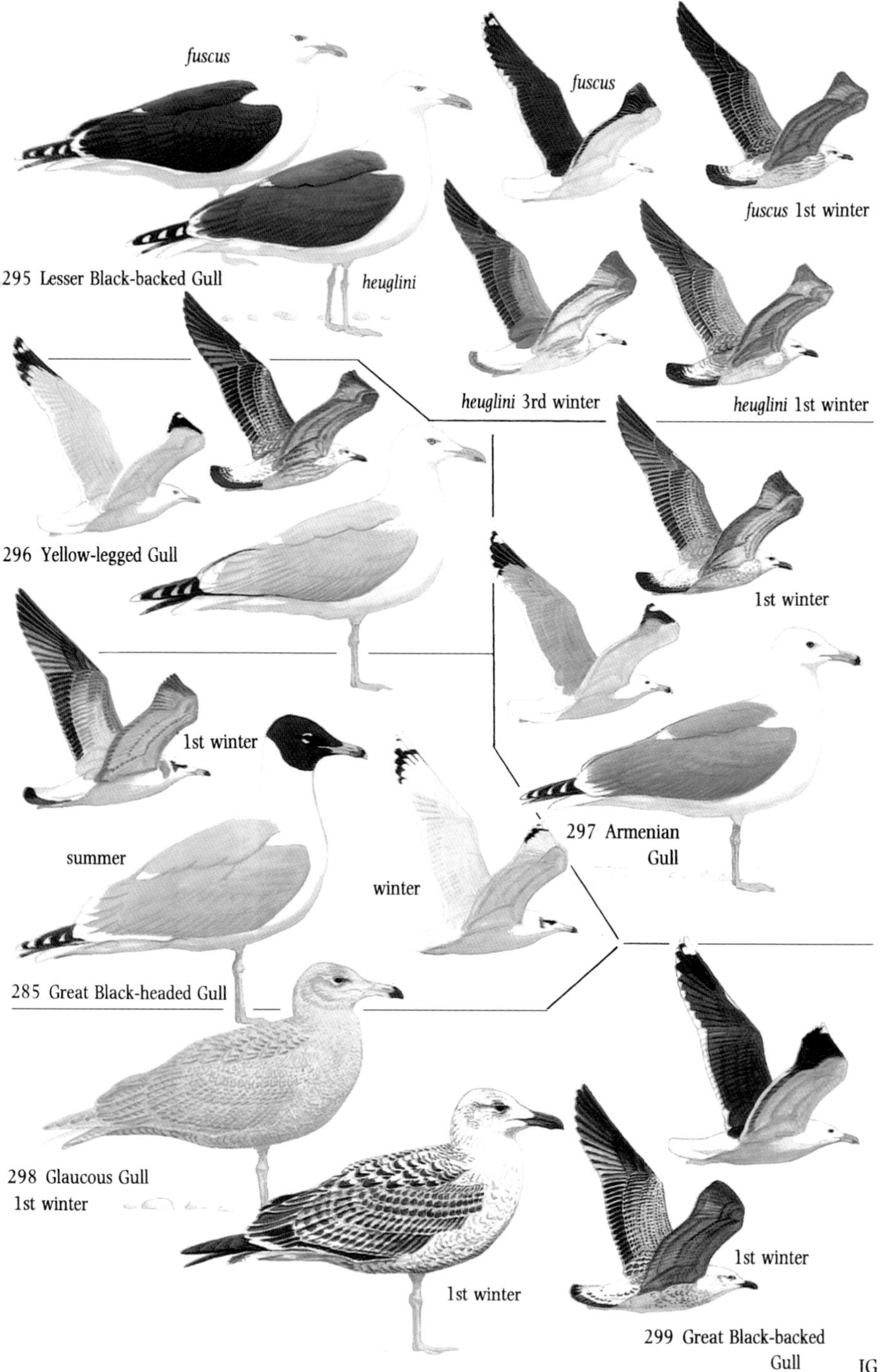
fuscus
fuscus
fuscus 1st winter
295 Lesser Black-backed Gull
heuglini
heuglini 3rd winter
heuglini 1st winter
296 Yellow-legged Gull
1st winter
1st winter
summer
winter
297 Armenian Gull
285 Great Black-headed Gull
298 Glaucous Gull 1st winter
1st winter
1st winter
299 Great Black-backed Gull
JG

301 **Gull-billed Tern** *Gelochelidon nilotica*

Status: first breeding Saudi Arabia (1994). Mainly summer visitor to breeding grounds. On passage E Mediterranean and coastal Arabia where some over-summer. Winters Iran and Arabia. May have bred Bahrain where large numbers also winter. **Habitat:** saltmarshes, sandy coasts and inland waters. Nests colonially on sandy shores and islets in saline lagoons.

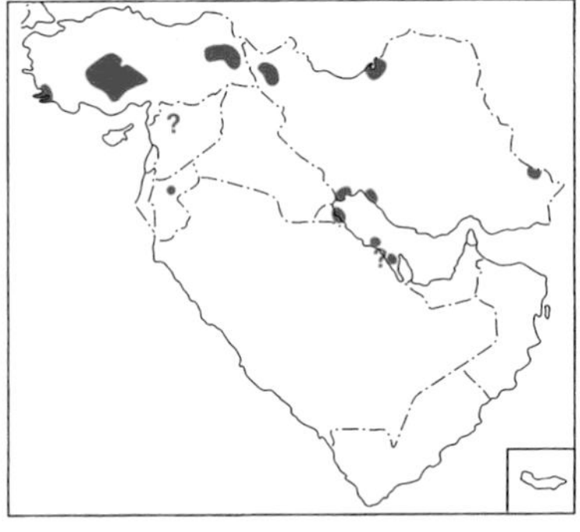

302 **Caspian Tern** *Sterna caspia*

Status: partial migrant; passage on most coasts though rare E Mediterranean and Cyprus. Winters coasts of S Iran and Arabia, very few in SW Turkey. Some over-summer S and SE Arabia. Scarce Jordan. **Habitat:** coastal, but also lakes and rivers on passage. Nests on ground on sandy coasts or islands, singly or colonially.

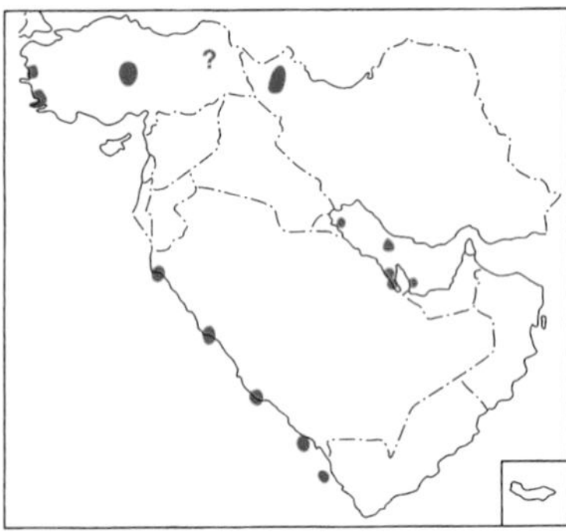

303 **Swift Tern** *Sterna bergii*

Status: resident but dispersive within breeding range; rare Israel. Vagrant Jordan (Aqaba). **Habitat:** coastal. Nests colonially on sandy or rocky shores and islands.

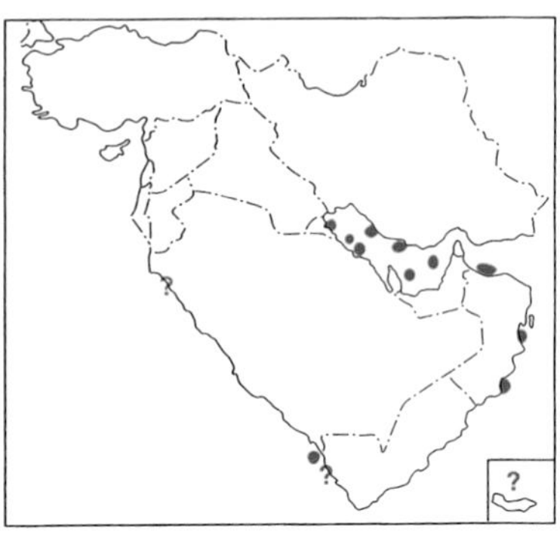

304 **Lesser Crested Tern** *Sterna bengalensis*

Status: resident in some areas, but dispersing; mainly passage Oman. Vagrant Turkey and Jordan (Aqaba). **Habitat:** coastal. Nests colonially on sandy or rocky shores or islands.

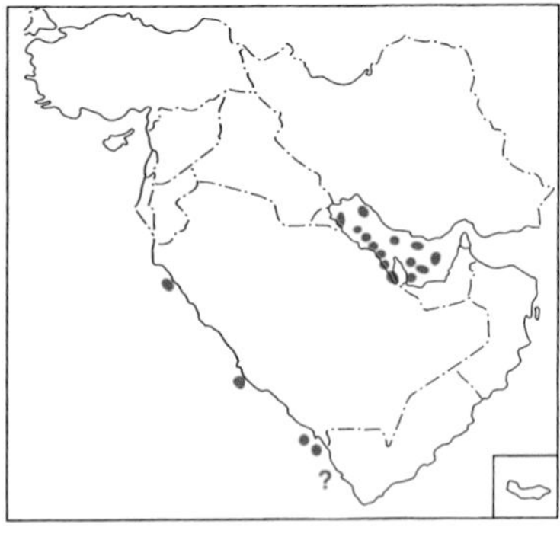

305 **Sandwich Tern** *Sterna sandvicensis*

Status: has bred recently Gulf coast of Saudi Arabia. Passage on most coasts but rare Cyprus. In winter, coasts of Arabia and Iran, S Caspian, Black Sea and E Mediterranean; over-summers Arabian Gulf and Oman. **Habitat:** coastal. Nests colonially on islands or sandy coasts.

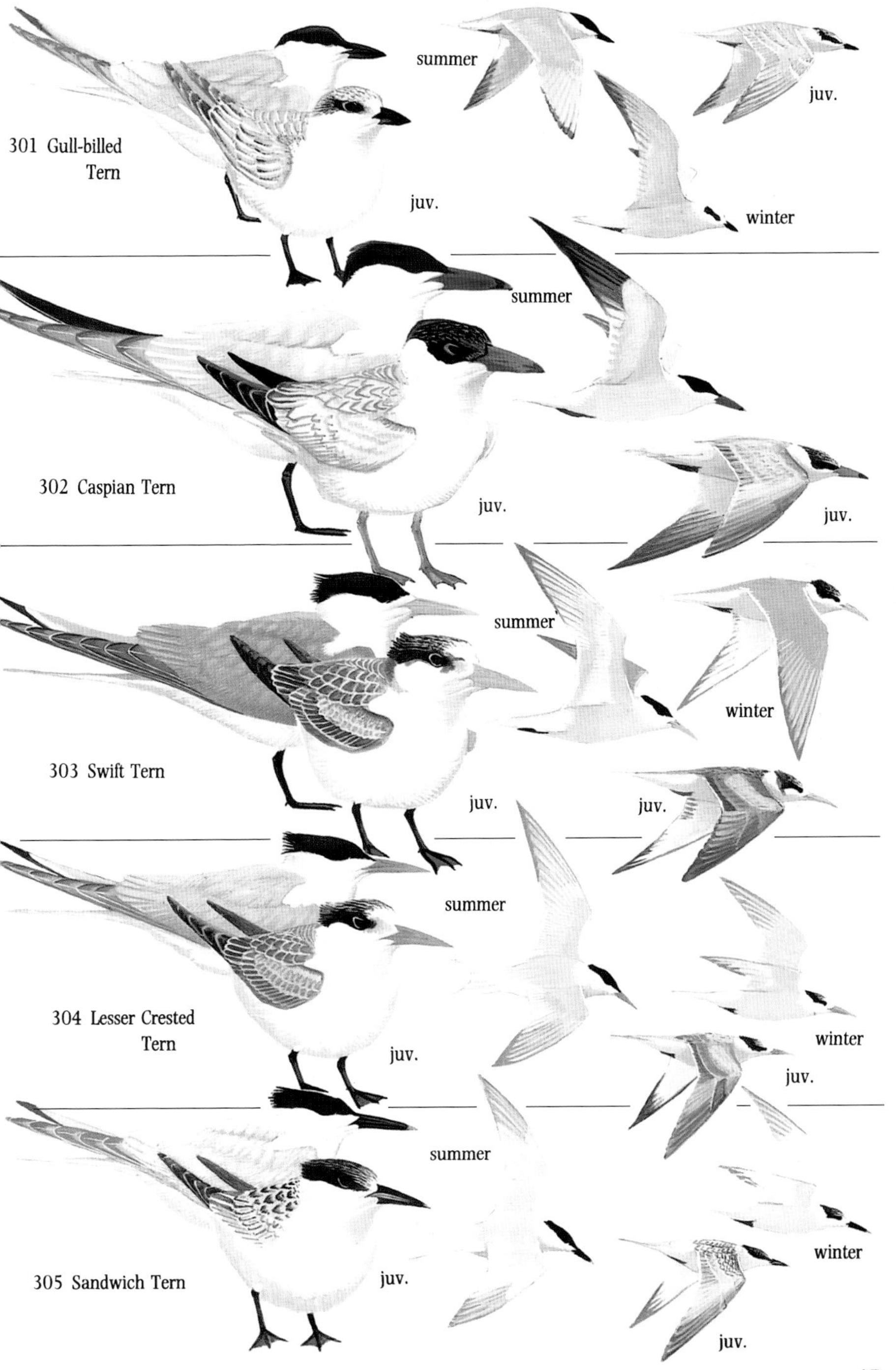
summer
juv.
301 Gull-billed Tern
juv.
winter
summer
302 Caspian Tern
juv.
juv.
summer
winter
303 Swift Tern
juv.
juv.
summer
304 Lesser Crested Tern
juv.
winter
juv.
summer
305 Sandwich Tern
juv.
winter
juv.

AB

306 **Roseate Tern** *Sterna dougallii*

Status: summer visitor. Vagrant Yemen, Arabian Gulf (e.g. Bahrain, Saudi Arabia, UAE) and Israel. **Habitat:** coastal; more maritime than Common Tern. Nests colonially on sandy/rocky islands, sometimes amongst White-cheeked Terns.

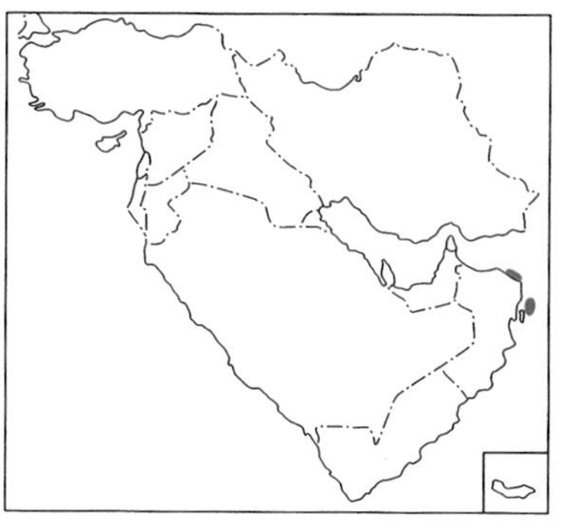

307 **Common Tern** *Sterna hirundo*

Status: may nest Euphrates valley, Syria. Mainly summer visitor, widespread on passage; a few winter Oman to Yemen; also over-summers S Arabia, Oman and UAE. **Habitat:** coastal and inland waters. Nests colonially on beaches, islands and lake-edges.

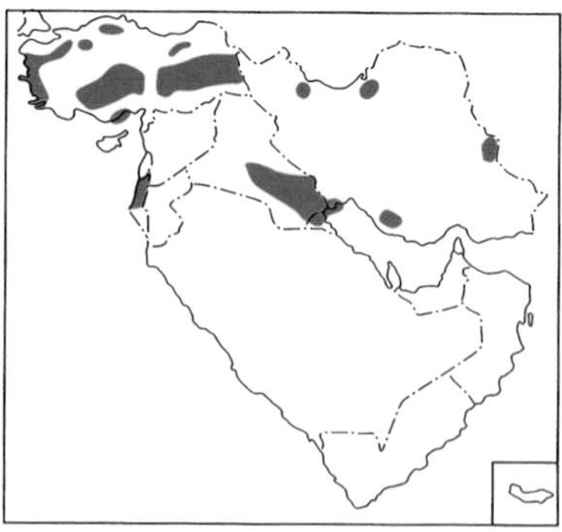

308 **Arctic Tern** *Sterna paradisaea*

Status: vagrant Cyprus, Turkey, Israel and Oman. **Habitat:** coastal; more maritime than Common Tern; occasional on inland waters.

309 **White-cheeked Tern** *Sterna repressa*

Status: mainly summer visitor; a few winter records Arabian Gulf, Oman; some may winter in S Red Sea. Rare Gulf of Aqaba. **Habitat:** coastal and maritime. Nests colonially on small sandy islands and beaches.

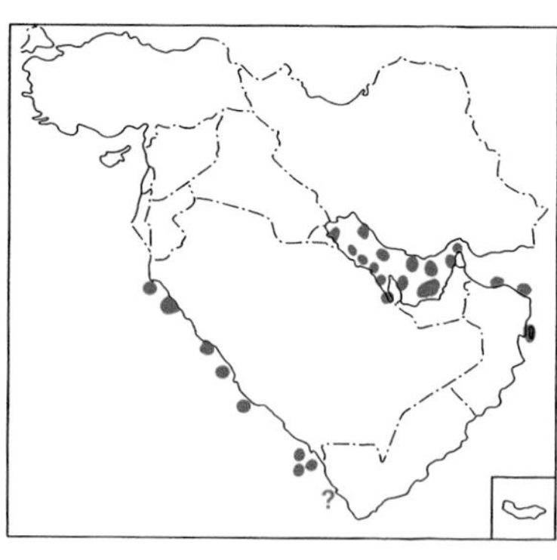

310 **Bridled Tern** *Sterna anaethetus*

Status: summer visitor to breeding grounds. In winter few remain SE and E Arabia, but mainly pelagic. Vagrant Gulf of Aqaba. **Habitat:** maritime. Breeds colonially on rocky or sandy islands with crevices and low vegetation below which it nests.

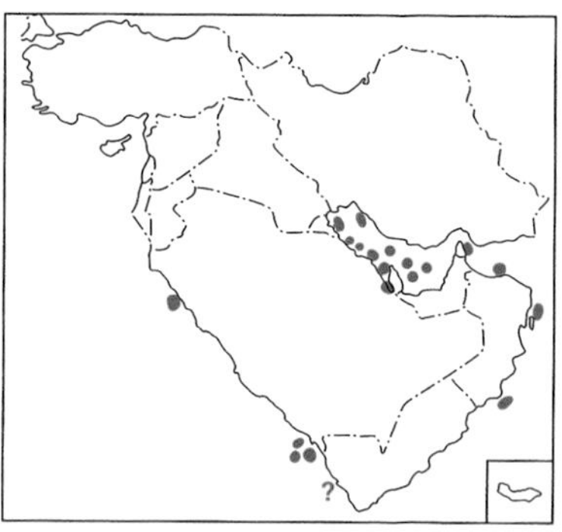

311 **Sooty Tern** *Sterna fuscata*

Status: summer visitor to breeding grounds. Pelagic in Arabian Sea, apparently absent mid-winter. Vagrant Israel, SW Saudi Arabia, Bahrain, UAE. **Habitat:** maritime. Nests colonially on sandy, stony or rocky islands, often with vegetation.

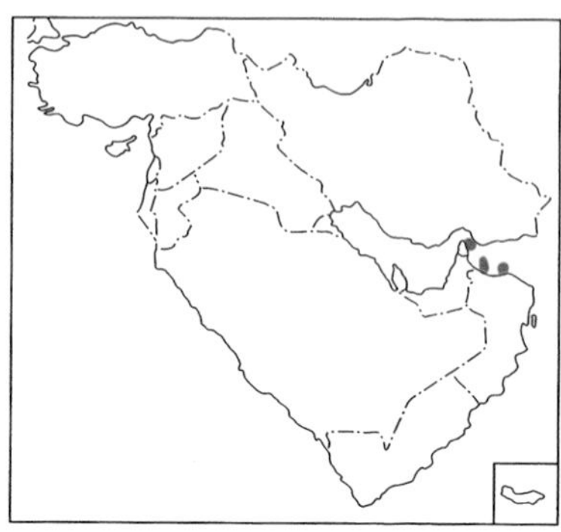

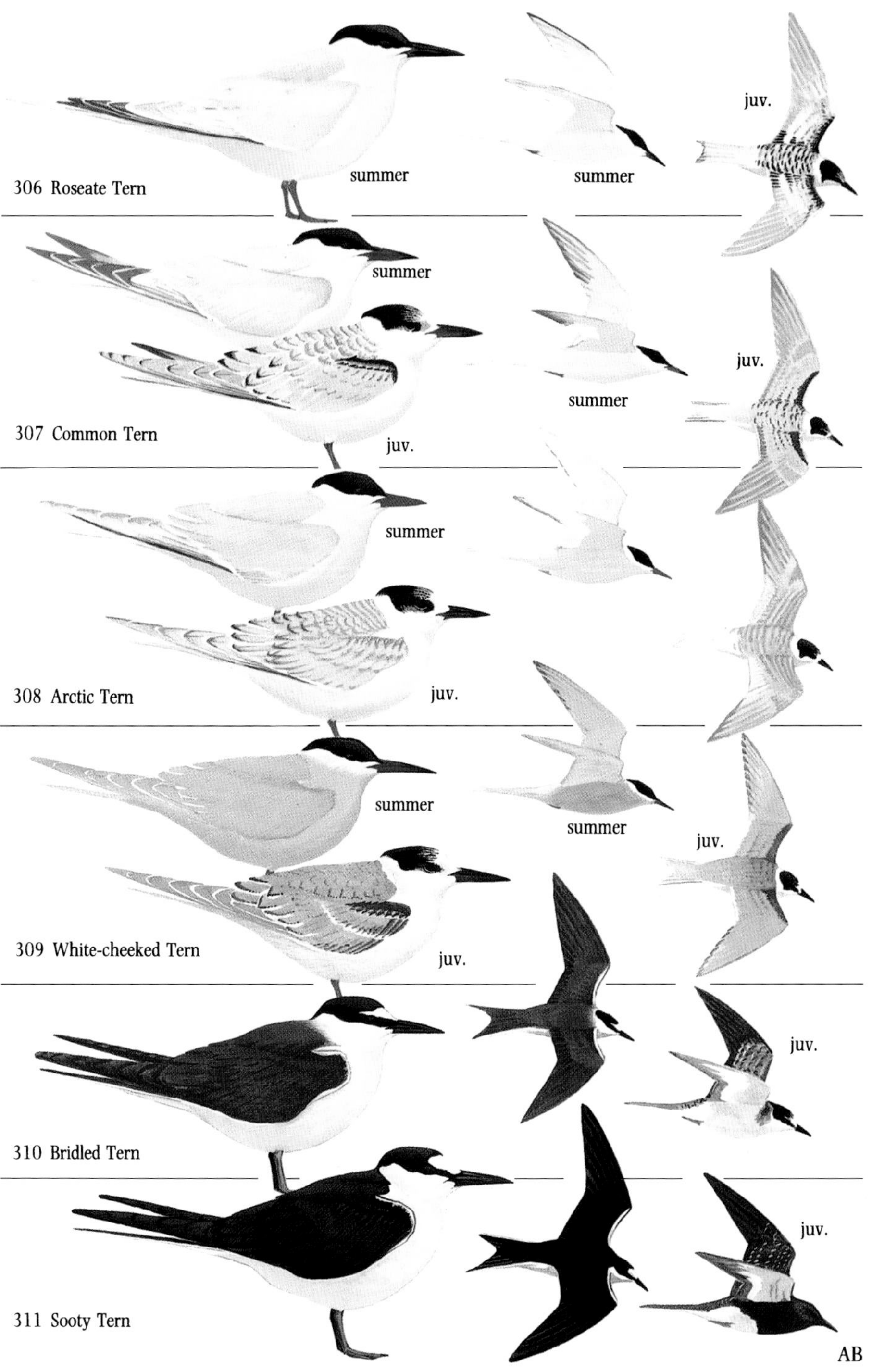
306 Roseate Tern
summer
summer
juv.
307 Common Tern
summer
juv.
summer
juv.
308 Arctic Tern
summer
juv.
309 White-cheeked Tern
summer
juv.
summer
juv.
310 Bridled Tern
juv.
311 Sooty Tern
juv.
AB

312 **Little Tern** *Sterna albifrons*

Status: has bred Cyprus and Syria. Summer visitor; on passage in small numbers Arabia; status on coasts of Iran and Arabia obscured by confusion with Saunders's Little Tern, but it does occur. **Habitat:** both coastal and inland waters, including wide rivers. Nests on sand or shingle beaches or sandbanks in rivers.

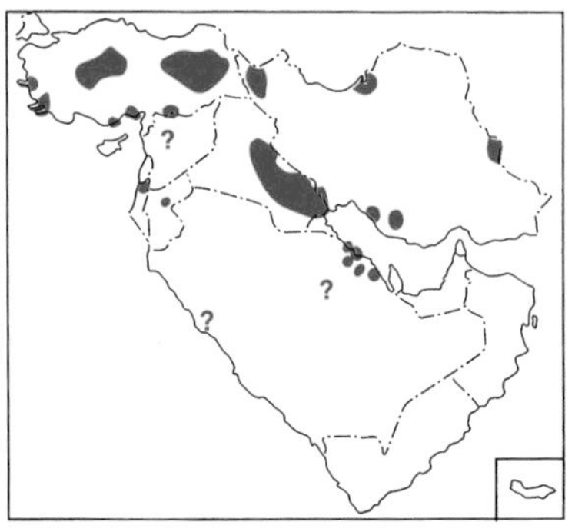

313 **Saunders's Tern** *Sterna saundersi*

Status: mainly summer visitor but recorded wintering E and S Arabia and S Iran. Vagrant Israel. **Habitat:** more maritime than Little Tern; coastal; occasionally inland. Nests on beaches.

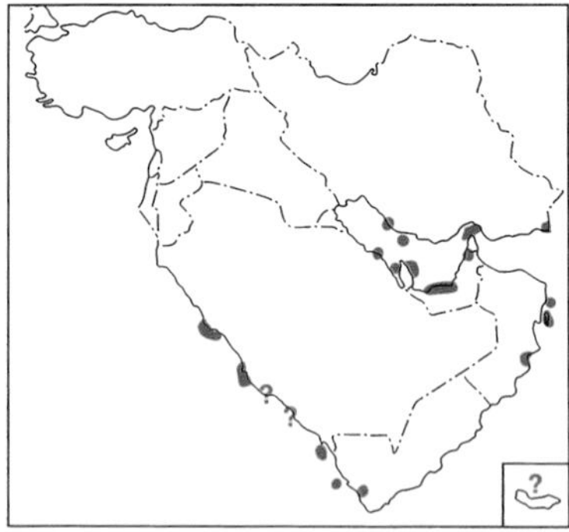

314 **Whiskered Tern** *Chlidonias hybridus*

Status: probably breeds more widely in Turkey. Mainly summer visitor; widespread on passage. Winters Iraq, Iran, Israel, E Arabia, rarely Turkey. **Habitat:** inland waters but also coastal areas on passage and in winter. Nests colonially building floating nest on vegetation in lakes and marshes.

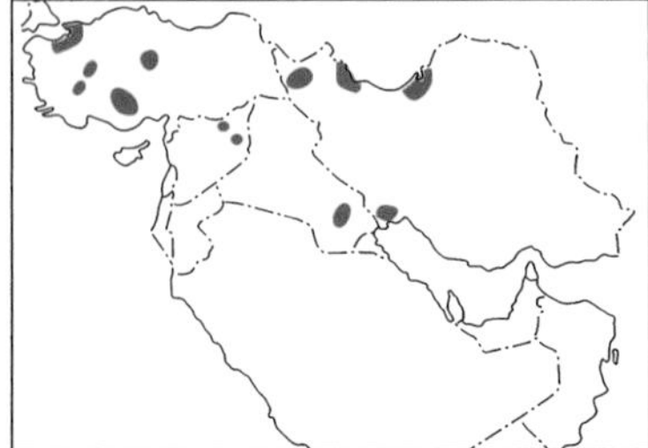

315 **Black Tern** *Chlidonias niger*

Status: summer visitor. Passage E Mediterranean, few N Iran; vagrant elsewhere. **Habitat:** inland waters, but also coastal areas on passage. Nests in scattered colonies building floating nest in shallow marshes and lagoons.

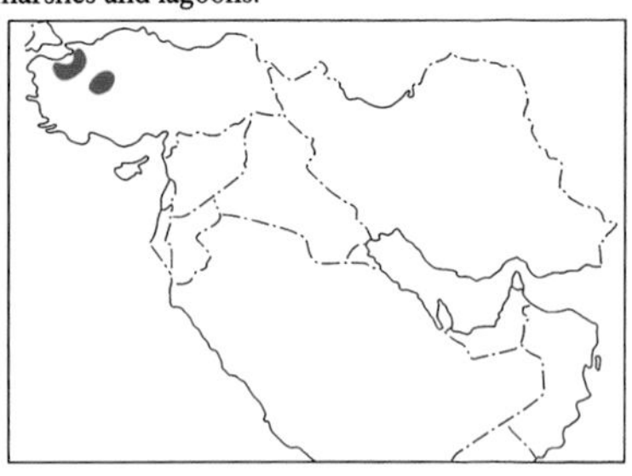

316 **White-winged Black Tern** *Chlidonias leucopterus*

Status: summer visitor. Widespread on passage. A few winter E and S Arabia where it may also oversummer. **Habitat:** inland waters but also coastal areas on passage. Nests in scattered colonies building floating nest in shallow marshes and lagoons.

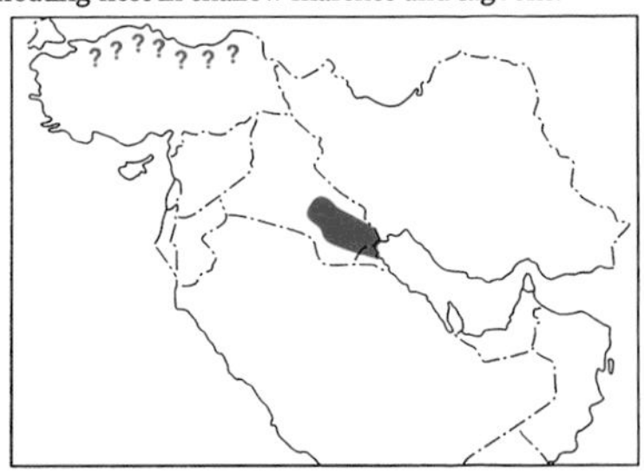

317 **Common Noddy** *Anous stolidus*

Status: summer visitor; occasional records Yemen; vagrant UAE. **Habitat:** maritime, rarely coming to land except to breed. Nests colonially on rocky islets and cliffs.

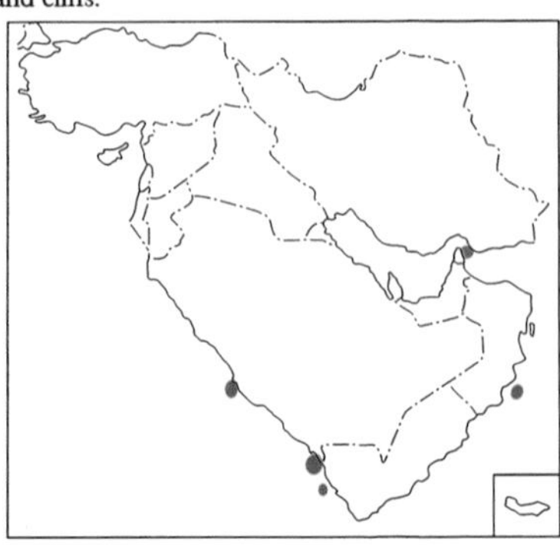

318 **Lesser Noddy** *Anous tenuirostris*

Status: a non-breeding visitor in small numbers to Masirah I. Oman. Vagrant UAE. Nearest range subtropical Indian Ocean. **Habitat:** maritime, rarely approaching land.

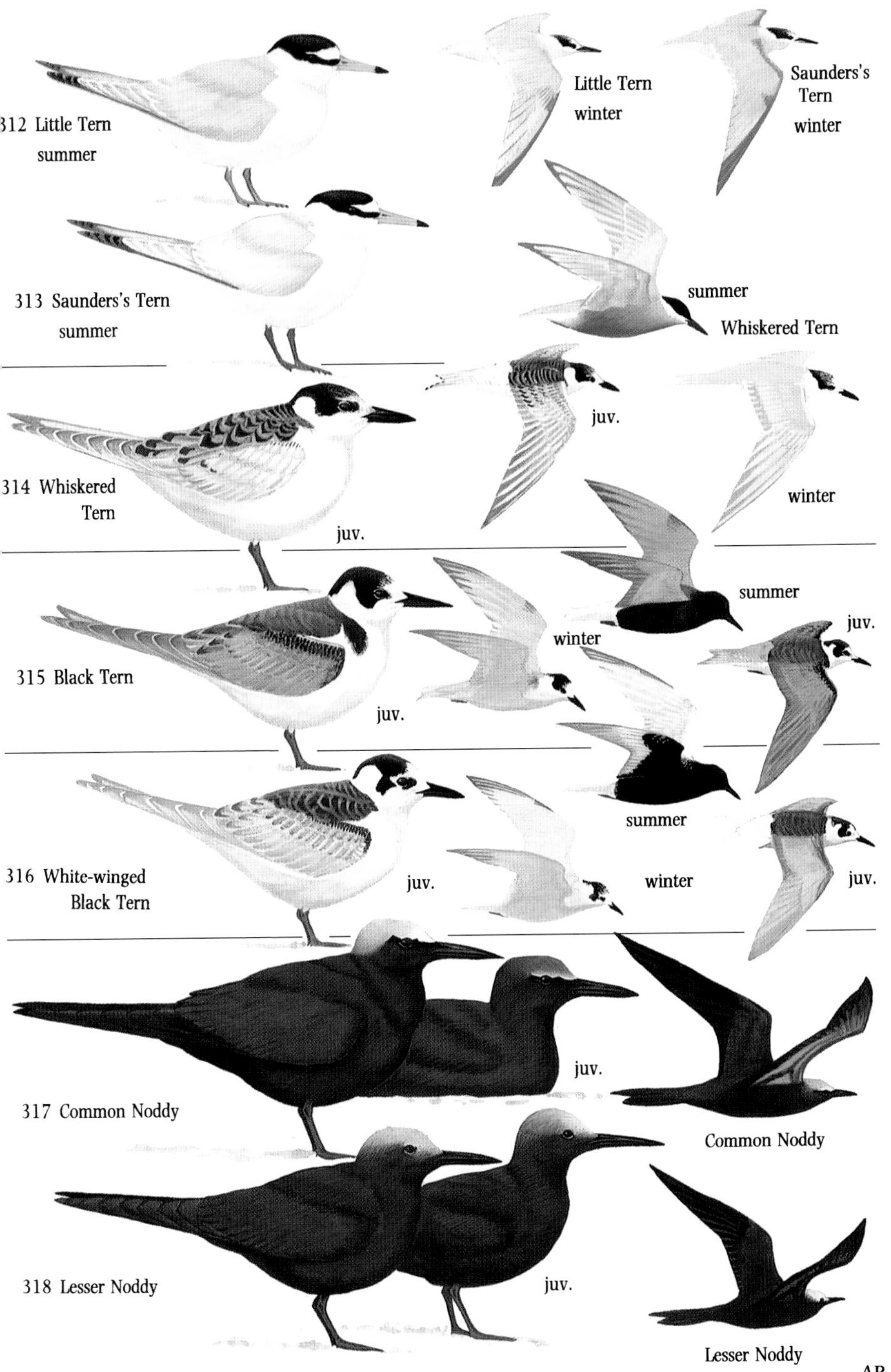
312 Little Tern
summer
Little Tern
winter
Saunders's
Tern
winter
313 Saunders's Tern
summer
summer
Whiskered Tern
314 Whiskered
Tern
juv.
juv.
winter
315 Black Tern
juv.
winter
summer
juv.
summer
316 White-winged
Black Tern
juv.
winter
juv.
317 Common Noddy
juv.
Common Noddy
318 Lesser Noddy
juv.
Lesser Noddy
AB

321 **Lichtenstein's Sandgrouse**
Pterocles lichtensteinii

Status: resident. Vagrant Iraq. **Habitat:** dry, rocky semi-deserts. Wadis and hillsides with scrub, usually acacia. Nests on ground.

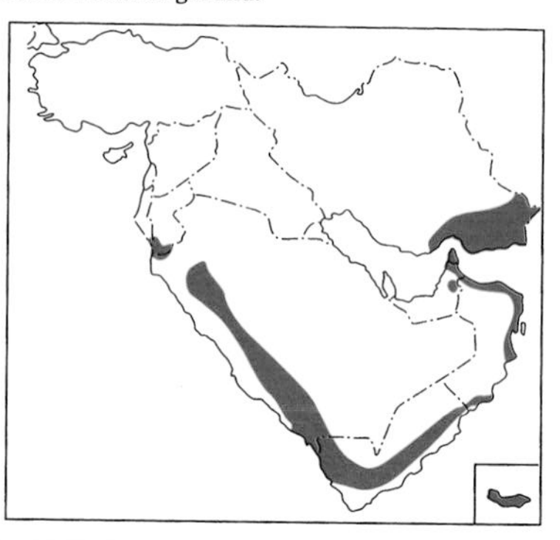

322 **Crowned Sandgrouse** *Pterocles coronatus*

Status: breeding distribution in Arabia mapped very tentatively. Mainly resident, somewhat nomadic. In winter some records in UAE where may breed. **Habitat:** stony and semi-deserts. Nests on ground.

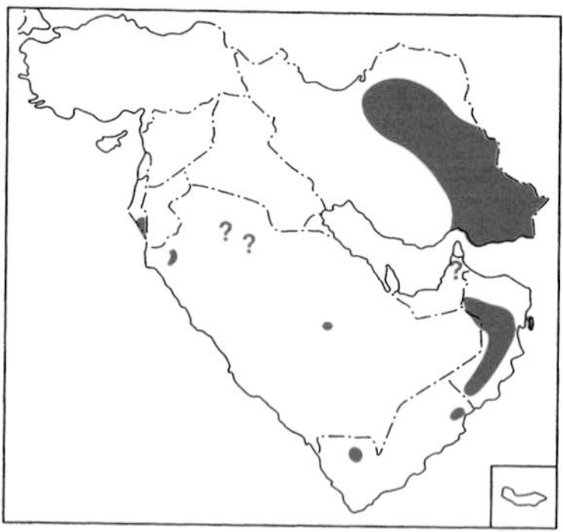

323 **Spotted Sandgrouse** *Pterocles senegallus*

Status: resident; occasional Syria. Vagrant Turkey. **Habitat:** mainly sandy deserts; also semi-deserts with sparse vegetation but sometimes fairly thick scrub. Nests on ground.

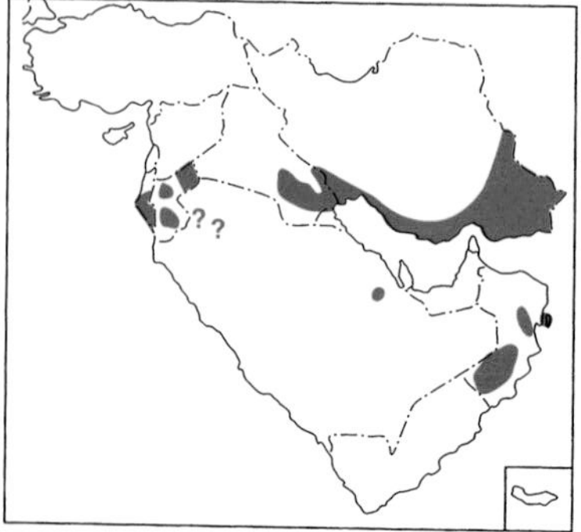

324 **Chestnut-bellied Sandgrouse**
Pterocles exustus

Status: resident and nomadic. Vagrant Israel and Jordan. **Habitat:** semi-deserts with sparse or moderate vegetation, though often near marginal agriculture. Nests on ground.

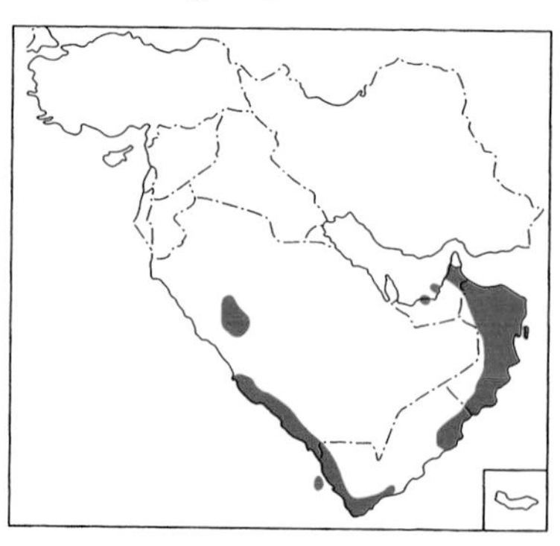

325 **Black-bellied Sandgrouse**
Pterocles orientalis

Status: partial migrant, expanding in winter to Iraq, Syria, Jordan, Kuwait (now scarce), occasionally central Arabia. Passage S Turkey, Cyprus. Vagrant UAE and Bahrain. **Habitat:** dry regions with fairly sparse vegetation and at the edges of cultivation. Nests on ground.

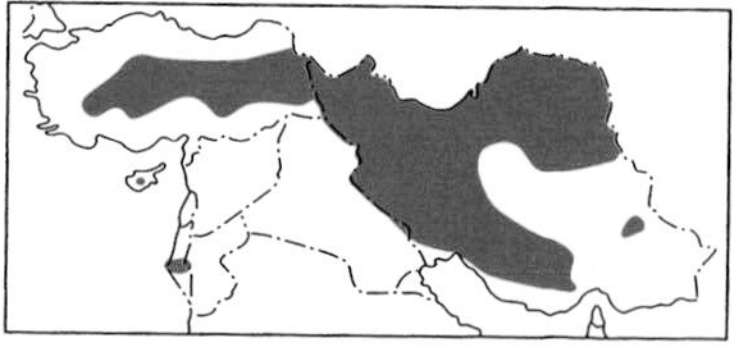

326 **Pin-tailed Sandgrouse** *Pterocles alchata*

Status: resident and nomadic or partial migrant. In winter expands in Saudi Arabia, abundant in NE Iran. Irregular Kuwait; vagrant Lebanon, Bahrain and Jordan. **Habitat:** dry plains and stony semi-deserts often near areas of cereal cultivation. Nests on ground.

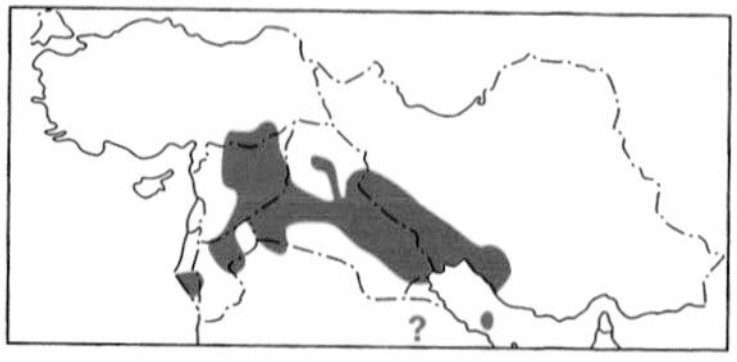

327 **Pallas's Sandgrouse** *Syrrhaptes paradoxus*

Status: probably irregular winter visitor to Iranian steppes east of Caspian; eruptions have reached Turkey in past. **Habitat:** during eruptions from its E Asian breeding grounds can inhabit any open country such as semi-deserts, steppe, waste land and cultivated areas.

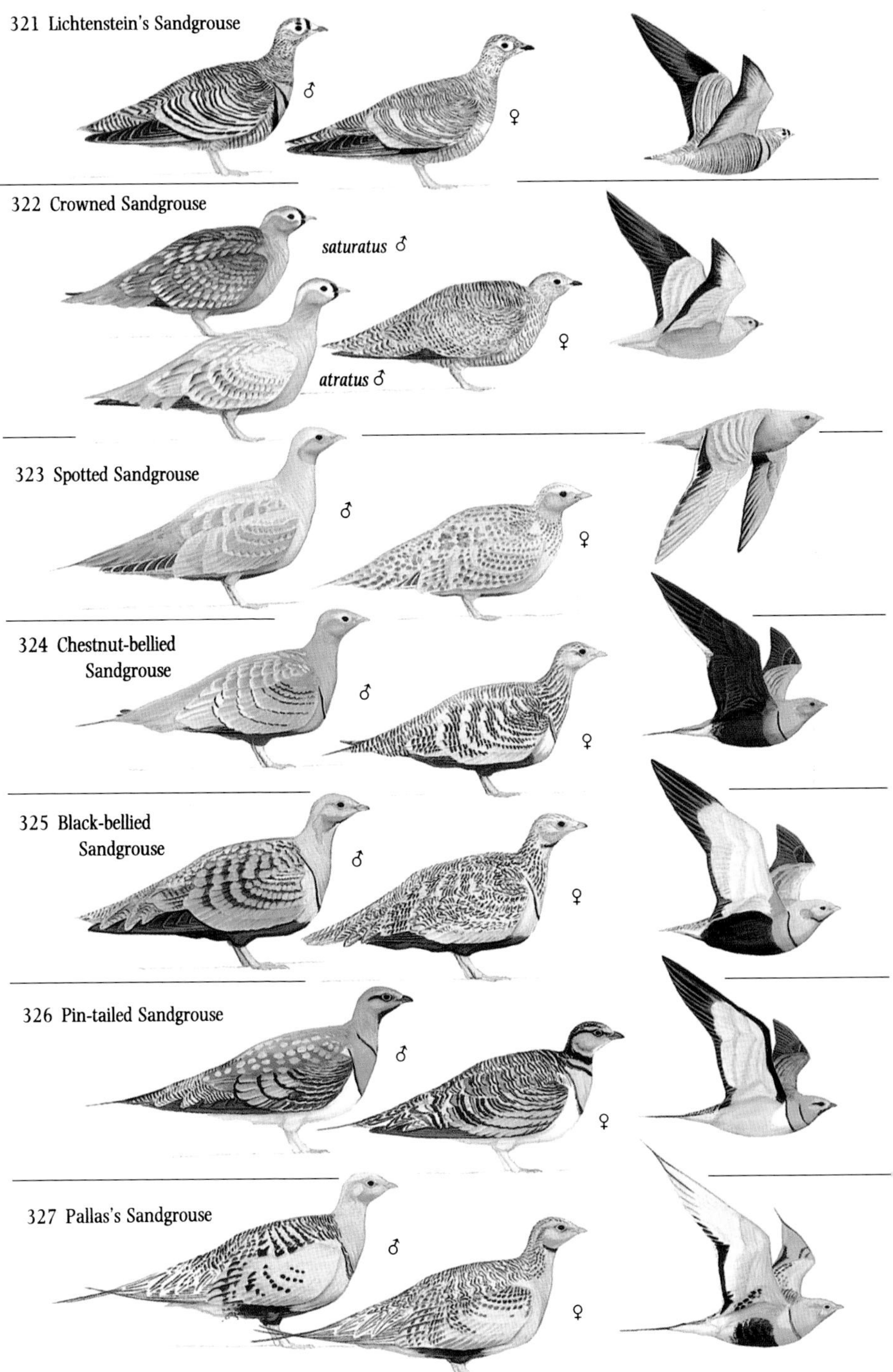
321 Lichtenstein's Sandgrouse
♂
♀
322 Crowned Sandgrouse
saturatus ♂
♀
atratus ♂
323 Spotted Sandgrouse
♂
♀
324 Chestnut-bellied Sandgrouse
♂
♀
325 Black-bellied Sandgrouse
♂
♀
326 Pin-tailed Sandgrouse
♂
♀
327 Pallas's Sandgrouse
♂
♀
JG

328 **Rock Dove (includes Feral Pigeon)** *Colomba livia*

Status: resident, with local movements. **Habitat:** rocky wadis, sea-cliffs, mountains, nesting in cave, rock-ledge or cliff. Feral Pigeon occurs anywhere, often in towns, nesting on buildings.

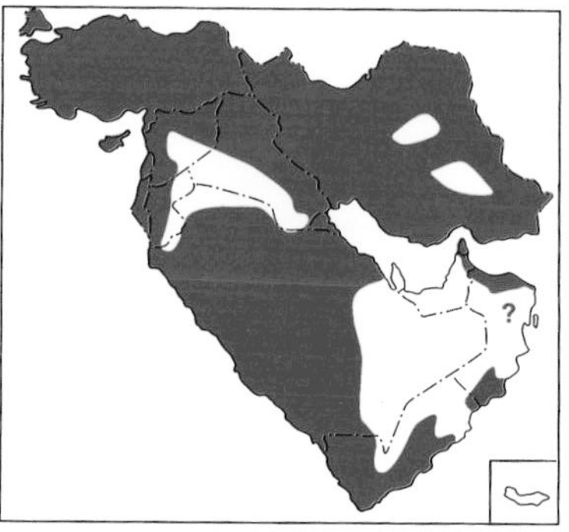

329 **Stock Dove** *Columba oenas*

Status: resident and partial migrant. Winter range W and SW Turkey, Cyprus, Israel, Iraq, Syria and NW Iran. Irregular Jordan. Vagrant Bahrain and Oman. **Habitat:** wooded open areas with old trees. More widespread and varied on passage, including fields. Nests in hole in tree, rock or building.

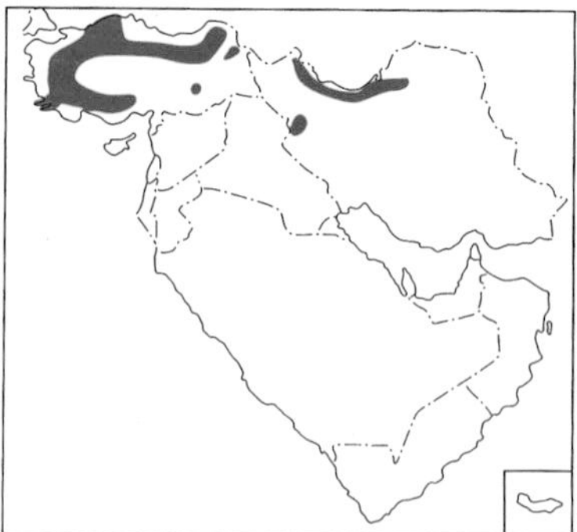

330 **Eastern Stock Dove** *Columba eversmanni*

Status: partial migrant; few present NE Iran all year (breeding not proved); winters apparently from Iranian Baluchistan eastwards. **Habitat:** lightly wooded areas in cultivation, river banks, cliffs or ruins. Nests in trees or hole in bank.

331 **Woodpigeon** *Columba palumbus*

Status: mainly resident, but enters W Turkey, S Iran and Israel in winter. Vagrant Jordan, Kuwait and UAE. **Habitat:** wooded areas, parks in towns, gardens; on passage also fields. Nests in tree, sometimes using old nest.

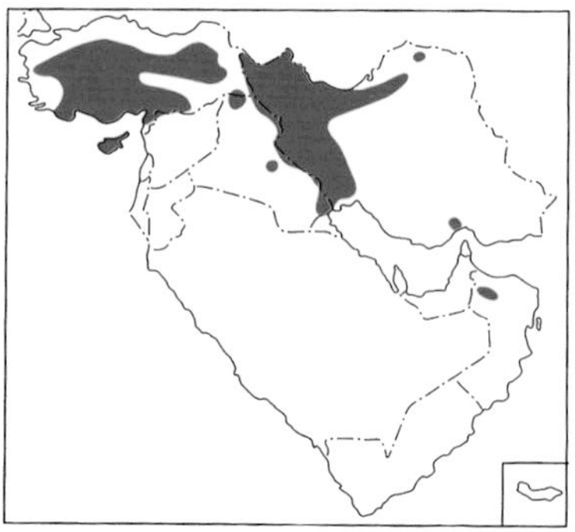

332 **Olive Pigeon** *Columba arquatrix*

Status: rare and local SW Saudi Arabia and Yemen where almost certainly breeds. Range Africa. **Habitat:** wooded slopes and highlands, with tall fruit-trees and wild olives.

328 Rock Dove
Feral
Pigeon (variations)
9 Stock Dove
330 Eastern
Stock Dove
juv.
331 Wood Pigeon
332 Olive Pigeon

ML

333 **African Collared Dove** *Streptopelia roseogrisea*

Status: resident, but recent range extension; vagrant Israel and Bahrain. **Habitat:** semi-desert and savanna with trees; also coastal mangroves, parks in towns. Nests in tree, sometimes colonially.

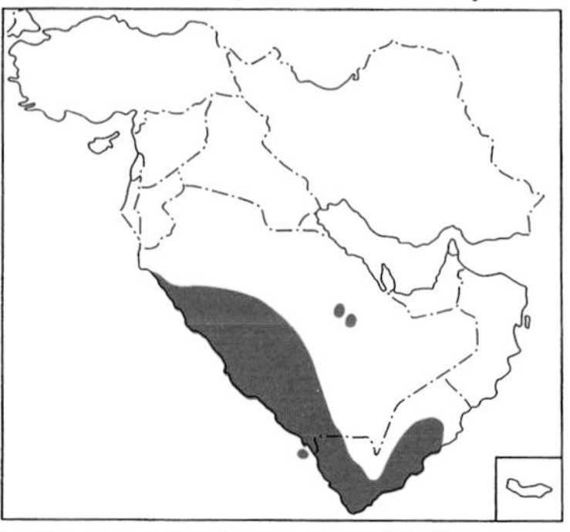

334 **Collared Dove** *Streptopelia decaocto*

Status: largely resident, but mainly summer visitor Iran, though wintering SW Iran and in Arabian Gulf area. Recent range extension in Arabia. **Habitat:** towns, villages, parks; often flocking in fields. Nests in tree or building.

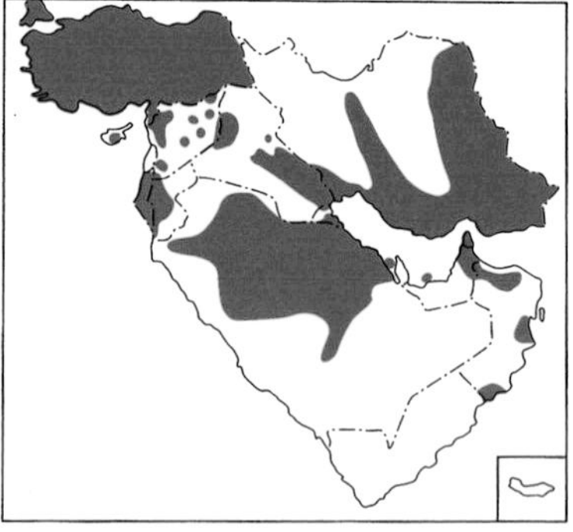

335 **Red-eyed Dove** *Streptopelia semitorquata*

Status: resident. **Habitat:** mainly between 500–1700 m in vegetated areas with tall trees and fertile fruit-gardens; often in wadis near water. Nests in trees.

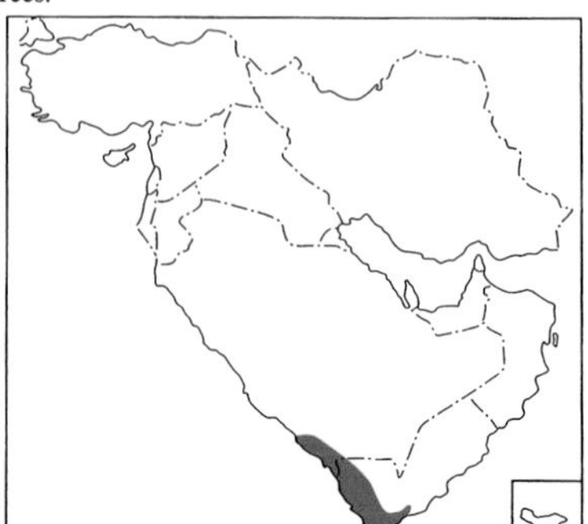

336 **Red Turtle Dove** *Streptopelia tranquebarica*

Status: vagrant Oman and Iran from Asia. **Habitat:** open wooded country.

337 **Turtle Dove** *Streptopelia turtur*

Status: summer visitor. Widespread on passage over most region. May winter rarely Yemen. **Habitat:** open, lightly wooded country with scrub, steppe and oases. Nests in tree, bush or thicket.

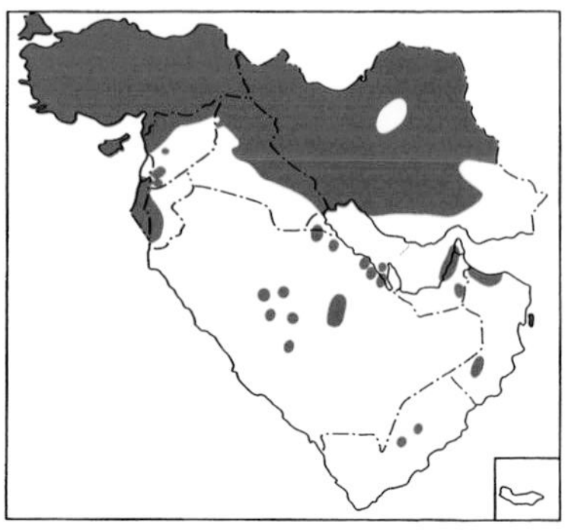

338 **Dusky Turtle Dove** *Streptopelia lugens*

Status: resident; some dispersal in winter. **Habitat:** mainly between 1000–2800 m, sometimes lower in winter; in wooded areas, acacia wadis, tree-lined agriculture. Nests in tree, rarely in bush.

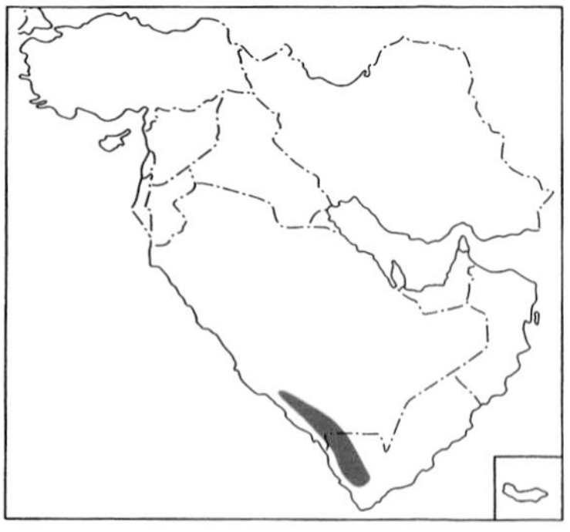

339 **Rufous Turtle Dove** *Streptopelia orientalis*

Status: scarce on passage in E Iran, rare Oman. Vagrant elsewhere in Arabian Gulf area (e.g. UAE), Israel. **Habitat:** open woodland, often near cultivation.

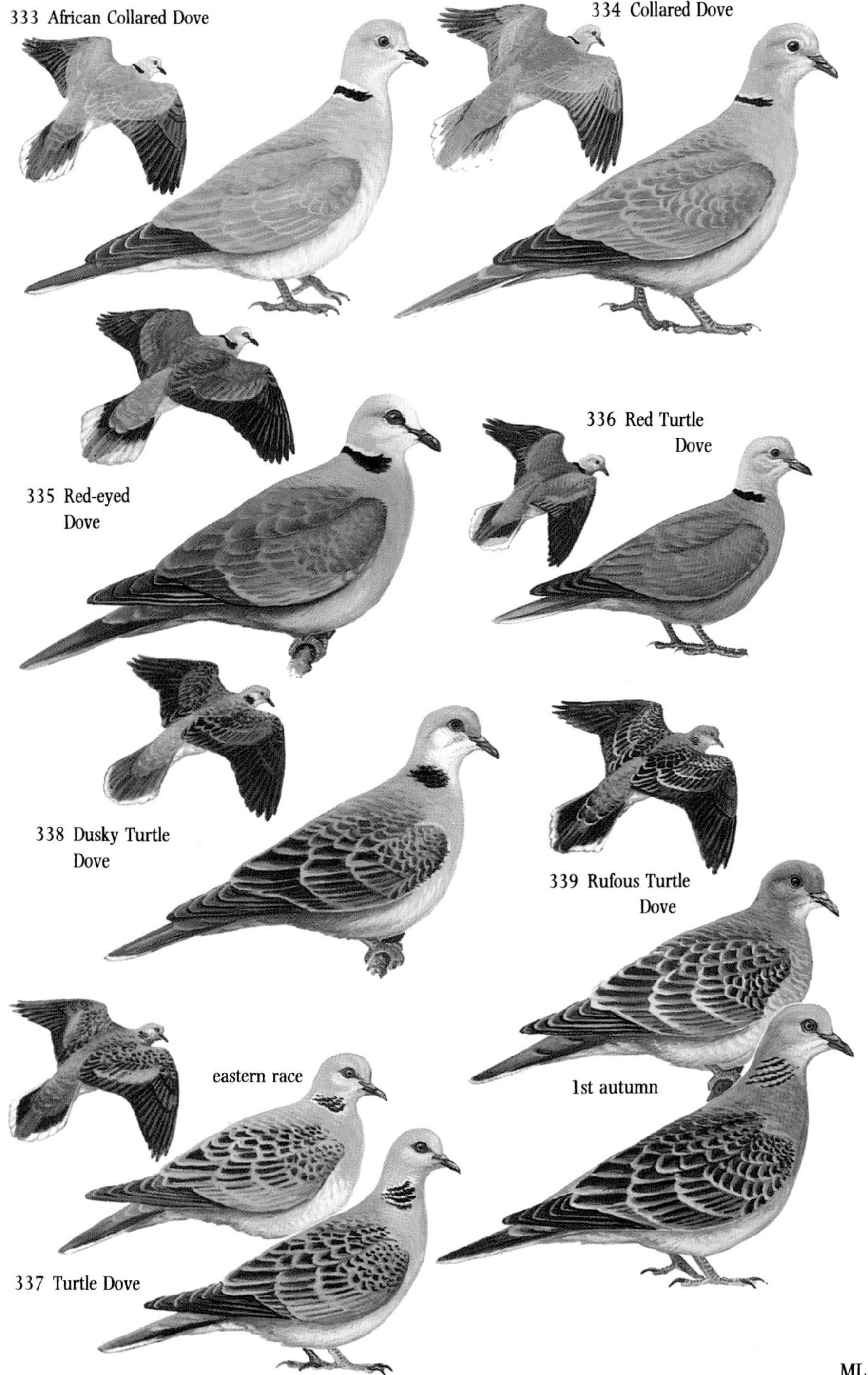
333 African Collared Dove
334 Collared Dove
336 Red Turtle
Dove
335 Red-eyed
Dove
338 Dusky Turtle
Dove
339 Rufous Turtle
Dove
eastern race
1st autumn
337 Turtle Dove
ML

340 **Laughing Dove** *Streptopelia senegalensis*

Status: mainly resident; recent extensive range expansion in Saudi Arabia and Turkey. Vagrant Cyprus. **Habitat:** towns, villages, gardens, oases, savanna with trees and agricultural areas. Nests in tree, bush or house.

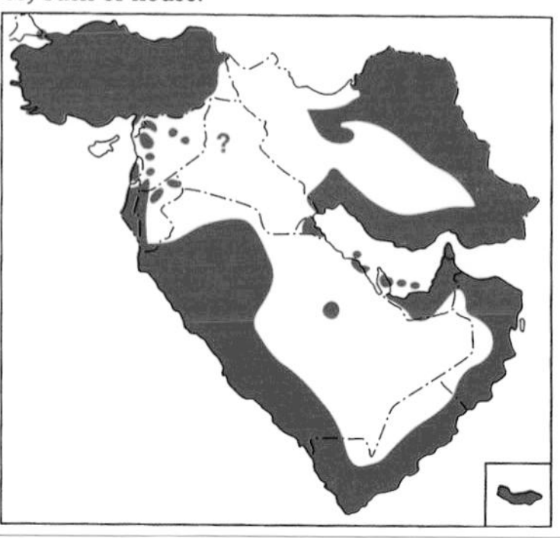

341 **Namaqua Dove** *Oena capensis*

Status: mainly resident, but nomadic movements. Spreading in Gulf States including UAE and Bahrain where rare but perhaps breeding. Scarce Kuwait. **Habitat:** savanna and semi-desert with thorn-bush or scrub. Nests in low bush or tree.

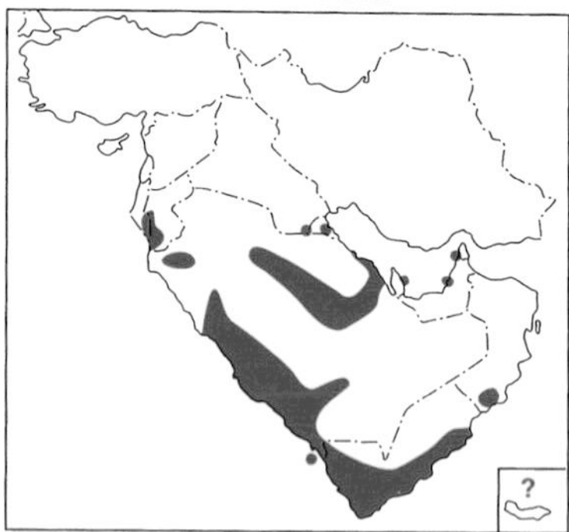

342 **Bruce's Green Pigeon** *Treron waalia*

Status: summer visitor, wintering Africa. **Habitat:** park-like open country, gardens with tall trees, especially fig, wooded wadis; chiefly between 700–2400 m. Nests in tree.

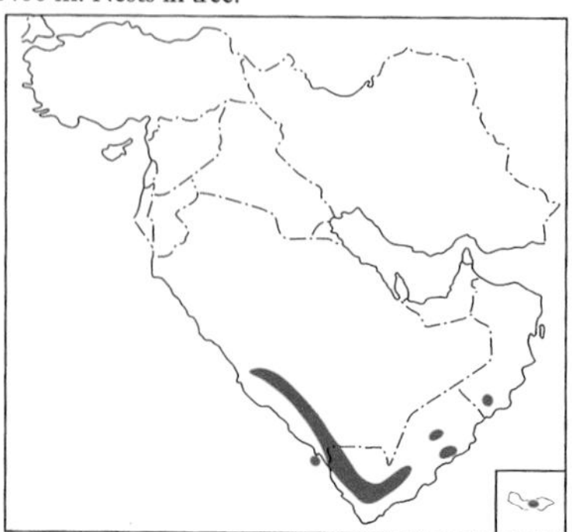

343 **Ring-necked Parakeet** *Psittacula krameri*

Status: mainly resident with some dispersal; perhaps some passage Oman: recorded Turkey. Probably originated from escapes. **Habitat:** gardens and open wooded areas; also near cultivation. Nests, often colonially, in holes in trees, walls or ruins.

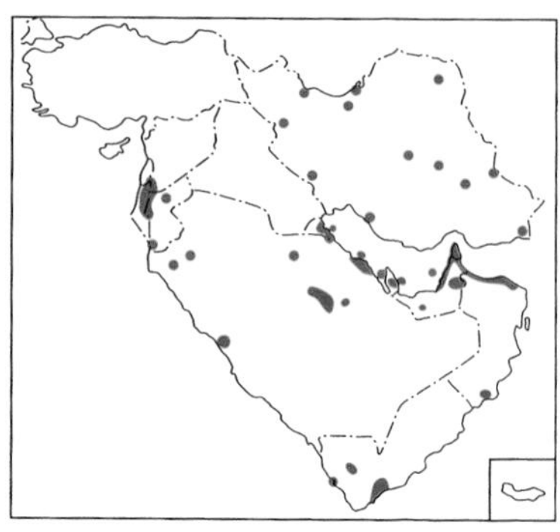

344 **Alexandrine Parakeet** *Psittacula eupatria*

Status: feral resident. Escapes recorded, and may breed, Gulf States, Oman and Yemen. **Habitat:** gardens and plantations. Nests in hole in tree.

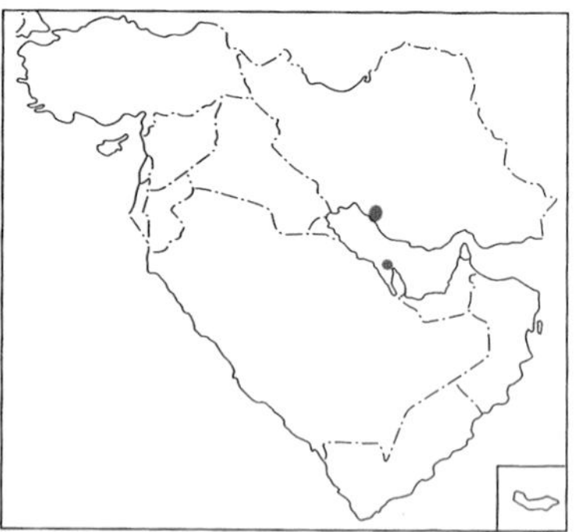

353 **Koel** *Eudynamys scolopacea*

Status: rare but regular winter visitor to Iran and Oman; vagrant UAE. Range Indian continent. **Habitat:** open scrub, gardens and woods, towns and villages, also coastal mangroves.

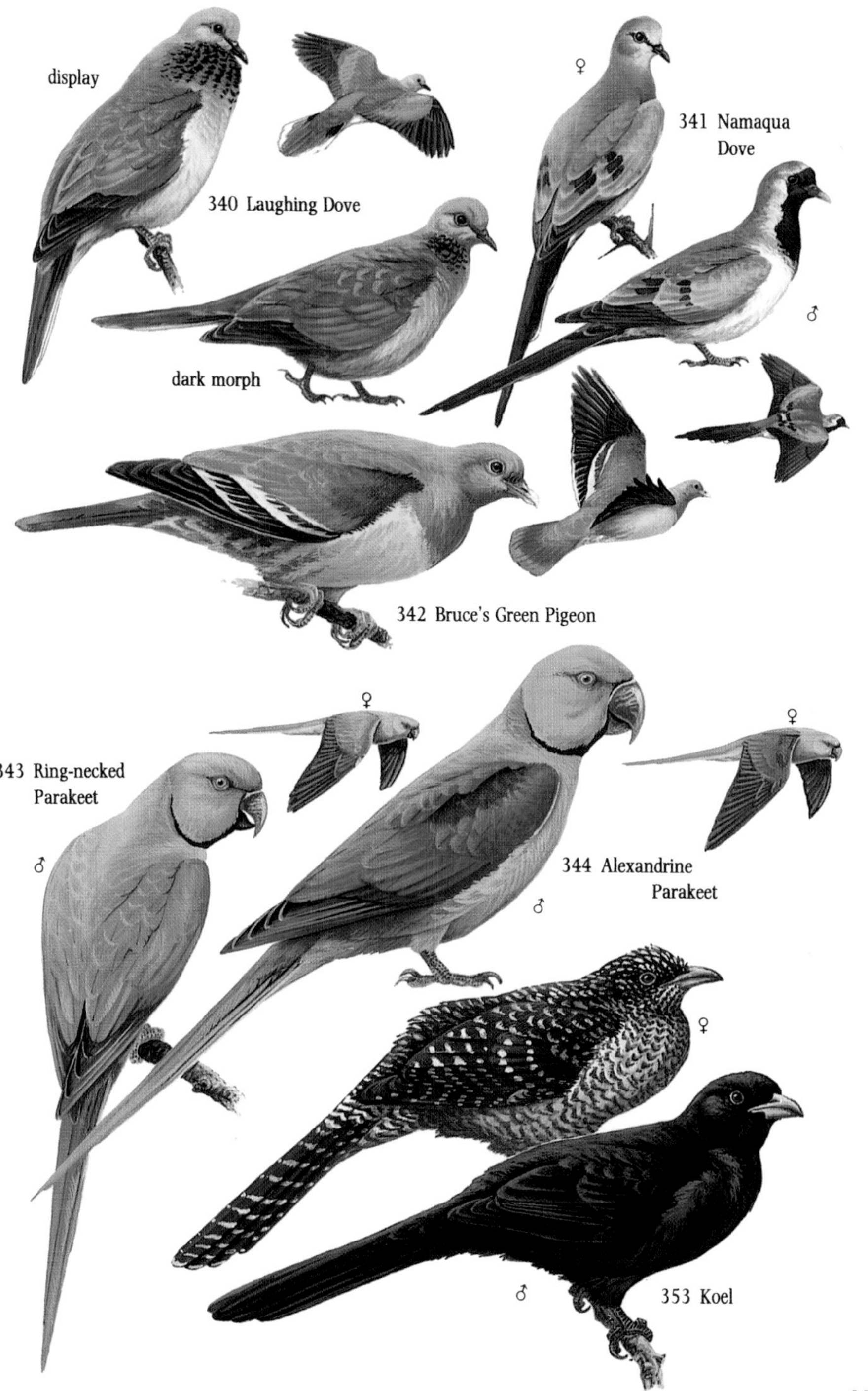
display
340 Laughing Dove
dark morph
♀
341 Namaqua Dove
♂
342 Bruce's Green Pigeon
♀
343 Ring-necked Parakeet
♂
♀
344 Alexandrine Parakeet
♂
♀
♂
353 Koel

ML

345 **Jacobin Cuckoo** *Clamator jacobinus*

Status: scarce summer visitor to Yemen and SW Saudi Arabia and Oman; formerly recorded in Iran. **Habitat:** thorn-scrub, woodlands and cultivated areas usually below 1000 m. Parasitizes nests of passerines, notably bulbuls and shrikes.

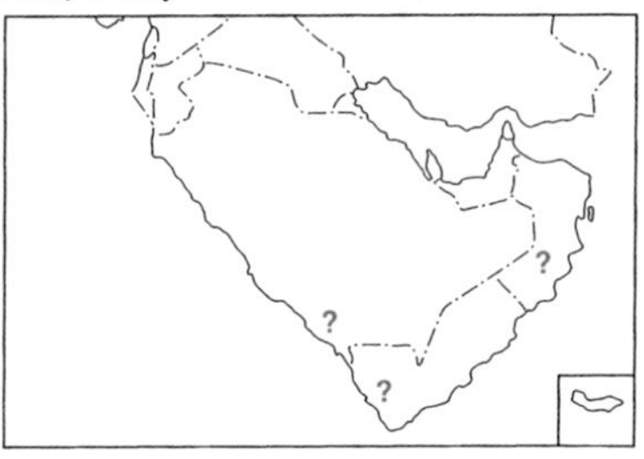

346 **Great Spotted Cuckoo** *Clamator glandarius*

Status: summer visitor. May breed Lebanon, Syria and Iran. Has wintered N Saudi Arabia. Scarce passage W Arabia, but vagrant elsewhere in Arabia. **Habitat:** woodland, olive groves, cultivation with bushes and trees. Parasitizes nests of crows and Magpies.

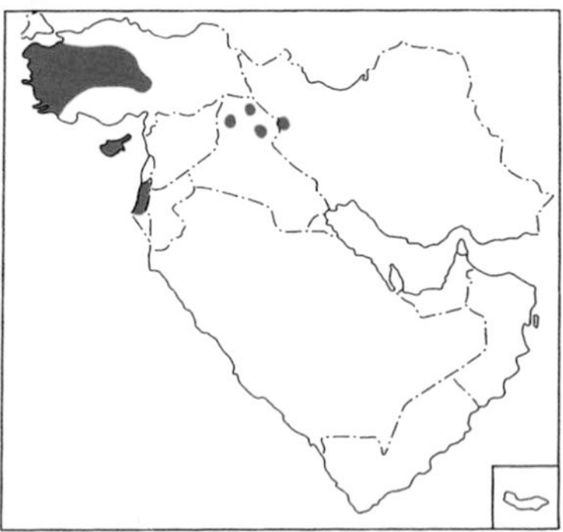

347 **Indian Hawk Cuckoo** *Cuculus varius*

Status: vagrant Oman (from Indian region). **Habitat:** usually in tall trees, but vagrants could occur anywhere with vegetation.

348 **Didric Cuckoo** *Chrysococcyx caprius*

Status: summer visitor. Vagrant Cyprus and Israel. **Habitat:** dry scrub and open woodland. Parasitizes nests of passerines.

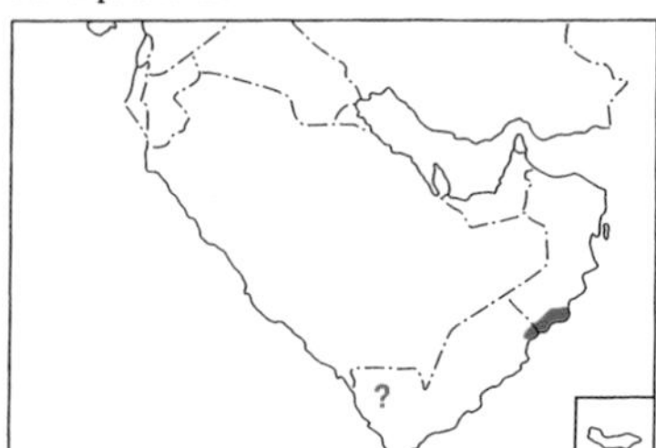

349 **Klaas's Cuckoo** *Chrysococcyx klaas*

Status: summer visitor. May breed Yemen and Saudi Arabia. **Habitat:** open woodland and bushland. Parasitizes nests of passerines.

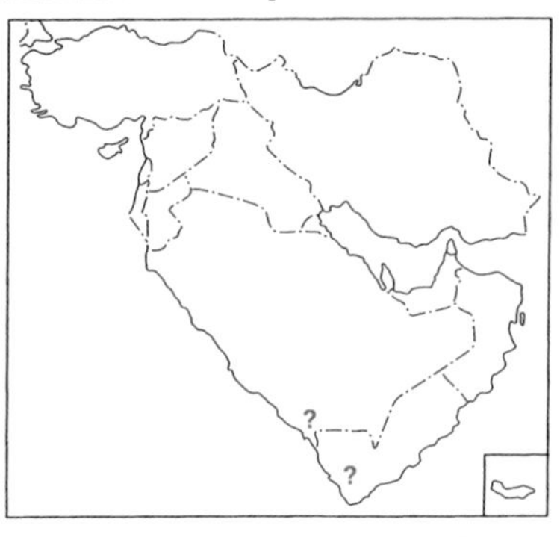

350 **Plaintive Cuckoo** *Cacomantis (merulinus) passerinus*

Status: vagrant Oman (from Indian Region), probably of race *passerinus*. **Habitat:** woodland, bushes and thickets.

351 **Common Cuckoo** *Cuculus canorus*

Status: may breed Syria, Lebanon and Cyprus. Has perhaps bred UAE. Summer visitor. Widespread on passage throughout the region. **Habitat:** open country with trees and bushes, also open woodland. Parasitizes passerine nests.

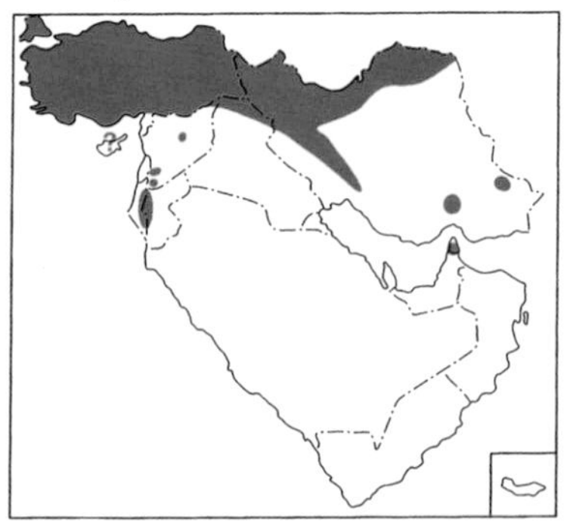

352 **White-browed Coucal** *Centropus superciliosus*

Status: resident. **Habitat:** dense scrub, palms, especially near water. Nests in thick cover.

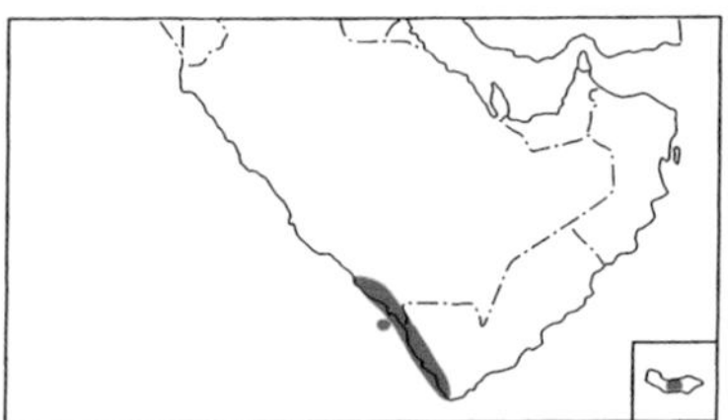

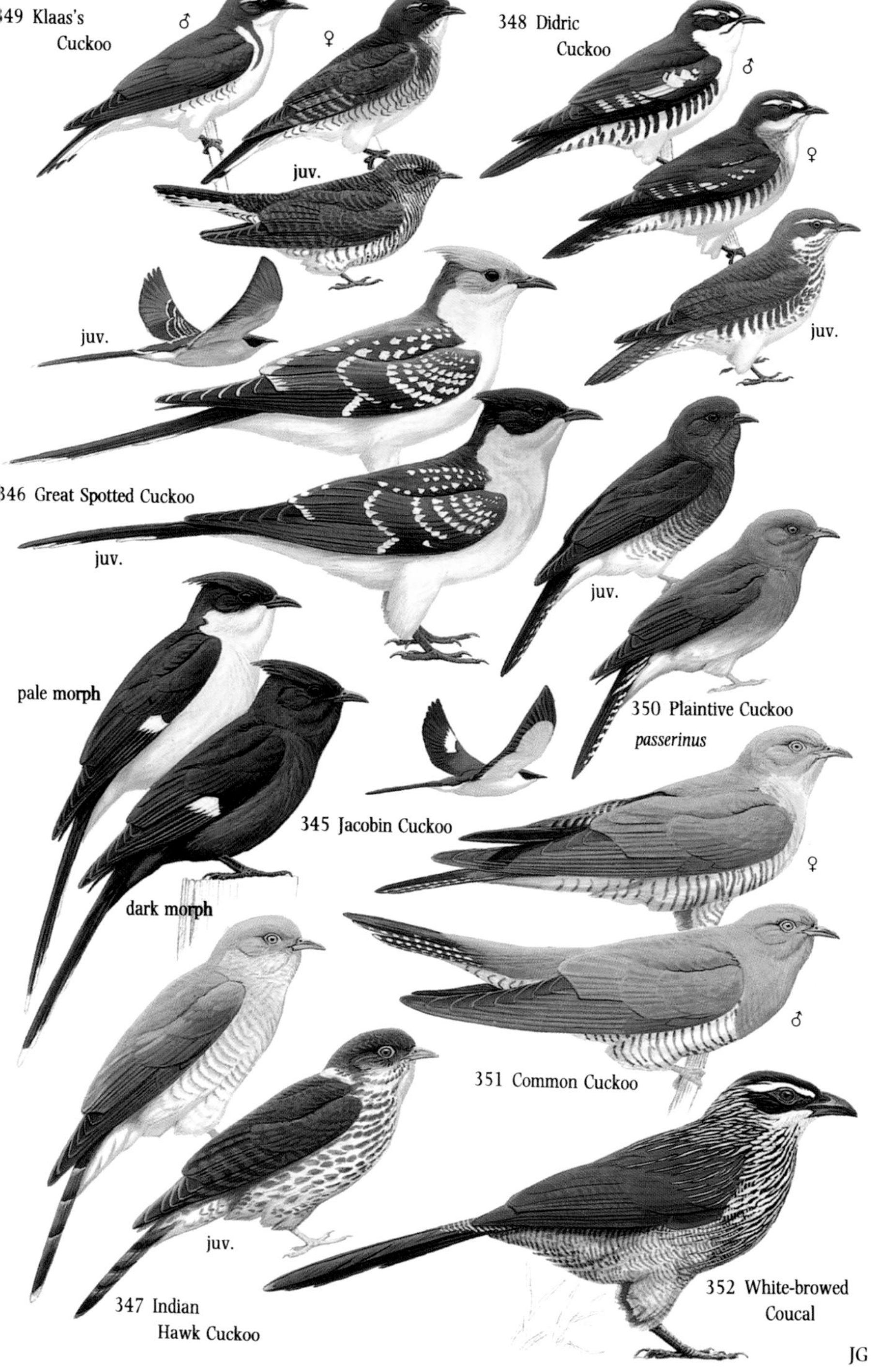
349 Klaas's Cuckoo
♂
♀
juv.
348 Didric Cuckoo
♂
♀
juv.
juv.
346 Great Spotted Cuckoo
juv.
juv.
pale morph
345 Jacobin Cuckoo
dark morph
350 Plaintive Cuckoo
passerinus
♀
♂
351 Common Cuckoo
juv.
347 Indian Hawk Cuckoo
352 White-browed Coucal
JG

355 **Indian Scops Owl** *Otus bakkamoena*

Status: uncertain. May perhaps breed in SE Iran but confirmation required; old record Oman may be erroneous. **Habitat:** woodlands in plains, hills and lower mountain slopes; also gardens.

356 **Striated Scops Owl** *Otus brucei*

Status: partial migrant, apparently present all year Iran; summer visitor Iraq, Oman; very scarce winter visitor S Israel; vagrant Bahrain and E province of Saudi Arabia. **Habitat:** arid foothills with preference for steep cliffs and gorges with small trees; also semi-deserts with tree cover; gardens in towns and palm groves. Nests in hole in tree.

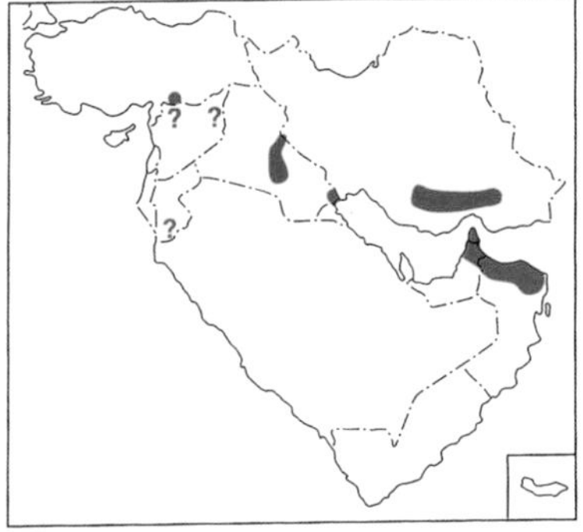

357 **European Scops Owl** *Otus scops*

Status: resident Cyprus. Otherwise mainly summer visitor and passage migrant, also occurring much of Arabia but vagrant Yemen. **Habitat:** trees near human habitation, plantations, oases, gardens. Nests in hole in tree or building, occasionally in old nest of other bird.

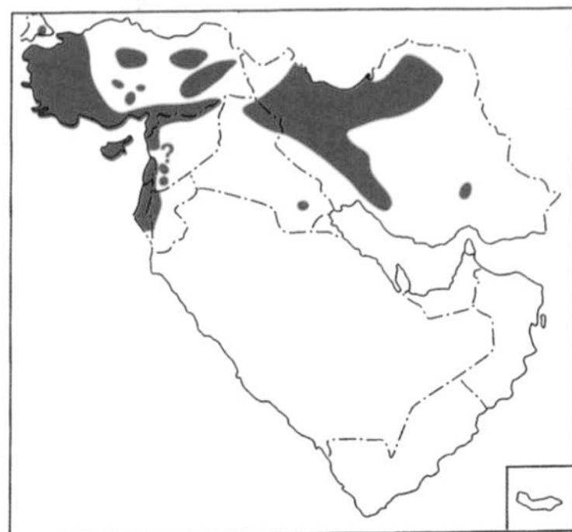

358 **African Scops Owl** *Otus senegalensis*

Status: resident. **Habitat:** hilly regions with trees. Nests in hole in tree.

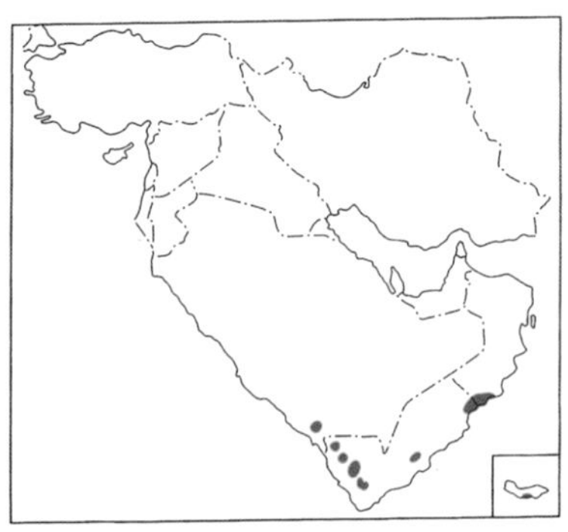

363 **Little Owl** *Athene noctua*

Status: resident. **Habitat:** fairly open country with trees, stony wasteland, wadis, rocky semi-deserts, cultivated areas. Nests in hole in tree and in rocks, buildings and burrows.

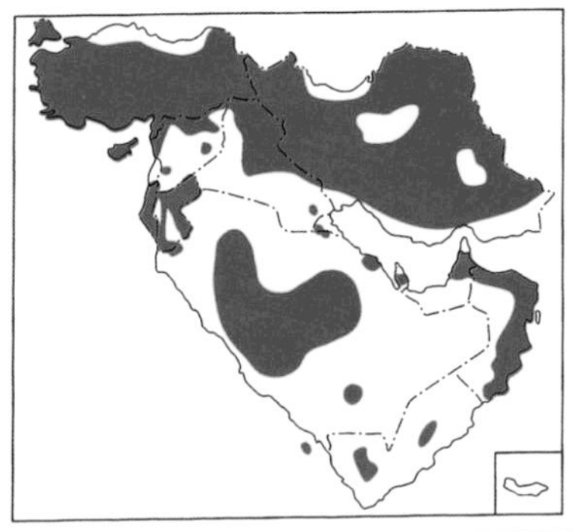

364 **Spotted Little Owl** *Athene brama*

Status: presumed resident. Apparently no overlap in range with Little Owl. **Habitat:** gardens, villages, cultivated areas, semi-deserts and earth cliffs. Nests in hole in tree or wall.

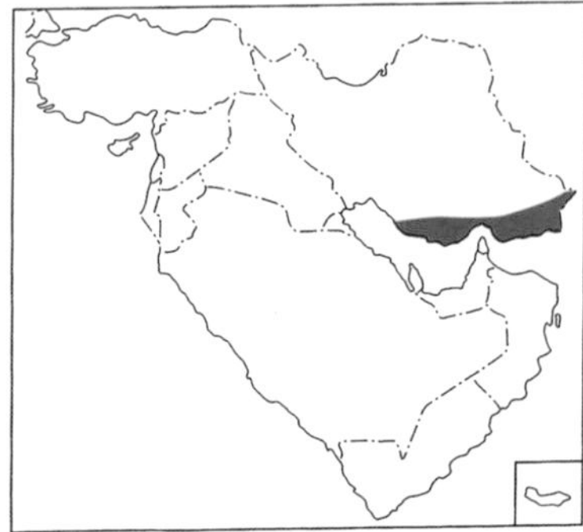

grey phase
rufous phase
355
Indian Scops
Owl
356
Striated Scops
Owl
357 European
Scops Owl
juv.
363 Little Owl
364 Spotted
Little
Owl
lilith
358 African Scops
Owl
Striated Scops Owl
European Scops Owl
Little Owl *lilith*
BS

354 **Barn Owl** *Tyto alba*

Status: resident. Distribution in Arabia uncertain. Recorded Kuwait and SW Iran. **Habitat:** open country (including semi-deserts) with trees, edges of woods, often near human habitation. Nests in holes in trees, buildings and ruins.

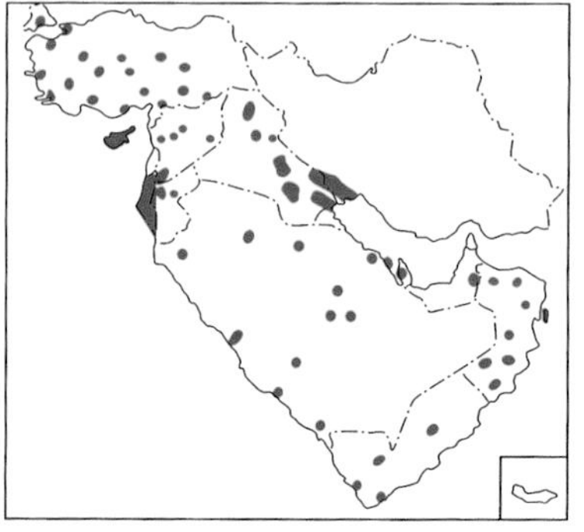

359 **Eagle Owl** *Bubo bubo*

Status: resident, sometimes wandering in winter. Vagrant Bahrain. **Habitat:** mountains and steppe regions, frequenting cliffs, crags and rocky outcrops. Nests on ledge or in crevice in rocks and down wells.

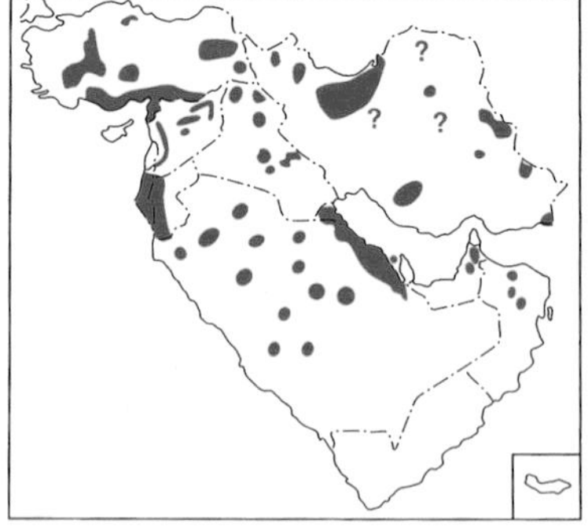

360 **Spotted Eagle Owl** *Bubo africanus*

Status: resident. **Habitat:** open woodland, rocky hills, ravines; sometimes in or near habitation. Nests on the ground under rock, cliff ledge or occasionally in tree.

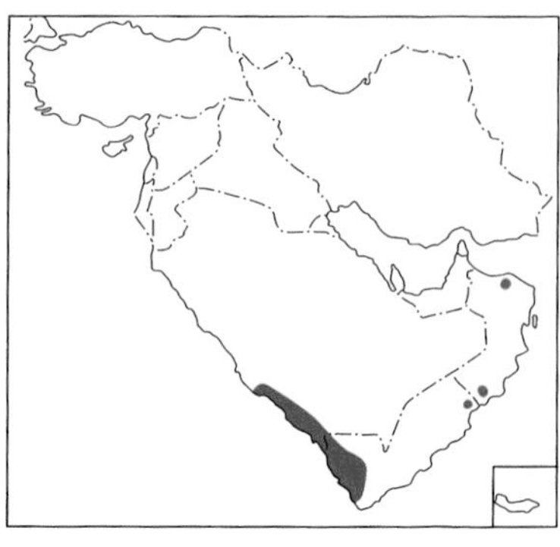

361 **Brown Fish Owl** *Ketupa zeylonensis*

Status: resident. Some old records SW Iran; also from S Turkey where recently recorded again after several decades. Vagrant Lebanon. **Habitat:** lowland tree-lined rivers or lakes. Nests in hole in cliff, dead tree or old raptor nest.

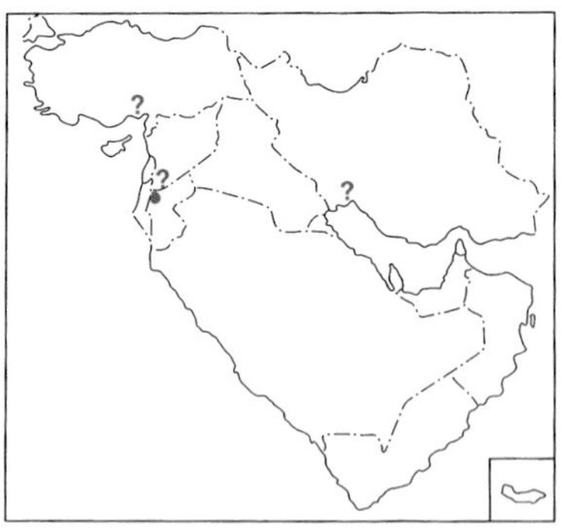

362 **Snowy Owl** *Nyctea scandiaca*

Status: rare winter visitor Turkoman steppes (NE Iran). **Habitat:** rolling plains and open steppe.

354 Barn Owl
desertorum
359 Eagle Owl
361 Brown Fish Owl
ascalaphus
Spotted Eagle Owl
360 Spotted Eagle Owl
362 Snowy Owl
1st year ♀
BS

365 **Tawny Owl** *Strix aluco*

Status: resident. **Habitat:** mature woods, parks, large gardens even in towns. Nests in hollow trees, old nests of large birds, also in buildings and sometimes on crags.

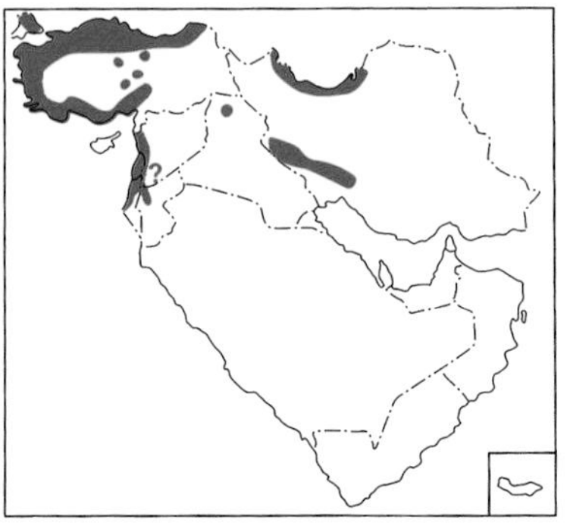

366 **Hume's Tawny Owl** *Strix butleri*

Status: isolated mapped spots in Yemen and Oman refer to records of single birds without proof of breeding. Range probably more extensive than shown and may include S Iran. Resident. **Habitat:** rocky deserts with gorges, rocky wadis, desert earth banks; often near palm groves, acacias and sometimes near springs and settlements. Nests in hole or crevice in rocks or cliff-face.

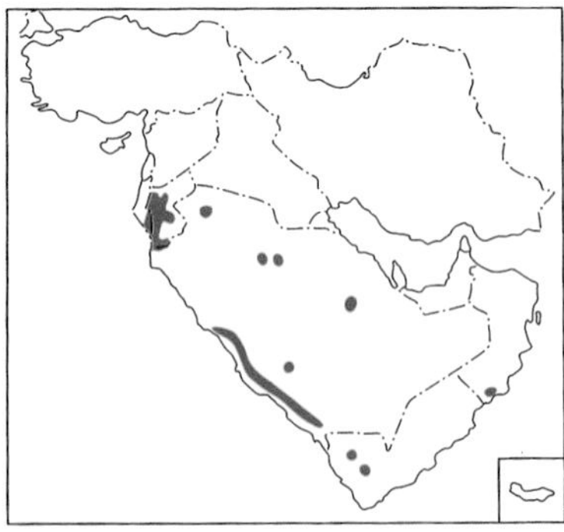

367 **Long-eared Owl** *Asio otus*

Status: partial migrant. In winter reaches Iraq, Israel and Jordan. Vagrant Arabian Gulf, Oman, SW Saudi Arabia. **Habitat:** deciduous woods and copses, also stands of conifers. Nests in old nest of raptor or crow, usually in a tree.

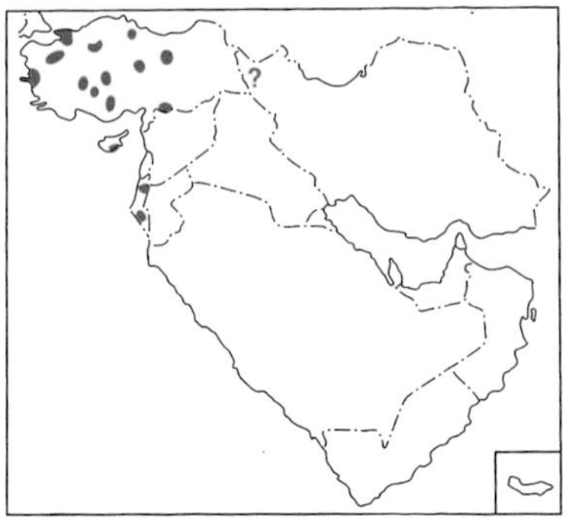

368 **Short-eared Owl** *Asio flammeus*

Status: has bred Turkey, Israel. Mainly winter visitor in most of the Asiatic countries south to central Saudi Arabia and Oman (where uncommon) vagrant farther south. **Habitat:** open, often marshy, country on coasts and inland. Nests on ground in vegetation near marsh.

369 **Tengmalm's Owl** *Aegolius funereus*

Status: recently recorded from N and NW Turkey where heard calling, so perhaps breeding. **Habitat:** coniferous (occasionally mixed) forest in mountains. Nests in hole in tree.

Tawny Owl
366 Hume's Tawny Owl
grey
phase
rufous
phase
365 Tawny Owl
368 Short-eared Owl
367 Long-eared Owl
369 Tengmalm's Owl
BS

370 **Plain Nightjar** *Caprimulgus inornatus*

Status: status uncertain. Probably a summer visitor (from Africa). **Habitat:** mountainous areas, often dry, with scattered vegetation up to 1500 m. Nests on ground.

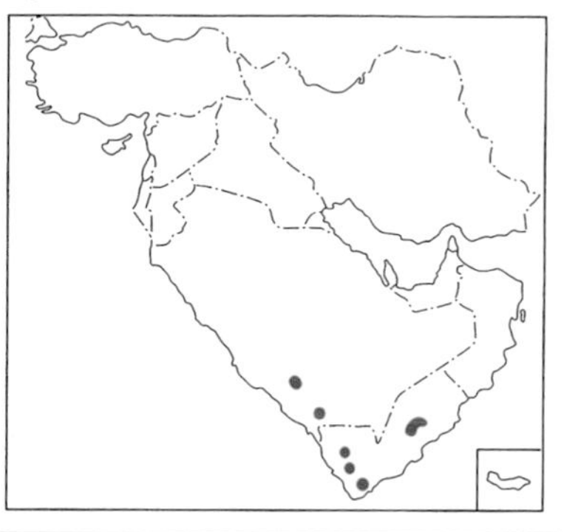

371 **Nubian Nightjar** *Caprimulgus nubicus*

Status: mainly resident but summer Israel; rare visitor Oman. **Habitat:** sand deserts with scattered vegetation, including palms and tamarisk. Nests on ground.

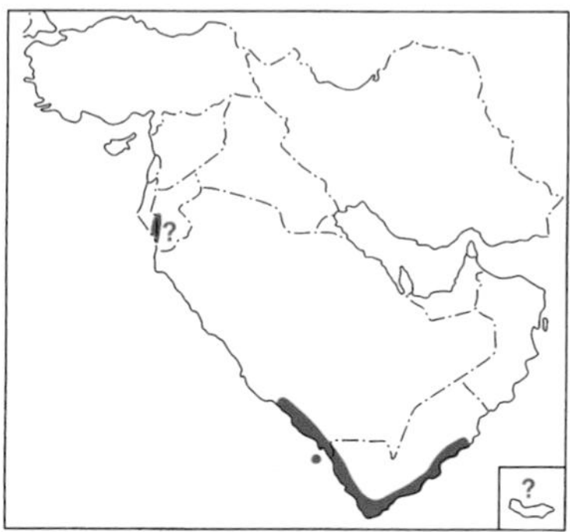

372 **Sykes's Nightjar** *Caprimulgus mahrattensis*

Status: resident. **Habitat:** deserts and semi-deserts, abandoned cultivation; often with tamarisks. Nests on ground.

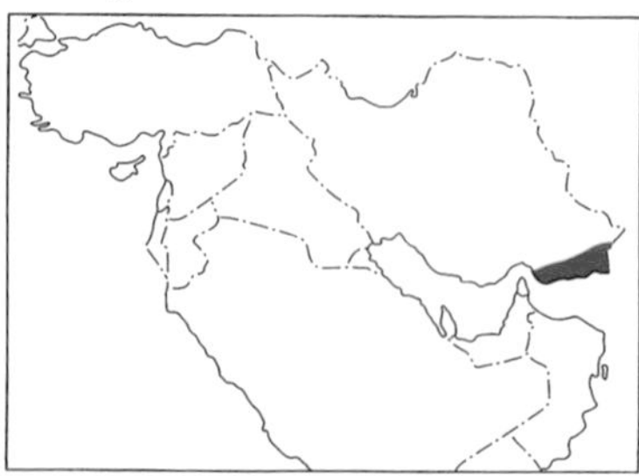

373 **Indian Nightjar** *Caprimulgus asiaticus*

Status: uncertain. May breed NE Iran; vagrant SE Iran. **Habitat:** open plains, including cultivation; also gardens. Nests on ground.

374 **European Nightjar** *Caprimulgus europaeus*

Status: summer visitor to breeding areas. Passage throughout region. Occasional in winter Oman. **Habitat:** edges of woods and heaths, steppes with sparse vegetation; any open areas on migration but often seeks shelter of trees.

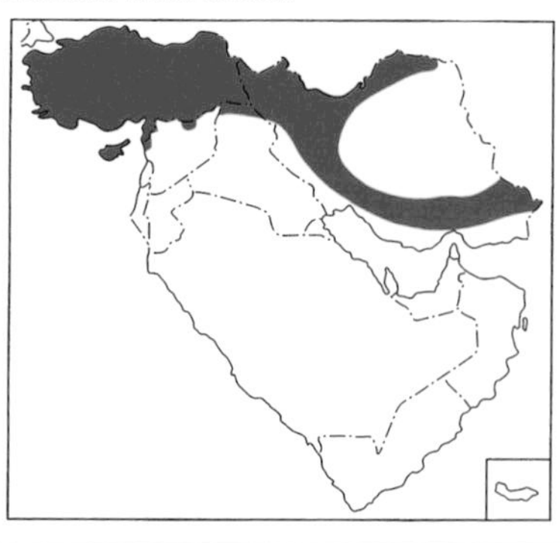

375 **Egyptian Nightjar** *Caprimulgus aegyptius*

Status: mainly summer visitor, also passage migrant in Iran and Arabia. Winter records from Oman, Saudi Arabia and Yemen. Vagrant Israel and Syria. **Habitat:** semi-deserts often with palms or scrub. Nests on ground.

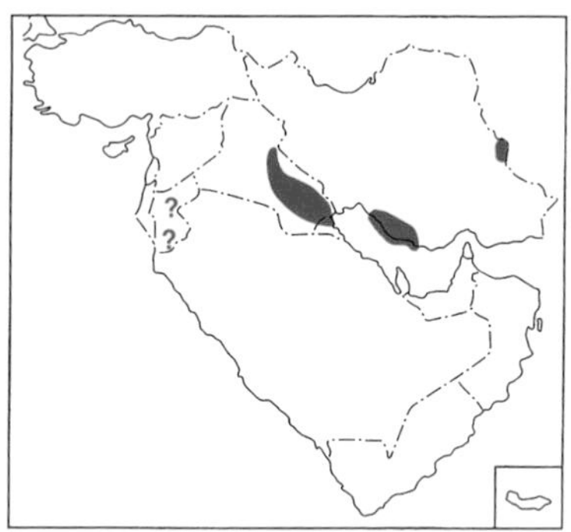

376 **Mountain Nightjar** *Caprimulgus poliocephalus*

Status: probably resident. **Habitat:** rocky mountains with junipers; 2000–3000 m, lower in winter.

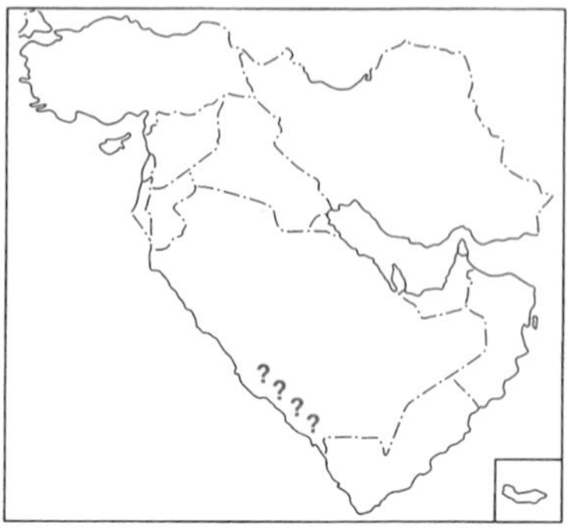

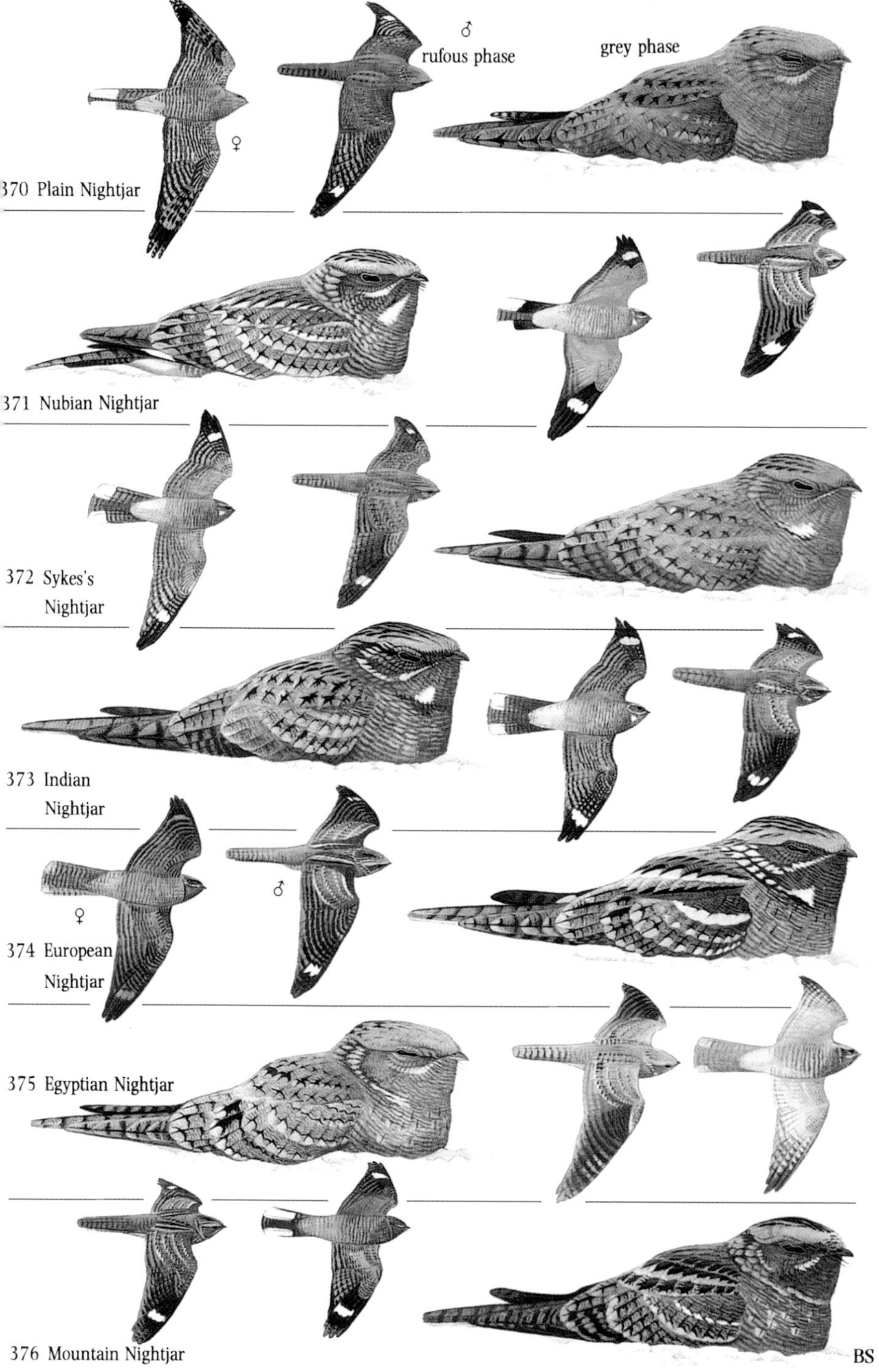
♂
rufous phase
grey phase
♀
370 Plain Nightjar
371 Nubian Nightjar
372 Sykes's Nightjar
373 Indian Nightjar
♂
♀
374 European Nightjar
375 Egyptian Nightjar
376 Mountain Nightjar
BS

377 **Common Swift** *Apus apus*

Status: summer visitor. Passage throughout region. Some winter records in S Arabia. **Habitat:** aerial; may occur anywhere but especially in areas, including towns, with suitable nesting sites or food abundance. Nests in buildings, under eaves, occasionally in cliffs.

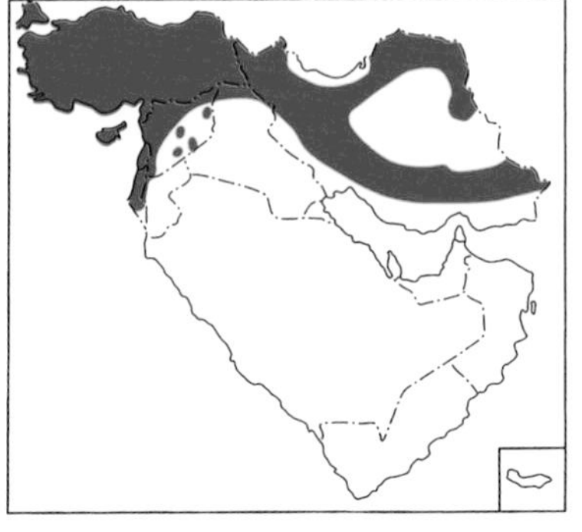

378 **Pallid Swift** *Apus pallidus*

Status: mainly summer visitor (can arrive in Arabian Gulf breeding localities in November), but locally resident. Occurs over most of region on passage. **Habitat:** as Swift, but greater preference for cliffs. Nests as Swift.

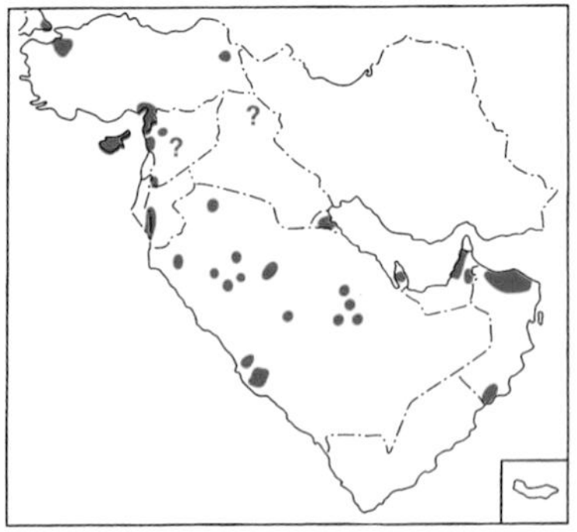

379 **Forbes-Watson's Swift** *Apus berliozi*

Status: probably summer visitor to Socotra where breeds. **Habitat:** mountains, foothills and plains.

380 **Alpine Swift** *Apus melba*

Status: mainly summer visitor and passage migrant, but probably resident in W Arabia; winter records Cyprus, Iran, Yemen. Irregular passage Oman, E Arabia and Gulf area. **Habitat:** rocky mountain regions, also along sea-cliffs and in old towns. Nests (usually colonially) in natural crevices and beneath rafters.

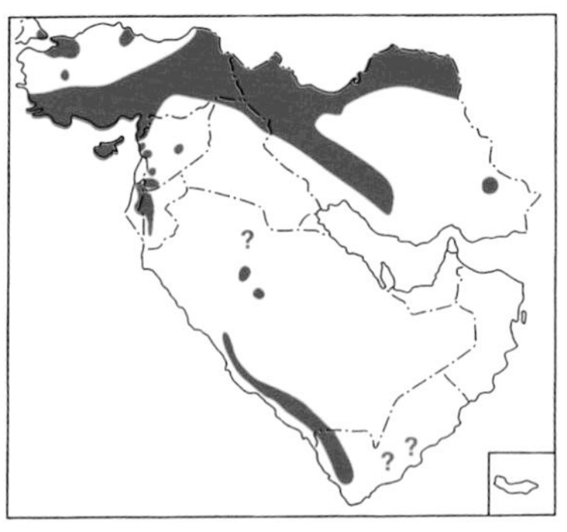

381 **White-rumped Swift** *Apus caffer*

Status: vagrant Yemen, from Africa. **Habitat:** mountainous regions and gorges, also towns and villages.

382 **Little Swift** *Apus affinis*

Status: occurs Syria in breeding season; partial migrant with few remaining in northern part of range in winter. Passage includes Yemen. Vagrant Iraq, UAE, Oman, Bahrain and Cyprus. **Habitat:** varied from rocky ravines and cliffs to large towns; often near water. Nests are clustered below eaves, rocky overhang or ceiling of open building.

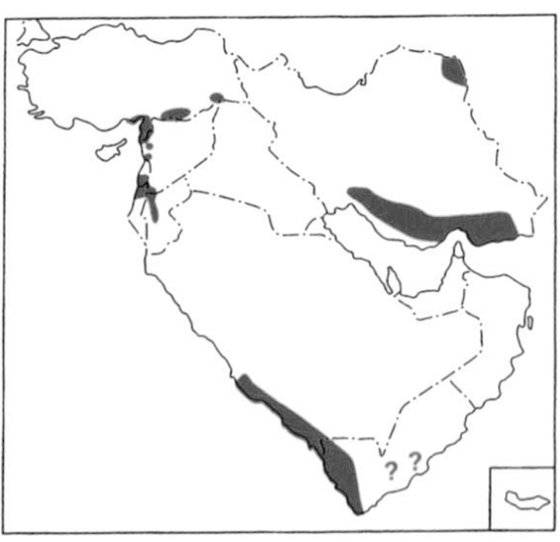

383 **Palm Swift** *Cypsiurus parvus*

Status: resident. **Habitat:** over open country, villages and towns, usually near palms in which it nests.

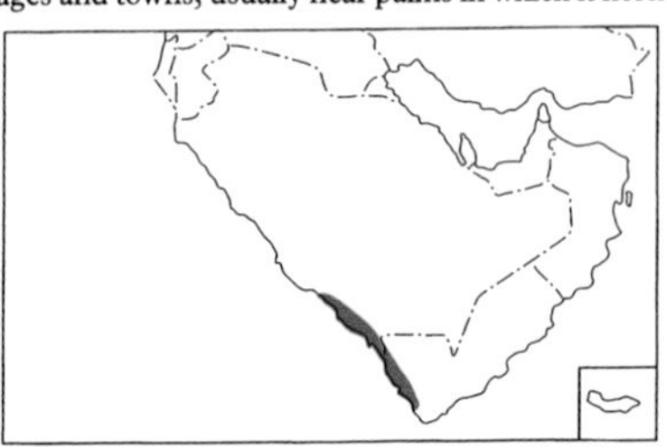

377 Common Swift
379 Forbes-Watson's Swift
378 Pallid Swift
380 Alpine Swift
381 White-rumped Swift
383 Palm Swift
382 Little Swift
ML

384 **White-breasted Kingfisher**
Halcyon smyrnensis

Status: mainly resident, but some movements in winter, when reaching Kuwait. Rare visitor central and E Saudi Arabia; vagrant Cyprus, Lebanon, Qatar, UAE. **Habitat:** wholly dry woodland glades and palm groves as well as tree-lined lakes, rivers or other wetlands. Nests in hole in bank.

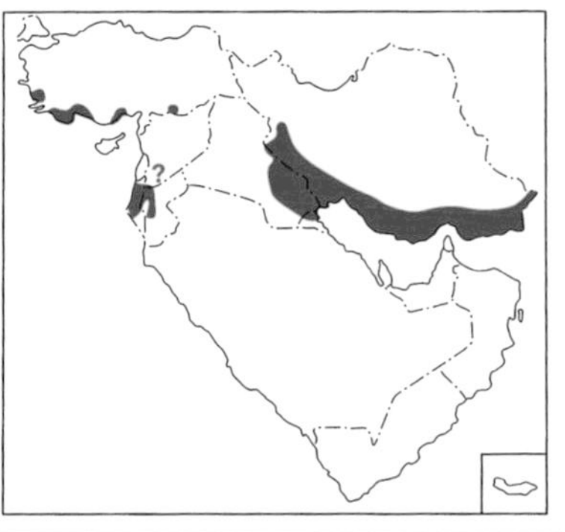

385 **Grey-headed Kingfisher**
Halcyon leucocephala

Status: summer visitor. Vagrant UAE. **Habitat:** wadis with trees (with or without water), particularly palms, usually between 250–1500 m, at lower altitudes on migration. Nests in hole in bank or cliff.

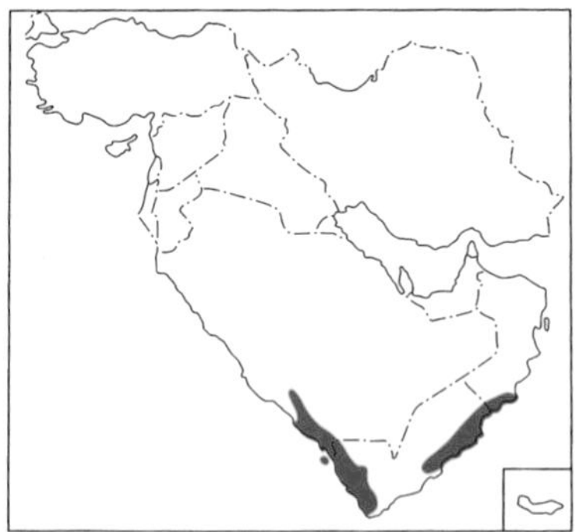

386 **White-collared Kingfisher**
Halcyon chloris

Status: resident; probably dispersive. **Habitat:** mangroves. Nests in hole in tree or bank.

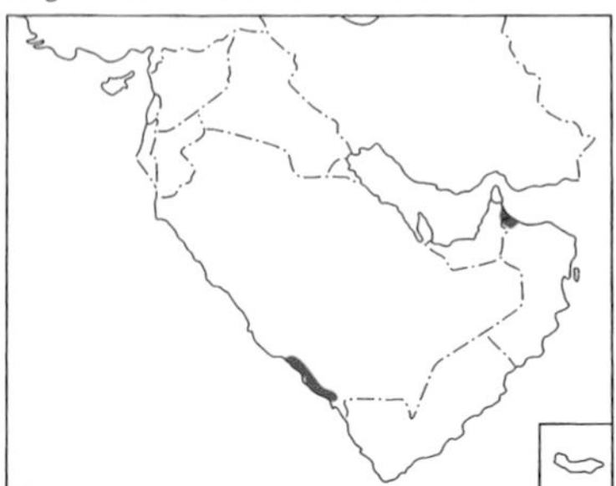

387 **Common Kingfisher** *Alcedo atthis*

Status: partial migrant though resident some places. Reaches coasts of E Mediterranean and Arabian peninsula, except southern part; vagrant Yemen. **Habitat:** rivers, calm streams, canals and lakes; in winter also coasts. Nests in hole excavated in bank.

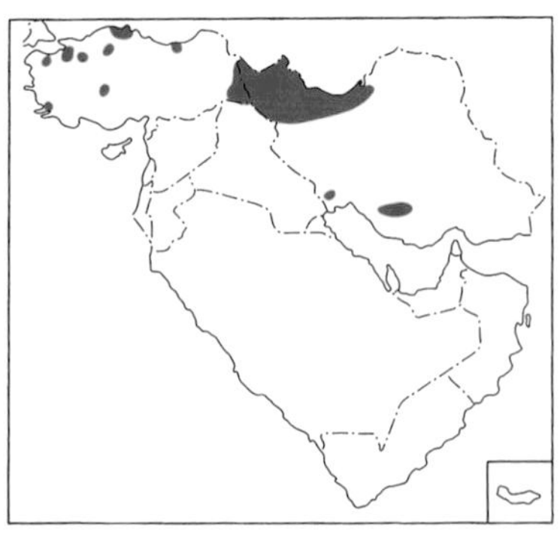

388 **Malachite Kingfisher** *Alcedo cristata*

Status: very rare Yemen where it may breed; from Africa. **Habitat:** permanent water with vegetation.

389 **Pied Kingfisher** *Ceryle rudis*

Status: mainly resident. Some dispersal in winter to Iran, Iraq; reaches Kuwait. Rare Cyprus and SE Arabian Gulf, vagrant N and central Saudi Arabia, Oman. **Habitat:** rivers, lakes, estuaries, salt pans and coasts. Nests in hole in bank.

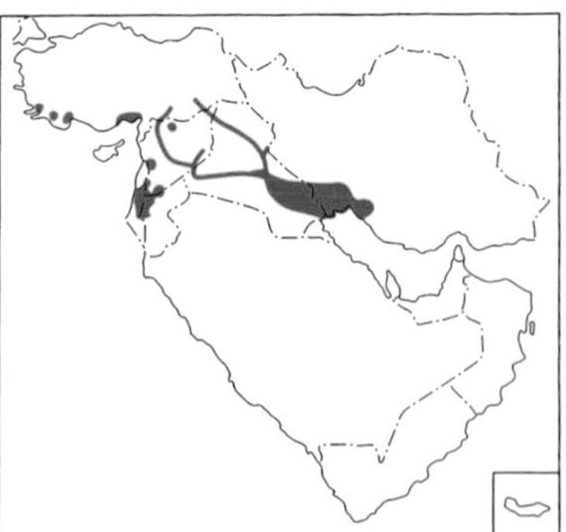

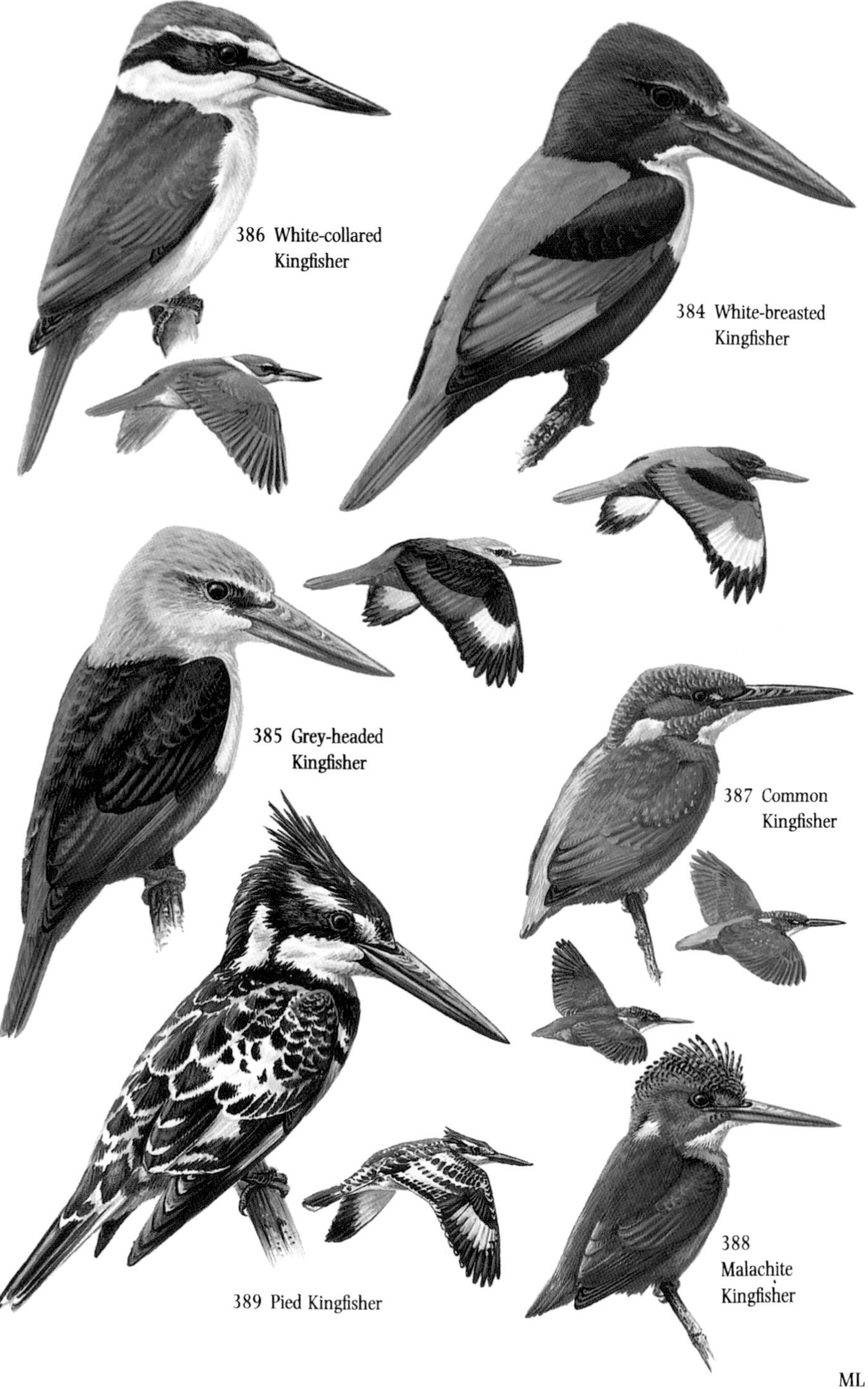
386 White-collared Kingfisher
384 White-breasted Kingfisher
385 Grey-headed Kingfisher
387 Common Kingfisher
388 Malachite Kingfisher
389 Pied Kingfisher
ML

390 **White-throated Bee-eater**
Merops albicollis

Status: summer visitor. Vagrant Oman. **Habitat:** hills, plains and wadis with bushes and trees; often in agriculture; from sea level to 1500 m. Nests in hole in bank.

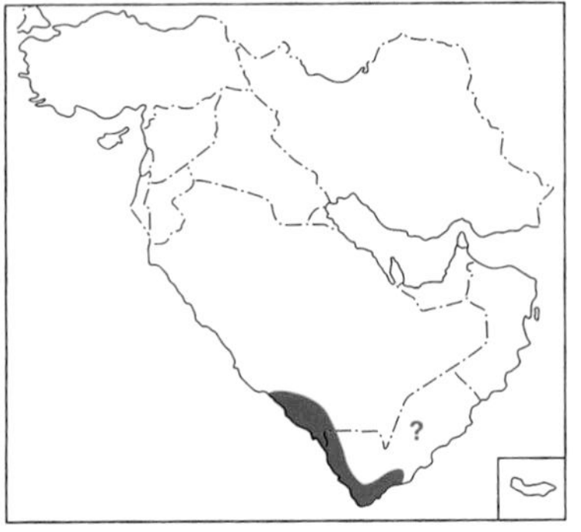

391 **Little Green Bee-eater** *Merops orientalis*

Status: mainly resident, but some seasonal movement. Vagrant southern Arabian Gulf, but spreading westward from UAE. **Habitat:** mainly lowland or open country with trees, cultivation. Nests in tunnel in bank or hole in ground.

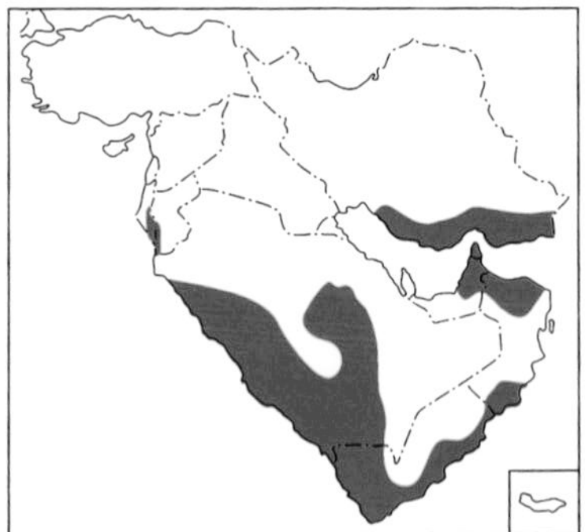

392 **Blue-cheeked Bee-eater**
Merops superciliosus

Status: mainly summer visitor. Passage Near East and Arabia. **Habitat:** dry sandy areas with scattered trees, often close to water. Nests colonially in holes in sandy bank.

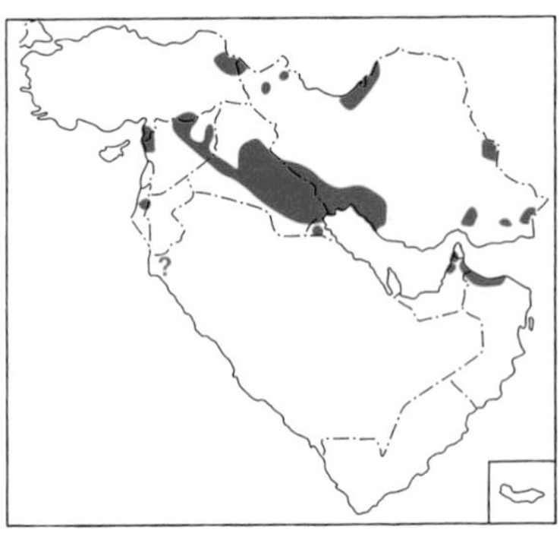

393 **European Bee-eater** *Merops apiaster*

Status: summer visitor. Broad fronted passage throughout region. **Habitat:** open bushy country with scattered trees and telegraph poles, river banks, also woodland glades. Nests colonially in holes in sand pits, river and roadside banks.

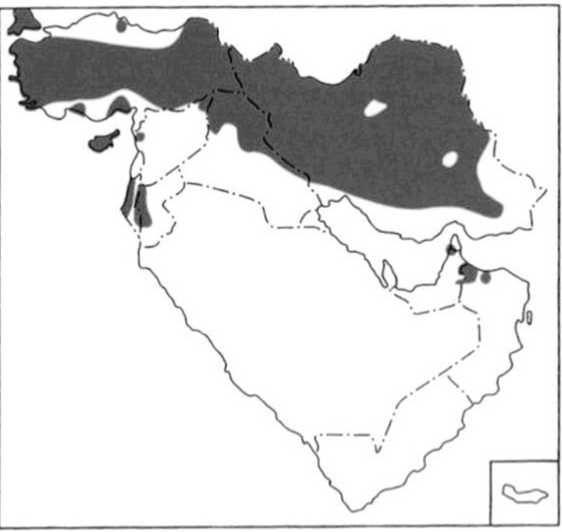

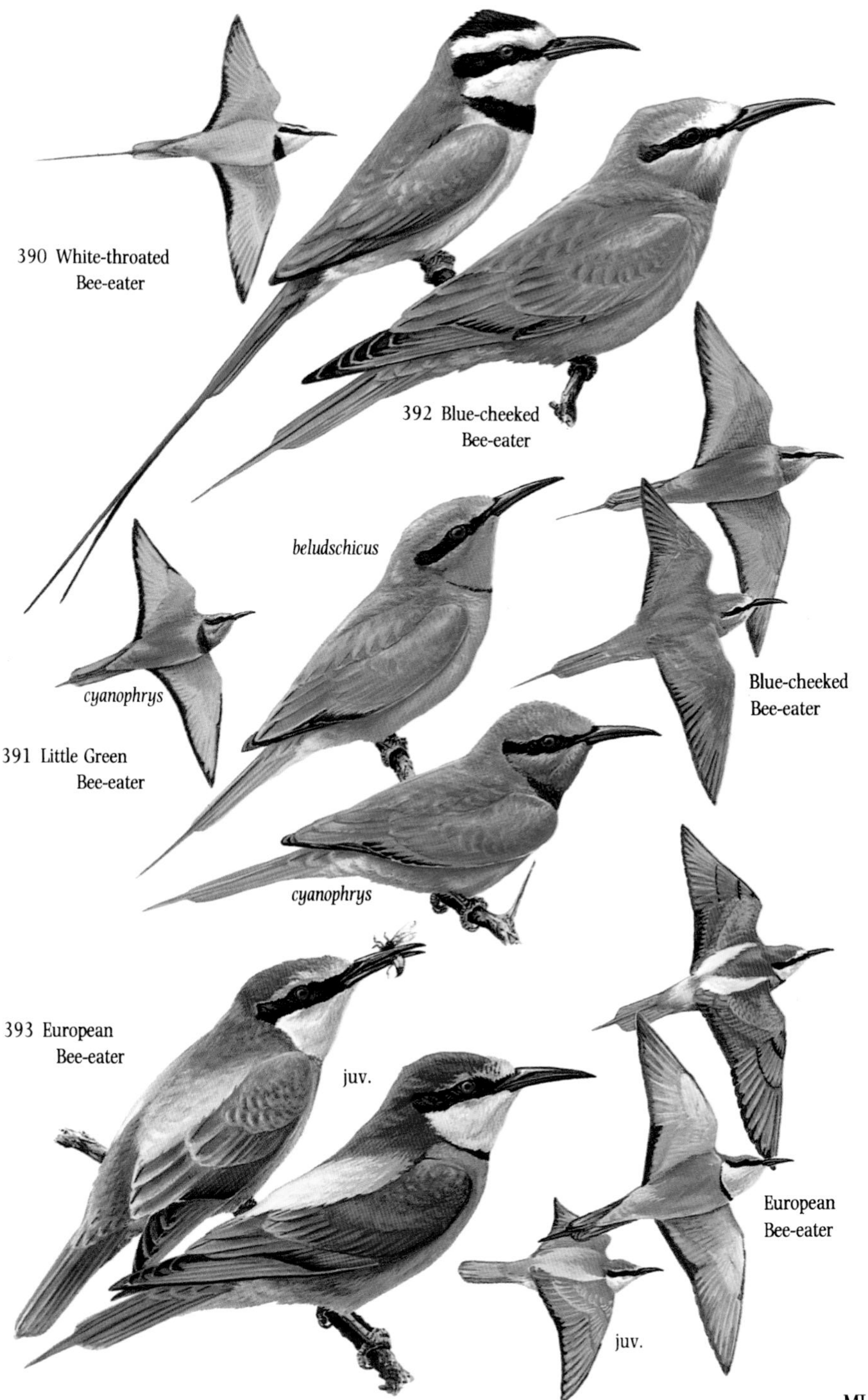
390 White-throated
Bee-eater
392 Blue-cheeked
Bee-eater
beludschicus
cyanophrys
391 Little Green
Bee-eater
Blue-cheeked
Bee-eater
cyanophrys
393 European
Bee-eater
juv.
European
Bee-eater
juv.
ML

394 **European Roller** *Coracias garrulus*

Status: summer visitor. Seen on passage over most of the region. **Habitat:** open country with clumps of large hollow trees; rarely up to 2000 m. More varied habitat on passage. Nests in hole in tree, ruin or rock.

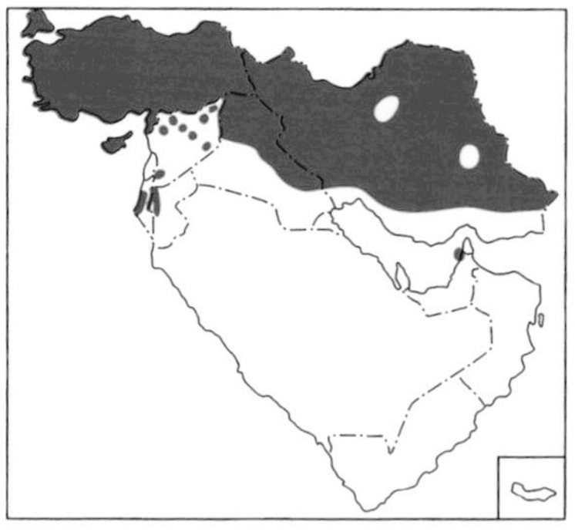

395 **Abyssinian Roller** *Coracias abyssinicus*

Status: mainly resident. **Habitat:** savanna and semi-desert with scattered trees; also tree-covered plains. Nests in hole in tree or building.

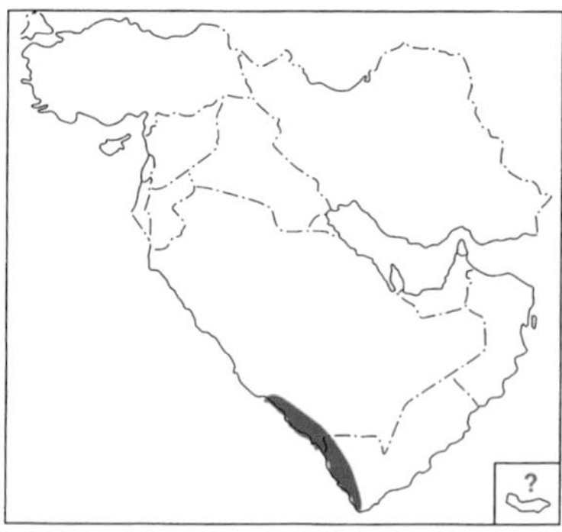

396 **Indian Roller** *Coracias benghalensis*

Status: mainly resident. Immigration to Oman in autumn. Vagrant Kuwait and Turkey. **Habitat:** open cultivated country with plantations or scattered trees, usually below 1000 m. Nests in hole in tree or wall.

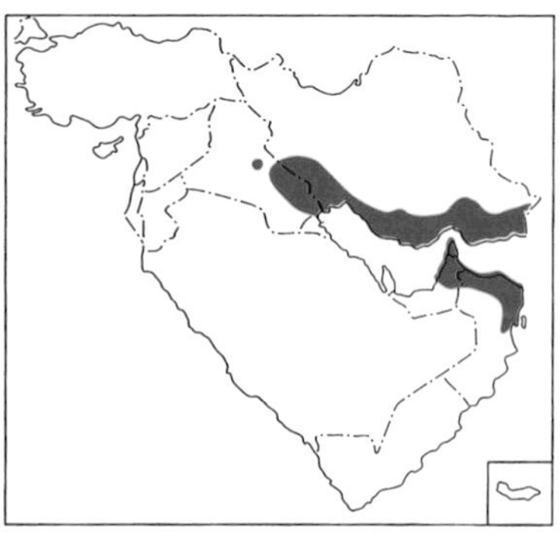

397 **Lilac-breasted Roller** *Coracias caudata*

Status: vagrant Yemen from Africa. **Habitat:** open woodland and plains with trees.

398 **Rufous-crowned Roller** *Coracias naevia*

Status: vagrant Yemen from Africa. **Habitat:** wooded areas.

399 **Hoopoe** *Upupa epops*

Status: partial migrant; does not winter Turkey, N and central Iran, but present farther south. Widespread on passage. **Habitat:** open woodland, olive and palm groves, oases; more in the open in winter. Nests in hole in tree or ruin.

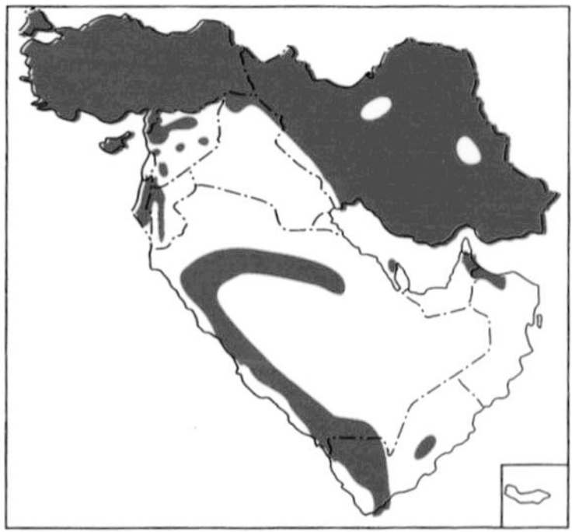

394 European
Roller
juv.
395 Abyssinian
Roller
juv.
juv.
397 Lilac-breasted
Roller
398 Rufous-
crowned Roller
396 Indian Roller
399 Hoopoe

ML

400 **African Grey Hornbill** *Tockus nasutus*

Status: resident. **Habitat:** acacia woodland, plains and hills with mature trees. Nests in hole in tree, where female is almost walled-in.

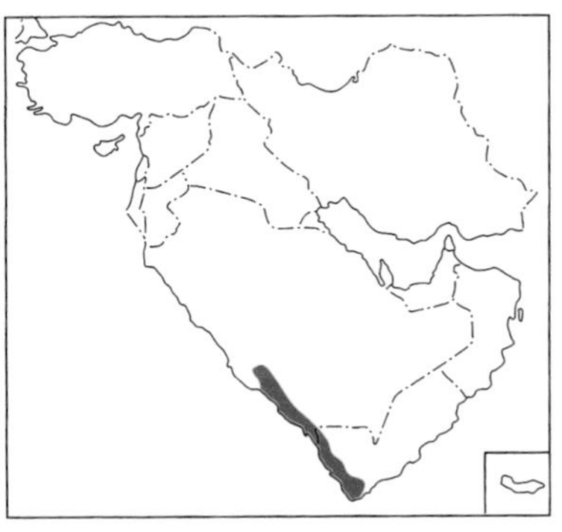

401 **Wryneck** *Jynx torquilla*

Status: may breed NW Iran. Summer visitor N Turkey otherwise widespread passage throughout the region, though uncommon. Has wintered SW Turkey, Cyprus and from E Arabian Gulf to Oman. **Habitat:** open woodland, orchards, parks; can occur in any area with scrub or bushes on migration. Nests mostly in natural holes in trees.

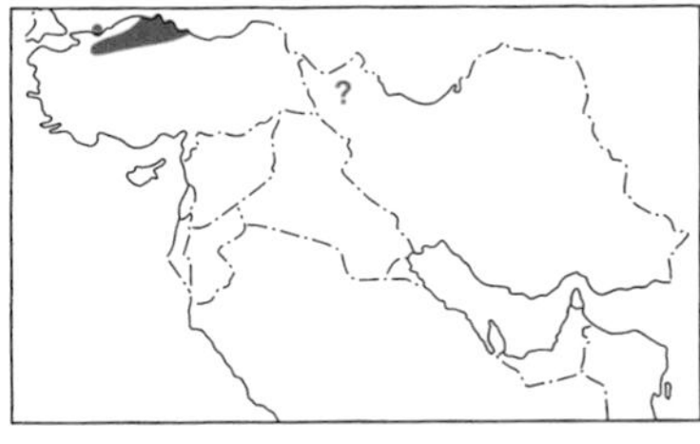

402 **Grey-headed Woodpecker** *Picus canus*

Status: recently proved breeding N Turkey, where potentially a widespread resident. **Habitat:** as Green Woodpecker but also deciduous mountain forests up to tree line; occasionally in coniferous woods. Nest as Green Woodpecker.

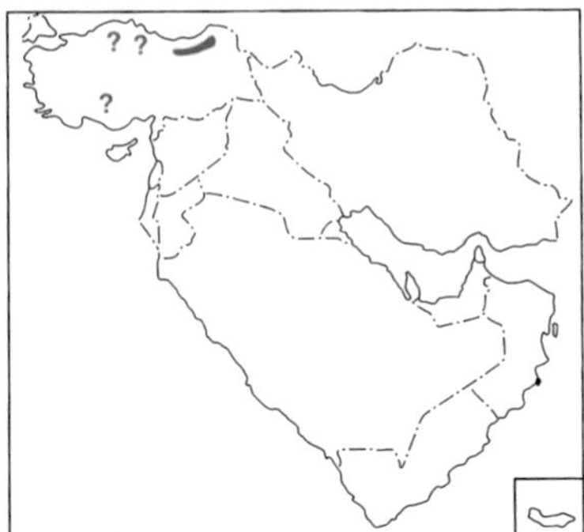

403 **Green Woodpecker** *Picus viridis*

Status: resident. **Habitat:** deciduous woods, parks and open areas with scattered trees. Nests in hole bored in tree.

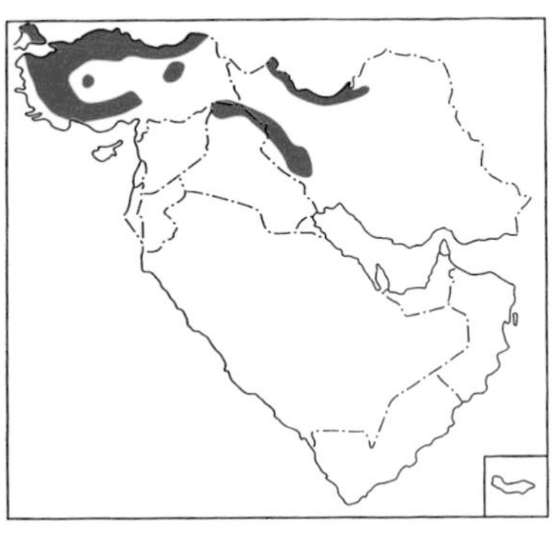

404 **Scaly-bellied Woodpecker** *Picus squamatus*

Status: uncertain. No recent records from Iran though typically resident in breeding areas; may vary altitude in winter. **Habitat:** mixed woodland and open country with scattered copses, orchards and junipers to 3000 m. Nests in hole bored in tree.

405 **Black Woodpecker** *Dryocopus martius*

Status: resident. **Habitat:** mature mountain forests particularly coniferous and beech. Nests in excavated hole often high in tree.

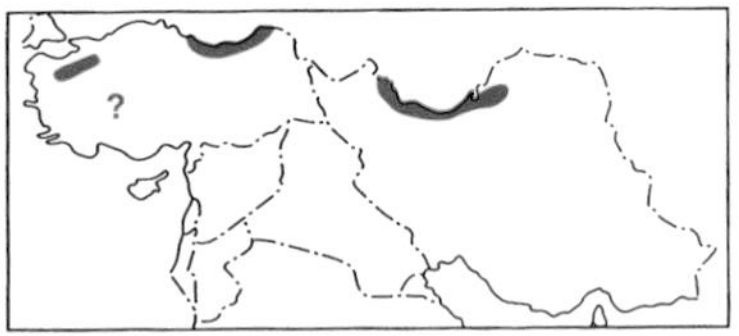

413 **Arabian Woodpecker** *Dendrocopos dorae*

Status: resident. **Habitat:** acacia woodland from near sea level to 2400 m, but most frequently above 1500 m; also in other trees including palms and often close to habitation. Nests in hole in tree.

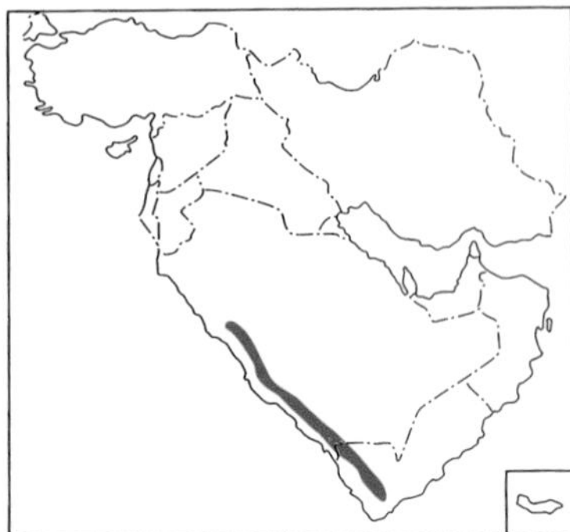

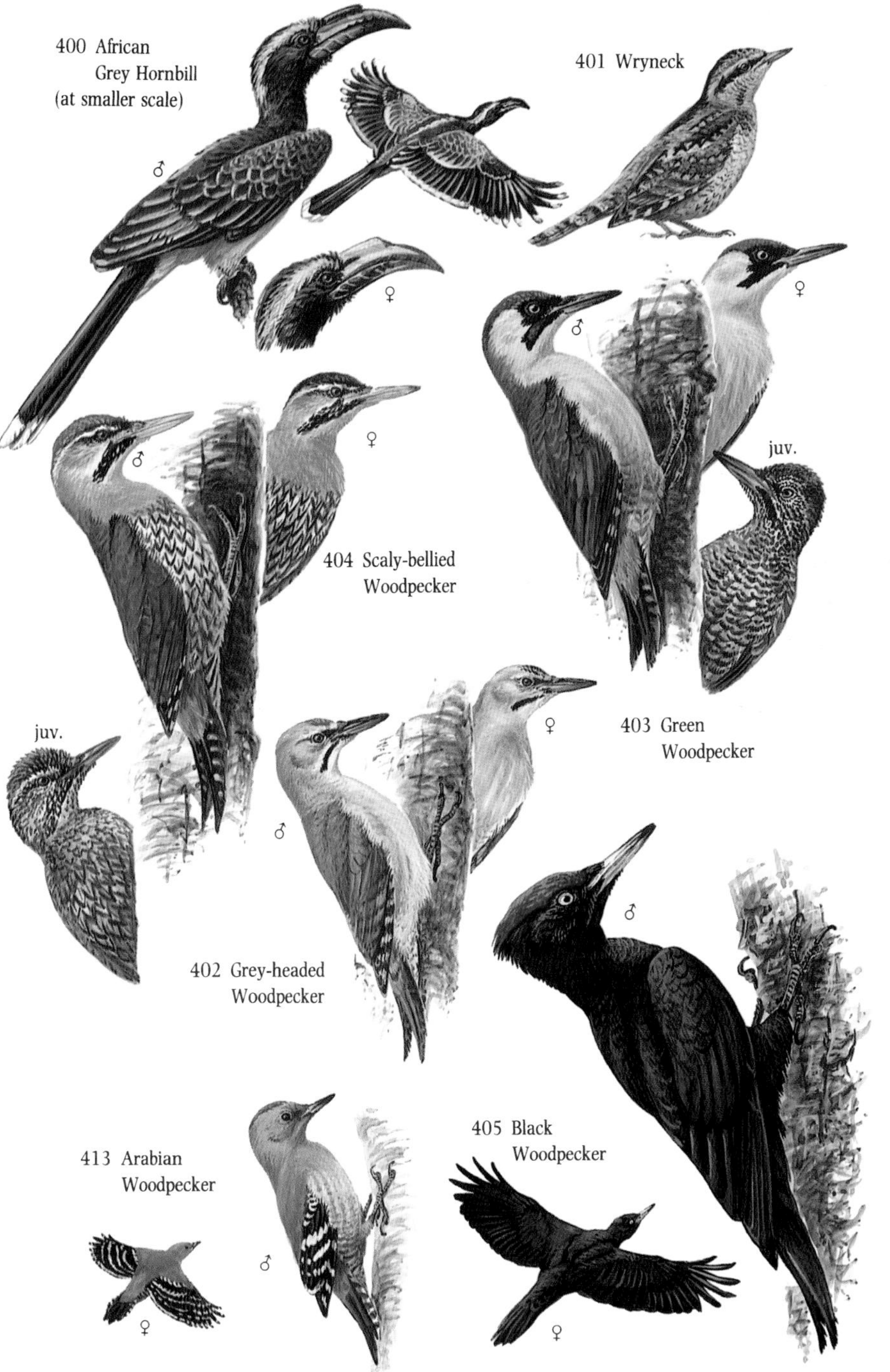
400 African
Grey Hornbill
(at smaller scale)
♂
♀
401 Wryneck
♂
♀
juv.
♂
♀
404 Scaly-bellied
Woodpecker
juv.
♀
403 Green
Woodpecker
♂
♂
402 Grey-headed
Woodpecker
405 Black
Woodpecker
413 Arabian
Woodpecker
♂
♀
♀

ML

406 **Great Spotted Woodpecker** *Dendrocopos major*

Status: resident. **Habitat:** deciduous and coniferous forests. Nests in hole bored in tree.

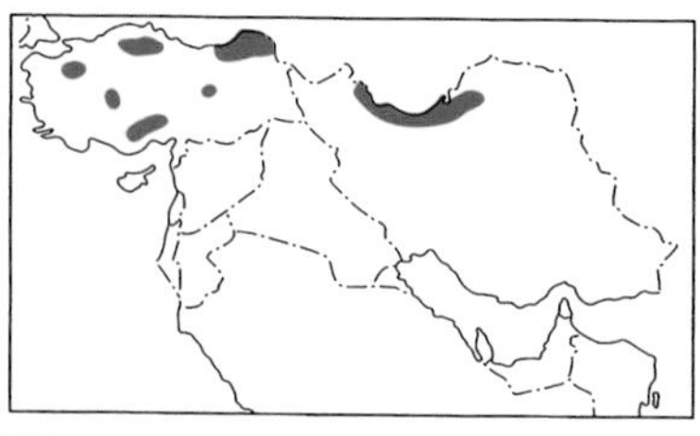

407 **White-winged Woodpecker** *Dendrocopos leucopterus*

Status: uncertain; may occur in extreme NE Iran. **Habitat:** sparsely wooded areas with deciduous trees particularly birch, poplars and willows in lowlands and lower slopes; also in gardens, orchards and oases in desert region.

408 **Syrian Woodpecker** *Dendrocopos syriacus*

Status: resident. **Habitat:** light woodland, gardens and olive groves, often near villages or cultivation. Nests in hole excavated in tree.

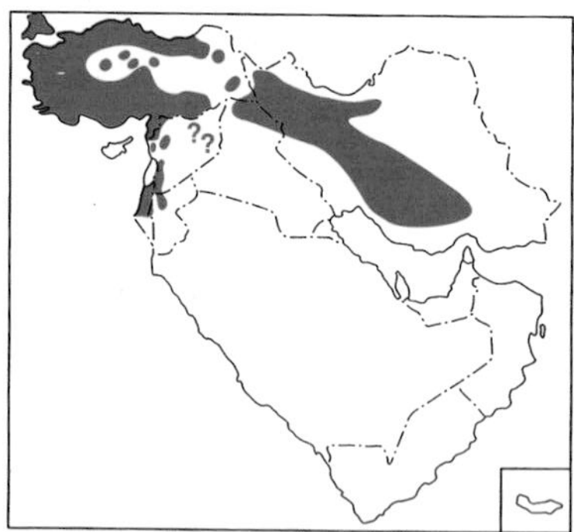

409 **Sind Pied Woodpecker** *Dendrocopos assimilis*

Status: resident. **Habitat:** trees, frequently palms and often in oases or gardens near water. Nests in hole bored in tree, with preference for tamarisk.

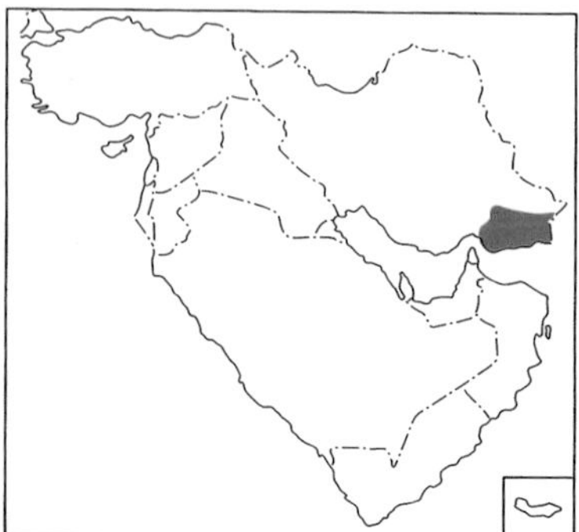

410 **Middle Spotted Woodpecker** *Dendrocopos medius*

Status: resident. Vagrant Iraq. **Habitat:** deciduous woods (especially oak and oak/juniper), olive groves. Nests in hole bored in tree.

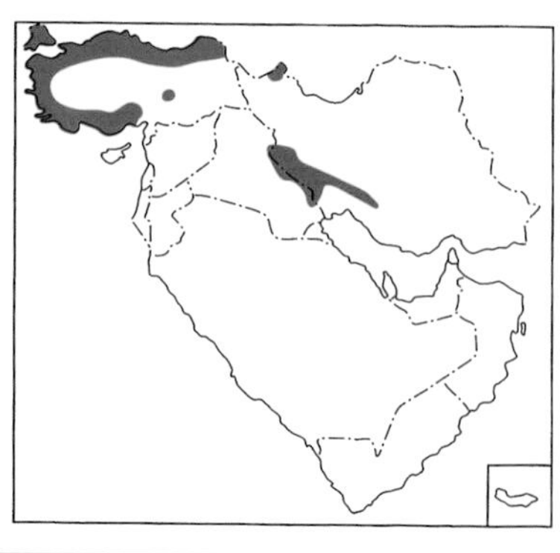

411 **White-backed Woodpecker** *Dendrocopos leucotos*

Status: resident. Probably more widespread in Turkey than has been recorded. **Habitat:** mature deciduous woodland with rotting trees; also old coniferous forests. Nests in hole bored in dead tree.

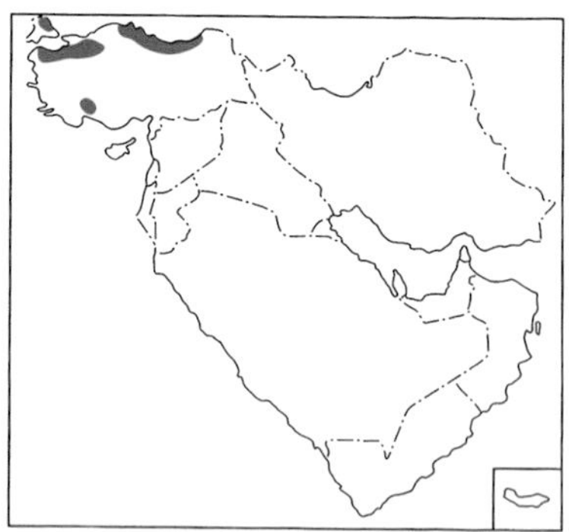

412 **Lesser Spotted Woodpecker** *Dendrocopos minor*

Status: resident. **Habitat:** fairly open woodland, old orchards. Nests in hole bored in tree.

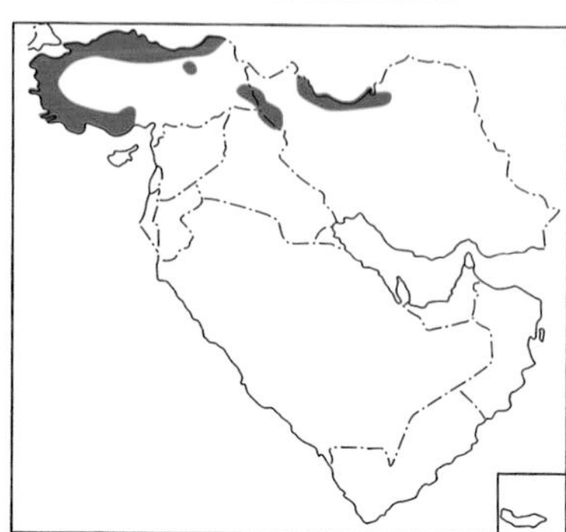

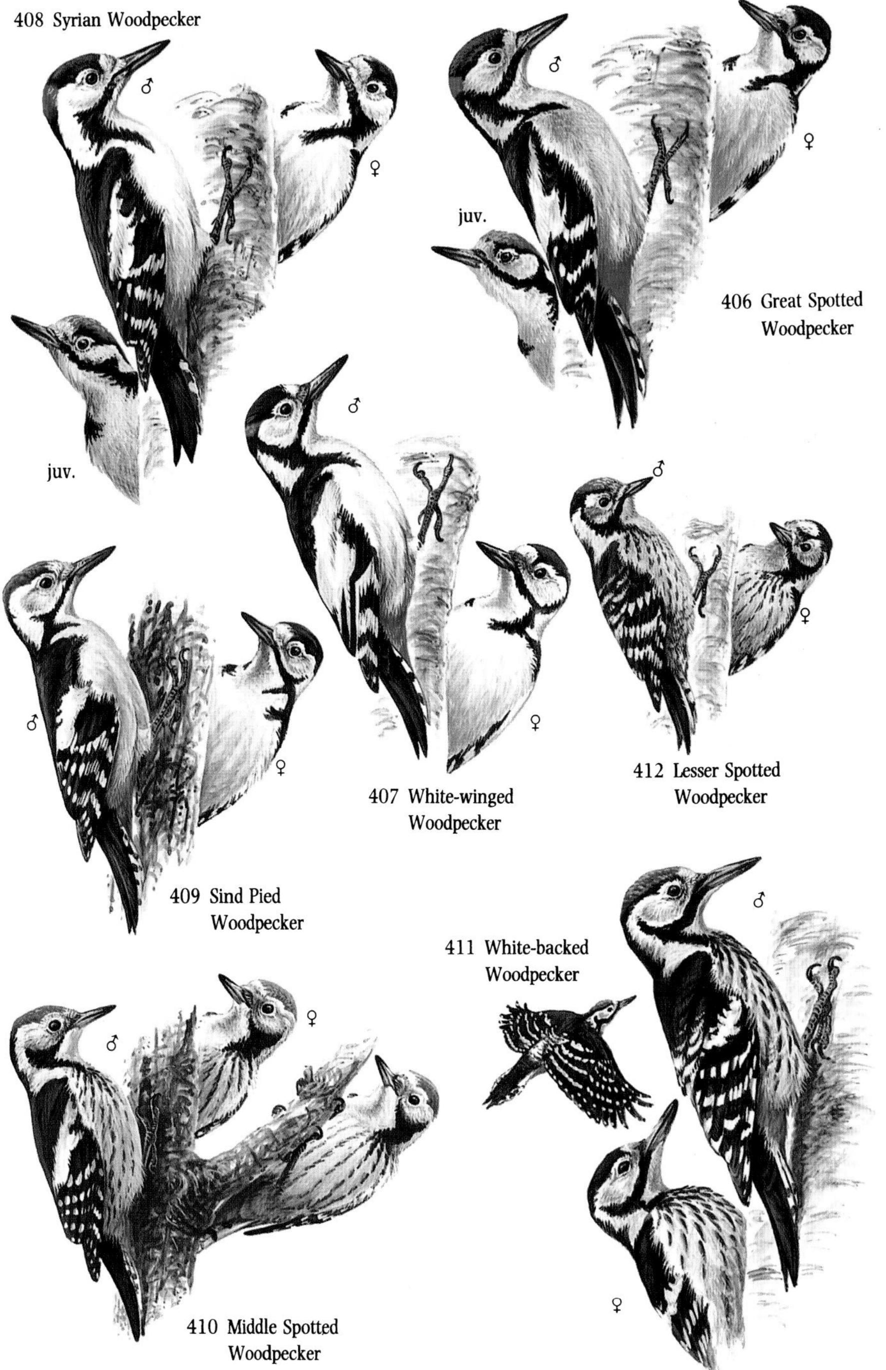
408 Syrian Woodpecker
♂
♀
juv.
♂
♀
juv.
406 Great Spotted Woodpecker
♂
♂
♀
♂
♀
407 White-winged Woodpecker
♀
412 Lesser Spotted Woodpecker
409 Sind Pied Woodpecker
♂
411 White-backed Woodpecker
♀
♂
410 Middle Spotted Woodpecker
♀

ML

414 **Singing Bush Lark** *Mirafra cantillans*

Status: apparently resident Yemen but mainly summer visitor to Oman. **Habitat:** dry grassy areas, undulating plains, foothills and sparsely vegetated semi-deserts; often near or in cultivation and with scattered bushes. Nests in grass or near tuft.

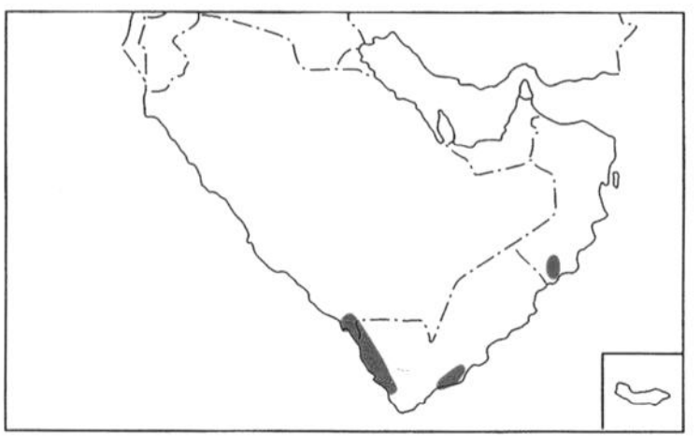

415 **Black-crowned Finch Lark**
Eremopterix nigriceps

Status: apparently breeds erratically outside mapped area. Partial migrant and wanderer. Mainly summer visitor S Iran, NE and central Arabia, but present throughout year E and S Arabia. Vagrant Israel. **Habitat:** flat semi-desert and sandy savanna, often with scattered low scrub, sometimes at edge of cultivation. Nests on ground.

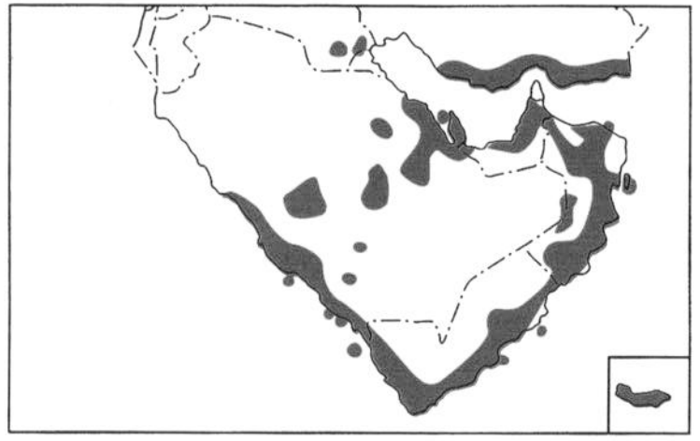

416 **Dunn's Lark** *Eremalauda dunni*

Status: mainly resident but sometimes nomadic movements, including Jordan and Israel where several pairs have bred once. Vagrant Lebanon, Qatar, Kuwait and Yemen. Possible old records from Turkey. **Habitat:** flat sandy or stony semi-desert with low scrub and grass. Nests on ground in grass or below scrubby vegetation.

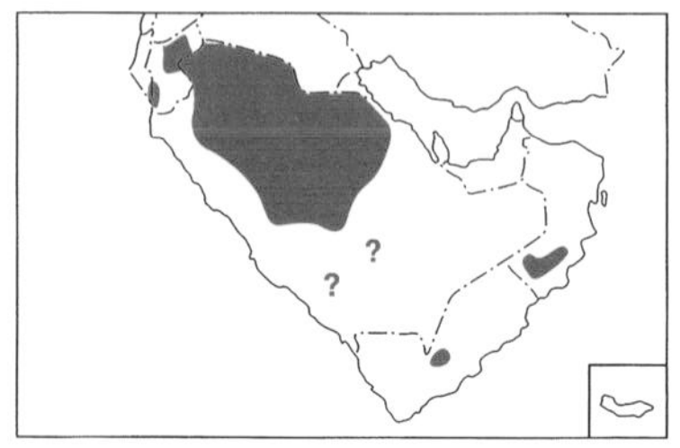

417 **Bar-tailed Desert Lark**
Ammomanes cincturus

Status: has bred Bahrain; mainly resident with local movements, but summer visitor to northern half of Iranian range. Irregular Kuwait and Qatar, vagrant Syria. **Habitat:** flat semi-desert with scattered vegetation. Nests on ground.

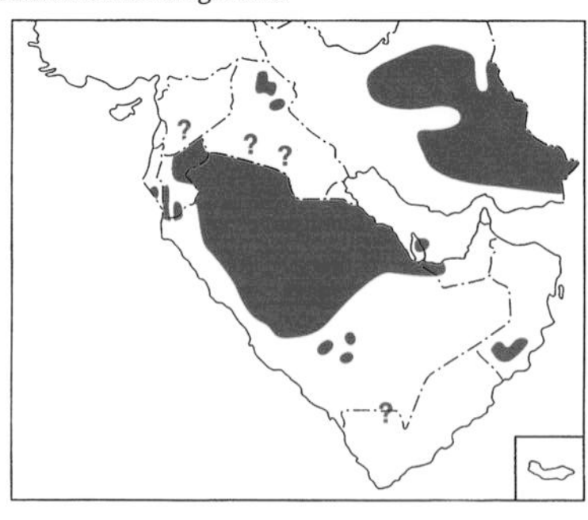

418 **Desert Lark** *Ammomanes deserti*

Status: recently discovered Turkey. Resident, but locally slight dispersal in winter. Rare Lebanon, vagrant Cyprus. **Habitat:** arid, broken hilly country, stony or rocky slopes, wadis, semi- and lava-deserts; often with scattered vegetation. Nests on ground beside stone or bush.

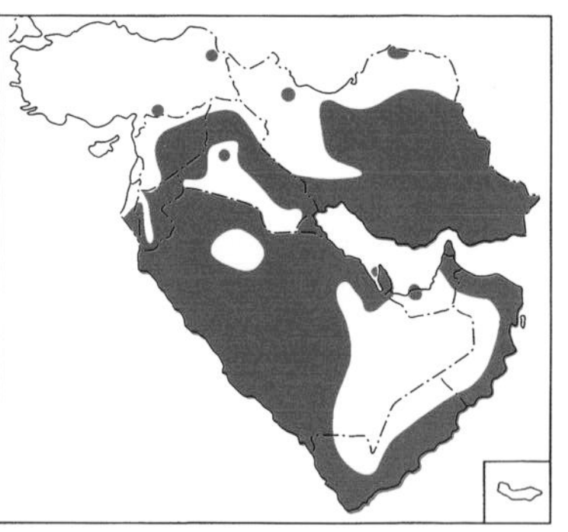

419 **Hoopoe Lark** *Alaemon alaudipes*

Status: mainly resident with some undefined seasonal movements. Vagrant Lebanon. **Habitat:** flat deserts and semi-deserts, rolling dunes with or without sparse vegetation/low bushes; also sandy shores. Nests under shelter of bush.

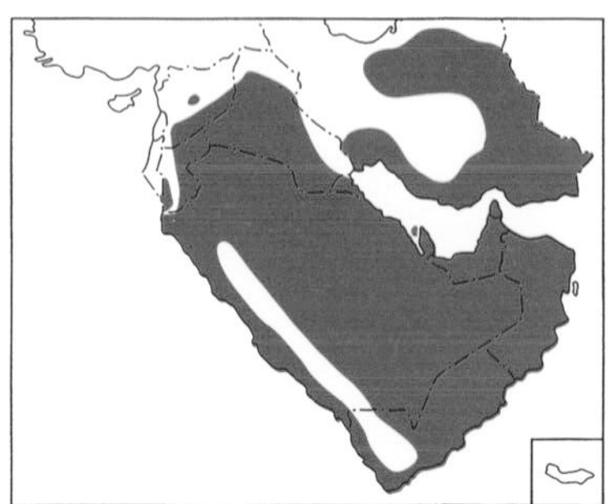

420 **Dupont's Lark** *Chersophilus duponti*

Status: vagrant Cyprus. **Habitat:** arid scrub; outside breeding season also agricultural fields.

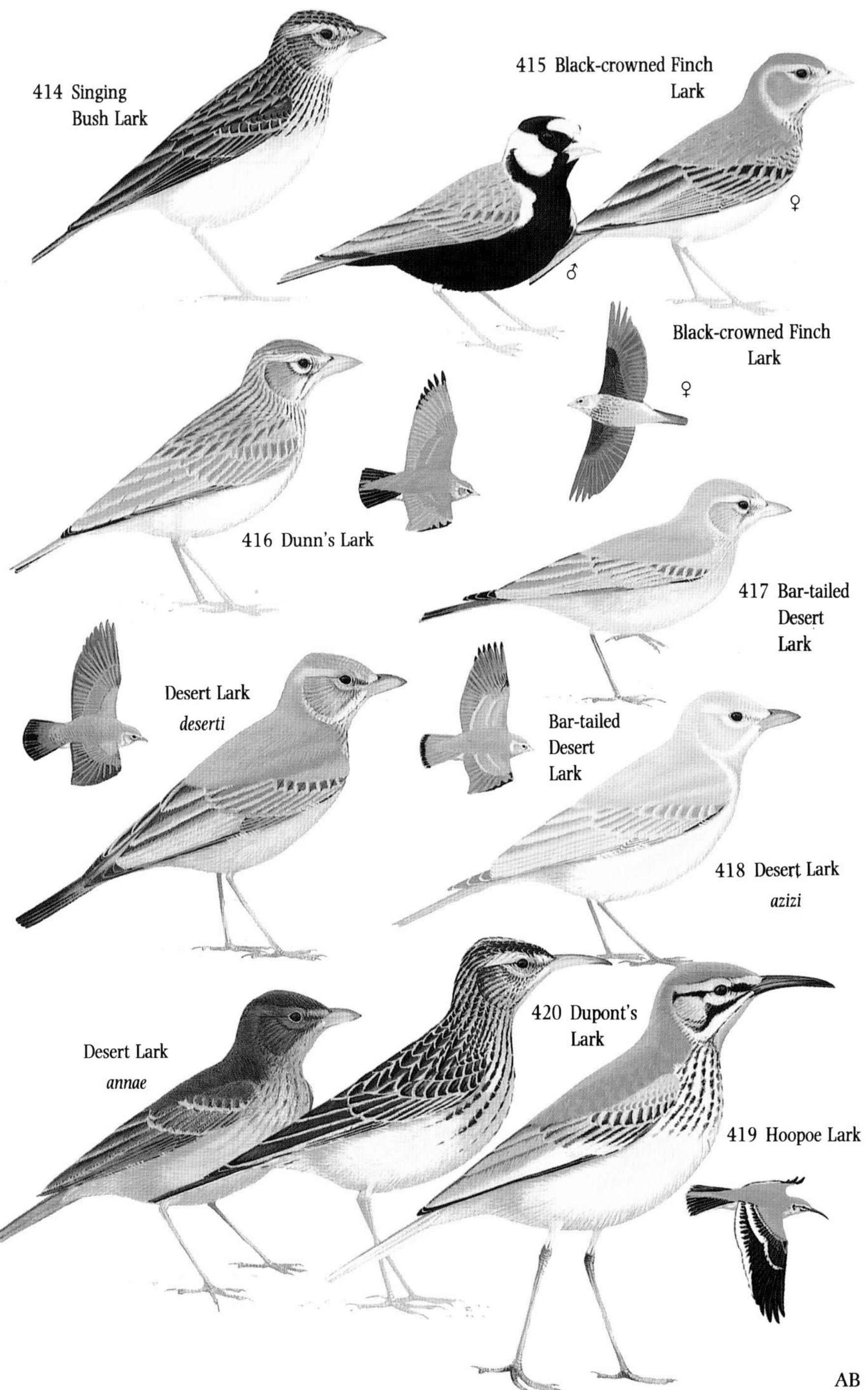
414 Singing
Bush Lark
415 Black-crowned Finch
Lark
♀
♂
Black-crowned Finch
Lark
♀
416 Dunn's Lark
417 Bar-tailed
Desert
Lark
Desert Lark
deserti
Bar-tailed
Desert
Lark
418 Desert Lark
azizi
420 Dupont's
Lark
Desert Lark
annae
419 Hoopoe Lark
AB

421 **Thick-billed Lark** *Ramphocoris clotbey*

Status: has bred Arabia. Mainly resident, some dispersal in winter; occurs irregularly S Israel and Kuwait and south to *c.* 25° in Saudi Arabia. Vagrant Yemen. **Habitat:** stony deserts; outside breeding season also in grassy wadis, rocky slopes and edges of cultivation. Nests in shallow depression on ground.

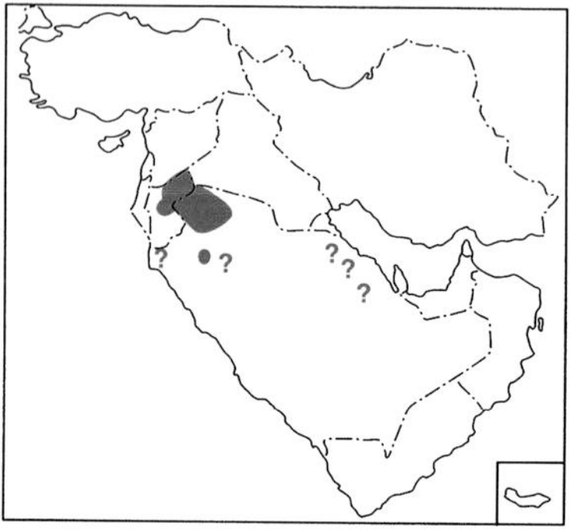

422 **Calandra Lark** *Melanocorypha calandra*

Status: partial migrant; present all year in most breeding range, extending in winter to parts of Iraq, SW and central Iran. Vagrant NE Arabia, Bahrain and UAE. **Habitat:** open cultivated plains, grass and cereal fields, steppes and wastelands. Nests on ground.

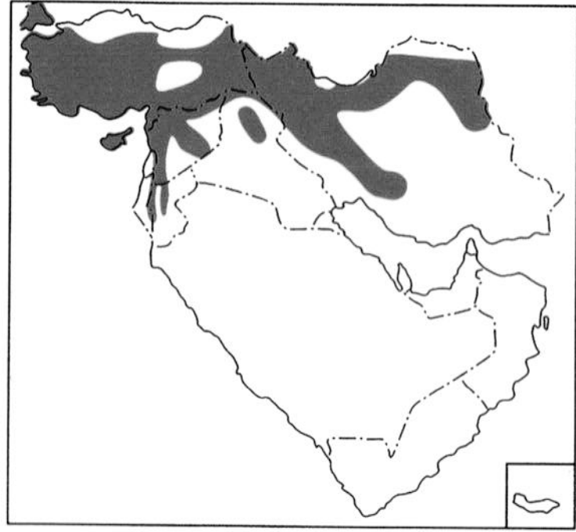

423 **Bimaculated Lark**
Melanocorypha bimaculata

Status: recent breeding Jordan. Has bred Kuwait and N Saudi Arabia. Mainly summer visitor. Generally scarce migrant or winter visitor Israel, Jordan, N Saudi Arabia, Arabian Gulf coasts, E Iran and Oman; scarce passage Cyprus. Vagrant Yemen. **Habitat:** thinly vegetated hills or marginal stony cultivation, 1200–2400 m. In winter, to sea level in agricultural areas. Nests on ground under tuft.

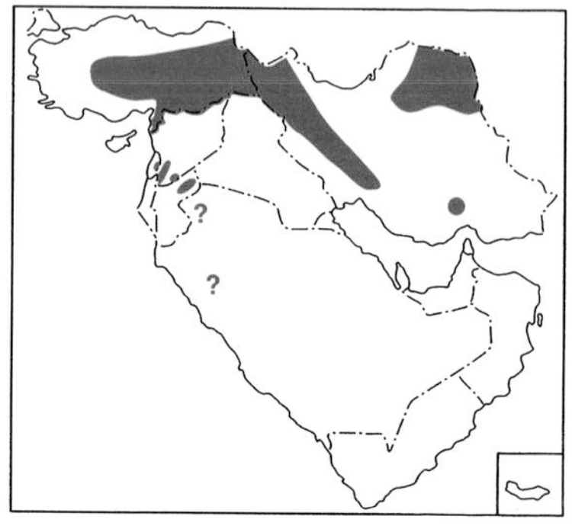

424 **White-winged Lark**
Melanocorypha leucoptera

Status: winter visitor N Iran on Turkoman Steppes, rarer elsewhere on S Caspian lowlands. Vagrant Turkey. **Habitat:** dry grasslands and steppe.

425 **Black Lark** *Melanocorypha yeltoniensis*

Status: very irregular sporadic winter visitor to N Iran, but can be numerous. Vagrant Turkey, Lebanon and Israel. **Habitat:** grassy steppes and agricultural regions.

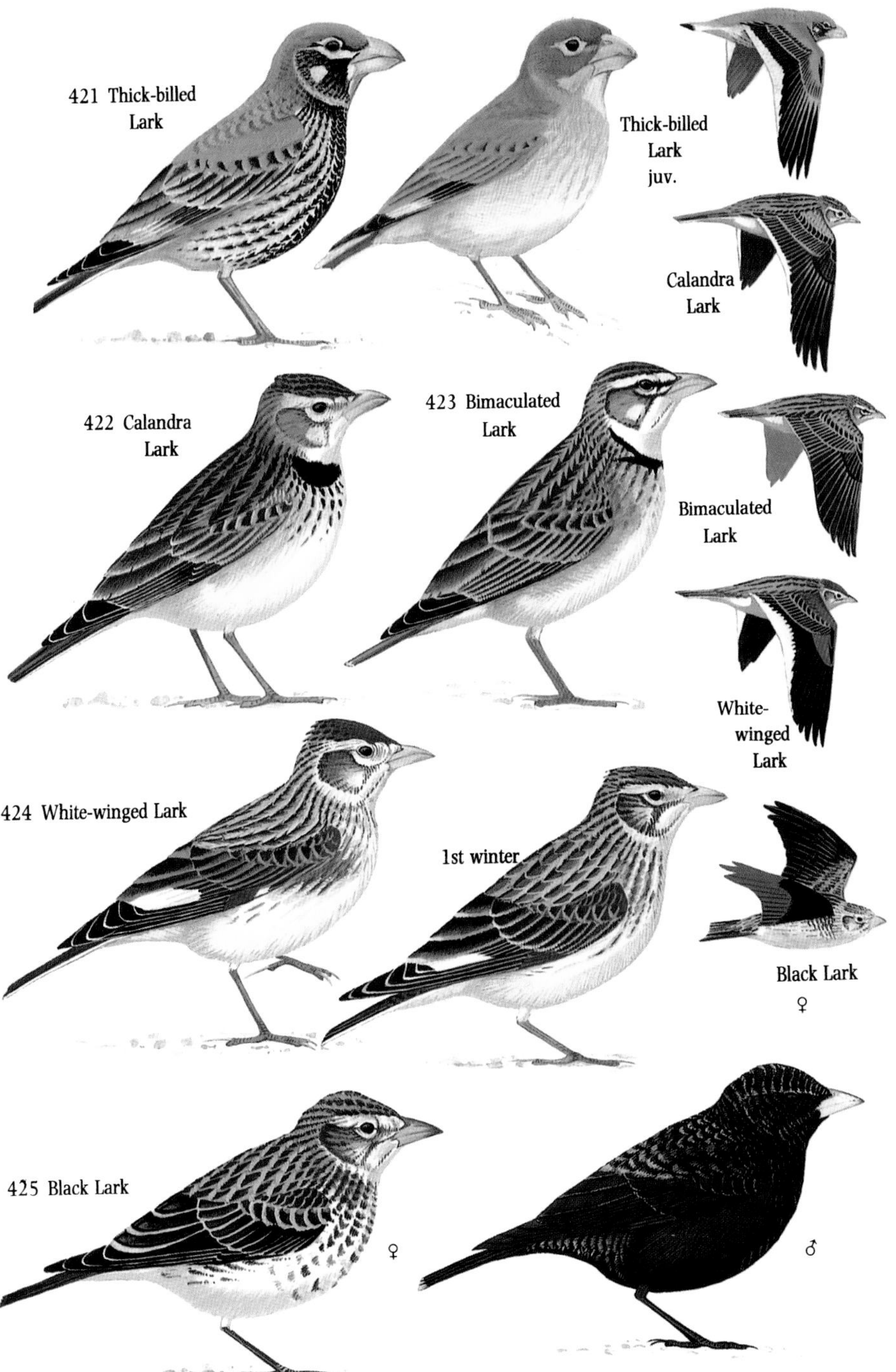

421 Thick-billed Lark
Thick-billed Lark juv.
Calandra Lark
422 Calandra Lark
423 Bimaculated Lark
Bimaculated Lark
White-winged Lark
424 White-winged Lark
1st winter
Black Lark ♀
425 Black Lark
♀
♂

A

426 **Red-capped Lark** *Calandrella cinerea*

Status: mainly resident; little dispersal in winter. Vagrant Oman. **Habitat:** open stony plateaux and fields often with scattered bushes, mainly 1800–2500 m but often lower in winter. Nests on ground.

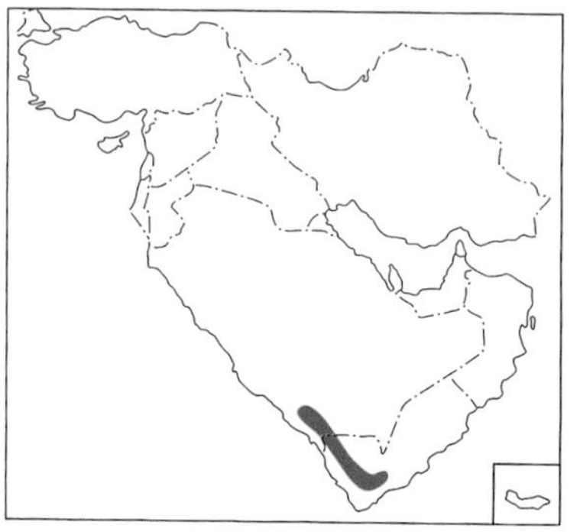

427 **Short-toed Lark** *Calandrella brachydactyla*

Status: has bred Oman. Mainly summer visitor. Common passage E Mediterranean and parts of Arabia including Oman; some winter N Saudi Arabia, few Israel, occasionally Oman, very rarely S Iran. **Habitat:** open country, wastelands, dried saltmarshes, steppes, semi-deserts, cultivated plains. Nests on ground.

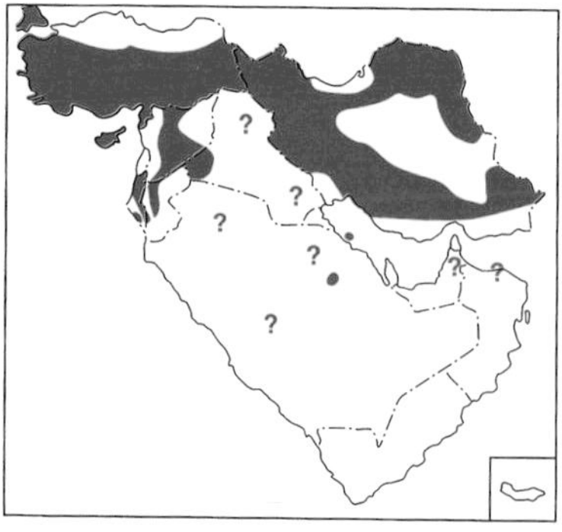

428 **Hume's Short-toed Lark** *Calandrella acutirostris*

Status: status uncertain, recorded NE Iran but perhaps only as vagrant; vagrant Israel. **Habitat:** grassy steppes and rocky uplands.

429 **Lesser Short-toed Lark** *Calandrella rufescens*

Status: mainly resident in breeding areas, but more widespread in winter and reaching Cyprus (irregularly), central and SW Saudi Arabia, Oman. **Habitat:** dry steppes and salt plains, bare wastelands or cultivated lands. Nests on ground.

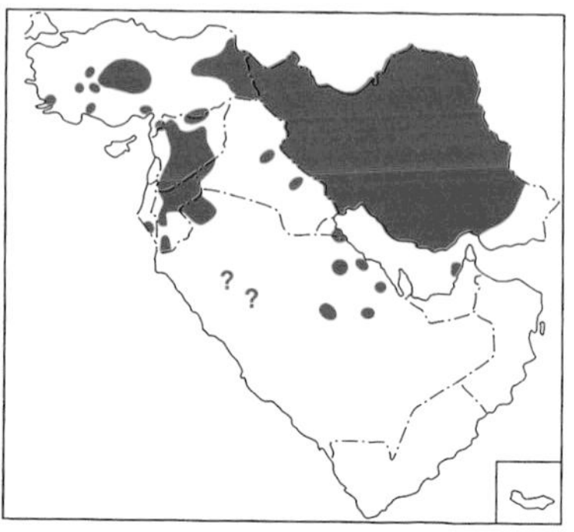

430 **Indian Sand Lark** *Calandrella raytal*

Status: resident. **Habitat:** dry open coastal and river sandbanks, flood-plains, salty mudflats. Nests at base of small bush or tuft in sand.

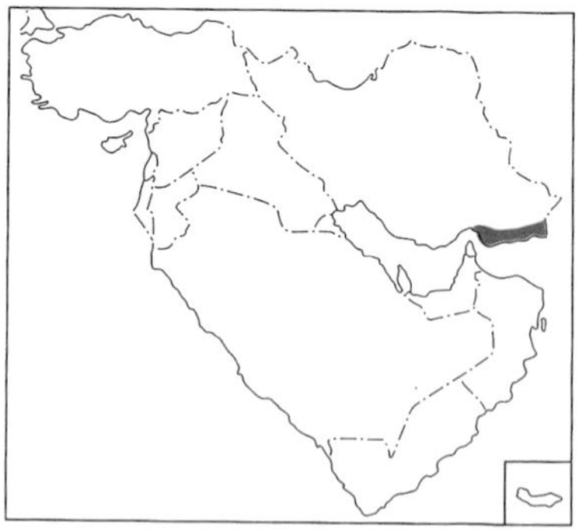

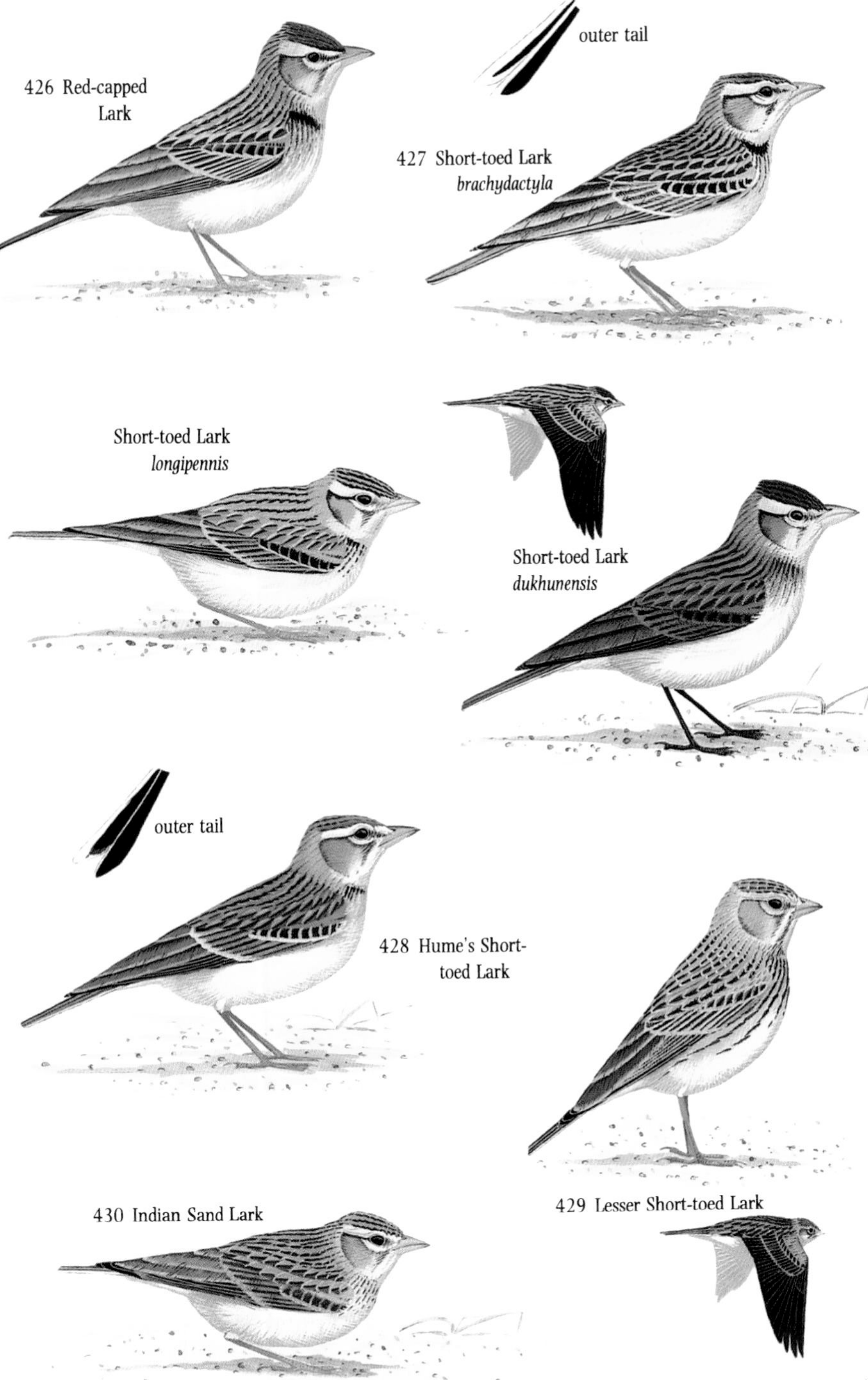
426 Red-capped Lark
outer tail
427 Short-toed Lark *brachydactyla*
Short-toed Lark *longipennis*
Short-toed Lark *dukhunensis*
outer tail
428 Hume's Short-toed Lark
430 Indian Sand Lark
429 Lesser Short-toed Lark
AB

431 **Crested Lark** *Galerida cristata*

Status: resident. Some slight dispersal, and immigration from north in winter. **Habitat:** generally flat grassy or arid country, cultivated plains and semi-deserts; often near habitation, dusty tracks and roadsides. Nests on ground.

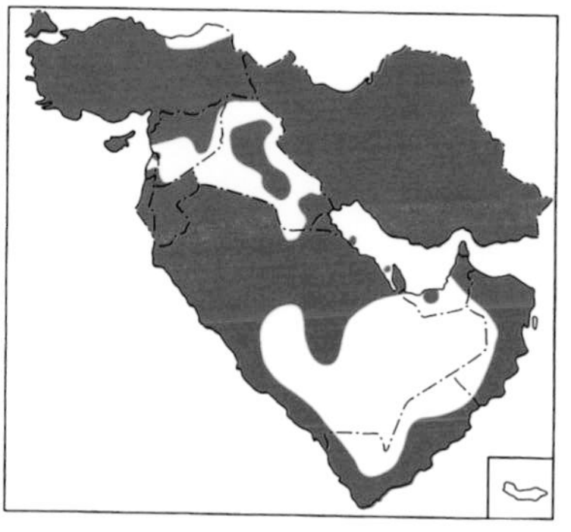

432 **Woodlark** *Lullula arborea*

Status: partial migrant, mainly resident in breeding areas. Winter immigrants reach Cyprus, Iraq, Near East. Vagrant E Arabia south to Bahrain. **Habitat:** open woodland edges, hillsides and heaths with scattered or occasional trees; in winter frequently in fields. Nests on ground.

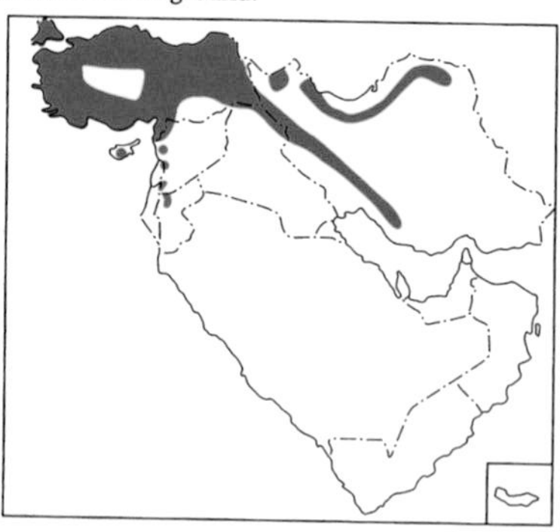

433 **Small Skylark** *Alauda gulgula*

Status: partial migrant, with movements mainly short distance, nomadic or altitudinal. Annual but very scarce in winter Israel. Regular Bahrain and UAE in small numbers. Vagrant NE Arabia and Kuwait. **Habitat:** open moist grassy and cultivated lowlands, but also in grassy hills. Nests on ground.

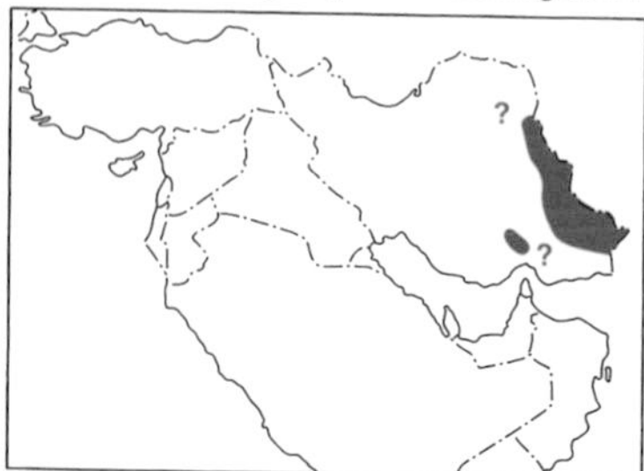

434 **Skylark** *Alauda arvensis*

Status: partial migrant and winter visitor. In most breeding areas regarded as resident, but summer visitor Iran. More numerous in winter, occurring in almost whole region but scarce Oman and absent SW Arabia. **Habitat:** high and low grasslands, cultivated fields; in winter more widely in open areas. Avoids deserts. Nests on ground.

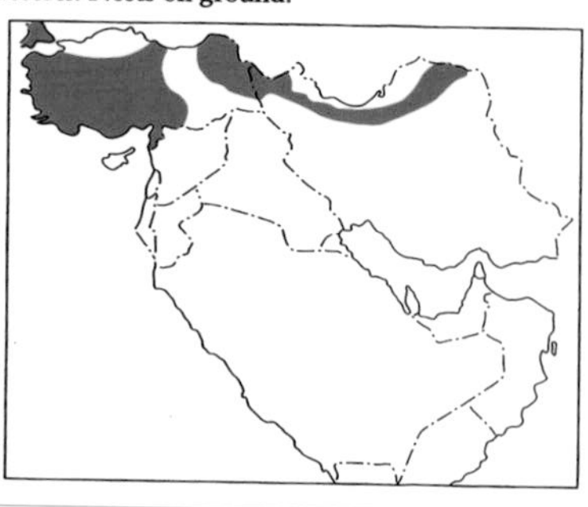

435 **Shore Lark** *Eremophila alpestris*

Status: mainly resident, dispersing to lower levels in winter when occasionally recorded SE Caspian. Vagrant Cyprus. **Habitat:** mountains above tree level (often over 3000 m, rarely below 2000 m); in winter reaches plains. Nests on ground under a tuft.

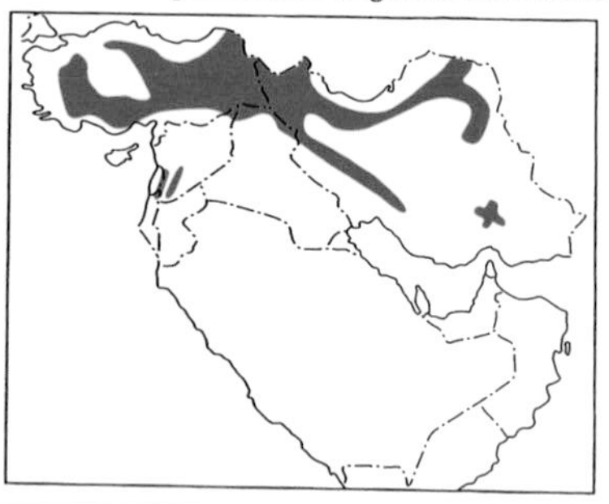

436 **Temminck's Horned Lark** *Eremophila bilopha*

Status: mainly resident; some dispersal in non-breeding season, occasionally in flocks. Vagrant Lebanon, Yemen and UAE. **Habitat:** open flat stony or sandy desert sometimes with sparse grassy vegetation, usually below 1500 m. Nests on ground.

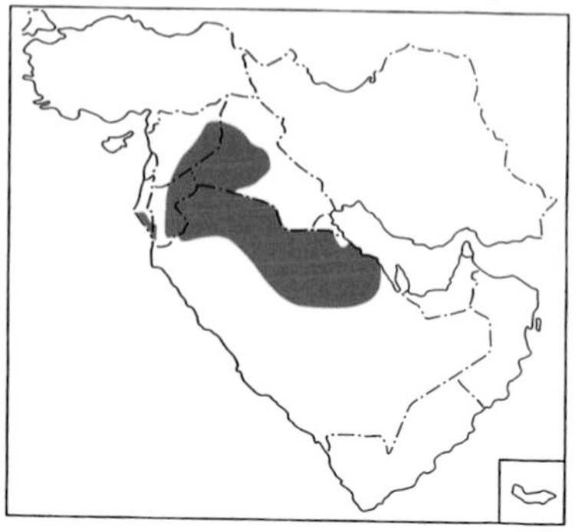

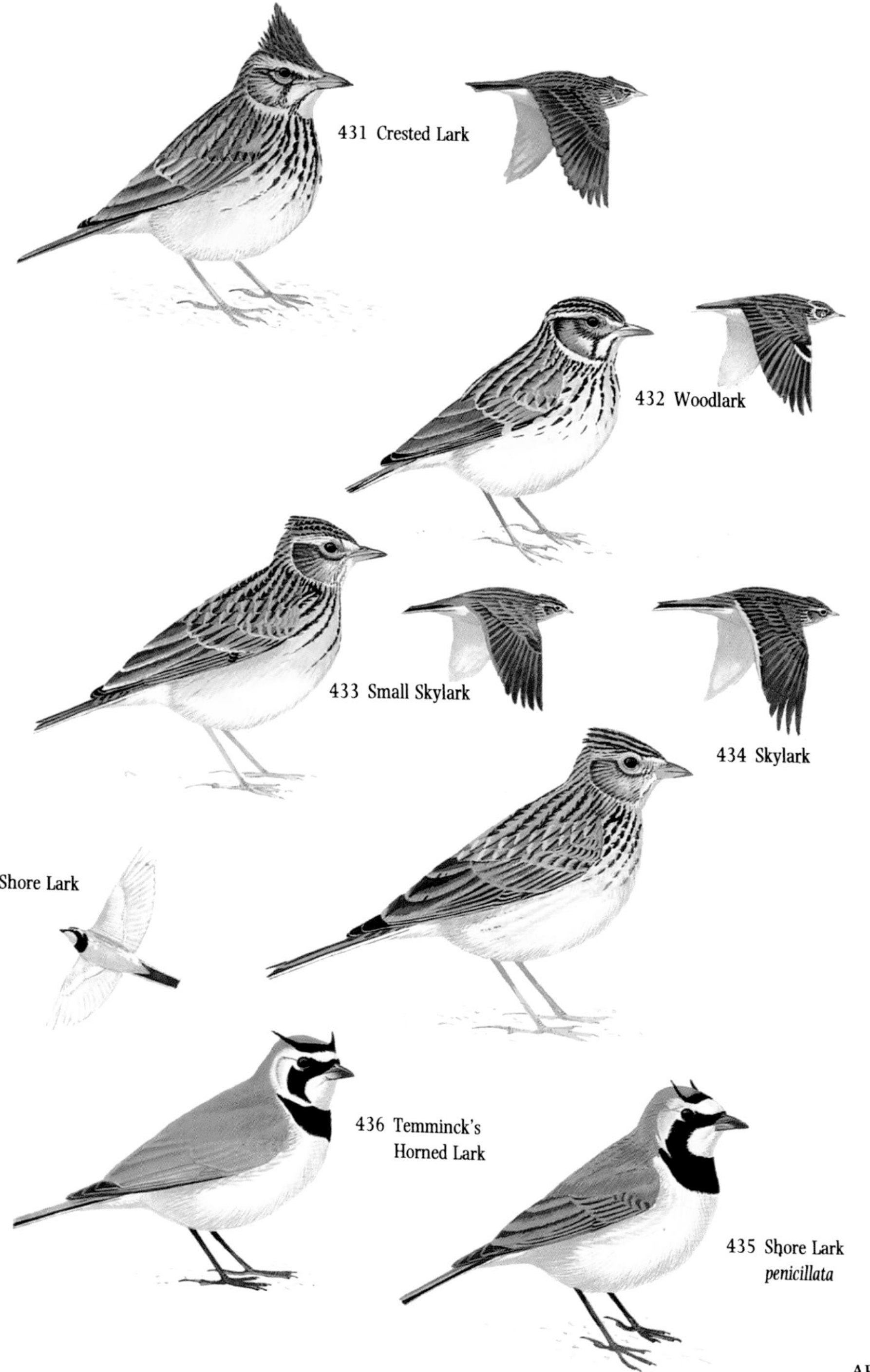
431 Crested Lark
432 Woodlark
433 Small Skylark
434 Skylark
Shore Lark
436 Temminck's Horned Lark
435 Shore Lark penicillata
AB

437 **Brown-throated Sand Martin** *Riparia paludicola*

Status: vagrant NW and E Saudi Arabia, Oman and Israel. **Habitat:** sandy river banks and their environs. Nests colonially in excavated holes.

438 **Sand Martin** *Riparia riparia*

Status: summer visitor. Passage throughout the region. Isolated winter records in Arabia, Iran. **Habitat:** open country with lakes and rivers. More widespread on migration, though usually over water. Nests colonially in tunnels bored in sandy banks.

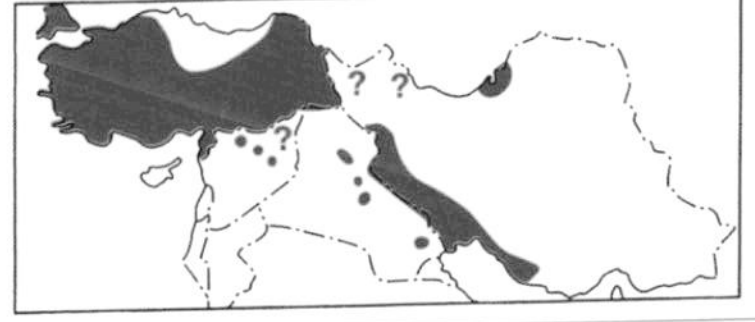

439 **Banded Martin** *Riparia cincta*

Status: vagrant Yemen. Range Africa. **Habitat:** over open grassland and bush country, often by water.

440 **African Rock Martin** *Ptyonoprogne fuligula*

Status: summer visitor central and E Iran, with wintering/passage Oman, but resident in S Iran and apparently all other breeding areas in our region. Vagrant Kuwait. **Habitat:** desert regions and dry hilly country with gorges and ravines often near human habitation. Nests like Crag Martin, in caves, on rock faces or buildings.

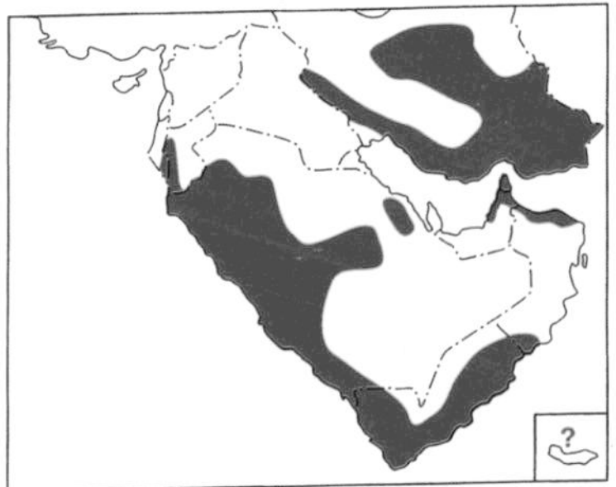

441 **Crag Martin** *Ptyonoprogne rupestris*

Status: may breed Yemen highlands. Northern populations, including Turkey and Iran, migratory; otherwise resident, or moving lower in winter as in Cyprus. Uncommon passage or winter visitor most other parts including Arabia. **Habitat:** mountain gorges and rocky inland and coastal cliffs. Builds open half-cup shaped nest in cleft rocks or caves, on cliff face or under eaves of building.

442 **Barn Swallow** *Hirundo rustica*

Status: has bred Bahrain; old breeding record for Yemen. Resident Israel, otherwise summer visitor with scattered winter records Arabia and S Iran. Passage throughout whole region. **Habitat:** open cultivated country with settlements, but can occur over any area on migration. Nests on ledge in building.

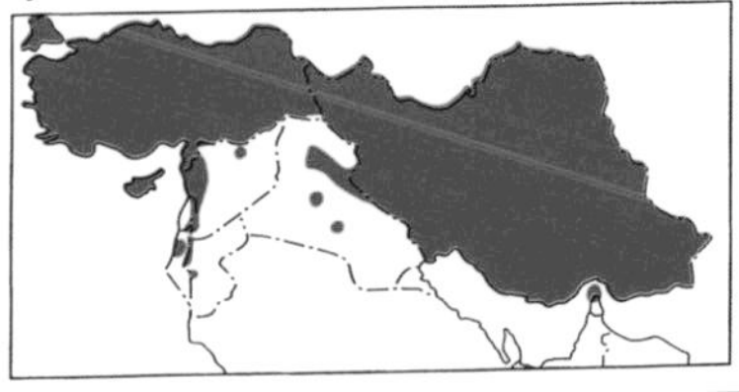

443 **Ethiopian Swallow** *Hirundo aethiopica*

Status: vagrant Israel. **Habitat:** open country; also frequents towns and villages.

444 **Wire-tailed Swallow** *Hirundo smithii*

Status: vagrant UAE and unconfirmed record from Oman; (range Africa). **Habitat:** near human habitation, bridges and along rivers and lakes.

446 **Red-rumped Swallow** *Hirundo daurica*

Status: summer visitor, but occasional winter records in SW Arabia. Light passage most parts of the region. **Habitat:** inland and sea-cliffs; less cultivated areas than Barn Swallow, but in flat country frequents bridges and buildings. Flask-shaped nest with spout-shaped entrance is made from mud in caves, under rocky overhangs, bridges or buildings.

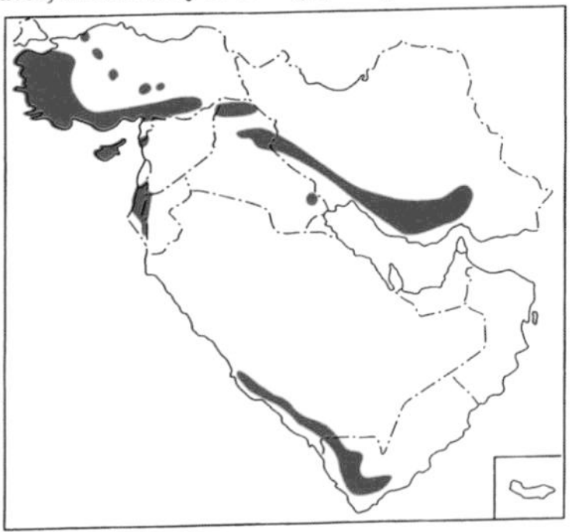

447 **House Martin** *Delichon urbica*

Status: summer visitor; also passage most areas but irregular in Oman. Winter records Arabian Gulf, Yemen. **Habitat:** like Barn Swallow but more often near human habitation. Cup-shaped mud nest is built under eaves of house, sometimes on cliff.

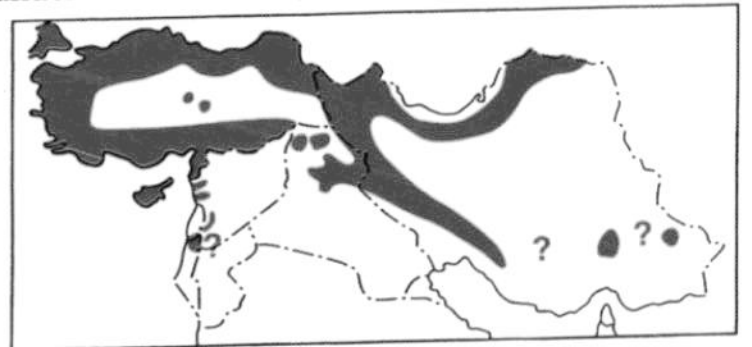

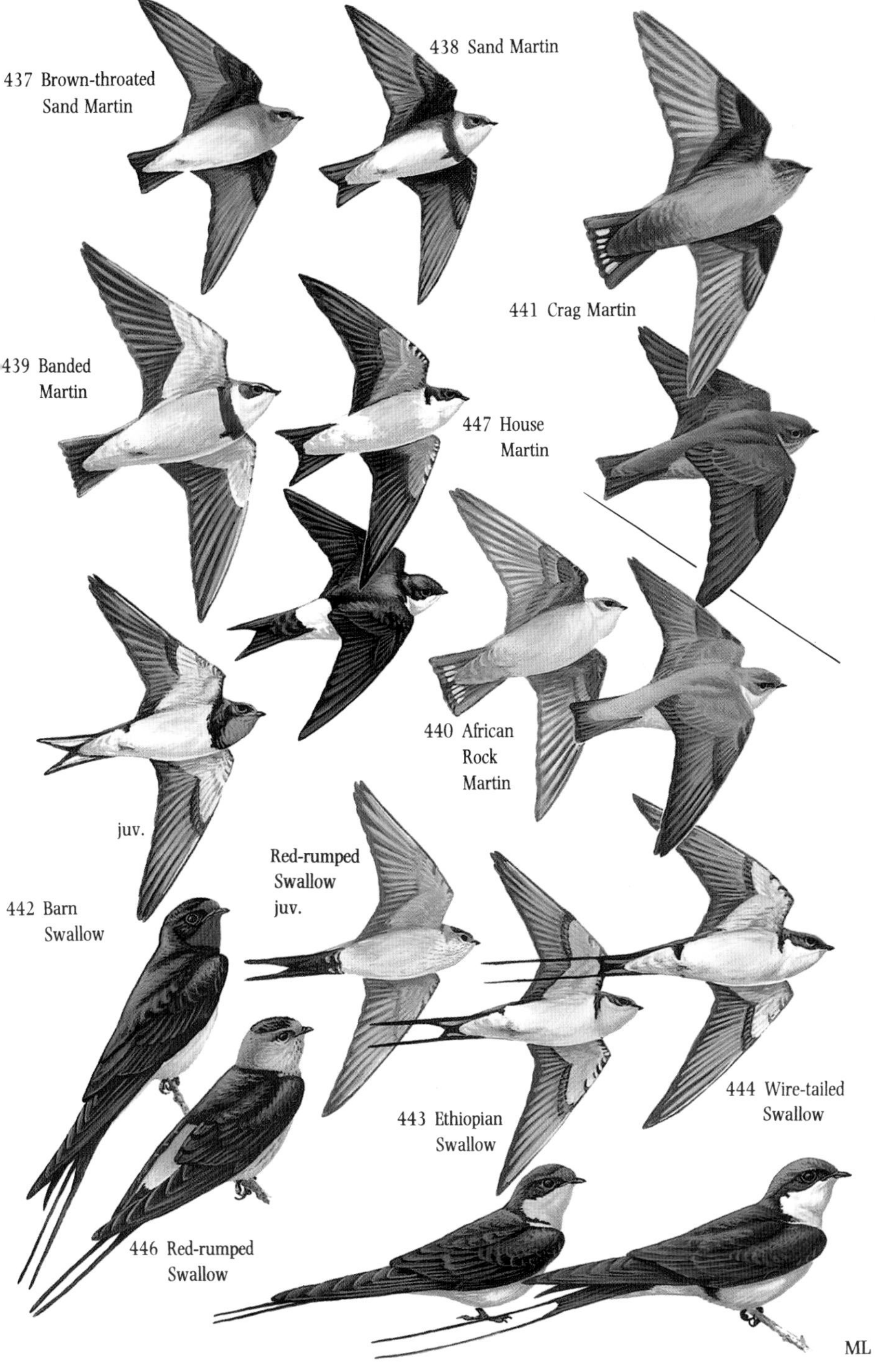
437 Brown-throated Sand Martin
438 Sand Martin
441 Crag Martin
439 Banded Martin
447 House Martin
440 African Rock Martin
juv.
Red-rumped Swallow juv.
442 Barn Swallow
444 Wire-tailed Swallow
443 Ethiopian Swallow
446 Red-rumped Swallow
ML

448 **Golden Pipit** *Tmetothylacus tenellus*

Status: vagrant Oman from Africa. **Habitat:** open dry bush country.

449 **Richard's Pipit** *Anthus richardi*

Status: rare resident SW Arabia. Scattered records of Asian migrants over most of region but uncommon winter visitor or passage migrant Israel, UAE and Oman, and south Arabian Gulf. Vagrant Kuwait and Syria. **Habitat:** breeds on grassy slopes between 1500–2400 m. On passage and winter in grassland, fields and marshes. Nests on ground.

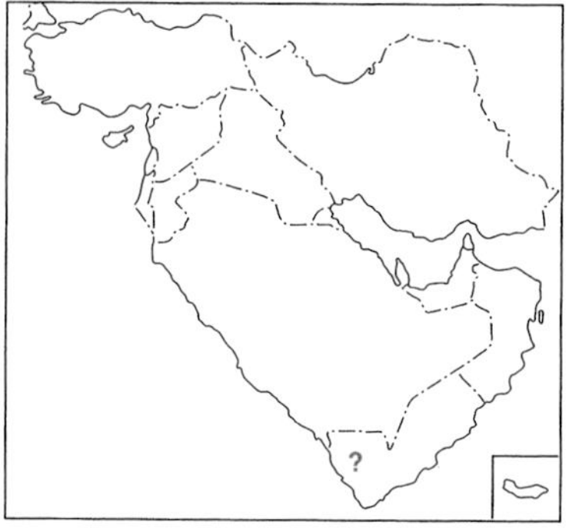

450 **Tawny Pipit** *Anthus campestris*

Status: summer visitor. Widely on passage; winters throughout southern part of the region including whole of Arabia, occasionally Cyprus and Iran. **Habitat:** sparsely vegetated wastelands, edges of cultivation, plateaux, but outside breeding season in almost any open bare country. Nests on ground.

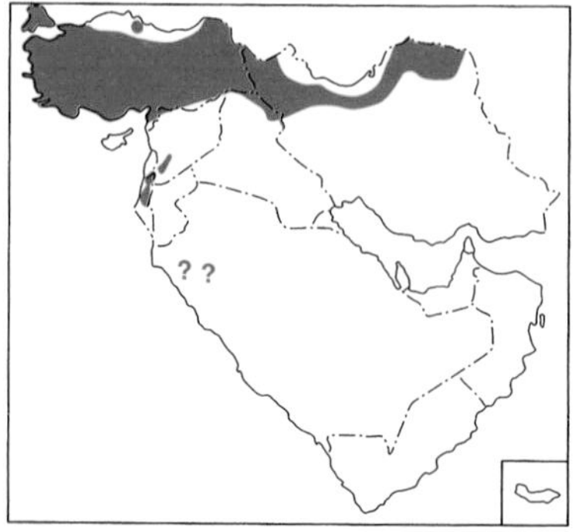

451 **Long-billed Pipit** *Anthus similis*

Status: Summer visitor Iran, otherwise mainly resident. Some winter coasts of S Iran and UAE (where bred recently). Vagrant Kuwait, Qatar, Iraq and Cyprus. **Habitat:** rocky mountain slopes with scattered vegetation up to 3000 m, descending in winter to plains. Nests on ground.

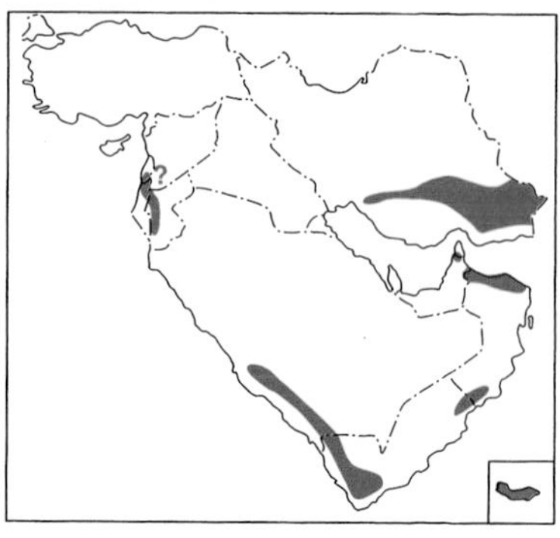

452 **Olive-backed Pipit** *Anthus hodgsoni*

Status: locally scarce winter visitor Oman; rare winter visitor Israel and UAE; vagrant Turkey, Cyprus, Iran, Kuwait and W and E Saudi Arabia. Range Asia. **Habitat:** similar to Tree Pipit.

453 **Tree Pipit** *Anthus trivialis*

Status: summer visitor. Widespread on passage. Regular winter Oman, UAE; few Yemen and Bahrain. **Habitat:** open woodland, fields with scattered trees or bushes. On migration more in the open. Nests on ground.

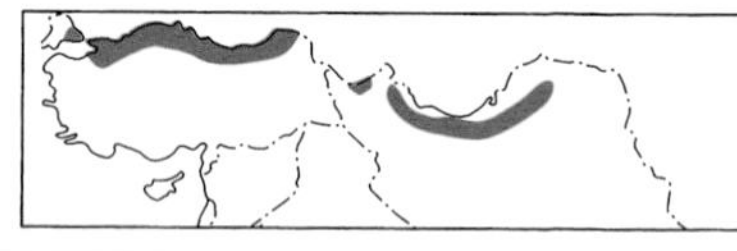

455 **Meadow Pipit** *Anthus pratensis*

Status: common winter visitor to Mediterranean part of the region east to central Iran and Bahrain, scarcer Iraq, Arabian Gulf and NW Saudi Arabia. Vagrant Oman. **Habitat:** open county, fields, wastelands, marshes and coasts.

456 **Red-throated Pipit** *Anthus cervinus*

Status: passage migrant over most of region. Small numbers winter from E Mediterranean to Iran and southwards. **Habitat:** marshes or moist grassland but also fields and wastelands, usually near water.

457 **Blyth's Pipit** *Anthus godlewskii*

Status: rare winter visitor UAE; vagrant Israel. Range Asia. **Habitat:** grassy and open ground.

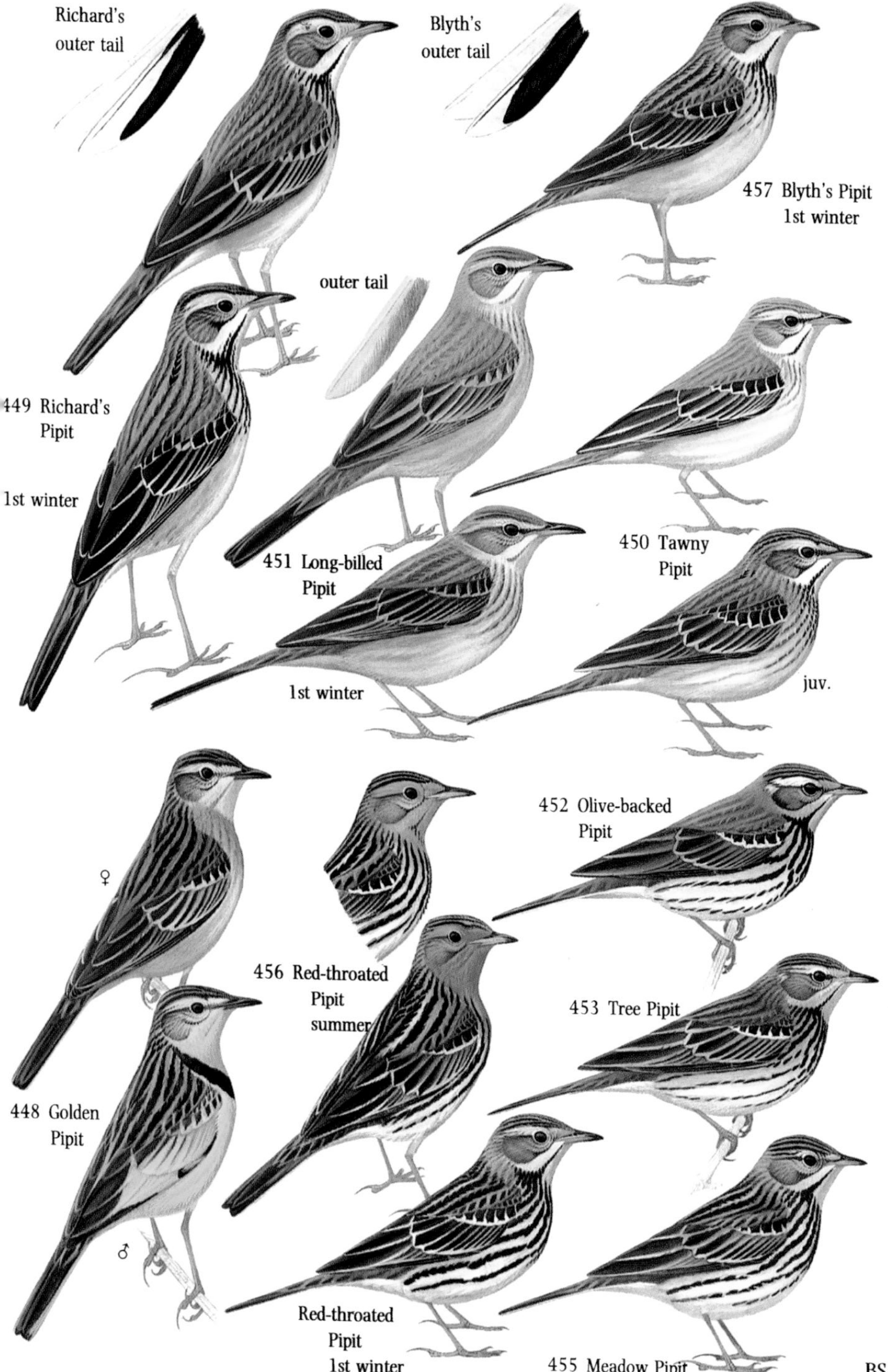
Richard's
outer tail
Blyth's
outer tail
457 Blyth's Pipit
1st winter
outer tail
449 Richard's
Pipit
1st winter
451 Long-billed
Pipit
450 Tawny
Pipit
1st winter
juv.
452 Olive-backed
Pipit
♀
456 Red-throated
Pipit
summer
453 Tree Pipit
448 Golden
Pipit
♂
Red-throated
Pipit
1st winter
455 Meadow Pipit
BS

458 **Buff-bellied Pipit** *Anthus rubescens*

Status: scarce but regular migrant and winter visitor to S Israel, vagrant UAE; may occur elsewhere; probably overlooked. **Habitat:** grassy fields near water.

459 **Water Pipit** *Anthus spinoletta*

Status: resident or partial migrant in Turkey and Iran. Winters Cyprus and Near East, south to NW Saudi Arabia, S Arabian Gulf to Oman. **Habitat:** mountains above tree-line. In winter in open wet lowlands, edges of lakes. Nests in crevice in rocks.

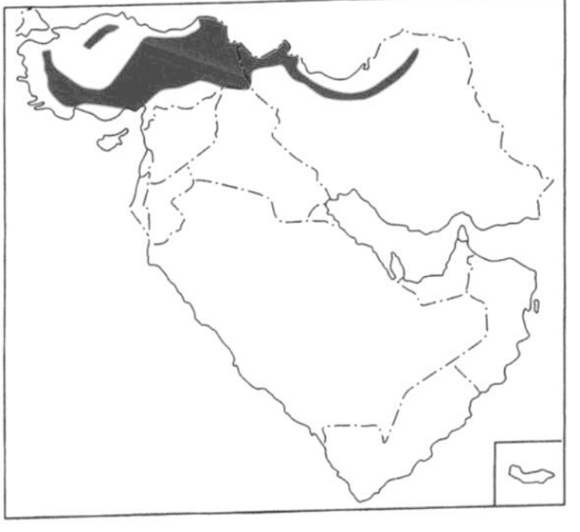

460 **Forest Wagtail** *Dendronanthus indicus*

Status: winter vagrant UAE and Oman. Range Asia. **Habitat:** in Asian habitat occurs in clearings in damp forest, often near water.

461 **Yellow Wagtail** *Motacilla flava*

Status: summer visitor (*feldegg* is the race breeding in the region); *feldegg*, *flava* and *thunbergi* occur widely on passage with few in winter E and S Arabia, rarely Iran and Cyprus; *lutea* recorded Near East and Arabia, vagrant Cyprus; *beema* on passage in Near East and Arabia; *melanogrisea* in E Iran; *leucocephala* vagrant NE Iran; *cinereocapilla* scarcely in E Mediterranean but recorded E and S Saudi Arabia. **Habitat:** flat, wet or dry grassland, and open fields. Nests on ground.

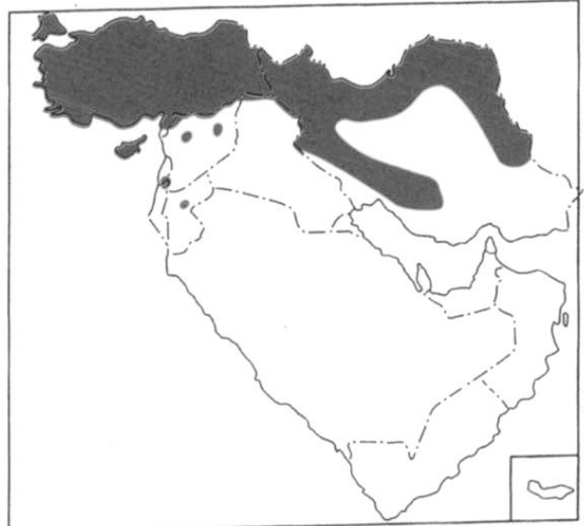

462 **Citrine Wagtail** *Motacilla citreola*

Status: summer visitor to breeding areas. On passage regular Israel, Turkey, S Arabian Gulf; scarce E Arabia and Yemen; rare or vagrant NW and SW Saudi Arabia and most Mediterranean countries including Cyprus. **Habitat:** between 1500–2500 m in swampy meadows, or near streams. Outside breeding season near water in open country at low altitudes. Nests in hollow on ground or in bank.

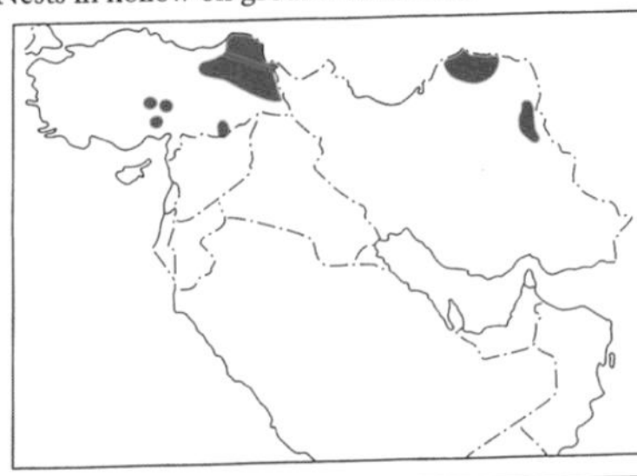

463 **Grey Wagtail** *Motacilla cinerea*

Status: has bred Lebanon. Partial migrant; widely on passage and in winter. **Habitat:** streams in hilly wooded country. In winter down to sea level, always near water. Nests in hole in bank or wall.

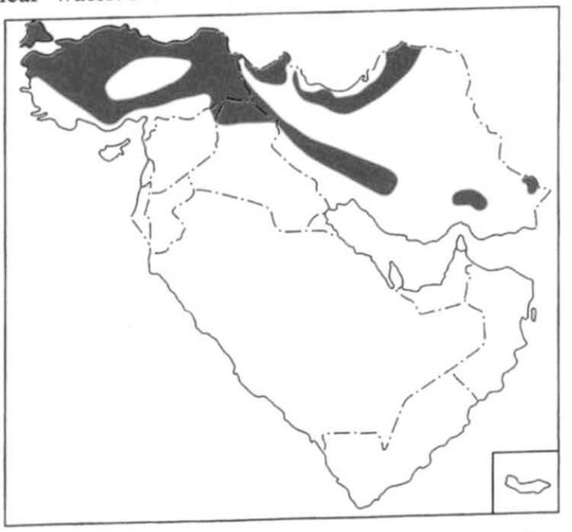

464 **White Wagtail** *Motacilla alba*

Status: has bred Cyprus. Some resident; widespread on passage and in winter throughout the region. Race *personata* winters south to Iraq; vagrant Bahrain, UAE and Cyprus. **Habitat:** open areas, often near habitation and cultivation. Nests in hole in wall or rock.

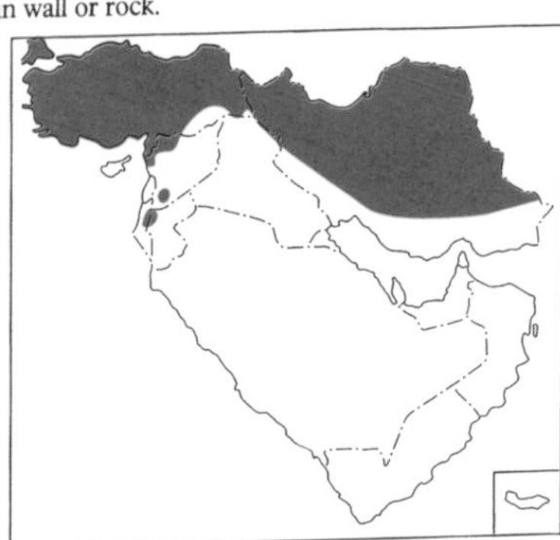

459 Water Pipit
summer
winter
summer
458 Buff-bellied Pipit
winter
feldegg ♀
1st winter
flava
beema
melanogrisea
Yellow Wagtail
thunbergi
461 Yellow Wagtail
feldegg ♂
cinereocapilla
lutea
leucocephala
462 Citrine Wagtail
citreola
calcarata
werae
♀
1st winter
460 Forest Wagtail
personata
♀
♀
♂
alba
1st winter
463 Grey Wagtail
464 White Wagtail
BS

465 **White-cheeked Bulbul**
Pycnonotus leucogenys

Status: resident; has bred Oman; may breed Syria. Apparently introduced Riyadh, Qatar, Bahrain and UAE. Vagrant (or escapes) Israel, Jordan. **Habitat:** as Yellow-vented Bulbul. Nests in bush, in verandahs, under eaves.

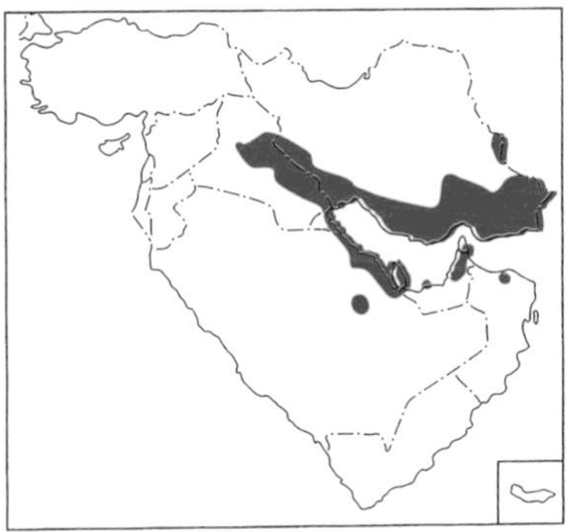

466 **Yellow-vented Bulbul**
Pycnonotus xanthopygos

Status: resident. Vagrant (perhaps breeding) Iraq. **Habitat:** trees and bushes, particularly berry-bearing; gardens, palm groves and wadis with cover. Nests in bush or trees.

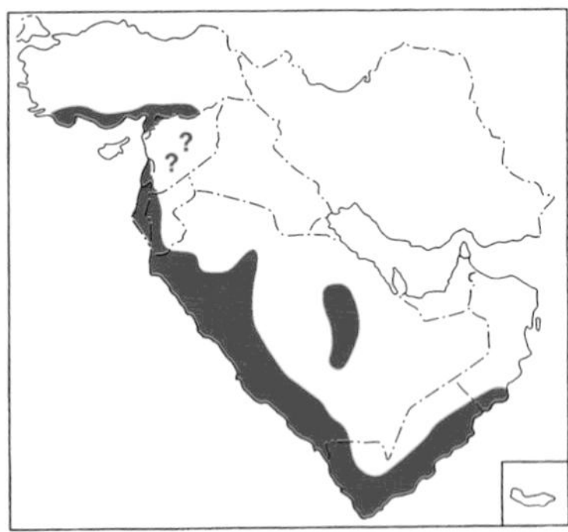

467 **Red-whiskered Bulbul**
Pycnonotus jocosus

Status: has bred UAE. Records in E Saudi Arabia and Gulf States are assumed to be escapes. **Habitat:** as other bulbuls.

468 **Red-vented Bulbul** *Pycnonotus cafer*

Status: resident, probably having originated from escapes. Also recorded elsewhere on southern Gulf coast away from breeding areas. **Habitat:** as other bulbuls but prefers evergreen trees. Nest: as other bulbuls.

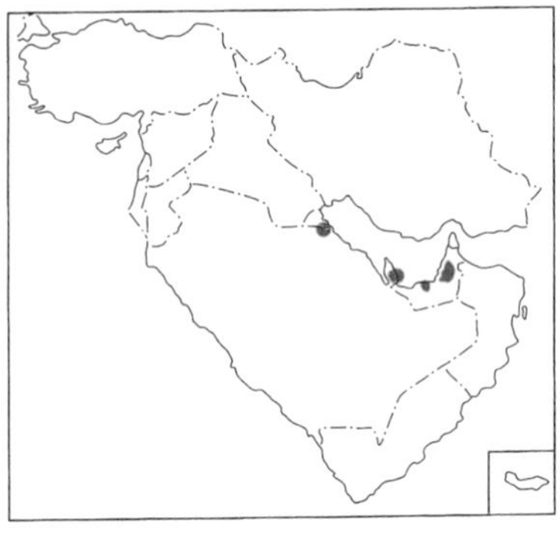

469 **Waxwing** *Bombycilla garrulus*

Status: vagrant Turkey, Iran, Cyprus and Israel. **Habitat:** berry-bearing trees and shrubs.

470 **Grey Hypocolius** *Hypocolius ampelinus*

Status: summer and winter range confined to Middle East. Summer visitor to breeding sites. Winters Bahrain, N and central Saudi Arabia and UAE; scarce W Arabia. Vagrant Oman, Qatar, Yemen and Israel. **Habitat:** bushes, scrub, date gardens. Bulky nest built in small tree.

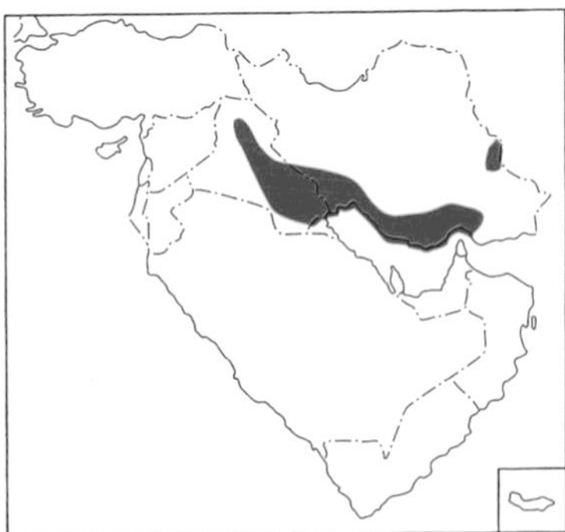

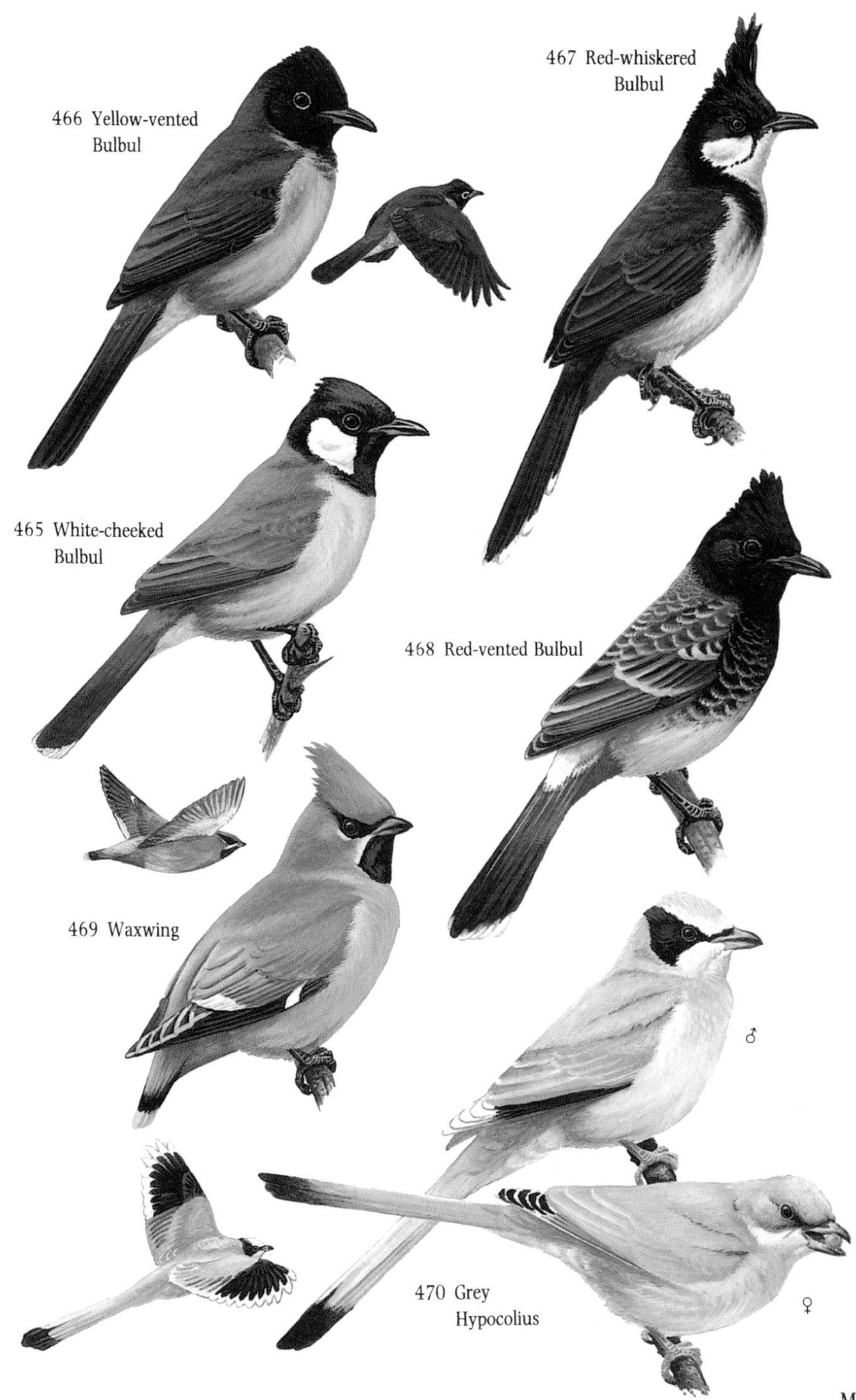
467 Red-whiskered
Bulbul
466 Yellow-vented
Bulbul
465 White-cheeked
Bulbul
468 Red-vented Bulbul
469 Waxwing
♂
470 Grey
Hypocolius
♀
ML

472 **Wren** *Troglodytes troglodytes*

Status: mainly resident, but in winter reaches SW Iran, central Iraq and most of Turkey. Vagrant Kuwait. **Habitat:** low shaded cover in gardens, thickets or woods. Builds domed nest in hedge, hole in tree, bank or building.

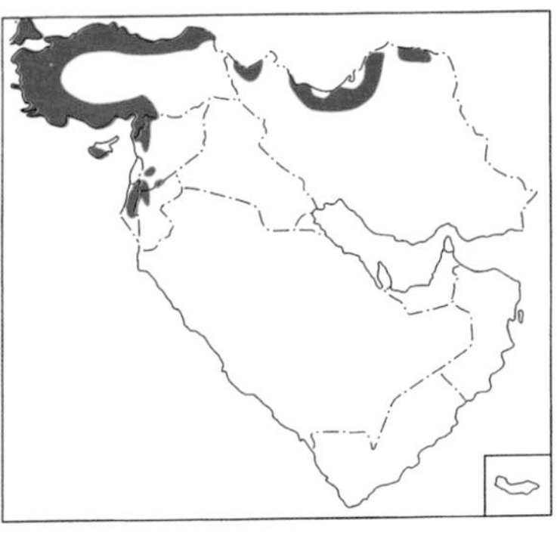

473 **Dunnock** *Prunella modularis*

Status: mainly resident. In winter reaches N and W Iran, N Iraq, Cyprus, Israel and Jordan. Vagrant Kuwait. **Habitat:** scrub and wooded areas on low and high ground.

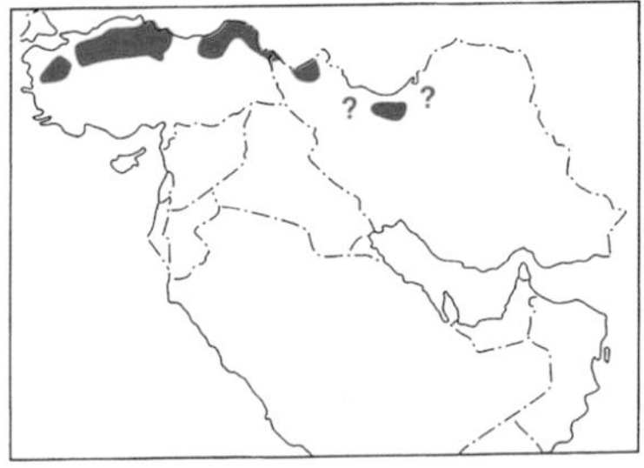

474 **Siberian Accentor** *Prunella montanella*

Status: vagrant Lebanon. Range Asia. **Habitat:** in winter prefers river bank thickets and streamside tangles but also forests and scrub.

475 **Radde's Accentor** *Prunella ocularis*

Status: summer visitor to breeding grounds; winters usually nearby at lower levels including Syria and N Israel, but winter distribution not fully known.

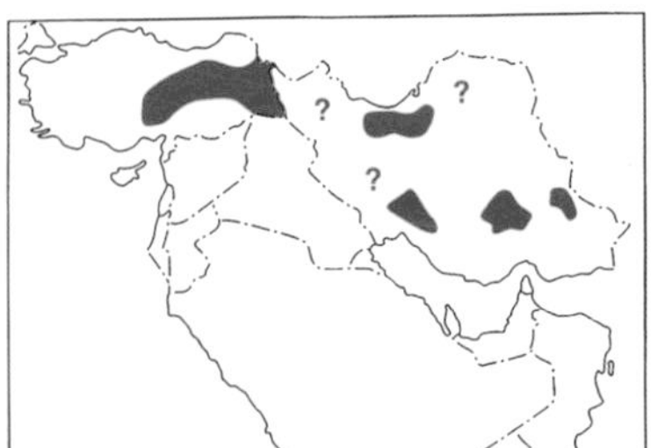

Habitat: low scrub and boulders in mountains 2500–3500 m; in winter down to 1000 m. Nests low in thick bush.

476 **Arabian Accentor** *Prunella fagani*

Status: resident. **Habitat:** open and bushy areas in rocky mountains above 2500 m; may straggle lower in winter. Nests in low vegetation.

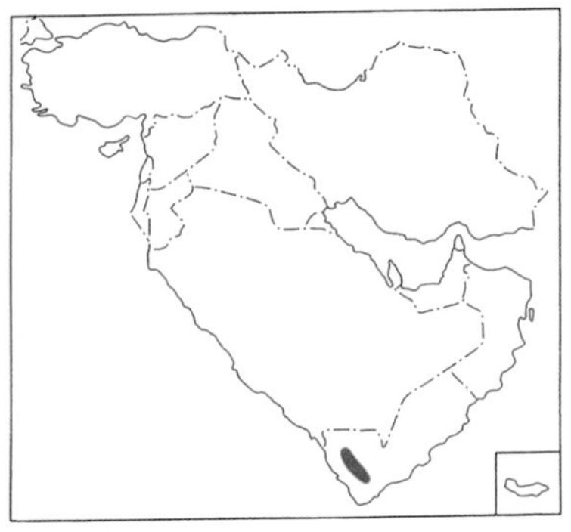

477 **Black-throated Accentor** *Prunella atrogularis*

Status: scarce winter visitor E Iran. Vagrant, perhaps annual, Israel, Kuwait. **Habitat:** in winter could occur in any area with cover including trees.

478 **Alpine Accentor** *Prunella collaris*

Status: mainly resident with some altitudinal movements and dispersal in winter reaching Iraq, Syria, Jordan and Israel. **Habitat:** mountains, above tree-line amongst boulders and rocks, stony alpine meadows, barren slopes, occasionally with scrub. Nests in hole among rocks or vegetation.

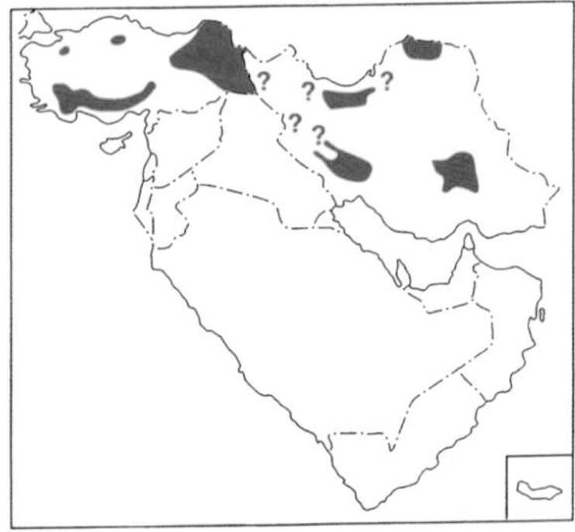

473 Dunnock
472 Wren
474 Siberian Accentor
475 Radde's Accentor
477 Black-throated Accentor
476 Arabian Accentor
winter
478 Alpine Accentor
montana

479 **Rufous Bush Robin** *Cercotrichas galactotes*

Status: summer visitor; occurs widely on passage. **Habitat:** a variety of habitats with low cover, from semi-desert scrub to gardens, palm groves and vineyards. Nests in prickly pear hedges, palm bushes and thickets.

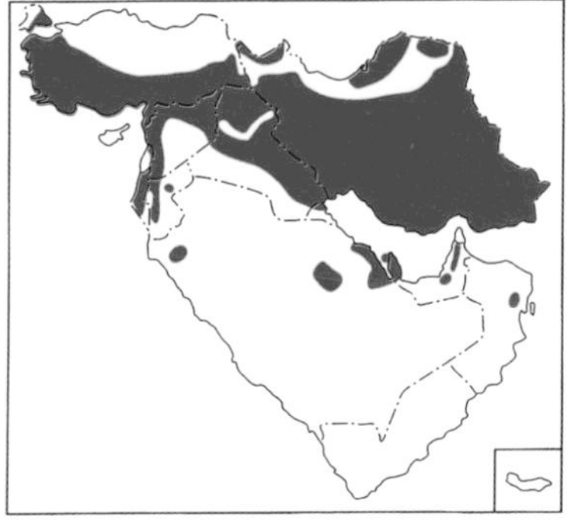

480 **Black Bush Robin** *Cercotrichas podobe*

Status: resident, but extending range north and east. Rare, but regular Israel. Vagrant UAE and Bahrain. **Habitat:** rolling or flat sandy plains with bushes, dry scrub and acacia wadis; also gardens. Nests in low bush.

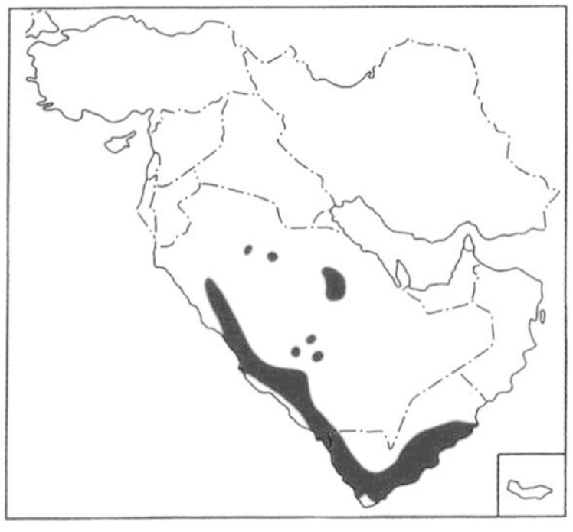

481 **Robin** *Erithacus rubecula*

Status: resident. Also winter visitor throughout the Mediterranean area, including Jordan, eastwards to

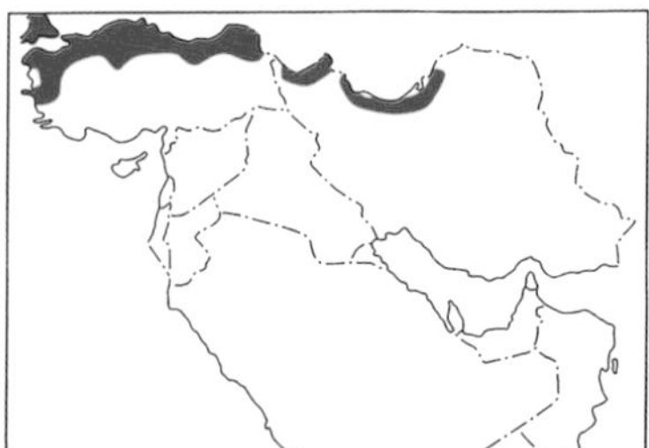

central Iran, Kuwait; rare E Saudi Arabia and UAE and vagrant Bahrain and Oman. **Habitat:** shady gardens, hedges, coppices, woods with undergrowth. Nests in holes or crannies in banks, trees and hedge-bottoms.

482 **Thrush Nightingale** *Luscinia luscinia*

Status: passage migrant Turkey, Iran; uncommon or rare Arabia and Near East. **Habitat:** on passage in thickets, swampy undergrowth and edges of reed-beds.

483 **Nightingale** *Luscinia megarhynchos*

Status: summer visitor. Occurs on passage in most countries south of breeding range although generally scarce, especially in Arabia, though common UAE and S Oman (in monsoon vegetation). **Habitat:** deciduous woods, gardens, thickets, sometimes on edge of wadis; on passage as Thrush Nightingale. Nests near ground well hidden in brambles, nettles, and other cover.

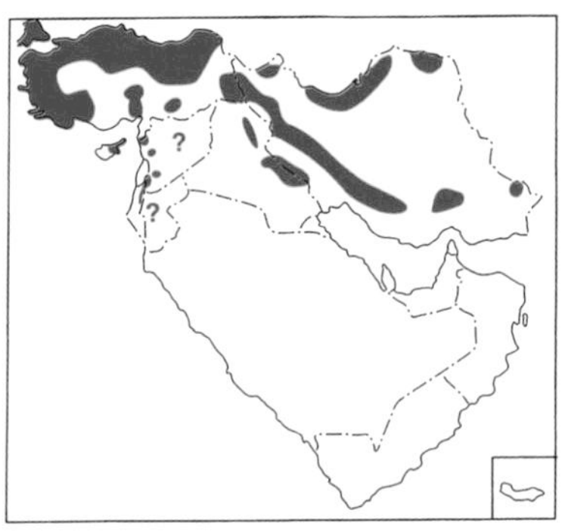

485 **Bluethroat** *Luscinia svecica*

Status: summer visitor to breeding area; occurs widely on passage throughout the region and at scattered localities in winter. **Habitat:** swampy thickets in mountains; in winter also in reedbeds and dense vegetation bordering swamps. Nests near ground within cover.

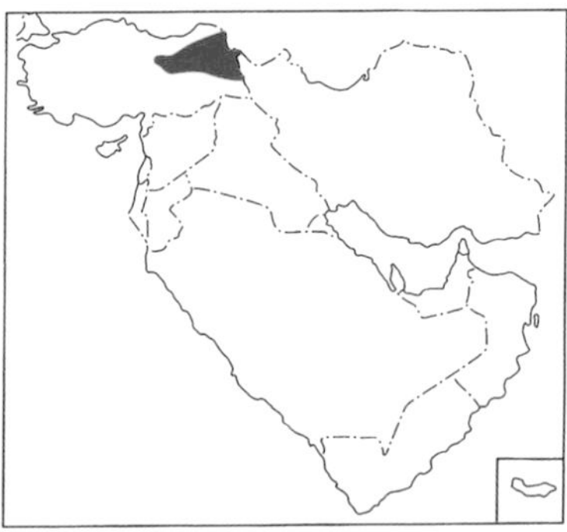

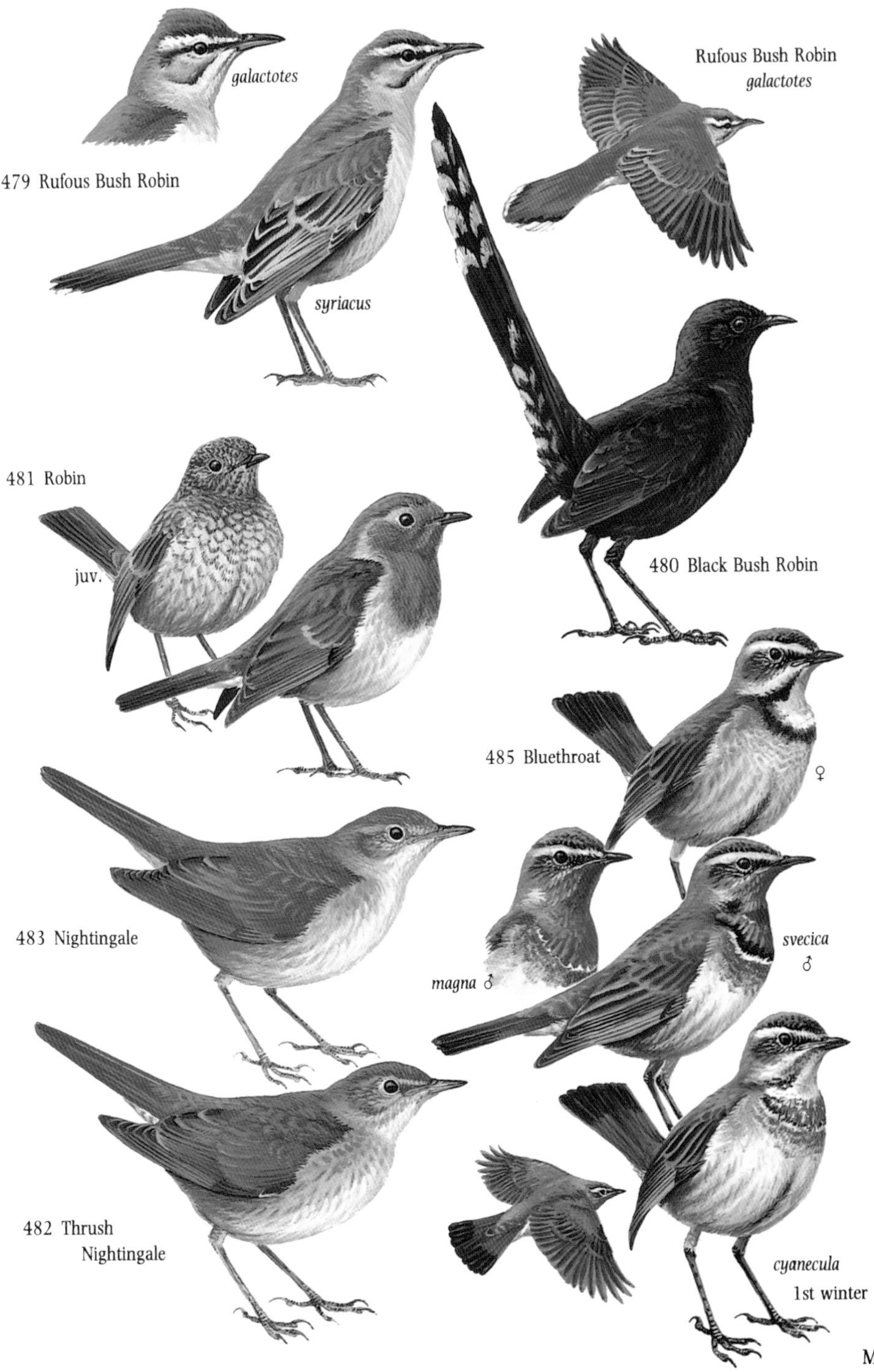
galactotes
479 Rufous Bush Robin
syriacus
Rufous Bush Robin
galactotes
481 Robin
juv.
480 Black Bush Robin
485 Bluethroat
♀
483 Nightingale
svecica
♂
magna ♂
482 Thrush
Nightingale
cyanecula
1st winter

486 **Red-flanked Bluetail** *Tarsiger cyanurus*

Status: vagrant Cyprus, Israel and Lebanon. Range Eurasia. **Habitat:** usually thick undergrowth.

487 **White-throated Robin** *Irania gutturalis*

Status: has bred Lebanon. Summer visitor. Migrant Near East, rare Arabia (but regular UAE, Oman, Yemen). Vagrant Cyprus. **Habitat:** stony hillsides and valleys with scrub, usually at 1000–2200 m. On passage in dense scrub. Nests in bush.

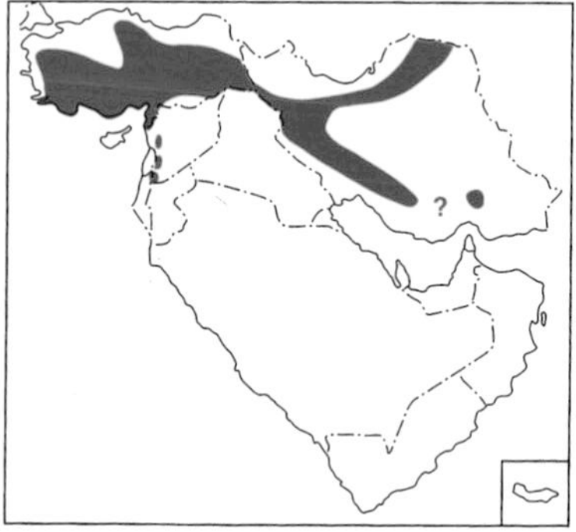

488 **Eversmann's Redstart**
Phoenicurus erythronotus

Status: thinly distributed winter visitor Iran (except NW), rare central Iraq and south side of Arabian Gulf to Oman; vagrant NW Saudi Arabia and Turkey. **Habitat:** oases, wetlands, coastal groves and dunes; also juniper woodland.

489 **Black Redstart** *Phoenicurus ochruros*

Status: largely summer visitor to mountain localities. Winter visitor throughout the region; some passage through central part of region. **Habitat:** between 2500–5000 m on stony slopes, rocks and cliffs. In winter to sea level in open rocky areas and wastelands. Nests in hole in crevice in rock or wall.

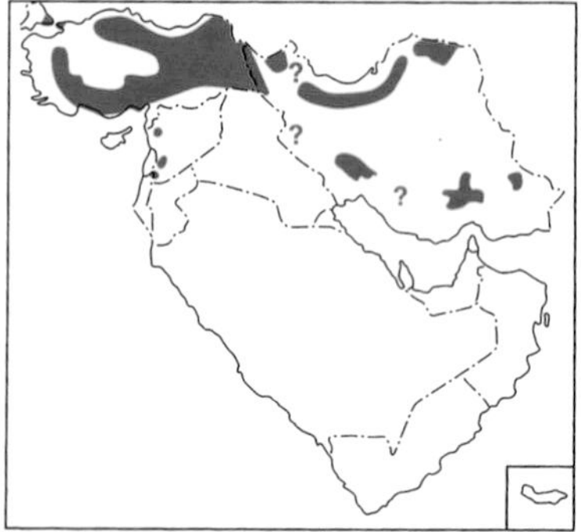

490 **Common Redstart**
Phoenicurus phoenicurus

Status: summer visitor. Passage migrant throughout the region; winter visitor E and SW Arabia (except Oman) and perhaps Iraq. **Habitat:** woodland and parks with old trees. On passage where bushes and trees. Nests in hole in tree or stone wall.

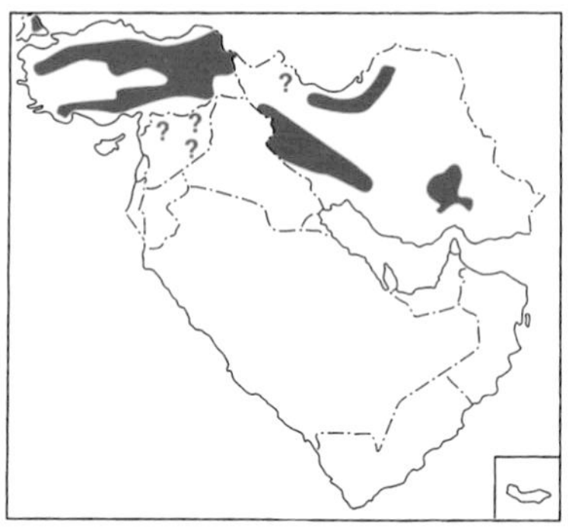

491 **Güldenstädt's Redstart**
Phoenicurus erythrogaster

Status: unsubstantiated claim from Turkey and Kuwait; range Central Asia. **Habitat:** rocky slopes, cultivation and gardens.

492 **Blackstart** *Cercomela melanura*

Status: resident, but local movements, e.g. into NE Israel in winter. Occasional Syria. **Habitat:** desert-fringed bushy or wild rocky wadis; sometimes near agriculture. Nests in crevice in or under rock or large stone.

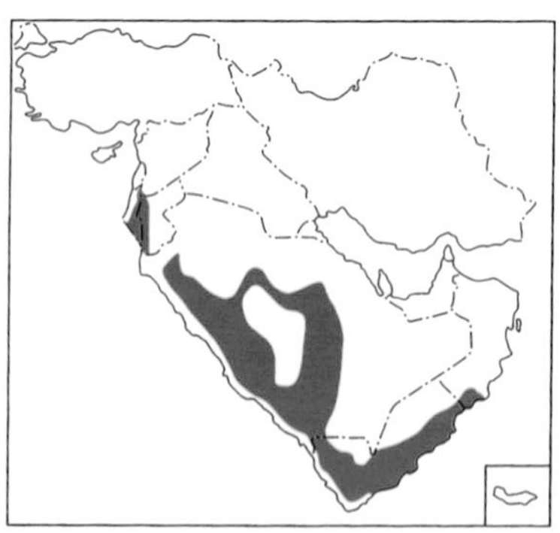

493 **Familiar Chat** *Cercomela familiaris*

Status: vagrant Yemen. Range Africa. **Habitat:** rocky ground and cultivated or lightly wooded areas.

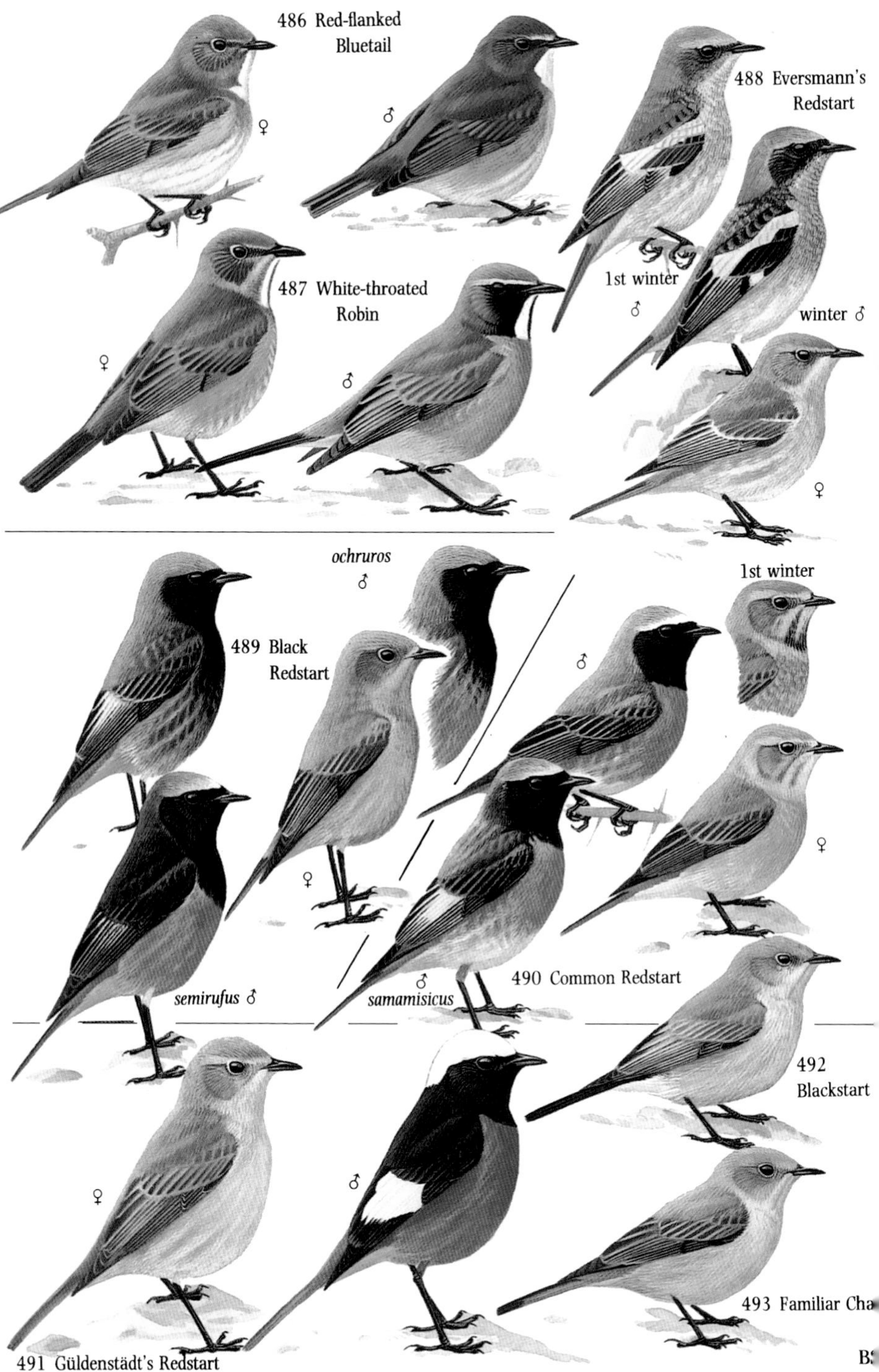
486 Red-flanked Bluetail
♀
♂
487 White-throated Robin
♀
♂
488 Eversmann's Redstart
1st winter ♂
winter ♂
♀
489 Black Redstart
ochruros ♂
♀
semirufus ♂
samamisicus ♂
490 Common Redstart
♂
1st winter
♀
491 Güldenstädt's Redstart
♀
♂
492 Blackstart
493 Familiar Cha

494 **Whinchat** *Saxicola rubetra*

Status: summer visitor. Widespread on passage, but uncommon Arabia. **Habitat:** wastelands, open bush country, marshes; on passage in open vegetated regions, often cultivated. Nests in grass at foot of bush.

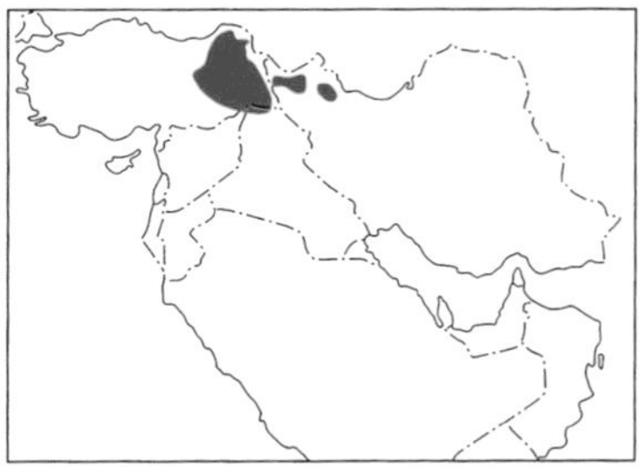

495 **Stonechat** *Saxicola torquata*

Status: resident or partial migrant; departs high ground in winter. Passage migrant or winter visitor over entire region though rare E Iran. S Arabian visitors are *maura, variegata* and *armenica*, which reach Near East on passage. European race mainly on passage in E Mediterranean and Near East. **Habitat:** open regions from sea level to 3000 m, in semi-cultivated areas or scrub-covered slopes; on passage/winter in any open area with vegetation. Nests on ground in vegetation.

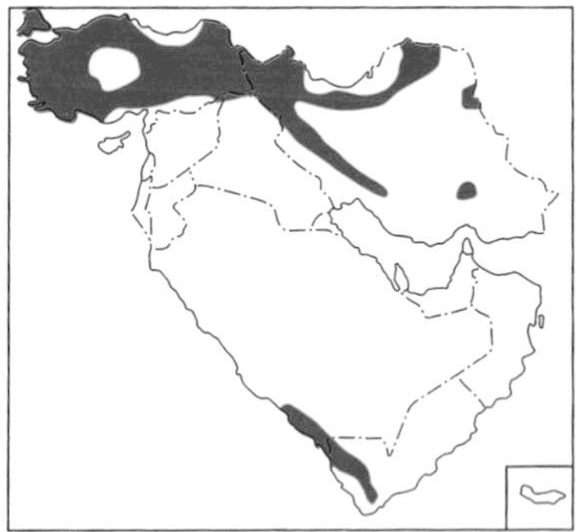

496 **Pied Stonechat** *Saxicola caprata*

Status: summer visitor. Vagrant Oman, Iraq, UAE and Saudi Arabia. **Habitat:** cultivated areas with open fields and scattered bushes, often near moist ground. Nests in hole in wall or rocks

497 **Isabelline Wheatear** *Oenanthe isabellina*

Status: summer visitor to breeding areas. Migrant and winter visitor from Arabian Gulf coasts to Iraq and Syria, southwards across Arabia; passage only in Turkey, Cyprus and Near East. **Habitat:** Stony, barren or grass-covered ground, also steppes; on passage/winter also in cultivation. Nests in hole in bank or burrow.

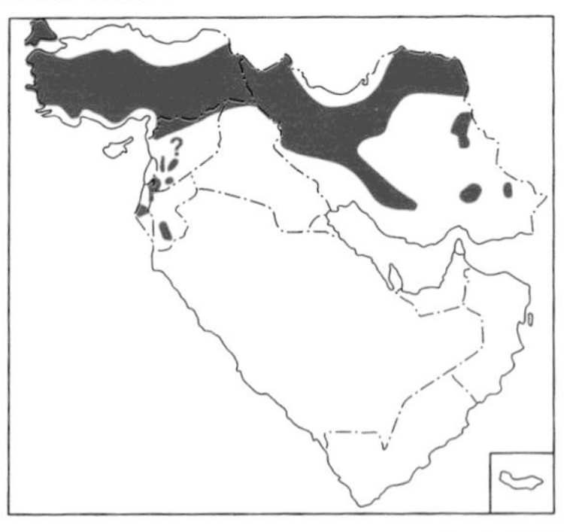

498 **Red-breasted Wheatear** *Oenanthe bottae*

Status: resident. **Habitat:** from 1600 m upwards on high plateaux, open hill-sides with sparse vegetation, cultivated terraces. Nests in earth-bank, stone wall or hole in ground.

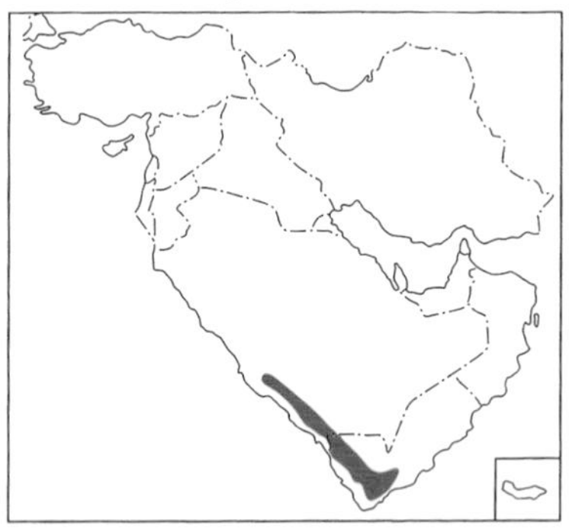

499 **Northern Wheatear** *Oenanthe oenanthe*

Status: summer visitor. Widely on passage throughout the region but scarce SE and SW Arabia, SE Iran. Occasional in winter. **Habitat:** open highlands with rocky or stony slopes often with bushes. On passage down to sea level in any open area, including cultivation. Nests in hole in rock or stones.

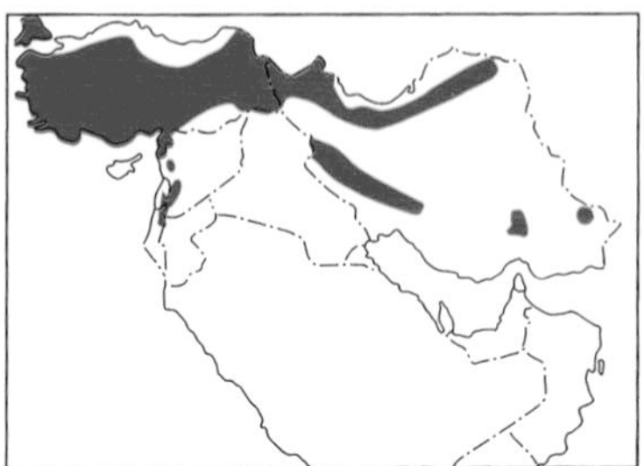

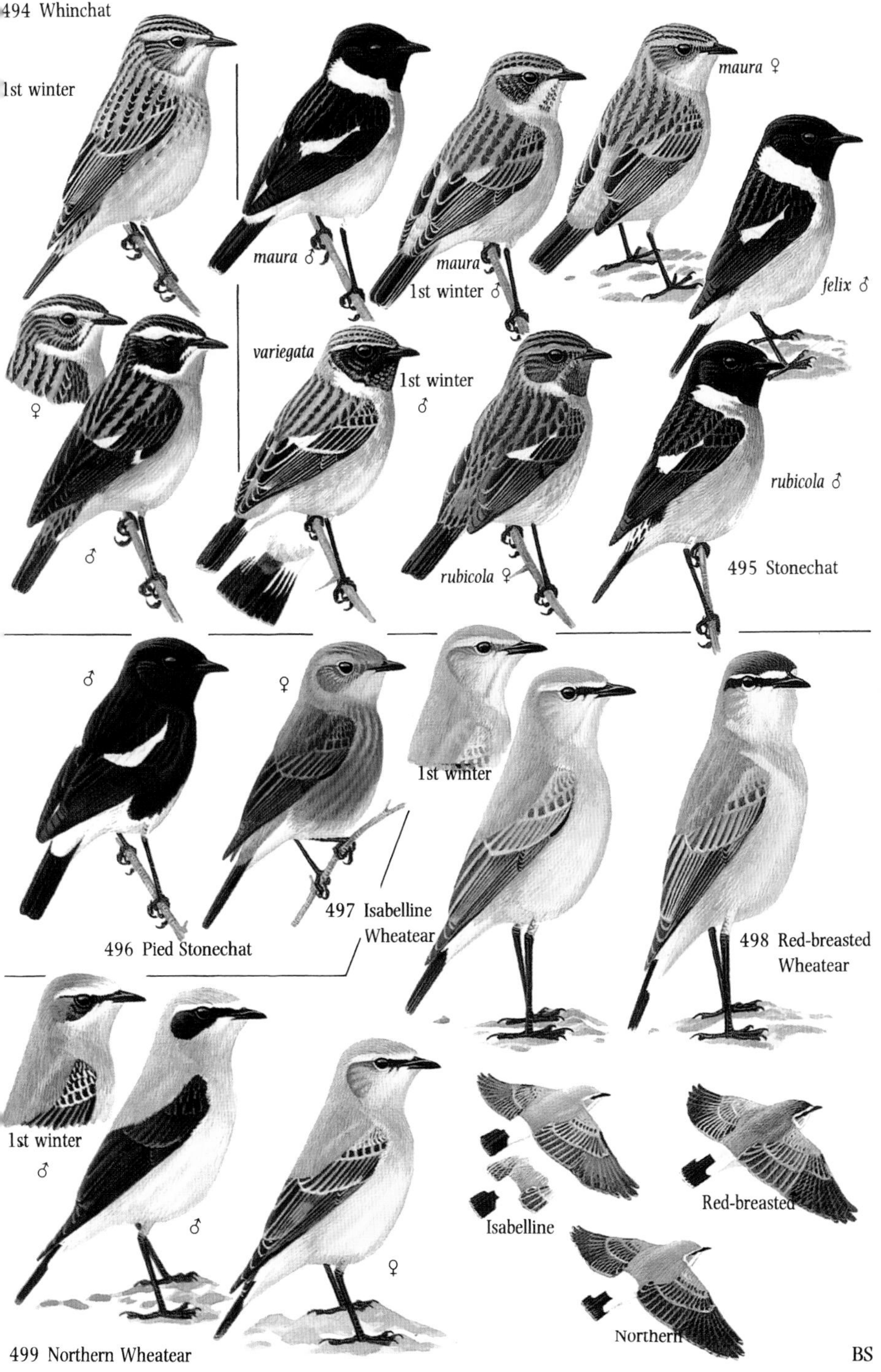
494 Whinchat
1st winter
♀
♂
maura ♂
maura
1st winter ♂
maura ♀
felix ♂
variegata
1st winter
♂
rubicola ♀
rubicola ♂
495 Stonechat
♂
♀
496 Pied Stonechat
1st winter
497 Isabelline
Wheatear
498 Red-breasted
Wheatear
1st winter
♂
♂
♀
Isabelline
Red-breasted
Northern
499 Northern Wheatear
BS

500 **Pied Wheatear** *Oenanthe pleschanka*

Status: summer visitor. Winters Yemen, occasionally Gulf States. On passage all Arabia, southern Iran, scarcer Near East, rarer Turkey. Vagrant Cyprus. **Habitat:** stony, barren, rocky hillsides with bushes. On passage/winter also bare fields, wastelands, wadis with vegetation and towns. Nests in hole, often in banks.

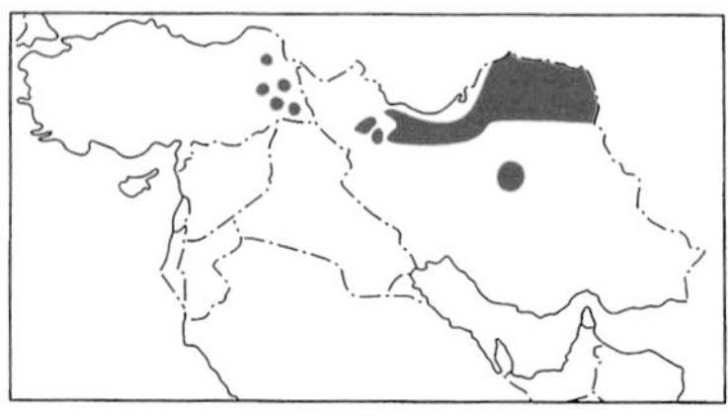

501 **Cyprus Pied Wheatear** *Oenanthe cypriaca*

Status: endemic summer visitor to Cyprus; winter range unclear (probably not S Arabia); passage Israel, Jordan and, in spring, nearby Turkey by birds overshooting Cyprus. Vagrant Saudi Arabia. **Habitat:** as Pied, but often on more forested slopes. On passage in vegetated wadis, tree-lined agriculture. Nests in hole in bank and even next-box.

502 **Black-eared Wheatear**
Oenanthe hispanica

Status: summer visitor. On passage in all the region, but rare Arabia, especially eastern half. **Habitat:** open, lightly vegetated and stoney slopes, dry river banks. On passage in many open areas with some vegetation. Nests in hole in bank or wall.

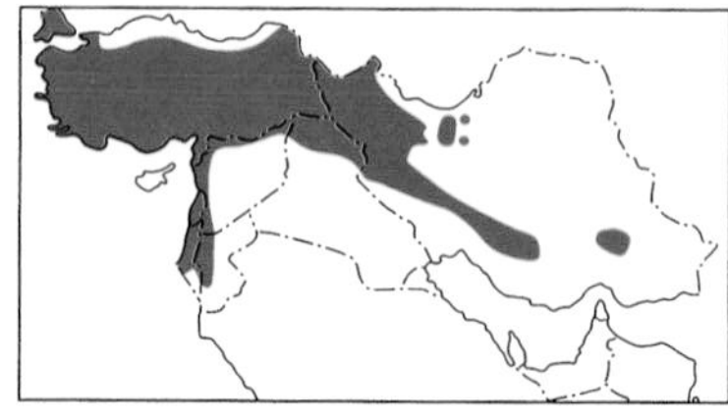

503 **Desert Wheatear** *Oenanthe deserti*

Status: has bred N Saudi Arabia. Summer visitor Iran, Iraq, perhaps resident Near East. On passage/winter in S Iran, Iraq and all Arabia; rare

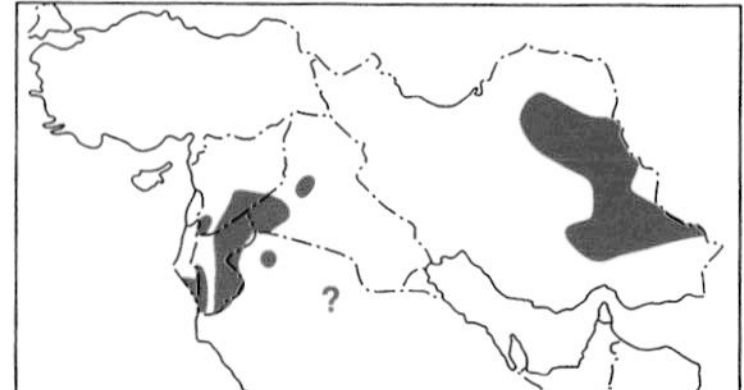

Cyprus, vagrant Turkey (where bred recently). **Habitat:** bushy desert plains and stony barren wastes. Nests in hole, burrow, amongst rocks.

504 **Finsch's Wheatear** *Oenanthe finschii*

Status: breeds occasionally Lebanon. Summer visitor to uplands, descending in winter, reaching Cyprus, Near East, N Arabia and Arabian Gulf coasts, but rare Bahrain and UAE. Vagrant Oman. **Habitat:** dry rocky, stony uplands and foothills, sparsely vegetated semi-deserts. Nests in hole or fissure in rock or bank.

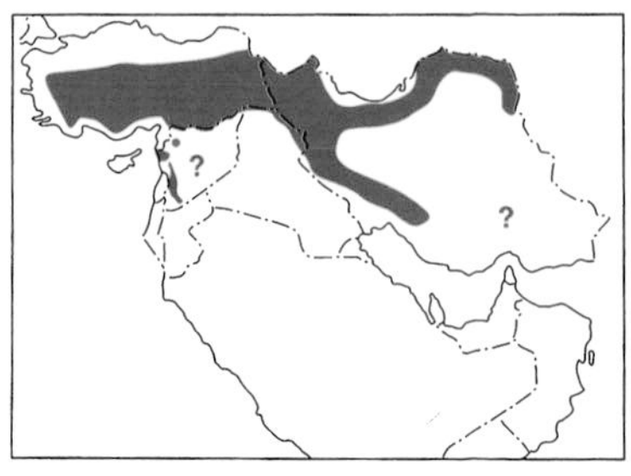

505 **Red-rumped Wheatear** *Oenanthe moesta*

Status: resident. Vagrant Saudi Arabia and Turkey. **Habitat:** flat saline regions and bush-clad desert-fringes. Nests in hole in earth bank.

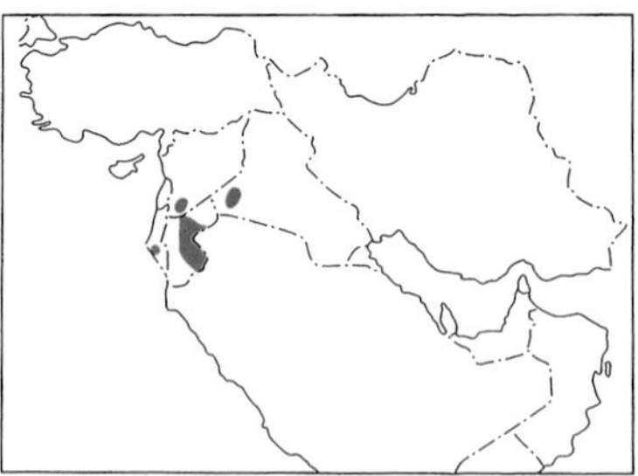

506 **Red-tailed Wheatear**
Oenanthe xanthoprymna

Status: summer visitor. Winters S Iran, Oman, UEA, uncommon elsewhere in Arabia, but probably regular Yemen; scarce passage Iraq; rare Israel, vagrant Syria, Jordan and Cyprus. **Habitat:** stony or barren hillsides with scree, scattered boulders, low scrubby vegetation. In winter includes agriculture, ruins in deserts. Nests in hole in bank or well.

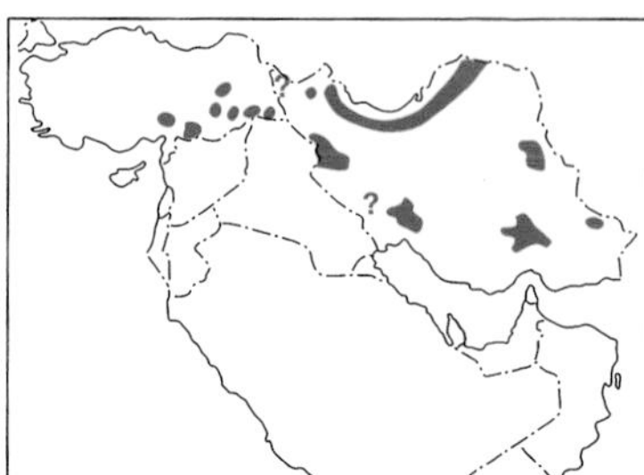

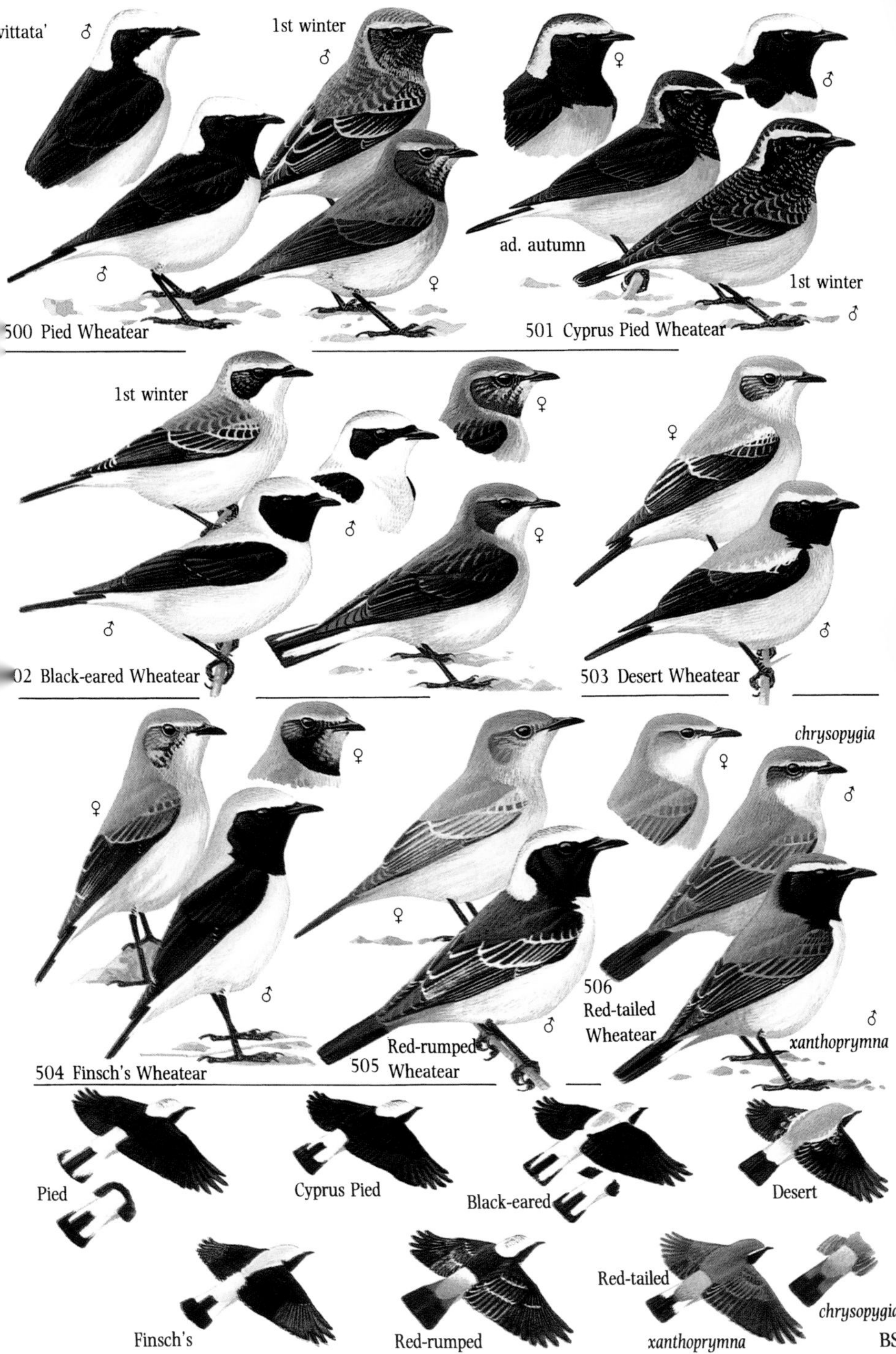
'vittata' ♂
1st winter ♂
♂
♀
♀
ad. autumn
♂
1st winter ♂
500 Pied Wheatear
501 Cyprus Pied Wheatear
1st winter
♀
♂
♀
♂
♀
♂
502 Black-eared Wheatear
503 Desert Wheatear
chrysopygia
♀
♀
♂
♀
♀
♂
♂
♂
504 Finsch's Wheatear
505 Red-rumped Wheatear
506 Red-tailed Wheatear
xanthoprymna
Pied
Cyprus Pied
Black-eared
Desert
Finsch's
Red-rumped
Red-tailed
xanthoprymna
chrysopygia
BS

507 **Eastern Pied Wheatear** *Oenanthe picata*

Status: summer visitor. Common winter visitor S Iran, including Mekran coast; scarce winter visitor UEA and N Oman (*picata*-morph). **Habitat:** barren, open, boulder-strewn country with vegetation, steep river banks; in winter also cultivation. Nests in hole in rocks or bank.

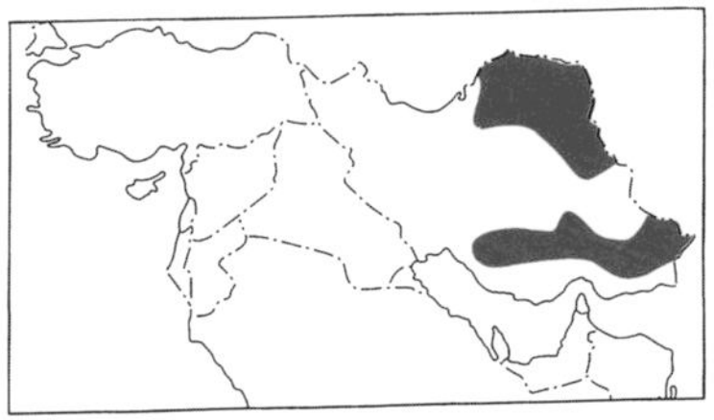

508 **Mourning Wheatear** *Oenanthe lugens*

Status: resident, but dispersive; summer visitor Iran; winters in small numbers Arabia. Occasional passage E Iran. Vagrant, Turkey, Lebanon, Cyprus and, dark morph, Israel and Saudi Arabia. **Habitat:** rather barren rocky mountain gorges and wadis, lava-plains, deserts and semi-deserts. Nests in hole in rock or in burrow.

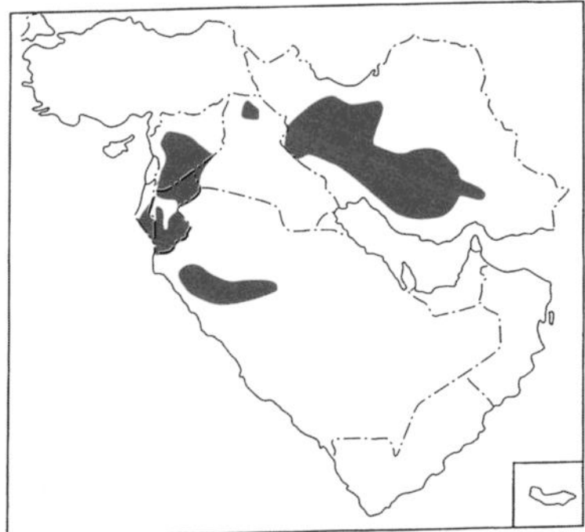

509 **South Arabian Wheatear** *Oenanthe lugentoides*

Status: resident. **Habitat:** rocky hillsides, mountains with sparse vegetation but often bushes and trees, including thick juniper scrub near cultivation; usually 1000–2500 m. Nests in hole in rocks, wall or bank.

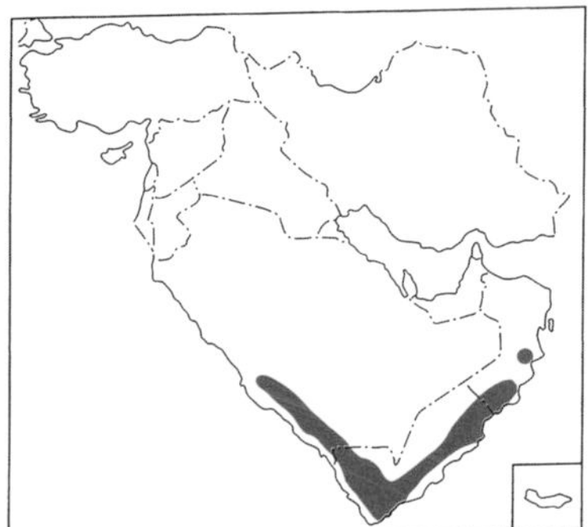

510 **Hooded Wheatear** *Oenanthe monacha*

Status: resident, probably some dispersal. Vagrant Bahrain, Iraq, Yemen and Cyprus. **Habitat:** desolate, barren rocky ravines, slopes, gorges and desert-regions, usually without vegetation. Nests in hole in rock.

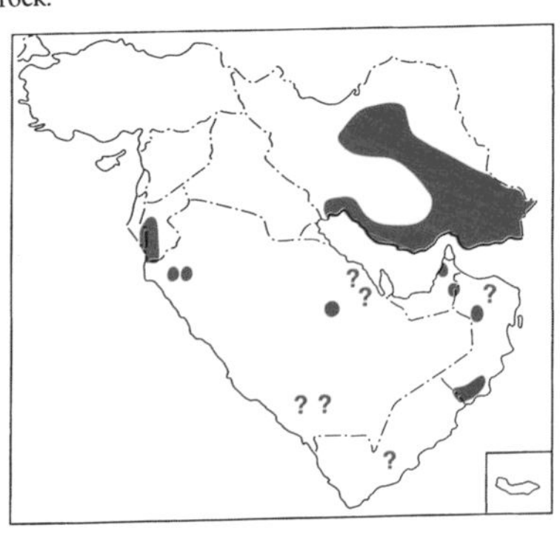

511 **Hume's Wheatear** *Oenanthe alboniger*

Status: resident. Vagrant Bahrain. **Habitat:** boulders at foot of barren hills with scanty vegetation. Nests in hole in rock.

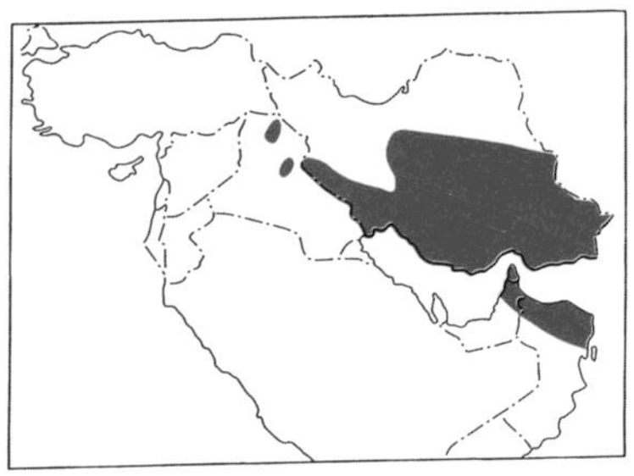

512 **White-crowned Black Wheatear** *Oenanthe leucopyga*

Status: resident with some dispersal. Rare Gulf States, Oman, Yemen and Cyprus. Vagrant Turkey, Bahrain and UEA. **Habitat:** rocky deserts, ravines in rocky mountains, usually without vegetation, often around human settlements. Nests in hole, sometimes in building.

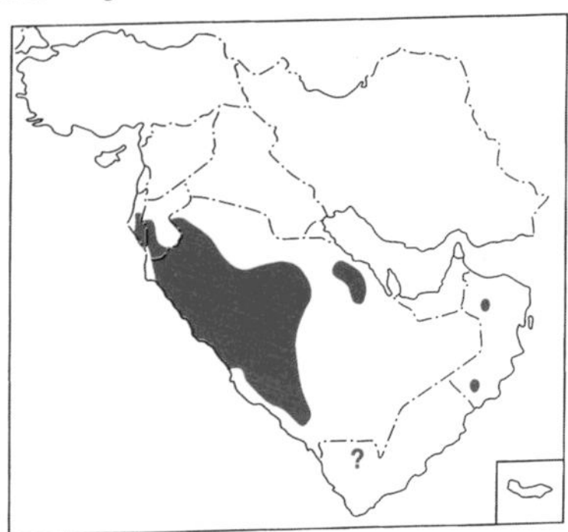

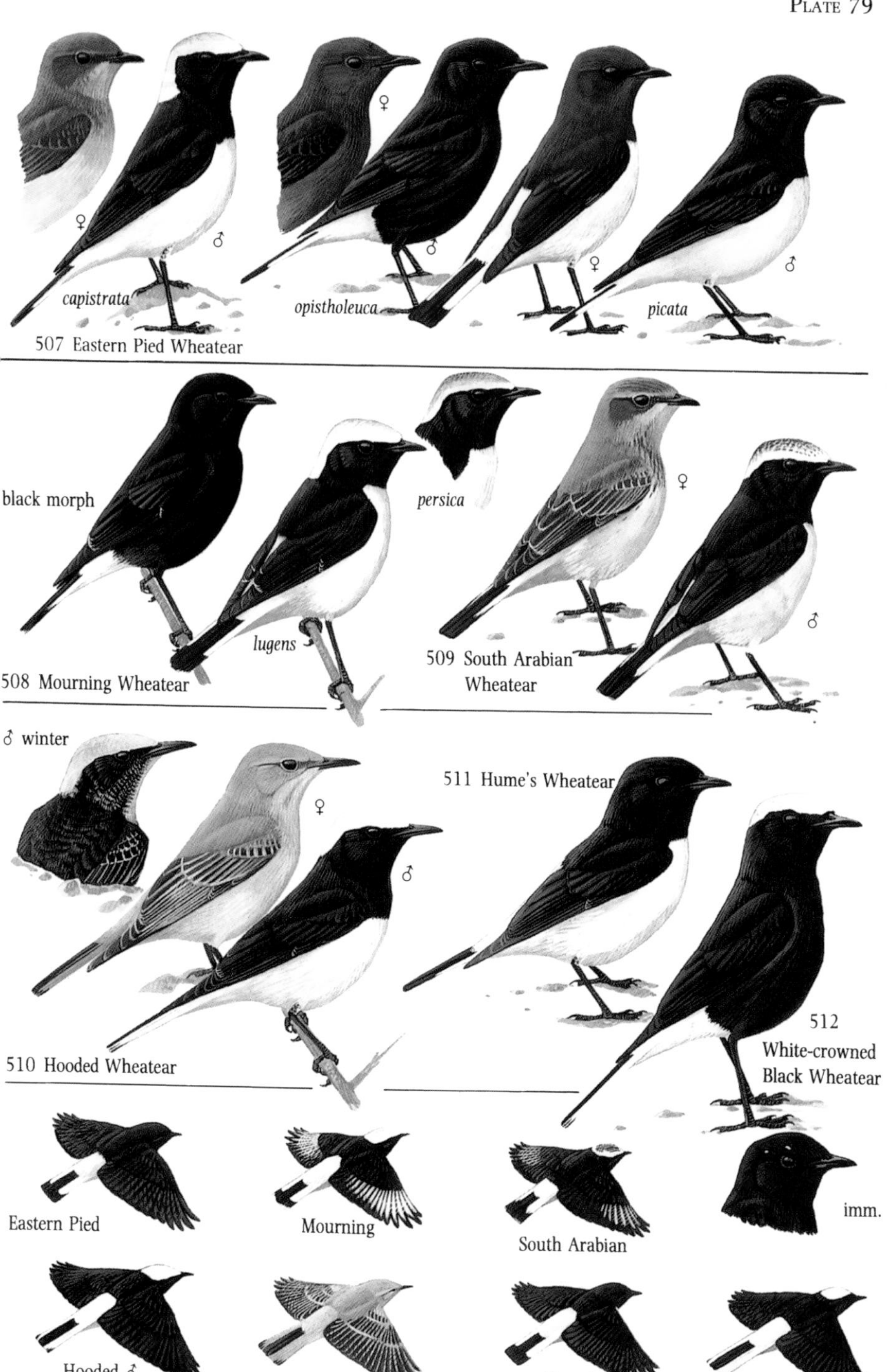
♀
♂
capistrata
♀
♂
opistholeuca
♀
♂
picata
507 Eastern Pied Wheatear
black morph
persica
lugens
508 Mourning Wheatear
♀
♂
509 South Arabian Wheatear
♂ winter
♀
♂
510 Hooded Wheatear
511 Hume's Wheatear
512 White-crowned Black Wheatear
Eastern Pied
Mourning
South Arabian
imm.
Hooded ♂
Hooded ♀
Hume's
White-crowned
BS

513 **Little Rock Thrush** *Monticola rufocinerea*

Status: resident. **Habitat:** rocky highlands, 1500–2500 m, with trees and bushes, typically Acacia; also in or near cultivation. Nests in tree.

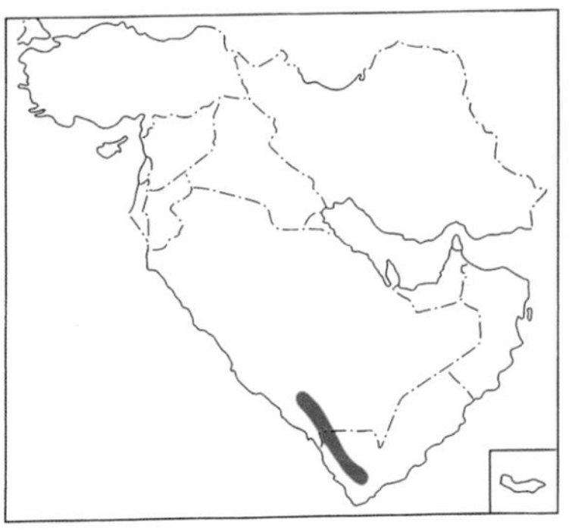

514 **Rock Thrush** *Monticola saxatilis*

Status: summer visitor. Passage throughout region but not common. Winter records from Arabian Gulf coasts, Oman and Yemen. **Habitat:** breeds in rocky sparsely vegetated areas, also ruins, 1000–2600 m. Can occur most areas on passage.

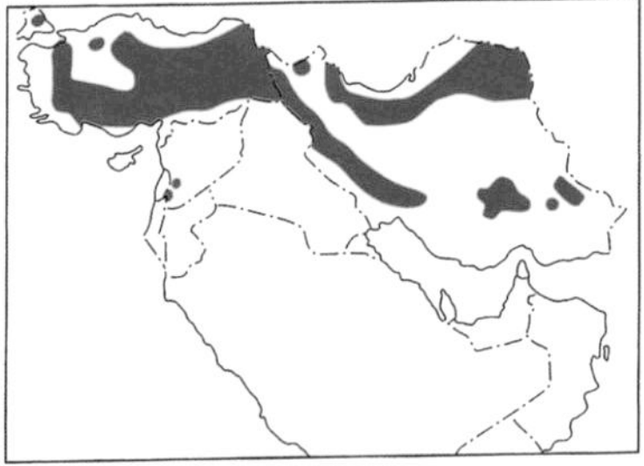

515 **Blue Rock Thrush** *Monticola solitarius*

Status: partial migrant. Summer visitor to mountains in Turkey and Iran but mainly resident in Cyprus and Near East. Winter and on passage in W Turkey, S Iran and most Arabia. **Habitat:** rocky deserts, mountains and cliffs, to sea level; in winter and passage also in cultivated areas with trees. Nests in rocks or buildings.

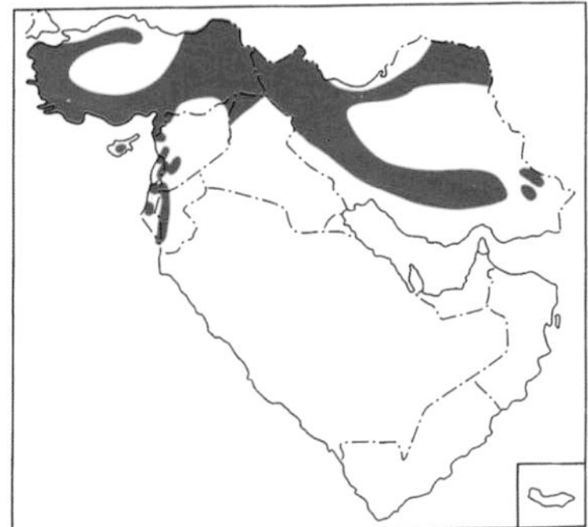

517 **Siberian Thrush** *Turdus/Zoothera sibirica*

Status: vagrant Israel. **Habitat:** usually near trees or bushes.

518 **Yemen Thrush** *Turdus menachensis*

Status: resident. **Habitat:** wooded areas (especially Acacia and Juniper) in mountains, on terraces and cultivated wadis 1700–3100 m. Nests in tree.

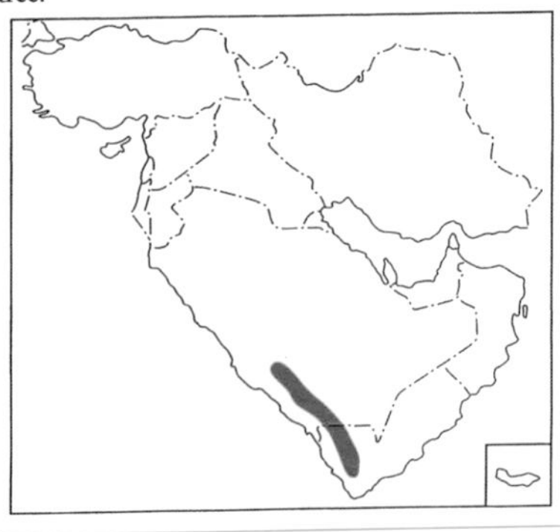

519 **Ring Ouzel** *Turdus torquatus*

Status: summer visitor to breeding areas. Winters W Iran, S Turkey, Cyprus; rare Near East. Vagrant Jordan, Syria, Gulf States and Oman. **Habitat:** mountains with alpine meadows and moorland, upper limit of open forest. In winter any area with trees or bushes. Nests in low vegetation or among rocks.

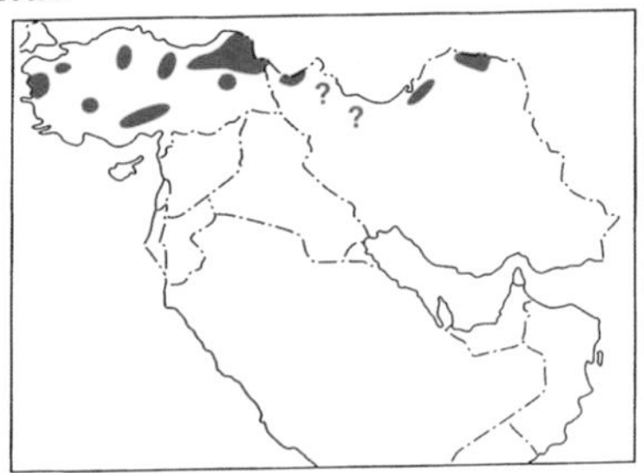

520 **Blackbird** *Turdus merula*

Status: present in winter through much of breeding range. Migrants from farther north also reach Iraq, N Saudi Arabia and Gulf States. **Habitat:** woodlands, gardens, orchards with rich undergrowth; often near cultivation. Nests in dense bush or tree.

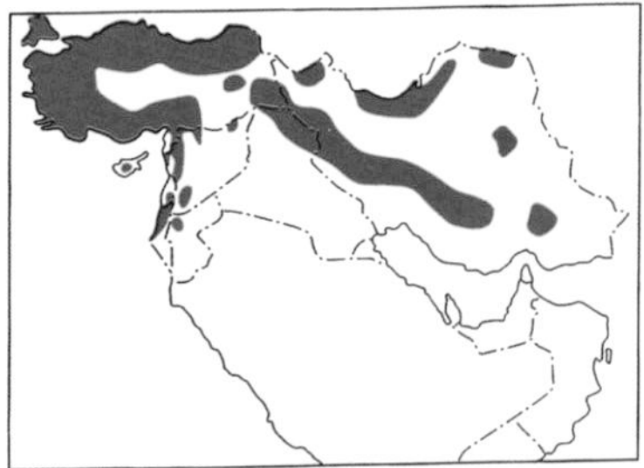

513 Little Rock Thrush
515 Blue Rock Thrush
♂
♀
14 Rock Thrush
♂
♀
520 Blackbird
♂
♀
♂
19 Ring Ouzel
♀
518 Yemen Thrush
♂
♀
1st winter
17 Siberian Thrush
1st winter ♂
ML

471 **Dipper** *Cinclus cinclus*

Status: formerly bred Cyprus. Resident; may disperse to lower levels in winter. Vagrant Iraq. **Habitat:** swift hill-streams. Nests besides running water, in crevice under waterfall, bridge, bank or cliff-face.

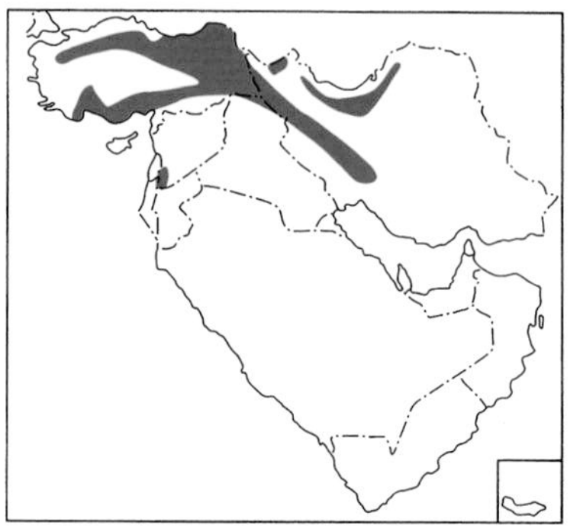

521 **Eyebrowed Thrush** *Turdus obscurus*

Status: vagrant Oman, UAE and Cyprus. Range Asia. **Habitat:** on passage/winter in wooded areas or open country with trees and bushes.

522 **Dusky/Naumann's Thrush** *Turdus naumanni*

Status: vagrant from Asia to Kuwait, Israel, Cyprus and Oman. **Habitat:** open wooded areas.

523 **Black-throated Thrush** *Turdus ruficollis*

Status: winter visitor, widespread Iran except NW; reaches Gulf and Mekran coasts, but scarce and irregular on Arabian side, from Iraq to Oman and central Saudi Arabia. Vagrant Yemen and Israel. Range Asia. **Habitat:** winters in open country with trees alternating with grassy areas.

524 **Fieldfare** *Turdus pilaris*

Status: winter visitor; fairly common NW Iran, Turkey, Cyprus, Lebanon; irregular Iraq, Syria, Arabian Gulf and NW Saudi Arabia but vagrant UAE. **Habitat:** open country; fields with groups of trees.

525 **Song Thrush** *Turdus philomelos*

Status: mainly resident. Winter visitor in Mediterranean part of the region, Iran, Iraq, Gulf States, E Saudi Arabia, scarcer Oman and irregular farther south. **Habitat:** woodland, parks, plantations; in winter also in open country with scattered trees and bushes. Nests in bush or tree.

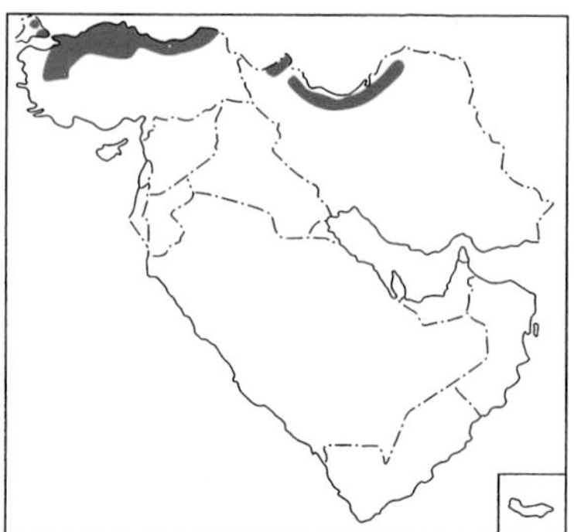

526 **Redwing** *Turdus iliacus*

Status: scarce or rare winter visitor to Near East though sometimes common N Turkey and N Iran; rare NW and E Arabia, Iraq and south side of Arabian Gulf. **Habitat:** in open country with scattered woods or trees, also bushes bordering cultivation.

527 **Mistle Thrush** *Turdus viscivorus*

Status: resident and dispersive; in winter reaching Cyprus and Near East but vagrant Jordan, Gulf States, Oman and N Saudi Arabia. **Habitat:** wooded regions with clearings. In winter in open country, fields with trees. Nests in tree.

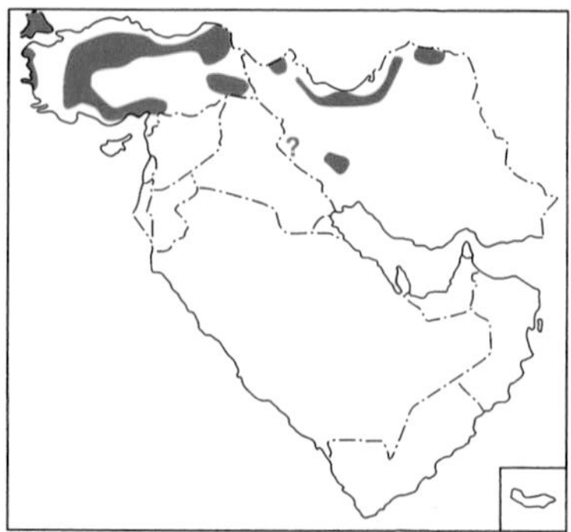

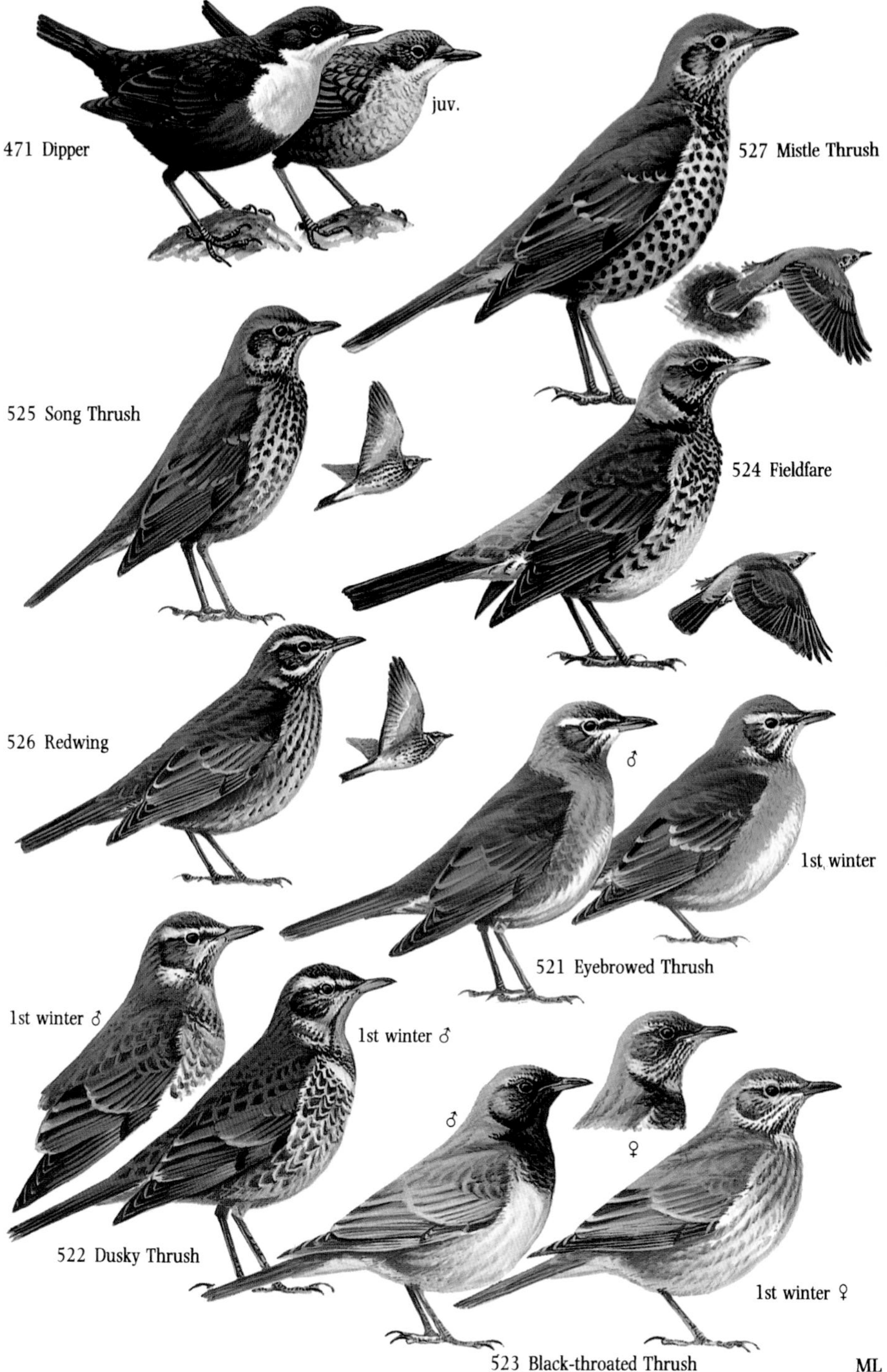

juv.
471 Dipper
527 Mistle Thrush
525 Song Thrush
524 Fieldfare
526 Redwing
♂
1st winter
521 Eyebrowed Thrush
1st winter ♂
1st winter ♂
♂
♀
522 Dusky Thrush
1st winter ♀
523 Black-throated Thrush
ML

528 **Cetti's Warbler** *Cettia cetti*

Status: resident, perhaps with some winter immigration in at least Iran and Kuwait. Vagrant Jordan and UAE. **Habitat:** low, dense, tangled vegetation, bushy thickets, often near swamps, streams, ditches and reedbeds, locally up to 2000 m. Nest well hidden in low vegetation.

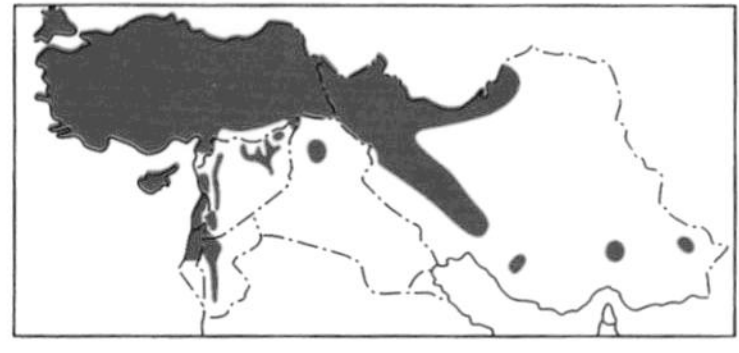

530 **Fan-tailed Cisticola** *Cisticola juncidis*

Status: resident with some dispersal; vagrant Kuwait and Oman. **Habitat:** wet and dry localities, grainfields, grassy plains, often along streams, marshes. Purse-shaped nest suspended in bush, long grass or corn.

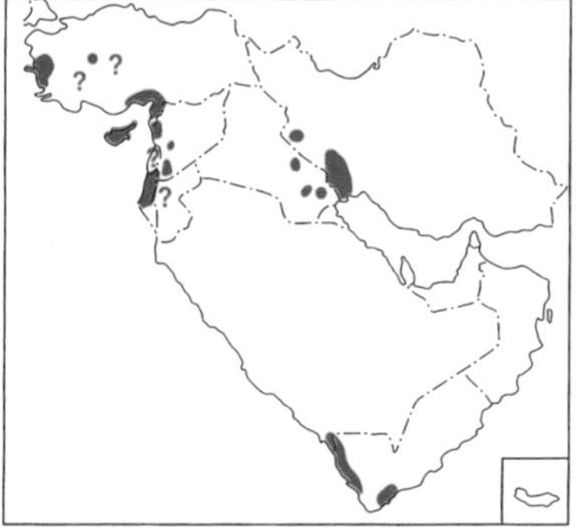

531 **Socotra Cisticola** *Cisticola haesitata*

Status: endemic to Socotra. **Habitat:** coastal sandy areas with tamarisk and other scrubby vegetation; has been recorded up to 750 m. Nest not described.

532 **Graceful Prinia** *Prinia gracilis*

Status: resident. Vagrant Cyprus. **Habitat:** scrub and low vegetation, in arid regions; cultivated areas, patches of wasteground in villages and towns, also near water. Domed nest in low bush or thick grass.

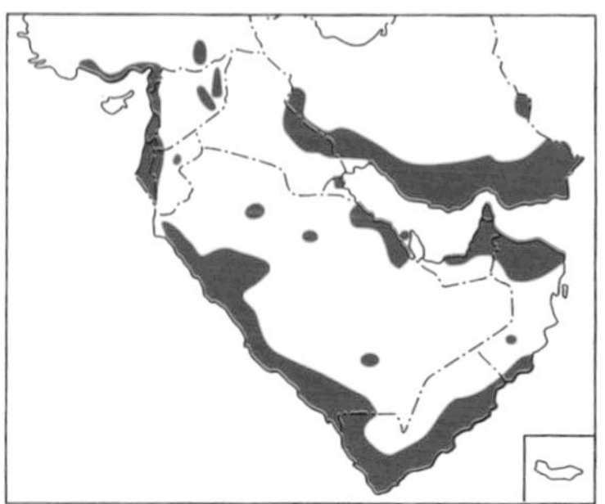

533 **Socotra Warbler** *Incana incana*

Status: endemic to Socotra. **Habitat:** tamarisk and other scrub from sea level to 800 m. Nest not described.

534 **Scrub Warbler** *Scotocerca inquieta*

Status: resident. Probably breeds S Syria. Possible record from Turkey. **Habitat:** rather barren, stony hillsides, semi-deserts and sandy plains with patches of low scrub, up to 2600 m. Nests in low bush or scrub.

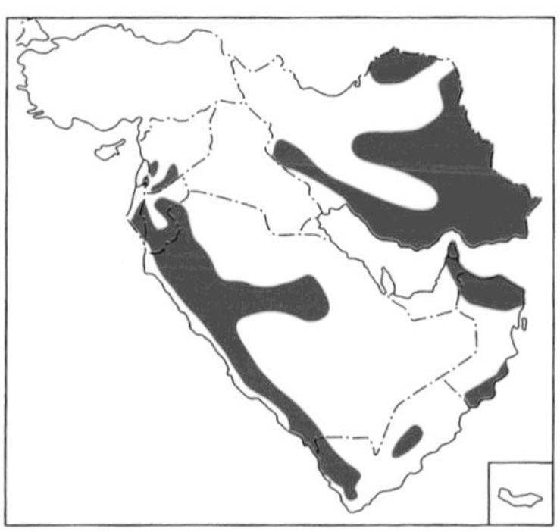

535 **Grasshopper Warbler** *Locustella naevia*

Status: may breed NW Iran. Uncommonly recorded or rare passage migrant in region generally, being especially rare Near East and Arabia, but probably overlooked. **Habitat:** thick moist vegetation, dense bushes, thickets near streams, reedbeds.

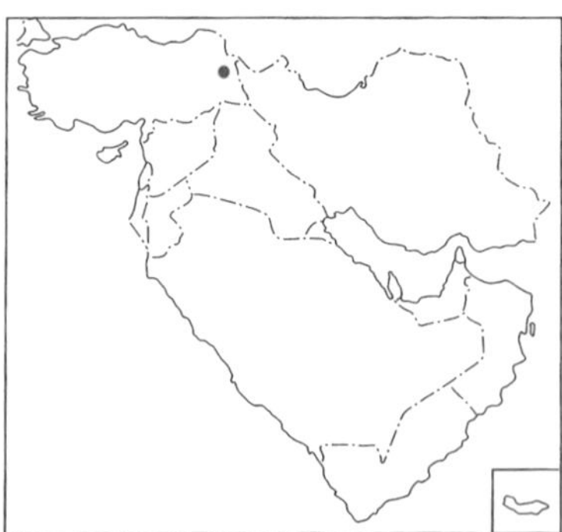

533 Socotra Warbler
531 Socotra Cisticola
532 Graceful Prinia
530 Fan-tailed Cisticola
song flight
28 Cetti's Warbler
buryi
inquieta
35 Grasshopper Warbler
534 Scrub Warbler
JG

536 **River Warbler** *Locustella fluviatilis*

Status: passage migrant throughout whole region but rarely seen because of secretive behaviour. **Habitat:** thick vegetation, dense bushes, usually near water.

537 **Savi's Warbler** *Locustella luscinioides*

Status: summer visitor. Passage migrant Turkey, Iran, Israel; scarce UAE, but vagrant (or probably under-recorded) elsewhere in Near East and Arabia. **Habitat:** reedbeds (in breeding season); at other times swamps, tall grass-fields, or anywhere with vegetation.

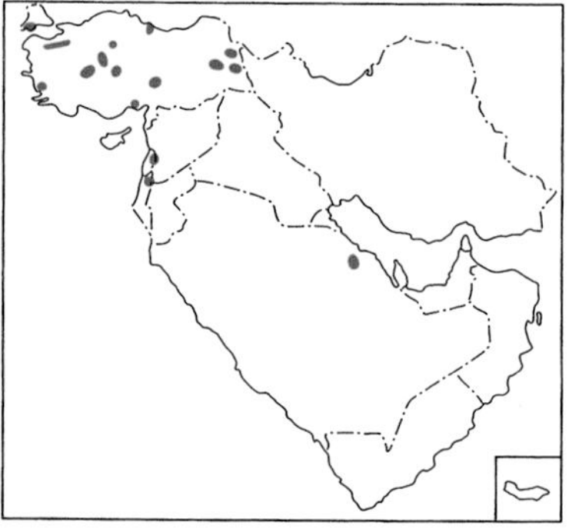

538 **Moustached Warbler** *Acrocephalus melanopogon*

Status: resident and partly migratory or dispersive; has been recorded away from breeding areas in most parts of the region (except SW Arabia) but generally rare. **Habitat:** reedbeds and swampy thickets. Nests in reeds or low bushes.

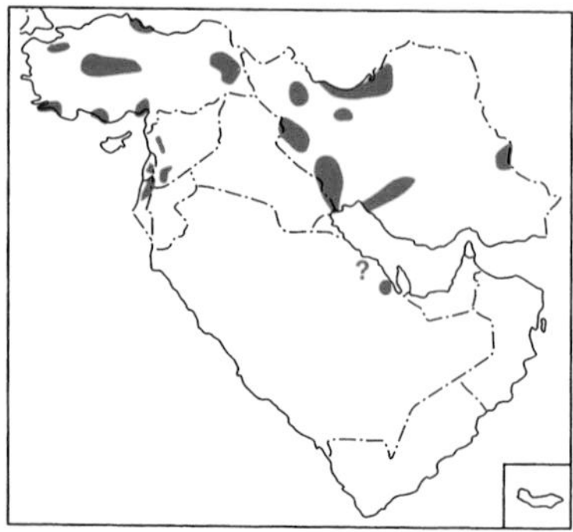

539 **Aquatic Warbler** *Acrocephalus paludicola*

Status: vagrant Turkey, Cyprus and Jordan, though probably much overlooked. **Habitat:** swampy thickets and marshes.

540 **Sedge Warbler** *Acrocephalus schoenobaenus*

Status: summer visitor. Passage across most of the region. Occasional winter Oman. **Habitat:** reedbeds, swampy thickets but will occur in drier habitat on passage. Nests in low, dense vegetation.

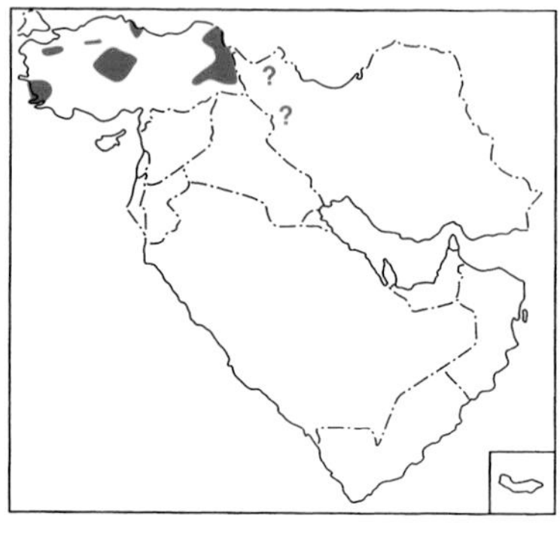

541 **Paddyfield Warbler** *Acrocephalus agricola*

Status: summer visitor; on passage in SE Iran and locally in Turkey outside breeding range. Vagrant Oman, Bahrain, Israel. **Habitat:** wetlands with reeds and scrub, particularly in the drier parts. On passage more widespread.

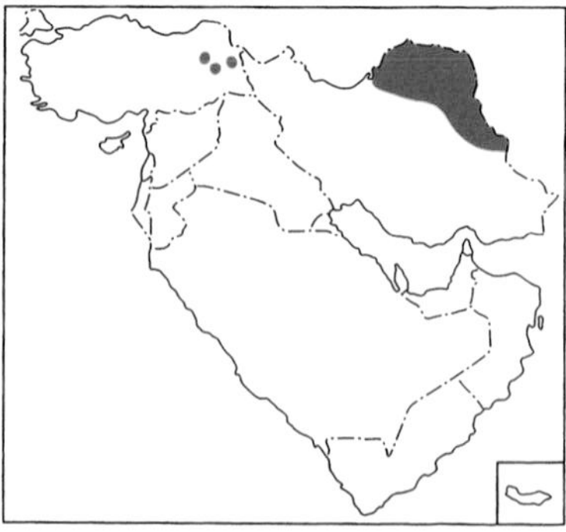

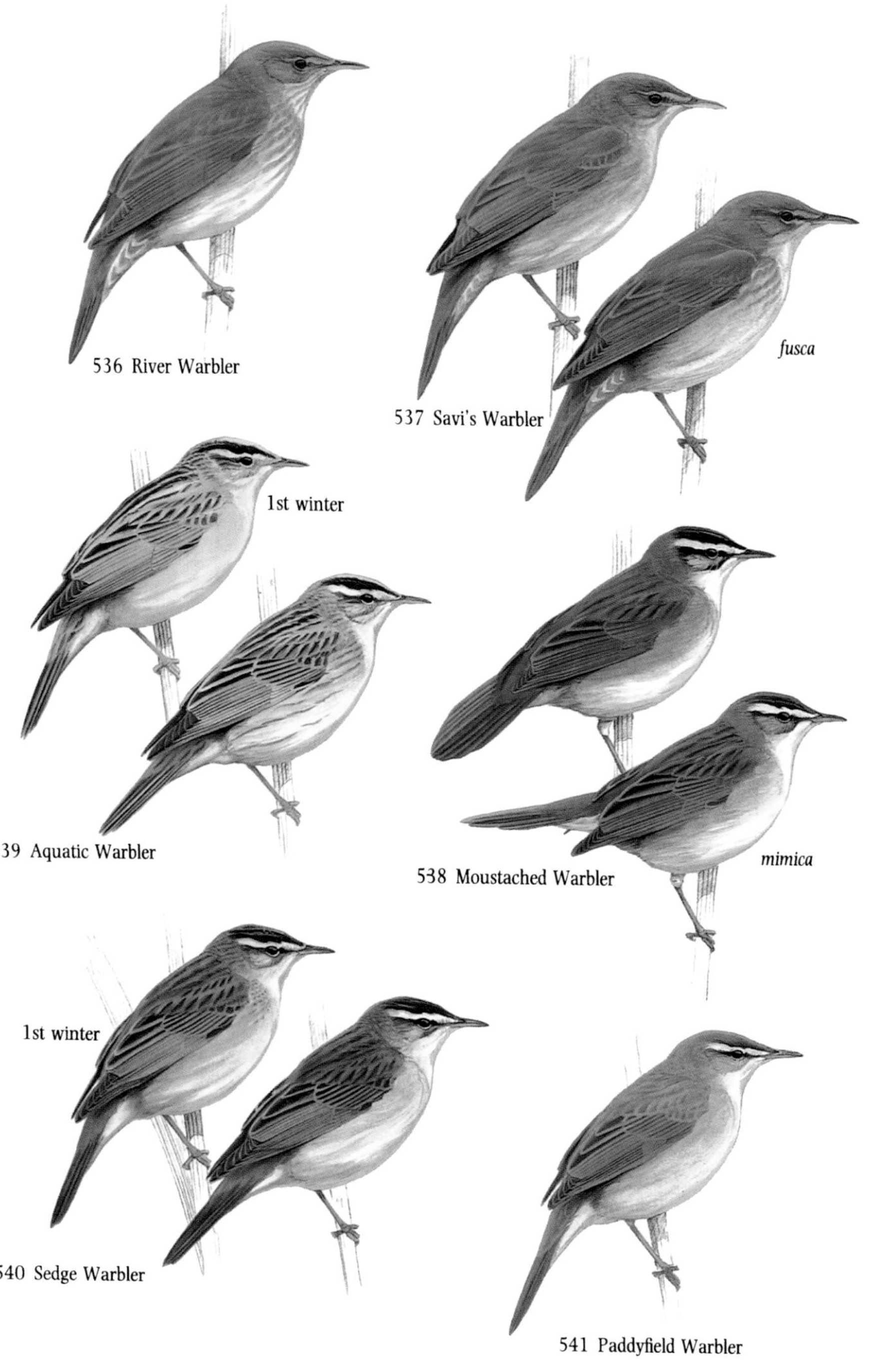
536 River Warbler
fusca
537 Savi's Warbler
1st winter
539 Aquatic Warbler
538 Moustached Warbler
mimica
1st winter
540 Sedge Warbler
541 Paddyfield Warbler
JG

542 **Blyth's Reed Warbler**
Acrocephalus dumetorum

Status: passage E Iran (where may breed in NE) and UAE; Vagrant Oman, Turkey, Syria, Israel, Cyprus, Bahrain, Syria. **Habitat:** thick bushy vegetation, hedges, light woodland, mostly near water or marshes.

543 **Marsh Warbler** *Acrocephalus palustris*

Status: summer visitor. Passage throughout most of region, but vagrant Qatar. **Habitat:** dense vegetation in ditches, thickets and stream-banks; in almost any thick cover on passage. Nests in low vegetation.

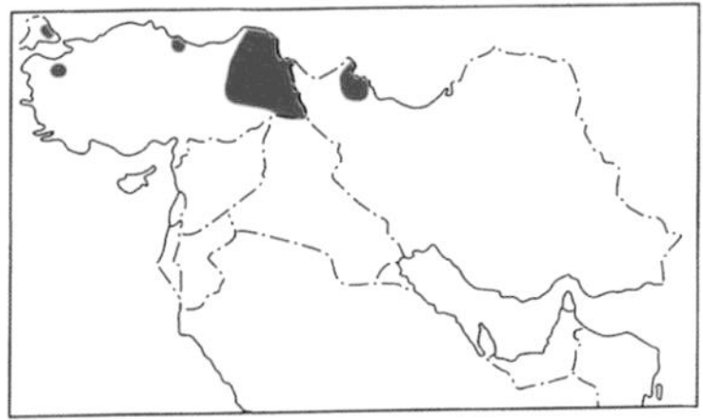

544 **European Reed Warbler**
Acrocephalus scirpaceus

Status: summer visitor. Passage throughout the region. **Habitat:** reedbeds and waterside vegetation. More widely on passage, in any area with cover. Nests in scattered colonies in reeds.

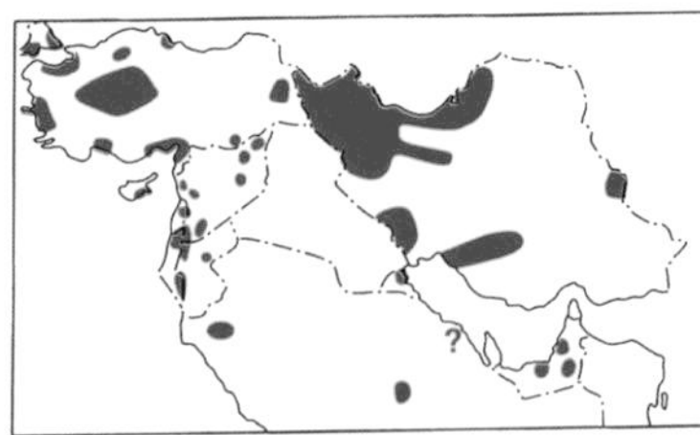

545 **African Reed Warbler**
Acrocephalus baeticatus

Status: apparently resident (on Red Sea coast and islands). **Habitat:** mangroves, in which it nests.

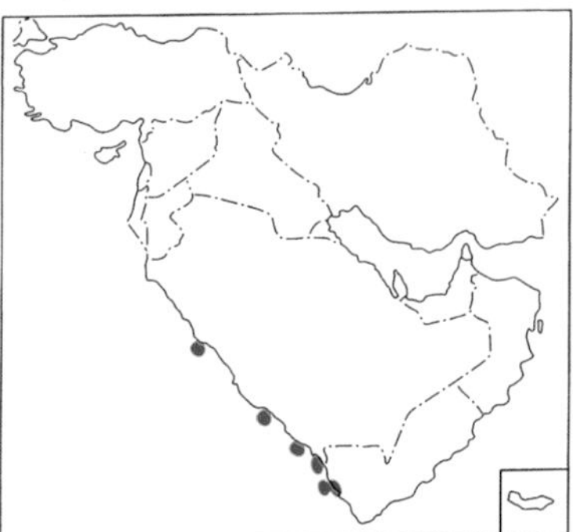

546 **Clamorous Reed Warbler**
Acrocephalus stentoreus

Status: resident and winter visitor in Near East. Summer visitor with passage Kuwait, UEA and Oman; some wintering in Iran, Iraq, Oman and Arabian Gulf. Probably resident in SW Arabia. **Habitat:** reeds, waterside vegetation including mangroves; also crops (e.g. beans) near water.

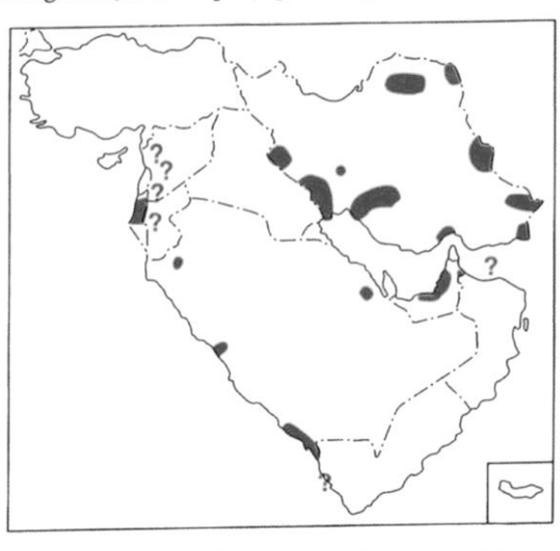

547 **Great Reed Warbler**
Acrocephalus arundinaceus

Status: has bred Cyprus. Summer visitor. Passage through most of region. **Habitat:** tall reeds bordering open water; on passage also drier habitats. Nests colonially in reeds.

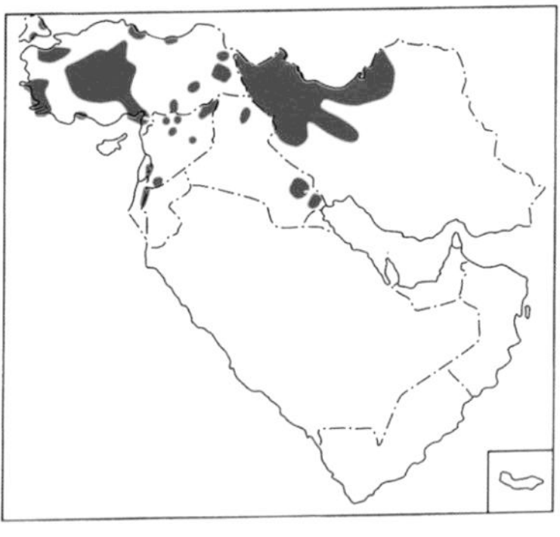

548 **Basra Reed Warbler** *Acrocephalus griseldis*

Status: summer visitor; passage Kuwait (where may breed) and E Saudi Arabia. Vagrant Israel. **Habitat:** expansive reedbeds; on migration in reeds, bushes and other cover.

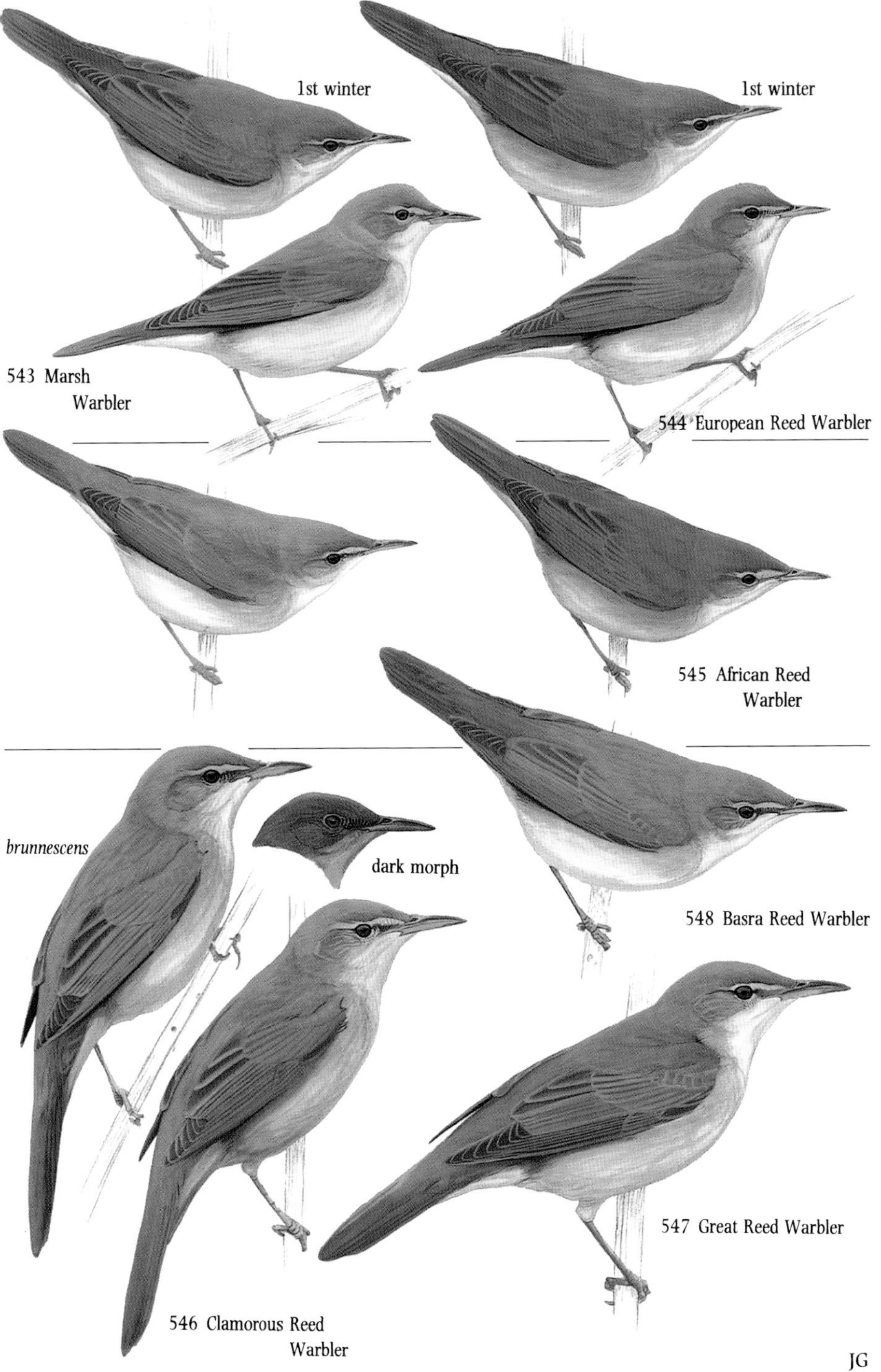
1st winter
1st winter
543 Marsh Warbler
544 European Reed Warbler
545 African Reed Warbler
brunnescens
dark morph
548 Basra Reed Warbler
547 Great Reed Warbler
546 Clamorous Reed Warbler
JG

549 **Olivaceous Warbler** *Hippolais pallida*

Status: summer visitor. Passage migrant throughout. Recorded rarely in winter in Turkey, Arabia (W and central Gulf). **Habitat:** scrub, gardens, many areas with trees and bushes, oases; on passage/winter also wadis with acacia and tamarisk. Nests in bush or small tree.

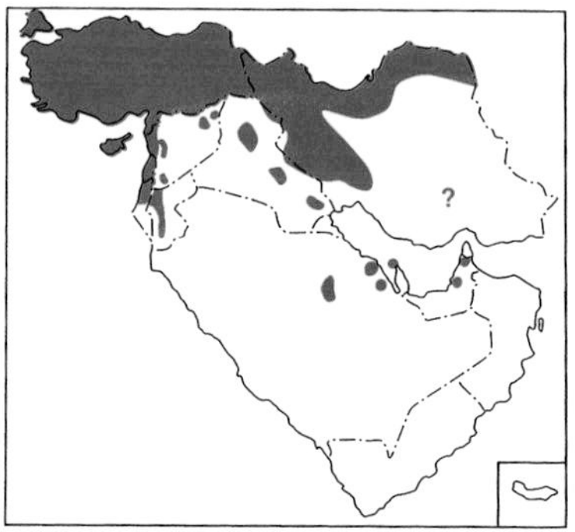

550 **Booted Warbler** *Hippolais caligata*

Status: summer visitor. Passage E Arabian Gulf and Oman, increasing regular to head of Gulf; rare Turkey, Israel. Vagrant W Arabia. **Habitat:** bushes, cultivated land with trees and cover; on passage in any bushy country, wadis, gardens, parks. Nests in tamarisk, mangroves and other bushes.

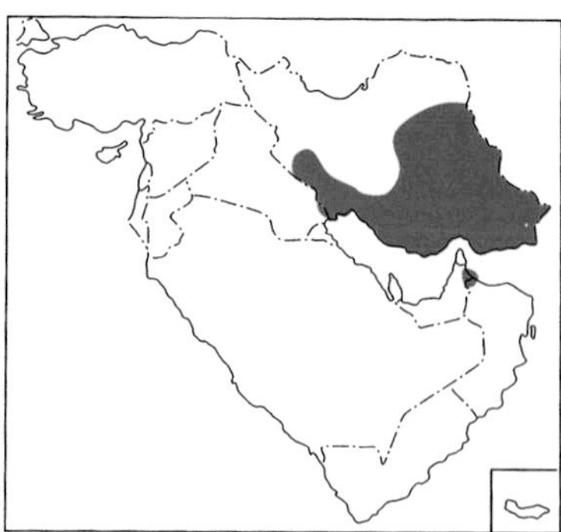

551 **Upcher's Warbler** *Hippolais languida*

Status: summer visitor. Passage migrant throughout Near East and Arabia south of breeding range, generally scarce (though common Oman and S. Yemen). **Habitat:** scrub from plains to mountain valleys, up to 2000 m; also in gardens and more wooded areas; on passage also wadis with acacia and tamarisk. Nests in bush.

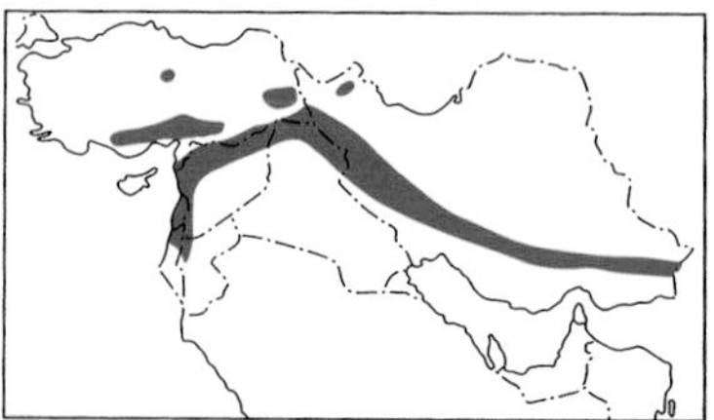

552 **Olive-tree Warbler** *Hippolais olivetorum*

Status: uncertain Syria and Lebanon but might breed. Summer visitor. Passage Jordan, Israel, occasional Cyprus; vagrant Saudi Arabia, Yemen. **Habitat:** patches of scrub, olive-groves, oak-woods, usually on edge of cultivation; on passage also acacia wadis. Nests in tree or tall bush.

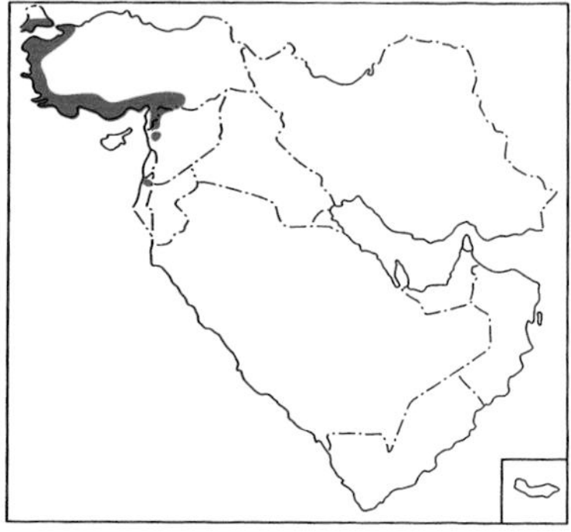

553 **Icterine Warbler** *Hippolais icterina*

Status: may occasionally breed NW Turkey. Summer visitor. Scarce passage migrant Turkey, Kuwait and W Iran, rare Near East and Arabia. **Habitat:** gardens, parks, wood thickets. On passage where trees and bushes exist. Nests in scrub, hedge or low in tree.

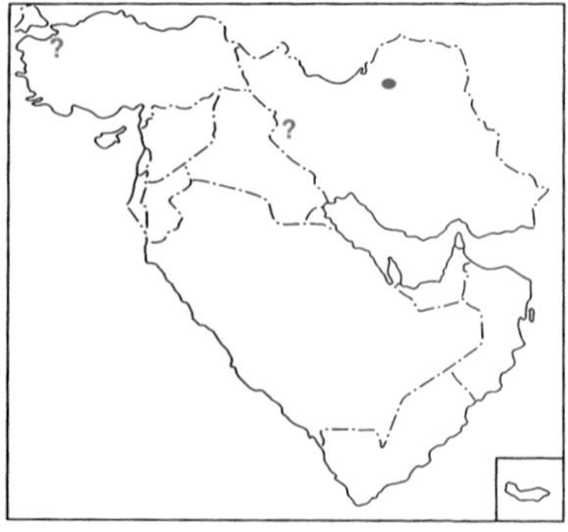

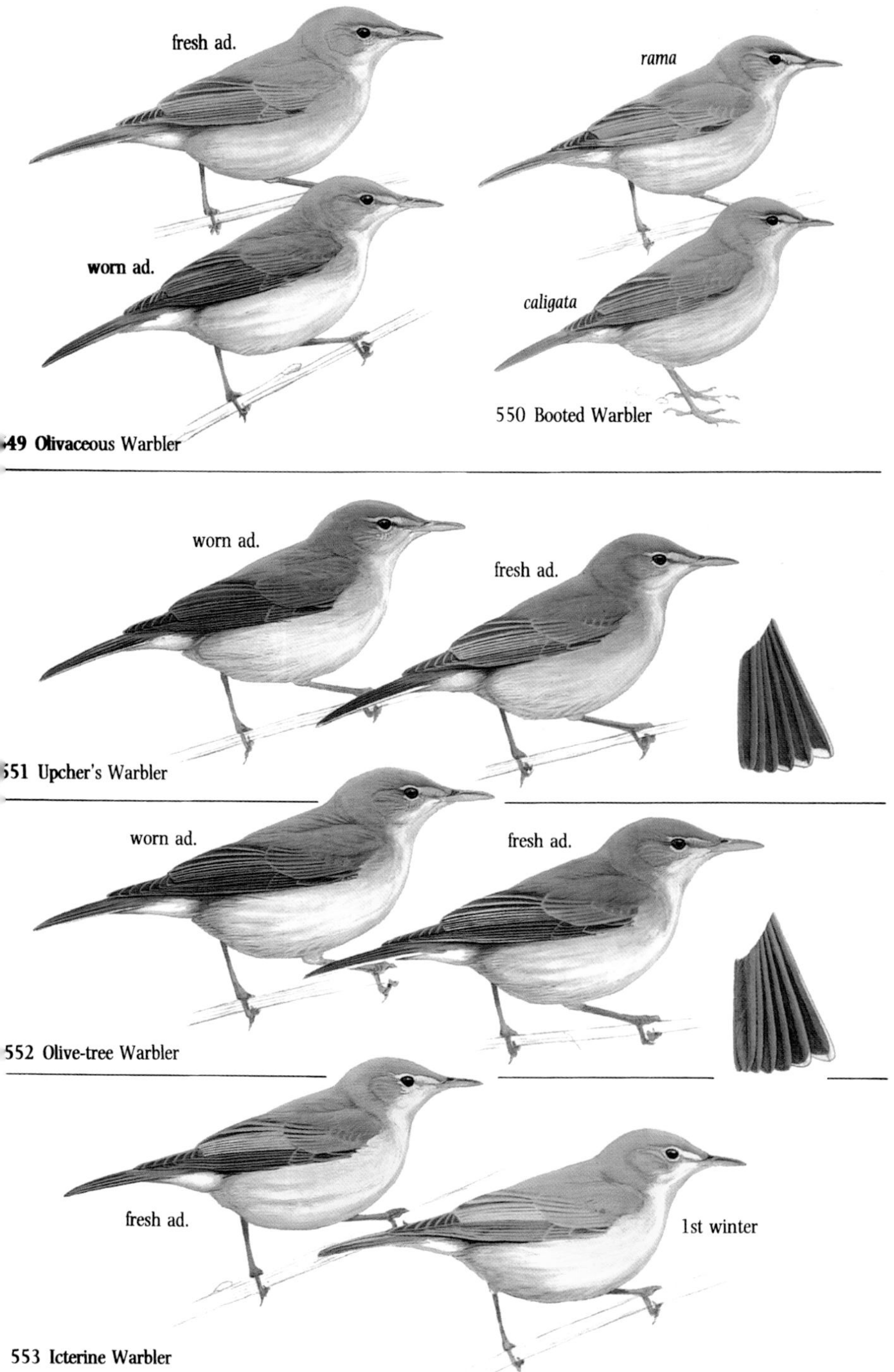
fresh ad.
worn ad.
rama
caligata
550 Booted Warbler
549 Olivaceous Warbler
worn ad.
fresh ad.
551 Upcher's Warbler
worn ad.
fresh ad.
552 Olive-tree Warbler
fresh ad.
1st winter
553 Icterine Warbler

JG

555 **Spectacled Warbler** *Sylvia conspicillata*

Status: resident or partial migrant. Recorded rarely in Lebanon. Vagrant Turkey, Iraq and Saudi Arabia. **Habitat:** chiefly in *Salicornia* on coastal flats often near lagoons, salty plains and semi-deserts with low scrub. Nests in low bush.

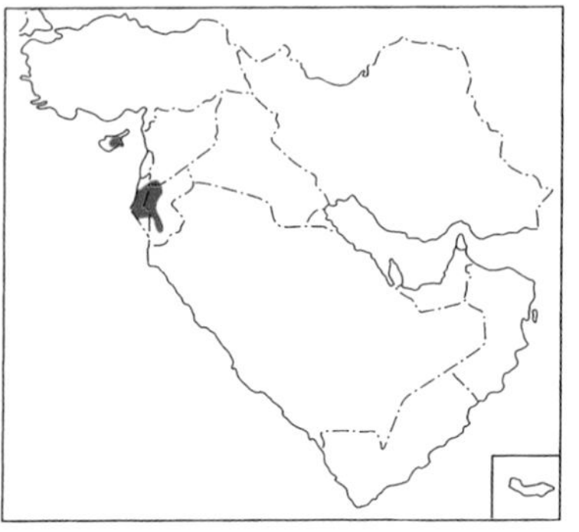

556 **Subalpine Warbler** *Sylvia cantillans*

Status: summer visitor. Uncommon spring migrant Cyprus, W Turkey, Near East and W Saudi Arabia. In autumn rare throughout. Vagrant Yemen. **Habitat:** scrub and thickets; also open woodland. On passage also wadis with acacia or dense scrub. Nests in thick bush.

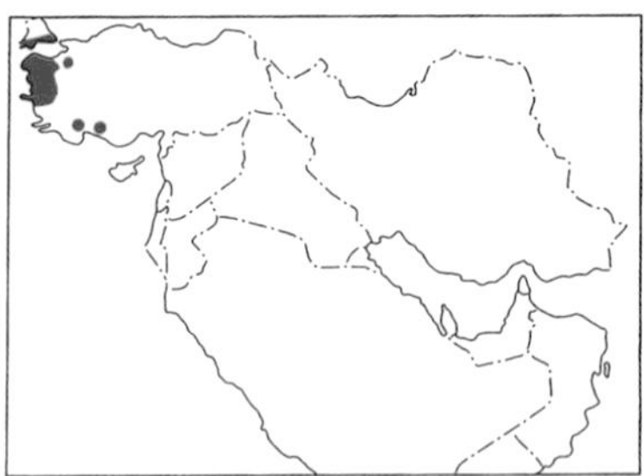

557 **Ménétries's Warbler** *Sylvia mystacea*

Status: mainly summer visitor. Winters Iran and Arabia, always in small numbers; scarcer Oman. Passage migrant Arabian Gulf area, scarcer rest of Arabia, rare Lebanon and Israel, vagrant Jordan. **Habitat:** scrub and small thickets, patches of cover along rivers, often in broken country. Nests in thick bush.

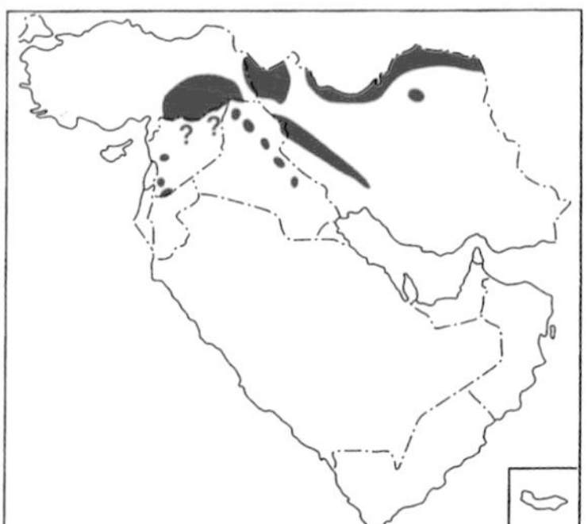

558 **Sardinian Warbler** *Sylvia melanocephala*

Status: resident or partial migrant. Passage noted in Turkey, Cyprus and Near East, rarely elsewhere. Recorded as winter visitor in Cyprus and Near East. Rare W Arabia. **Habitat:** dry, fairly open bushy scrub, thickets, pine and evergreen oak forests. In winter, wherever scrub exists, including acacia-wadis and desert-edges, oases. Nests in low bush or undergrowth.

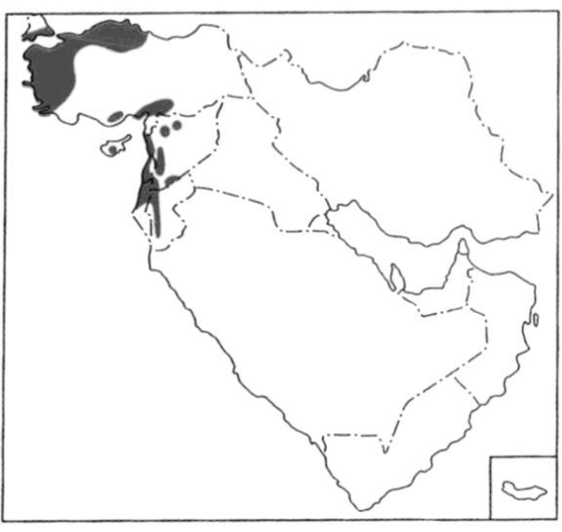

559 **Cyprus Warbler** *Sylvia melanothorax*

Status: Cyprus endemic; resident and partial migrant. Passage and winter visitor Israel, uncommon Lebanon and Jordan. Vagrant Turkey and NW Saudi Arabia. **Habitat:** well vegetated rocky slopes and foothills, dense scrub country. In winter, also acacia wadis. Nests in low bush.

560 **Rüppell's Warbler** *Sylvia rueppelli*

Status: summer visitor. Passage migrant Cyprus, Near East, NW Saudi Arabia. Vagrant Iran and Kuwait. **Habitat:** low scrub with bushes and often rocky outcrops; on passage also acacia wadis. Nests in bush.

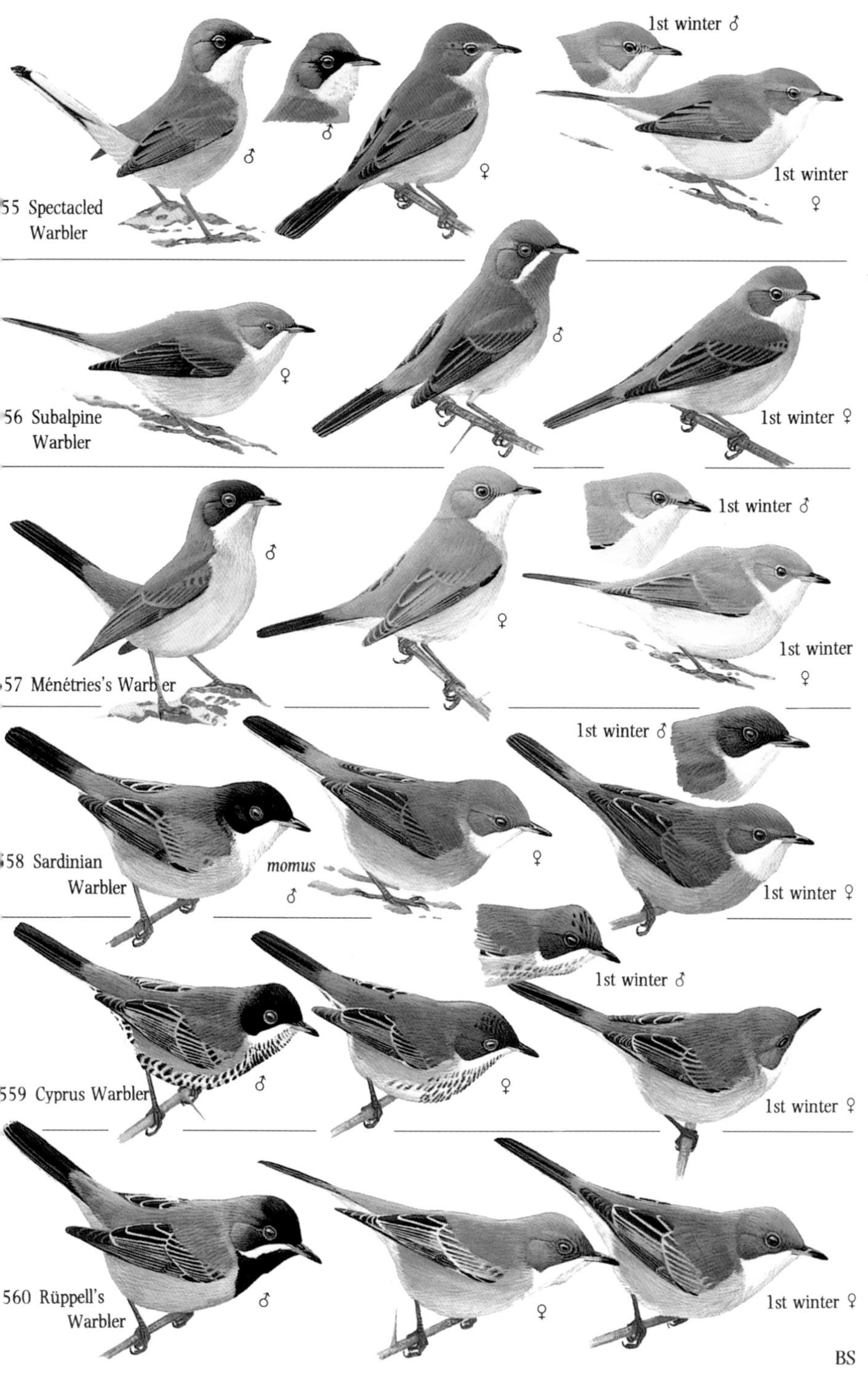
1st winter ♂
♂
♂
♀
1st winter
♀
55 Spectacled
Warbler
♀
♂
56 Subalpine
Warbler
1st winter ♀
1st winter ♂
♂
♀
1st winter
♀
57 Ménétries's Warbler
1st winter ♂
558 Sardinian
Warbler
momus
♂
♀
1st winter ♀
1st winter ♂
559 Cyprus Warbler
♂
♀
1st winter ♀
560 Rüppell's
Warbler
♂
♀
1st winter ♀
BS

529 **Yemen Warbler** *Parisoma buryi*

Status: resident. **Habitat:** mountains with acacia (especially those with flakey bark) and juniper scrub, lightly wooded cultivated valleys, usually 1500–2900 m. A nest found in low branch of acacia.

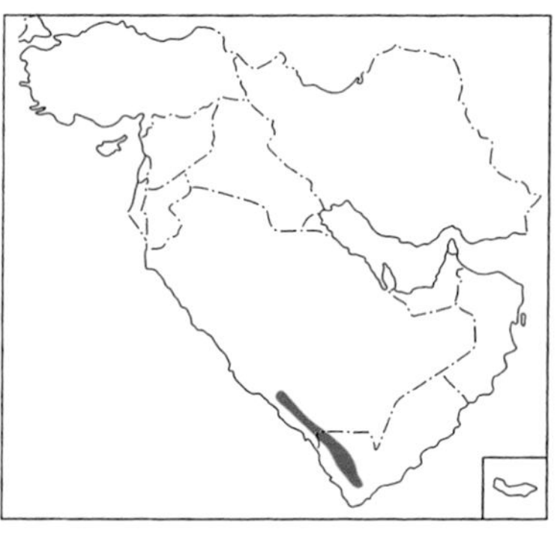

561 **Desert Warbler** *Sylvia nana*

Status: migratory; on passage and in winter S Iran and Arabia, including Kuwait, Oman and Yemen; scarce Israel. Vagrant Jordan, Turkey and Cyprus. **Habitat:** open semi-desert, hill-sides with low scattered dense bushes. On passage/winter flat semi-desert with low saline bushes. Nests in bush.

562 **Arabian Warbler** *Sylvia leucomelaena*

Status: resident. **Habitat:** acacias in semi-deserts and wadis. In SW Arabia, rocky hills with trees up to 1500 m. Nests in tree.

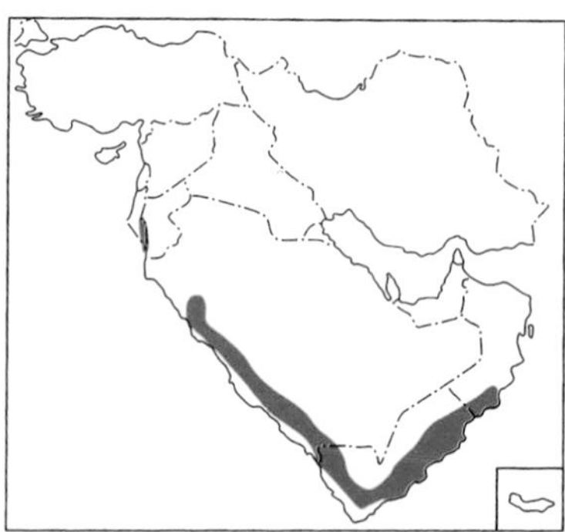

563 **Orphean Warbler** *Sylvia hortensis*

Status: summer visitor. On passage throughout the region, but scarce E and S Arabia. Scarce winter visitor UAE and Oman, fewer Iran and Yemen. **Habitat:** bushy hillsides, deciduous thickets, orchards, parkland, olive-groves. On passage also acacia. Nests in bush or low tree.

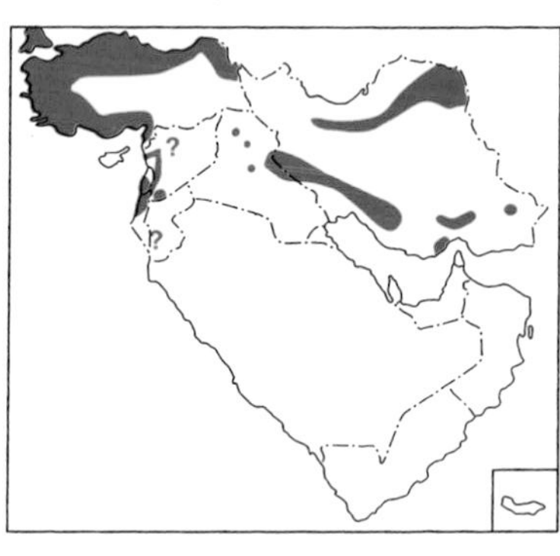

564 **Barred Warbler** *Sylvia nisoria*

Status: summer visitor. Passage most parts of SW Arabia, UAE (spring), Oman (where occasional in winter) and W Arabian Gulf; less common in Near East. **Habitat:** thorny thickets with scattered trees, hedges or bushy wood clearings. Nests in bush.

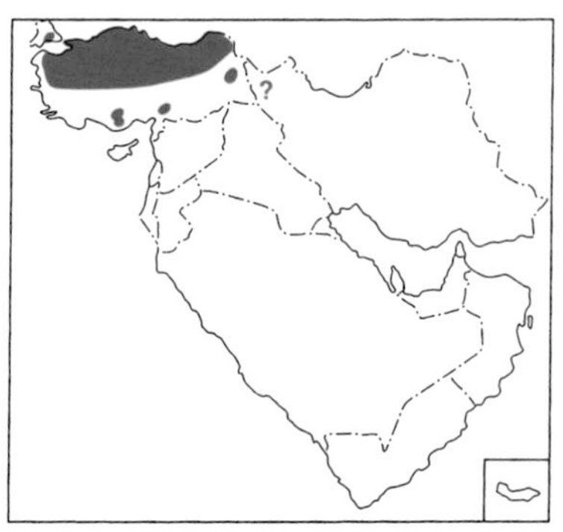

569 **Garden Warbler** *Sylvia borin*

Status: summer visitor. Passage migrant throughout the region, but scarce most of Arabia; vagrant Yemen. **Habitat:** open woodland with thick undergrowth, overgrown hedges. On passage where trees and bushes. Nests in low bush.

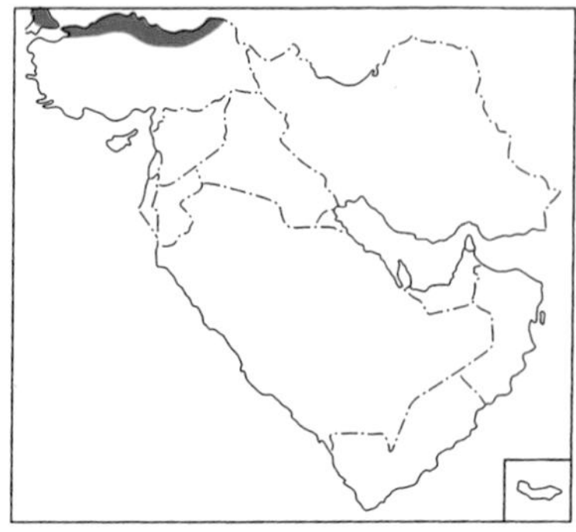

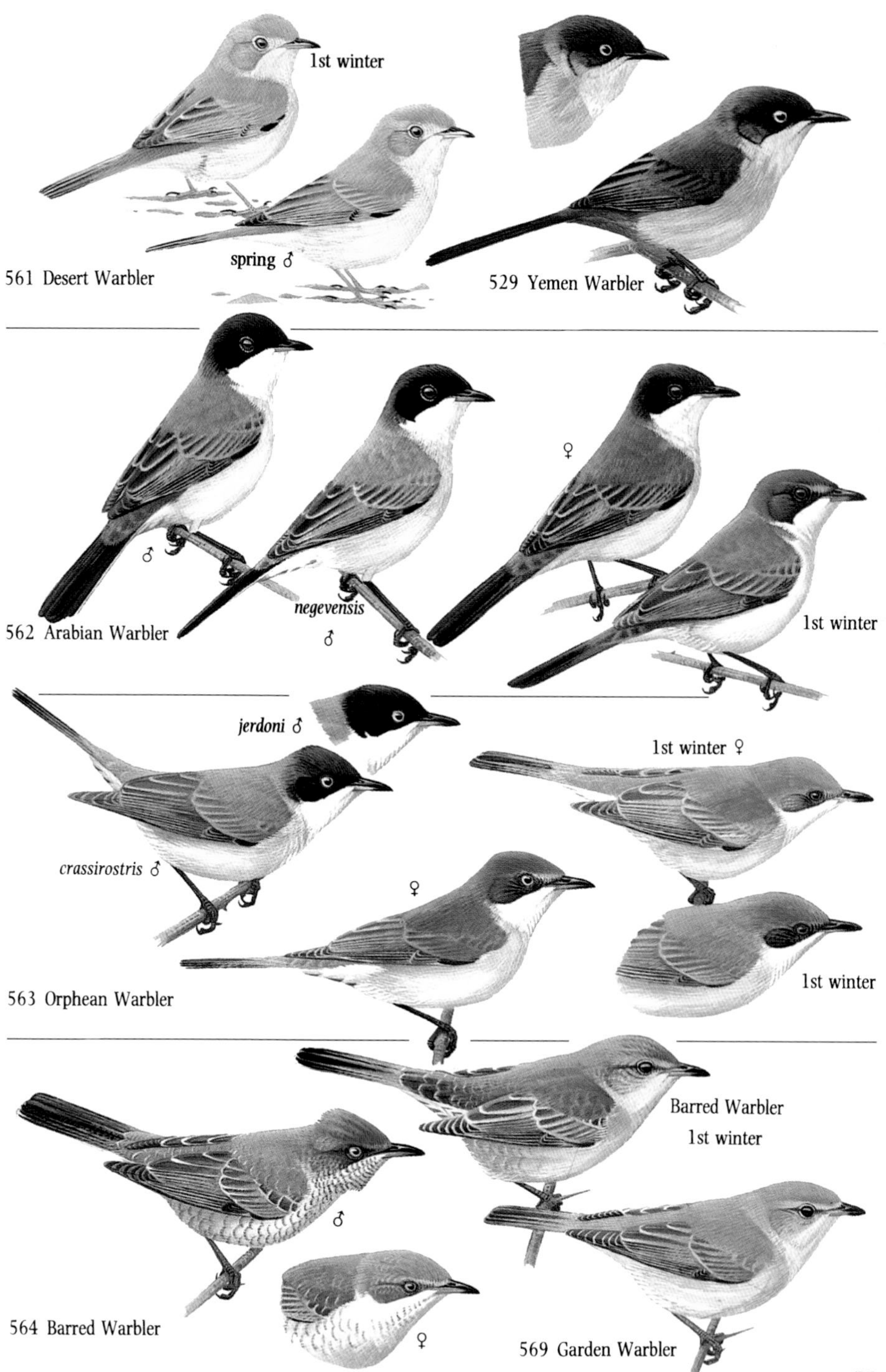
1st winter
spring ♂
561 Desert Warbler
529 Yemen Warbler
♂
562 Arabian Warbler
negevensis
♂
♀
1st winter
jerdoni ♂
1st winter ♀
crassirostris ♂
♀
1st winter
563 Orphean Warbler
Barred Warbler
1st winter
♂
564 Barred Warbler
♀
569 Garden Warbler

BS

565 **Lesser Whitethroat** *Sylvia curruca*

Status: summer visitor. Passage migrant throughout region but uncommon Oman. A few in winter in E and SW Arabian, but rarely Near East. **Habitat:** dense undergrowth, trees. Nests in bush or tangled vegetation.

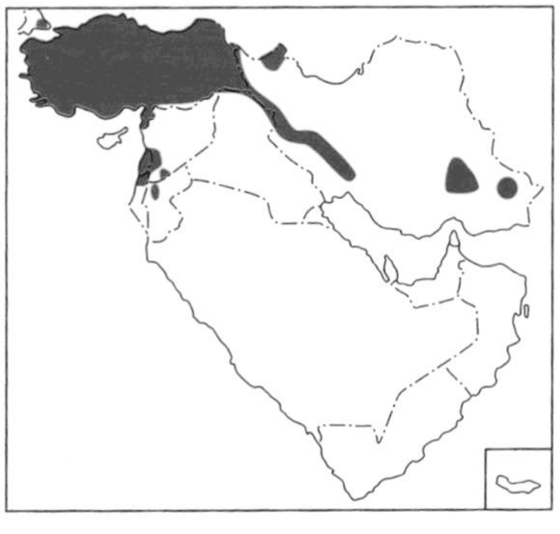

566 **Hume's Lesser Whitethroat** *Sylvia curruca althaea*

Status: summer visitor with some wintering in S Iran. Uncommon UAE, Bahrain and Israel. **Habitat:** sub-alpine juniper forest and small bushes, above 2500 m. Can occur in any trees and bushes on passage/winter. Nests in low bush.

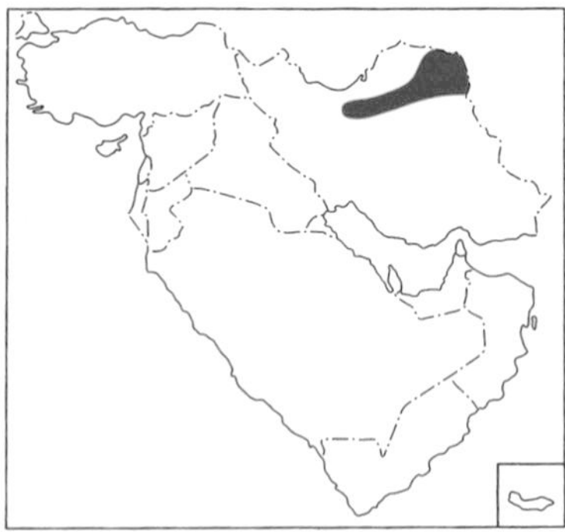

567 **Desert Lesser Whitethroat** *Sylvia curruca minula*

Status: winter visitor and passage migrant, occurring in much of Arabia, including SW also S Iran. Vagrant Israel. **Habitat:** trees, especially acacia in semi-deserts and rolling hills.

568 **Whitethroat** *Sylvia communis*

Status: summer visitor. Passage migrant throughout region; in winter uncommon Oman and rare Arabian Gulf. **Habitat:** scrub and other patches of dense vegetation usually in open country. Nests low in cover.

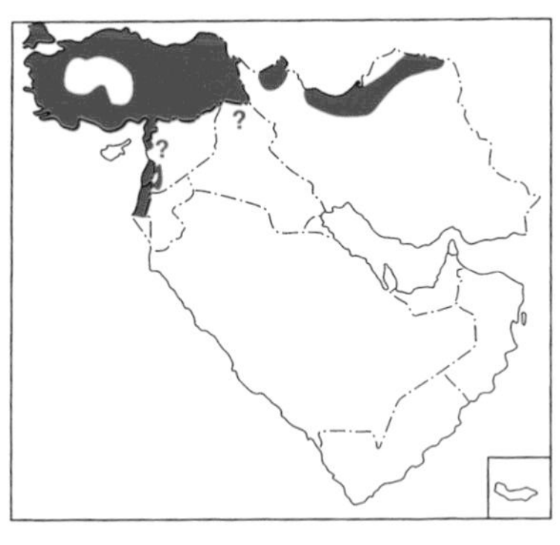

570 **Blackcap** *Sylvia atricapilla*

Status: summer visitor. Passage throughout the region; few winter in Near East, Arabian Gulf, S Iran, Oman and Yemen. **Habitat:** woodland with undergrowth. On passage anywhere with trees or scrub. Nests in bush.

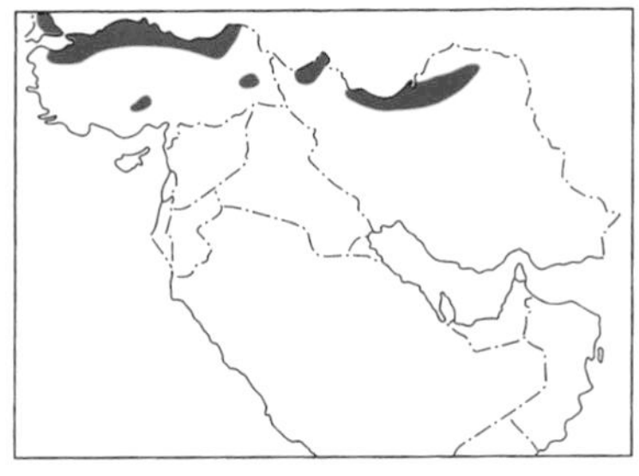

621 **White-breasted White-eye** *Zosterops abyssinica*

Status: resident, with seasonal movements. **Habitat:** trees in wooded mountains through to gardens from *c.* 300–3000 m. Builds cup-shaped nest in outer-branches of tree.

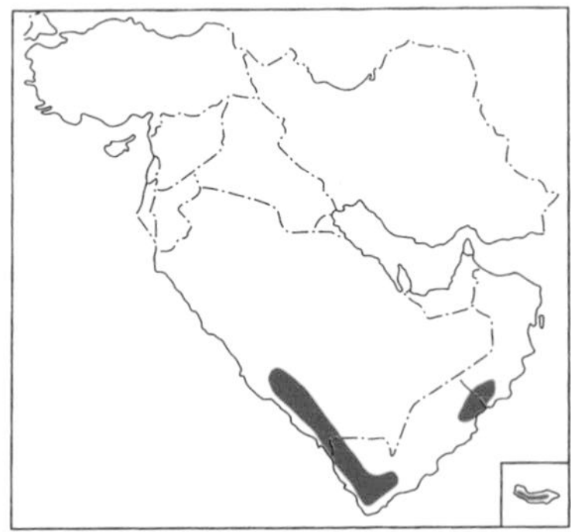

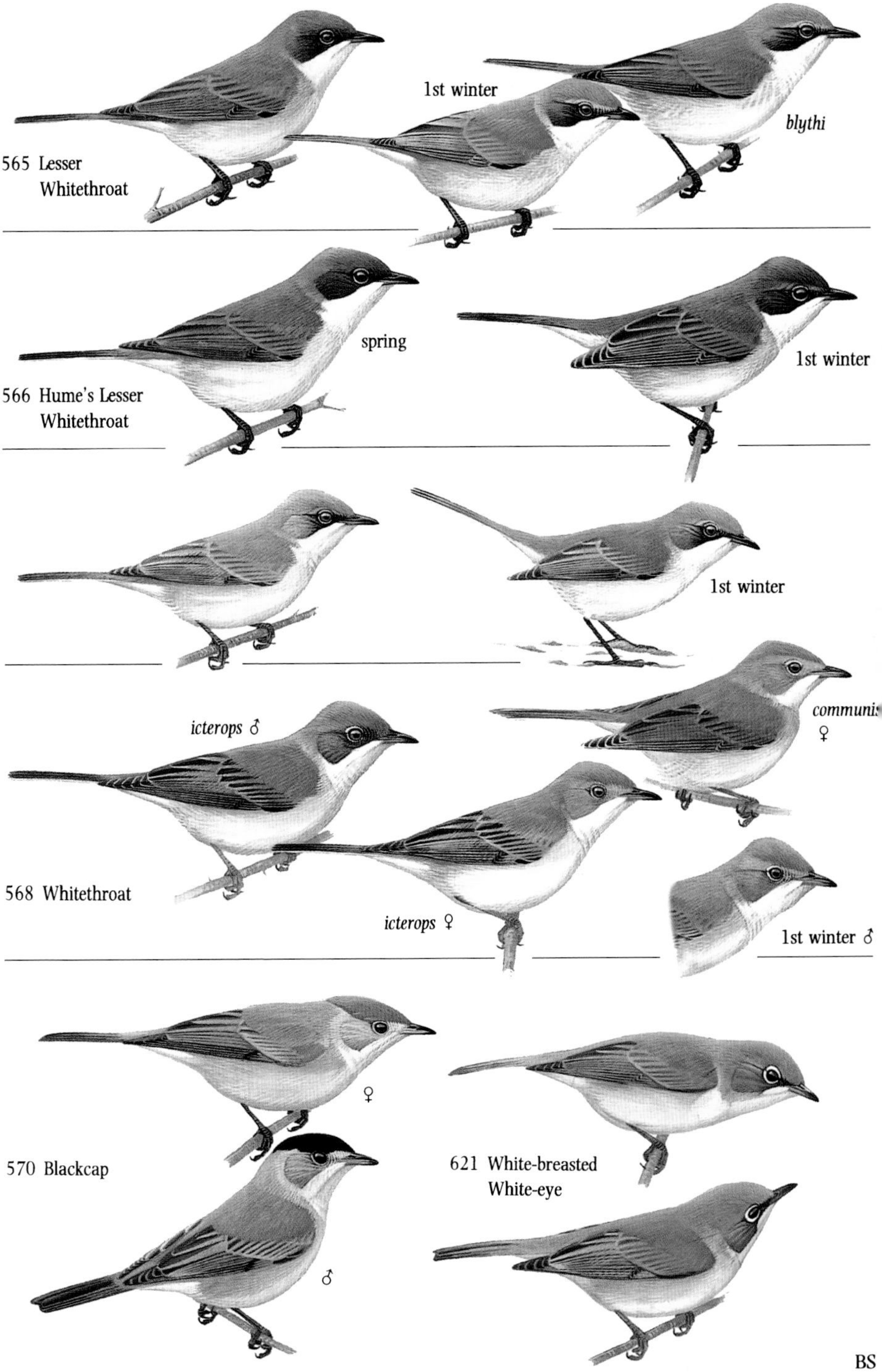

1st winter
blythi
565 Lesser
Whitethroat
spring
1st winter
566 Hume's Lesser
Whitethroat
1st winter
communis
♀
icterops ♂
568 Whitethroat
icterops ♀
1st winter ♂
♀
570 Blackcap
♂
621 White-breasted
White-eye
BS

571 **Brown Woodland Warbler** *Phylloscopus umbrovirens*

Status: resident. **Habitat:** hillsides and wadis with trees and bushes (especially acacia), junipers, lush gorges, 1600–2600 m. Domed nest built near ground.

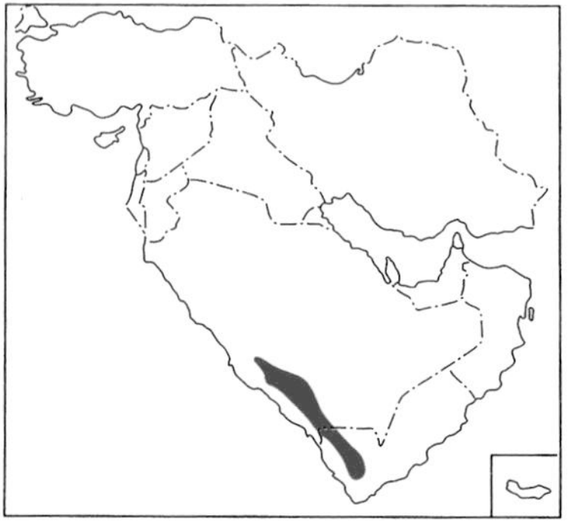

572 **Green Warbler** *Phylloscopus nitidus*

Status: summer visitor; also passage in Iran, scarce Oman. Vagrant Lebanon, Bahrain, Israel and UAE. **Habitat:** mixed coniferous and deciduous woods, usually with scrub layer on mountain slopes 1500–2000 m. Nests on ground.

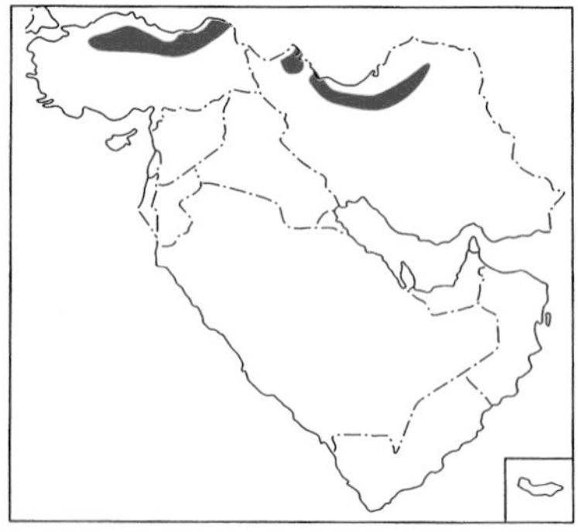

573 **Greenish Warbler** *Phylloscopus trochiloides*

Status: vagrant or rare passage Turkey, Iran, E Arabia and Israel. **Habitat:** on passage in bushes and trees.

574 **Arctic Warbler** *Phylloscopus borealis*

Status: vagrant E Saudi Arabia, Oman. **Habitat:** on passage in bushes and trees.

575 **Pallas's Warbler** *Phylloscopus proregulus*

Status: vagrant Israel and UAE. **Habitat:** trees and bushes.

576 **Yellow-browed Warbler** *Phylloscopus inornatus*

Status: scarce winter visitor SW Turkey, SE Iran, Oman, UAE and Israel; rare migrant Cyprus; vagrant Kuwait, Bahrain. Many of these records may refer to Hume's Yellow-browed, which see. **Habitat:** on passage/winter in trees and bushes.

577 **Hume's Yellow-browed Warbler** *Phylloscopus humei*

Status: because until recently Hume's has been regarded as a race of Yellow-browed, the only definite records are vagrants in Israel, Turkey and UAE, but probably commoner (see Yellow-browed). **Habitat:** on passage/winter in bushes and trees.

580 **Bonelli's Warbler** *Phylloscopus bonelli*

Status: also apparently breeds Lebanon. Summer visitor. Passage Near East, Cyprus and Turkey, scarce NW Saudi Arabia. Vagrant Iran, Oman, UAE and Yemen. **Habitat:** deciduous or mixed woodland, pinewoods, evergreen oaks, or even scattered trees up to tree-limit. On passage in any trees and bushes. Nests on ground in undergrowth under tree.

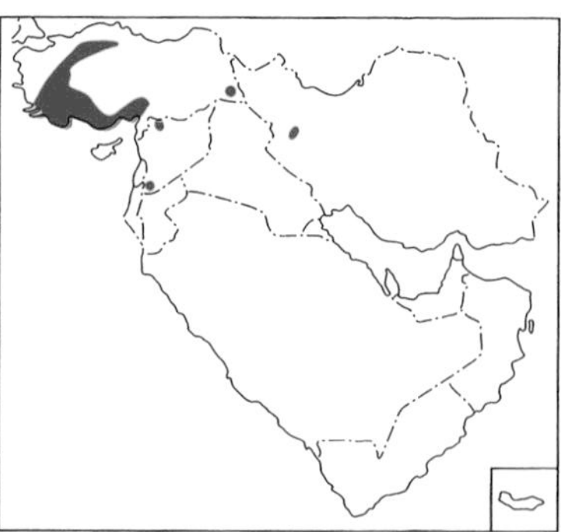

581 **Wood Warbler** *Phylloscopus sibilatrix*

Status: passage mainly spring in whole region east to W Iran, but scarce in Jordan and Arabia. Vagrant Qatar. **Habitat:** on passage in woodland and tall trees, also scrub.

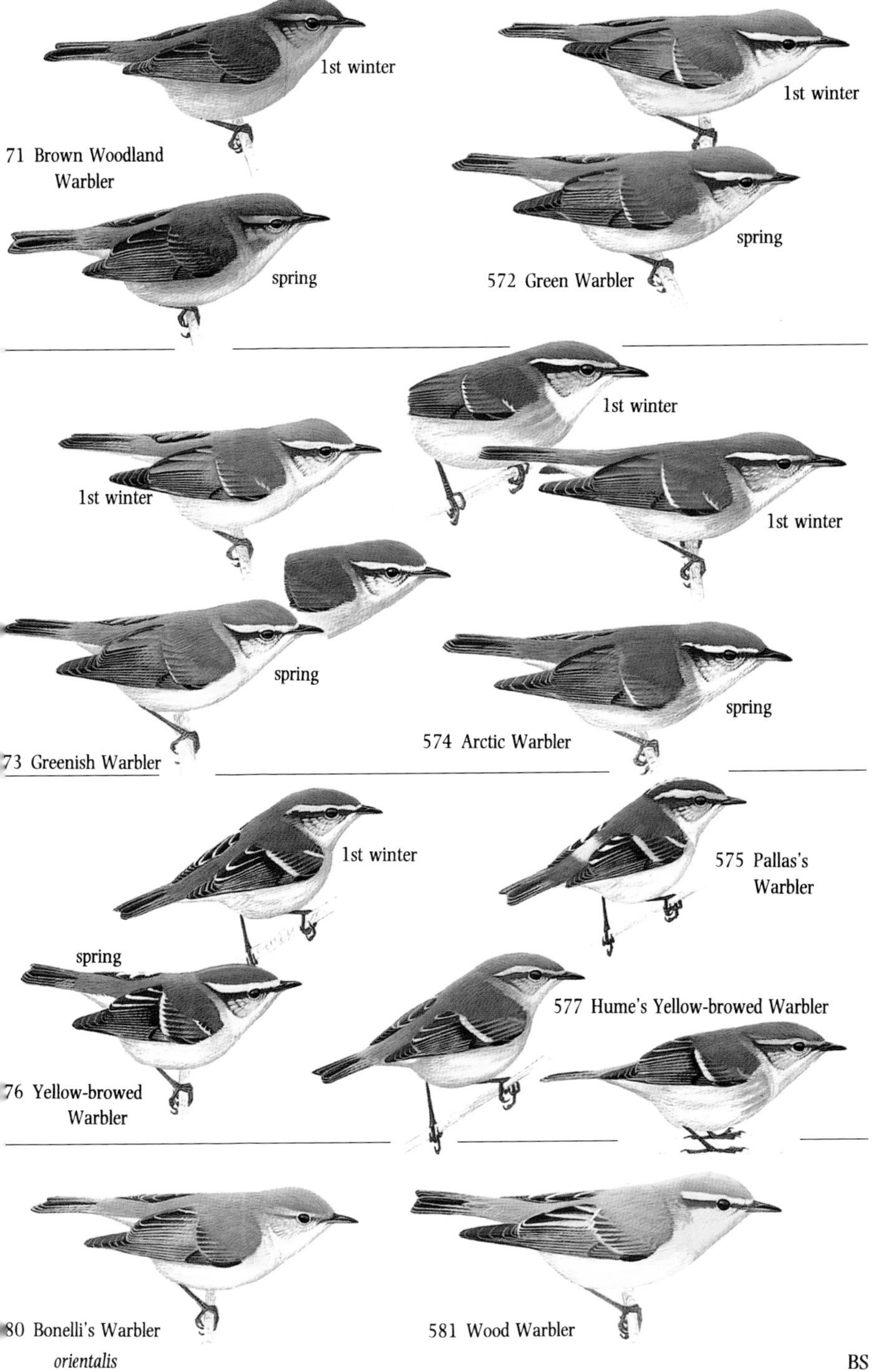
1st winter
71 Brown Woodland Warbler
spring
1st winter
spring
572 Green Warbler
1st winter
1st winter
1st winter
spring
spring
73 Greenish Warbler
574 Arctic Warbler
1st winter
575 Pallas's Warbler
spring
577 Hume's Yellow-browed Warbler
76 Yellow-browed Warbler
80 Bonelli's Warbler
orientalis
581 Wood Warbler
BS

578 **Radde's Warbler** *Phylloscopus schwarzi*

Status: vagrant Israel and UAE. **Habitat:** in winter, in low vegetation with bushes and tall grasses on river banks or along roadside ditches.

579 **Dusky Warbler** *Phylloscopus fuscatus*

Status: vagrant Yemen and UAE. **Habitat:** in winter, on low ground cover, long grass or bushes in swampy places, along rivers or in reed or scrub.

582 **Plain Leaf Warbler** *Phylloscopus neglectus*

Status: short-distance migrant. Winters Iran south of breeding range; also UAE and Oman. Vagrant Jordan, Israel, Bahrain, and possibly Yemen. **Habitat:** sparse low mountain scrub (deciduous or evergreen) with occasional trees, 2000–3000 m; scrub down to sea level in winter. Nests low in bush.

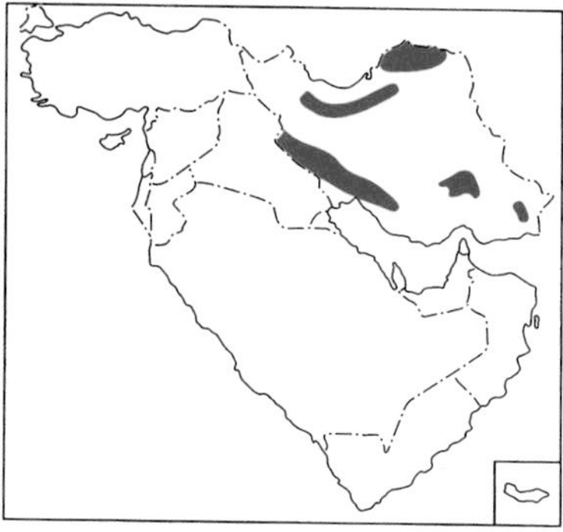

583 **Mountain Chiffchaff** *Phylloscopus sindianus*

Status: summer visitor. Breeding range and wintering/passage areas not fully understood owing to identification difficulties. Vagrant Iraq. **Habitat:** in breeding season, in mountain forest with scrub in subalpine zone, and in willow scrub along rivers.

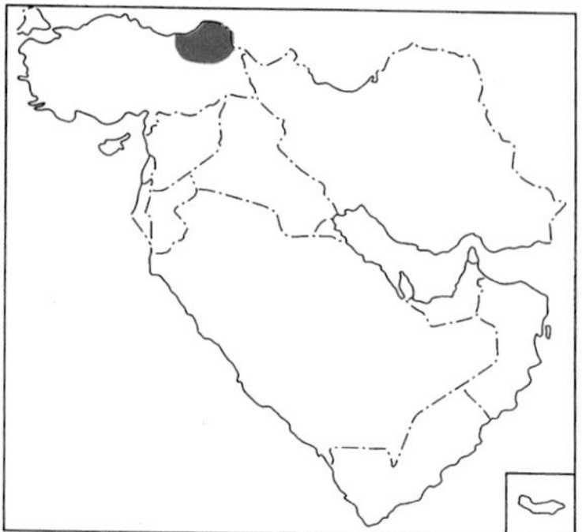

584 **Chiffchaff** *Phylloscopus collybita*

Status: partial migrant, also occurs on passage or as winter visitor almost throughout the region. **Habitat:** deciduous, mixed or coniferous woodland. On passage/winter almost anywhere with bushes and trees. Nests usually just above ground in vegetation.

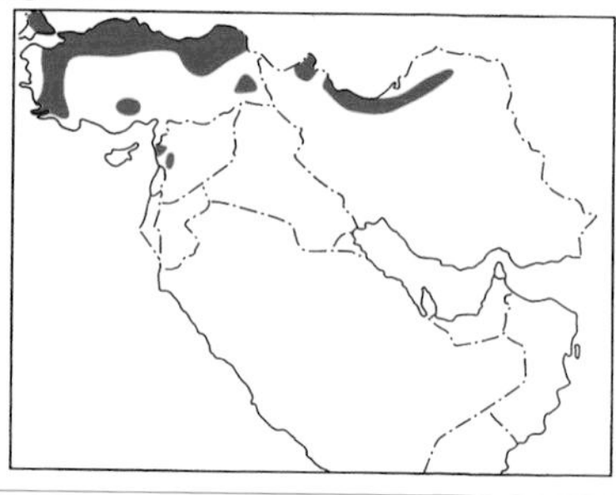

585 **Willow Warbler** *Phylloscopus trochilus*

Status: passage migrant throughout region. **Habitat:** bushy areas, woodland edges, gardens, parks.

586 **Goldcrest** *Regulus regulus*

Status: resident. Also winter visitor to Turkey, Cyprus, and uncommonly to N Iran, Near East. Vagrant Jordan. **Habitat:** coniferous or mixed woodland; in winter also hedges and undergrowth. Nests in conifer.

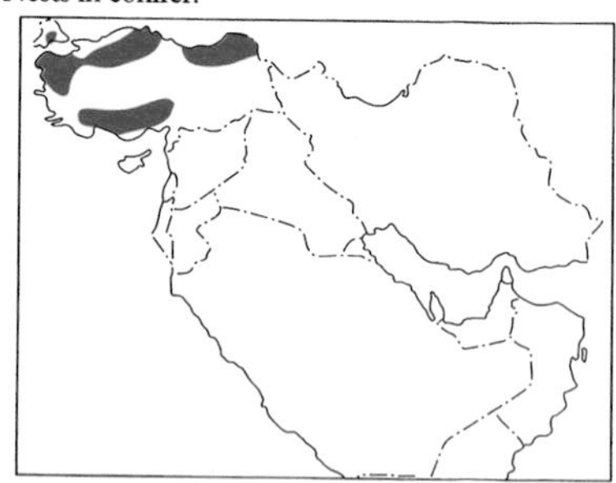

587 **Firecrest** *Regulus ignicapillus*

Status: resident. Immigration to Turkey in winter. Rare winter visitor to Lebanon. Vagrant Cyprus. **Habitat:** as Goldcrest, but less partial to coniferous woods; more often in low undergrowth, but also in higher mixed forest. Nests in tree.

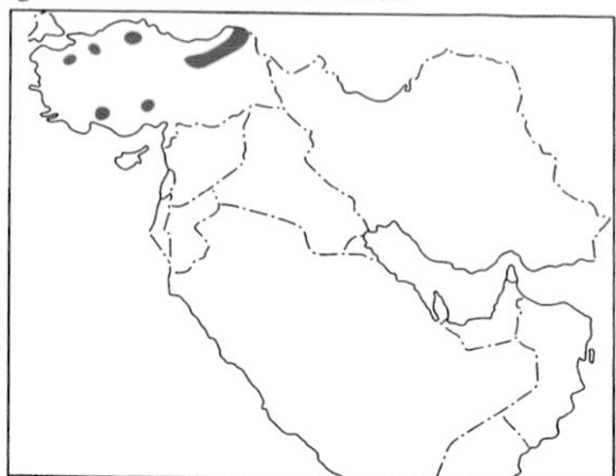

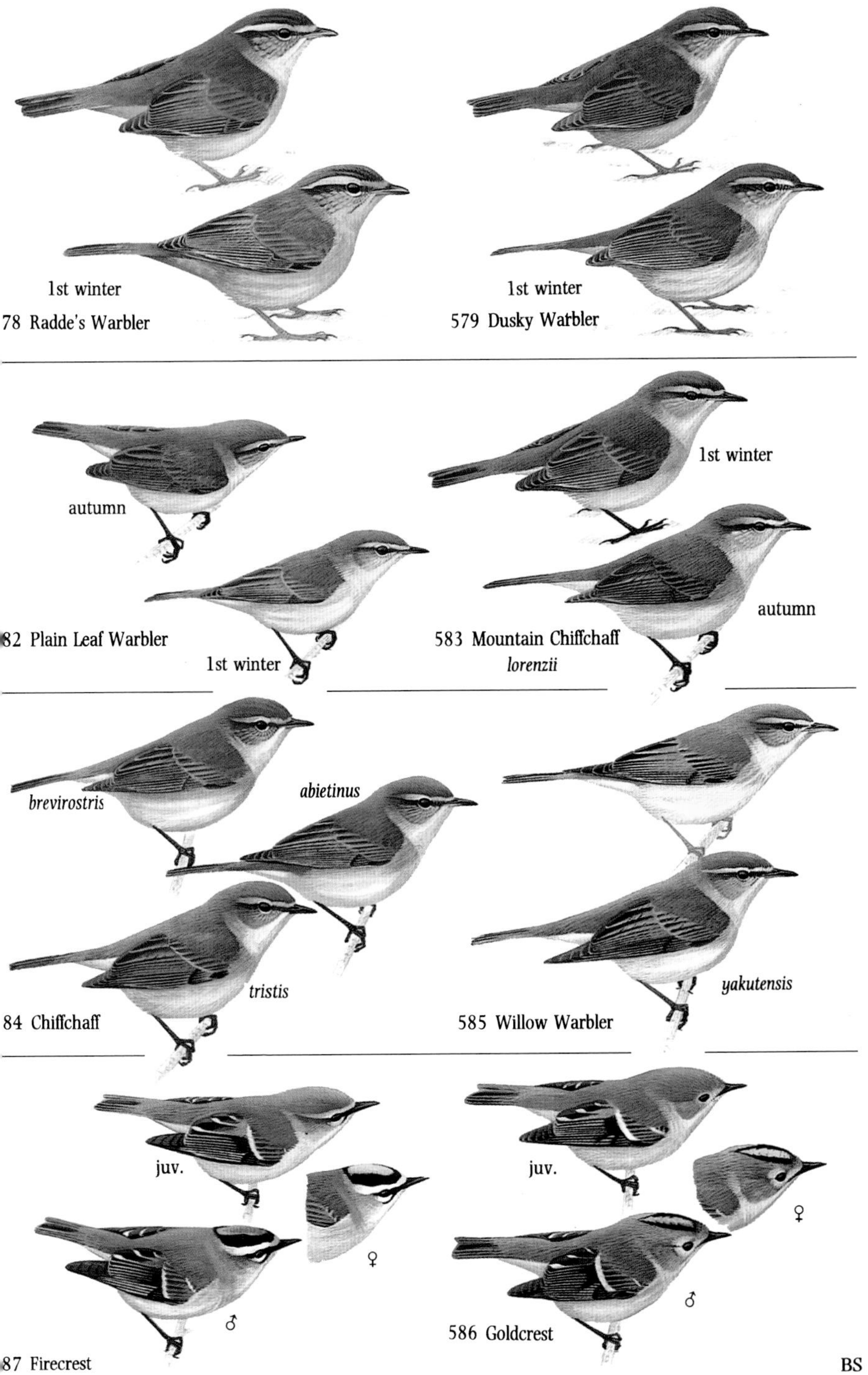
1st winter
78 Radde's Warbler
1st winter
579 Dusky Warbler
autumn
1st winter
82 Plain Leaf Warbler
1st winter
autumn
583 Mountain Chiffchaff
lorenzii
brevirostris
abietinus
tristis
84 Chiffchaff
yakutensis
585 Willow Warbler
juv.
♀
♂
87 Firecrest
juv.
♀
♂
586 Goldcrest

BS

588 **Blue-and-white Flycatcher**
Cyanoptila cyanomelana

Status: vagrant from Asia to Oman and UAE. **Habitat:** on passage in areas with trees.

589 **Spotted Flycatcher** *Muscicapa striata*

Status: summer visitor. On passage throughout the region. **Habitat:** gardens, parks, edges of woods, but anywhere with trees or scrub on passage. Nests on or in building, walls, against tree-trunk.

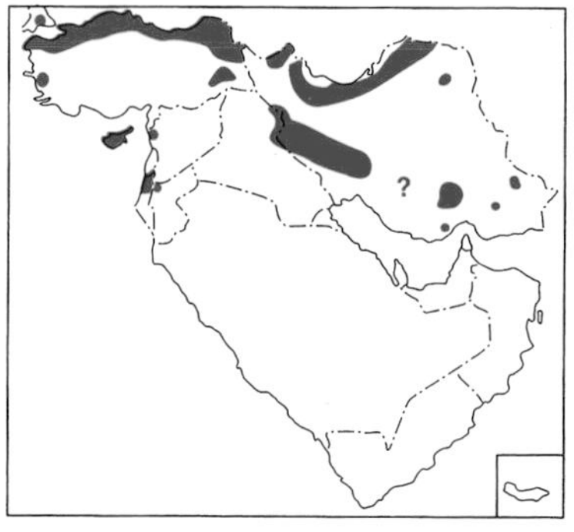

590 **Gambaga Flycatcher** *Muscicapa gambagae*

Status: summer visitor. **Habitat:** wooded highlands, bushy acacia scrub, up to 2500 m. Nests on branch, usually in acacia.

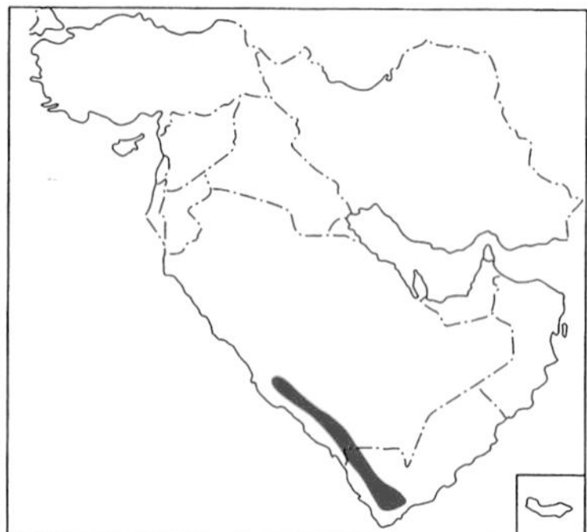

591 **Red-breasted Flycatcher** *Ficedula parva*

Status: summer visitor. Regular migrant Turkey (autumn) and Iran, rather scarce Cyprus, Near East, N and E Arabia and Gulf States; rare or vagrant Lebanon and Jordan. Occasional UAE in winter. **Habitat:** deciduous or mixed shady damp forest. On passage anywhere with trees. Nests in tree-crevice, hole or against trunk.

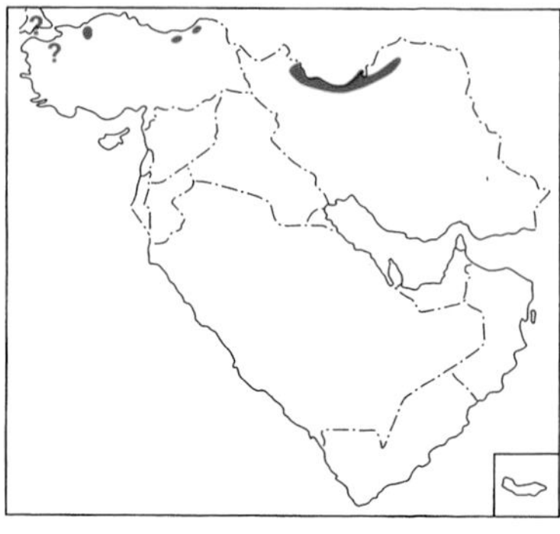

592 **Semi-collared Flycatcher**
Ficedula semitorquata

Status: summer visitor. On passage through Asian area and Near East; scarce or rare Arabia though some spring passage UAE. **Habitat:** woods, parks, large gardens. On passage in areas with trees. Nests in hole in tree.

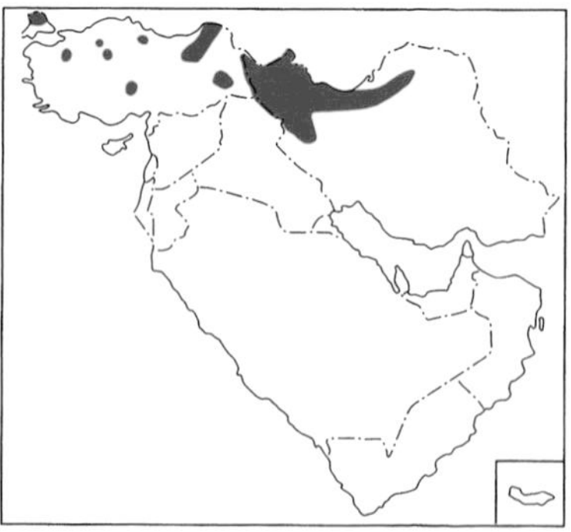

593 **Collared Flycatcher** *Ficedula albicollis*

Status: passage migrant Near East, Cyprus, Turkey. Vagrant Bahrain. **Habitat:** on passage, in areas with trees.

594 **Pied Flycatcher** *Ficedula hypoleuca*

Status: passage Turkey, Cyprus, Near East, Syria and Iran (where uncommon). Vagrant UAE. **Habitat:** on passage, where trees exist.

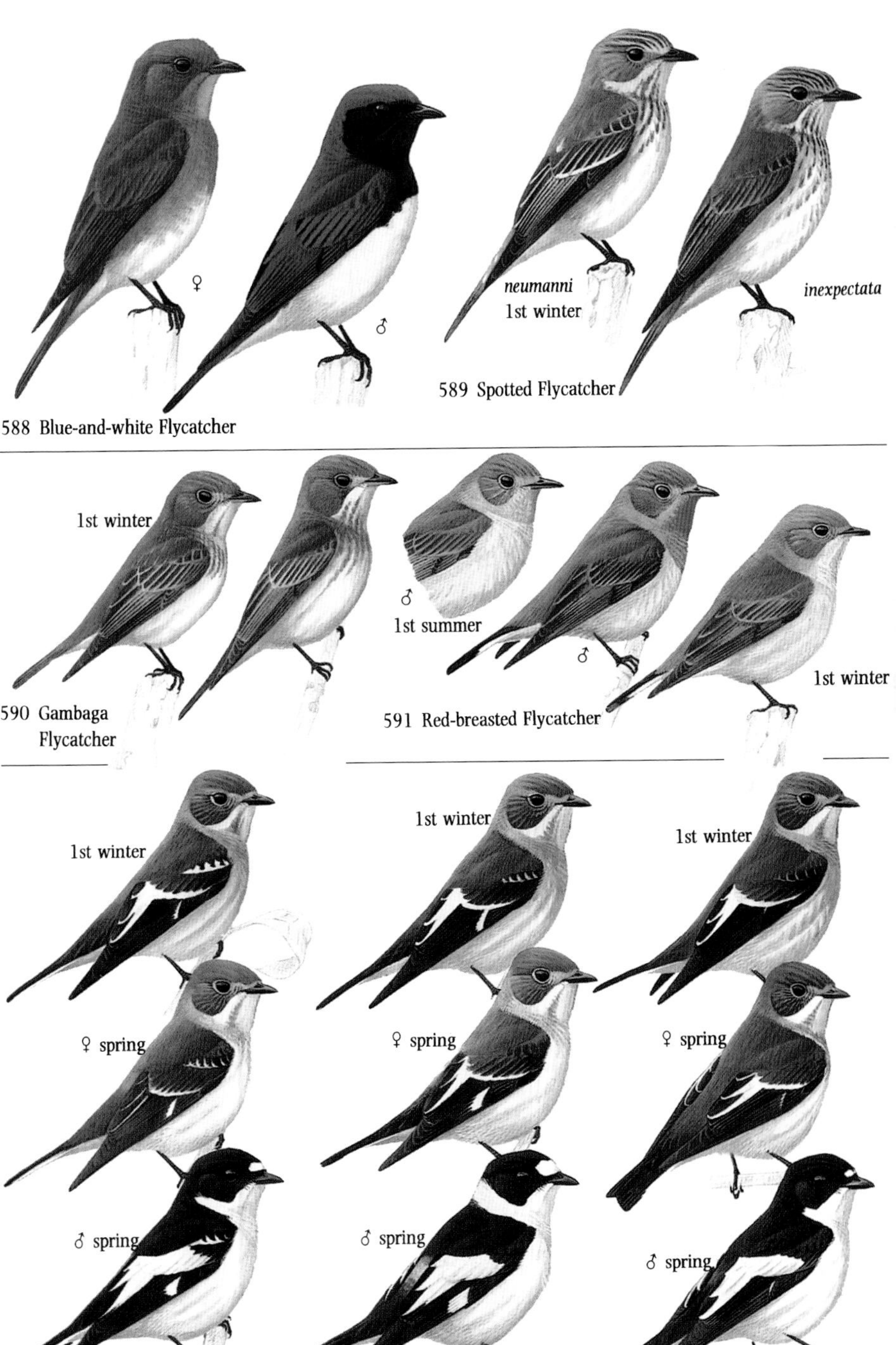
♀
♂
neumanni
1st winter
inexpectata
589 Spotted Flycatcher
588 Blue-and-white Flycatcher
1st winter
♂
1st summer
♂
1st winter
590 Gambaga
Flycatcher
591 Red-breasted Flycatcher
1st winter
1st winter
1st winter
♀ spring
♀ spring
♀ spring
♂ spring
♂ spring
♂ spring
592 Semi-collared Flycatcher
593 Collared Flycatcher
594 Pied Flycatcher
BS

595 **African Paradise Flycatcher** *Terpsiphone viridis*

Status: resident. **Habitat:** semi-tropical woodlands, 600–2400 m; usually near water. Nests in fork of tree or bush.

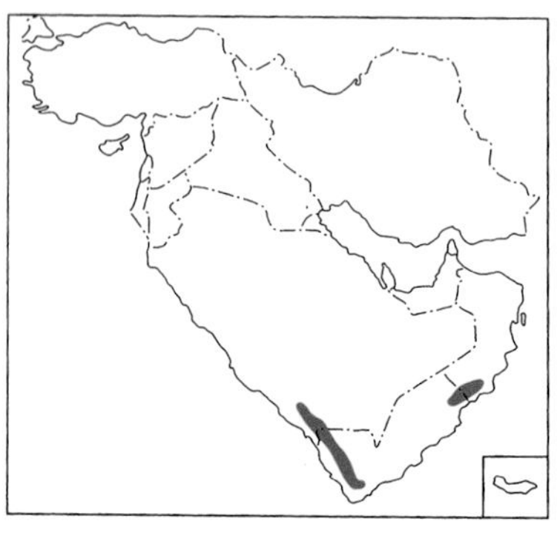

596 **Bearded Tit** *Panurus biarmicus*

Status: resident or partial migrant. Rare or irregular winter visitor Iran, Cyprus, Israel. Vagrant Kuwait. **Habitat:** extensive reedbeds. Nests low down in reeds.

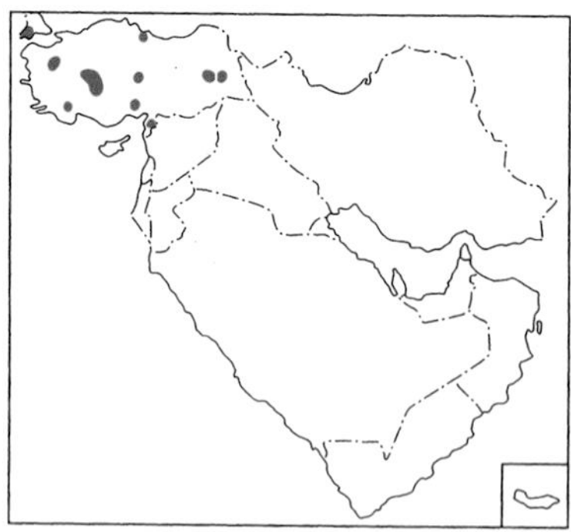

597 **Iraq Babbler** *Turdoides altirostris*

Status: resident. **Habitat:** chiefly river beds, riverine thickets, palm-groves, cultivated fields.

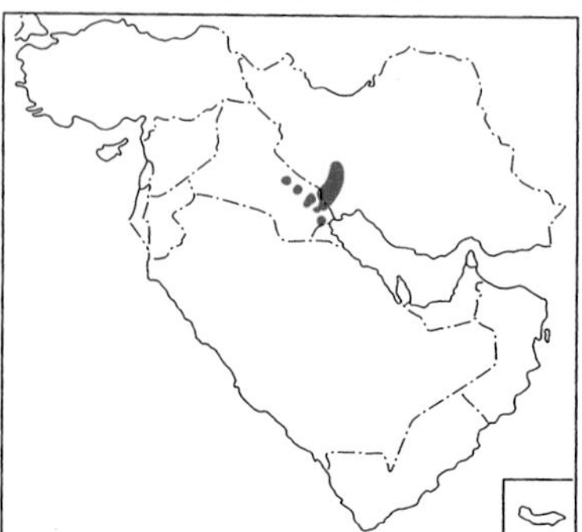

598 **Common Babbler** *Turdoides caudatus*

Status: resident. **Habitat:** cultivated and dry regions with scattered bushes and trees. Nests in tree.

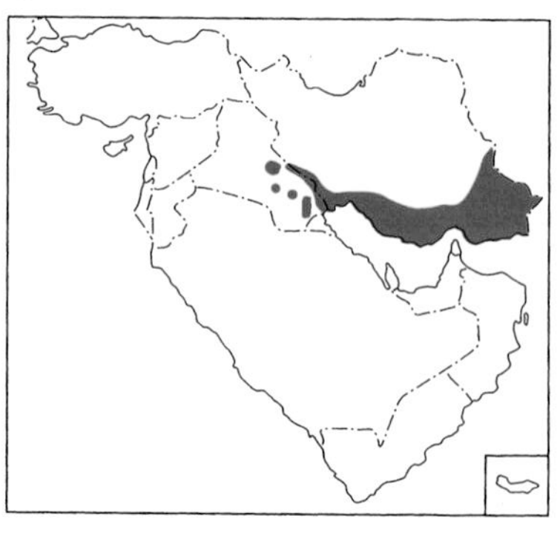

599 **Arabian Babbler** *Turdoides squamiceps*

Status: resident. **Habitat:** dry areas with scrub, scattered acacias, wadis, palm groves and savanna from sea level to 2400 m. Nests usually in acacia.

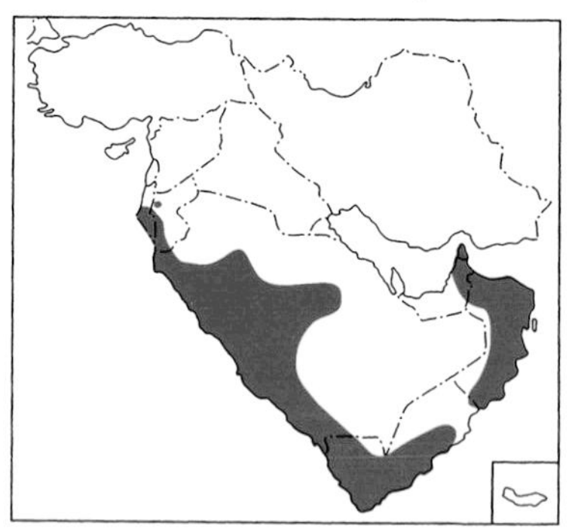

600 **Long-tailed Tit** *Aegithalos caudatus*

Status: resident; some winter dispersal within Turkey and Iran. **Habitat:** mixed woods with bushes and other undergrowth, open country with bushes. Nests in bush or tree-fork.

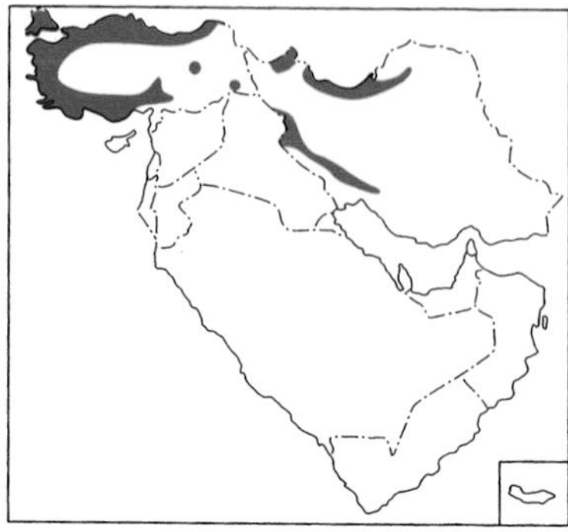

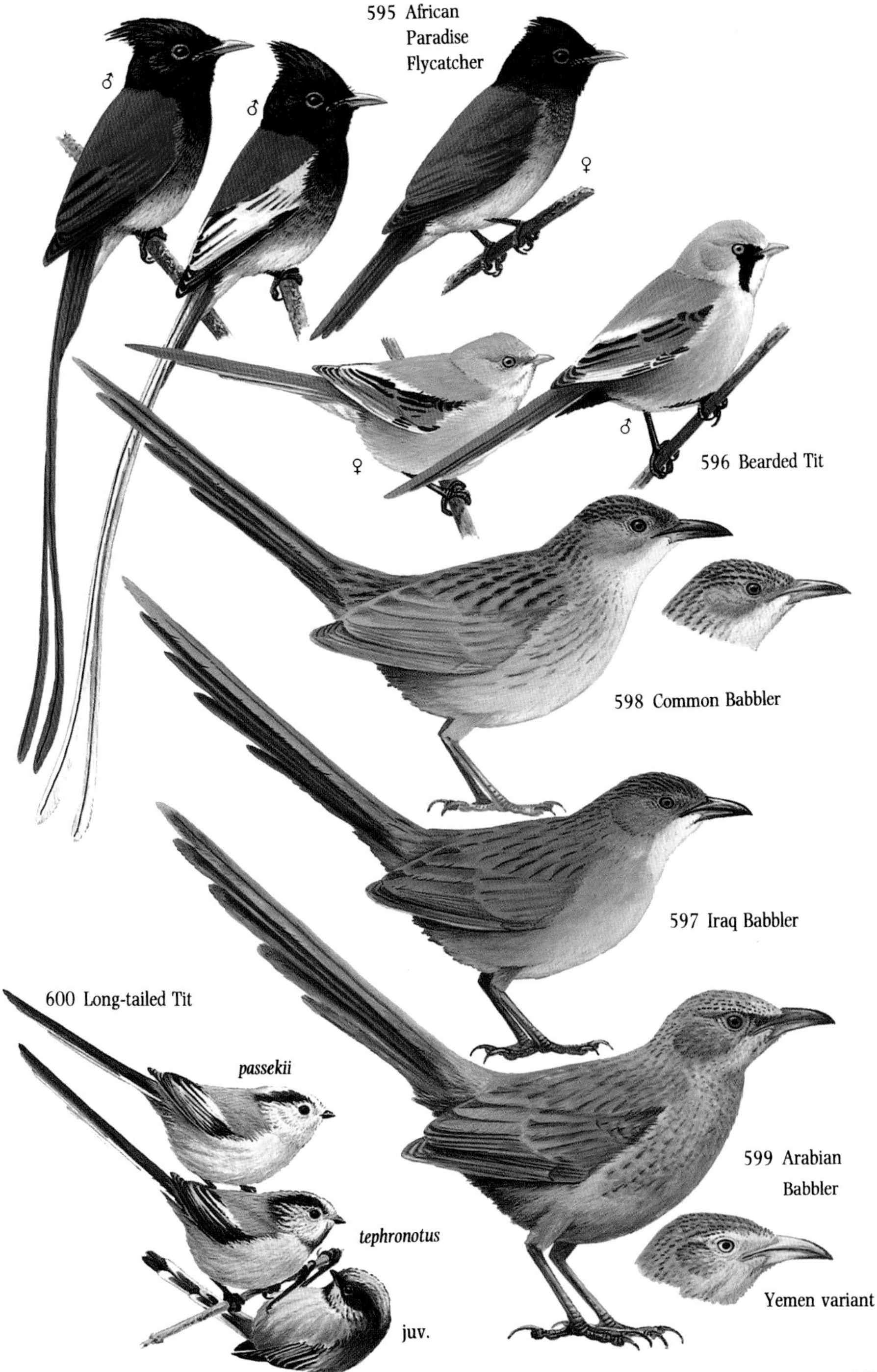
595 African Paradise Flycatcher
♂
♂
♀
596 Bearded Tit
♀
♂
598 Common Babbler
597 Iraq Babbler
600 Long-tailed Tit
passekii
tephronotus
juv.
599 Arabian Babbler
Yemen variant
ML

601 **Marsh Tit** *Parus palustris*

Status: resident. **Habitat:** deciduous woods, thickets. Nests in hole in tree.

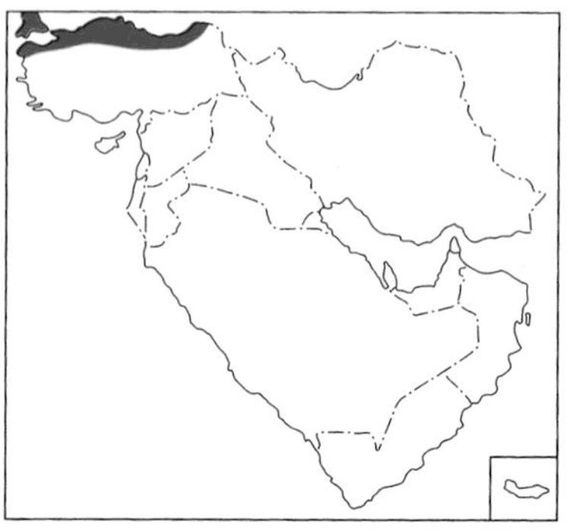

602 **Sombre Tit** *Parus lugubris*

Status: resident. **Habitat:** lowland plains and mountain slopes with mixed woodland, scrub, and rocky outcrops. Nests in hole in tree or other structure, occasionally among rocks.

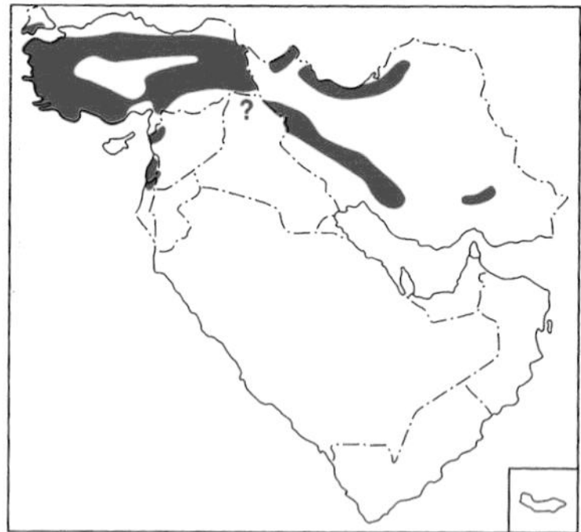

603 **Coal Tit** *Parus ater*

Status: resident, slight dispersal. **Habitat:** mainly coniferous woodland but also oak and cork-oak; mainly in mountainous areas. Nests low in stump or hole in tree, bank or ground.

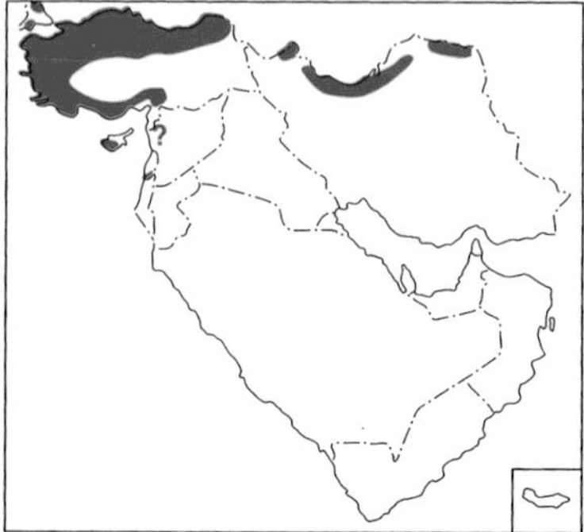

604 **Blue Tit** *Parus caeruleus*

Status: resident, breeds regularly Jordan now. Vagrant Syria (where has bred), Lebanon and Israel. **Habitat:** woodland, gardens, thickets, plantations, sometimes also oases. Nests in hole in tree, wall or other structure.

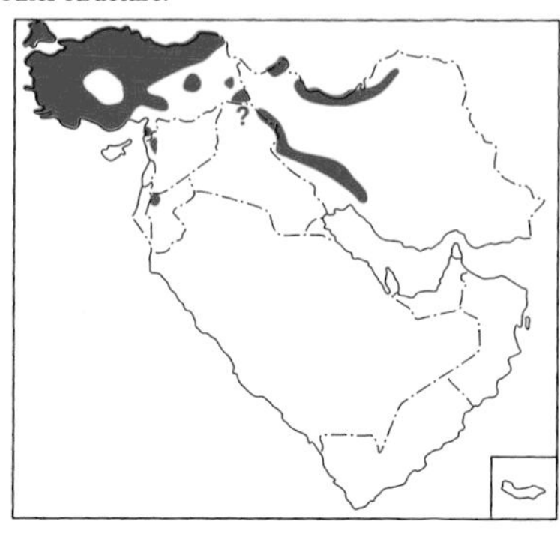

605 **Azure Tit** *Parus cyanus*

Status: vagrant Iran. **Habitat:** trees and bushes (notably willow), invariably near water.

606 **Great Tit** *Parus major*

Status: resident. **Habitat:** anywhere with trees (mixed woods, gardens) but not at high altitudes. Nests in hole in tree or wall.

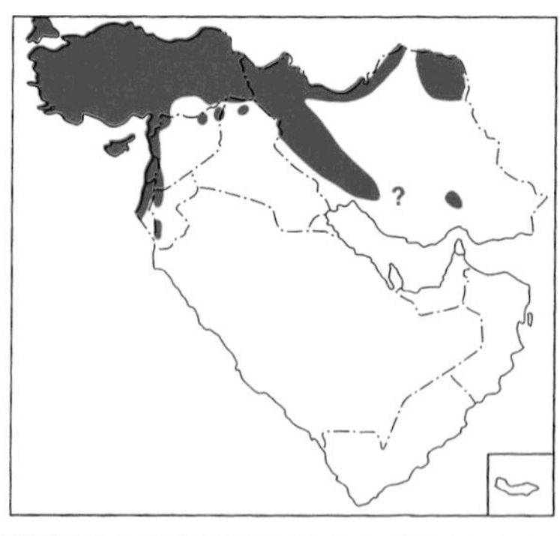

607 **Turkestan Tit** *Parus bokharensis*

Status: resident. **Habitat:** riverine thickets in lowlands. Nests in hole in tree or wall.

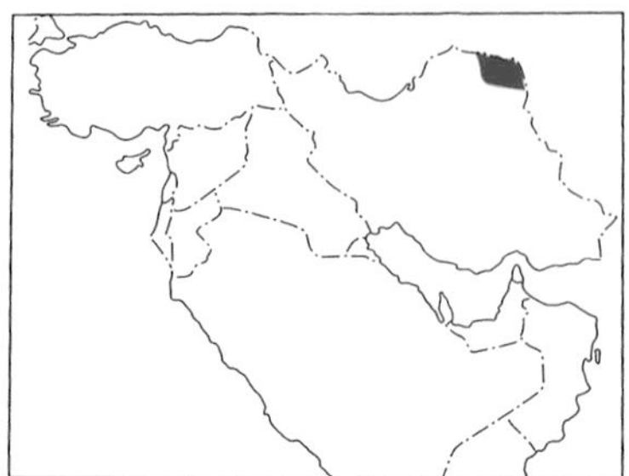

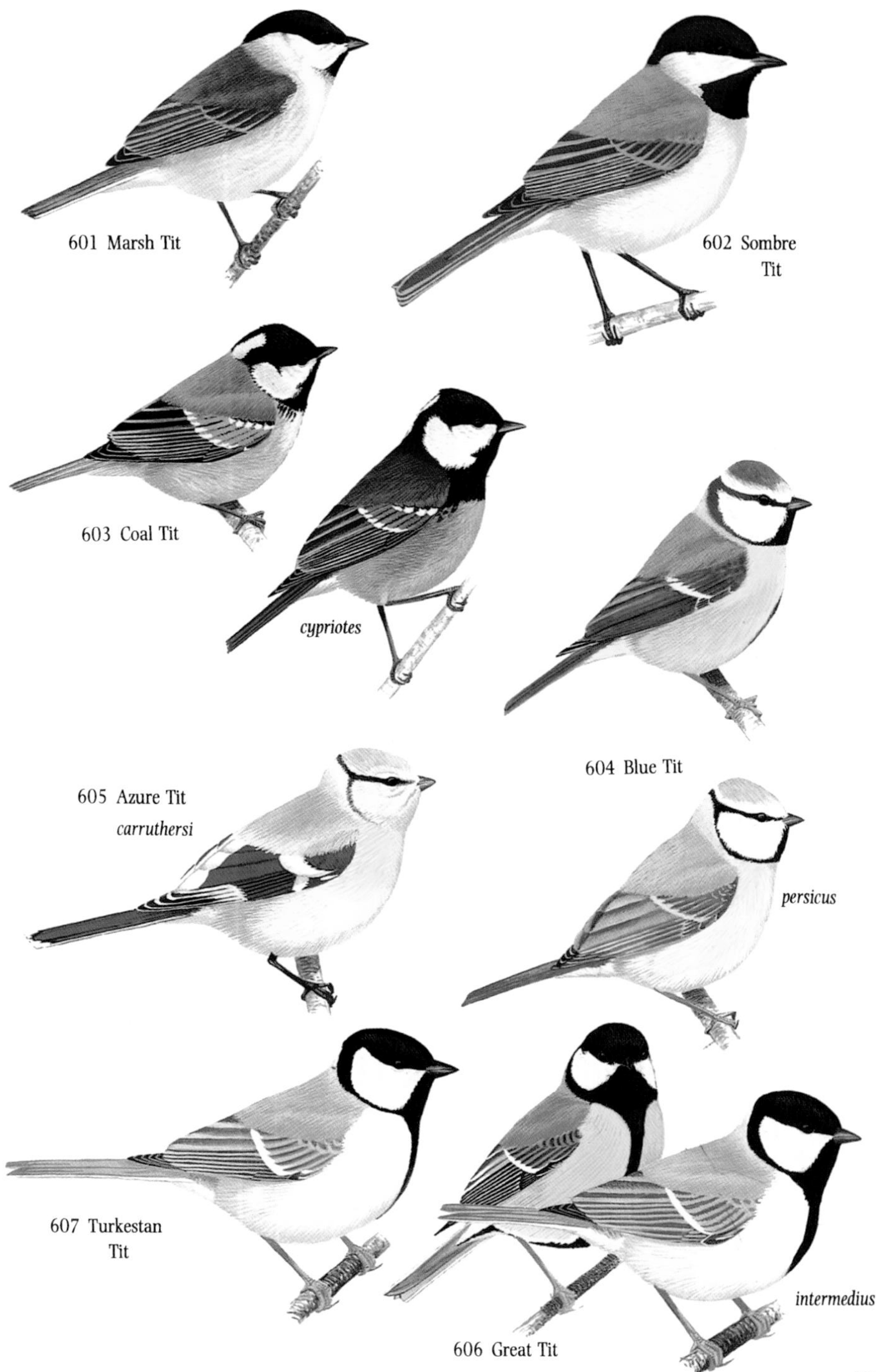
601 Marsh Tit
602 Sombre Tit
603 Coal Tit
cypriotes
604 Blue Tit
605 Azure Tit
carruthersi
persicus
607 Turkestan Tit
606 Great Tit
intermedius

AB

608 **Krüper's Nuthatch** *Sitta krueperi*

Status: resident. **Habitat:** pine forest, cedars and junipers, from sea level to tree-limit, but typically between 1200–1700 m. Nests in hole in tree without mud-plastering.

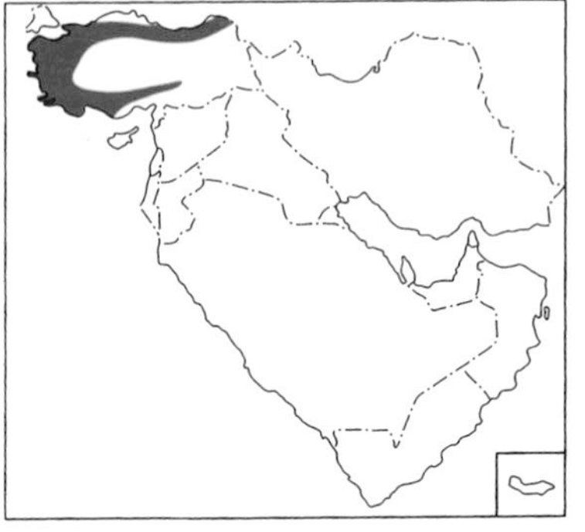

609 **Nuthatch** *Sitta europaea*

Status: resident. **Habitat:** large deciduous trees in woods and parks. Nests in hole in tree, plastering entrance hole with mud.

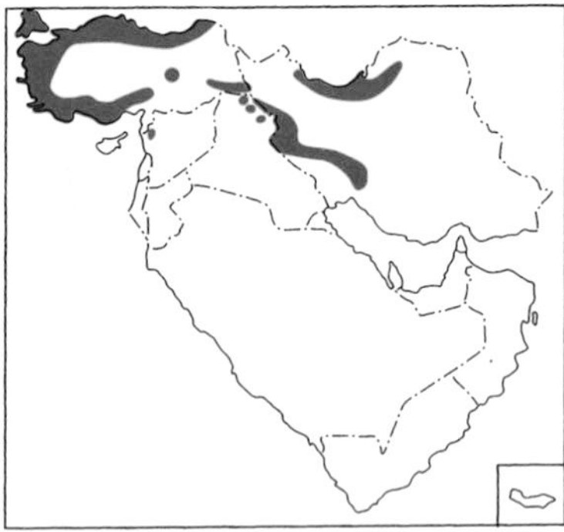

610 **Eastern Rock Nuthatch** *Sitta tephronota*

Status: resident. **Habitat:** rocky gorges, mountain sides, cliffs and semi-desert edges, sometimes with scattered trees. Nests on rock surface, often overhung, mud construction with hole near top.

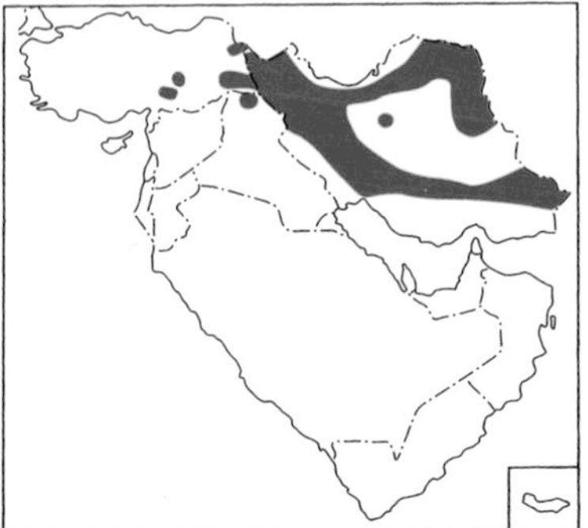

611 **Western Rock Nuthatch** *Sitta neumayer*

Status: resident. **Habitat:** rocky gorges, mountain sides, cliffs. Nests in protected cranny in rock; built of mud, with protruding tunnel entrance.

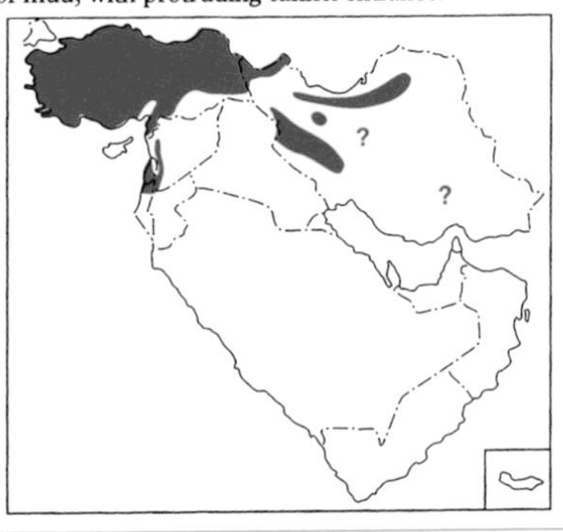

612 **Wallcreeper** *Tichodroma muraria*

Status: resident with some winter dispersal outside breeding range. Winters also in Cyprus and Israel (where may have bred), perhaps also Lebanon. Vagrant Syria and Jordan. **Habitat:** rocky ravines, earth cliffs, ruins, from 2000 m to snowline. In winter, rocky valleys and foothills, but also coastal cliffs. Nests in deep crevice on cliff-face, rocks, occasionally buildings.

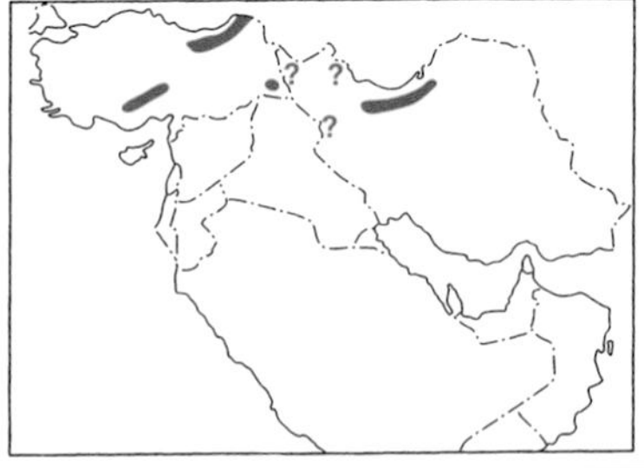

615 **Penduline Tit** *Remiz pendulinus*

Status: resident and dispersive with winter records from Israel, Jordan and Cyprus (where may breed), Lebanon, NW, central and E Arabia. Vagrant Oman. **Habitat:** trees (especially willows, poplars and tamarisks) and reedbeds, always near water, including coastal marshes and oases. The oval-shaped nest with tube entrance at side near top is suspended (usually over water) from outermost branch of tree.

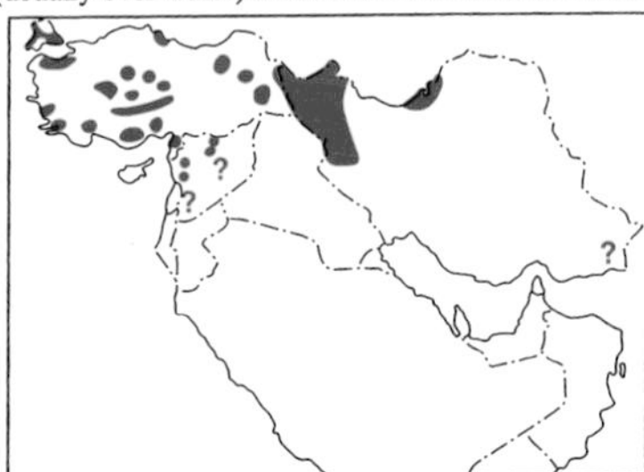

608 Krüper's Nuthatch
♂
♀
609 Nuthatch
610 Eastern Rock Nuthatch
611 Western Rock Nuthatch
tschitscherini
macronyx ♂
612 Wallcreeper
juv.
615 Penduline Tit
coronatus ♂
AE

613 **Treecreeper** *Certhia familiaris*

Status: resident, but with some winter dispersal, e.g. to S Turkey. **Habitat:** mixed mountain woodland up to 3000 m. Nests behind loose bark or in split trunk.

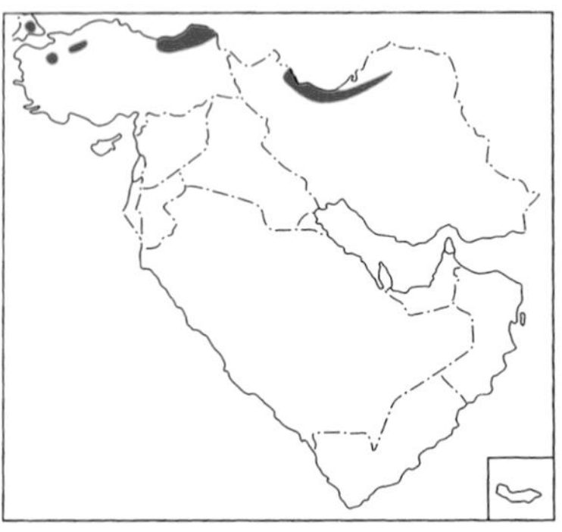

614 **Short-toed Treecreeper** *Certhia brachydactyla*

Status: resident. **Habitat:** large gardens or parks, deciduous woods, pines or cedars, from sea level to 1500 m (seldom higher). Nests in slit, hole or loose bark on trunk, or in wall.

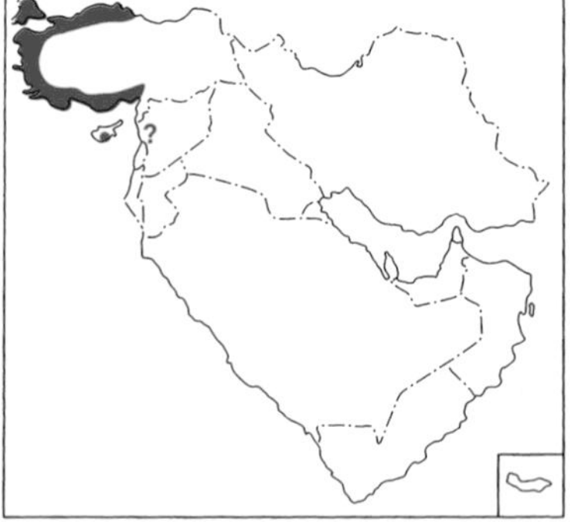

616 **Nile Valley Sunbird** *Anthreptes metallicus*

Status: resident. **Habitat:** gardens, savanna with acacia or dry scrub, wadis, from sea level to 2500 m, commonest at low altitudes. Nest is an oval structure suspended from twig.

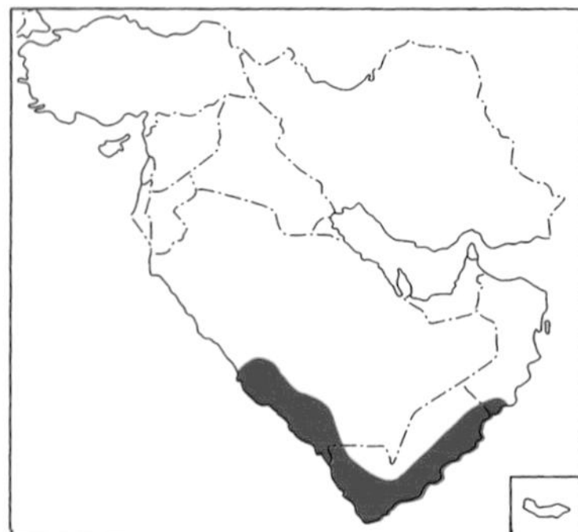

617 **Purple Sunbird** *Nectarinia asiatica*

Status: mainly resident; some dispersal in autumn. **Habitat:** large gardens, cultivation, tamarisk along rivers, thorn scrub and dry forest. Nest pear-shaped, suspended on bush.

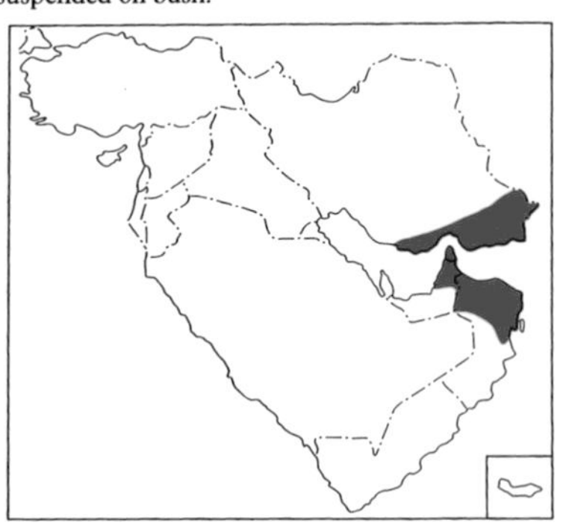

618 **Shining Sunbird** *Nectarinia habessinica*

Status: resident, with seasonal movements. **Habitat:** luxuriant vegetation in wadis; dry savanna scrub, 250–2500 m but mostly lower and middle altitudes. Bottle-shaped nest suspended in tree or bush.

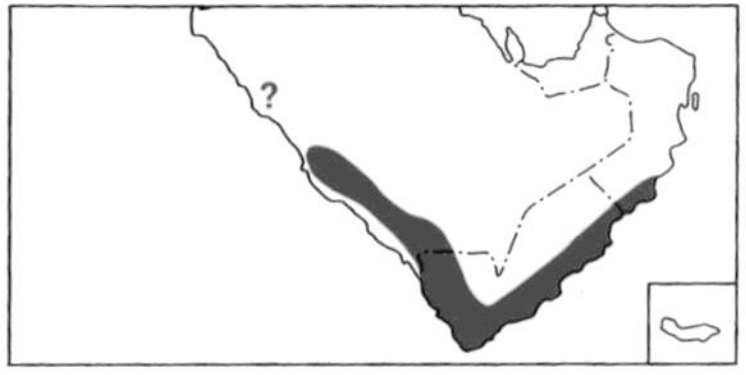

619 **Palestine Sunbird** *Nectarinia osea*

Status: resident; slight dispersal in winter, reaching Lebanon and Syria. **Habitat:** well vegetated areas, rocky wadis with acacia, acacia steppe, savanna, river banks, gardens in towns, from sea level up to 3200 m. Nest suspended on tree or bush.

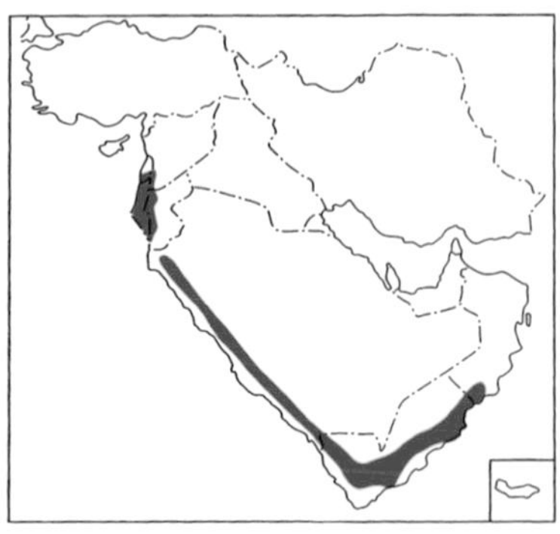

620 **Socotra Sunbird** *Nectarinia balfouri*

Status: resident endemic on Socotra. **Habitat:** wooded hillsides, up to 1300 m. Domed nest built in tree.

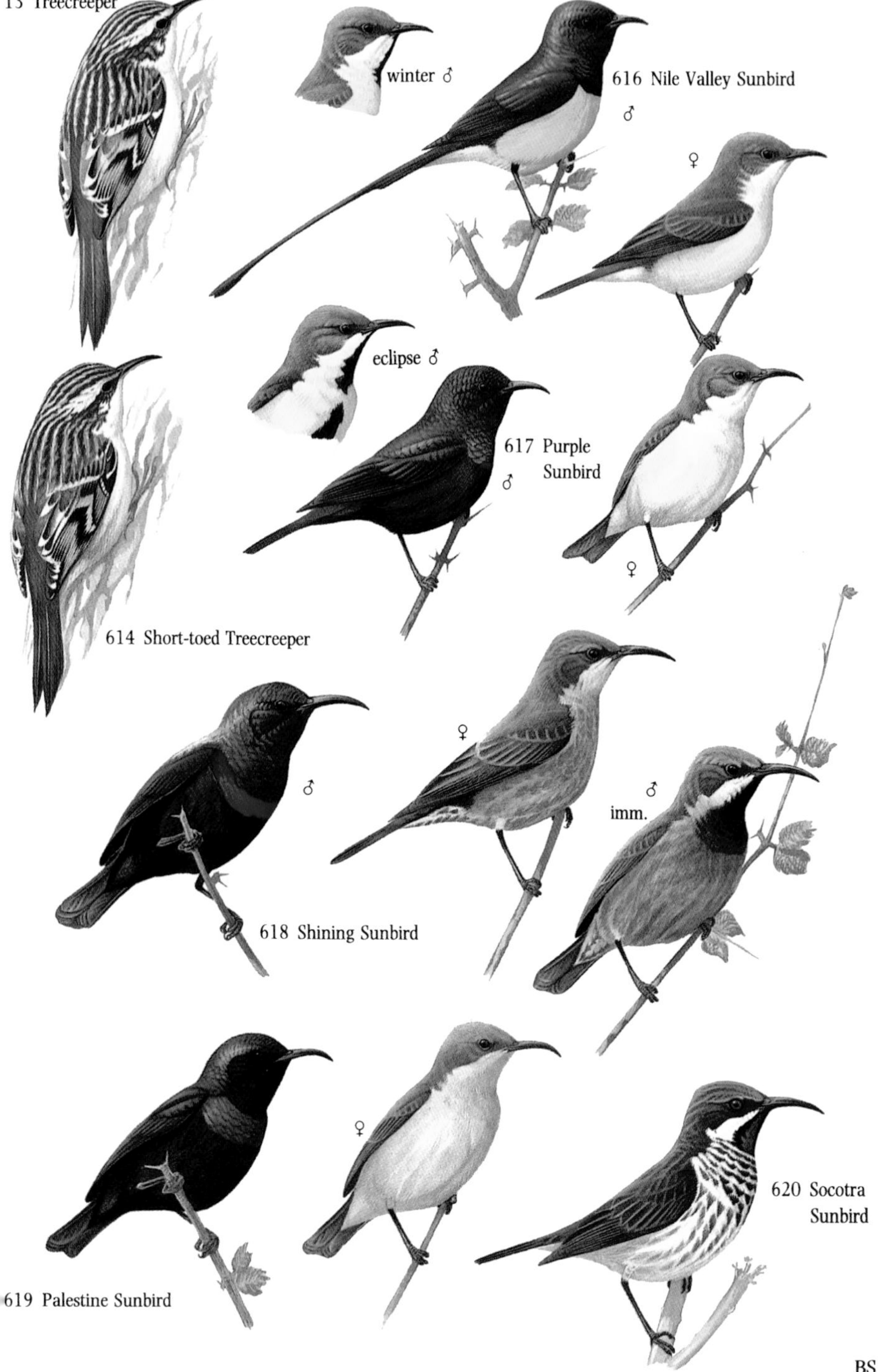
13 Treecreeper
winter ♂
616 Nile Valley Sunbird
♂
♀
eclipse ♂
617 Purple
Sunbird
♂
♀
614 Short-toed Treecreeper
♀
♂
imm.
♂
618 Shining Sunbird
♀
620 Socotra
Sunbird
619 Palestine Sunbird
BS

622 **Golden Oriole** *Oriolus oriolus*

Status: has bred Israel and NW Saudi Arabia and has attempted in E Saudi Arabia. Summer visitor; on passage throughout the region. Occasional in winter Oman. **Habitat:** parks with tall trees, broad-leaved woods. Nest usually slung between horizontally forked branches.

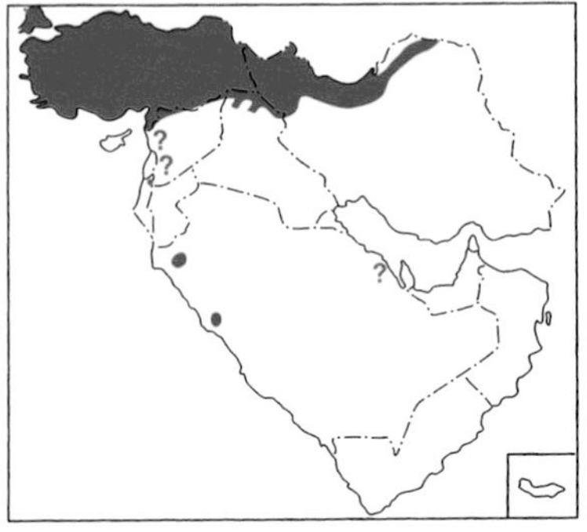

623 **Black-crowned Tchagra** *Tchagra senegala*

Status: resident. **Habitat:** dense dry scrub with scattered trees. Nests in low tree or bush.

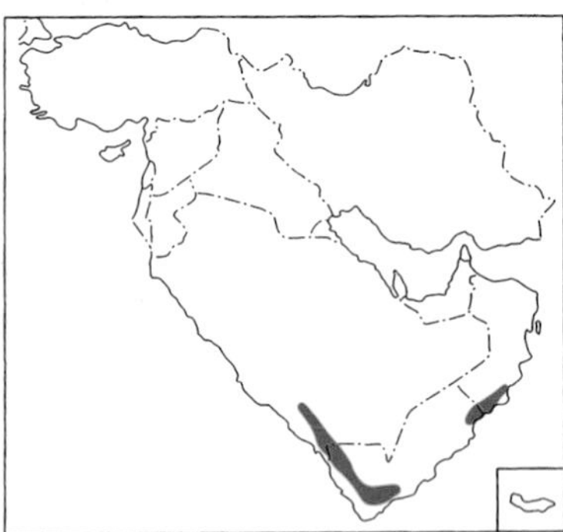

624 **Isabelline Shrike** *Lanius isabellinus*

Status: summer visitor. Passage Iran, Iraq, Arabian peninsula; scarce winter S Iran, Iraq and Arabia; rare Turkey and Near East. Vagrant Cyprus. **Habitat:** dry steppes with bushes and scattered acacias, semi-desert or barren mountains. On passage/winter also in cultivation. Nests in bush.

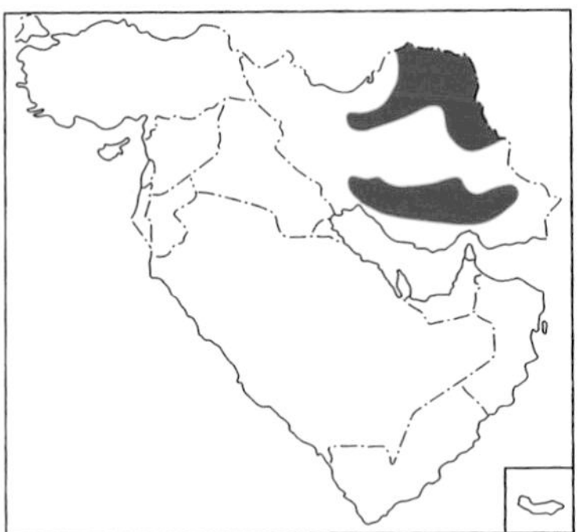

625 **Red-backed Shrike** *Lanius collurio*

Status: has bred Lebanon. Summer visitor; on passage in the whole region. **Habitat:** bushy waste-lands, thickets and lightly wooded areas with thorny clearings. Nests in dense bush or small tree.

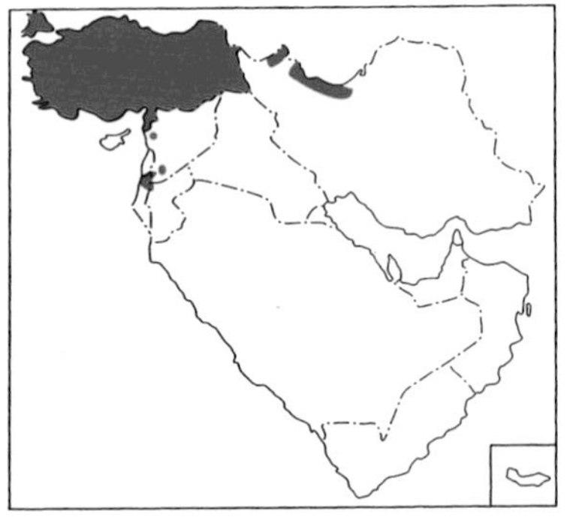

626 **Bay-backed Shrike** *Lanius vittatus*

Status: resident. Vagrant Qatar, UAE and Oman. **Habitat:** open, often rocky country with scattered trees and low bushes. Nests in bush.

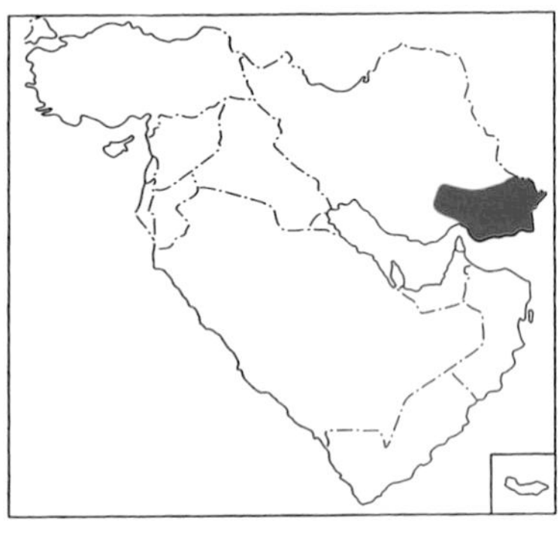

627 **Long-tailed Shrike** *Lanius schach*

Status: formerly bred in NE Iran. Vagrant Israel, Turkey and Oman. **Habitat:** open scrubland and cultivated regions. Nests in bush.

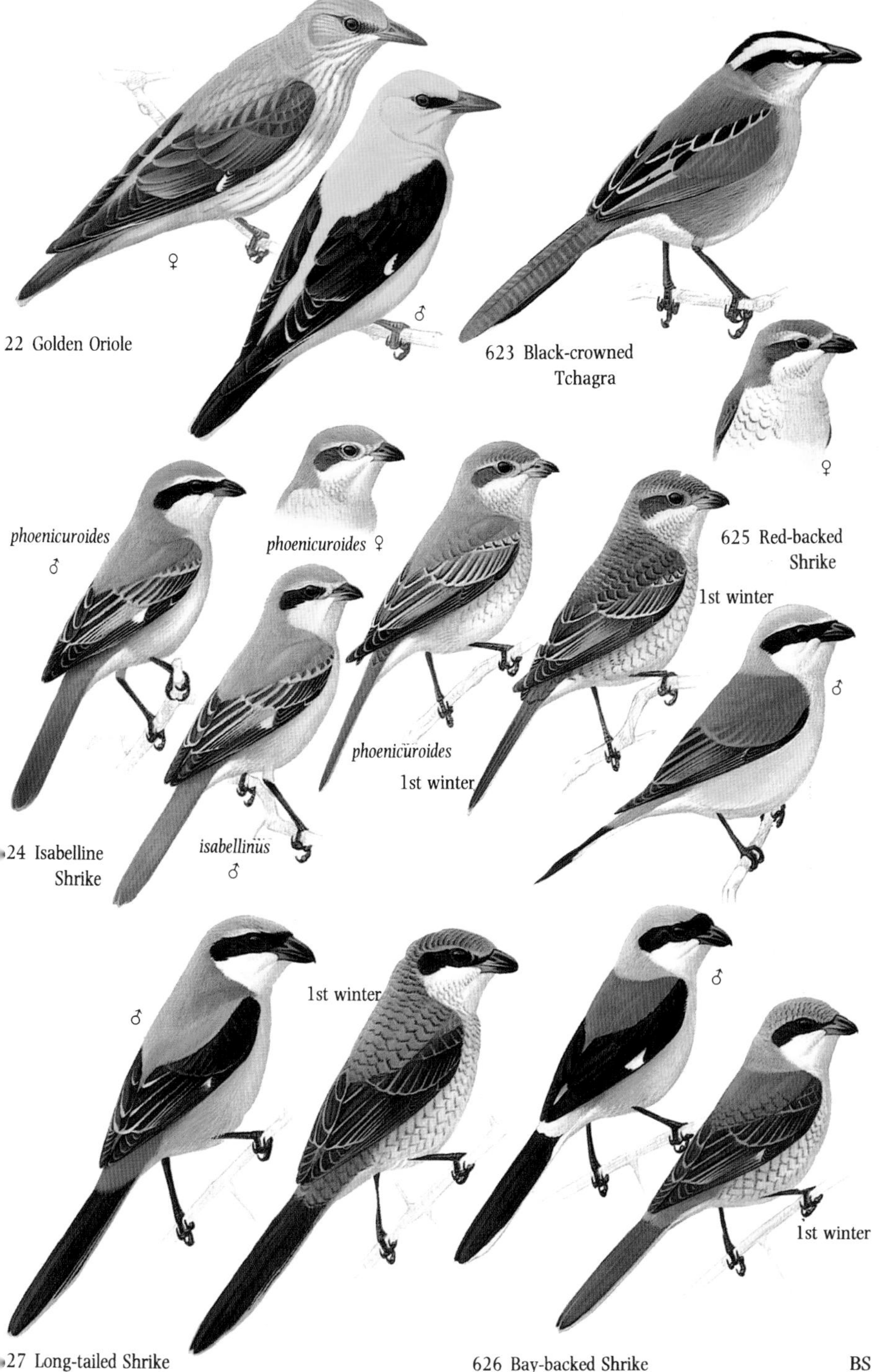

♀
♂
22 Golden Oriole
623 Black-crowned Tchagra
♀
phoenicuroides ♂
phoenicuroides ♀
625 Red-backed Shrike
1st winter
♂
phoenicuroides 1st winter
24 Isabelline Shrike
isabellinus ♂
♂
1st winter
♂
1st winter
27 Long-tailed Shrike
626 Bay-backed Shrike

BS

628 **Lesser Grey Shrike** *Lanius minor*

Status: summer visitor. Passage through Turkey, Cyprus, Iran, Near East and, more scarcely, Arabian peninsula. **Habitat:** fairly open cultivated country with scattered trees and bushes. Nests fairly high in tree.

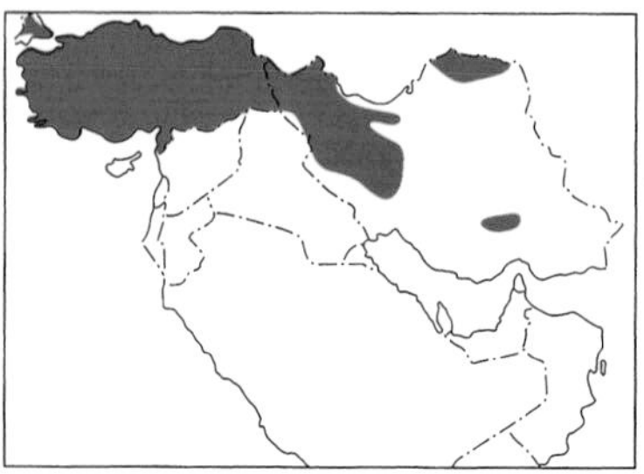

629 **Great Grey Shrike** *Lanius excubitor*

Status: may breed Lebanon and probably more widespread in N Saudi Arabia; also passage migrant and winter visitor throughout much of the region but generally rather scarce in Turkey. Vagrant Cyprus. **Habitat:** anywhere with patches of scrub and trees near open areas. Nests in bush or small tree.

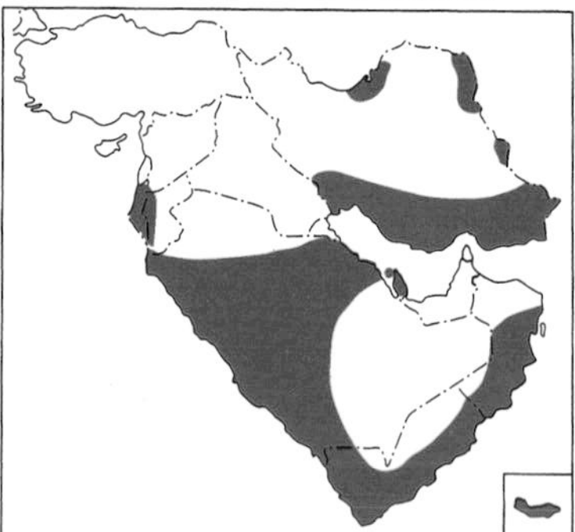

630 **Woodchat Shrike** *Lanius senator*

Status: occasionally breeds Cyprus, summer visitor. Passage through most of the region but scarce in Oman. Wintering recorded in Kuwait and SW Arabia. **Habitat:** dry open country, olive groves, orchards, bushy wasteland, oases, hedges in cultivated areas, occasionally open woods. Nests in tree.

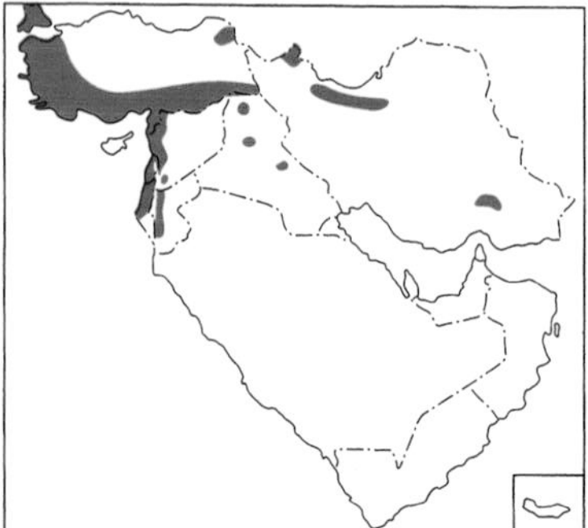

631 **Masked Shrike** *Lanius nubicus*

Status: summer visitor. Passage Turkey and Iran and all countries south including Arabia, though generally scarce. Winters in small numbers in SW Arabia. **Habitat:** olive-groves, parkland, cultivation with scattered trees and scrub, in lowlands and foothills; on passage may occur in almost any group of low trees and bushes. Nests in tree.

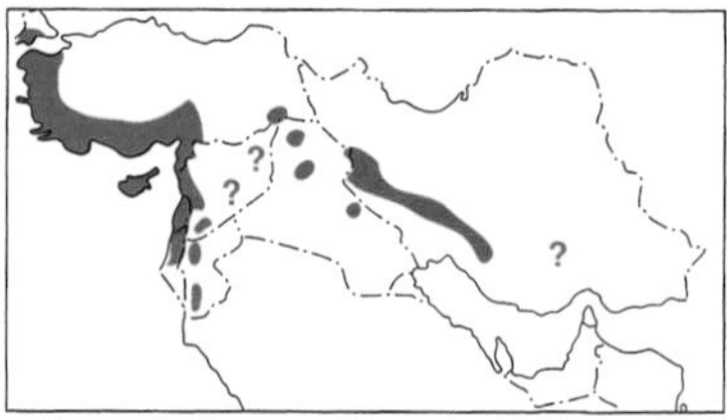

632 **Black Drongo** *Dicrurus macrocercus*

Status: formerly bred in SE Iran but no records this century; rare UAE. Vagrant Oman. **Habitat:** cultivated lowlands, particularly abundant near water; frequently near habitation and even edges of towns.

633 **Jay** *Garrulus glandarius*

Status: resident, slight dispersal outside breeding range. **Habitat:** coniferous, oak or mixed woodland. Nests in tree.

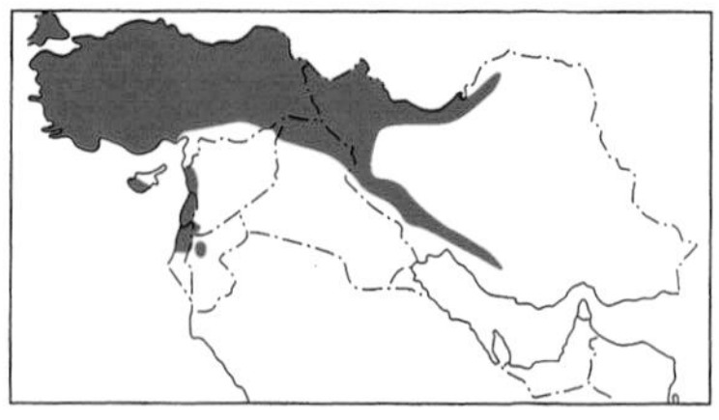

634 **Magpie** *Pica pica*

Status: resident with some dispersal outside breeding range in winter, e.g. to S Iraq, also occasionally S Lebanon. **Habitat:** bushy or open country with tall trees, scattered woods, pine and juniper-covered slopes, up to 2900 m. Builds domed nest, in tree or large bush.

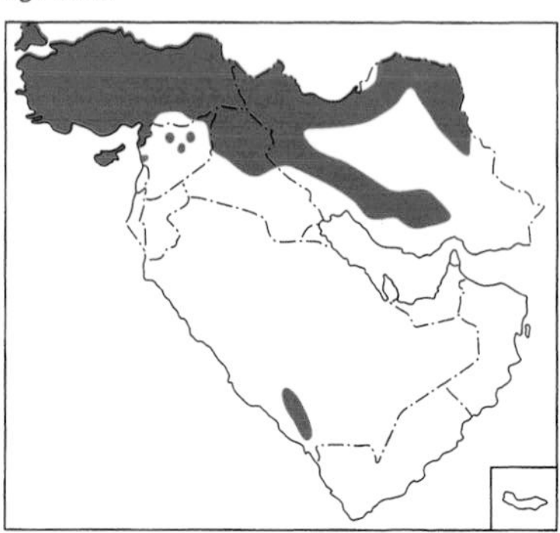

juv.
pallidirostris
1st winter
buryi
628 Lesser Grey Shrike
1st winter
♂
629 Great Grey Shrike
aucheri
630 Woodchat Shrike
1st winter
♂
632 Black Drongo
631 Masked Shrike
krynicki
633 Jay
634 Magpie
asirensis
BS

635 **Pleske's Ground Jay** *Podoces pleskei*

Status: resident. **Habitat:** steppe with scattered low bushes. Nest undescribed.

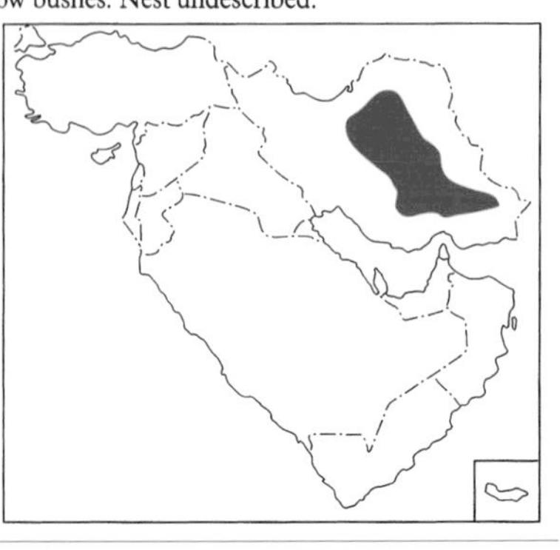

636 **Nutcracker** *Nucifraga caryocatactes*

Status: vagrant Turkey and N Iran. **Habitat:** mainly coniferous woods.

637 **Alpine Chough** *Pyrrhocorax graculus*

Status: resident; has occurred Syria in winter. **Habitat:** high mountains, often around habitation, occasionally descending in winter. Nests colonially on cliffs or ruins.

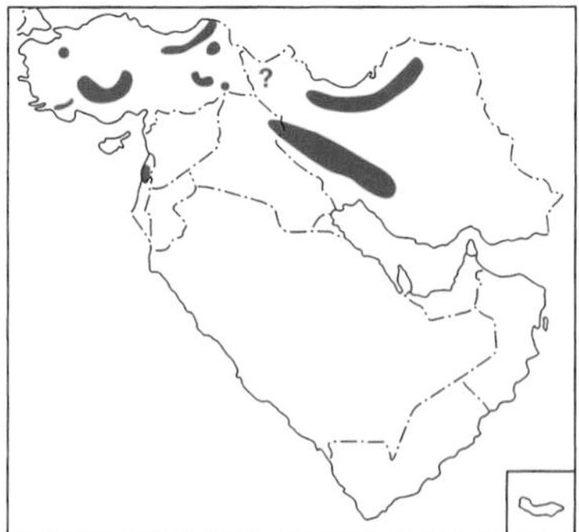

638 **Chough** *Pyrrhocorax pyrrhocorax*

Status: resident. Rare Syria in winter. **Habitat:** mountains, sometimes near habitation. Nests colonially in rocks, cliffs or caves.

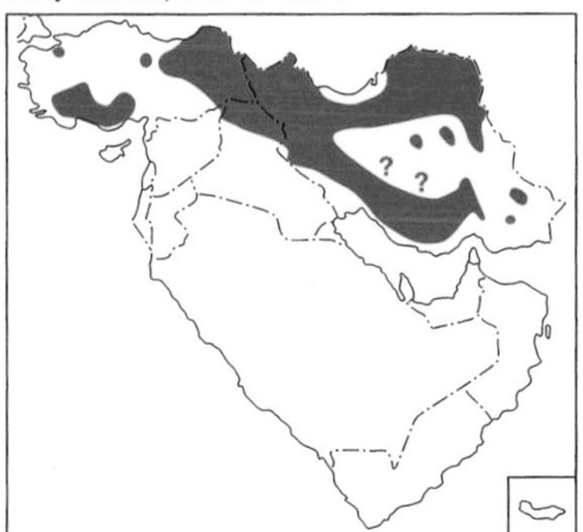

639 **Jackdaw** *Corvus monedula*

Status: has bred Jordan. Resident, but some winter dispersal in Syria, Jordan (where scarce) and Iraq. **Habitat:** often near habitation; also farmland and mountain cliffs. Nests colonially in holes in cliffs, buildings or trees.

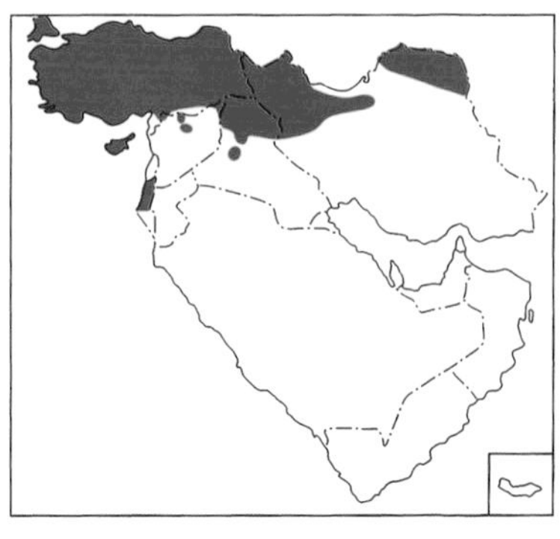

640 **House Crow** *Corvus splendens*

Status: recently colonized many sea port areas, probably by introduction. **Habitat:** coastal towns and villages and their environs. Nests in tree or other structure.

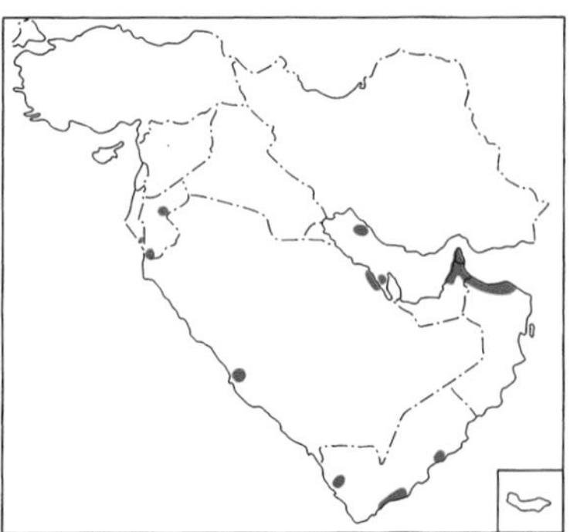

638 Chough
juv.
637 Alpine
Chough
639 Jackdaw
635 Pleske's Ground
Jay
640 House Crow
imm.
636 Nutcracker
ML

641 **Rook** *Corvus frugilegus*

Status: mainly resident. Passage through Turkey. In winter, reaches SW Iran and S Iraq, Syria and irregularly Cyprus and Israel. Rare Kuwait and Jordan. **Habitat:** farmland with trees. Nests and roosts in colonies in trees.

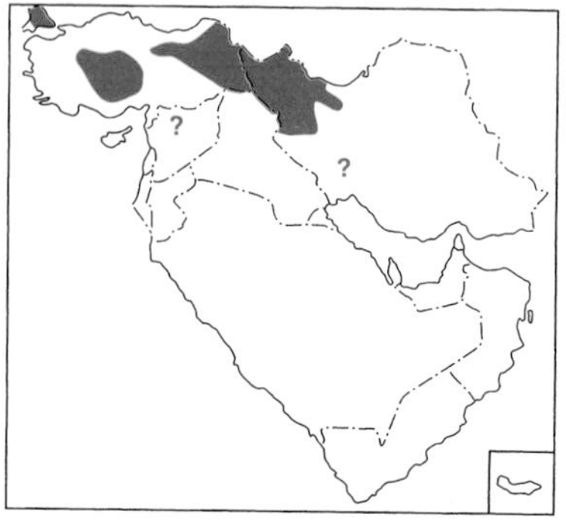

642 **Carrion Crow** *Corvus corone corone*

Status: vagrant Turkey. **Habitat:** as Hooded Crow.

643 **Hooded Crow** *Corvus corone cornix*

Status: resident, but some reach S Iraq in winter. **Habitat:** wide range of habitats; open country with scattered trees.

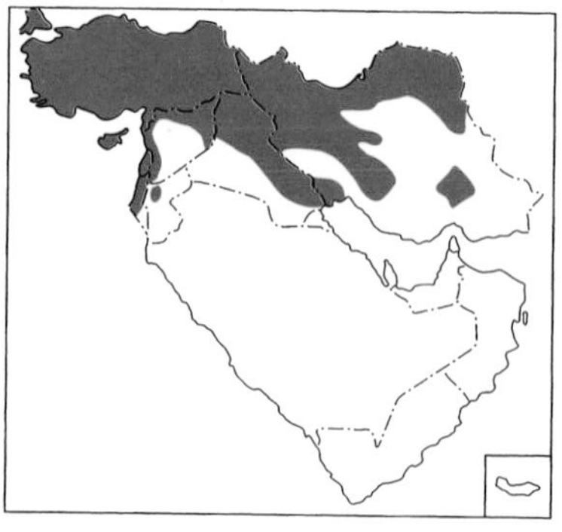

644 **Jungle Crow** *Corvus macrorhynchos*

Status: bred last century in SE Iran, but apparently no recent records. **Habitat:** wooded countryside, also edges of towns and villages. Nests in tree.

645 **Brown-necked Raven** *Corvus ruficollis*

Status: resident. Vagrant Syria and SE Turkey. **Habitat:** mountains, deserts, semi-deserts, usually near habitation. Nests on cliffs, ruin or outcrops.

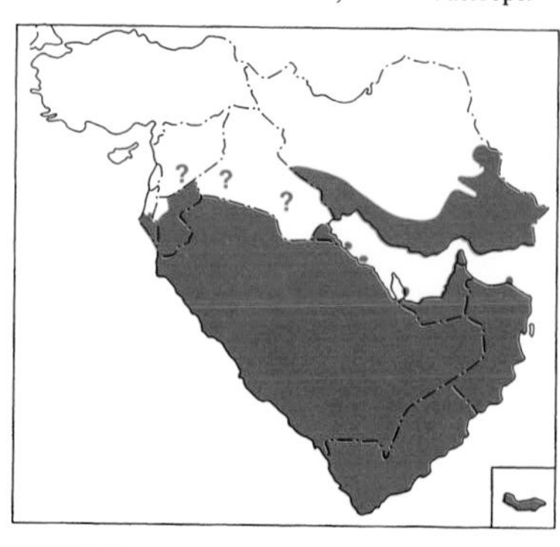

646 **Raven** *Corvus corax*

Status: resident. Reaches S Iraq in winter. Rare Lebanon. **Habitat:** mountains. Nests on cliff, or sometimes tree.

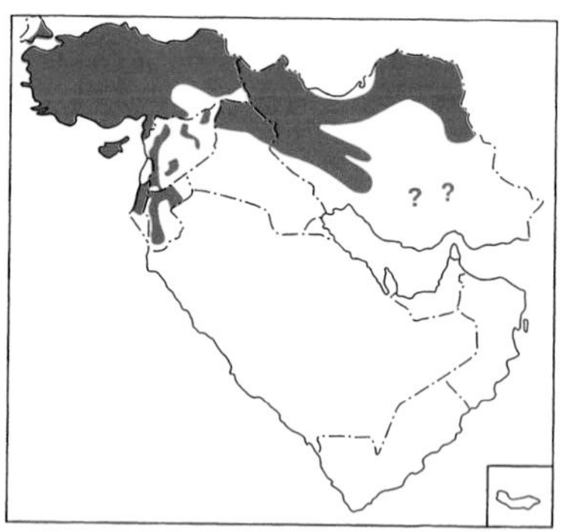

647 **Fan-tailed Raven** *Corvus rhipidurus*

Status: resident. **Habitat:** a wide variety of habitats from sea level to over 3000 m; often close to human habitation. Nests on ledge or hole in rock face.

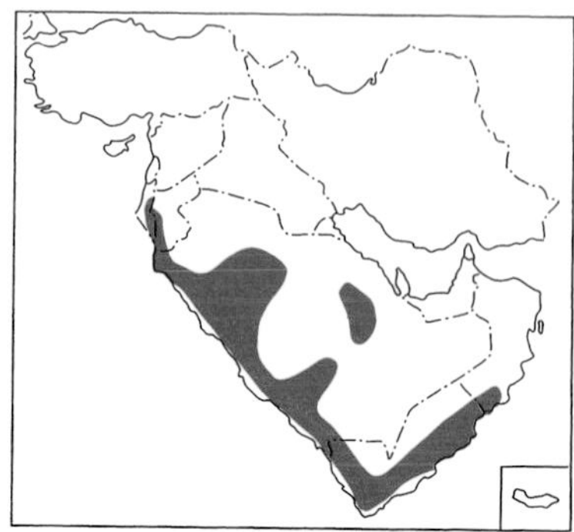

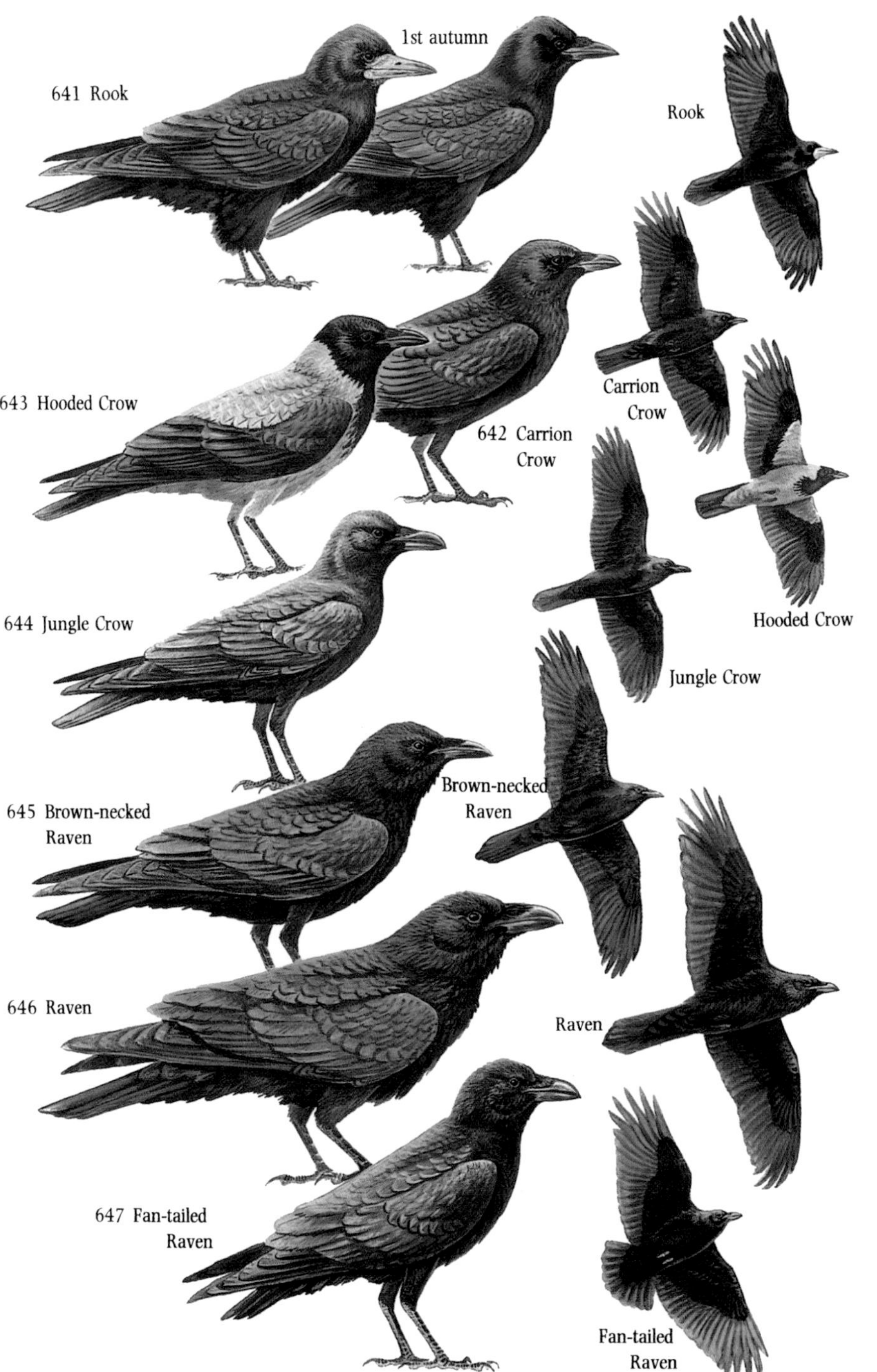
641 Rook
1st autumn
Rook
643 Hooded Crow
642 Carrion Crow
Carrion Crow
Hooded Crow
644 Jungle Crow
Jungle Crow
645 Brown-necked Raven
Brown-necked Raven
646 Raven
Raven
647 Fan-tailed Raven
Fan-tailed Raven
ML

648 **Socotra Starling** *Onychognathus frater*

Status: resident endemic on Socotra. **Habitat:** woodland and open plains with scattered trees.

649 **Somali Starling** *Onychognathus blythii*

Status: resident on Socotra. **Habitat:** open areas with cattle, scattered trees and patches of woodland.

650 **Tristram's Grackle** *Onychognathus tristramii*

Status: resident, with slight winter dispersal. **Habitat:** rocky hills and ravines, semi-deserts with acacia, also towns and villages. From sea level up to 3000 m. Nests among rocks.

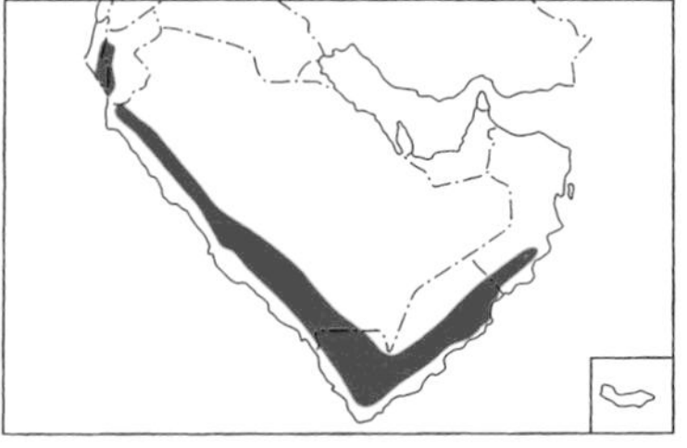

651 **Amethyst Starling** *Cinnyricinclus leucogaster*

Status: probably more widespread in Yemen. Summer visitor. Vagrant Israel. **Habitat:** plains, hills and wadis with trees and other vegetation, mainly between 500–2000 m. Nests in hole in tree.

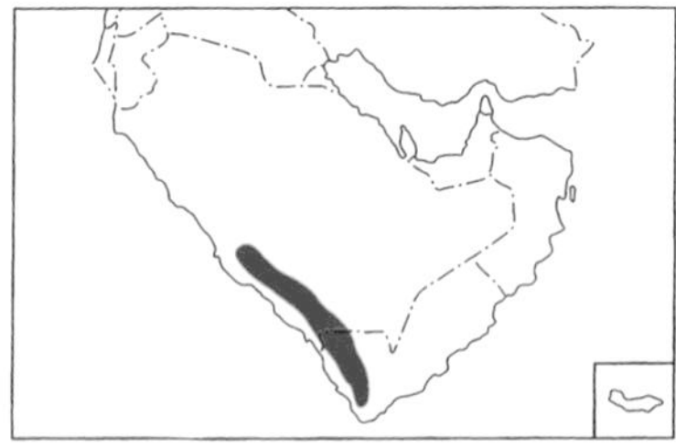

652 **Starling** *Sturnus vulgaris*

Status: resident and winter visitor; winters over most of the region but irregular S Iran and vagrant Yemen. **Habitat:** towns and villages, woods or old parks. In winter, also farmland; outside breeding season often roosts in reeds. Nests in hole in tree or building, or in nest-box.

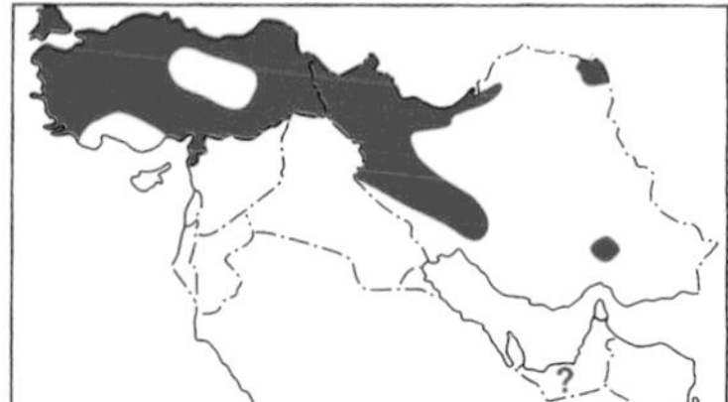

653 **Rose-coloured Starling** *Sturnus roseus*

Status: breeds erratically central and E Turkey, exceptionally west to Istanbul. Summer visitor. Passage Turkey, N and W Iran Syria and Iraq; scarce and irregular Cyprus, Near East, Arabian Gulf and Oman where also irregular in winter; vagrant rest of Arabia. **Habitat:** open regions, near agriculture, cliffs, slopes. Nests colonially in holes among stones on open ground, in walls or woodstacks.

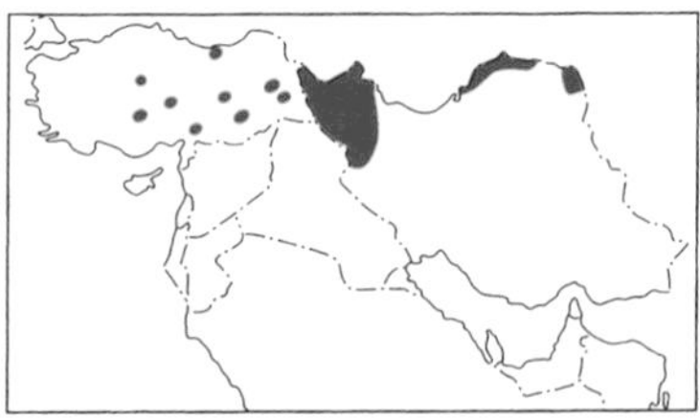

654 **Wattled Starling** *Creatophora cinerea*

Status: very scarce, irregular visitor to Oman and Yemen (mid-summer to winter); nomadic from Africa. **Habitat:** open bush and savanna.

655 **Pied Mynah** *Sturnus contra*

Status: recorded recently from several places in UAE; range Indian continent. **Habitat:** open, cultivated moist ground near human habitation, dumps, grassy fields.

656 **Bank Mynah** *Acridotheres ginginianus*

Status: breeding in UAE, probably from escaped birds. **Habitat:** towns, villages, fields, grassy areas; roosts in trees or reeds. Nests in hole in bank or masonry.

657 **Common Mynah** *Acridotheres tristis*

Status: resident Iran. Most colonies in Arabia suspected of originating from escapes. **Habitat:** as Bank Mynah. Nests in hole in tree, under eaves or similar.

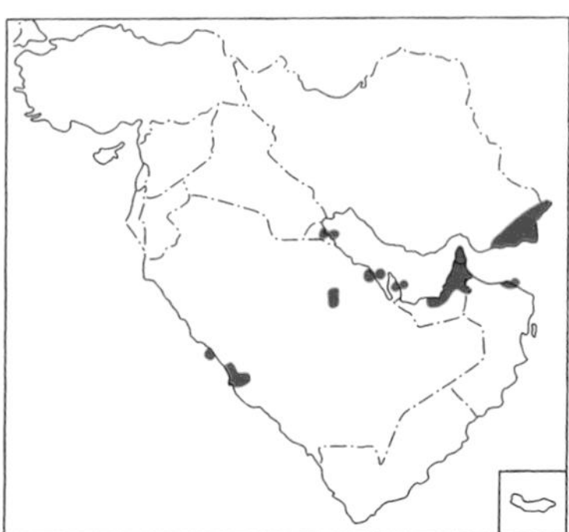

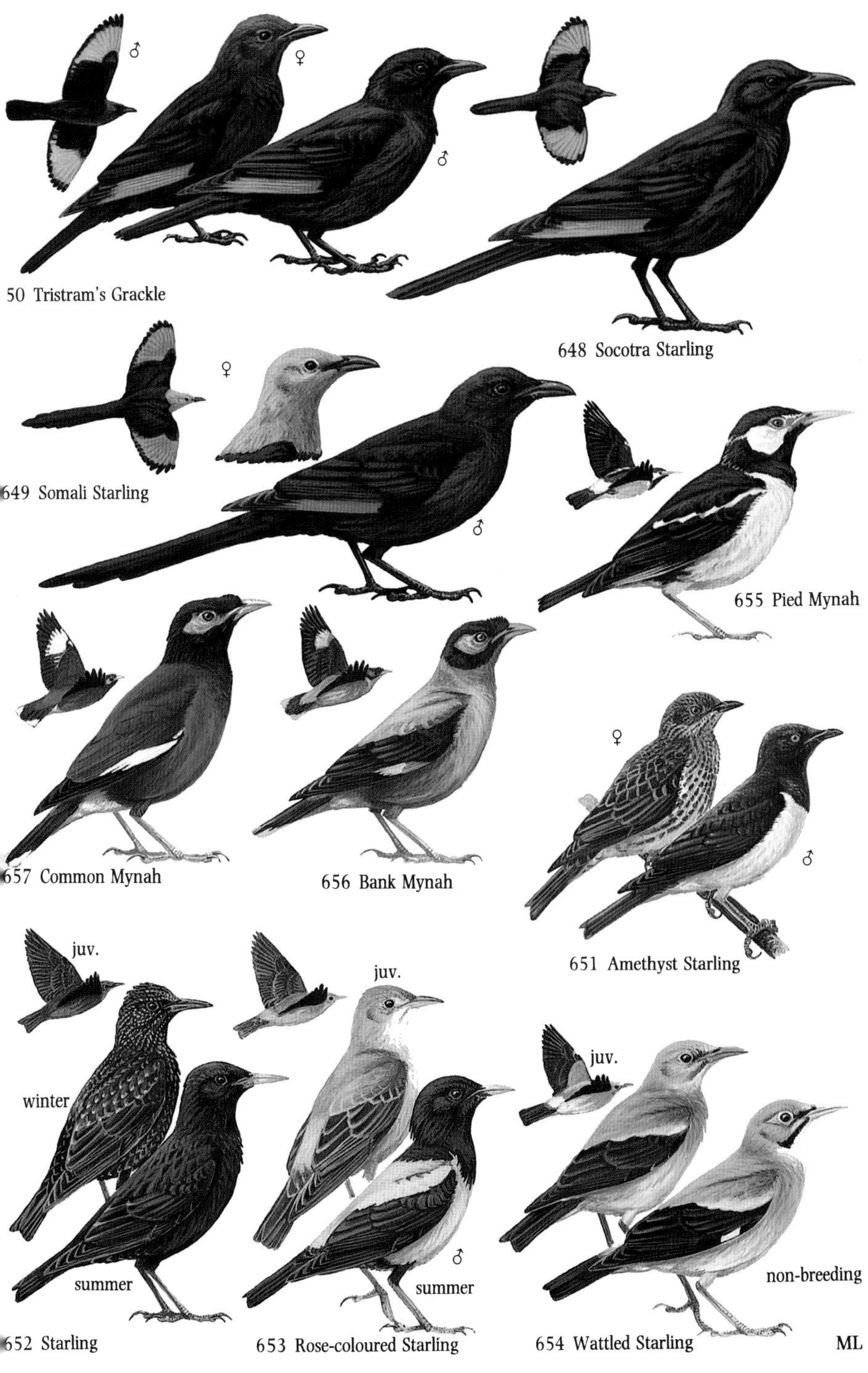
♂
♀
♂
50 Tristram's Grackle
648 Socotra Starling
♀
649 Somali Starling
♂
655 Pied Mynah
657 Common Mynah
656 Bank Mynah
♀
♂
651 Amethyst Starling
juv.
juv.
winter
juv.
♂
summer
summer
non-breeding
652 Starling
653 Rose-coloured Starling
654 Wattled Starling
ML

658 **Saxaul Sparrow** *Passer ammodendri*

Status: an old record from Iran. Range central Asia. **Habitat:** sandy or stoney desert with banks and rocky outcrops, river valleys, oases and large bushes, sometimes near human settlements.

659 **House Sparrow** *Passer domesticus*

Status: resident. **Habitat:** in or near human habitation, towns and villages, but also farmland. Nests in holes, crevice in building, but sometimes builds domed nest in tree.

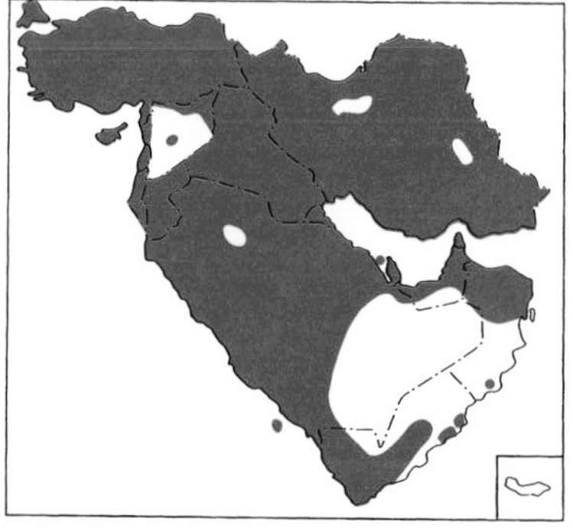

660 **Spanish Sparrow** *Passer hispaniolensis*

Status: may breed some years N Saudi Arabia. Summer visitor and resident Turkey, mainly resident elsewhere. Passage Turkey, Cyprus, N and E Iran; winters Turkey, S Iran, Iraq, Near East, NW Saudi Arabia; rare or vagrant in the Gulf States (where now breeding sporadically) and Oman. **Habitat:** like House Sparrow but prefers villages to towns; also woodland and palm groves. In winter, in open cultivation. Nests in foundation of large nests, hole in wall or builds domed nest in tree, often colonially.

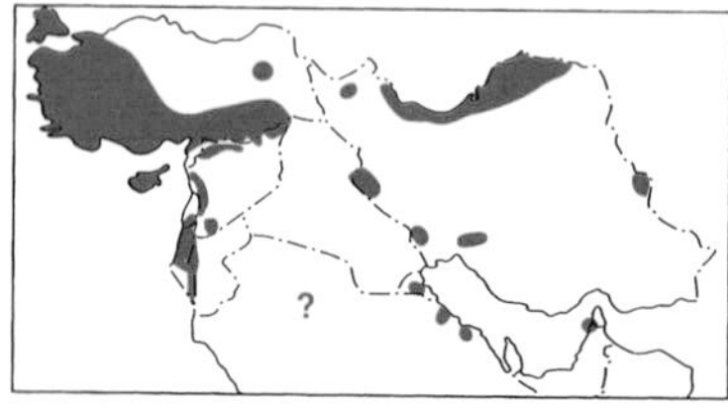

661 **Sind Jungle Sparrow** *Passer pyrrhonotus*

Status: resident. **Habitat:** rivers and swamps with tamarisk, acacia and thorn bush in which it nests.

662 **Dead Sea Sparrow** *Passer moabiticus*

Status: largely resident Iran and Iraq, largely migratory Turkey and Cyprus; Israeli birds partly dispersive or migratory; recent winter records on south side of Arabian Gulf, including UAE. **Habitat:** near water in thick scrub, tamarisk or poplars. Outside breeding season also in farmland. Nests in scrub, colonially.

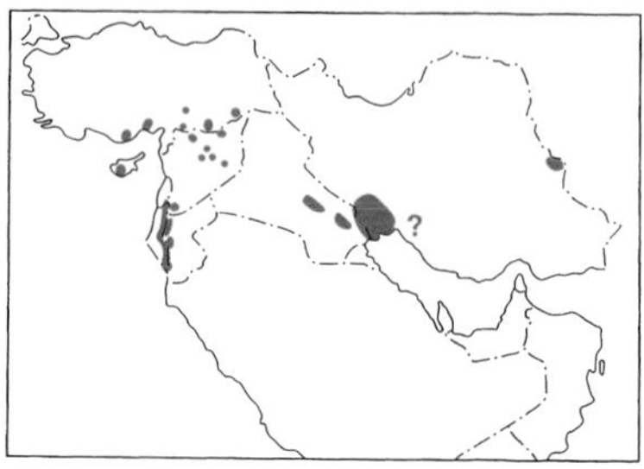

663 **Socotra Sparrow** *Passer insularis*

Status: resident endemic on Socotra. **Habitat:** trees, scrub, rock faces and human settlements.

664 **Tree Sparrow** *Passer montanus*

Status: resident, occasionally wandering to Iraq, Syria, Cyprus and Israel in winter; vagrant Gulf States. **Habitat:** parks, gardens or woodland often near habitation. Nests in hole in tree, crevice in house but sometimes builds domed nest in tree.

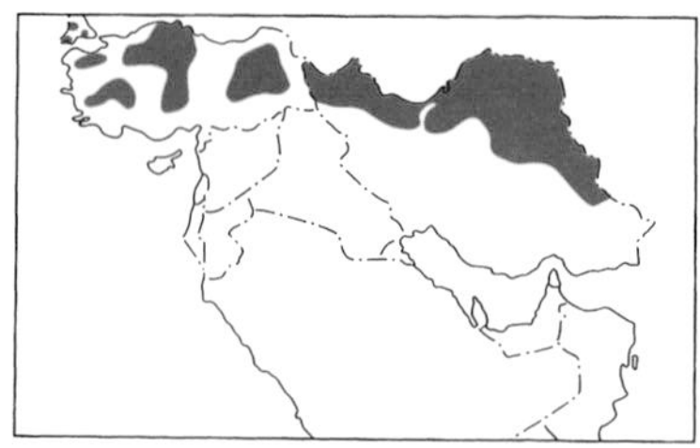

665 **Arabian Golden Sparrow** *Passer euchlorus*

Status: resident, but some local movements. **Habitat:** arid thorn-savanna and nearby cereal crops. Nests colonially in tree, typically acacia.

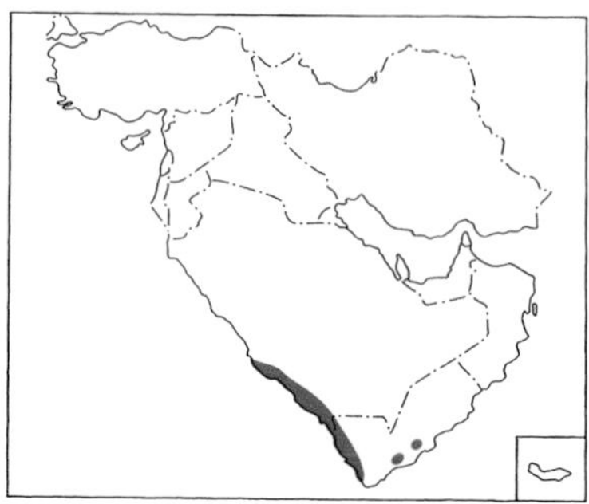

♀
♂ winter
♀
♂
658 Saxaul Sparrow
659 House Sparrow
♂
♀
hybrid ♂
Spanish x House
♀
660 Spanish Sparrow
♂
♂ winter
♂
661 Sind Jungle Sparrow
yatii ♀
♀
♂ winter
664 Tree Sparrow
♂
662 Dead Sea Sparrow
yatii ♂
♂ 1st winter
♀
♀
♂
♂
663 Socotra
Sparrow
665 Arabian Golden Sparrow
BS

666 **Pale Rock Sparrow**
Petronia brachydactyla

Status: recent breeding Jordan and UAE. Summer visitor; passage in variable numbers in SW and W Saudi Arabia, Jordan, UAE and Oman. Winter visitor W Saudi Arabia. Vagrant Yemen and Cyprus. **Habitat:** rocky and scrubby areas in moderate altitudes; in winter also plains, semi-deserts and cultivation. Nests in low bush.

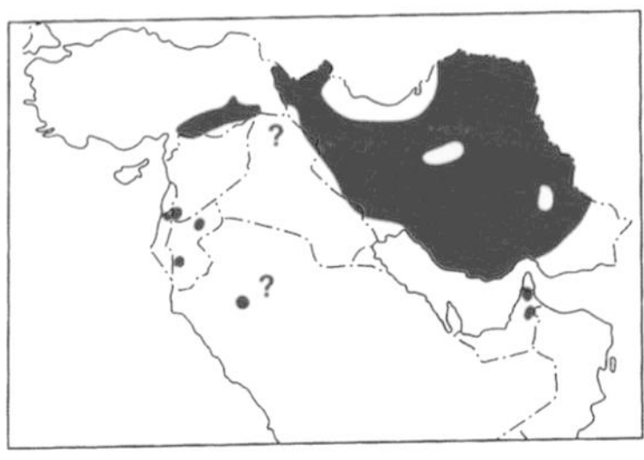

667 **Yellow-throated Sparrow**
Petronia xanthocollis

Status: summer visitor. Passage Kuwait and S Oman. Vagrant E Saudi Arabia, Bahrain and Israel. Occasional in winter Oman. **Habitat:** open dry woodland, date groves, villages, river bottoms with trees, cultivated areas. Nests in hole or crevice in tree.

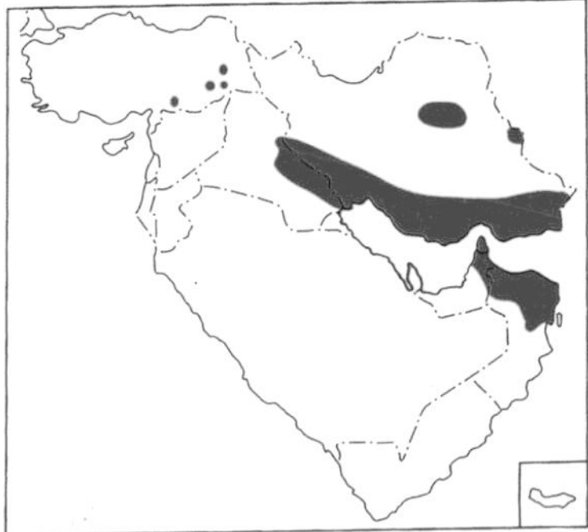

668 **Bush Petronia** *Petronia dentata*

Status: resident. **Habitat:** wadis and terrace slopes with scattered trees, 650–1700 m. Nests in hole or crevice in tree.

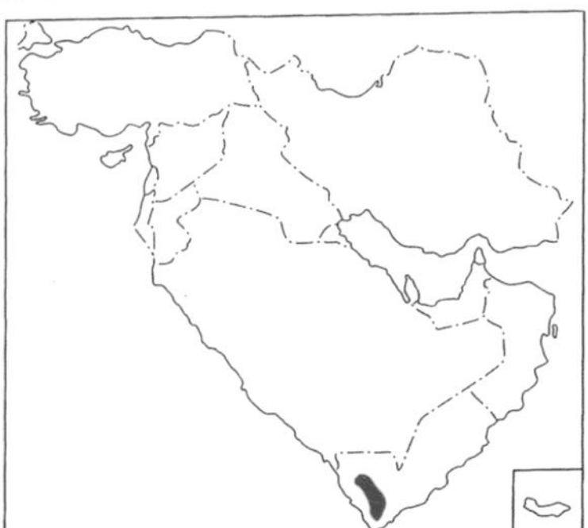

669 **Rock Sparrow** *Petronia petronia*

Status: resident but summer visitor to extreme NW Iran; in winter recorded E, SW Iran, Iraq. Vagrant Cyprus. **Habitat:** rocky mountain slopes, stony ground, ruins; also dry river beds and farmland; sometimes among trees. Nests in crevice in rock, building or tree.

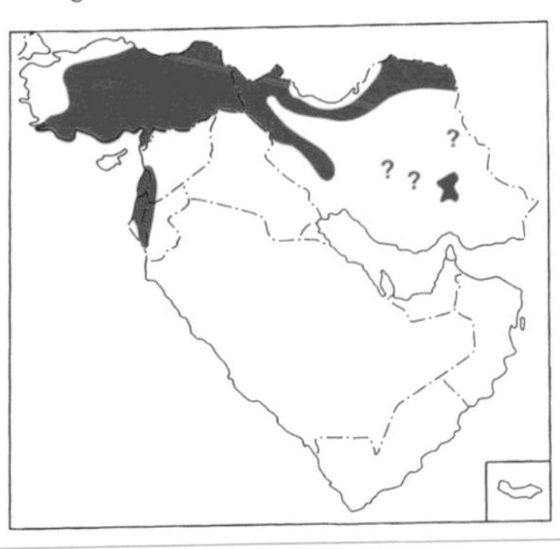

670 **Snow Finch** *Montifringilla nivalis*

Status: resident, with altitudinal movements in winter. **Habitat:** bare mountains above 2000 m, lower in winter; visits mountain huts and camps. Nests in rock crevice or wall.

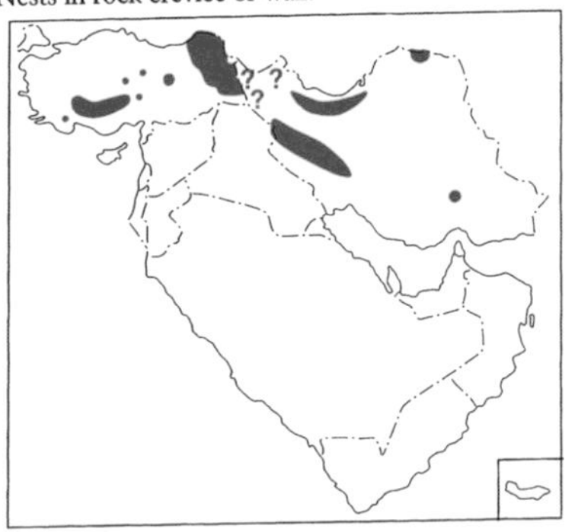

671 **Rüppell's Weaver** *Ploceus galbula*

Status: resident. **Habitat:** crops, palm groves, savanna, wadis with acacia and other bushes up to 2500 m, commonest at lower altitudes. Nest suspended from acacia or other large tree; colonial.

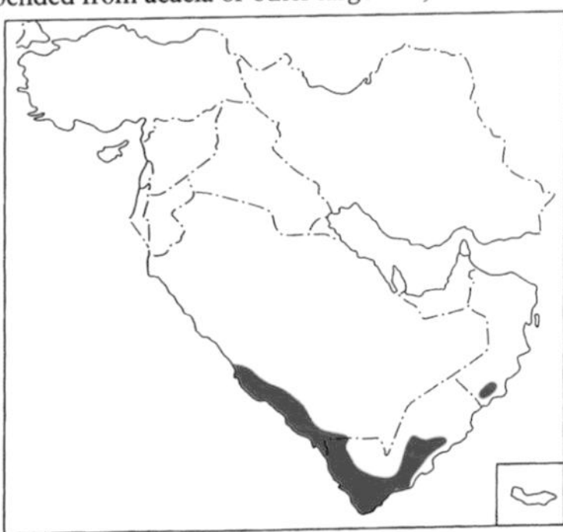

666 Pale Rock Sparrow
667 Yellow-throated Sparrow
♀
♂
♂
Pale Rock Sparrow
Yellow-throated Sparrow
669 Rock Sparrow
668 Bush Petronia
♂
♀
imm.
♀
♂
Bush Petronia ♀
Rock Sparrow
alpicola
♂ winter
alpicola ♂
♀
670 Snow Finch
♂
♂
671 Rüppell's Weaver
BS

672 **Arabian Waxbill** *Estrilda rufibarba*

Status: resident. **Habitat:** wadis and hillsides with scrub and often cultivation especially grain crops; 250–2400 m. Nest not described.

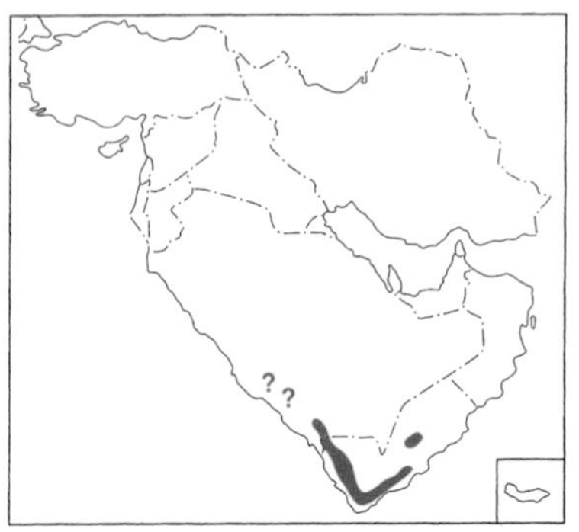

673 **Avadavat** *Amandava amandava*

Status: recorded Lebanon., Iran, Bahrain, UAE, Saudi Arabia, Oman and Israel: all originating from escapes. **Habitat:** varied, with trees or scrub; often near human habitation.

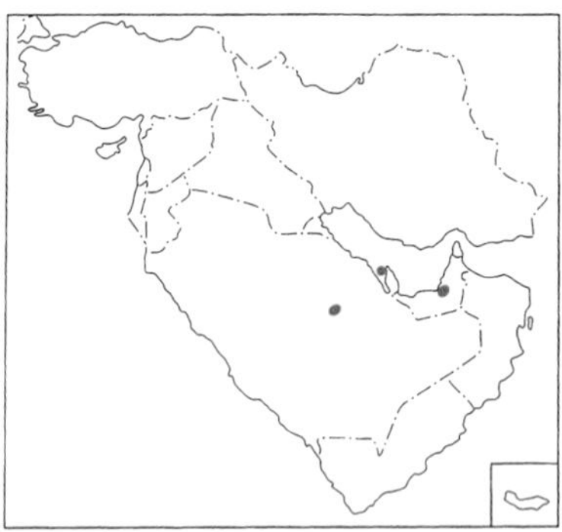

674 **Zebra Waxbill** *Amandava subflava*

Status: resident; has been recorded in central Saudi Arabia. **Habitat:** cereal fields, trees, scrub. Nests low in vegetation or in nest of other small bird.

675 **Indian Silverbill** *Euodice malabarica*

Status: resident; some dispersal away from breeding areas which will probably lead to further colonization and even hybridization with African Silverbill if the ranges merge. Recorded Jordan. **Habitat:** dry open grassland with scrub, cultivation or palm groves. Nest as African Silverbill.

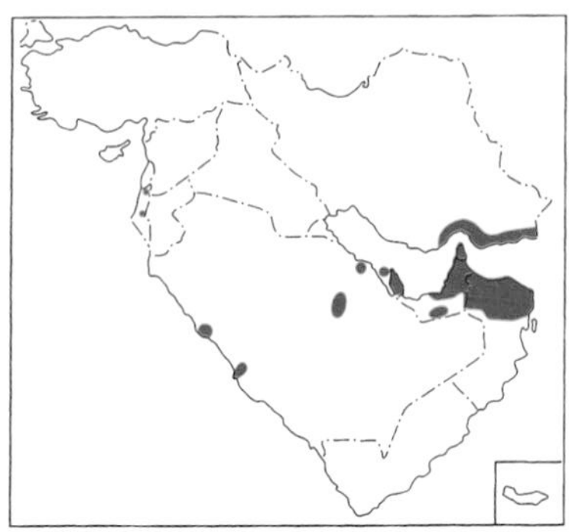

676 **African Silverbill** *Euodice cantans*

Status: resident. See also Indian Silverbill. Occasional records in Saudi Arabia and Gulf States may be from escapes. **Habitat:** as Indian Silverbill. Nests in bush, crevice or old nest of weaver; will also build suspended nest.

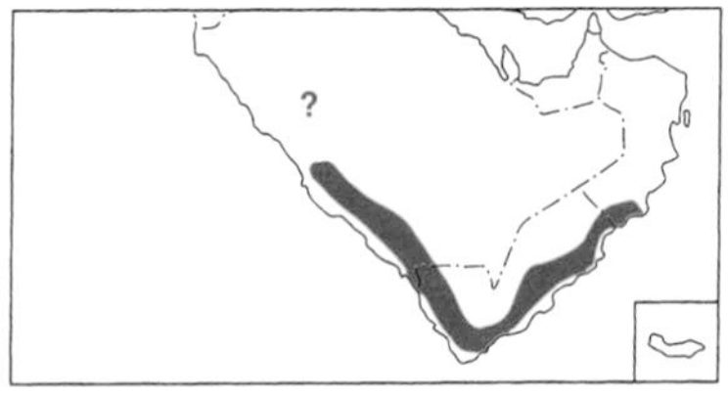

677 **Chaffinch** *Fringilla coelebs*

Status: resident; also winter visitor to Near East, Iraq and SW Iran; irregular Kuwait, has become regular E Saudi Arabia. Vagrant UAE. **Habitat:** forests, orchards, thickets, etc. In winter, often open fields. Nests in tree or bush.

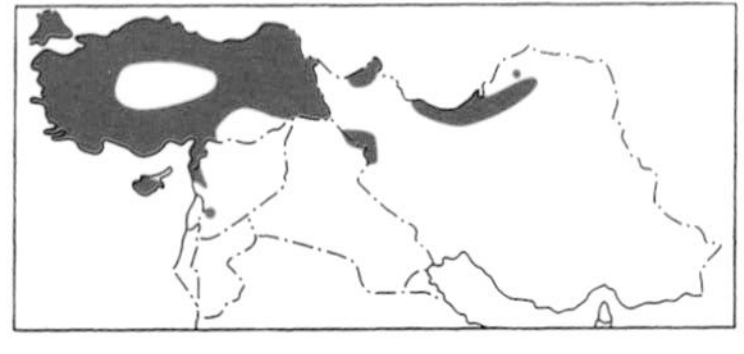

678 **Brambling** *Fringilla montifringilla*

Status: winter visitor to Turkey, Syria, Iraq, W and N Iran, Near East, Cyprus (scarce), becoming regular E province Saudi Arabia; rare UAE, Jordan and Oman. **Habitat:** fields or other cultivation and often amongst trees.

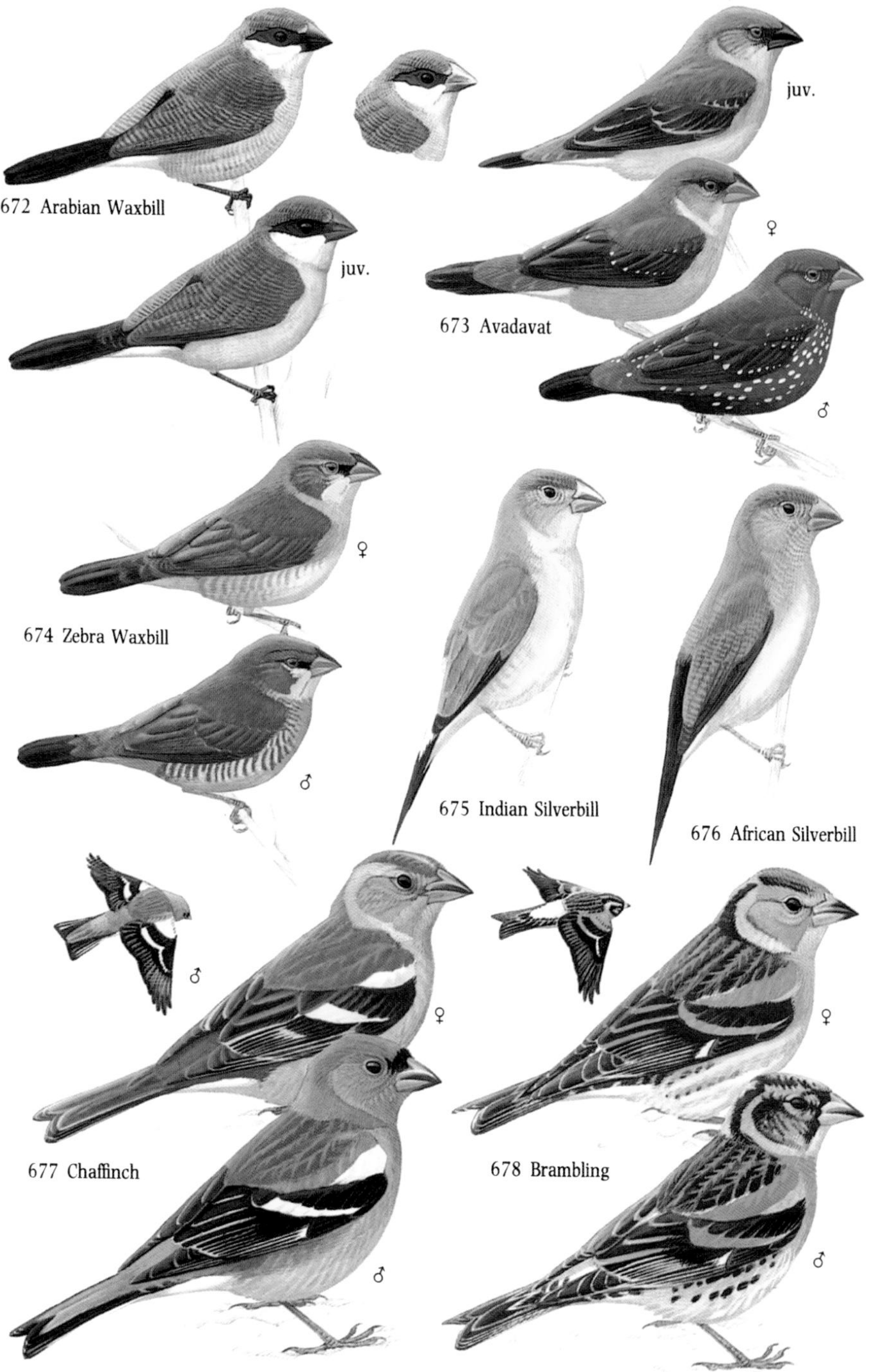
672 Arabian Waxbill
juv.
673 Avadavat
juv.
♀
♂
674 Zebra Waxbill
♀
♂
675 Indian Silverbill
676 African Silverbill
677 Chaffinch
♂
♀
♂
678 Brambling
♀
♂

BS

679 **Red-fronted Serin** *Serinus pusillus*

Status: resident, but descending to lower altitudes in winter and occasionally reaching Iraq, Syria, Lebanon, Israel and, rarely, Cyprus. Vagrant Jordan. **Habitat:** rocky mountains (usually 2000–4000 m) with open grassy areas and junipers, conifers, willows and birches; lower in winter. Nests in tree.

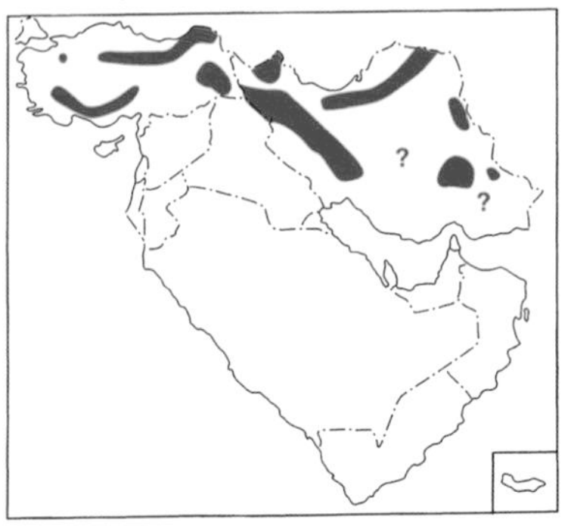

680 **European Serin** *Serinus serinus*

Status: mainly resident but dispersal in winter to Cyprus and Near East. Vagrant W Iran. **Habitat:** woodland edges, thickets, gardens, orchards; in more open country in winter. Nests in tree or bush.

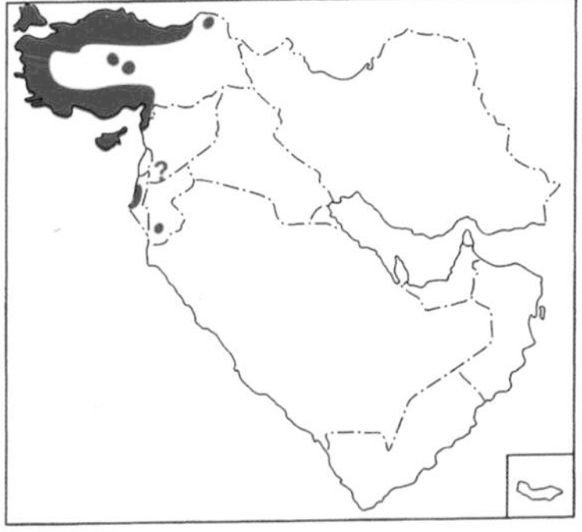

681 **Syrian Serin** *Serinus syriacus*

Status: mainly resident but extends to S Israel in winter; probably breeds W Syrian coast; has also been recorded in N Iraq. **Habitat:** mountains with trees (deciduous, cedars, junipers); more open areas outside breeding season. Nests in tree.

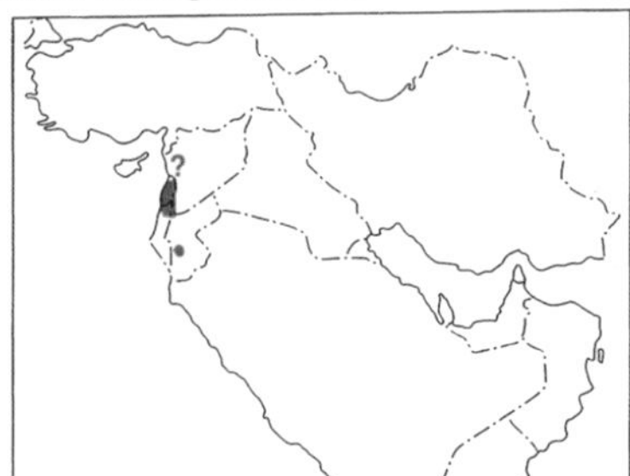

682 **Arabian Serin** *Serinus rothschildi*

Status: resident. **Habitat:** rocky hills and wadis with trees and bushes; 1500–2600 m. Nests in tree.

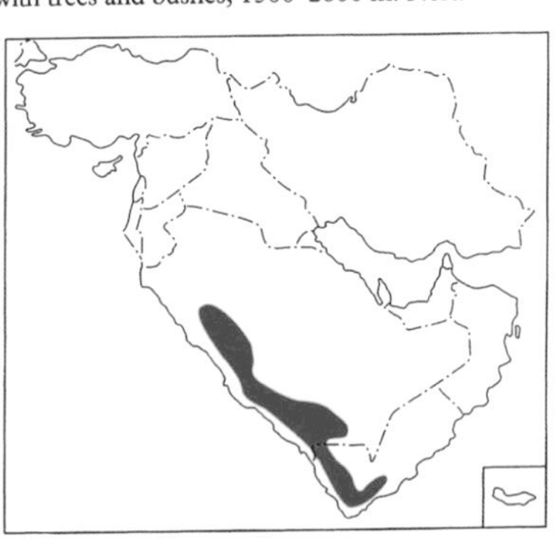

683 **Yemen Serin** *Serinus menachensis*

Status: resident. **Habitat:** high plateaux and rocky hillsides; villages, towns; trees not essential; 2000–3200 m. Nests in hole in rock or wall.

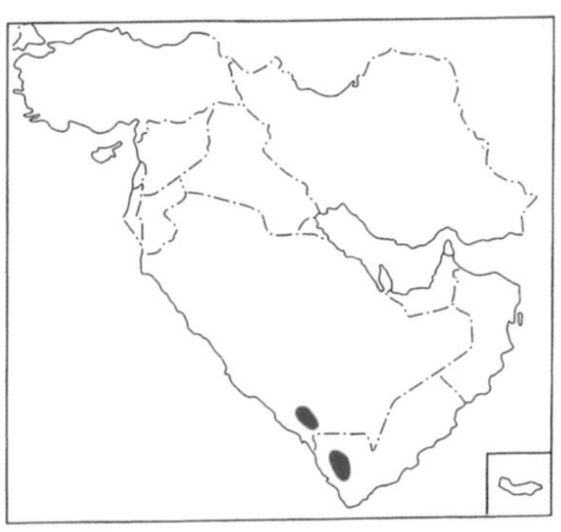

684 **Golden-winged Grosbeak**
Rhynchostruthus socotranus

Status: resident. **Habitat:** hills and wadis with euphorbias, acacias, junipers and other fruit-bearing trees, 250–2000 m. Nest undescribed.

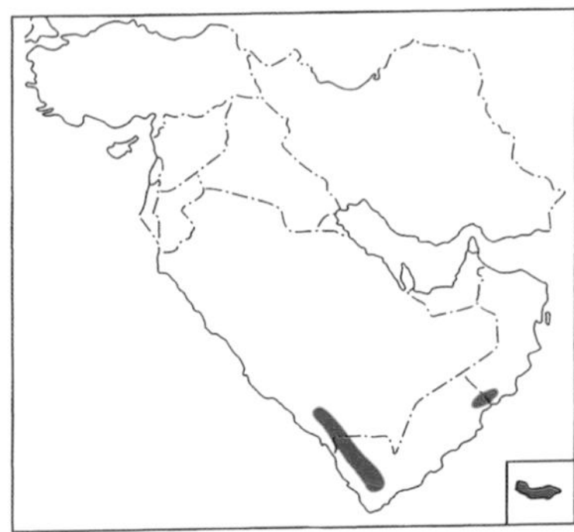

1st summer
1st winter
679 Red-fronted Serin
♀
juv.
680 European Serin
♂
♀
1st winter
682 Arabian Serin
681 Syrian Serin
♂
683 Yemen Serin
♀
juv.
♂
socotranus
♂
684 Golden-winged Grosbeak
BS

685 **Greenfinch** *Carduelis chloris*

Status: mainly resident but dispersive; winter visitor from Near East to SW Iran; uncommon NW Saudi Arabia, unrecorded rest of Arabia. **Habitat:** gardens, parks, pine woods, olive plantations, oases; in winter also open farmland and coastal areas. Nests in bush or small tree, particularly evergreen.

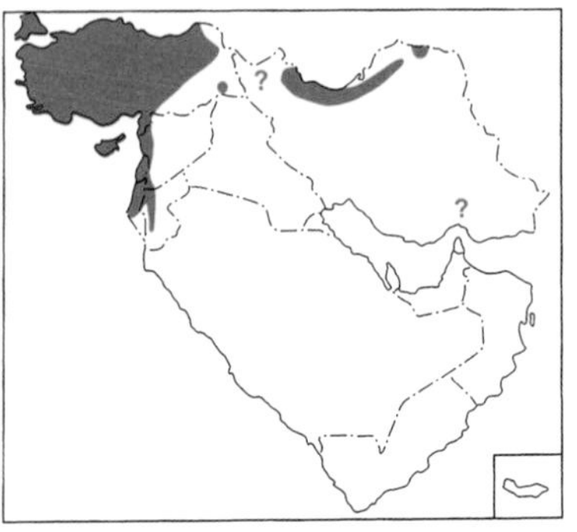

686 **Goldfinch** *Carduelis carduelis*

Status: mainly resident but dispersive; also winter visitor Cyprus, Near East to S Iraq; reaches NW Saudi Arabia; rare Kuwait; vagrant Bahrain and E Arabia. **Habitat:** gardens, orchards, cultivated areas, scrub or tree-covered hills and valleys, up to 2200 m. In winter often open country with herb vegetation, wastelands. Nests in tree.

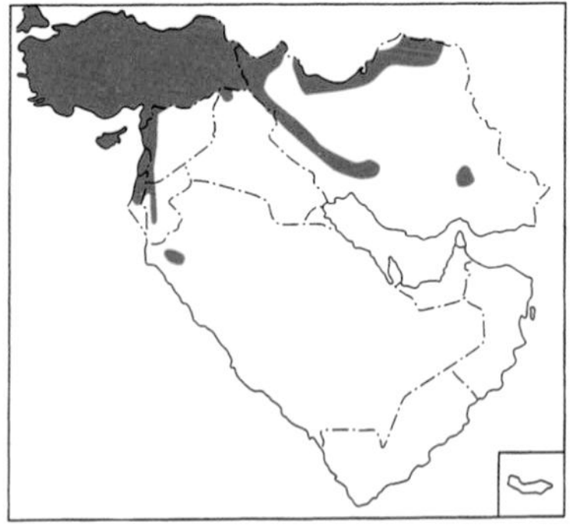

687 **Siskin** *Carduelis spinus*

Status: resident; also winter visitor Cyprus, Near East to N Iran; rare S Iran, Jordan, E, N and central Arabia; vagrant Syria, Qatar and Oman. **Habitat:** coniferous woods, birches with alder thickets; in winter often in open cultivated areas with trees. Nests high in conifer.

688 **Linnet** *Carduelis cannabina*

Status: resident; also winter visitor to Near East and SW Iran; rarely Kuwait and E Saudi Arabia; vagrant UAE. **Habitat:** open country with thickets and plantations; in winter in wastelands, cultivation, marshes and coastal areas. Nests sociably in bushy thickets, vines or low scrub.

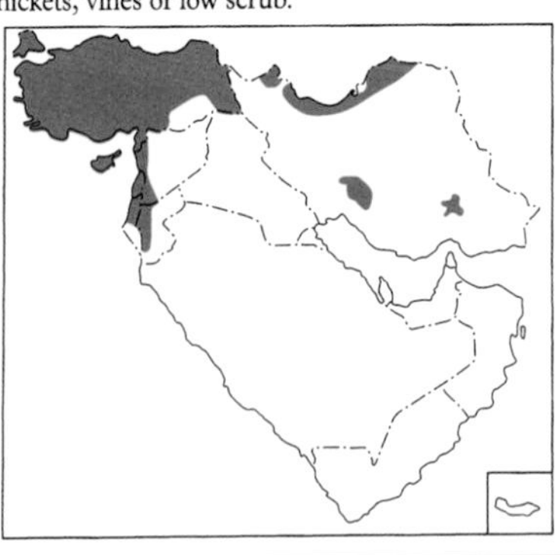

689 **Yemen Linnet** *Carduelis yemenensis*

Status: resident. **Habitat:** lightly wooded slopes and wadis, cultivation with scattered trees (e.g. juniper, acacia) from 1700 to over 3000 m. Nests in tree, semi-colonial.

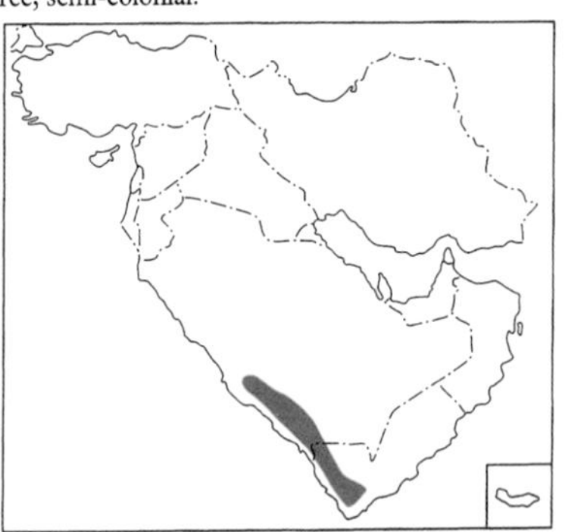

690 **Twite** *Carduelis flavirostris*

Status: descends and disperses from high breeding grounds in winter. Vagrant SE Iran. **Habitat:** mountain slopes; also plains in winter. Nests on ground under cover of vegetation or rock.

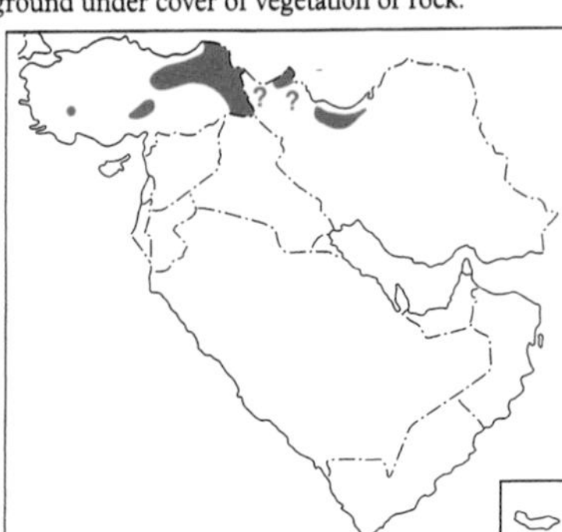

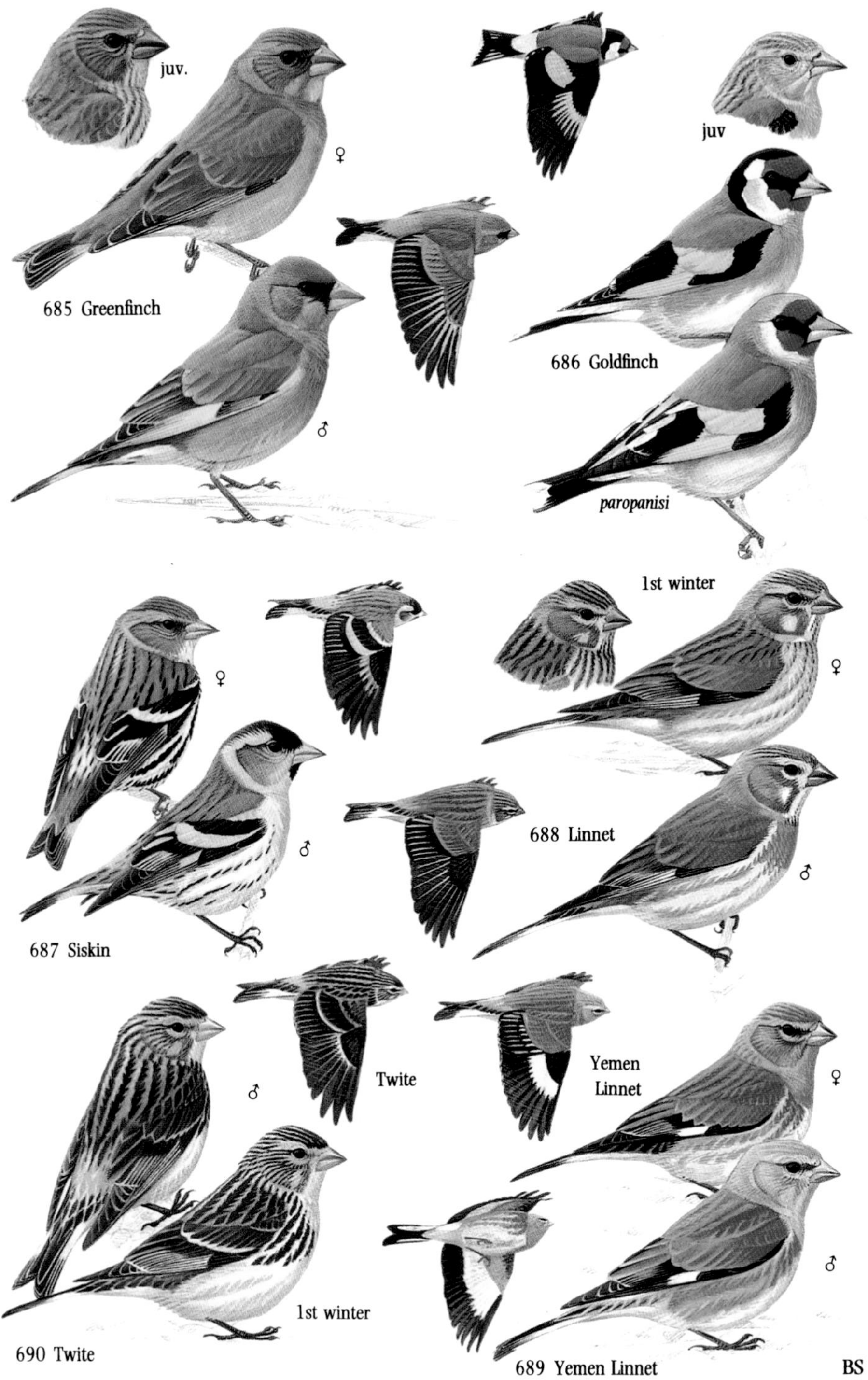
juv.
♀
685 Greenfinch
♂
juv
686 Goldfinch
paropanisi
1st winter
♀
688 Linnet
♂
♀
♂
687 Siskin
Twite
Yemen
Linnet
♂
♀
1st winter
♂
690 Twite
689 Yemen Linnet
BS

691 **Redpoll** *Carduelis flammea*

Status: vagrant Turkey, Cyprus and perhaps Israel. **Habitat:** lightly wooded regions, near cultivation and wasteland.

692 **Crossbill** *Loxia curvirostra*

Status: has bred Israel in recent years. Resident and dispersive; in Turkey also irruptive, and recorded on passage. Vagrant Iran. **Habitat:** coniferous woods in which it nests.

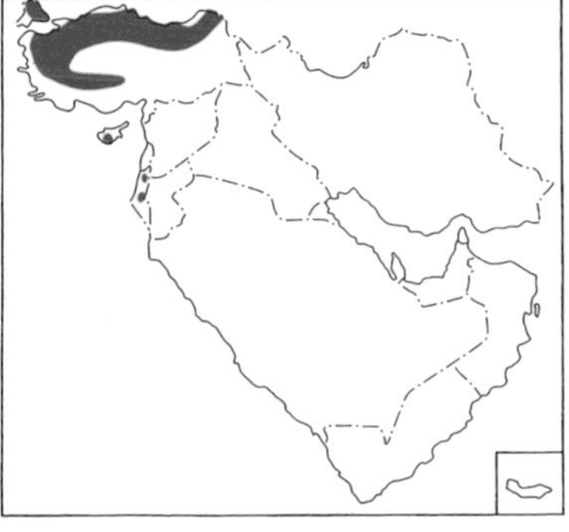

693 **Crimson-winged Finch**
Rhodopechys sanguinea

Status: resident, but moving to lower altitudes in winter. **Habitat:** rocky mountains and slopes, sometimes with scrubby areas usually above 2000 m; at lower altitudes in winter, including cultivated fields or stony grassland. Nests under stone or bush.

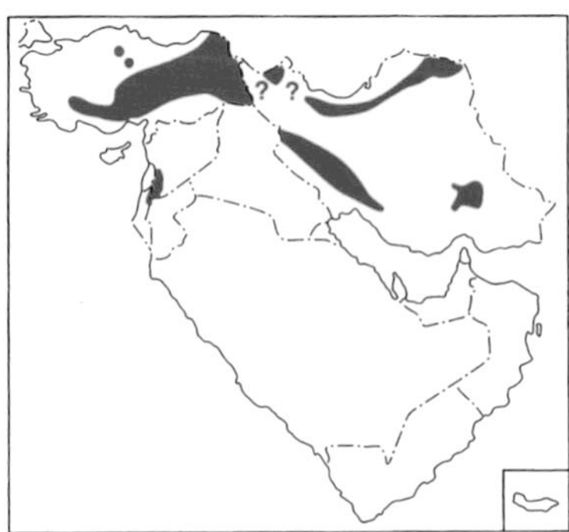

694 **Desert Finch** *Rhodospiza obsoleta*

Status: recent colonization of central Arabia. Mainly resident, some wandering in winter when recorded in E Iraq. **Habitat:** open country with trees and bushes, orchards and vineyards; in winter, also in cultivation with trees nearby. Nests in small tree; loosely colonial.

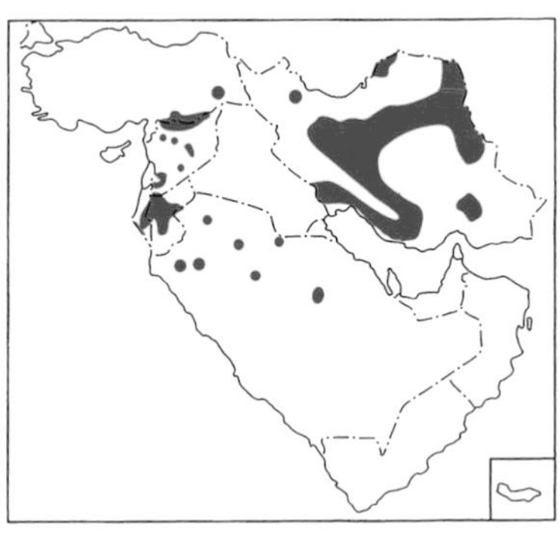

695 **Mongolian Trumpeter Finch**
Bucanetes mongolicus

Status: resident or partial migrant with altitudinal movements in winter. Vagrant Bahrain. **Habitat:** mountains, from 2700 m upwards, in steep rocky sites; in winter at lower altitudes to 1500 m, including desert plains. Nests on ground, in tussock or bush.

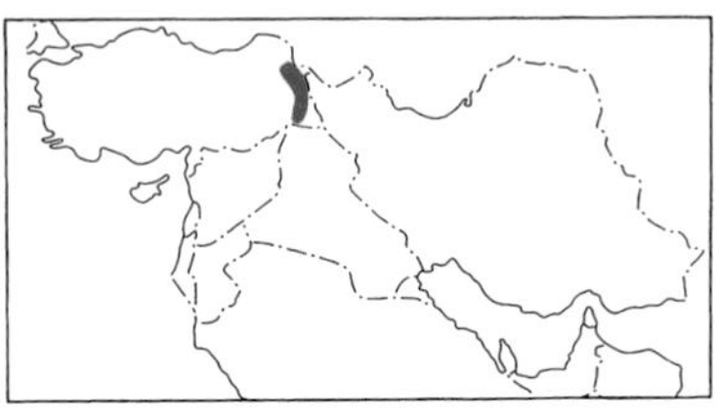

696 **Trumpeter Finch** *Bucanetes githagineus*

Status: mainly resident, but dispersive outside breeding season. Rare Cyprus. **Habitat:** bare rocky and stony hill-sides, plains and wild arid wadis, stony deserts and semi-deserts. Nests amongst rocks or under clump.

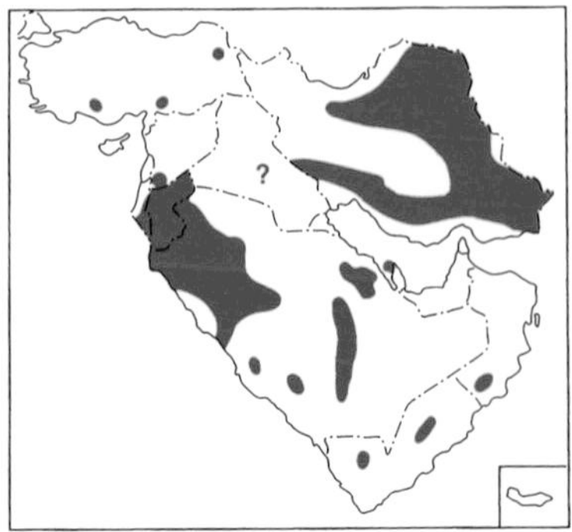

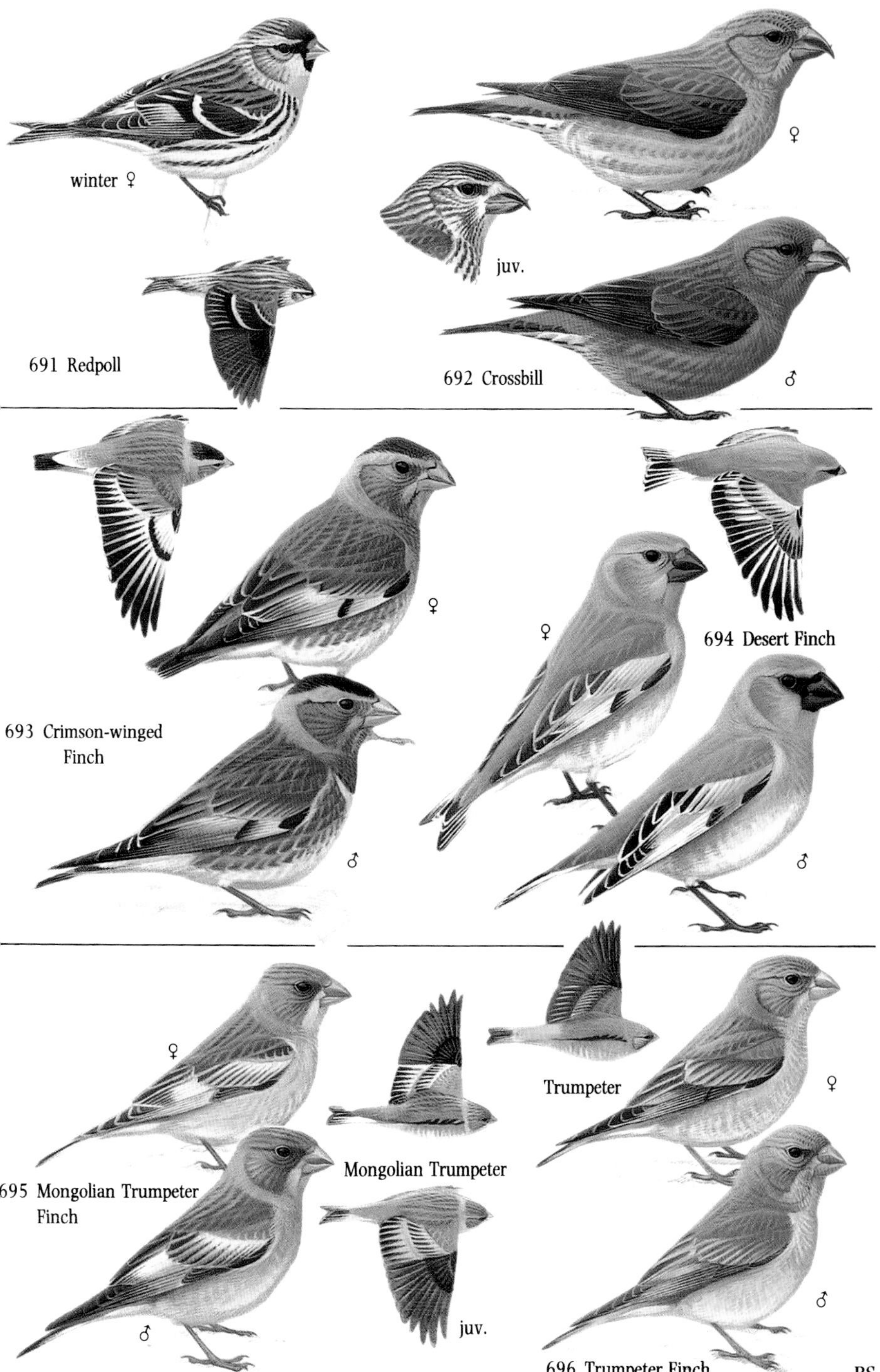
winter ♀
691 Redpoll
♀
juv.
692 Crossbill
♂
♀
693 Crimson-winged
Finch
♂
♀
694 Desert Finch
♂
♀
Trumpeter
♀
Mongolian Trumpeter
695 Mongolian Trumpeter
Finch
♂
juv.
♂
696 Trumpeter Finch
BS

697 **Common Rosefinch**
Carpodacus erythrinus

Status: summer visitor. On passage occurs Iran and Oman but scarce UAE and Bahrain. Vagrant E Arabia, Cyprus and Israel. **Habitat:** woodlands in mountains, and valley bottoms, especially near water. On passage anywhere with trees. Nests low in bush.

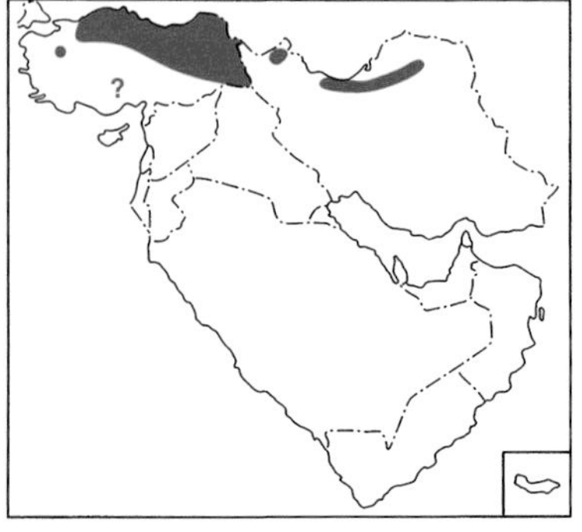

698 **Sinai Rosefinch** *Carpodacus synoicus*

Status: resident, but moves to lower ground in winter. **Habitat:** rocky hills and cliffs up to 2000 m. Nests in hole in rock face.

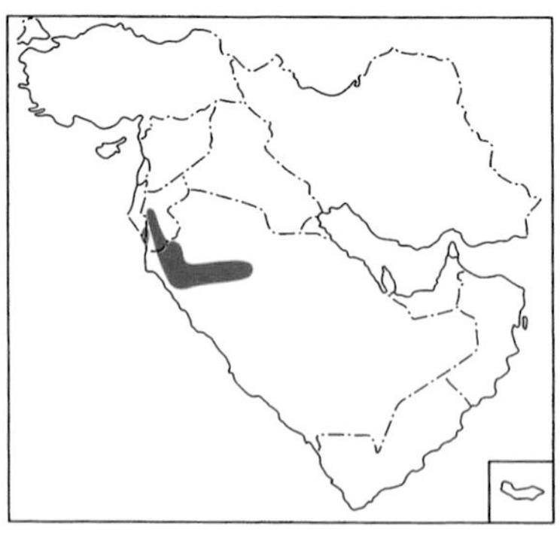

699 **Great Rosefinch** *Carpodacus rubicilla*

Status: status uncertain; may occur N Turkey and N Iran but records unconfirmed. **Habitat:** descends in winter from high alpine zones to lower altitudes.

700 **Bullfinch** *Pyrrhula pyrrhula*

Status: mainly resident. **Habitat:** woodlands particularly coniferous and orchards. Nests in bush or similar vegetation.

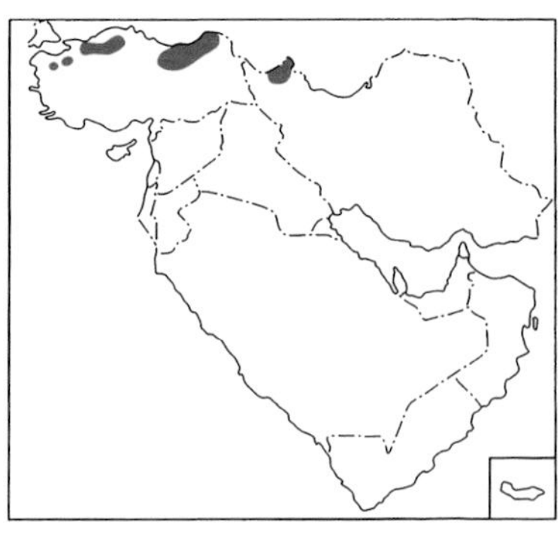

701 **White-winged Grosbeak**
Mycerobas carnipes

Status: resident. **Habitat:** mountains with juniper woodland. Nests in juniper.

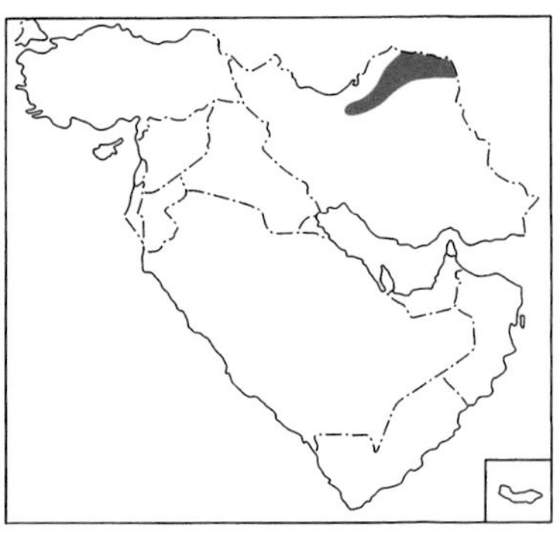

702 **Hawfinch** *Coccothraustes coccothraustes*

Status: mainly resident, but dispersive in winter when occasionally occurs Cyprus, Lebanon, Israel, Jordan. Vagrant Saudi Arabia and Syria. **Habitat:** dense and light mixed woodland. Nests in tree often in loose associations.

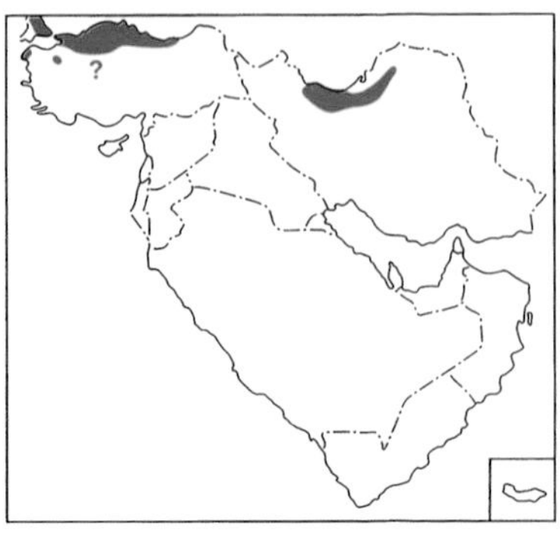

♀
♀
697 Common Rosefinch
698 Sinai Rosefinch
♂
♂
♀
♀
699 Great Rosefinch
700 Bullfinch
♂
♂
♀
♂
701 White-winged Grosbeak
♂
702 Hawfinch

704 **Pine Bunting** *Emberiza leucocephalos*

Status: winter visitor Iran; scarce Israel. Vagrant Cyprus, Iraq, Jordan, Syria, Turkey, UAE and E Saudi Arabia. **Habitat:** as for Yellowhammer.

705 **Yellowhammer** *Emberiza citrinella*

Status: may breed W and extreme NE Turkey. Winter visitor Iran, Turkey, Syria; scarce Cyprus, Near East, Iraq, Jordan. Vagrant UAE and Bahrain. **Habitat:** farmland, open hilly country with bushes; often along roads.

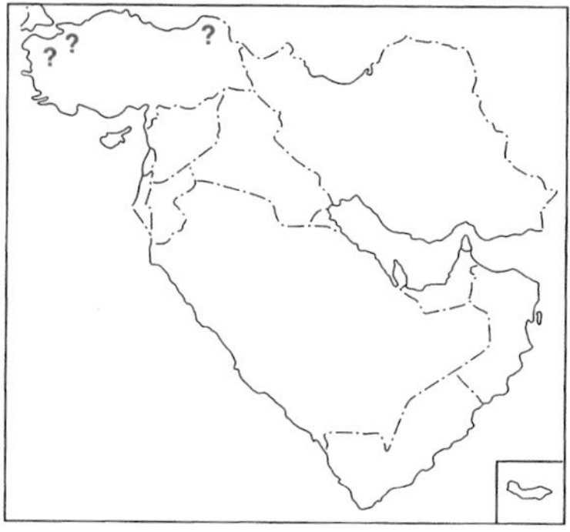

706 **Cirl Bunting** *Emberiza cirlus*

Status: mainly resident, wandering slightly in winter. Vagrant SW Iran. **Habitat:** scrubby broken country with scattered trees and scrub; in winter also farmland. Nests in low bush or tree.

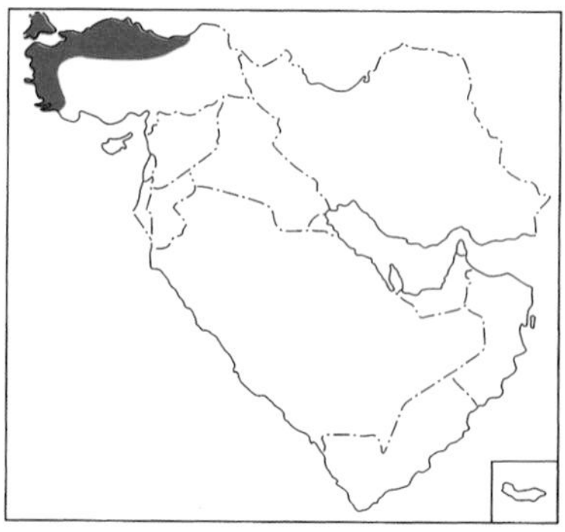

707 **White-capped Bunting** *Emberiza stewarti*

Status: vagrant from Asia to E Iran and UAE. **Habitat:** in winter in open dry hilly country with scattered bushes.

708 **Rock Bunting** *Emberiza cia*

Status: resident, but wanders to lower levels in winter. Rare winter visitor Cyprus, N Jordan, Iraq. Vagrant Kuwait and Saudi Arabia. **Habitat:** rocky and bushy slopes in mountains with or without scattered trees; in winter at more sheltered areas. Nests on or near ground.

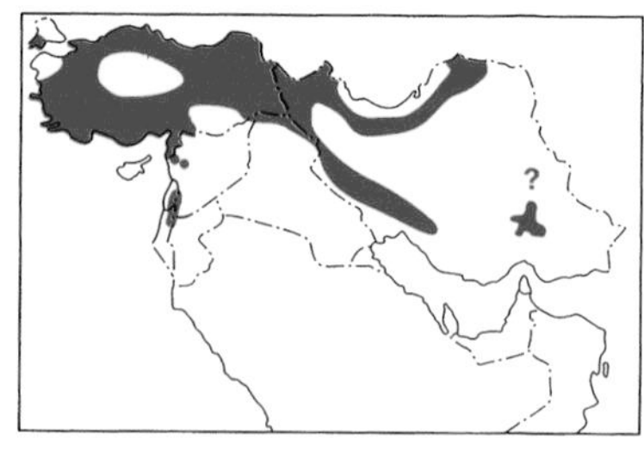

709 **House Bunting** *Emberiza striolata*

Status: resident; some seasonal movements occur. Vagrant Kuwait, and possible record from Turkey. **Habitat:** desert oases, wild desolate rocky wadis with little vegetation. Nests in hole in building, crack in wall or rock crevice.

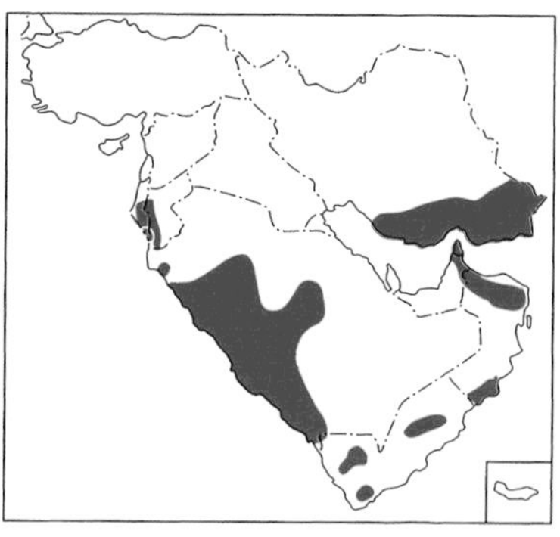

710 **African Rock Bunting** *Emberiza tahapisi*

Status: resident, but seasonal movements Oman and SW Saudi Arabia. **Habitat:** dry rocky and stony hill-sides with scattered vegetation: 300–2500 m. Nests on ground at base of stone or bush.

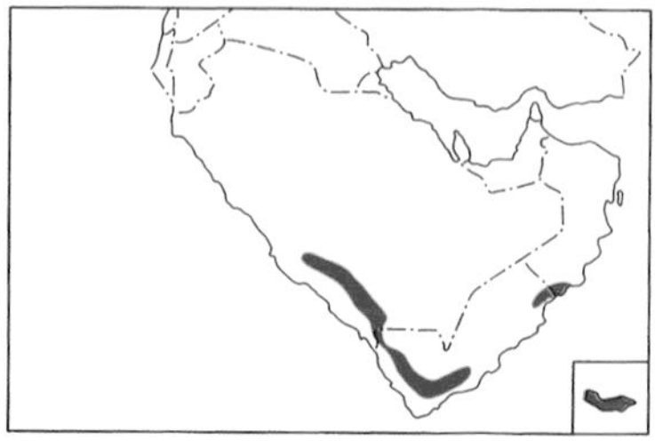

711 **Socotra Bunting** *Emberiza socotrana*

Status: resident endemic on Socotra, but scarce. **Habitat:** mountains descending at times to high grassy plateaux. Nest undescribed.

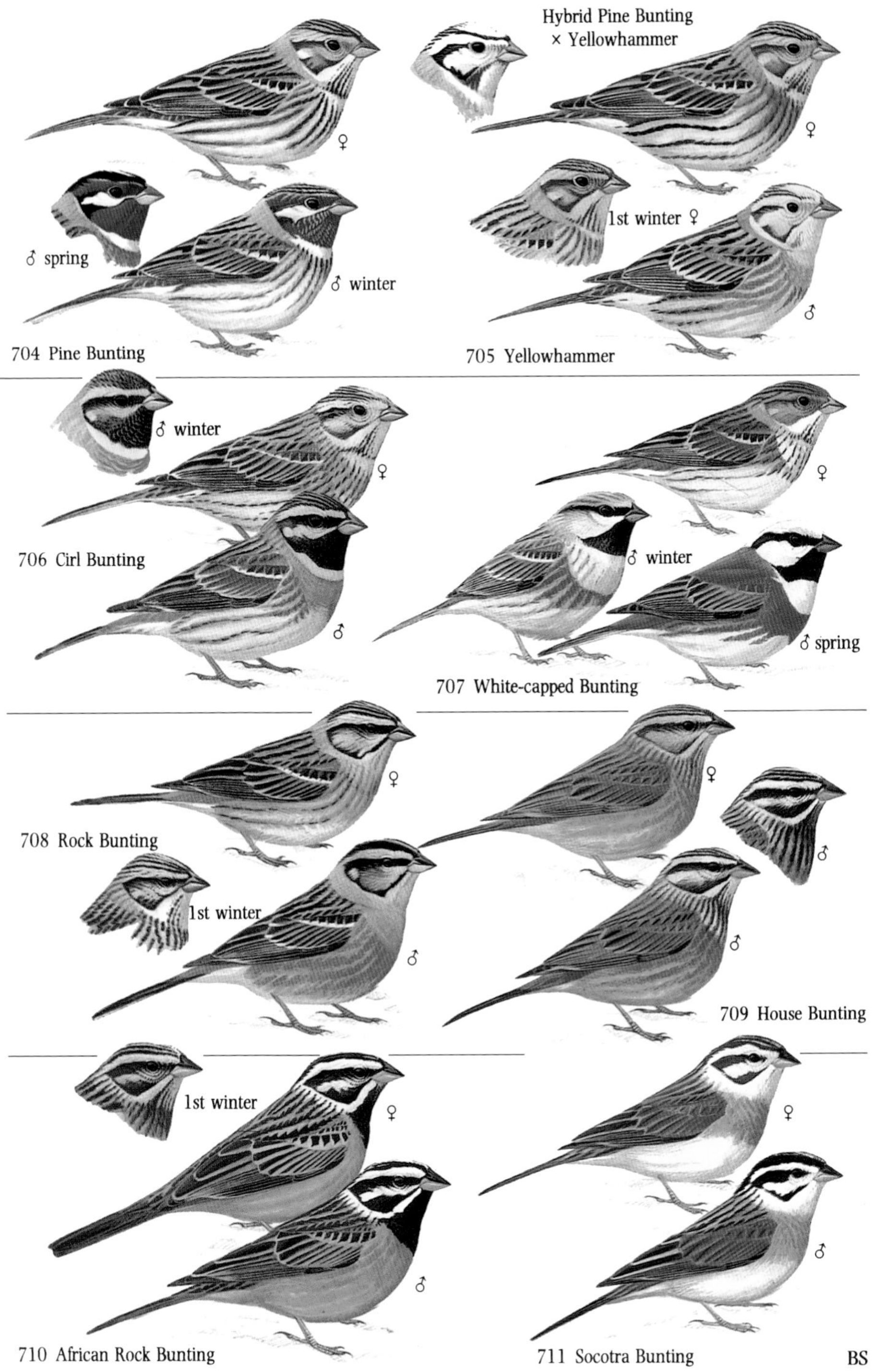
Hybrid Pine Bunting
× Yellowhammer
♀
♂ spring
♂ winter
1st winter ♀
♂
704 Pine Bunting
705 Yellowhammer
♂ winter
♀
706 Cirl Bunting
♂ winter
♂
♂ spring
707 White-capped Bunting
♀
708 Rock Bunting
1st winter
♂
709 House Bunting
1st winter
710 African Rock Bunting
711 Socotra Bunting
BS

712 **Cinereous Bunting** *Emberiza cineracea*

Status: summer visitor. Very scarce or rare migrant in Arabia, Cyprus, Israel, Jordan, Syria and Iraq. Vagrant Lebanon, Oman and Qatar. **Habitat:** dry rocky stony slopes with sparse vegetation up to tree-limit; on passage/winter also semi-deserts or bushy wadis. Nests on ground.

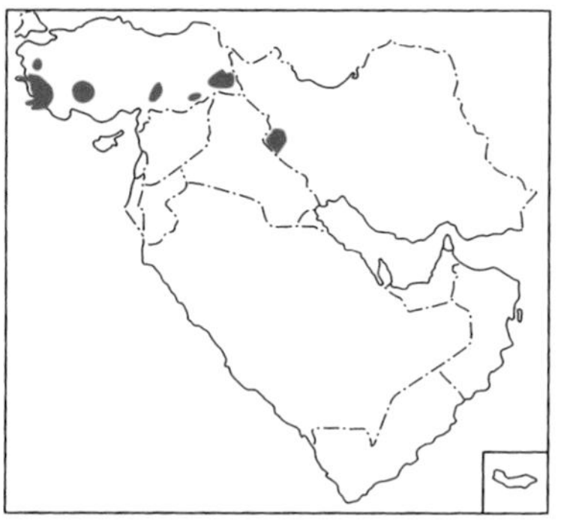

713 **Ortolan Bunting** *Emberiza hortulana*

Status: summer visitor. Widespread on passage, scarcer NW and central Arabia, S Iran and Oman; rarely in winter Oman and Yemen. **Habitat:** broken lowland or hilly country, bare cultivation with scattered trees and scrub. On passage also semi-deserts. Nests on or near ground in vegetation.

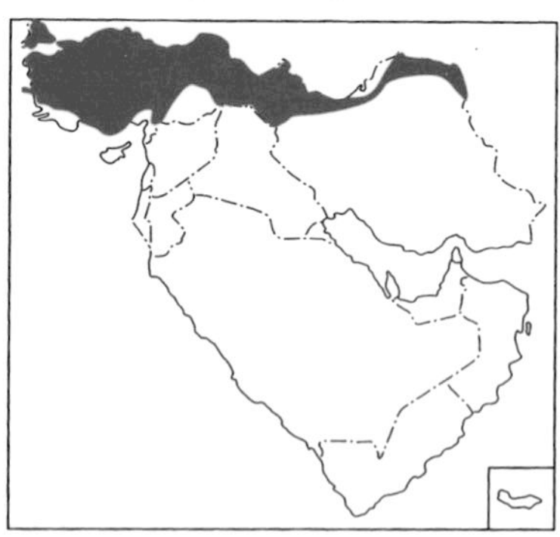

714 **Grey-necked Bunting** *Emberiza buchanani*

Status: summer visitor; passage SE Iran. Vagrant Syria. **Habitat:** rocky barren mountain hillsides and plateaux with sparse vegetation, mainly above 2000 m. Nests on ground.

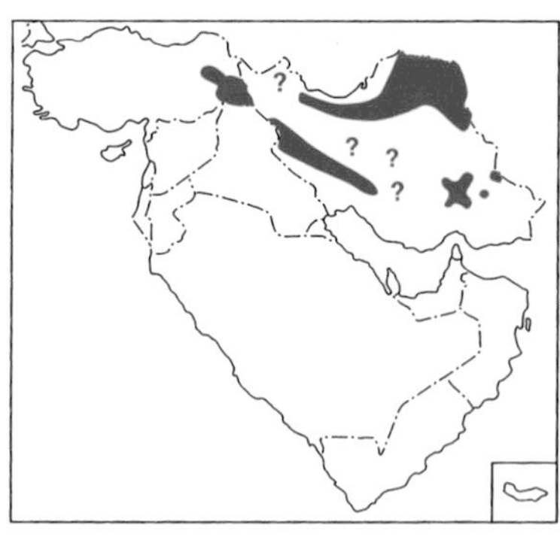

715 **Cretzschmar's Bunting** *Emberiza caesia*

Status: summer visitor; passage Cyprus, Near East and NW Saudi Arabia, rarer SW Saudi Arabia; winter records from W Saudi Arabia. Vagrant Iran, Oman and Kuwait. **Habitat:** bare rocky hill-sides, sometimes with bushes. On passage also cultivation, semi-deserts, etc. Nests on ground.

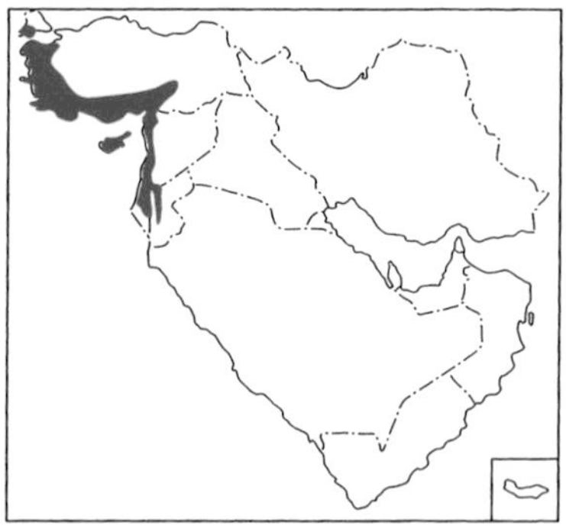

716 **Rustic Bunting** *Emberiza rustica*

Status: rare migrant or vagrant Iran, Israel, Syria, Kuwait, E Arabia and Turkey. **Habitat:** on passage generally in damp, bushy areas.

717 **Little Bunting** *Emberiza pusilla*

Status: recorded rarely in Lebanon, Israel, Jordan, Turkey, Iran, Kuwait, E Arabia. **Habitat:** on passage generally in damp bushy areas.

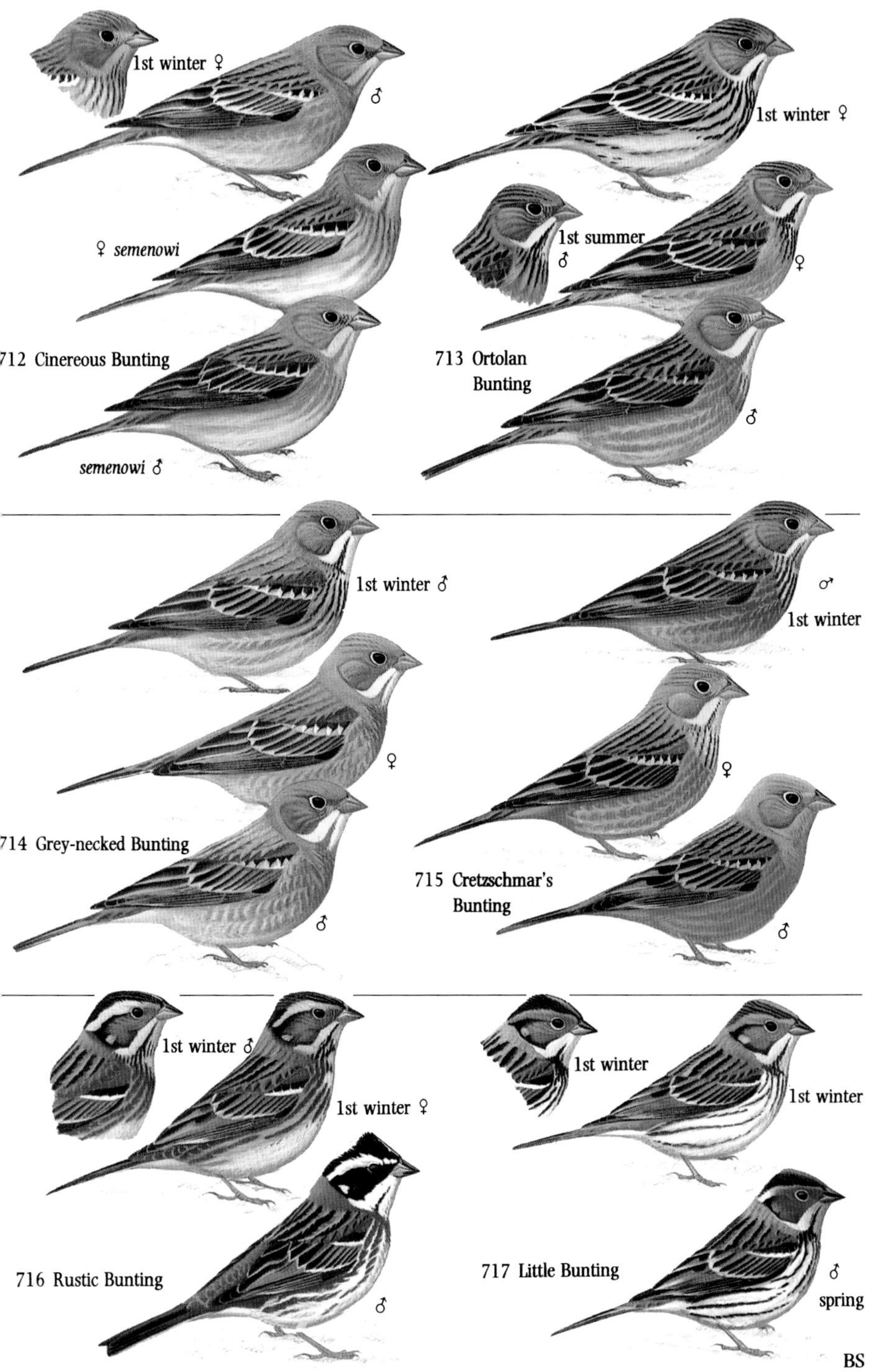
1st winter ♀
♂
♀ semenowi
712 Cinereous Bunting
semenowi ♂
1st winter ♀
1st summer
♂
♀
713 Ortolan
Bunting
♂
1st winter ♂
♀
714 Grey-necked Bunting
♂
♂
1st winter
♀
715 Cretzschmar's
Bunting
♂
1st winter ♂
1st winter ♀
716 Rustic Bunting
♂
1st winter
1st winter
717 Little Bunting
♂
spring
BS

703 **Snow Bunting** *Plectrophenax nivalis*

Status: winter vagrant to N Turkey and Israel. **Habitat:** in winter, mostly in open coastal fields or marshes.

718 **Yellow-breasted Bunting** *Emberiza aureola*

Status: vagrant or rare migrant; recorded Cyprus, Israel, N Iran, Bahrain and E Arabia. **Habitat:** on passage in open country with trees or bushes.

719 **Reed Bunting** *Emberiza schoeniclus*

Status: probably breeds more widely in Turkey. Resident and partial migrant. Scarce winter visitor to Near East, Iraq and Iran (common in north); rare E Arabia (e.g. Bahrain). Vagrant Oman and UAE. **Habitat:** reedbeds, swampy thickets. Nests in reeds and bushes.

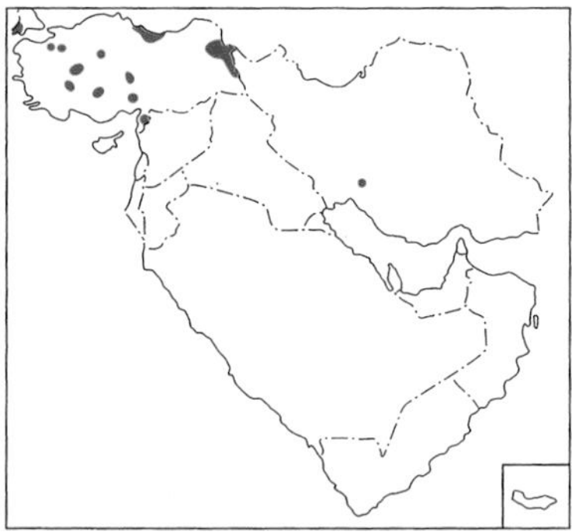

720 **Red-headed Bunting** *Emberiza bruniceps*

Status: summer visitor; recorded as vagrant Turkey, Israel, NW Saudi Arabia, Kuwait and UAE. **Habitat:** irrigated farmland and other open country with bushes, mainly in upland areas. Nests in bush close to ground.

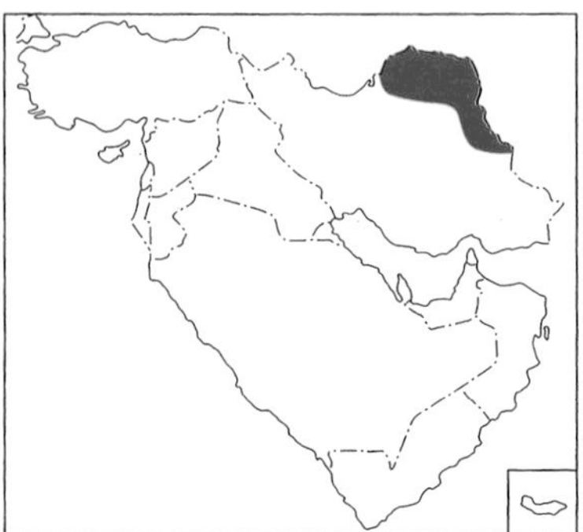

721 **Black-headed Bunting** *Emberiza melanocephala*

Status: summer visitor; widely on passage Iran, regular Oman (where recorded in winter), E Saudi Arabia and S Israel; scarce or irregular UAE, Bahrain and central Saudi Arabia. **Habitat:** bushy hilly or rolling country with or without cultivation; also fairly open farmland. Nests in vegetation close to ground.

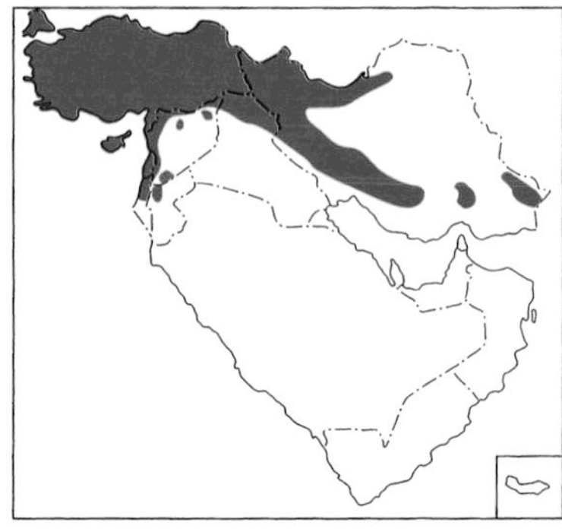

722 **Corn Bunting** *Miliaria calandra*

Status: mainly resident; dispersive in winter to S Iran, S Iraq, Arabian Gulf, E Arabia, Oman but rare Saudi Arabia. **Habitat:** open farmland, hillsides with scattered bushes. Nests in grass or other vegetation near or on ground.

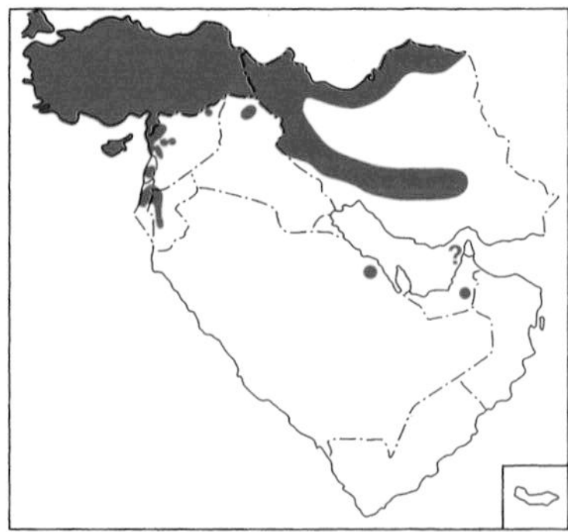

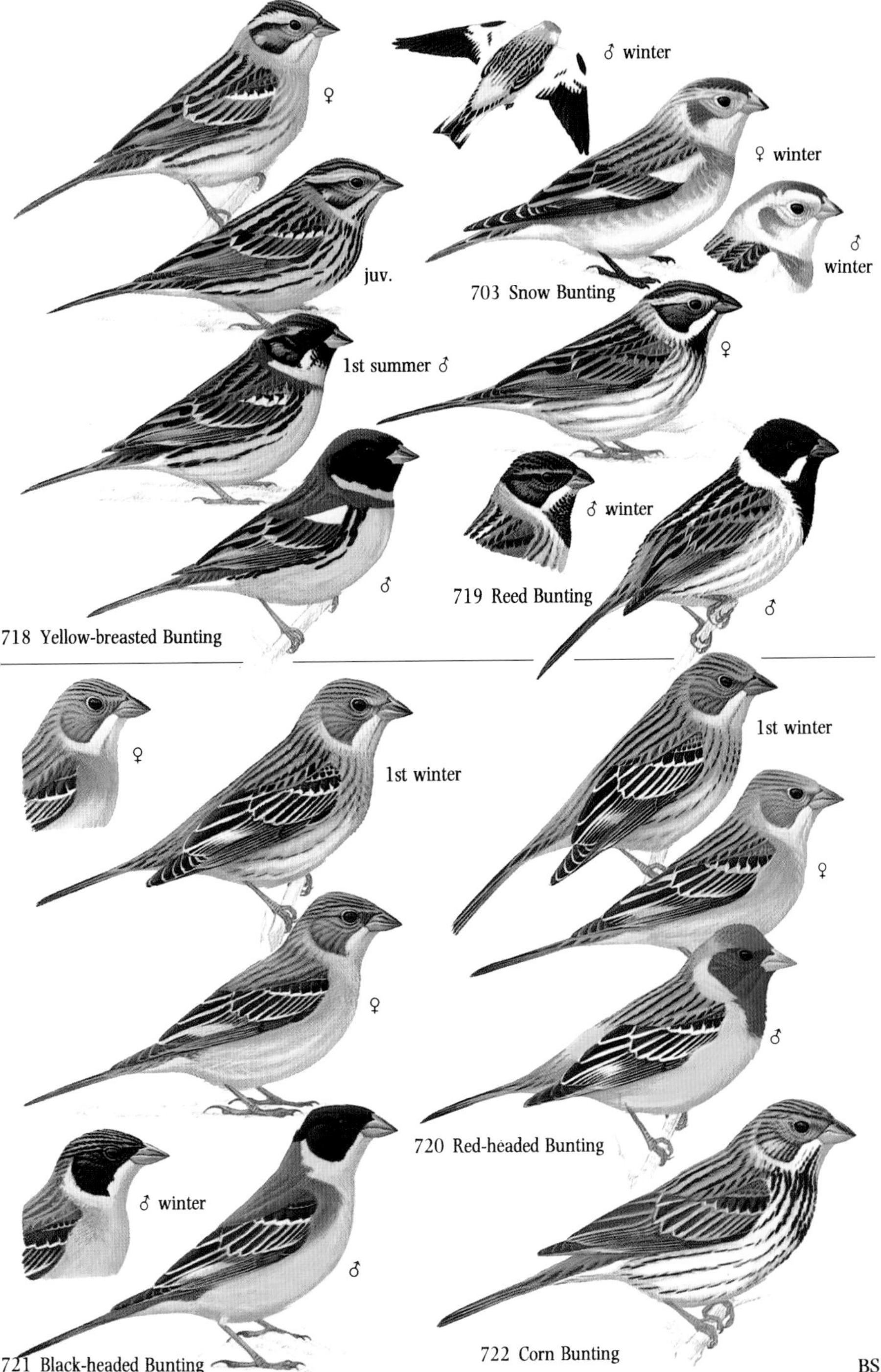

♂ winter
♀
♀ winter
♂ winter
juv.
703 Snow Bunting
1st summer ♂
♀
♂ winter
♂
719 Reed Bunting
♂
718 Yellow-breasted Bunting
♀
1st winter
1st winter
♀
♀
♂
♂ winter
720 Red-headed Bunting
♂
721 Black-headed Bunting
722 Corn Bunting
BS

88 **Comb Duck** *Sarkidiornis melanotos*

Status: vagrant Oman. **Habitat:** reedy coastal areas.

---

104 **Eider** *Somateria mollissima*

Status: vagrant Turkey. **Habitat:** coastal, seldom on inland waters.

---

119 **Brahminy Kite** *Haliastur indus*

Status: vagrant UAE. **Habitat:** wastelands and marshes.

---

190 **Lesser Moorhen** *Gallinula angulata*

Status: vagrant Oman. Range Africa. **Habitat:** similar to Moorhen but also often occurs in areas of temporary water.

193 **Watercock** *Gallicrex cinerea*

Status: vagrant Oman. **Habitat:** swamps, marshes and, on migration, along coasts.

---

207 **Egyptian Plover** *Pluvianus aegyptius*

Status: an old record from Jordan. Range Africa. **Habitat:** rivers and lakes with sandy edges.

---

239 **Red-necked Stint** *Calidris ruficollis*

Status: vagrant UAE. **Habitat:** sandy coasts and edges of wetlands.

---

243 **Baird's Sandpiper** *Calidris bairdii*

Status: vagrant Oman. **Habitat:** sandy shores and wetland margins.

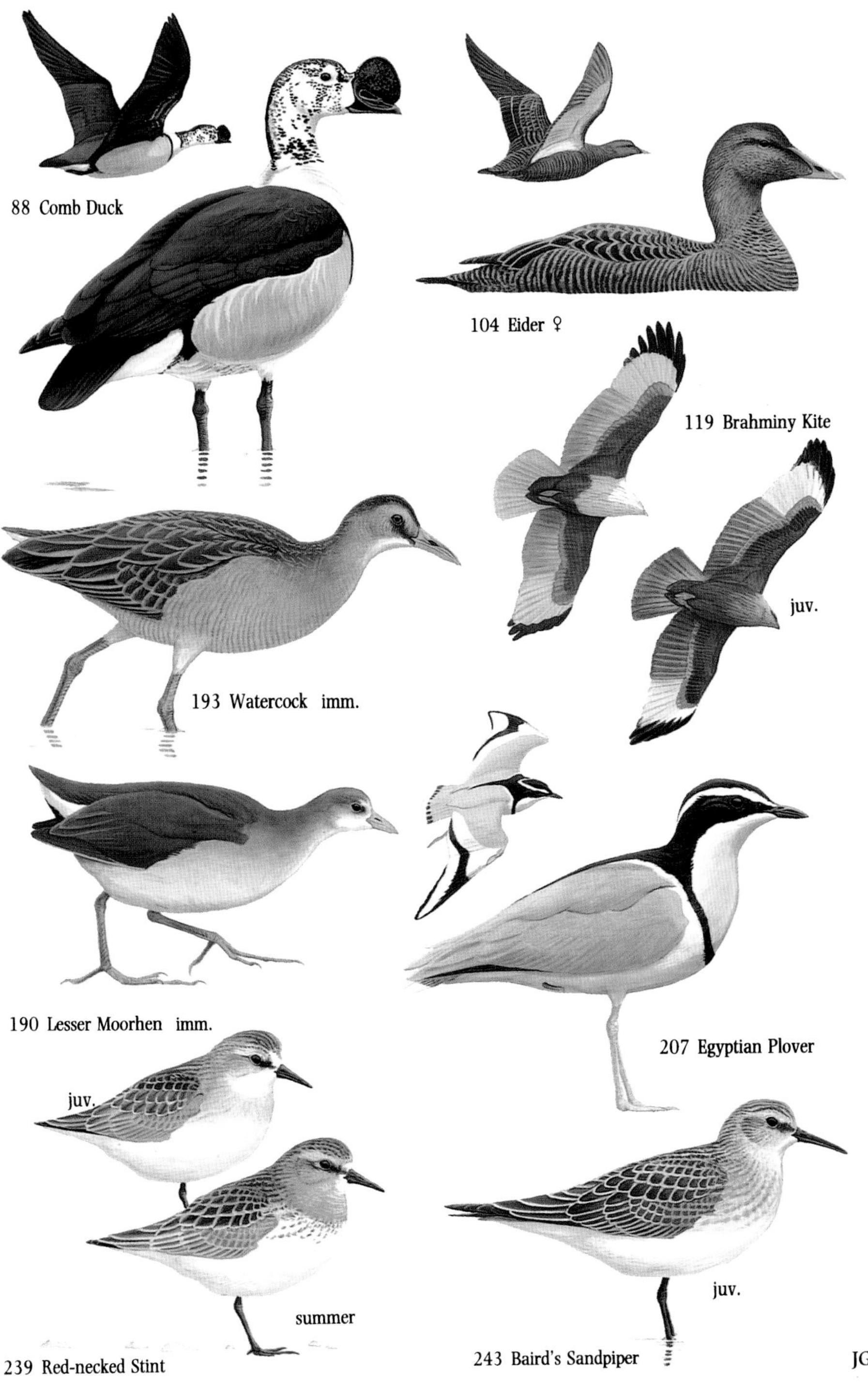
88 Comb Duck
104 Eider ♀
119 Brahminy Kite
juv.
193 Watercock imm.
190 Lesser Moorhen imm.
207 Egyptian Plover
juv.
summer
239 Red-necked Stint
juv.
243 Baird's Sandpiper
JG

445 **Lesser Striped Swallow** *Hirundo abyssinica*

Status: vagrant Oman. **Habitat:** open areas, as other swallows.

---

454 **Pechora Pipit** *Anthus gustavi*

Status: unconfirmed record UAE; this Asian species could occur in the region. **Habitat:** often in thick vegetation or tall grass, crops or stubble-fields.

---

484 **Siberian Rubythroat** *Luscinia calliope*

Status: vagrant Israel. **Habitat:** dense scrub or thickets near water, hedges near villages, ditches, underbrush along sides of country roads, also long grass or reeds.

516 **White's Thrush** *Zoothera dauma*

Status: vagrant Oman. **Habitat:** always near dense undergrowth or woods in which it skulks.

---

554 **Melodious Warbler** *Hippolais polyglotta*

Status: vagrant Turkey. Range SW Europe. **Habitat:** as Olivaceous Warbler.

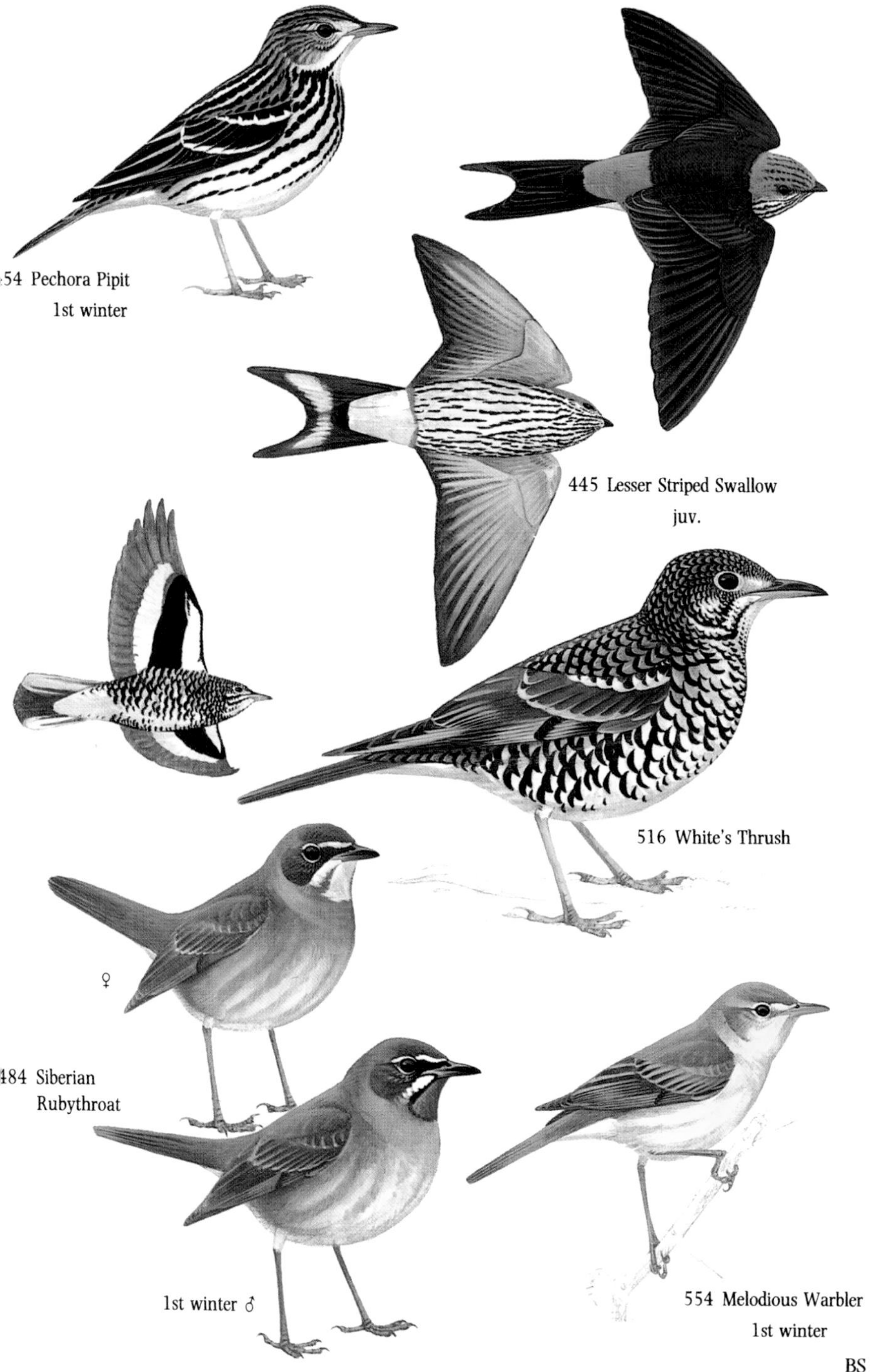
454 Pechora Pipit
1st winter
445 Lesser Striped Swallow
juv.
516 White's Thrush
♀
484 Siberian
Rubythroat
1st winter ♂
554 Melodious Warbler
1st winter
BS

1 **Red-throated Diver** *Gavia stellata*

(Red-throated Loon) **PLATE 1**

L: 61. W: 111. Smallest diver. Long, slender body and bill recalling Great Crested Grebe, but has shorter, thicker neck and ***often upturned head and bill*** (unlike that usually shown by Black-throated). In non-breeding plumage easily confused with Black-throated, but upperparts speckled white giving a lighter, greyish appearance; ***grey crown and hindneck contrast much less with clean white face and fore-neck***; bill is light grey, and dark reddish eye is often fairly conspicuous, surrounded by white. ***Lacks white patch on rear flanks***, but has often continuous whitish flank-line. In flight best separated from Black-throated by lack of contrast on head and neck; also by less protruding legs, more hunch-backed appearance, thinner neck and often higher upstrokes of wings. Distinguished from grebes by size and plain wings. Juvenile is duller and darker grey on upperparts, head and neck until moult in late winter. Adult in summer plumage (not likely to occur in the Middle East) has dark reddish throat (looks black at distance) with pale chin, black and white striped hindneck, and without white patches on shoulders.

2 **Black-throated Diver** *Gavia arctica*

(Arctic Loon) **PLATE 1**

L: 66. W: 120. Intermediate in size between Red-throated and Great Northern, but overlaps with both. Non-breeding plumage close to Great Northern, and more contrasting black and white than Red-throated, ***especially head and neck with sharp demarcation below eye and down side of neck***. No visible eye-ring makes eye invisible at distance unlike other divers. Head and nape often greyer than nearly black upperparts unlike Great Northern. ***Often shows white patch to rear of flanks***. Straight bill held horizontally like Great Northern (unlike Red-throated), but is shorter, more fine-pointed and not so heavy; ***bill often conspicuously two-toned with blackish tip, unlike Red-throated***. Sometimes angular head with steep forehead like Great Northern. Juvenile has narrow pale edgings to upperparts, but looks blackish at distance. Black chin and throat in summer plumage bordered by diagnostic vertical white bars and squarish white patches on scapulars. In flight, size closest to Red-throated and ***best separated by head contrast and black and white appearance***.

3 **Great Northern Diver** *Gavia immer*

(Common Loon) **PLATE 1**

L: 80. W: 137. Largest diver, heavy looking, low-lying with long, deep-based bill often held horizontally; head mostly angular with steep forehead (but not always). Large size, heavy-billed and large-headed appearance compared with Black-throated. In non-breeding plumage black upperparts and white underparts with extensive black and white on head and neck like Black-throated, but demarcation line is less clear-cut and ***often shows a conspicuous dark half or full collar on fore-neck. Upper head and hindneck often darker than upperparts*** (the reverse of pattern in Black-throated); pale eye-ring makes eye more obvious than Black-throated; bill pale grey with dark culmen; no white patch on rear flanks, but sometimes a long, paler flank-line. In juvenile/1st winter ***upperparts show distinct pale feather edgings***, more obvious than juvenile Black-throated. Flight strong and heavy, reminiscent of a goose or cormorant; ***large feet project conspicuously behind tail***, and neck appears thicker than in Black-throated.

4 **Little Grebe** *Tachybaptus ruficollis* **PLATE 1**

(Dabchick)

L: 27. W: 43. Smallest grebe in the region; identified by small size, ***blunt-ended body***, short neck and bill, and absence of white in upperwing. Rich dark brown adult has ***bright chestnut throat and cheeks*** and ***conspicuous yellow gape patch***. In winter, adult and young paler brown above, pale buff below, variably mixed with dull chestnut on fore-neck (adult); light gape often reduced or absent. Juvenile has white striped head and hindneck. Dives with fast jump and splash; when alarmed dives rather than flies. Flight is only for a short distance with very rapid wing-beats low over water.
Voice: in breeding area more often heard than seen. Commonest call is a variable, high-pitched trilling recalling ***whinny of a horse***.

5 **Great Crested Grebe** *Podiceps cristatus* **PLATE 1**

L: 49. W: 88. ***Largest grebe*** in the region. Swims with long body low, and ***long, white slender neck erect***; bill pink, dagger-like, held horizontally. In flight, extended neck and feet are held below line of body, with hunch-backed appearance recalling a diver, but smaller with more rapid wing-beats; ***large white patch on secondaries, white border to front of innerwing and on shoulders along body***. Easily visible on open water and usually submerges smoothly, without leaping. Breeding adult unmistakable with ***black crest and black and chestnut tippets***. Winter plumage lacks tippets, but retains short, dark grey crest; ***black lores and narrow white line over eye*** making eye clearly visible at distance; fore-neck and underparts gleaming white. Juvenile largely like winter adult but with striped black and white head and neck pattern.
Voice: most common sound, in the breeding season, is a loud, harsh far-carrying 'rah-rah-rah . . .'.

6 **Red-necked Grebe** *Podiceps grisegena* **PLATE 1**

L: 45. W: 81. Close in size and shape to Great Crested, but head more rounded, ***neck shorter and thicker. Bill black with diagnostic yellow base (adult)*** or with yellow extending diffusely towards the tip on lower mandible (juvenile/1st winter). In all plumages told from Great Crested by ***black cap down to eye, pale greyish cheeks contrasting with dark crown and dusky neck***. Breeding adult unmistakable with black cap, grey cheeks, and ***chestnut neck and breast***. Distinguished from smaller grebes in winter plumage by size, long bill and oblong headshape. Flight silhouette less attenuated than Great Crested; ***white patch to secondaries and front of innerwing smaller and not connected along the body***; wing pattern thus like Slavonian Grebe, but has more white at base of forewing.
Voice: commonest sound of breeding season (often heard by night) resembles Water Rail's squealing pig-like call.

7 **Slavonian Grebe** *Podiceps auritus* **PLATE 1**

(Horned Grebe)

L: 35. W: 62. Similar size to Black-necked. Distinctly smaller and with relatively shorter bill than Great Crested and Red-necked. Breeding plumage (unlikely to occur in the region) unmistakable: black head with gold crest, chestnut neck, upper breast and flanks with upperparts glossy brown-black. ***Winter plumage strikingly black and white; white*** *cheeks and fore-neck contrasting with black cap down to eye and black upperparts.* Distinguished from Black-necked in winter plumage by *gleaming white fore-neck, white cheeks and straight bill.* In flight, distinctly smaller, with more rapid wing-beats than Great Crested and Red-necked; upperwing shows white wing-patch confined to secondaries (includes inner primaries in Black-necked) and some white on inner forewing (like Red-necked, but unlike Black-necked).

8 **Black-necked Grebe** *Podiceps nigricollis* — PLATE 1

(Eared Grebe)

L: 31. W: 58. Small grebe compared with Great Crested and Red-necked. ***Short, uptilted bill; steep forecrown with peaked head***; sometimes puff-backed, recalling Little Grebe. Breeding plumage ***mostly black (including fore-neck)*** with golden ear-coverts and chestnut flanks. Winter plumage basically black and white, mostly recalling Red-necked and Slavonian. Distinguished from Red-necked by size, short bill, steep forehead and more rounded back. From Slavonian (with which most easily confused) by ***head and bill-shape, cheek pattern and less clean black and white appearance***. Uptilted bill difficult to see at distance. In flight, shows long, white wing-patch extending to inner primaries (unlike Slavonian); also lacks white on inner forewing (present in the other grebes).
Voice: commonest call in breeding area a plaintive whistle 'ooo-eep'.

9 **Shy Albatross** *Diomedea cauta* — PLATE 2

(White-capped Albatross, White-capped Mollymawk)

L: 91. W: 255. Large seabird, ***over twice the size of Cory's Shearwater***. Effortless flight with long periods of gliding on stiffly held wings. ***Underwing with narrow black border and dark mark at base of leading edge***. The dark mark through eye and greyish bill with yellow tip are visible only at close range. Juvenile similar to adult but bill tipped blackish and often has dark half-collar. Although this is the only species of albatross recorded in the Middle East, others could occur as vagrants. Several are similar to Shy (notably Grey-headed Albatross *D. chrysostoma* and Black-browed Albatross *D. melanophris*) but all lack the characteristic dark mark at base of underwing and have broader black margins on underwings.

10 **Cape Petrel** *Daption capensis* — PLATE 3

(Cape Pigeon, Pintado Petrel)

L: 39. W: 85. Very distinctive medium-sized stocky petrel with ***black and white patterned upperparts***, black head and clean white underparts except for well defined black rim to underwing and ***black tail band***. Flight alternate bursts of fairly fast shallow wing-beats with short glides on stiffly held wings, low over sea or rising higher in strong winds.

11 **Atlantic Petrel** *Pterodroma incerta* — PLATE 2

(Schlegel's Petrel)

L: 43. W: 104. Near to Cory's in size, thus larger than Soft-plumaged Petrel and appears longer winged and longer tailed. Identified by dark brown upperparts, with darker primaries and also a darker shade on rump and tail. Usually very distinctive below where ***white breast and belly-patch contrast with brown head and upper breast, underwings,***

lower belly and tail. In worn plumage the brown on head and throat can fade to a much paler shade, when it often shows dark eye-mask. Gliding flight can be low over the sea in calm conditions, but in strong winds (when wing-beats are less frequent) can glide very high above the waves at great speed, on angled wings. Atlantic Petrel may flap at the peak of a glide, unlike Soft-plumaged.

12 **Soft-plumaged Petrel** *Pterodroma mollis* PLATE 2

L: 35. W: 90. Larger than Mediterranean Shearwater but noticeably smaller than Cory's. Best told by combination of ***dark and light patterned underwing contrasting with white underparts***; grey or brownish upperparts with dark primaries and upperwing coverts, ***forming an 'M' pattern***; grey breast band (which is characteristic of the southern races likely to occur in the Middle East) and ***dark eye-mask***. The complicated pattern of this species can vary according to distance and light conditions. At a distance the underwing can look all-dark but close views reveal a variable pale stripe down the centre and white wedge from the body along the leading coverts. In addition, the 'M' marking on the upperwing can be difficult to see and then the upperparts look all dark. In light winds rises only about 5 m above the waves, with five or six fairly fast wing-beats followed by a long, low glide on slightly angled and bowed wings. In stronger winds can rise in arcs high above the waves with long glides, twisting from side to side on a zig-zag course: in such conditions the wings become more angled. The only possible confusion species is Atlantic Petrel in worn plumage, but Soft-plumaged can always be told by ***white lower belly and undertail coverts*** (blackish in Atlantic), ***white patterning on underwing*** (all-dark on Atlantic) and shorter tailed appearance.

13 **Jouanin's Petrel** *Bulweria fallax* PLATE 2

L: 30. W: 75. Noticeably larger and broader winged than its close relative, Bulwer's Petrel *B. bulwerii* (an Atlantic species which has not been recorded in the Middle East). The general impression is of a ***small all-dark shearwater type*** and care is needed for separation from dark morph of Wedge-tailed, particularly by those with no experience of the two. Jouanin's is smaller than Wedge-tailed with less languid flight, ***smaller all-dark bill*** (greyish with dark tip in Wedge-tailed) ***which is held down at 45° angle*** (Wedge-tailed holds bill almost horizontally). Tail is fairly long, though not obviously so and the graduation in tail is not visible unless close. In windy conditions will rise 2–5 m above the waves in long banking arcs, interspersed with spells of four or five rather leisurely flaps, which tend to be at the peak of the arcs. Otherwise flies fairly close to the sea, progressing through troughs with a mixture of wing-beats and long glides. Plumage entirely sooty brown with diffuse pale edge to upperwing coverts, though this may not be obvious.

14 **Cory's Shearwater** *Calonectris diomedea* PLATE 3

L: 46. W: 110. Large shearwater with greyish brown upperparts and white underparts; lacking any distinctive markings. Underwing white with narrow brownish margins and may show diffuse whitish band at base of uppertail. Note ***lack of clear demarcation between grey-brown head and white chin and throat***. At close range ***yellow bill*** is an important field mark. Flight relaxed with fairly slow wing-beats interspersed with long glides, often in an arc above the waves, on characteristic bowed wings. At a distance confusable with

Great Shearwater but Cory's is featureless, lacking clearly demarcated black cap and clear white band at base of tail. See also Streaked Shearwater.

15 **Streaked Shearwater** *Calonectris leucomelas* PLATE 3

(White-faced Shearwater)

L: 48. W: 120. Slightly smaller than Cory's Shearwater, which it somewhat resembles in general plumage and flight, but distinguished by ***white face*** and ***off-white head, streaked with brown***. Upperparts dark brown, darker than Cory's and usually lacking any pale on uppertail-coverts (which Cory's can often show). Underparts white with dark patches by the shoulders, more restricted and contrasting than in Cory's. Underwing white with dark flight feathers (like Cory's) but with ***diagnostic dark patch*** on ***primary coverts***. Bill more slender than Cory's, grey (never yellow like most Cory's) with darker tip. Flight similar to Cory's but less heavy, with slower wing-beats and less shearing. Wings flexed less than Cory's, and raised higher, giving a less stiff, almost elastic flight.

16 **Pale-footed Shearwater** *Puffinus carneipes* PLATE 3

(Flesh-footed Shearwater)

L: 43. W: 100. Slightly smaller than Cory's but uniformly blackish brown with often slightly paler coverts and ***silvery flash on primaries below***, visible even at a distance but only in good light. It is this feature, if seen, that distinguishes it from Wedge-tailed Shearwater, but if close, also note Pale-footed's larger and paler bill which can contrast markedly with dark plumage. Like Wedge-tailed, the bill has a darker tip. In addition, Pale-footed is slightly larger, broader winged and ***shorter tailed than Wedge-tailed***. Lazy flight, generally more steady than Wedge-tailed with straighter wings: a series of slow, heavy flaps followed by a long glide, rising well above the waves in windy conditions.

17 **Great Shearwater** *Puffinus gravis* PLATE 3

L: 46. W: 105. A large shearwater, close to Cory's in size, but immediately ***identifiable from this and all other large dark and white shearwaters by combination of clear-cut black cap, white collar and white uppertail coverts***. Other useful features are the white underwings with patterning of dark lines on the coverts, the all-dark bill (yellowish and more obvious in Cory's) and a smudgy dark patch in centre of lower belly seen in good conditions. Flight powerful with long glides on stiffly held, fairly straight wings, unlike the more languid flight of Cory's, which glides on bowed wings.

18 **Wedge-tailed Shearwater** *Puffinus pacificus* PLATE 3

L: 43. W: 100. Fairly large sooty brown shearwater, smaller than Cory's and with longish tail, ***the wedge shape being difficult to observe until the tail is spread*** when manoeuvring in flight. May show slightly paler bar on greater coverts above and, on close views, light edging to some coverts, forming narrow bars. ***Bill grey with dark tip***. In light winds flight lazy and low over the sea with three or four fairly quick wing-beats followed by a short glide on wings bowed forwards and downwards. In stronger winds flight is erratic, often changing direction and soaring in low arcs between short bursts of wing-beats. A rare pale phase occurs (with white underparts and underwing-coverts) but this has not been recorded in the Middle East. Three other all-dark shearwater types occur in the Arabian and Red Sea: Jouanin's Petrel (the commonest), Sooty Shearwater and

Pale-footed Shearwater. Told from Jouanin's by larger size, ***broader-based wings, longer grey bill with dark tip which is held horizontally***, and pale feet; from Sooty by longer tail, broader wings, ***all-dark underwings*** and pale feet; from Pale-footed by slightly smaller size, all dark underwing (frequently pale underside to primaries in Pale-footed) and thinner bill, which is grey with dark tip (flesh with dark tip in Pale-footed).

19 **Sooty Shearwater** *Puffinus griseus* **PLATE 3**

L: 42. W: 100. A dark brown shearwater, similar in size to Wedge-tailed but with shorter tail, narrower wings and ***variable white on underwing coverts***, which can often flash in contrast to dark plumage; feet dark. Flight faster than Wedge-tailed, more direct with faster wing-beats between glides on stiffly held wings.

20 **Mediterranean Shearwater** *Puffinus yelkouan* **PLATE 2**

(Levantine Shearwater, Yelkouan Shearwater)

(For ease of presentation we have treated this as a separate species from Manx Shearwater *Puffinus puffinus* (of the North Atlantic) with which it has until recently been regarded as a subspecies).

L: 33. W: 82. A fairly small, dark and white shearwater, noticeably smaller than Cory's, with much faster wing-beats and relatively short periods of gliding. Stouter than Manx and shorter tailed with slightly projecting feet, this structure leading to a more fluttery flight often low over the water, with usually only short spells of gliding. ***Upperparts brown*** (but can look black at a distance, especially in fresh plumage); ***white below though in worn plumage can have rather dirty underparts being smudgy on breast, flanks and undertail, thus without the clean 'black and white' appearance of Manx***. Underwing sullied white with flight feathers variably brownish. For separation from Little Shearwater and Persian Shearwater see those species.

21 **Little Shearwater** *Puffinus assimilis* **PLATE 2**

L: 27. W: 62. A distinctly small shearwater, black above and white below, which is most easily confusable with Persian Shearwater (the commonest shearwater in the Arabian Sea). It differs from Persian (which see) by its slightly smaller size, ***white on face extending above eye*** (which can make whole face appear white at a distance) and ***white extending from undertail-coverts onto sides of rump***. Care, however, is needed as there are several races of Little Shearwater and one of those in the South Atlantic (which could reach the Middle East) has less white-faced appearance—the dark cap extending to eye level or just below it. At close range upperwing of at least some birds in autumn ***shows silvery-grey secondaries*** causing whole wing to look silvery against strong sunlight. In calm conditions wing-beats are fairly ***shallow, rapid and fluttering***, interspersed with short, flat glides low over sea with little banking; in strong winds periods of gliding are longer with banking glides several metres high. In characteristic feeding flight, patters over surface with outstretched wings.

22 **Persian Shearwater** *Puffinus (lherminieri) persicus* **Plate 2**

(Persian race of Audubon's Shearwater)

L: 30. W: 70. ***A small dark and white shearwater, most easily confused with Little Shearwater***, with white underparts contrasting with dark brown upperparts (which look nearly black at a distance) extending to just below level of eye, unlike most Little Shearwaters. Apart from colour of upperparts, which are black in Little Shearwater, it also differs from that species by ***lacking white extending on to rump above undertail-coverts and darker underwing with broad dark margins*** (in Little white underwings have narrow dark margins) and flesh-coloured legs (grey in Little). Note that the ***undertail coverts are white*** (not black) ***in the Persian race of Audubon's***). At least one record has occurred in the Arabian Sea of a bird showing dark brown head and throat, extending onto breast, thus resembling Heinroth's Shearwater *P. heinrothi* (breeds Solomon Is.). Fairly fast wing-beats interspersed with short glides, low over sea, though often rising in low arcs in strong winds. Flocks at fish shoals with other seabirds.

23 **Wilson's Storm-petrel** *Oceanites oceanicus* **Plate 4**

(Wilson's Petrel)

L: 18. W: 40. Petrels are scarce in the Middle East but Wilson's is the most likely to occur; identified (with care) from other similarly marked white-rumped petrels by ***long legs which protrude beyond the tip of square-ended tail*** in flight, and ***bold white rump which extends on to sides of uppertail-coverts***. Wilson's is larger and longer winged than European Storm-petrel with more pointed wings and fairly conspicuous pale brownish panel on the upperwing-coverts; unlike European Storm (which see) ***the underwings are wholly dark***. Flight is less fluttery than European Storm with series of flaps interspersed with glides, which are fairly horizontal to the sea. For separation from Leach's and Madeiran see those species.

24 **Black-bellied Storm-petrel** *Fregetta tropica* **Plate 4**

(Black-bellied Petrel)

L: 20. W: 45. Small square-tailed sooty petrel with white rump, ***white underwing-coverts*** and ***underbody***, the latter with inconspicuous dark line extending up middle to ***sooty brown head and upper breast, and feet projecting slightly beyond end of tail***. The upperwing-coverts show obscure pale band. Erratic flight, zig-zagging from side to side but usually close to water. Similar to White-bellied Storm-petrel, which see.

25 **White-bellied Storm-petrel** *Fregetta grallaria* **Plate 4**

(White-bellied Petrel)

L: 19. W: 46. Similar to Black-bellied Storm-petrel (which see) but ***lacks dark line through centre of white underparts***, giving a more white appearance below. The upperparts are generally paler sooty brown, but there appears to be little difference in the paler upperwing band on the coverts. Close views show that the ***feet project very slightly beyond tip of tail***. Flight less erratic than Black-bellied, which often dangles feet and 'bounces' onto water.

26 **White-faced Storm-petrel** *Pelagodroma marina* **PLATE 4**

(Frigate Petrel, White-faced Petrel)

L: 20. W: 45. Unmistakable small petrel with patterned blackish-brown upperparts, ***white underparts, underwing-coverts, forehead and cheeks with dark patch through eye***. Close views show black crown and nape, with brownish grey upperwing-coverts contrasting with blackish flight-feathers and ***grey rump, accentuating blackish forked tail***. Strong and erratic flight with much banking, and often 'bounces' off the sea in 'yo-yo' fashion, seemingly pushed off the surface by its ***long legs which project distinctly beyond the tail-end*** and are occasionally dangled.

27 **European Storm-petrel** *Hydrobates pelagicus* **PLATE 4**

(British Storm-petrel, Storm Petrel)

L: 15. W: 38. A small black petrel with square-ended tail and white rump. Most easily confused with Wilson's (which see) but told by characteristic ***white stripe on underwing***, smaller size, shorter wings, fairly uniform dark upperwing and less extensive white rump. Unlike Wilson's, ***feet do not project beyond end of tail***. Flight is the weakest of all the small petrels with nearly continuously fluttering wing-beats interspersed with short glides when white stripe on underwing observable. Differs from Leach's and Madeiran in smaller size and shorter wings, shape of rump patch, darker upperwing, white on underwing-coverts and flight.

28 **Leach's Storm-petrel** *Oceanodroma leucorhoa* **PLATE 4**

(Leach's Petrel)

L: 20. W: 46. Larger and longer winged than Wilson's (the commonest petrel to occur in Middle East waters) with ***forked tail***, which can be difficult to see unless close. The plumage is sooty brown, often looking black, with a fairly noticeable pale bar across the upperwing-coverts and rather dirty white rump with a dark central band—noticeable only at close range; the white rump is the least conspicuous of all the small petrels. ***Flight is a good feature: erratic and bounding and at times often reminiscent of a small Black Tern*** with its long rather angled wings and deep wing-beats. Much larger than European Storm-petrel and differs in longer wings, noticeable covert bar, forked tail, ***lack of white on underwing*** and flight. For differences from Madeiran see that species.

29 **Swinhoe's Storm-petrel** *Oceanodroma monorhis* **PLATE 4**

(Swinhoe's Petrel)

L: 20. W: 45. Similar in size and shape to Leach's (including forked tail) but with ***all-dark rump***. Plumage blackish brown with paler bar across upperwing-coverts, usually less obvious than that of Leach's. Flight bounding and Black Tern-like, and thus similar to Leach's. Although this is the only dark-rumped petrel to have occurred in Middle East waters, care must be taken to ensure other possible species have been eliminated. Matsudaira's Petrel *Oceanodroma matsudairae* is a species that could occur: it is larger with longer, broader based wings, deeper tail-fork and has a whitish patch at base of outer-primaries.

30 **Madeiran Storm-petrel** *Oceanodroma castro* **PLATE 4**

(Band-rumped Storm-petrel, Harcourt's Storm-petrel, Madeiran Petrel)

L: 20. W: 43. Fairly long-winged petrel with ***slightly forked tail*** (very difficult to see unless in hand), which most closely resembles Leach's. As both species are vagrants to Middle East waters great care must be taken over their separation. ***The best distinguishing features from Leach's are the clear-cut narrow white rump*** (no dark central line) ***which extends onto sides of uppertail-coverts***, less distinct pale panel on upperwing-coverts and ***flight which is much steadier and usually fairly low over the sea*** with quick wing-beats interspersed with shearwater-like glides.

31 **Red-billed Tropicbird** *Phaethon aethereus* **PLATE 5**

L: 48 (plus 50 cm tail). W: 105. Unlikely to be confused with any other seabird in the Middle East. Plump-bodied, white with ***long white tail-streamers and conspicuous red bill***; the white plumage is relieved by black eye-stripe, black outer primaries and narrow black barring on upperparts and coverts. Juvenile (which lacks tail-streamers), has black-tipped tail, yellow bill and blackish collar. Flight a useful character: direct with ***fast wing-beats and interspersed with glides on horizontally held wings***, usually fairly high. **Voice:** shrill, rapid rasping notes.

32 **Red-footed Booby** *Sula sula* **PLATE 5**

L: 75. W: 100. This vagrant to the Arabian seas can be similar, depending on plumage, to Masked Booby and Brown Booby. It is the smallest booby to occur in the region and has two main phases: white and brown. Those likely to be seen are from the Indian Ocean which in ***adult plumage, irrespective of phase, have all-white tails***. All other adult boobies have black tails, except Gannet which is likely to be seen in the Mediterranean only where Red-footed Booby has never occurred. ***The white phase adult resembles Masked Booby but is smaller, has white tail, black carpal patches below***, often yellowish wash to head and ***red feet***; it lacks a black mask. Brown phase adult is grey-brown with darker back and wings (both above and below) and white tail. Juvenile is rather featureless, all brown with noticeably dark underwing lacking white or pale on coverts (as shown by juvenile Brown Booby).

33 **Masked Booby** *Sula dactylatra* **PLATE 5**

(Blue-faced Booby)

L: 85. W: 150. Slightly smaller than Gannet which adult superficially resembles but differs in ***entire broad black trailing-edge of wings*** (above and below), ***black tail and black face mask; head lacks yellow wash often seen in Gannet***. See also Red-footed Booby for differences. Juvenile Masked has ***brown head and neck separated from paler brown back by white collar*** which broadens with age, while upperparts become mottled with white especially, initially, on the scapulars; white underparts and ***white underwing-coverts with black line through centre*** and black flight-feathers. Juvenile differs from Brown Booby by larger size, variable white mottling on upperparts, white neck-collar and more prominent band on underwing-coverts.

34 **Brown Booby** *Sula leucogaster* **PLATE 5**

L: 70. W: 145. The commonest booby in the Red and Arabian Sea and readily identified in adult plumage by ***uniform dark chocolate-brown upperparts, head and neck and conspicuously white underbody and underwing-coverts.*** The pale greenish yellow bill contrasts with the dark head even at a distance. Juvenile pattern similar to adult but underparts buffish brown (thus less contrast with brown head, neck and upper breast); white underwing-coverts appear at an early age but in some very young birds can look quite brownish. Flight similar to other boobies, with a series of wing-beats followed by a long glide with the wings held fairly horizontal. Catches fish by diving, with folded wings, often at a shallow angle from a short height above the sea. See Masked Booby for separation from juvenile.
Voice: raucous, high-pitched calls heard on breeding grounds.

35 **Gannet** *Sula bassana* **PLATE 5**
(Northern Gannet)

L: 92. W: 170. A large seabird with long, pointed bill, cigar-shaped body and long narrow wings. Adult is unmistakable (but see Masked Booby which superficially resembles it) ***with white body and tail, black wing-tips and yellowish wash to head.*** Adult plumage is reached after about five years during which time the young bird moults from its grey-brown juvenile form. Juvenile most closely resembles juvenile Brown Booby but always told by white crescent on uppertail-coverts, pale speckling especially on upperparts (at close range) and larger size. As bird grows older white feathers start to appear producing a patchy black and white plumage with a ***yellow wash on white head from an early stage.*** Flight as Brown Booby but banks much more during glides and plunges into sea for fish more vertically.

36 **Cormorant** *Phalacrocorax carbo* **PLATE 6**
(Great Cormorant)

L: 90. W: 140. Large waterbird with powerful bill; swims low in the water and frequently perches with wings outstretched. In breeding plumage glossy black with ***white feathering on nape and neck*** and ***white thigh-patch. In winter adult still retains white patch on chin and throat*** unlike Socotra Cormorant or Shag (which see). Juvenile browner than adult with brownish-white or dirty-white underparts; ***identified from juvenile Shag (which see) by larger size, thicker neck, larger, deeper bill and lack of white fringes to wing-coverts.*** The distinctive African race, *lucidus* (which has been reported in Israel and Yemen) has a clear-cut white fore-neck and breast.

37 **Shag** *Phalacrocorax aristotelis* **PLATE 6**
(European Shag)

L: 75. W: 100. Similar to Cormorant but smaller and slimmer with thinner neck and bill. Careful examination of head shape shows it to be more rounded in Shag with ***steeper forehead*** (in Cormorant rather flat crown merges into bill). Entirely black with green sheen (more blue sheen in Cormorant) ***tuft on forehead in breeding season*** and ***pronounced yellow gape-flange*** which also shows in winter when plumage duller (lacks white face-patch of Cormorant). Juvenile similar to juvenile Cormorant but differs in structure, browner upperparts, whitish-fringed upperwing-coverts which contrast with dark flight-feathers and whiter underparts with ***contrasting dark thigh-patch***, a characteristic of *desmarestii*, the race that occurs in the Middle East.

38 **Socotra Cormorant** *Phalacrocorax nigrogularis* **PLATE 6**

L: 80. W: 130. Slightly smaller than Cormorant, with slimmer head and neck, resembling Shag in structure, though wing length in flight more Cormorant-like. ***Adult has sooty black plumage with bronze-green wings and back*** (only noticeable when close) ***and without white face and chin patch*** of Cormorant; ***bill much slimmer than Cormorant's*** and although greyish or even greeny grey (greyish yellow in Cormorant) usually shows up as pale against dark head. In breeding plumage, more glossy and with white eye-streak, white flecks on neck and fine whitish streaks on rump. Immature grey-brown above with pale fringes to coverts and brownish white below with brownish spotting on breast and belly. Juvenile, which has less obvious pale fringes to coverts and lacks spotting on breast and belly, can be easily confused with young Cormorant; best separated on structure, especially bill shape, pale fringes to coverts and, where present, dark spotting below. Often congregates in large flocks in and out of breeding season.

39 **Pygmy Cormorant** *Phalacrocorax pygmeus* **PLATE 6**

L: 48. Small cormorant with short neck, rounded head, stubby bill and long tail, these features being useful when seen in flight silhouette. Bronze-brown head and neck (with fine white streaks in breeding season) and glossy black body and wings. In winter throat becomes white. Juvenile whitish below with brownish wash on breast. ***Separated from Long-tailed Cormorant by smaller size and shorter, more stubby bill and, in adult, by black-fringed scapulars and wing-coverts*** (tipped black in Long-tailed) and dark eyes (red in both adult and immature Long-tailed).

40 **Long-tailed Cormorant** *Phalacrocorax africanus* **PLATE 6**
(Reed Cormorant)

L: 55. Slightly larger and slimmer than Pygmy Cormorant. In non-breeding plumage blackish brown with paler underparts and lacking the forehead crest, green gloss on head and orange-red bill of blackish breeding plumage; the ***silvery grey wing-coverts*** of breeding plumage are still largely retained, and form a ***pale patch in flight***; at close range the ***silvery grey wing-coverts and scapulars are broadly tipped black*** (fringed black in the other cormorants). Immature similar to non-breeding adult, but silvery coverts less pronounced and underparts whiter, particularly about the head. From Cormorant by much smaller size, distinctly longer tail and pale patch on wing-coverts. Long-tailed and Pygmy Cormorants both have long tails and size difference is not obvious. Note Long-tailed has pale upperwing-patch and longer, less stubby bill, often with some red on it (never shown in Pygmy); at close range note also different pattern of scapulars and wing-coverts (see above) and red eyes (dark in Pygmy).

41 **Darter** *Anhinga rufa* **PLATE 6**
(African Darter, Anhinga)

L: 95. W. 125. Similar to Cormorant but immediately identified by extremely ***thin, kinked, snake-like neck, long pointed bill and long tail*** which is especially obvious in flight. Sooty brown plumage with ***chestnut-brown head and neck with pale line down side;***

long, silvery feathers on coverts and scapulars making back (and wings, if held out) pale streaked. Immature paler and browner, palest on neck and underparts. Often swims with only thin neck showing; sits on perch with wings outstretched. In flight, long tail is slightly fanned and slightly kinked neck held forward (thinness can make it difficult to observe at distance). Wing-beats interspersed with short glides and can recall large *Accipiter* at a glance. Frequently soars high.

42 **White Pelican** *Pelecanus onocrotalus* **PLATE 7**

(Great White Pelican)

L: 140–175. W: 270–330. Large pelican with huge wing-span. Shape of bill prevents confusion with other birds except other pelicans. Adult unmistakable ***being white with contrasting, solid black flight-feathers below*** recalling White Stork's pattern (body tinged yellowish rosy in breeding plumage; Dalmatian has greyish underwing, and body appears greyish white). At close range (on the water) ***nape has short shaggy crest*** (breeding season), ***dark eye set in large area of naked rosy skin, legs are fleshy yellow and feathers of forehead are pointed where meeting culmen***; these latter characters also useful when separating immature from similar Dalmatian; immature has also ***clearly darker grey-brown upperparts than the grey-buff Dalmatian***. See Pink-backed for separation. Flight consists of a few slow wing-beats followed by a glide; flocks often fly in regular lines, or circle in formation.

43 **Dalmatian Pelican** *Pelecanus crispus* **PLATE 7**

L: 150. W: 310–345. Resembles White Pelican in general appearance, build and behaviour, ***but easily told in flight by greyish underwing with pale band through centre and greyish white body*** (in White, flight-feathers below solid black and white body tinged with yellowish rosy). At close range (on the water) ***the nape feathers curl upwards*** (drooping in White). Combination of colour of eyes, legs, shape of bare skin round eye and of feathers where meeting culmen—differences described under White Pelican (which see)—useful at all ages. Immature dirty white below, ***pale grey-buff above*** (similar White is dark grey-brown above). Told from Pink-backed by larger size (but hard to see without comparison), paler secondaries with less obvious pale band through centre; at close range also by darker upper mandible, whitish eye, lead-grey legs; lacks the bold white and black eye-surroundings of Pink-backed.

44 **Pink-backed Pelican** *Pelecanus rufescens* **PLATE 7**

L: 130. W: 265–290. Resembles the other pelicans but smaller (hard to see without comparison). ***Adult duller and greyer than other pelicans with darkish curly crest on nape. In flight, shows dull greyish flight-feathers below*** (bleaching paler), ***separated from whitish grey or pink-rufous underwing-coverts by whitish translucent band through centre of wing*** (pattern close to Dalmatian's); ***flanks, back and rump have pink tinge in breeding season***, otherwise whiter. At close range told from Dalmatian ***by black markings round dark eye and pale legs***. Immature brownish above and on tail, whitish below; in flight, broad white rump narrows into white band up centre of mantle, almost to base of neck, framed by brownish shoulders and wing-coverts. Never shows solid black flight-feathers below of White Pelican.

Voice: clacking guttural conversations in breeding colonies; otherwise silent.

45 **Great Frigatebird** *Fregata minor* — Plate 5

L: 93. W: 218. Large, dark, piratical seabird with long, narrow wings and long, deeply forked tail (often held closed). Very difficult to separate from Lesser Frigatebird but note ***absence of white on axillaries, at all ages; adult male is all black*** and thus most easily distinguished. See Lesser for further discussion. Soars and glides majestically, with only an occasional deep wing-beat.

46 **Lesser Frigatebird** *Fregata ariel* — Plate 5

L: 75. W: 185. Smaller but otherwise very similar in structure and flight to Great Frigatebird and unless good views of the underwing are obtained, most frigatebirds in the field will not be identified. ***White extending onto axillaries is the most useful field feature at all ages.*** Female has black head which when seen from below contrasts with white on breast (white chin, throat and breast in Great). Juvenile often shows black mottling on lower white breast whereas in Great the white breast-patch is neater.

47 **Bittern** *Botaurus stellaris* — Plate 8

(Eurasian Bittern)

L: 75. W: 130. The most skulking heron, more often heard than seen. Smaller than Grey Heron with ***stocky neck and entirely dark brown and golden buff streaked plumage,*** providing excellent camouflage in its reedy habitat. In low, owl-like flight (but with quick wing-beats) neck often extended in front, before being retracted. Juvenile Little Bittern is similar in plumage but is very much smaller with contrasting whitish coverts in flight; juvenile Night Heron is smaller and shows white spots on wings; juvenile Purple Heron lacks streaks on back and has a long, thin neck.
Voice: in breeding season a loud boom like blowing across the mouth of an empty bottle: 'upwhoom', usually repeated many times. Flight call a hoarse 'kaau'.

48 **Little Bittern** *Ixobrychus minutus* — Plate 8

L: 35. W: 55. The smallest heron, identified by ***tiny size and, at all ages, conspicuous pale covert-panels contrasting with black flight-feathers and dark back.*** Female resembles male but duller with more rufous tinge and buff streaks on upperparts and brownish streaks on underparts. Juvenile more boldly streaked buff and brown with duller, streaked (but still conspicuous) covert panels. Most often seen in flight, when flushed, usually flying a short distance on rather jerky wing-beats before diving into cover.
Voice: a short 'kek', usually at dusk, which can also be insistently repeated 'kek-kek-kek-kek'. Male in breeding season has loud croaking 'khok' repeated at intervals of about 2 s.

49 **Night Heron** *Nycticorax nycticorax* — Plate 8

(Black-crowned Night Heron)

L: 60. W: 110. Stocky, about half the size of Grey Heron and most active at dusk. ***Adult's grey plumage with black back and crown,*** unmistakable. The ***brownish juvenile is prominently spotted with white on back and coverts*** and this feature, also noticeable in flight, immediately identifies it from other brown heron-types, notably the larger Bittern. By 2nd calendar year the spotting is reduced in size and the mantle and scapulars become

grey-brown (mirroring the black of the adult). During the day more likely to be flushed than seen in the open.
Voice: a harsh deep 'kwark', frequently heard in flight, especially at dusk.

50 **Striated Heron** *Butorides striatus* **PLATE 8**

(Green-backed Heron, Little Green Heron)

L: 43. W: 60. ***Small, rather dark, thick-set heron*** found almost exclusively on coastal shores. Adult readily identified by black crown with elongated nape-plumes, ***bluish grey upperparts*** (which can show olive gloss), ***buff-fringed coverts and greyish neck and underparts***; yellow patch on lores and dark moustachial streak give marked facial pattern. ***Rosy, pinkish or yellowish legs*** extend just beyond tail in flight. ***Immature brownish with white spots on tips of wing-coverts***, brown and white streaked upper breast and yellowish green legs; may recall young Night Heron, but is smaller with dark crown and lacks white spots on mantle. In flight (low with fast wing-beats), both adult and immature show dark upperwings without any obvious pattern. Solitary and often skulking, adopting horizontal crouching position if disturbed; most active at dusk.
Voice: occasional alarm 'chook-chook-chook'; when flushed a croaking 'kweuw'.

51 **Squacco Heron** *Ardeola ralloides* **PLATE 8**

L: 45. W: 85. A small heron, which in adult breeding plumage is a delicate golden buff with purple sheen on mantle and ***long streaked nape-plumes***; the bill has a greenish blue base. Nape-plumes lost in winter plumage when neck becomes streaked and bill has yellowish base. Juvenile is brownish buff with streaked neck and upper breast, making it well camouflaged. ***In fast flight, at all ages, white wings are revealed making the bird look predominantly white.*** Closely resembles Indian Pond Heron and great care needed to separate the two in SE Iran, Oman/E Arabia, where both can occur. Differences given under Indian Pond.
Voice: flight note a short, harsh 'kar'.

52 **Indian Pond Heron** *Ardeola grayii* **PLATE 8**

L: 45. W: 85. ***Very similar to Squacco Heron and only readily separable in breeding plumage when has unstreaked yellowish buff to buff-brown crown, hindneck and nape feathers*** (crown, hindneck and nape feathers streaked blackish in Squacco), ***blackish maroon back*** (paler, wine-coloured in Squacco) and pale buff breast (pale ochre in Squacco). In winter, loses head plumes and others are reduced in length; head, neck and upper breast become streaked brownish. Juvenile similar to winter adult but primaries with dark tips. Legs greenish to yellowish (red in some breeding birds). Winter adult and juvenile virtually impossible to separate from Squacco.

53 **Cattle Egret** *Bubulcus ibis* **PLATE 8**

L: 50. W: 85. A small, white, meadow-dwelling heron, usually seen in flocks and frequently associates with cattle. Separated from Little Egret by stockier build, ***shorter pale bill, neck and legs, extended 'jowl' under bill*** and faster flight; also, in ***breeding season, orange-buff wash on crown, back and breast, and reddish bill.*** Usually seen feeding in flocks or flying in groups to and from roost sites.
Voice: a short 'ark' and duck-like 'og-ag-ag'.

54 **Western Reef Heron** *Egretta gularis* **PLATE 9**

L: 60. W: 90. White, dark and intermediate phases occur. White phase similar to Little Egret but less elegant with thicker bill with ***curve to culmen giving slightly drooping look***. Bill in all plumages pale brown to yellowish usually with reddish flush in breeding season. Legs brownish green. Dark phase is ***slate-grey with white chin and throat*** and occasionally a few white flight-feathers. Juvenile in white phase often has grey feathers in plumage; dark phase is paler grey than adult with whitish on fore-neck, breast and belly; like adult can show some white feathers in wing. Outside breeding season facial skin yellow or greenish yellow whereas blue/grey in Little. See also Intermediate Egret. Typical feeding behaviour is slow wading in shallow coastal waters with occasional sudden rushes for prey.

55 **Little Egret** *Egretta garzetta* **PLATE 9**

L: 60. W: 90. A graceful, all-white egret, which in adult plumage has ***all-black bill and black legs with yellow feet***. In breeding season, shows long, delicate plumes on nape and mantle. Immature birds have brownish green legs and pinkish base to lower mandible. May be confused with white phase of Western Reef Heron and for differences, see that species. Told from Cattle Egret by larger size, bill colour and lack of buffish wash to plumage.

Voice: occasionally a short, grunting 'raaak'.

56 **Intermediate Egret** *Egretta intermedia* **PLATE 9**

(Yellow-billed Egret)

L: 65. W: 110. An all-white egret between Great White and Little in size. Told from Little by ***yellowish facial skin, yellow or orange-yellow bill*** (sometimes with black on tip) and blackish legs with brownish grey joints and tibia but ***without yellow feet***. Less easily told from Great White but smaller in size and with ***shorter, stouter bill***; best feature is ***thin black gape-line not extending behind eye*** as does bold line in Great White. Can appear similar to white phase of Western Reef Heron but larger with longer legs and feet blackish, brighter yellow straighter bill and always lacking nape-plumes.

57 **Great White Egret** *Egretta alba* **PLATE 9**

(Great Egret)

L: 95. W: 155. The ***largest white heron with long, angular neck*** often stretched to full extent. Adult in breeding plumage told by large size, scapular plumes, black bill with yellow base and black legs, yellowish above joint. In winter adult and juvenile, bill is yellow and lower legs blackish green or brownish. Then told from Intermediate Egret by larger size, ***longer thinner bill and black gape-line extending behind eye***. Sedate in flight and on the ground.

58 **Black-headed Heron** *Ardea melanocephala* **PLATE 9**

L: 85. W: 160. Slightly smaller than Grey Heron with ***black crown and hindneck contrasting with white throat and fore-neck*** in adult plumage. Grey upperparts darker than Grey Heron and legs blackish; bill has grey upper mandible (yellowish in Grey Heron). In juvenile, crown and hindneck dark grey contrasting less with whitish foreneck which is often tinged rusty. ***At all ages has characteristic white underwing-coverts contrasting with black flight feathers.***

59 **Grey Heron** *Ardea cinerea* **PLATE 9**

L: 95. W: 185. Large heron with long neck and legs and powerful bill. Told in all plumages by ***grey upperparts and white underparts and neck***. Adult has ***characteristic black crest and markings down front of neck***. Juvenile has darker grey (sometimes browner) upperparts and crown with rest of plumage white. Bill yellowish and legs brownish yellow in adult but in younger plumages bill browner and legs greyish. Juvenile can superficially resemble Purple Heron, but is much paler with thicker neck, stouter bill and never shows buffish brown on neck and breast.
Voice: a raucous 'waak' in flight.

60 **Purple Heron** *Ardea purpurea* **PLATE 9**

L: 80. W: 135. Slightly smaller and much darker than Grey Heron, with ***angular, snake-like neck***. Adult dark grey with black belly, ***chestnut on neck edged by black streak, black crown and line across cheeks***; dull yellow bill. Juvenile has sandy brown upperparts and hindneck with diffuse dark streaking on neck and blackish crown. ***In flight, has more angular neck than Grey Heron with prominent spread feet; note especially brownish upperwing-coverts***. For differences from Goliath Heron see that species. Mostly seen in, or flushed from vegetation, unlike Grey Heron and Goliath Heron, which feed in more open areas.
Voice: when flushed often gives a harsh 'kroonk'.

61 **Goliath Heron** *Ardea goliath* **PLATE 9**

L: 145. W: 220. ***Very large with long legs and heavy bill. Resembles Purple Heron in plumage but twice the size***. Distinctive features of adult are ***rich chestnut head and hindneck*** (contrasting with white throat), bluish grey upperparts and ***rich chestnut underparts and underwing-coverts. Immature birds are paler rufous-orange on head*** merging into grey neck and upperparts, which have some rufous edgings to feathers; white underparts are dark-streaked. ***Told from Purple Heron by large size, stouter head and neck, lack of any black on crown, absence of black line across cheeks, greyish bill*** (yellowish in Purple), ***dark legs*** (yellow/yellowish green in Purple) and upperwing more uniform grey. In flight, wing-beats slow and heavy with legs protruding more beyond tail than in Grey or Purple.

62 **Hamerkop** *Scopus umbretta* — PLATE 10

L: 60. W: 85. Unmistakable when perched with ***all-brown plumage, large 'hammer-shaped' head with heavy bill and blunt crest on nape***. Slow wing-beats in flight often interspersed with glides on slightly bowed wings; then plumage shows slight orange hue with paler orange-brown bases to primaries and tail. May be seen standing still, hunched or actively feeding with running and jabbing movements.
Voice: far-carrying laughing cry with rising inflexion, but usually silent when alone; in flight a high-pitched, nasal 'yip' or 'kek'.

63 **Yellow-billed Stork** *Mycteria ibis* — PLATE 10

L: 100. W: 160. Unmistakable in adult plumage with ***slightly drooping orange-yellow bill, bare red face and long orange-red legs***. Plumage resembles White Stork's but ***tail black*** (visible in flight) and ***mantle and tips of wing-coverts tinged pink***. In subadult plumage sandy buff with ***some pinkish on underwing-coverts***, much duller bill and legs and greyish to pale orange facial skin. In 1st-winter plumage shows greyish wash, brownish underwing-coverts, yellowish grey bill and greyish brown legs.

64 **Black Stork** *Ciconia nigra* — PLATE 10

L: 95. W: 150. Glossy black stork with white lower underparts. Only confusable with Abdim's (which see) but told by ***all-black upperparts (no white on lower back and rump) and small white axillary-patch on black underwing*** (larger white area in Abdim's). Adult has red bill and legs, whereas browner, less glossy juvenile has greyish green bill and legs.

65 **Abdim's Stork** *Ciconia abdimii* — PLATE 10

L: 80. W: 140. Very similar to Black Stork but smaller with ***white on lower back and rump*** noticeable in flight when also shows ***more extensive white on underwing-coverts, less protruding greenish legs and shorter neck***. At rest often shows white extension above bend of wing; bill which is shorter than Black Stork's is greyish green, tipped reddish; there is a crimson surround to the eye, a small white forehead-spot and the bare cheeks are bluish. Juvenile is browner, lacking the green-purple gloss of adult; bill is dirty flesh and bare skin of cheeks whitish blue. Often in flocks; circles or glides at height.

66 **Woolly-necked Stork** *Ciconia episcopus* — PLATE 10
(White-necked Stork)

L: 85. W: 160. Similar to White Stork in shape and size but easily told from it and all other storks by ***black upperparts, wings and breast with white, woolly neck*** and black cap. ***In flight, all-black wings and underparts with long white undertail-coverts and white neck give unmistakable pattern***. Immature browner than adult with longer, fluffier neck feathers. Often stands hunched; soars and glides at height.

67 **White Stork** *Ciconia ciconia* PLATE 10

L: 100. W:170. Easily told both on ground and in flight by ***combination of large size, white plumage with black flight-feathers, straight red bill and long red legs***. In flight, the neck is extended and legs protrude behind end of tail. Juvenile has duller white plumage and duller red bill and legs. Told from adult Yellow-billed Stork (a vagrant to the region) by straight red bill and all-white tail (black in Yellow-billed).
Voice: clatters bill when greeting mate at nest; otherwise silent.

68 **Marabou Stork** *Leptoptilos crumeniferus* PLATE 10

L: 150. W: 240. Unmistakable large stork with long, very stout bill and long legs. In adult, ***naked head, neck and chest-pouch, slate-grey upperparts and whitish underparts***. Juvenile has downy cover to bare head and neck. In flight shows ***blackish underwing with conspicuous white armpits***. Soars like a vulture on broad, flattish wings with curved trailing-edge and deep fingers; looks thick-necked with head sunk between shoulders.

69 **Glossy Ibis** *Plegadis falcinellus* PLATE 11

L: 60. W: 90. ***Blackish with long, decurved bill*** but close views show plumage to be ***a deep purple-chestnut***, glossed green on wings; in breeding season has white marks on face at base of dull pink bill; in winter, bill brownish with fine pale streaks on head and neck. Juvenile plumage much duller. Fast wing-beats in flight often interspersed with long glides; frequently in long-line formation. For differences from Bald Ibis see that species.
Voice: crow-like 'kraa kra kra'; grunting notes often given by feeding flocks.

70 **Bald Ibis** *Geronticus eremita* PLATE 11
(Hermit Ibis, Waldrapp)

L: 75. W: 125. Similar to Glossy Ibis but stouter with ***bare reddish vulturine head with shaggy mane, shorter legs and longer, more drooping tail***. Black plumage has an oily bronze-green sheen and a purple-bronze forewing at rest; long, decurved, dull red bill. 1st winter brownish with glossy green tail and some green feathers in coverts; lacks adult's shaggy mane, and partly bald head is pinkish-grey. In flight, wing-beats much slower than Glossy Ibis, frequently interspersed with glides and ***legs do not protrude beyond tail***.

71 **Sacred Ibis** *Threskiornis aethiopicus* PLATE 11

L: 65. W: 112. Unmistakable with black and white plumage, ***long black scapulars drooping over rear-end and long, decurved bill***. Juvenile has mottled head and neck. ***In flight, shows diagnostic black line on rear-edge of wings***. Rather heavy in flight with longish neck outstretched and legs protruding just beyond tail.

72 **Spoonbill** *Platalea leucorodia* PLATE 11
(Eurasian Spoonbill)

L: 85. W: 120. Heron-sized with ***all-white plumage, characteristic black spatulate bill with yellow tip and black legs***, these being the main differences from African Spoonbill (which see). Adult has nape-plumes and yellowish neck band, which are lost in winter. Immature has dull flesh-coloured bill and legs and black wing-tips. In flight, neck extended and wing-beats fast and shallow, interspersed with short glides, usually in small groups in line-formation. When feeding keeps in close groups, unlike egrets and herons.

73 **African Spoonbill** *Platalea alba* PLATE 11

L: 90. W: 130. Similar to Spoonbill in structure and white plumage but differs in having ***bare skin of face red and, in adult, grey bill, edged red and pinkish or red legs. No nape-plumes or yellow on neck.*** Juvenile has reduced red skin on face (no red skin on face of Spoonbill), black wing-tips (like young Spoonbill) and dusky yellow bill which overlaps in colour with that of Spoonbill; ***legs of juvenile are blackish*** (becoming red with age) whereas young Spoonbill has dull flesh-coloured legs. Actions like Spoonbill.

74 **Greater Flamingo** *Phoenicopterus ruber* PLATE 11

L: 130. W: 155. Flamingos are unmistakable, large, long-legged, long-necked pinkish wading birds with characteristic bill-shape. Greater is the commoner of the two species in the region; the white plumage of adults gradually acquires a pink hue and red wing-coverts. Juveniles, which are half the size of adults, are greyish with brown markings. Most readily told from Lesser Flamingo (which see for full differences) by larger size (but beware much size variation in Greater) and ***pink (adult) or greyish (juvenile) bill with black tip*** (all blackish in Lesser).

Voice: a short nasal grunt; feeding flocks may give goose-like sound.

75 **Lesser Flamingo** *Phoenicopterus minor* PLATE 11

L: 85. W: 130. ***Smaller than Greater and generally deeper and brighter pink with all blackish bill*** (bi-coloured in Greater) though very close views show dark carmine near tip; in full adult plumage has deep rose-pink on face bordering bill (lacking in Greater) and on the long scapulars (pinkish white/white in Greater). In flight, shows rose-pink patch across centre of upperwing-coverts, bordered pinkish-white (in Greater all secondary-coverts are rose-pink, but can bleach to white). Iris red (pale yellow in Greater). ***Juvenile similar to Greater in plumage but blackish bill readily identifies it***; Lesser's smaller size can often be difficult to establish in lone birds; also immatures of both species much smaller than adults. ***In flight, which is similar to Greater, note shorter neck and legs, faster wing-beats and more uptilted chin.***

76 **Mute Swan** *Cygnus olor* PLATE 12

L: 153. W: 223. Heavy, all-white swan; ***orange bill with black knob*** (largest in male) and base. Graceful curve to long neck with bill often pointing down. Wings often arched when swimming. Juvenile dingy brown; ***bill grey with black base***, gradually becoming pink and then orange. Plumage grows increasingly white during 1st winter and spring, sometimes still partly brownish above until 2nd winter. ***In flight, wing-beats produce loud, characteristic, rhythmic, singing sound*** 'vaou-vaou-vaou—'.

Voice: mostly silent. In breeding season sometimes a faint, hoarse snoring.

77 **Bewick's Swan** *Cygnus columbianus* **Plate 12**

(Tundra Swan)

L: 121. W: 195. ***Smaller with shorter neck and bill*** than Mute and Whooper. Adult basically similar to Whooper, but ***yellow on bill reduced to base, ending well behind nostril.*** Told from Mute by shape and colour of bill, erect neck, raised head, shorter tail and voice. Immature more greyish than Mute, and ***pink bill lacks black base***, becoming white and later yellow during 1st winter. From immature Whooper by ***shorter neck and bill and more rounded head.*** Walks more easily than Mute, and in flight, wing-beats slightly faster, more goose-like and with no humming sound.

Voice: calls similar to Whooper (and very musical in flight and on water) but obviously higher pitched, monosyllabic or disyllabic (Whooper often has three syllables), sometimes recalling distant barking dogs.

78 **Whooper Swan** *Cygnus cygnus* **Plate 12**

L: 153. W: 231. Basically similar to Bewick's, but larger (size of Mute) with proportionally longer neck and bill; flattened forehead and bi-coloured bill: ***black with yellow base extending in wedge usually to below nostril or beyond.*** Distinguished from Mute by erect neck, shape, colour of bill, short tail, raised head, and voice. Walks well, not with awkward, waddling action of Mute. Silent flight similar to Bewick's but with powerful wing-beats like Mute, but note flattened head profile. ***Juvenile/immature similar to Bewick's, but note longer neck and head profile*** of Whooper. Also, young birds are often in company with parents during 1st winter.

Voice: mostly calls in flight and on water. Similar to Bewick's but deeper and stronger and with musical, trumpet-like quality 'ahng-ha' or 'ko-ko-ko-'. Often trisyllabic.

79 **Bean Goose** *Anser fabalis* **Plate 12**

L: 75. W: 158. Smaller and slimmer than Greylag, slightly larger than White-fronted. Variable often wedge-shaped head and bill; ***dark neck and head (almost blackish forehead)*** and pale brown body; ***bill mixture of black and orange***, sometimes mostly orange (in tundra race, *rossicus*, bill mostly dark and deeper based), occasionally small, white base to bill; ***bright orange legs***, but exact colour can be difficult to discern. Strong white feather-edges to dark, brownish upperparts. In flight, upperwing is dull brownish with slightly greyish outerwing; coverts similar in tone to upperparts, dark head and neck conspicuous. Greylag is heavier, more broad-winged, has grey head, pink bill and legs and striking silvery forewing. White-fronted lacks dark brown head, has pink bill and, if adult, white facial blaze and black belly-barring.

Voice: least vocal of grey geese; flight-call, repeated disyllabic 'ung-ank' or 'ow-ow'.

80 **White-fronted Goose** *Anser albifrons* **Plate 12**

(Greater White fronted Goose)

L: 72. W: 148. Smaller than Greylag, and normally larger than Lesser White-fronted. Warmer grey-brown than Greylag, with slightly darker head and hindneck; has orange legs and lacks silvery forewing; ***nearly plain upperparts with less defined pale edges to coverts and tertials than Bean.*** Wing pattern similar to Bean and Lesser White-fronted. Bill not heavy, but quite long, ***pink with white nail***; no eye-ring visible in field. ***Adult has large white area surrounding base of bill*** (rarely also forecrown) and ***black bars***

on underparts. Juvenile warmer brownish, lacks white forehead and black bars on underparts; ***forehead and area round bill very dark, also dark nail on bill***. See also Lesser White-fronted.
Voice: musical and much higher pitched than other grey geese except Lesser White-fronted. Repeated disyllabic, sometimes trisyllabic with metallic, laughing quality 'kow-lyow' or 'lyo-lyck'.

81 **Lesser White-fronted Goose** *Anser erythropus* PLATE 12

L: 59. W: 127. ***Smallest grey goose***. Resembles White-fronted, but smaller (though a few small White-fronted overlap), with proportionally ***shorter neck and bill, steep forehead, yellow eye-ring*** and slightly longer wings often extending past tail. Adult has ***extensive white on forehead and forecrown to above eye***; upperparts dusky brown with only dull transverse lines; underparts with only a few black blotches. Juvenile similar to juvenile White-fronted, but told by size, head and neck proportions and yellow eye-ring. Flight similar to White-fronted, but faster wing-beats and silhouette more compact. On ground noticeably faster walk.
Voice: higher pitched, squeakier and more rapid than White-fronted. A di-or trisyllabic yelp, 'kow-yow', 'kyu-yu-yu' or piping 'yi-yi-yi'.

82 **Greylag Goose** *Anser anser* PLATE 13

L: 83. W: 164. Largest goose in the region. Thick-neck with ***heavy, pink bill, greyish head and neck and pink legs***. In flight, shows distinctive, ***pale silvery forewing***. Upperparts more greyish than other grey geese; belly with small, black blotches, not forming bars. Juvenile similar to adult but lacks sharply defined transverse lines of upperparts and dark belly marks. From all other geese in the region by combination of size, bill and leg colour and pale forewing.
Voice: similar to domestic geese. Flight call loud, most characteristic 'ang-ang-ang—', rather deep and sonorous.

83 **Brent Goose** *Branta bernicla* PLATE 13

L: 59. W: 115. ***The only sooty black goose to occur in the region***. Small, stocky, with short neck, ***white neck-patch, black bill and legs***; vent and undertail-coverts gleaming white. Juvenile lacks neck-patch (starts to show in 1st winter), and has prominent pale-edged wing-coverts. Strictly coastal and more aquatic than other geese. Flight fast with quick wing-beats.
Voice: rather silent except in flight: gargling or growling 'r-rot'.

84 **Red-breasted Goose** *Branta ruficollis* PLATE 13

L: 55. W: 125. Small, compact, very dark goose with ***extremely short, black bill***. Unmistakable pattern of ***black body, chestnut cheeks and breast*** (looks dark at distance) ***bordered by white***; broad shaggy ***white line along flanks*** and ***white patch in front of eye***. Juvenile has smaller chestnut cheek-patch. Often mixes with grey geese. Flies fast with agile actions.
Voice: flight call a shrill, staccato 'ki-kwi'.

85 **Egyptian Goose** *Alopochen aegyptiacus* **Plate 13**

L: 68. W: 144. Slightly larger than Ruddy Shelduck, and like that species has black flight-feathers, green speculum and striking white forewing and underwing-coverts. Otherwise ***basically brownish buff*** with variable brownish or greyish upperparts, pale fore-neck and face with ***prominent dark eye-patch***; dark collar and belly-patch; proportionally long, pink legs. Juvenile/immature browner, lacking eye patch, collar and belly patch, and with yellowish-grey legs. Distinguished from Ruddy Shelduck by longer legs, dark eye-patch and body colour.

Voice: rather silent except in social situations: flocks calling with hoarse, rasping 'kvaa-kvaa-kvaa'. Male a harsh, wheezy hiss. Female a guttural, strident cackle 'honk-haah-haah-haah'.

86 **Ruddy Shelduck** *Tadorna ferruginea* **Plate 13**

L: 64. W: 133. Shelduck size. Short-necked, small-billed, goose-like duck with ***orange-chestnut body*** and paler cinnamon-buff head; ***black flight-feathers*** with green speculum, ***striking white forewing*** and underwing-coverts similar to Egyptian Goose; black bill and legs. In flight, told from other ducks or geese by size, striking black and white wing pattern and orange-chestnut body. Female and non-breeding male lack black neck-collar, and female has paler head. Juvenile resembles adult female, but has browner back. From Egyptian Goose by shape, head pattern and darker chestnut underparts.

Voice: from flocks on water or in flight: nasal, trumpeting or whooping rather penetrating 'ang—ang—'.

87 **Shelduck** *Tadorna tadorna* **Plate 13**

(Common Shelduck)

L: 63. W: 122. A large, ***mainly white, goose-like duck***. Black head with green gloss, ***broad chestnut belt round forepart of body and black belly-stripe***; prominent ***red bill*** with fleshy knob in breeding male. In flight, shows ***black flight-feathers contrasting strongly with white upper- and underwing-coverts*** and black shoulders. Flight is goose-like with slow wing-beats. Juvenile lacks chestnut belt, and head, hindneck and upperparts are grey-brown contrasting with white underparts, foreneck and face; bill and legs grey or pale flesh.

Voice: mostly silent outside breeding season. Male's call is a high whistle 'siss-siss-siss—', answered by the female's whinnying 'geheheheheheh'; sometimes nasal 'ah-hang'.

88 **Comb Duck** *Sarkidiornis melanotos* **Plate 111**

(Knob-billed Goose)

L: 76. W: 150. Large duck, with strong flight recalling goose. ***Male has large fleshy knob on base of bill and forehead***, absent in female and juvenile. ***White head and neck speckled blackish***, underparts largely white, in male with narrow black half-collar down sides of breast (absent in rather smaller female which has also more mottled underparts). Upper back blackish, glossed green and purple, ***lower back conspicuously grey***. In flight, upperwing, rump and tail appear all-dark except for bronze speculum. Immature duller, less glossy than female. Perches freely on trees and branches and feeds largely by grazing.

89 **Cotton Teal** *Nettapus coromandelianus* **PLATE 13**

(Cotton Pygmy-goose)

L: 33. W: 55. ***Smallest duck to occur in the region.*** Short-neck and short, ***stubby, goose-like bill***; round body with high rear on water. Male unmistakable; ***white head, neck and underparts with black cap, eye and band across white breast***; ***black back*** with greenish and purple gloss; flight is fast; looks white with green-tinged ***black wings with conspicuous white trailing edge across full length of wing, very broad on primaries.*** Female a drab version of male, browner with dark eye-line and greyish flanks; in flight brown wings and narrow white trailing edge restricted to secondaries.
Voice: a sharp, staccato cackle of male 'car-car-carawak' or, chiefly in flight, 'quack, quack-quackyduck'. Female a weak 'quack'.

90 **Wigeon** *Anas penelope* **PLATE 13**

(Eurasian Wigeon)

L: 48. W: 80. Medium-sized ***rather rufous duck*** with short neck, rounded, ***steep forehead*** and ***small grey bill***, tipped black; looks high at rear end on water. ***In all plumages has white belly-patch***, obvious in flight and when grazing on land. Adult ***male has creamy yellow forehead*** and forecrown contrasting with chestnut head and neck; pinkish breast, grey back and flanks and ***black undertail-coverts***; conspicuous white forewing in flight showing as white band along side at rest. Adult female variable in plumage from rufous to greyish brown lacking white forewing. Juvenile resembles adult female (immature male does not usually assume white forewing until 2nd winter). In flight, shows narrow wings, pointed tail, and contrasting, white belly-patch; ***white forewing contrasting with blackish green speculum is very distinctive in adult male.*** Female and juvenile have grey-brown forewing, darker flight-feathers and black speculum; for differences from female Falcated see that species.
Voice: male utters characteristic clear whistle 'wheeooo', both in flight and when on water.

91 **Falcated Duck** *Anas falcata* **PLATE 14**

(Falcated Teal)

L: 50. W: 79. Dabbling duck, between Mallard and Teal in size; looks high at rear end on water. Male basically grey with large, glossy green and bronzy chestnut head and ***drooping crest on nape***; steep forehead and black bill; characteristic ***long drooping tertials at rear end***; conspicuous Teal-like tail pattern. Distinguished from Mallard by black bill and grey breast; from Teal by size and head-shape. Female and juvenile resemble female Gadwall or Wigeon, but have ***plain, dark head*** (not rufous like Wigeon) with full nape, ***black bill and legs*** (orange in Gadwall), different wing pattern and buffish (not white) belly. Male in flight has grey forewing, black and green speculum bordered in front by broad, white wing-bar. Female in flight resembles male, but has grey-brown forewing and buff wing-bar.
Voice: generally silent outside breeding ground. Male a low trilling whistle 'tyu-vit ... tyu-tyu-vit'. Female a quacking call like female Gadwall.

92 **Gadwall** *Anas strepera* **Plate 14**

L: 51. W: 89. Medium-sized, slightly-built dabbling duck. Swims high on water with elevated rear end like Wigeon. Flies with rapid wing-beats and pointed wings like Wigeon. ***Male mainly dark grey with black around tail***; in flight, ***white speculum bordered black in front,*** and ***white belly.*** Female and juvenile resemble larger and heavier built female Mallard, but tail is grey-brown (not white), and thin dark bill is orange along sides; ***in flight, shows white speculum patch close to body*** (often smaller than male) and ***white belly*** bordered by dark flanks.

Voice: male has a rasping, low 'rrep' call, and a shrill 'pyee' in courtship. Female's call resembles Mallard's but is higher.

93 **Teal** *Anas crecca* **Plate 14**

(Common Teal)

L: 36. W: 61. ***Smallest of the common ducks*** in the Middle East; readily taking to the wing with swift flight. At distance, male looks dark-headed with greyish body; ***yellow patches on side of rump and thin white horizontal stripe on side of body.*** Female told from Garganey female (which see) by ***evenly coloured head and throat,*** small light patch on side of tail base and darker upperwing with white stripe through centre; from all other ducks by small size. In flight, both sexes show white belly and ***greyish underwing with light band through middle. Upperwing-pattern of both sexes is dark with prominent white or yellow streak in middle of wing in front of greenish speculum.*** Male told from Garganey by head- and wing pattern.

Voice: male has far-carrying ringing whistle 'kreek-kreek'. Female a higher pitched, more nasal quacking than female Mallard's.

94 **Mallard** *Anas platyrhynchos* **Plate 14**

L: 56. W: 95. Large dabbling duck. Male easily told by ***dark greenish head, yellow bill, brown breast and grey body***; in flight by ***grey wings, dark blue speculum distinctly bordered by white, and white underwing-coverts.*** Female mottled brown like other female ducks and told from most by size, bill shape and bill colour, and wing pattern; in addition, told from Gadwall by mottled brown belly, from Pintail by structure and from Teal and Garganey by size. Male in eclipse resembles female, ***but has greenish yellow bill.***

Voice: male rather silent. Has a soft nasal 'raehb' and, during courtship, a weak, high-pitched whistle 'piu'. Female has a well known deep quacking.

95 **Pintail** *Anas acuta* **Plate 14**

(Northern Pintail)

L: 56 (excluding central tail-feathers of larger male). W: 88. Large, very slim and elegant dabbling duck; ***long neck and tail, also obvious in flight.*** Male unmistakable: ***white neck and underparts,*** grey upperparts contrasting with ***dark head and tail and dark slender bill.*** Female recalling some other female dabbling ducks, but note ***dark bill, greyer plumage and much slimmer appearance.*** In flight, both sexes show distinctly longer neck and tail, and slender, more pointed wings than other dabbling ducks. Male has grey upperwing with ***green speculum and conspicuous white border to rear.*** Female has generally brownish upperwing with distinctive ***white border at rear of***

secondaries (two white bars and blue speculum in Mallard; blue-grey forewing in Shoveler; white wing-patch in Gadwall).
Voice: male has a Teal-like whistle, but lower pitched. Female a hoarse quack.

96 **Garganey** *Anas querquedula* PLATE 14

L: 39. W: 63. Small, Teal-sized dabbling duck. Male unmistakable: broad, ***white stripe above eye to hindneck***, black-mottled breast contrasting with grey sides and white belly; ***in flight, striking blue-grey forewing*** contrasts with dark head, breast and upperparts. Female similar to female Teal, but note ***longer bill and more contrasting dark and light head-stripes widening in front into pale patch at bill base***, and whiter throat. In flight, forewing is slightly lighter and white border along secondaries is distinctive (Teal has white wing-bar in middle of wing in front of speculum). Male in eclipse like female, but wing pattern as male adult. Juvenile similar to female, but lighter forewing in male.
Voice: ***male's call is a dry, drawn-out rattling*** like a nail upon a comb 'knerreck'. Female has a short, sharp quack.

97 **Shoveler** *Anas clypeata* PLATE 14
(Northern Shoveler)

L: 51. W: 78. Medium-sized, rather heavily-built and short-necked dabbling duck. Told from all other ducks by ***huge, spatulate bill.*** Swims with front end low and bill often dabbling in water. Shape and proportions characteristic. In flight, wings appear set far back. Male unmistakable: ***dark bill and glossy green head (looks black at distance) contrast markedly with white breast and shoulder; flanks and belly strikingly chestnut.*** In flight, looks black, white and chestnut with ***distinctive blue forewing.*** Female and juvenile on water resemble female Mallard and Gadwall, but enormous bill is distinctive; in flight, has bluish forewing, somewhat resembling smaller Garganey male; from below, brown belly contrasts with white underwing. Male eclipse largely resembles female, but brighter blue forewing.
Voice: rather quiet. Male's call in spring is a hollow, double-note 'g-dunk—g-dunk', often in flight. Female quacks at the same time 'pe-ett'.

98 **Marbled Teal** *Marmaronetta angustirostris* PLATE 15
(Marbled Duck)

L: 41. In flight resembles small female Pintail having fairly long neck, wings and tail. On the water identified by ***pale plumage, dappled dark and cream*** ('marbled') with ***dark oval patch around eye*** in otherwise pale head and long ***black bill***; head appears ***large and rounded*** with steep forehead and, in adult, bulky crest on lower nape. In flight, wings are fairly pale with slightly paler secondaries but ***no speculum***; tail creamy white. Often difficult to see, hiding in vegetation.
Voice: low nasal wheezing 'jeak' or double whistling note.

99 **Red-crested Pochard** *Netta rufina* PLATE 15

L: 56. Plump, Mallard-sized, diving duck but often behaving as a surface-feeder; sits high on the water. Male diagnostic with ***large red head*** and paler erectile crest, ***red bill***, black breast and ***white flanks***; in eclipse resembles female but bill red. Female drab brown with ***pale grey cheeks contrasting with dark crown*** (superficially like female Common

Scoter); dark bill has ***pink band near tip***. Appears large in flight in which it shows ***broad white translucent wing-bar throughout wing***.
Voice: male has a double 'weep-weep', female a grating 'keerr'.

100 **Pochard** *Aythya ferina* — Plate 15
(Common Pochard)

L: 46. W: 80. ***Sloping forehead grading into long bill*** characteristic; no crest. Male has ***chestnut head and neck, contrasting with black breast*** and ***pale grey body***; black bill has pale blue-grey band in centre; in eclipse resembles female but greyer above. Female fairly nondescript with dull brownish head and breast, chin and eye-stripe paler; dark bill becomes paler towards broad black tip. In flight, appears longer and plumper than Tufted; greyish-brown wings (palest in male) have ***indistinct pale grey wing-bar***.
Voice: courting male has low whistle; also hoarse wheezing note; female has rough harsh 'krra-krra'.

101 **Ferruginous Duck** *Aythya nyroca* — Plate 15

L: 40. W: 66. Slightly smaller than Tufted; shape of head close to Pochard's; ***no crest; compared with Tufted has higher crown, flatter sloping forehead and longer dark grey bill with black nail only*** (compare Tufted). Male is ***rich dark chestnut-brown*** (particularly head and breast) with ***white eye, sharply defined pure white undertail-coverts*** and smaller white belly area than Tufted. Female and immature have dark eyes (but white in one-year-old male); they lack whitish at base of bill of some Tufted; dark grey-brown ***female has warmer brown head than similar Tufted*** (which has yellow eye); pure white and sharply defined white undertail-coverts (sometimes seen in Tufted). ***First-year birds identified by shape of head, length and pattern of bill and dark eye***. In flight separated from Tufted by ***broader, more conspicuous white wing-bar extending onto outermost primaries*** (in Tufted wing-bar not pure white on outermost primaries).
Voice: female has repeated 'karr' with high rising tone. In flight a high-pitched 'err-err'.

102 **Tufted Duck** *Aythya fuligula* — Plate 15

L: 42. W: 70. Small diving duck with roundish head but fairly steep forehead; ***crest at nape***, long and drooping in male, minute in female; ***bill blue-grey*** (darker in female) ***with broad black tip. Black and white male separated from Scaup by black upperparts*** and purple head (green-headed male Scaup has pale grey back, more rounded head and lacks crest). Female and male in eclipse have brownish sides to body; female sometimes with whitish band round base of bill (but seldom so conspicuous as in female Scaup). In summer female has ***darker back*** (washed greyish in female Scaup), ***short crest*** (absent Scaup) ***and broad black tip to bill*** (compare Scaup). Sometimes shows white undertail-coverts (see Ferruginous). In rapid flight has conspicuous white wing-bar, almost the full length of wing. Gregarious in winter.
Voice: a 'kerr' seldom heard outside breeding season.

103 **Scaup** *Aythya marila* — **Plate 15**

(Greater Scaup)

L: 46. W: 79. Larger than Tufted with larger, more rounded head, ***no crest and only black nail to longer blue-grey bill*** (compare Tufted); male also told from Tufted by ***pale grey back and greenish sheen to head.*** Female (and male in eclipse) has ***dark back washed greyish***; usually has ***bold white band at base of bill*** (though sometimes approached by female Tufted); in spring a ***pale patch on ear-coverts not seen in Tufted***; in winter this patch is absent or indistinct. Some 1st winter birds may lack white at base of bill but told at close range by shape of head, bill pattern and some greyish mottling on back. Though larger and stockier than Tufted in flight not easily told by silhouette without comparison.

104 **Eider** *Somateria mollissima* — **Plate 111**

L: 60. W: 100. Large, heavy diving duck, told from other ducks by ***elongated, flat profile of forehead and wedge-shaped bill.*** Male has ***black belly but white breast*** (tinged pink), ***forewing and mantle***; white head, black crown and pale green nape-patch are other features. Brown female ***closely barred black.*** In flight, which has heavy wing-beats, neck is thicker and head held lower than in other ducks; female has purple-blue speculum bordered by narrow white stripes (recalling female Mallard's); immature female lacks white borders to speculum.

105 **Long-tailed Duck** *Clangula hyemalis* — **Plate 16**

(Oldsquaw)

L: 42 (excluding central tail-feathers of male). W: 76. Small, short-billed, distinctive seaduck. ***Male in winter black and white with very long tail; white, steep-fronted head with black cheek patch*** and broad, black breast band contrasting with otherwise generally white plumage. Female variable, more brownish, washed out, often with pale head and dark cheek markings; lacks long tail. Juvenile generally like female, but male shows bi-coloured bill. ***In all plumages shows plain, dark brown wings in flight,*** which is swift and close to sea-level.

106 **Common Scoter** *Melanitta nigra* — **Plate 16**

(Black Scoter)

L: 49. W: 84. Medium-sized dark seaduck. ***Male all black with yellow on part of upper mandible and black wings in flight,*** though flight-feathers paler. Female has dark brown body and crown with ***contrasting greyish buff cheeks*** and dark wings in flight. Similar coloured Velvet Scoter separated by head and wing pattern.

107 **Velvet Scoter** *Melanitta fusca* — **Plate 16**

(White-winged Scoter)

L: 55. W: 94. Rather large, very ***dark seaduck with distinctive white wing-patch*** covering all secondaries, very obvious in flight at long range. ***Male black with partly orange bill and white fleck below eye.*** Female dark brown with variable white patches on head. Both sexes resemble Common Scoter, but note different head and wing pattern and sloping forehead. White wing-patch sometimes hidden when on water. Flight is heavy and low over the waves.

Voice: rather silent. When pair circling over breeding area, female gives nasal rolling 'arr-ha', and then a whistling call can be heard from male.

108 **Goldeneye** *Bucephala clangula* (Common Goldeneye) — PLATE 16

L: 46. W: 73. Fairly small, easily identified seaduck with ***big rounded head***, yellow eye (adult) and much white in wing, obvious in flight and often on folded wing too. ***Male largely white*** with black back and rump and black head, glossed green; ***large white circular patch between eye and bill.*** Female greyish with ***contrasting brown head and white belly.*** In flight, both show characteristic shape with round head, short neck and ***nearly white innerwing*** and dark underwing. Flight with rapid wing-beats, male producing characteristic ***whistling wing-noise***, audible at long range.

109 **Smew** *Mergus albellus* — PLATE 16

L: 41. W: 62. Small, very pale and distinctive duck. ***Male almost white*** in appearance at distance; ***white head patterned by black face mask***, thin breast and side marks and black back; slightly crested and steep forehead. Female resembles Goldeneye female, but ***white cheeks are distinctive***; from Red-crested Pochard by size and wing pattern; and much paler than female Common Scoter. Flight is fast and agile, and shape is slender like other sawbills. Wings are black in male with white patch on forewing as in Wigeon. Female similar, but with grey-white forewing and brown flight feathers. Eclipse male (resembles female) until Nov.

110 **Red-breasted Merganser** *Mergus serrator* — PLATE 16

L: 56. W: 78. Large, rather dark sawbill with ***conspicuous wispy crest in both sexes.*** In flight, appears elongated with long head and neck, and shows much white in innerwing, (no visible feet beyond tail as in grebes). ***Males greenish black head, thin red bill, white neck collar and rusty black-spotted breast***, easily separates it from other ducks. Female and juvenile closely resemble female Goosander, but smaller, and more brownish (less greyish) upperparts, ***less contrasting head pattern, and without sharp demarcation from brown head to greyish buff neck and breast.*** Flight is fast and direct.

111 **Goosander** *Mergus merganser* (Common Merganser) — PLATE 16

L: 65. W: 90. Largest sawbill. Looks very elongated both on water and in flight. Much more contrasting plumage than Red-breasted Merganser. Flight is strong, fast and direct, and sometimes high over land. Male unmistakable: ***clean-coloured, salmon-pink or white neck and underparts contrasting with clear-cut greenish black head without crest***, but sometimes very steep forehead with ruff on hind neck; ***nearly all white innerwing*** very striking in flight. ***When overhead looks white with black head.*** Female and juvenile easily confused with female Red-breasted Merganser, but note greyish upperparts, ***chestnut-brown head with sharp demarcation to white chin and to whitish neck.*** In flight, larger, paler, more contrasting brown head and with more white on innerwing than female Red-breasted Merganser.

112 **White-headed Duck** *Oxyura leucocephala* **Plate 16**

L: 46. W: 66. Medium-sized strange-looking diving duck with ***short neck and large bill, and long stiff tail often held cocked*** at rest on water. Male unmistakable: ***white head, grotesquely swollen base of blue bill***, black crown and neck, and largely chestnut body. Female buffish-brown with dark brown head and variable ***grey- and dark-striped cheek pattern***; bill grey, swollen at base. Male eclipse more female-like, but usually with white cheeks. Juvenile duller and paler than female. Swims low on water. In flight, which is seldom, looks elongated with short uniform wings appearing far forward on body. Rather silent most of year.

113 **Honey Buzzard** *Pernis apivorus* **Plate 24**

(European Honey Buzzard)

L: 55. W: 135–150. ***Recalls Common Buzzard but tail longer and narrower with rounded corners, head and neck narrower, protruding in cuckoo-like manner; wing-beats more flexible.*** Soars on flattish- and glides on slightly lowered wings (more up-lifted in Common Buzzard). Male has greyish head and upperparts; female browner. Below, some are blackish brown, others virtually white on body and coverts, ***but most are barred, with large black carpals; tail has dark band near tip and 2 at base.*** Cere grey, eyes yellow. Juvenile dark brown, rufous-brown or creamy-white with streaked breast; ***often, but not always, with dark carpal-patches; head often whitish with dark eye-mask***; uppertail-coverts whitish (unlike Common Buzzard); pale band on greater underwing-coverts often separates secondaries from dark forewing (unlike Common Buzzard). Slenderer, more curved wings and shorter tail than adult give more Common Buzzard-like outline, but shape of tail and head, wing position and wing action separate it.

Voice: musical, drawn-out 'plee-oo-lee' heard in mid-summer.

114 **Crested Honey Buzzard** *Pernis ptilorhynchus* **Plate 21**

L: 65. W: 160. Resembles Honey Buzzard in shape and wing position but wing-span slightly larger, wings broader, wing-tip more ample (6 long 'fingers' against 5 in Honey) and tail broader and shorter. Adult identified by ***dark gorget across throat, lack of bold black carpal-patches and pattern on flight and tail feathers below.*** Adult male (dark red eyes, yellow in male Honey) ***has distinctive black undertail with broad pale band in centre*** and longer black flight-feather band below than male Honey, reaching body (in male Honey, inner part of band hidden beneath greater coverts). Adult female has undertail pattern as in male Honey, but inner bar (at tips of tail-coverts) averages broader; compared with female Honey ***it has 3 dark evenly spaced bars across secondaries*** (female Honey has usually 2 bars with wider gap between dark trailing edge and next bar). Juvenile like dark morph of young Honey with similarly dark eyes and yellow cere; best told by shape and structure (see above). Adult Crested usually migrate in autumn with new inner 4–5 primaries (in adult Honey at best 1–3).

115 **Black-winged Kite** *Elanus caeruleus* **Plate 26**

(Black-shouldered Kite)

L: 33. W: 76. Size of large Kestrel, but appears larger in flight. Easily identified by combination of ***black primaries on white under-surface and large black patch on forewing*** in otherwise pale grey upper-surface, but tail whitish. Juvenile darker and dirtier above with thin white line on greater coverts. Well protruding, broad head, shortish tail with slight notch when closed, long pointed but relatively broad-based wings; almost owl-like flight with soft wing-beats, raised wings like harrier when soaring, interspersed with persistent hovering.
Voice: in display, thin whistling 'wee-oo, wee-oo'.

116 **African Swallow-tailed Kite** *Chelictinia riocourii* **Plate 26**

L: 37. W: 75–80. Small raptor with ***long, narrow, pointed wings*** and ***long, deeply forked tail*** (adult) and ***graceful, tern-like flight***. Adult is grey above, white below with ***large oval black patch near carpal***; flight- and tail-feathers often translucent. ***Browner juvenile lacks elongated tail-feathers***, tail-shape being like Black-winged Kite, but ***lacks black forewing above***. Hovers persistently with great skill and grace with little wing-movement.

117 **Black Kite** *Milvus migrans* **Plate 17**

L: 50–63. W: 135–150. Resembles Red Kite in flight and silhouette but ***less forked tail shorter*** (slightly rounded when fully spread), wings little shorter and broader and ***flight less buoyant***. Adult of European *migrans* (smaller than Red Kite) darker and browner than Red, with ***hardly any white on primaries below*** and ***dark brown, not reddish uppertail***; head darker and pale band across upperwing-coverts less conspicuous. Adult *lineatus* (winter visitor Iran, Iraq) differs from *migrans* by ***paler belly, without contrast to pale undertail*** and has larger white patch on primaries. In yellow-billed *aegyptius* (SW Arabia) most adults have yellow bill (some are dark-billed like juvenile); under surface is brighter rufous-brown and white patch on primaries large (approaching Red). Young Black Kite (all races) has dark eye-mask, pale feather tips on mantle and shoulders, ***boldly dark-spotted breast but pale belly with no contrast to undertail***, dark band near tip of tail and ***whitish line on tips of greater upperwing-coverts***. Flies with elastic wing-beats, soars and glides on slightly arched wings; manoeuvres tail when scanning for food. Gregarious at rubbish dumps and at night-roosts outside breeding season. Can be confused with distant dark phase Booted Eagle, which see.
Voice: gull-like, high-pitched, whinnying, descending, drawn-out 'yiieerr' is most frequent call.

118 **Red Kite** *Milvus milvus* **Plate 17**

L: 61. W: 145–155. Larger and slimmer than Black Kite with ***longer, deeply forked rusty red tail, narrower wings with bold white patch on primaries below*** and ***more buoyant flight with more elastic wing-beats***. In adults, underparts largely dark rusty red, contrasting with pale undertail; broad buffish brown band across innerwing above. Juvenile has dark breast with bold pale streaks, rest of underparts and undertail pale; duller red-brown uppertail has slight, dark subterminal-band, broader, paler band on innerwing above and ***greater upperwing-coverts show thin white line***. Light and graceful soaring and gliding on arched wings and constantly manoeuvred tail (slightly forked when fully spread). Feeds largely on carrion but takes prey up to size of partridge and small hare.

119 **Brahminy Kite** *Haliastur indus* — **Plate 111**

L: 48. W: 135. Size of Black Kite, with similar wing position though sometimes soars with wings in shallow 'V'; tip of tail slightly rounded. Distinctive adult ***has white head, neck and breast*** (at close range thinly dark-streaked); otherwise red-brown above and on underbody with paler rufous underwing-coverts; flight-feathers and undertail creamy buff; blackish wing-tips most conspicuous from below. Juvenile lacks white head and neck, is darker brown above and below, with greyish secondaries below but ***conspicuous whitish primary-patch*** and more extensive black wing-tip than adult. Sometimes recalls Marsh Harrier when foraging low over ground.

120 **Pallas's Fish Eagle** *Haliaeetus leucoryphus* — **Plate 17**

(Pallas's Sea Eagle)

L: 80. W: 190–220. Smaller than White-tailed with narrower wings and longer, ***rounded tail***. Adult has creamy-white head and neck merging into chestnut-brown body and ***white tail with broad black terminal band***; bill dark grey (yellow in White-tailed). Juvenile told from similar White-tailed by ***paler head with dark patch behind eye, uniform pale brown underparts***, often darker foreneck but yellowish throat (White-tailed has blackish head and streaked or mottled underparts), ***broad pale band through under-wing-coverts, contrasting with dark brown leading coverts***, (white lines in well-marked White-tailed), ***distinct white patch or white streaks on primaries below*** (absent in White-tailed) and ***blackish tail*** (sometimes faint whitish mottling at base of spread undertail); legs grey (yellow in young White-tailed). Up to three years old, paler head and underparts emphasize dark eye-patch and, often, band round foreneck; under-wing like juvenile but ***tail-centre distinctly white-mottled, forming pale band in some***; differs from young Golden by pale axillary patch and centre of underwing, paler underparts, more ***parallel-edged wings held flattish*** and by longer, more protruding neck. When perched, bare tarsus and pale loral patch separates it from *Aquila* eagles.

121 **White-tailed Eagle** *Haliaeetus albicilla* — **Plate 17**

L: 80–90. W: 190–250. Large, bulky eagle, adult with broad parallel-edged wings dark underwings, ***short white wedge-shaped tail***, whitish brown head and upper breast, ***large yellow bill and long protruding neck***. Juvenile has curved trailing edge to wings, arm broader, hand narrower than adult; appears dark at distance, but at close range ***head and neck blackish***, contrasting with rusty brown dark-streaked underparts; spread tail has ***pale feather centres approaching*** tip of tail; whitish axillary-patch and lines along central underwing-coverts; pale loral patch. Immature variable but rather whitish mottled below and on mantle and shoulders; still dark-headed. Soars and glides on flattish or slightly arched wings; active flight with long series of heavy wing-beats.
Voice: a loud 'klee-klee-klee'.

122 **Lammergeier** *Gypaetus barbatus* **PLATE 17**
(Bearded Vulture)

L: 102–115. W: 250–280. Identified by ***unique outline with long, narrow wings, long wedge-shaped tail and large size***. Adult has ***orange-buff underparts contrasting with dark underwings and tail***; upperparts metallic slate-blue (look black at distance); at close range golden head has black eye-mask and feathers drooping below bill. Juvenile has blackish-brown head and neck, contrasting with dirty brownish grey underparts and underwings; largely dark brown above but mantle whitish. Soars and glides effortlessly along mountain ridges on slightly lowered wings. Solitary.
Voice: has whistling and querulous sounds at breeding site.

123 **Egyptian Vulture** *Neophron percnopterus* **PLATE 18**

L: 62. W: 155. Small vulture, size of Lesser Spotted Eagle. Adult has ***white, wedge-shaped tail***, white underparts with black flight-feathers (secondaries greyish white above), ***small pointed head and thin bill***; colour-pattern resembles pale phase Booted Eagle or White Stork below but shape quite different. Juvenile: mid-brown below, ruff blackish, dark brown above with creamy bars on wing-coverts, pale rump, whitish uppertail-coverts; ***wedge-shaped tail grey-brown, tipped paler***. Soars much on slightly arched wings; active flight has many deep wing-beats before a glide. Scavenges for offal.
Voice: occasional grunts and whistles but mostly silent.

124 **Indian White-backed Vulture** *Gyps bengalensis* **PLATE 18**
(White-rumped Vulture)

L: 87. W: 208. Resembles Griffon in outline but smaller. Adult ***white on underwing-coverts contrasting*** with dark flight-feathers; underparts slate (much darker than Griffon's); with ***conspicuous white patch on lower back and rump framed by slate-grey upperparts and wing-coverts which are darker than flight-feathers*** (Griffon pale brown above with contrasting dark flight-feathers). Juvenile ***lacks white on rump and underwing-coverts*** (except usual white 'vulture-streak' near leading forewing); resembles Griffon but darker buff-brown underwing-coverts show less contrast with darker flight-feathers (Griffon has paler, warmer wing-coverts); warm dark brown upperwing-coverts contrast less with flight-feathers than in Griffon (but coverts bleach paler); pale streaks on body and upperwing-coverts broader than in Griffon. Immature develops creamy buff bands in middle of underwing, creating contrast like Griffon, but ***diagnostic white rump-patch starts to appear***. Gregarious, often allowing close approach.
Voice: hisses and squeals at breeding site.

125 **Griffon Vulture** *Gyps fulvus* **PLATE 18**
(Eurasian Griffon Vulture)

L: 95–105. W: 245–270. Large, heavy vulture with long, broad, deeply-fingered ***wings with curved trailing edge***; short, broad tail almost square-cut and thin head/neck protrudes only slightly. ***Soars effortlessly for long periods on raised, but glides on kinked, wings***; active flight with very slow, deep wing-beats. Adult is ***gingery-buff above and below contrasting with dark flight-feathers***. Juvenile even paler brownish yellow

on rear underwing-coverts (greater contrast with flight-feathers). Gregarious when feeding at carcasses and when roosting.
Voice: croaking and whistling notes at breeding site.

126 **Rüppell's Vulture** *Gyps rueppellii* — **Plate 18**

(Rüppell's Griffon Vulture)

L: 95. W: 240. Resembles Griffon in flight but often soars on less raised wings. In adult of Ethiopian race (most likely race in Arabia) ***broadly scaled underparts appear almost creamy in flight in contrast to blackish brown underwing-coverts***; at close range there are ***narrow white lines on rear underwing-coverts and the usual 'vulture-streak' on leading forewing is pure white***. From above, ***broadly scaled wing-coverts form creamy forewing in contrast to flight-feathers***; at close range usually 2 whitish lines on rear-coverts. ***Juvenile lacks scaly appearance of adult***; told with difficulty from Griffon by darker wing-coverts (both surfaces) contrasting less with flight-feathers and accentuating white 'vulture-streak' near leading edge of underwing.

127 **Lappet-faced Vulture** *Torgos tracheliotos* — **Plate 18**

L: 105. W: 255–290. Very large, heavy vulture, ***paler than Black, darker than Griffon***. Very long, deeply-fingered wings and short tail with distinctly pointed tips; wings less parallel-edged than in Black but less curved than in Griffon. ***From above, dull grey-brown wing-coverts contrast much less with flight-feathers than in Griffon*** (but bleach paler; in Black, virtually no contrast). From below, dark wing-coverts have variable whitish 'vulture streak' near leading edge and ***flight-feathers and their coverts are clearly paler greyish, but wing-tip blackish*** (in Black, almost unmarked black under-wing-coverts contrast with paler flight-feathers, and wing-tip not clearly darker). Dark brown underparts have whitish brown mottling on breast, creamy upper-flanks, browner upper thighs but paler lower thighs and ventral region, ***giving variegated appearance to underparts***. Young birds more uniform with less developed 'vulture streak' and less variegated underparts but usually have some pale on vent. When perched, adult (Arabian birds) identified by very heavy bill, feathered hindneck, pinkish white head and fore-neck and long lanceolated breast feathers; lappets often inconspicuous. Solitary or in pairs but small parties at carcasses.
Voice: usually silent, but has growling sounds and grunts at breeding site.

128 **Black Vulture** *Aegypius monachus* — **Plate 18**

(Cinereous Vulture, Monk Vulture, Eurasian Black Vulture)

L: 105. W: 255–295. Very large, size as Lappet-faced which it resembles in outline. ***Told from Griffon by all blackish plumage without any large contrasts***, almost parallel-edged wings, held flattish or slightly downcurved, particularly when gliding; tail also slightly longer and less square-cut than Griffon's. Young birds blacker than adults but in both ***pale legs stand out against black undertail-coverts***. At close range adult has contrasting black and whitish head pattern; head blackish brown in young. Plumage blacker throughout than Lappet-faced (which see). Told from dark eagles by larger size, longer and more deeply fingered wings and less protruding head. The occasional wing-beat is slow and deep (like Lappet-faced).
Voice: rather silent.

129 **Short-toed Eagle** *Circaetus gallicus* **PLATE 24**

L: 64–73. W: 165–180. ***Large long-winged eagle with very pale underparts, square-cut tail with about 3 evenly spaced dark bands*** and broad head. Whitish underparts variably spotted and barred; ***some are nearly all whitish, others with contrasting dark head and upperbreast***; lacks dark carpal-patches. Flies with slow, flexible wing-beats, soars on flat or slightly lifted wings and hovers regularly. Separated from Osprey by broader wings, lack of carpal-patch and different flight. Pale Common Buzzards and Honey Buzzards usually have dark carpals, blacker wing-tips, different spacing of tail-bands and are much smaller with quicker wing-beats.
Voice: whistling, disyllabic 'kee-yo' with long ascending start and short descending finish.

130 **Bateleur** *Terathopius ecaudatus* **PLATE 19**

L: 55. W: 175. Unmistakable. Adult in flight has ***very short red tail with protruding legs, obvious 'S'-curved trailing edge to wings, black head and underparts, chestnut-red back and white underwing-coverts***; male has black, female has white, secondaries, tipped black. Juvenile dark brown with paler head; flight-feathers below paler than coverts; dark brown tail longer than adult's (feet do not project). Subadult sooty black on both surfaces but flight-feathers a mixture of dark and white feathers (all-dark in male). Glides fast, 'tilting', on raised wings. When leaving perch wing-beats fast and light. Displaying birds make flapping sound with wings.
Voice: in display, harsh 'chaaa-aw'; at nest a barking 'kak-kak-kak'.

131 **Marsh Harrier** *Circus aeruginosus* **PLATE 19**
(Eurasian Marsh Harrier, Western Marsh Harrier)

L: 48–55. W: 115–130. Larger, heavier and more broad-winged than other harriers; wavering low glides on raised wings when hunting; otherwise wings with bend at carpals (recalling Rough-legged Buzzard) which, together with more slender build, separates from distant Common Buzzard. ***Male tri-coloured on upperwing*** (silver-grey, brown and black); uppertail silver-grey; under surface white but wing-tip black and ***rear-body red-brown***, latter separating it from high-flying male Hen. Female dark brown, tail rusty brown, ***crown, throat and breast-spot yellow-white***. Juvenile blackish brown usually with rusty yellow on head. Immature male has dirty grey areas on upperwing and tail, rusty brown body and underwing-coverts, and more extensive black wing-tips than adult. Rare dark morph solidly blackish brown but adult male has flight-feathers below distinctly white at base. Display flight comprises steep up and down swoops with slow, deep wing-beats, performed at great height.
Voice: loud Lapwing-like 'vay-ee' from displaying male.

132 **Hen Harrier** *Circus cyaneus* **PLATE 19**
(Northern Harrier)

L: 45–56. W: 100–120. Smaller, slimmer and slightly narrower-winged than Marsh, with similar flight, but male more buoyant. Male has ***clear-cut white uppertail-coverts, uniform pale blue-grey upperparts, head and upper breast, rest of under surface white but wing-tip black*** (both surfaces). 2nd autumn male can show black wedge on wing-tip like male Pallid (as result of primary-moult). Female and juvenile brownish with distinct white uppertail-coverts; streaked underparts whitish or rusty-yellow (warmest in juvenile); banding on secondaries below generally most distinct in female; juvenile has pale line along tips of greater upperwing-coverts. ***Young birds separated from similar Montagu's and Pallid by proportionally shorter, broader wings with more ample rounded wing-tip*** (formed by 4 outermost primaries against 3 in the other two species), ***less buoyant flight and streaked breast*** (unstreaked rusty yellow-brown in juveniles of the other two species).

133 **Pallid Harrier** *Circus macrourus* **PLATE 19**

L: 40–48. W: 95–117. Flight and proportions almost identical to Montagu's. Male pale grey above ***without clear-cut white rump; whitish head and undersurface with black wedge on wing-tip.*** Female separated from similar Montagu's by ***pale, dark-streaked collar*** (like female Hen; virtually absent in Montagu's), ***less spacing between dark bands on secondaries below, with pale bands becoming darker towards body*** and absence of dark band on secondaries above. Primaries below often pale, contrasting with darker secondaries and lacking distinct dark trailing edge of hand; heaviest barring confined to central primaries ***with bases often unbarred, creating pale 'boomerang' surrounding darkish coverts***; distal primaries with faint or no barring, except for narrow dark 'finger-tips' of longest primaries (unlike Montagu's). Except for pale leading arm, ***most underwing-coverts and axillaries rather dark-streaked*** and lacking distinct pattern (not bold rufous-barred as Montagu's); ***they merge into dark secondaries*** (unlike Montagu's). Streaks of underparts largely confined to upperbreast which contrasts more with paler rear-body than in Montagu's. Juvenile like Montagu's ***but has broad yellow-white collar bordered behind by uniform dark brown neck.*** Primaries below rather evenly barred from base to tip ***though often with the pale 'boomerang' at primary-bases, 'fingers' never all-dark as in most young Montagu's.*** Male (9–12 months old) has paler head and breast than Montagu's and lacks contrast with pale belly; new central tail-feathers show diffuse dark bands near tip (similar Montagu's has grey neck and breast, contrasting with paler belly and new central tail-feathers plain grey).

134 **Montagu's Harrier** *Circus pygargus* **PLATE 19**

L: 43–47. W: 97–115. Slender build, narrower, longer more pointed wings, longer narrower tail and more buoyant flight helps separation from Hen Harrier. In adult, rear-body slimmer than in Pallid. Male has ***dark grey back and innerwing, silvery grey outerwing but extensive black wing-tips; one black band on secondaries above and two below; red-brown streaks below dark grey upperbreast.*** For 1st summer male, see Pallid. 2nd autumn male can show black wedge at wing-tip as result of primary-moult (thus recalling male Pallid). Grey-brown female has rufous-streaked underparts with ***male-like well spaced dark bands across pale secondaries below*** (and one above). ***Primaries below evenly barred from base to tip and trailing edge of hand dark***; at close range ***pale underwing-coverts and axillaries have uniform bold rufous barred pattern*** (unlike Pallid). Juvenile dark rufous to yellowish ochre below, ***largely unstreaked*** (few streaks confined to breast and flanks, absent in juvenile Pallid). Generally, ***'fingers' and trailing hand below dark, but hand otherwise pale*** with fine regular barring from base to tip (unlike most young Pallid, which see); ***lacks distinct pale collar of young Pallid.*** Rare melanistic phase (both sexes) largely sooty black with pale primary-bases below.
Voice: male's display-call a shrill, quick, 'kek, kek, kek'; female's alarm a fast 'jick-jick-jick'.

135 **Dark Chanting Goshawk** *Melierax metabates* **PLATE 20**

L: 38–48. W: 95–110. Medium-sized, ***blunt-winged raptor with long, graduated tail*** and *Accipiter*-like wing-beats; flies fast, low over ground, ascending to perch in upright position when ***long orange-red legs*** conspicuous. Soars on raised wings. Adult ash-grey but belly white, finely barred grey; ***in flight, tri-coloured above with grey back and innerwing, silvery outerwing but black wing-tip***; bold black and white bands on outer tail-feathers; ***white uppertail-coverts thinly barred grey***; whitish underwing has blackish wing-tip. Juvenile brown where adult grey but ***belly boldly barred***; in flight dark brown above with barred tail-coverts; underwing buffish, tail boldly banded. Spends much time perching in tree-tops or on telegraph-poles.

Voice: loud series of fluty whistles 'kjiu-kjiu-kjiu' with upward inflection; also high-pitched 'kleee-u'.

136 **Gabar Goshawk** *Micronisus gabar* **PLATE 20**

L: 30. W: 63. Superficially like Sparrowhawk but ***striking white uppertail-coverts at all ages; head and upper breast grey***, rest of underparts white, finely barred darker; tail boldly barred above and below. At close range ***white trailing edge to secondaries at all ages***, which is absent in other small hawks in region. Juvenile dark brown, ***breast boldly streaked, belly boldly barred rufous-brown; white edges to tertials conspicuous when perched.*** A melanistic phase (black uppertail-coverts) as yet unrecorded in SW Arabia, has boldly barred flight- and tail-feathers above but white tips to secondaries absent. Recalls Sparrowhawk in flight and silhouette but ***tail rounded with clearly shorter outer feathers*** (square-cut in Sparrowhawk); glides and soars on upcurved wings (unusual in Sparrowhawk).

Voice: varied 'kew-ke, kew-ke, kew-ke' during courtship; also high-pitched 'twee-twit-twee-twee-twee-twit'.

137 **Goshawk** *Accipiter gentilis* **PLATE 20**

(Northern Goshawk)

L: 48–60. W: 90–125. ***Female distinctly larger than male, with wing-span of Common Buzzard. Male Goshawk has deeper belly, slower, stronger, stiffer wing-beats and longer, broader-based but more pointed wings than female Sparrowhawk***; shorter, broader-based tail usually with rounded tip (thinner tail more square-cut in Sparrowhawk) and ***more protruding head and neck.*** Stronger, straighter glides than Sparrowhawk; soars more often on upturned wings. Adult dark grey above, darker head appears 'hooded' but supercilium white; underparts finely barred. Juvenile dark brown above with pale mottling on ear-coverts; rusty yellow ***underparts boldly streaked darker***; lacks 'hooded' appearance of adult. Female told from large falcons by more rounded wings, bold tail-bands and *Accipiter*-like flight. Hunts like Sparrowhawk but also runs down prey on the ground. Display flight with soft harrier-like wing-beats in shallow waves. Goshawks, unlike sparrowhawks, are treated cautiously by crows.

Voice: alarm a loud 'gi-ak, gi-ak, gi-ak'.

138 **Sparrowhawk** *Accipiter nisus* **PLATE 20**

(Eurasian Sparrowhawk)

L: 28–38. W: 60–80. Female much larger than male, approaching male Goshawk, but wing-beats faster and lighter; rear-body thinner; wing-tips blunter; tail thinner, longer and more square-cut and head protrudes less. Series of quick wing-beats interspersed with short descending glides (stronger, straighter glides in Goshawk). Adult dark ash-grey above, barred rufous or brown below (rufous most in male); dark bands on uppertail can be inconspicuous in male; whitish supercilium in female (infrequent in male); ***whitish underwing without dark tip***. Juvenile browner above, streaked or blotched throat and upper breast, otherwise barred below; clear white supercilium. Hunts swiftly low over ground behind cover before sudden attack or stoop on prey. Display flight with slow harrier-like wing-beats, sometimes also adopted when hunting.
Voice: alarm call 'kyi, kyi, kyi, kyi' faster, but in shorter series than in Goshawk.

139 **Shikra** *Accipiter badius* **PLATE 20**

L: 33. W: 60–70. Resembles Sparrowhawk in outline (blunt wing-tip) but ***slightly shorter tail is tapered with rounded corners*** (square-cut in Sparrowhawk). Two races: male of larger Iranian *cenchroides* ***pale dove-grey above with darker wing-tip; creamy-white under surface with narrow black wing-tip***; female pale brown above with blackish band near tip of tail; wing-tip almost without black below. Perched adult has ***dark central throat stripe, grey cheeks but no white supercilium***; male also has obscure buff neck-collar. In smaller SW Arabian *sphenurus* both sexes darker, soft blue-grey above, male with ***contrasting black wing-tip***; female less pure white below, less black on wing-tip; lacks buffish hind-neck collar of Iranian male. Juvenile has ***dark longitudinal spots on underparts*** (largely barred in young Sparrowhawk), banded uppertail and hardly any black on wing-tip. Sparrowhawk shows white supercilium (female), more or less rufous cheeks and no dark throat-stripe. Habits and flight recall Sparrowhawk's.
Voice: unhurried 'ch-wick, ch-wick' like quiet Tawny Owl's call; also loud 'kik-kooi' repeated at nest; in display a whistling 'piu-piu-piu'.

140 **Levant Sparrowhawk** *Accipiter brevipes* **PLATE 20**

L: 32–39. W: 65–80. Resembles Shikra in plumage, but ***longer, narrower, less curved wings have more pointed tip, giving falcon-like appearance when gliding*** (separating also from blunt-winged Sparrowhawk). Male dull blue-grey above with ***blackish wing-tip*** and plain central tail-feathers; female browner, less contrasting and closed uppertail has dark band near tip. From below, ***male has white underwing with contrasting black tip, extending to inner primaries***; female less white, less contrasting. When perched, ***black throat-streak, greyish cheeks but no supercilium separates from Sparrowhawk***. Juvenile dark grey-brown above with closed tail banded throughout (sometimes indistinct); ***underparts with dark longitudinal spots*** and dark throat-streak but ***told from young Shikra by shape of wing-tip***. Adult male has whiter underwing and more extensive black wing-tip than male Iranian Shikra and darker upperparts. Always told from Sparrowhawk by shape of wing-tip (4 free outermost primaries against 5 in Sparrowhawk), and often by more rounded tail-corners; adult female has dark outermost primaries (absent Sparrowhawk) and lacks female Sparrowhawk's fully banded central

tail-feathers above. Flight typical *Accipiter*. Unlike Sparrowhawk, forms dense flocks on migration.
Voice: shrill 'ke-wik'; at start of breeding also 'kéwek, kéwek, kéwek'; calls unlike Sparrowhawk.

141 **White-eyed Buzzard** *Butastur teesa* **PLATE 21**

L: 45. W: 100. Resembles cross between Honey Buzzard and harrier when soaring, with relatively narrow wings held flattish; when gliding wings angled and tail relatively long and narrow, usually half-spread when soaring; active flight *Accipiter*-like. From above, brown adult has ***cinnamon-rufous tail with black band near tip*** (otherwise unbarred or faintly barred), warm brown primary-patch and buffish panel across wing-coverts. From below, forebody rufous or brown grading into whitish rear-body; wing-coverts either dark brown or whitish, mottled darker, but ***wing tips blackish; half-spread undertail reveals unmarked whitish outer feathers***. At distance appears rather white below with darker breast. Young has paler head, whitish underparts (streaked at close range) and brown iris. Perches erect for long periods; ***then white throat with dark central streak and dark cheek-streak visible*** (latter narrow or absent in young).
Voice: plaintive, mewing 'pit-weer, pit-weer'.

142 **Common Buzzard (includes Steppe Buzzard)** **PLATE 21**
Buteo buteo
(Buzzard)

L: 50. W: 115. The main race that occurs in the region is the highly migratory Steppe Buzzard, *B. b. vulpinus*, which is slightly smaller with narrower, more pointed wings and longer tail compared with Western European birds of the race *B. b. buteo*, which are rare.
Steppe Buzzard: Three types occur: fox-red, grey-brown and blackish (scarce); intermediates are frequent. Fox-red and blackish types are very similar to Long-legged Buzzard but have ***shorter, narrower wings, shorter tail, stiffer wing-beats*** and ***glide on flattish wings*** (Long-legged has flexible wing-beats and kinked gliding profile); also wings are less up-lifted when soaring. Fox-red type generally has more variegated rusty orange underwing-coverts framed by dark greater coverts, with narrow blackish comma-shaped carpal-patch (but patch *can* be large as in Long-legged); rufous-brown uppertail more barred (but *can* be plain like Long-legged), head dark and pale primary-patch above is usually small. Blackish type best told from Long-legged by flight and silhouette. Grey-brown type resembles Western European birds. Juvenile (streaked breast, diffuse band on trailing edge of underwing, no broad dark band at tip of tail) often pale-headed, with pale upperwing-coverts and prominent primary-patch above; told from Long-legged by structure. Young blackish type has fine regular barring on secondaries below and on uppertail (coarser, more irregular in similar Long-legged). Soars on raised wings and hovers when hunting.
Socotra Buzzard: The resident buzzard on Socotra has not been assigned to a race and indeed may be a separate species. It is shorter-winged and shorter-tailed than *vulpinus*, with rather pale plumage below, prominent dark carpal-patches, heavily streaked belly and 'trousers' and variable mottling on breast.
Voice: high mewing 'peeeoo'.

143 **Long-legged Buzzard** *Buteo rufinus* — Plate 21

L: 60–66. W: 130–155. Size of Black Kite with ***long wings and tail***, kinked wing position when gliding and flexible wing-beats; ***soars on highly raised wings***. Wide plumage variation: creamy white, rufous-brown and blackish forms on account of colour on body and underwing-coverts. The blackish form can show coarse dark bars on flight- and tail-feathers. Many Long-legged (except dark birds) have ***very pale head and breast darkening towards belly***, pale sandy or rufous-brown upperwing-coverts contrasting with flight-feathers, ***unbarred pale rusty-orange uppertail*** (appearing very pale at distance); ***large black carpal-patches*** usually distinctive. Juvenile has finely barred outer tail and diffuse dark trailing edge to underwings (usually clear-cut in adult); some are almost white below with bold carpal-patches and dark rusty brown belly (sometimes sides only), but absence of broad blackish band at tip of uppertail separates it from young Rough-legged. See Common Buzzard for separation from Steppe Buzzard. Soars and hovers in search of prey.

Voice: recalls Common Buzzard's but less vocal.

144 **Rough-legged Buzzard** *Buteo lagopus* — Plate 21

(Rough-legged Hawk)

L: 50–61. W: 125–145. Larger and heavier than Common Buzzard, wings and tail longer, ***wing-beats flexible and wings kinked when gliding***, thus like Long-legged (Common Buzzard has stiff wing-beats and lacks obvious kink). ***Separated from other Buteos by pure white tail with broad blackish band near tip***; in adult, tail-band narrowest and most clear-cut, also conspicuous from below (in male with 2–4, in female with 0–1 extra bars). In juvenile tail-band less clear-cut and inconspicuous from below. Adult typically pale-headed with dark upperwings; pale 'U' on breast separates dark belly from dark fore-neck. ***Large dark carpal-patches at all ages***. Juvenile has pale band across paler upperwing-coverts, larger pale primary-patch above and is whiter below (but with bold dark belly and carpal-patches); lacks dark clear-cut trailing edge of adults underwing. Some pale Common Buzzards show whitish base to uppertail (mostly in centre) but upperwing-coverts usually show some white (absent in Rough-legged); also in Common Buzzard carpal-patches usually small and 'comma'-shaped. Tarsus feathered to toes. Recalls Common Buzzard in behaviour, but hovers more.

145 **Lesser Spotted Eagle** *Aquila pomarina* — Plate 22

L: 62. W: 145–165. ***Never looks blackish except in silhouette. Golden-brown adult has contrast between pale upperwing-coverts and darker brown mantle***, a ***small well defined pale-patch at base of primaries above and underwing-coverts paler than flight-feathers***; also a small creamy yellow area on uppertail-coverts. Juvenile darker, warm mid-brown below with ***flight-feathers of either same shade or little darker (coverts never darker than flight-feathers)***; unlike adult sometimes no contrast between mantle and wing-coverts; darker head has ***rusty yellow spot on nape*** (absent in Greater Spotted, present in some adult Steppe); narrow white bar on greater upperwing-coverts and distinct whitish inner primary-patch. Some birds (mostly non-juvenile) are very similar to odd-coloured young Greater Spotted (which see). ***Short or minute 4th primary, less deeply fingered wings and smaller hand separates from Steppe at all ages***; lacks dark band on trailing edge of underwing and tail of many adult Steppe. Wings relatively

narrow, tail medium-long; active flight less heavy than in larger species of *Aquila*. Soars and glides on arched wings with primaries more or less lowered. On ground lacks heavy 'trousers'.
Voice: series of about 10 'kyeep-kyeep' cries with upward inflection on second note, lasting 4 s.

146 **Greater Spotted Eagle** *Aquila clanga* **Plate 22**
(Spotted Eagle)

L: 65. W: 155–180. A difficult species. ***Normally darker than Lesser Spotted.*** Adult dark brown below, ***flight-feathers similarly dark or a shade paler*** (reverse of Lesser); ***leading underwing-coverts sometimes blackish brown*** (never in Lesser); on upper surface, mid to dark brown coverts sometimes contrast with darker mantle (like typical Lesser); no conspicuous pale primary-patch above (unlike Lesser and Steppe). Adult Greater Spotted usually lacks trailing edge band of underwing, seen on many adult Steppe; also lacks pale nape-spot of young Lesser and many adult Steppe. ***Juvenile is blackish brown below with paler flight-feathers*** (in Lesser, coverts are brown but flight-feathers never paler); ***blackish brown upperwing has 1–3 white covert-bars, often creating pale mid-wing panel***; large, diffuse primary-patch formed by whitish primary shafts and pale inner primaries (patch smaller, neater, more conspicuous in Lesser). Infrequently, young Greater Spotted is abnormally coloured on body and wing-coverts: (1) 'fulvescens' type (see Plate 22), (2) yellow-brown above and below, (3) yellow-brown above, normal below, (4) yellow-brown below, normal above, (5) under-wing-coverts greyish, or dark, mottled paler, underbody darker and (6) normal upper-and underwing, but contrasting paler underbody (. combination yet unrecorded in other *Aquila*). Irrespective of age, secondaries below may have ***thin dense bars*** (broader in Lesser, more well spaced in Steppe). Adult has relatively broad and parallel wings with slightly broader hand and deeper fingers than Lesser. Juvenile has narrower hand than adult with trailing edge fairly strongly incurved at body (visible when tail closed). Hand slightly shorter, less ample than Steppe (Greater Spotted has shortish 4th primary) and bill generally smaller. Wings held as Lesser in flight.
Voice: recalls Lesser Spotted.

147 **Tawny Eagle** *Aquila rapax* **Plate 22**

L: 70. W: 165–185. A difficult species. Slightly smaller than Steppe; ***plumage often creamy or rufous*** (both surfaces) unlike brownish Steppe; ***lacks white band through underwing of young Steppe; creamy white lower back and pale wedge on inner primaries below typical of many birds.*** Some adults are grey- or black-brown and hard to separate from adult Steppe (but rare in the Middle East). Juvenile, and often adult, have pale plumage which, particularly in juvenile, bleaches to creamy white (recalling 'fulvescens' Greater Spotted, quite unlike Steppe). Adult often has small primary-patch above but in juvenile patch is larger. ***Flight-feathers below at all ages are dark to pale grey, mostly with fine dense bars*** (bold, well spaced in Steppe) ***or no bars at all***; diffuse dark trailing edge band seen in adult only (usually more clear-cut in adult Steppe). Some pale birds (all ages) have pale primaries below with defined dark 'fingers', thus lacking pale wedge on inner primaries. Poorly known immature ***often rufous or blackish brown on head and/or forebody, contrasting with buffish rear-body.*** Underwing-coverts very pale, in SE Iranian *vindhiana* often with blackish bands. Adult has ***yellow***

iris (usually dark in Steppe) and ***shorter gape flange than Steppe***; heavy 'trousers' and heavy bill separates from spotted eagles. Fairly broad-winged with ample, deep-fingered hand (long 4th primary) and well protruding head. Wing position like Steppe.
Voice: during courtship a barking 'kowk-kowk'.

148 **Steppe Eagle** *Aquila nipalensis* **Plate 22**

L: 75. W: 175–210. Dark brown adult has variably coloured underwing-coverts, either similar to body or paler; ***underwing uniform, or with paler or darker flight-feathers with well spaced broad dark bars and clear-cut band on trailing edge*** (pattern sometimes diffuse); ***large dark carpal-patch typical, except in darkest birds***. From above, coverts often palest part of wing; ***usually large, dark-barred, pale primary-patch*** (patch virtually absent in adult Imperial, Tawny and Spotted); ***grey-brown tail often boldly barred with broader dark band at tip*** (absent in spotted eagles). Juvenile pale brown with ***diagnostic broad white band through middle of underwing***; from above, large primary-patch and dark rump separates it from most young Imperials. Subadult usually has darker body than underwing-coverts, very like some immature Lesser Spotteds, but told by ***remains of white underwing-band or well spaced flight-feather barring, long deeply fingered wings, ample hand*** (long 4th primary) ***and longer, heavier bill***. Flight heavy; often soars on flexed, flattish wings but can soar and particularly glide on arched wings with lowered hand. When perched, large heavy 'trousers' unlike the spotted eagles; ***long yellow gape flange to rear-end of eye***, separates it from other *Aquila*. See Tawny for separation.

149 **Imperial Eagle** *Aquila heliaca* **Plate 23**

L: 72–83. W: 190–210. In adult, combination of ***blackish-brown plumage, contrasting yellow-white hindneck, pale uppertail with broad black band, white 'shoulder-braces'*** (hard to see), pale ventral region, darkish primaries above, ***parallel-edged wings held flattish*** and, often, ***closed narrow tail when soaring***, identify it from Golden. Juvenile has ***dark-streaked breast forming pectoral band which contrasts with unstreaked yellow-buff rear-body***, and ***distinct pale wedge on inner primaries below***; yellow-brown upperparts show 1–2 complete whitish bars on coverts and ***lower back and rump usually creamy-white***; lacks white band through underwing of young Steppe. Immature from below, ***mottled blackish-brown and yellowish*** with rear-body clearly paler; adult head- and tail-pattern start to show early. Rather long-winged with ample hand, deep-fingered wing-tip (long 4th primary), well-protruding head and relatively long tail; juvenile has broader, more 'S'-curved wings. Wings sometimes slightly lifted when soaring, but arched during fast glides; wing-beats slightly more flexible than Steppe's. Pale Tawny (our region) lacks streaks below of young Imperial (but seen in some African birds, which could occur in Arabia).
Voice: deep 'kraw-kraw-kraw' in display.

150 **Golden Eagle** *Aquila chrysaetos* **PLATE 23**

L: 78. W: 190–230. Powerful flight with ***flexible wing-beats, lifted wings*** (both when soaring and gliding), ***fairly long tail and slightly 'S'-curved rear-edge of wings*** (more pronounced in juveniles) characteristic. Dark brown adult has ***rusty yellow hindneck, pale panel across upperwing-coverts***, dark-barred, black-tipped greyish flight-feathers showing up as ***greyish area on outerwing above*** and greyish tail with blackish band at tip (adult Imperial is blacker, including outerwing above, has flatter wings more parallel-edged and tail often looks narrow). ***Juvenile and immature have white patches in primaries and inner tail, latter with broad black band at tip***, unique in the *Aquila* eagles. Birds older than one year show pale panel across upperwing-coverts. Young White-tailed has shorter tail, more protruding head and broader wings, held flattish when soaring, and lacks white wing-patches.
Voice: loud barking 'kliak-kliak-kliak' at breeding site.

151 **Verreaux's Eagle** *Aquila verreauxii* **PLATE 23**
(Black Eagle)

L: 80–95. W: 225–245. ***Unique wing shape separates it at all ages*** from other eagles: long with ***distinct 'S'-curved trailing edge, and very narrow base***. Adult ***black with conspicuous white primary-patches above and below, pure white lower back and white 'V' on shoulders*** (hard to see at distance). Juvenile has pale primary-patches, creamy panel on upperwing-coverts, ***blackish brown throat and breast which contrast with buffish white rear-body*** and yellow-buff crown/hindneck. ***Soars gracefully on lifted wings*** for long periods without wing-beats; when gliding on half-closed wings, narrow wingbase 'disappears'.
Voice: male has a 'chorr-chorr-chorr', female a thinner 'che-che-che'; also a tremulous ascending 'whace-whace-whace'.

152 **Booted Eagle** *Hieraaetus pennatus* **PLATE 23**

L: 43–53. W: 110–130. Two phases. Size of Common Buzzard, but outline and wing-position close to Black Kite; ***tail square-cut***. More ample deep-fingered wings than Common Buzzard. Pale phase identified by ***creamy white underparts with contrasting blackish flight-feathers***, kite-like band on innerwing above, ***pale scapulars and uppertail-coverts and diagnostic small white spot at base of neck***, seen on front-view. Dark phase similar above but under surface dark brown, or rufous with black band through centre of underwing. ***Lacks dark carpal-patch of most pale Common Buzzards and Honey Buzzards***; also has darker flight-feather bases, paler inner primaries, different upperpart pattern and tail shape. When perched, ***feathered tarsus also separates from Common Buzzard***. Has deeper, more powerful wing-beats and steadier glides than Common Buzzard. Soars on flat wings and does not hover.
Voice: shrill, chattering, often three-toned whistling near nest.

153 **Bonelli's Eagle** *Hieraaetus fasciatus* PLATE 23

L: 60–70. W: 150–165. Size of Lesser Spotted but recalls thick-set Honey Buzzard. ***Adult identified by dark underwings contrasting with whitish underbody, pale tail with black band at tip*** and variable white patch on mantle; also at close range ***white on leading edge of wing.*** Pale rusty buff juvenile lacks black tail-band, and flight-feathers are pale with fine dark barring; ***paler translucent primaries contrast with blackish wing-tip***; when present, ***narrow dark bar on rear underwing-coverts diagnostic***, but in others confined to dark 'comma' on primary-coverts; upper surface cinnamon-brown with large pale primary-patch. Soars on flat or slightly arched wings, often with long, almost square-cut tail held closed; glides with carpals pressed forward and almost straight trailing edge of wings (recalling Honey Buzzard). Often in pairs; may stoop at great speed.
Voice: shouting 'juii-juii-gi-gi'.

154 **Osprey** *Pandion haliaetus* PLATE 24

L: 56–61. W: 145–165. ***Easily told by long, narrow wings, distinctly angled when gliding, white under-surface with black carpals and band through centre of underwing, white crown and dark eye-mask.*** Variable dusky band across fore-neck (usually boldest in female). Juvenile has whitish scales and white line on greater coverts above. Flies with very steady shallow wing-beats, glides on smoothly curved wings; may recall large soaring gull at distance. Hovers over water persistently for fish and dives with splash, almost disappearing.
Voice: mournful 'yeelp-yeelp' and a loud whistle 'pyip'.

155 **Lesser Kestrel** *Falco naumanni* PLATE 25

L: 28–33. W: 63–74. Very like Kestrel and not always separable. Slightly smaller and slimmer, with slightly narrower wings, slightly shorter, more wedge-shaped tail and quicker wing-beats. ***Male is unmarked rufous above but greater coverts often blue-grey*** (hard to see). ***Whole head ash-grey without moustache or pale area on cheeks.*** From below, ***white underwing contrasts with dark wing-tip and creamy buff underbody,*** both of which have small black spots; in some, underwing-coverts virtually unmarked (unlike Kestrel). Female very like Kestrel but on average has slightly whiter, less barred flight-feathers below, sometimes fewer and finer spots on underwing-coverts and greyer uppertail-coverts. ***With practice, female and juvenile identified by wing-tip formulae and pale claws*** (see Kestrel). Hovers less persistently than Kestrel and takes insects in flight. Gregarious, both at breeding site and on passage.
Voice: rasping, trisyllabic 'chae-chae-chae', very different from Kestrel's.

156 **Kestrel** *Falco tinnunculus* PLATE 25

(Common Kestrel)

L: 32–38. W: 70–78. Long, narrow, fairly pointed wings, long ***rounded or tapering tail, shallow loose wing-beats, persistent hovering and rufous upperparts, contrasting with darker flight-feathers***, characteristic. Compared to Lesser, ***male has black spots on back and wing-coverts*** (absent in male Lesser), ***lacks blue-grey greater upperwing-coverts***, has different head-pattern and underwing. Female has dark bars on red-brown upperparts; some approach male in greyness on head, tail-base and uppertail-coverts. Juvenile paler brown with thin white line along greater coverts above. Active flight

alternates with slow glides, some soaring and frequent hovering. ***At all ages told from Lesser by black claws and wing-tip formulae, useful when soaring at close range outside autumn period of primary-moult*** (p.10 shorter than p.8 and equalling p.7; in Lesser p.10 longer than p.8 and clearly longer than p.7).
Voice: a shrill 'kee-kee-kee' in short series, heard mostly in breeding season.

157 **Red-headed Merlin** *Falco chicquera* PLATE 25

L: 31–36. W: 65–80. Proportions like Merlin with short wings and long tail, but larger. From above, adult resembles male Merlin but ***crown pale rufous-chestnut; underparts white, finely barred darker from mid-breast down***; at distance under-surface appears greyish, but upper breast white; a short, diffuse, ***pale rufous moustache at close range***. Juvenile has buffier forehead and hindneck, crown finely streaked; ***creamy underparts boldly barred except in centre where diffusely streaked***. Hobby differs by dark upperparts, bold black moustache and streaked underparts. When perched, wing-tip of Hobby extends to tip of tail; in Red-headed wings are clearly shorter. Flight bold, dashing, and often low over ground; stoops on prey. Usually occurs in pairs.
Voice: shrill screaming calls heard at breeding grounds.

158 **Red-footed Falcon** *Falco vespertinus* PLATE 25

L: 30. W: 73. Has a variety of plumages. Distinctive ***male slate-grey with silvery primaries above, red 'trousers' and undertail-coverts***, blackish underwing-coverts and unbarred tail. ***Female has rusty yellow underparts and head***, but with ***dark eye-mask***; greyish upperparts barred darker, ***tail densely barred***; yellow-buff underwing has dark trailing edge. ***Juvenile, in which dark trailing edge is more conspicuous***, has streaked breast and spotted underwing-coverts; often brownish above, contrasting with buff cheeks and hindneck collar; ***moustache short and pointed***. 1st spring male has slate-brown upperparts, with rufous and slate underparts, but underwing-coverts, flight- and most tail-feathers as juvenile; in 2nd autumn some adults' ***central primaries show variegated pattern***. Proportionally longer and slightly more pointed wings than Kestrel but tail shorter, less rounded. At times Hobby-like in flight (but less stocky), but often flies with loose Kestrel-like wing-beats which, together with frequent hovering, can cause confusion. Hobby's combination of dark upperparts, plain uppertail, conspicuous moustache and streaks below prevent confusion with 1st spring male Red-footed. Gregarious, often hunting in flocks for insects, alternating with periods of hovering.

159 **Amur Falcon** *Falco amurensis* PLATE 25

(Eastern Red-footed Falcon, Manchurian Red-footed Falcon)

L: 26–32. W: 65–75. Resembles Red-footed, but ***male has white underwing-coverts*** (hard to see in poor light). ***Female white below except for warm buff thighs and streaked breast***, boldly barred flanks and ***lightly spotted underwing-coverts*** (female Red-footed rufous-yellow below with unmarked underwing-coverts); ***dark crown and short moustache contrast with white cheeks*** (Red-footed has golden head with dark eye-mask). ***Juvenile resembles Red-footed but white ground-colour below and darker crown separate it*** (young Red-footed usually has brownish buff crown with isolated dark eye-mask). 2nd year male like similar Red-footed until white underwing-coverts appear at about one year. Flight and silhouette like Red-footed. Female and young separated from Hobby by densely barred uppertail, smaller moustache, paler undersurface, flight and hovering.

160 **Merlin** *Falco columbarius* PLATE 25

L: 25–30. W: 55–65. (Female clearly larger.) Smallest falcon in the region; has ***short pointed wings*** and ***relatively long tail***, speedy flight with fast wing-beats, interspersed with short quick glides. ***Male told by blue-grey upperparts with blackish primaries, broad black tail-band*** and ***ill-defined head pattern***. Female and juvenile are brownish above, creamy below with dark streaks, ***diffuse moustache, barred primaries above*** and ***five pale/dark bands of equal width on uppertail***. In Siberian race, *pallidus*, the male is distinctly paler blue-grey above with some rusty on neck, shoulders and mantle and underparts are whiter; female and juvenile *pallidus* are rufous above with Kestrel-like dark bars (but Kestrel's flight, proportions and denser tail-barring prevent confusion). Hunts usually low over ground, attacking with thrush-like undulating flight, changing direction, followed by a straight attack. When perched, wings fall clearly short of tip of tail.

161 **Hobby** *Falco subbuteo* PLATE 26

(Eurasian Hobby)

L: 32–36. W: 74–92. Has ***long, scythe-like, pointed wings*** (recalling large swift) and relatively short tail. In adult, combination of ***dark slate-grey upperparts, uniform tail, contrast between prominent moustache and white cheeks/throat***, densely streaked underparts with ***red thighs and undertail-coverts***, and shape and length of wings and tail separate it from Red-footed, Merlin and Kestrel. Juvenile browner above with pale feather fringes and absence of red thighs and undertail-coverts. Flight swift and agile; has strong steady wing-beats, short fast glides; accelerates when hunting birds but more relaxed when catching insects, recalling Red-footed, but does not hover. See Eleonora's and Sooty Falcons for separation.

Voice: long, rapid 'kikikiki', recalling Kestrel but less sharp.

162 **Eleonora's Falcon** *Falco eleonorae* PLATE 26

L: 39. W: 97. ***Long-winged and long-tailed falcon***; apart from longer tail and narrower wings recalls Hobby in outline, but wing-span of male Peregrine. Flight swift and agile, Hobby-like when hunting but ***relaxed flight has soft, surprisingly slow wing-beats***. Two phases: pale recalling Hobby but separated by ***darker*** underparts (red-brown) and ***very dark, unmarked underwing-coverts contrasting with pale-based, unbarred flight-feathers***; at distance appears dark below except for pale throat and cheeks. Dark phase (25 per cent of population) uniform blackish brown; told from male Red-footed by size, proportions, flight, underwing-pattern, dark primaries above and lack of red thighs. Juvenile (both phases) paler below than adult pale phase; ***told from Hobby by dark underwing-coverts*** (spotted and barred) ***contrasting with paler flight-feathers (barred), which have dark band on trailing edge***; also thinner moustache. See Sooty Falcon for separation. Often hunts in flocks or pairs, especially at dusk. Catches insects like Hobby, but sometimes hovers; capable of tremendous stoops. Breeds late summer; feeds young on autumn migrants.

Voice: loud hoarse 'kjie-kjie-kjie' heard over breeding cliffs.

163 **Sooty Falcon** *Falco concolor* **Plate 26**

L: 32–38. W: 85. Size and shape between Hobby and Eleonora's. Long-winged like latter, but tail slightly shorter; elongated central tail feathers (sometimes seen in Hobby) but closed tail-tip pointed like Eleonora's. Adult told from dark-phase Eleonora's by ***slaty blue-grey upperparts with darker primaries and outer uppertail*** (male Red-footed has paler, silvery primaries), ***blue-grey underparts*** (underwings appear paler) ***without Eleonora's contrasting underwing pattern***. Female darker, less contrasting above. Young Sooty told from similar Hobby and Eleonora's (pale adult and juvenile) by ***greyer upperparts with darker wing-tip, less clearly streaked yellow-brown underparts, the spot-streaks almost merging on upper breast as a diffuse dark band; lightly marked dusky underwing has dark wing-tip and trailing edge*** (Eleonora's has dark coverts contrasting with pale-based flight-feathers; Hobby has uniform underwing); undertail finely barred except near end (Hobby and young Eleonora's have undertail barred to tip). Wing-beats faster than Eleonora's; flight recalls Hobby but glides on level wings (lowered in Hobby). When perched, wing-tip as long as, or slightly longer than tip of tail. Breeds in late summer, feeding young on autumn migrants.
Voice: Hobby-like, but slower, 'kee-kee-kee'.

164 **Lanner Falcon** *Falco biarmicus* **Plate 27**

L: 42–52. W: 95–115. Size of Peregrine but resembles Saker in plumage and shape; long wings with ample hand, slightly blunt-ended when soaring; tail relatively long. ***Adult separated from most Sakers by barred, greyish upperparts, including distinctly barred uppertail***, more contrasting head-pattern of ***black upper-forehead band, clear-cut narrow black eye-stripe, always conspicuous moustache*** and with ***crown unstreaked creamy buff*** (Near Eastern *tanypterus*), ***chestnut*** (SW Arabian *abyssinicus*) or ***pale rufous rear crown and nape*** (European *feldeggi*); also some ***spot-bars on flanks***. Juvenile very like young Saker (dark brown above, including head, boldly streaked underparts and rear underwing-coverts, contrasting with pale flight-feathers) but has ***unbarred closed uppertail*** (unlike most Sakers). The contrasting underwing pattern and, sometimes more densely streaked underparts, separates it from similar Peregrine and Barbary. See also other large falcons. Active flight with moderately slow, stiff wing-beats, faster when hunting; stoops or runs down prey. Soars with wings level or slightly upcurved.
Voice: slow scolding 'kraee-kraee-kraee' at breeding site.

165 **Saker Falcon** *Falco cherrug* **Plate 27**

L: 47–55. W: 105–125. Largest falcon of the region with wing-span of female Common Buzzard. Like Lanner in outline, but larger, and heavier-chested. Separated from Lanner by combination of ***whitish crown*** (sometimes confined to nape), ***unbarred Kestrel-like contrast above, less distinctly barred uppertail, poorly developed moustache*** and ***less contrasting head pattern*** (diffuse dark eye-stripe, no dark forehead-band); ***whitish supercilium*** often more conspicuous and belly more spotted, but lacks Lanner's spot-bars on flanks (except in NE Iranian *milvipes* which has pinky-rufous crown, dark bars above and boldly barred uppertail, like Lanner). Other Saker subspecies (*saceroides*-type) are greyish above with dark bars, including uppertail; such are ***best told by head pattern and size***. Juvenile almost identical to young Lanner, but outer tail-feathers generally conspicuously spotted buff on outer webs (seen well in half-spread tail) and dark stripe

behind eye less clear-cut; usually best told by size. When perched, wing-tip falls short of tail-tip (unlike most Lanners) and 'trousers' generally heavier, covering larger parts of tarsus. See Peregrine for separation. Flight with slow, flattish wing-beats; when soaring, wings flat or slightly up-curved.
Voice: like Peregrine, but harsher; also thin, querulous note like cross between Curlew and Herring Gull of Western Europe.

166 **Peregrine Falcon** *Falco peregrinus* — PLATE 27

L: 40–52. W: 85–120. Large, stocky falcon with relatively short tail and ***broad-based, sharply tapering wings*** which, when soaring, are ***more pointed than in other large falcons*** (except Barbary). Adult separated from Lanner and Saker by ***black crown and bold moustache, contrasting strongly with white throat and small cheek-patch***, darker upperparts, ***barred underparts***, but gleaming white upperbreast, and ***plain greyish, barred underwing***. Dark brown ***juvenile told by uniform underwing*** (distinctly bi-coloured in young Lanner and Saker) and ***smaller whitish cheek-patch*** (not reaching eye, unlike Lanner and Saker). Some young of the race *calidus* are tricky, showing Saker-like head pattern and large size, but underwing pattern and wing shape identifies them. See Barbary for separation, and note separation from the rare *brookei* (illustrated) can cause problems as its distribution is not fully understood. Normal flight with fairly quick, shallow, stiff wing-beats; impressive when hunting, with long fast stoops.
Voice: alarm note a scolding 'keck-keck-keck'; also 'kyok' and 'keek-keek'.

167 **Barbary Falcon** *Falco pelegrinoides* — PLATE 27

L: 32–45. W: 80–100. Resembles Peregrine in outline and flight but slightly narrower based wings (male) give impression of longer tail. Adult separated from Peregrine by ***rufous nape and rear eyebrow, narrower moustache and larger pale cheek-patch, almost reaching eye; more creamy, less barred underparts***, in some confined to flanks, particularly in E Iranian *babylonicus* which is also more red-crowned; ***underwings whiter with more extensive dark wing-tips*** than Peregrine (greyer underwing in Peregrine) ***and often with dark 'comma' on greater primary coverts*** (absent Peregrine); upperparts paler blue-grey, more often with darker end to tail. Told from Lanner by ***dark top of head, lack of dark forehead-band and distinct eye-stripe, barred, not streaked, underparts and wing-coverts*** and narrower hand with more pointed wing-tip. Juvenile like Peregrine, ***but moustache narrower, cheek-patch larger and has tawny supercilium and rusty nape***; rustier underparts with thinner, more restricted streaks (bolder in most Peregrines). Some juveniles have yellow cere and legs soon after fledgling (in Peregrine blue-grey, usually becoming yellow in 1st winter). From young Lanner (which see) by pattern of underwing and underparts.
Voice: like Peregrine's harsh 'keck-keck-keck'.

168 **Caucasian Black Grouse** *Tetrao mlokosiewiczi* — PLATE 28
(Caucasian Grouse)

L: 40–48. Robust gamebird. ***Male black with white underwing-coverts and long, downwardly curved, forked tail***; close views show plumage glossed blue-green, small white patch at carpal and red eye-wattle. Female smaller than male, brown, finely

vermiculated black with greyish underparts; tail much shorter than male's and square-cut. Flight rapid, short and direct with whistling wings. In communal displays (morning and evening) males make fluttering jumps about 1.5 m in air.
Voice: males utter a quiet guttural 'chr-chr' and a short chirping whistle when displaying; female has a fairly loud cackle. In flight, male utters a high-pitched prolonged, slightly falling whistle.

169 **Caspian Snowcock** *Tetraogallus caspius* — PLATE 28

L: 56. ***A large greyish buff gamebird with conspicuous white flight-feathers*** (in flight), occurring in high mountain regions. Close views reveal a darker belly, dark moustachial stripe and necklace on white cheeks and throat and (again in flight) ***dark brown tail with rufous tips***. Female is smaller, duller and buffer on crown and throat. SW Iranian race is buffer, less grey. Very wary and usually difficult to approach, running from the observer, but when alert adopts an upright posture. Voice important.
Voice: distinctive and often indicating presence: a far-carrying Curlew-like ascending whistle with last note very high pitched. Also a bubbling of variable length which accelerates and then descends slightly in a downward curl. Alarm-note a cackling 'chok-chok-chok' both on ground and in flight.

170 **Chukar** *Alectoris chukar* — PLATE 28

L: 33. Medium-sized, plump ground-dwelling gamebird, which closely resembles Arabian Partridge (which see) but does not overlap with it in range. Plumage grey above with conspicuous black, white and chestnut barred flanks and white face encircled by black necklace; differs from Arabian in having ***grey crown*** (black in Arabian) ***and narrow white supercilium*** (broad in Arabian). The SW Iranian race is buffer, less grey. Wary, runs with agility and will often adopt an upright posture when alert.
Voice: male utters an accelerating fast series 'ka-ka-kaka-kaka-kaka …' followed by nasal and hoarse chuckling 'chukara-chukar-chukara-chukar …'.

171 **Philby's Partridge** *Alectoris philbyi* — PLATE 28
(Philby's Rock Partridge)

L: 33. Similar to Chukar, but does not overlap with its range; differs from it and Arabian Partridge (which does occur in parts of its range) ***by large black throat-patch***, the only gamebird to show such a feature; further differs from Arabian in having a grey crown (black in Arabian) greyish-white supercilium extending behind eye and bordering black throat; ***in flight, chestnut on outer tail-feathers a useful feature*** (blackish in Arabian). Like Arabian, appears quite long-tailed. Juvenile brownish, finely barred above and with a shadow of adult's black throat-patch and pale pinkish-white legs.
Voice: a rapid repeated 'chuk chuk chuk kar' or 'chuk chuck-a-chuk kar' with accent on the last note, mostly heard at dawn and dusk. An explosive 'chork chork chork', uttered in flight; also a squealing 'chuck-a-chuck-a-chuck', often repeated several times.

172 **Arabian Partridge** *Alectoris melanocephala* **PLATE 28**

(Arabian Red-legged Partridge)

L: 36. Slightly larger than Philby's and Chukar, and most resembles latter. Immediately told from Philby's, with which its range coincides, by ***white throat*** (solid black in Philby's), ***long, broad, white supercilium and black crown*** (black crown and long, broad supercilium also separate it from Chukar). Juvenile, which lacks flank-bars and gorget, is grey-brown above, grey-buff below, lightly pale-spotted all over.
Voice: a variety of calls, most commonly 'kok, kok, kok, kok, kok, chok-chok-chok chook' beginning as separate notes, then accelerating and descending; also a rapid 'chuk-chuk chuk chuk chuk'; a loud throaty 'crook' and conversational, soft 'croo, croo, croo, croo'. More often heard during day than Philby's.

173 **See-see Partridge** *Ammoperdix griseogularis* **PLATE 29**

L: 24. W: 41. An almost ***sandy brown short-tailed partridge,*** smaller than Chukar, which runs fast among rocks and slopes, when disturbed, or flies low with whirr of wings. In flight, shows chestnut outer tail-feathers in both sexes. Male has ***grey head, chin and throat with conspicuous black forehead and supercilium bordered below by striking white eye-stripe,*** being very broad behind eye; orange bill; flanks strongly striped with 5–7 curving black and chestnut lines. Female and juvenile paler sandy, generally uniform, lacking head- and flank marks and resembling female Sand Partridge, although confusion unlikely due to different distribution.
Voice: more often heard than seen. A far-carrying repeated 'who-it' or a Spotted Crake-like 'wuid—wuid—'; also heard 'wuii-div' (stress on first syllable). Flight- call a fast 'bit-bit-bit'.

174 **Sand Partridge** *Ammoperdix heyi* **PLATE 29**

L: 24. W: 40. Similar to See-see (which see) in size, behaviour and general appearance, but sandier and more uniform. ***Male has white patch on forehead and behind eye, but lacks black head-marks of See-see.*** Flanks heavily banded black and chestnut as in See-see. Both sexes show chestnut outer tail-feathers in flight. Female and immature are uniform sandy-grey and are probably inseparable from female See-see in field, but distributions do not overlap. Subspecies *intermedia* (SW, S and NE Arabia) darker vinous-cinnamon.
Voice: commonest call repeated 'qwei-qwei-qwei-'. Alarm call in flight 'wit-wit'.

175 **Black Francolin** *Francolinus francolinus* **PLATE 29**

L: 35. W: 53. Large, plump, stub-tailed partridge. More often heard than seen, and very hard to flush. Male distinctive, ***mostly black with chestnut collar***; white spots on mantle and breast sides merging into ***white chevrons on flanks; obvious white patch behind eye.*** In flight, black- and brown-barred wings and blackish tail. Female lacks strikingly dark pattern of male, is ***warm brown*** with paler head and ***arrowhead-shaped feather-centres***, most obvious on lower breast and flanks; ***indistinct chestnut patch on hind-neck.*** In flight, dark tail with blackish outer tail-feathers.
Voice: male's call, most intense sunrise and evening, is ***loud, penetrating and very distinctive*** and with a grating character, mostly from a mound or a bush, sometimes

hidden on ground 'gldlri—*djjii*—djji-djji-*djjii*—djji-djjii', first note lower and 2nd and 5th stressed.

176 **Grey Francolin** *Francolinus pondicerianus* PLATE 29

L: 30. A rather drab, stub-tailed greyish brown gamebird separated from slightly larger female Black Francolin by much ***finer mottling and barring on upper- and underparts with chestnut forehead and cheeks***; lacks chestnut patch on hindneck and has ***pale throat-patch bordered below by black 'U'***. In flight, shows chestnut tail. Hard to flush, but if pressed rises with explosive whirr of wings. Normally runs off very swiftly.
Voice: a commonly heard call is a loud far-carrying series of 9–15 notes: 'kik-kjyw-ku, kik-kjyw-ku, kik-kjyw-ku . . .'.

177 **Grey Partridge** *Perdix perdix* PLATE 30

L: 30. W: 46. Adult much larger than Quail, but smaller, more round-bodied and thickset compared with Black Francolin and *Alectoris* partridges. Easily recognized by ***orange forehead and throat, grey breast and underparts with large dark horseshoe-like patch*** (smaller in female), and broad chestnut flank-bars. In flight, shows mottled grey-brown wings and upperparts contrasting with grey neck, orange head pattern and conspicuous rusty outer tail-feathers. In pairs or close together in families. Completely terrestrial. Rises with whirr and much wing noise.
Voice: song, mainly heard in evening or night, is a creaking slightly drawnout 'girrr-reck, girrr-reck'. Flushed birds rise with a grating 'grrree-grrree—'.

178 **Quail** *Coturnix coturnix* PLATE 30
(Common Quail)

L: 17. W: 33. Common over much of the region, but highly secretive; ***more often heard than seen***, and very hard to flush. In flight *size of Starling*; rather pointed narrow wings, fast shallow wing-beats, ***plain wings and striped back***; creeps on ground and is a rather compact, very short-tailed gamebird, with noticeably pale ***buffish white under-body, streaked flanks, and dark- and yellow-striped upperparts***. Head pattern black, chestnut and white with a ***neck-band in male***. Female has paler head pattern and lacks neck-band. Confusion possible in SW Arabia with smaller sized, but very rare Harlequin Quail and smaller Little Button Quail (which see).
Voice: male's call is a ***characteristic sound in open country*** over much of the region especially in farmland. An explosive, rhythmically repeated ***trisyllabic whistle*** 'pit, pil-it' (rendered as 'wet my lips').

179 **Harlequin Quail** *Coturnix delegorguei* PLATE 30

L: 17. Slightly smaller, with plainer upperparts and ***much darker underparts (male) than Quail***. Rarely seen unless flushed. In flight, less narrow-winged than Quail. Male separated from Quail by ***broad black centre of breast, chestnut flanks with black streaking***, and black and white head pattern. Female distinguished from Quail by ***unmarked buff throat and warmer rufous lower breast and belly without blackish spots*** (though present in juvenile). Little Button Quail is much smaller with short, rounded wings, contrasting upperwing in flight, absence of strong head pattern, and has golden-orange underparts with distinct black markings on breast-sides and flanks.

Voice: male's call resembles that of Quail but more metallic; an explosive 'whit-whit-whit' or rising and falling 'wit, wit-wit, wit-wit' repeated at intervals of 1–2 s.

180 **Pheasant** *Phasianus colchicus* **PLATE 30**

(Common Pheasant)

L: male 85, female 60. W: male 85, female 75. Unmistakable ***large gamebird with long, pointed tail*** and strutting gait. Rises with loud burst of whirring wing-beats in short flight. Mainly ground-dwelling, but perches and roosts in trees; runs fast with tail held high. ***Male has red facial skin contrasting with glossy green head and neck, deep chestnut body*** marked black and yellow. Female and juvenile pale brown with dark markings. Note half-grown fully fledged chick with short tail confusable with Grey Partridge, but lacks rusty tail.
Voice: male displays with a disyllabic crowing 'korrk-kok' (stress on 1st note) followed by rapid noisy wing-beats. Alarm call in flight a metallic loud clucking 'egg-egg-egg—'.

181 **Helmeted Guineafowl** *Numida meleagris* **PLATE 30**

L: 63. W: 97. Adult unmistakable, large, ***short-tailed, slate-grey gamebird***, the size of domestic hen. Tiny bluish white and red head with pale ***brown horny protuberance on crown. Plumage completely covered by fine white spots***. Juvenile duller and browner with buff or white spotting and less pronounced head pattern. Gregarious, runs fast with head held high or flies rapidly with occasional glides.
Voice: commonest call from alarmed flock or at roosting time, a far-carrying, raucous, nasal staccato 'kek, kek, kek, kek, kaaaaaa, ka, ka, ka, ka, kaaaaa, ka ka'.

182 **Little Button Quail** *Turnix sylvatica* **PLATE 30**

(*Andalusian Hemipode*, Small Button Quail)

L: 15. W: 27. Resembles Quail in buffish plumage, shape and gait, but smaller with ***distinctly uniform orange breast with bold black spots on breast-sides; lacks solid dark head marks of Quail***; female brighter than male. Extremely retiring and very hard to flush. ***In flight, looks lark-sized with short tail, short, rounded wings*** (quite unlike Quail, which has narrow wings), ***dark flight-feathers contrasting with buffish wing-coverts and body*** (Quail lacks contrast); rapid, whirring wing-beats for just a short distance before settling.
Voice: advertising call of female a curious deep, hollow, resonant 'crooo—' or 'hooo—' usually lasting 1 s and repeated every 1–2 s for 30 s or more; often by night, and resembling distant lowing of cattle.

183 **Water Rail** *Rallus aquaticus* **PLATE 31**

L: 26. W: 41. Mostly heard, seldom briefly seen walking in cover along edge of reeds or other aquatic vegetation. Obviously smaller than Moorhen, but ***bigger than the three smaller crakes*** and easily told from them by ***long, slender bill***. Adult has dark, ***mottled brown upperparts***, uniform slate-blue sides of head and underparts with ***heavy black and white bars on flanks***, conspicuous ***white undertail and red bill***. Juvenile has browner face, and ***mottled, greyer, less blue underparts*** and blackish bill. Tail often carried up and flicked in alarm; can run very fast. When flushed, flies only a short distance on fluttering, rounded wings with long legs dangling, before diving into cover.

Voice: several grunting, groaning, whining and ***stomach-churning sounds*** from within vegetation; sometimes ***like a squealing pig***. In spring, male (and female) utters for hours a rhythmic 'trüt—trüt—trüt—' sometimes ending with a trill.

184 **Spotted Crake** *Porzana porzana* — PLATE 31

L: 23. W: 40. A small-headed, round-bodied crake, much shorter and a little smaller than Water Rail, and slightly bigger than Little. Otherwise similar to Little, but ***heavily white-spotted plumage***. Short, red-based, yellow bill; flanks and vent strongly barred black and white, but has ***buff undertail-coverts*** (barred in Little and Baillon's), latter visible, when walking with tail cocked. Juvenile lacks grey head pattern, has whitish throat and bright brownish underparts with whitish spots. Rarely seen; moves with slow, stalking steps and sudden crouching run. Like other crakes very hard to flush. Flight action like Water Rail. Can also recall a small Moorhen with dangling legs.
Voice: song (both sexes) ***a rhythmical, far-carrying whistle*** 'whitt—whitt—' repeated each second, mainly from late dusk onward through night.

185 **Little Crake** *Porzana parva* — PLATE 31

L: 19. W: 37. Smaller than Spotted and separated by ***uniform grey underparts (adult male) and heavily barred undertail-coverts***. Size similar to Baillon's, but less compact with longer legs and neck, and ***much longer wing-projection*** (as long as exposed tertials; less than half the length of tertials in Baillon's). Tertial pattern differs markedly: pale buff fringes to inner webs form ***broad creamy line along inner aspect of folded wing*** in Little (may be worn off in spring/summer), in Baillon's paler edges to tertials never form broad continuous line. Male also told from Baillon's by ***less barring on flanks***, less spotted upperparts (some show pale bands, not fine irregularly scattered ring-shaped spots as in Baillon's), ***red base to bill***, and green legs. ***Female has brown-buff underparts***, white chin and throat and some grey on cheeks and supercilium. Juvenile lacks grey head pattern and has stronger flank-barring than adult; from juvenile Baillon's mostly by ***structure, tertial pattern*** and less barred underparts. ***Any small crake with buff underparts seen in the region in mid-winter and spring must be a female Little***. Hard to see, but often walks on floating vegetation.
Voice: male's song a loud, accelerating croaking or barking 'kwak——kwak——kwak—kwak—kwak-kwak-kva-kva-kva-kva-kva—'. Female's call is short, rapidly accelerating, with a vibrant terminal trill 'kwek—kwek-kverrrrr'.

186 **Baillon's Crake** *Porzana pusilla* — PLATE 31

L: 18. W: 35. Closely resembles Little (which see) but slightly smaller, more compact and with ***very short primary projection***. Upperparts warmer rufous with distinct, but smaller, irregularly scattered ring-shaped white spots; ***tertials do not form continuous pale line along inner aspect of folded wing*** as in Little; underparts bluish grey in both sexes with ***heavier black and white barring on flanks. Uniform green bill without red base***, and dirty olive legs and feet (sometimes look greenish). ***Juvenile more strongly barred below than Little, and best told by short primary projection and absence of broad, pale tertial-line***. Most skulking of crakes, but sometimes seen on floating vegetation.
Voice: male's song a series of dry, creaking or rattling frog-like sounds lasting 1–2 s and

repeated at intervals of 1–2 s, not so loud as, and very different from, Little, sometimes recalling the sound of male Garganey 'trrrrr—trrrrr—trrrrr—'.

187 **Corncrake** *Crex crex* — PLATE 31

L: 29. W: 50. Skulking land rail, larger than Quail, but recalling *Porzana* crakes in shape and general character. Rarely seen out of cover unless flushed. Flight, which is seldom seen, can suggest young gamebird, but note ***chestnut wing-coverts*** and often dangling legs. Otherwise only noticeable by ***characteristic breeding call.***
Voice: male's breeding call a far-carrying disyllabic rasping 'arrp-arrp, arrp-arrp ...' like grating a comb with a nail, repeated frequently, especially at night.

188 **White-breasted Waterhen** *Amaurornis phoenicurus* — PLATE 32

L: 32. W: 49. Slightly smaller and slimmer than Moorhen with longer neck and bill. Distinctive ***slaty brown and white, with rufous-chestnut vent and undertail-coverts,*** and yellowish green bill and feet; in breeding season reddish base to upper mandible; the crimson eye set in white face is very conspicuous. Juvenile has white face obscured by slate-brown. Skulking but sometimes easy to see in open, walking with jerking tail and displaying undertail-coverts. Flight with dangling legs like other rails. Occasionally swims.
Voice: very vocal in breeding season. Calls a prolonged ululating 'kaargh-kaargh' or breaking into 'kurrwah-kurrwagh-kurrwagh—', 'krrr-kwok-kwok-krr-oowark-oowark'.

189 **Moorhen** *Gallinula chloropus* — PLATE 32
(Common Moorhen)

L: 33. W: 52. A dark waterhen with prominent ***red bill and shield, white flank line and constantly flirting tail, exhibiting white undertail-coverts*** both when walking and swimming. ***Swims with vigorous nodding movements*** and body stooped forward unlike Coot. Juvenile paler, grey-brown with dark bill, and distinguished from young Coot by ***white flank line and undertail pattern***. See also Allen's Gallinule and Lesser Moorhen. Swims often but seeks cover more than Coot. Often runs over water surface when disturbed.
Voice: sings at night with persistent clucking 'kreck-kreck-kreck—'; sometimes a short variable 'kek' or 'kr-r-eck'. Many calls can be confused with Coot's. Most characteristic calls are a sudden, loud, gurgling, 'grrll' and a two- or three-note 'kwett—kwette-wett'.

190 **Lesser Moorhen** *Gallinula angulata* — PLATE 111

L: 28. Smaller than Moorhen and told from it in adult plumage by ***yellow bill*** (red with yellow tip in Moorhen) and absence of red at top of legs. Juvenile very similar to juvenile Moorhen but differs in having ***dull yellow bill*** (dark in Moorhen), ***paler head and underparts, the latter contrasting with brown upperparts*** (more uniform plumage in juvenile Moorhen). Habits similar to Moorhen.

191 **Allen's Gallinule** *Porphyrula alleni* **PLATE 32**

L: 23. W: 50. Smaller and more elegant than Moorhen, with bluish head and underparts, and ***iridescent greenish wings. Lacks white flank line***, but has white undertail-coverts like Moorhen. ***Note red legs and greenish upperparts.*** Much larger Purple Gallinule is almost uniform bluish with very heavy red bill and frontal shield (the latter greenish blue in Allen's). Juvenile/1st winter warm brown upperparts, distinctly pale-fringed tertials and greenish wash to flight-feathers; underparts buffish with white belly undertail-coverts; frontal shield brownish; legs brownish turning red. Much smaller female Little Crake (which see) also has long wing projection but note leg colour and undertail markings. Jerks tail when moving, and walks easily over floating plants; swims well. Long legs dangle in short flight. Secretive.

192 **Purple Gallinule** *Porphyrio porphyrio* **PLATE 32**

(Purple Swamphen)

L: 47. W: 95. ***Almost twice the size of Moorhen; huge red bill and frontal shield; red legs prominent. Plumage mostly uniform bluish with greyish head and neck.*** Swims with body stooped forward like Moorhen; ***white undertail-coverts often striking***. In flight, shape like huge Moorhen with blue wings and long red legs. Juvenile drabber with greyish underparts, dull red legs and greyish bill. Generally shy and often in cover.
Voice: rich and variable, often deep and mooing; contact call a low 'chock-chock', when feeding a 'tschak-tschak', flight-call a metallic 'krr'. Song is a long series of nasal notes 'quin quin krrkrr, quin quin krrkrr'.

193 **Watercock** *Gallicrex cinerea* **PLATE 111**

L: male 43, female 36. Larger than Moorhen. Male in breeding plumage blackish with pale feather-edging and ***bright red frontal shield and legs***. Female, non-breeding male and immature buff-brown ***streaked darker above*** and narrowly banded below; frontal shield and bill yellowish; legs brownish green. Mainly seen at dawn and dusk, never far from cover. Jerks tail; in flight legs dangle.

194 **Coot** *Fulica atra* **PLATE 32**

(Common Coot, Eurasian Coot)

L: 37. W: 75. Unmistakable. ***Sooty black with white bill and frontal shield***; hunch-backed appearance on water. Upright stance on long green-grey legs and ***long lobed feet*** when out of water. Flight is stronger and heavier than other rails, more duck-like, on rounded wings and with long pattering run across water before take-off. Note long toes trail behind tail-tip. Juvenile duller and paler with nearly white underparts and smaller frontal shield. Dives well, but only for a short time. Markedly gregarious especially in winter concentrations.
Voice: commonest call throughout the year is a short staccato 'kewk', also an explosive high 'pitts'.

195 **Common Crane** *Grus grus* PLATE 33

(Crane)

L: 115. W: 233. Big and majestic when walking slowly on ground. ***Huge in flight; shape resembles White Stork*** (though slightly bigger and shorter-billed), but ***grey plumage contrasting with black flight-feathers, black head and upper neck, and white stripe from eye down along neck,*** are obvious features. Red band above eye visible only at closer range. ***Looks 'bushy' at rear*** on ground. Juvenile has brownish head without contrasting head pattern. Adult separated from Demoiselle mainly on size, shape and ***absence of black breast*** (see that species); ***from herons by size and projecting neck in flight.*** In pairs in breeding area, gregarious on migration and in winter. Flight action regular and powerful interspersed with long glides and soaring; often in 'V'-formation or lines.

Voice: while on migration flocks often noticed by very far-carrying, guttural trumpeting calls 'krrllaa—krrllaa . . .' or 'krrlll— krrlll . . .'

196 **Siberian White Crane** *Grus leucogeranus* PLATE 33

(Siberian Crane)

L: 135. W: 245. Size and flight similar to Common Crane. ***Adult all white with bare red skin on face and with flesh-coloured legs. Black primaries conspicuous in flight,*** but concealed by elongated white scapulars and tertials when perched. Juvenile lacks naked red face and shows rusty orange head and neck and rusty buff wing-coverts. From White Stork in flight by ***lack of black secondaries.***

Voice: less vocal and higher pitched than Common Crane, a rather rapid alternating of two calls with a yodelling effect, the first high-pitched, the second more falsetto: 'kerrh-khoon' or 'khee-khoon-khee-khoon'.

197 **Demoiselle Crane** *Anthropoides virgo* PLATE 33

L: 95. W: 175. Markedly ***smaller than Common Crane when standing.*** In flight, however, appears more similar in size especially when alone, but note ***black on head extending onto breast,*** shorter neck and shorter bill. Adult is pale grey with largely black head and neck and ***elongated breast feathers hanging down in a narrow black fringe.*** Behind eye white, down-curving ear tufts; very long tertials curve down neatly over rear end concealing wing-tips and tail. Juvenile mostly grey on head and neck with short whitish band behind eye. Immature gradually acquires adult plumage, but the general coloration is browner with duller black parts, and both juvenile and immature have shorter feathers on fore-neck and tertials.

Voice: like Common Crane, but slightly higher pitched.

198 **Little Bustard** *Tetrax tetrax* PLATE 33

L: 43. W: 110. Pheasant-sized ground bird with ***small head and rather straight long neck.*** Male breeding has ***grey, black and white throat and neck pattern,*** upperparts vermiculated sandy brown and underparts white; male non-breeding, female and juvenile lack the striking neck marks. Flight fast and gamebird-like with hunched and rigid wings, then reveals ***almost completely white wings with black mainly confined to outermost 4 primaries.*** Male displays in spring with inflated neck and brief leaps in the air. Often

in flocks, especially in winter; rather shy with slow movements and often in cover of grass or low vegetation.
Voice: male displays with a short 'prrrt' call. In flight, male's wings make whistling noise.

199 **Houbara Bustard** *Chlamydotis undulata* PLATE 33

L: 60. W: 150. Nearly size of female Great Bustard; much larger and more attenuated than Little, with longer neck and rather pale sandy appearance; a ***black frill down side of neck in all plumages is unique***. Flight rather slow on shallow wing-beats, with a ***long-tailed and narrow-winged appearance*** and with a ***white patch confined to outer primaries only***. Shy, and prefers sneaking away without flying. Solitary when breeding; small flocks in winter.
Voice: virtually silent.

200 **Arabian Bustard** *Ardeotis arabs* PLATE 33

L: 82. W: 228. Large, nearly size of Great Bustard, but rather uniform, more drab upperparts (not cinnamon); ***head has small black crest bordered by white supercilium***; mostly ***grey neck without black frill*** of Houbara. In flight, shows ***broad wings and short tail***, and wing-beats are crane-like with accent on upward stroke; wings dark with a whitish band across innerwing similar to Houbara's but more conspicuous; ***dark hand has pale barring on inner primaries, lacking pure white patch on outer primaries of Houbara***. Usually in pairs or small groups.
Voice: a rasping croak 'pah pah' or a hollow 'puk-puk' during display (when neck balloons out, wings droop and tail is raised).

201 **Great Bustard** *Otis tarda* PLATE 33

L: 80–100. W: 190–260. ***The largest bustard of the region***; adult male twice as big as female. Huge, robust and stately with upright neck, ***tail sometimes cocked***; tri-coloured: ***grey head, cinnamon upperparts with black bars, and white underparts. Male develops brown-red breast band with age***, and breeding male has white whiskers (lost in non-breeders). Female and juvenile have slimmer neck without breast band. ***In flight, looks powerful with deep, eagle-like wing action***; broad-winged and short-tailed and ***reminiscent of a giant goose; large white wing-patches*** are very conspicuous. Usually in flocks but very shy.
Voice: usually silent, but can make deep, hollow territorial sounds.

202 **Pheasant-tailed Jacana** *Hydrophasianus chirurgus* PLATE 34

L: 31 (48 when breeding). Rail-like and mostly seen walking ***on floating vegetation on very long toes***. Adult breeding has long ***black down-curved tail***, chocolate-brown body, striking head marks with ***white face and fore-neck, golden-yellow hindneck, edged black; white wings with black tips are very conspicuous in flight***. In non-breeding plumage tail is short, ***underparts turn white, but a dark breastband remains, running up the neck and joining the eye-stripe***. Juvenile resembles adult non-breeding but breast band flecked white, and head and neck pattern are duller. Low, rapid flight with dangling legs, and landing with raised wings.
Voice: flight-call a distinctive plaintive 'queeear'; territorial song is a mewing 'mee-ooph' accelerating to 'ooph-ooph-ooph-' or 'meyu-meyu-meyu-'.

203 **Painted Snipe** *Rostratula benghalensis* — **PLATE 34**

(Greater Painted Snipe)

L: 25. W: 52. Snipe-sized skulking wader. When flushed flies with dangling legs like a rail, short tail and rather long, rounded wings; ***bill is shorter than Snipe's and slightly decurved at tip***; white flank feathers can give ***Ruff-like white rump pattern in flight***. Distinctive head marks, buff 'V' on mantle and ***white line in front of wing***; dark forepart contrasts sharply with white belly. ***Male has prominent buff eye-ring and streak behind eye***, buff crown-stripe, grey-brown wings spotted golden; in flight, shows golden-buff forewings and black wing-bar. ***Larger and brighter female is mainly greeny bronze*** with chestnut head and neck with ***distinctive white marks similar to male's***. ***Rises suddenly and silently***, and with slower wing-beats than Common Snipe.
Voice: usually silent, but sometimes a 'kek' from flushed birds.

204 **Oystercatcher** *Haematopus ostralegus* — **PLATE 34**

(Eurasian Oystercatcher)

L: 43. W: 83. Large unmistakable ***black and white wader with long, red bill and rather short, red-pink legs (adult)***. Flight is strong, direct and duck-like and shows ***conspicuous white wing-bar***, white rump and ***terminal black tail band***; non-breeding adult has white neck collar and duller bill-tip. Juvenile and immature have duller black upperparts, dark tip to bill, greyish pink legs and white neck collar. Gregarious after breeding.
Voice: noisy; common calls are a far-carrying 'kleep-kleep' and a disyllabic 'pick-pick'.

205 **Black-winged Stilt** *Himantopus himantopus* — **PLATE 34**

L: 37. W: 75. Unmistakable. ***Exceptionally long, red legs*** and rather slim black and white body; walks with high, graceful carriage. ***In flight, white with uniform black wings and noticeable trailing legs***. Head and neck pattern varies from pure white to dusky grey. Female has slightly browner mantle and scapulars. Non-breeding adults have dusky head- and neck marks. Juvenile and immature are brownish above with greyish crown and hindneck and white trailing edge to wings in flight.
Voice: noisy in breeding area; a variable sharp 'kek', a high-pitched continuous 'kikikikik' or 'kee-ack'.

206 **Avocet** *Recurvirostra avosetta* — **PLATE 34**

(Pied Avocet)

L: 44. W: 79. Rather large, slender and pied wader with mainly white appearance; at distance flocks on water somewhat recalling small gulls. ***Black crown and hindneck and long, thin upcurved bill*** are distinctive features; in flight, shows ***white wings with prominent black markings***; from below pure white with black wing-tips and long trailing greyish blue legs. Flight is stiff-winged, giving flickering appearance. Juvenile tinged brown on black parts, and white is mottled buff in upperparts. Walks steadily and delicately with head down when feeding in shallow water. Gregarious throughout the year.
Voice: noisy on breeding ground; most common call a repeated 'bluit-bluit'.

207 **Egyptian Plover** *Pluvianus aegyptius* **PLATE 111**

L: 22. W: 55. Unmistakable with unique ***black and white head pattern, black pectoral band*** and ***black band down back***; plumage otherwise grey above and pale buff below. In flight (fast wing-beats) ***black diagonal band across wings*** diagnostic.
Voice: a short, trilling 'trrrt'.

208 **Crab Plover** *Dromas ardeola* **PLATE 34**

L: 39. W: 77. Long-legged pied wader with ***large head and straight massive bill.*** Adult breeding has black mantle, white head and neck and ***distinctive black spot around eye***; adult winter and immature have dark streaks on crown and nape. Juvenile shows streaking on rear crown, silvery grey mantle, scapulars and tail, and at distance could be mistaken for a gull. Feeding action like a plover. ***In flight, shows white wings with blackish flight-feathers and long trailing legs***; flight slow with stiff wing-beats and head sunk back into shoulders. Often gregarious.
Voice: noisy; shrill 'tchuk-tchuk-tchuk—' near nest; alarm call a sharp 'kjep' or 'kiep'; in flight 'chee-rruk' often by night.

209 **Stone Curlew** *Burhinus oedicnemus* **PLATE 35**
(Eurasian Thick-knee)

L: 42. W: 81. Large, streaked greyish, Curlew-coloured wader with rather long tail and ***short bi-coloured bill.*** Usually in dry habitat, 'frozen' motionless with a prominent staring yellow eye; when slowly moving, often shows a hunched posture. Flight with ***stiff wing-beats recalling a huge Ruff; flight-feathers black with two small 'windows' in primaries and a paler mid-wing panel, bordered in front with a dark and a white line.*** Underwing pattern white with black tip and rear; white belly contrasts with dark streaked breast. Spotted Thick-knee (which see) is darker and boldly spotted in adult (more streaked in juvenile).
Voice: vocal mainly at night, reminiscent of Curlew: 'cur-lee' with accent on higher pitched second syllable; also an Oystercatcher-like 'ku-beek, ku-beek'.

210 **Senegal Thick-knee** *Burhinus senegalensis* **PLATE 35**

L: 37. Resembles Stone Curlew but the large whitish grey wing panel which is bordered by black, lacks the white band above (so typical in Stone Curlew); ***closed uppertail plain rufous or grey-brown*** (Stone Curlew has buff and dark brown markings on central tail-feathers). Bill larger than Stone Curlew's with less yellow at base. Juvenile resembles adult. Flight and general behaviour like Stone Curlew. Spotted Thick-knee (which see) has barred tertials and uppertail at all ages.

211 **Spotted Thick-knee** *Burhinus capensis* **PLATE 35**
(Dikkop)

L: 43. Resembles Stone Curlew in shape and appearance, but upperparts more brown-buff, ***spotted black in adult*** (but streaked in juvenile like Stone Curlew), the spotting being particularly obvious on the paler coverts; ***lacks black and white bars on coverts shown by Stone Curlew. In all ages tertials and tail are diagnostically barred*** (mainly plain in Stone Curlew). Flight pattern shows 2 prominent white patches on black

primaries, and underwing usually shows a strong dark bar along central wing, unlike Stone Curlew. Prefers to stay nearer to cover of bushes than Stone Curlew.
Voice: usually at night: a whistled 'ti-ti-ti-tee-tee-tee ti ti ti' growing to crescendo, then dying away.

212 Great Stone Plover *Esacus recurvirostris* PLATE 35

(Great Thick-knee)

L: 50. W: 95. Larger than Stone Curlew with ***plain sandy-grey upperparts***, white underparts with ***unstreaked grey neck and breast***. Head distinctive: ***long, very heavy upturned black, yellow-based bill*** and ***striking black and white head marks***, conspicuous at long distance; big yellow eye. In strong flight, short-tailed appearance recalling a small bustard or goose; flight-feathers black with striking white patches in primaries; mid-wing panel greyish, contrasting with dark band on lesser coverts, also visible on closed wing. Runs fast.
Voice: territorial note a wailing whistle with a rising inflection, mostly at night: 'see' or 'see-ey', alarm call a harsh 'see-eck'.

213 Cream-coloured Courser *Cursorius cursor* PLATE 35

L: 21. W: 54. Adapted to desert habitat, well camouflaged; runs quickly with plover-like behaviour and sudden stops, prefers running away rather than flying. ***Mainly sandy buff with distinctive black and white head markings joining in 'V' on nape, bill short and down-curved.*** When standing has an erect posture with long pale legs. Flight rapid with long, slightly rounded wings and short tail, legs protrude well beyond tail. Wing pattern very distinctive, ***black outerwing above, and black underwing contrast sharply with rest of plumage***. Juvenile lacks grey and black on crown; upperparts, head and breast have faint brown spots or irregular dark subterminal lines, and primaries are fringed buff.
Voice: rather silent on ground. In flight, a short 'kwit-kwit'.

214 Collared Pratincole *Glareola pratincola* PLATE 35

L: 25. W: 63. Distinctive, ***highly aerial wader***, resembling marsh terns in ***graceful fast flight, long pointed wings, deeply forked tail and short bill.*** Usually in loose flocks chasing winged insects. On ground, plover-like with quick tripping actions on rather short legs, often with upright stance, head held high. ***Adult has creamy buff throat bordered black***; tail and wing-tips are equal; in flight, uniform dark olive-brown above (slightly darker flight-feathers) with ***narrow, but distinctly white trailing edge to secondaries***, contrastingly, white rump and belly; ***underwing-coverts reddish brown***, but often look shadowy black. Black-winged Pratincole (which see) has generally darker upperparts, lacks white trailing edge to wings, and has black underwing-coverts. Adult non-breeding and juvenile lack distinct black throat line, and juvenile brown feathers have pale tips and fringes; outer tail-feathers are shorter than in adult.
Voice: most characteristic call a tern-like, sharp, chattering 'kikki-kirrik', and a short 'check' or 'che-keck'.

215 **Oriental Pratincole** *Glareola maldivarum* **PLATE 35**

L: 23. W: 63. Has distinctive elements of both Collared and Black-winged Pratincole, both of which it resembles. Colour of upperparts and ***lack of white trailing edge to secondaries are similar to Black-winged but shares reddish brown underwing-coverts with Collared***; from both by ***obviously shorter tail*** in adult plumage.

216 **Black-winged Pratincole** *Glareola nordmanni* **PLATE 35**

L: 24. W: 64. Structure and behaviour as Collared (which see); difficult to distinguish between the two. Adult slightly darker above than Collared, with no contrast between flight-feathers and coverts, ***lacks white trailing edge to secondaries, and underwing-coverts are jet black*** (but reddish brown in Collared can be hard to see, often appearing dark), tail fork shallower; when perched wings project beyond tail tip (in Collared approximately equal), and shows less red at base of bill. Non-breeding adult and juvenile resemble corresponding Collared but identifiable in flight by wing-pattern.
Voice: generally resembles Collared's, though a little sharper and in breeding area alarm call is a shorter 'pwik' or 'pwik-kik-kik'.

217 **Little Pratincole** *Glareola lactea* **PLATE 35**
(Small Pratincole)

L: 17. W: 45. Small, ***pied, swallow-like pratincole.*** In flight, appears sandy grey, black and white with diagnostic wing pattern: striking, broad ***white secondaries with narrow black trailing edge and black primaries; underwing black with broad striking white panel on secondaries and inner primaries***; white belly with no black line encircling upper breast; slightly forked black and white tail. Flight is swift, bouncy and rather swallow-like, on long and pointed wings. Juvenile resembles adult.
Voice: harsh notes recalling marsh tern, also a Little Tern-like 'tuck-tuck-tuck' or 'ke-terrick-ke-terrick'; when feeding in flocks high-pitched rolled 'prrip' or 'tiririt'.

218 **Little Ringed Plover** *Charadrius dubius* **PLATE 36**

L: 15. W: 45. Small, ***slim and long-winged plover***; horizontal stance when perched. In breeding plumage similar to Ringed Plover, but breast band narrower and ***lacks white wing-bar in flight***; note also ***yellow orbital ring***, dark bill, ***white line behind black forecrown***, and muddy coloured or pinkish legs. Adult non-breeding and juvenile have almost plain brown forehead with ill-defined (or absent) pale patch behind or above eye and duller orbital ring; breast band is broken or absent; additionally juvenile has yellow-buff tinge to face and throat. Distinguished from paler Kentish by head marks, leg colour and lack of wing-bar; larger Greater and Lesser Sand Plover have heavier bill, longer legs, narrow, but obvious white wing-bar, and lack white hindneck collar. ***Separated from all plovers by call and lack of wing-bar.***
Voice: flight and alarm call a loud, plaintive almost monosyllabic 'diu'; in wavering display flight with slow wing-beats gives 'pree-pree-pree—' and tern-like 'krre-u krre-u'.

219 **Ringed Plover** *Charadrius hiaticula* — PLATE 36

(Common Ringed Plover)

L: 19. W: 52. Dunlin-sized plover with ***orange legs and bill-base***, strong black breast band and white hindneck collar; in flight, shows ***conspicuous white wing-bar***, note also characteristic call. In adult non-breeding black is replaced by dark grey-brown, supercilium and forehead are tinged brown, and bill is nearly black. Juvenile similar to adult non-breeding, but is paler and duller, often with broken breast band, and upperparts slightly scaly with buff fringes. Distinguished from Little Ringed (which see) by white wing-bar, lack of pale orbital ring, and ***obvious pale supercilium in juvenile***. Paler Kentish has longer blackish legs, thin black bill, in flight more white in tail, and never has breast band. ***Separated from other plovers by obvious white hindneck-collar and call.***
Voice: commonest call a soft, rising, disyllabic whistle 'tooip'.

220 **Kittlitz's Plover** *Charadrius pecuarius* — PLATE 36

L: 13. W: 42. Structure and actions mostly resemble Kentish, but is smaller with rather ***brown, buff-fringed, dark-centred upperpart feathers, long legs, striking head markings and characteristic creamy buff wash on breast and belly*** (depth of buff colour variable). In adult breeding, white supercilium and black eye-stripe both meet at base of hindneck. In non-breeding and juvenile, head markings are much duller, lacking black, with white tinged buff. In flight, rather dark with blackish forewing and prominent, but short, white wing-bar, especially visible on inner primaries; tail pattern similar to Kentish and toes project well beyond tail-tip.
Voice: in flight, a plaintive 'pipeep'.

221 **Kentish Plover** *Charadrius alexandrinus* — PLATE 36

L: 16. W: 44. Small, sandy plover with white underparts, rather long, blackish legs, ***conspicuous white hindneck-collar and never complete breast band***; in flight, a clear, white wing-bar and ***broad, white sides to tail***. Adult breeding male has variable, ***rufous cap, black frontal bar, and lateral black breast-patches***. Breeding female, adult non-breeding and juvenile are much duller, lacking black in plumage, and resemble Ringed and Little Ringed non-breeding, but separated by rather long, blackish legs, white breast, more white in tail, and call.
Voice: flight call a soft 'kip' or 'twit' recalling Little Stint; sometimes drawn-out, more rolling; song a rattled repetition of 'tjekke-tjekke'.

222 **Lesser Sand Plover** *Charadrius mongolus* — PLATE 36

(Mongolian Plover)

L: 20. W: 51. Very similar to Greater Sand in all plumages, behaviour and general outline, and isolated individuals can be impossible to distinguish. Best separated by: size (***body only slightly larger than Ringed***, Greater Sand distinctly larger); ***shorter bill***, less pointed, more swollen at tip (but overlap with Greater Sand occurs in western birds); head looks smaller, more rounded, less angular; ***legs are proportionally shorter, darker, more greyish, not yellowish green***; stance in relaxed birds often more upright than in Greater Sand. Breeding male of *atrifrons*, the most common race in the region, has ***all-black forehead and face-mask***, and has ***broad reddish chestnut breast band*** (Greater Sand never shows completely black forehead, and breast band is usually narrower); female

has black of head reduced (like Greater Sand); the eastern race, *mongolus*, has almost white forehead. Non-breeding and immature plumage is very like Greater Sand. ***Separated from 'ringed' plovers by lack of white hindneck collar***, and from Caspian by shorter legs and wings, bolder white wing-bar, white underwing, and less bold supercilium. In flight, legs reach to or slightly beyond tip of tail (in Greater Sand legs show fairly prominently beyond), and both Sand Plovers show clear, but variable, white wing-bar: in Greater Sand often most prominent on inner primaries, of more even width in Lesser Sand.
Voice: quieter than Greater Sand: short, sharp and less trilling 'chitik', 'chi-chi-chi', 'chik-tik', etc. also 'kruit-kruit' or 'drriiiit'.

223 **Greater Sand Plover** *Charadrius leschenaultii* **Plate 36**

L: 24. W: 56. Resembles larger edition of Lesser Sand (which see for further details), and isolated individuals can be almost impossible to separate. Note especially: larger size (***obviously larger than Ringed***); ***longer, more pointed bill***, and larger, more angular head and big eye; sometimes reminiscent in shape of much larger Grey Plover; ***longer, usually yellowish green legs*** (especially thighs) with ***toes projecting well beyond tail-tip in flight***; call often more trilling. From most other plovers by lack of white hindneck-collar or by size; Caspian is slimmer with proportionally longer legs, and wings protrude well beyond tail when perched; head is more rounded with broader supercilium, underwing is dusky, and it has only a faint white wing-bar.
Voice: more vocal than Lesser Sand: when flushed and in alarm a trilling 'kyrrr', 'kirr' or 'trrr'; in song-flight 'huit-huit-huit' or an ascending 'dui-dui-tui-dit'.

224 **Caspian Plover** *Charadrius asiaticus* **Plate 36**

L: 19. W: 58. ***A slim and delicate plover*** mostly recalling sand plovers (which see) in size and general appearance, but separated by ***long attenuated body, with wings projecting well beyond tail-tip, when perched; proportionally longer legs***, long neck and smaller head with rather fine tapering bill; ***broader white supercilium gives a 'capped' effect***. Male breeding shows distinct ***blackish lower border to rufous breast band***. In flight appears long-winged with ***faint wing-bar***, only visible on inner primaries, and toes clearly projecting beyond tail; ***underwings are dusky*** (not white), and tail is rather dark with less white at sides and tip.
Voice: in flight, a short, sharp, single 'tyup' or 'tup', sometimes repeated, and occasionally combined to a rapid series of rattling notes 'tptptptptp'.

225 **Dotterel** *Charadrius morinellus* **Plate 37**
(Eurasian Dotterel)

L: 21. W: 60. Rather dark, brownish plover recalling non-breeding European Golden in general outline, but smaller; additionally ***white or buffish supercilia, meeting in a 'V' on nape, and narrow white or pale upper breast band*** are distinctive features in all plumages. Adult breeding female (male duller) has striking ***white supercilium and throat contrasting with blackish cap*** and greyish neck, ***upper breast bordered white with chestnut and blackish below***. Adult non-breeding and immature rather similar with ***blackish brown scaly upperparts and buff underparts***, lacking black and chestnut. In flight, ***no visible wing-bar***, but shows white tip to dark tail, and buffish grey underwing; white shaft on outer primary sometimes rather distinctive. Sand Plovers and Caspian have whitish underparts and longer legs, perched Sociable is larger and lacks pale breast band. **Voice:** rises with a trilling, rather dry Dunlin-like 'dryrrr'.

226 **American Golden Plover** *Pluvialis dominica* — PLATE 37

L: 26. W: 68. ***Almost identical to Pacific Golden*** (which see) in size, slim structure, underwing colour and voice (see also European Golden). Separated in all plumages from Pacific by slightly shorter legs (though still proportionally longer than in European Golden) with toes not projecting beyond tail-tip in flight (which they do obviously in Pacific but not in European Golden); ***tertials fall much shorter than tail-tip producing very long primary projection*** (tertials longer, only slightly shorter than tail-tip, with short primary projection in Pacific). In breeding plumage has darker upperparts (smaller yellow spots), ***all-black underparts including flanks and vent*** (whitish flank-line usually visible on ground and in flight in both European Golden and Pacific). Non-breeding and immature generally greyer than European Golden and Pacific with darker crown and upperparts, and ***lacking yellow/greenish tone on neck and breast***, with supercilium and forehead rather distinct whitish (lacking yellow tone of Pacific), underparts grey-vermiculated (whitish belly in Pacific).
Voice: ***flight-call resembles Pacific's more than European Golden's***, being di- or trisyllabic but softer and more variable 'kluilip', 'kuee-eep', 'chu-eep' or shorter 'hyyd'.

227 **Pacific Golden Plover** *Pluvialis fulva* — PLATE 37

L: 24. W: 66. Very similar to and easily confused with slightly bigger European Golden in all plumages. ***Best separated by generally slimmer structure, voice and underwing.*** Compared with European Golden, wings often (but not always) protrude 1–2 cm beyond tail-tip (most obvious in adult), ***bill is finer and longer, legs are clearly longer***, particularly thighs, making it appear more elegant; ***in flight, toes extend beyond tip of tail*** (never in European Golden), ***wings appear longer and narrower than European Golden's***, and ***underwing and axillaries are greyish brown in all plumages*** (white in European Golden). Adult breeding similar to European Golden; non-breeding and immature have rather distinct yellow-buff supercilium (more indistinct in European Golden). Flight pattern above uniform blackish brown speckled golden, faint, nearly absent, wing-bar (slighter than in European Golden) most obvious in primaries. Larger Grey Plover has grey appearance, stronger bill and, in flight, white rump and black axillaries. For separation from American Golden Plover see that species.
Voice: most typical call a soft disyllabic 'gru-it' ***resembling Spotted Redshank's*** (European Golden's is barely disyllabic, and vowel sounds 'y'-like).

228 **European Golden Plover** *Pluvialis apricaria* — PLATE 37
(Golden Plover)

L: 28. W: 71. Resembles Pacific and American Golden in all plumages but ***separated by heavier, short-necked and pot-bellied appearance, comparatively shorter legs and bill, white underwing coverts and axillaries*** (greyish brown in Pacific/American), ***and voice***. Adult breeding is strikingly coloured black, white and golden; very contrasting in flight from below with white underwings and black belly; ***feet do not extend beyond tip of tail*** (unlike Pacific). Flight pattern above is rather uniform, dark brown spotted golden with faint wing-bar mainly shown in primaries, no white in tail. Non-breeding and immature similar to Pacific, but yellow-buff supercilium less distinct. Larger Grey Plover has black axillaries, stronger bill, white rump and grey appearance. From American Golden see that species.

Voice: call resembles Pacific's in character, but is ***barely disyllabic*** and has an obvious 'y'-sound: a melancholy whistle 'pyüh' or repeated 'pyü-pü' (Pacific/American call resembles Spotted Redshank's).

229 **Grey Plover** *Pluvialis squatarola* **PLATE 37**

L: 29. W: 77. Large, rather robust-looking plover, bigger than European Golden, with ***heavier head and longer, stouter bill***; lacks obvious yellow or greenish tones in upperparts, except for faint yellow-buff tinge in juvenile. ***Black axillaries diagnostic in all plumages in flight*** (most obvious and contrasting in juvenile and non-breeding). Adult breeding recalls European Golden, but more white on head and nape, larger white breast-side patch, ***coarsely speckled black and white upperparts, and lacks white flank-line***. Non-breeding and winter similar to European Golden, but distinguished by size, structure, and ***mainly greyish appearance***. Wing-beats distinctly slower with longer, slimmer wings than in European Golden; note also obvious ***white wing-bar and white rump***. Behaviour like European Golden but more coastal.

Voice: flight-call a characteristic mournful tri-syllabic whistle, somewhat recalling Curlew's in tone, 'dee-oo-wee' (second note lower-pitched), often repeated.

230 **Spur-winged Plover** *Hoplopterus spinosus* **PLATE 37**
(Spur-winged Lapwing)

L: 26. W: 75. Elegant, long-legged plover, mainly sandy brown above, ***black head and underparts and contrasting white cheeks and sides of neck***. In flight, tri-coloured: conspicuous white band between black flight-feathers and sandy wing-coverts; broad black tail-band; from below black belly and flight feathers contrast with white underwing-coverts. Flight action rather jerky. Juvenile similar to adult. From White-tailed and Red-wattled by mainly black underparts and black legs.

Voice: noisy in breeding area; alarm call a shrill Oystercatcher-like 'dwitt–dwitt' or 'kwitt–kwitt'.

231 **Black-headed Plover** *Hoplopterus tectus* **PLATE 37**
(Black-headed Lapwing)

L: 25. Unmistakable: striking black and white head and breast markings with ***black crown, short up-turned crest, and prominent black 'tie' down middle of breast***; distinctly red bill with black tip, yellow eyes and rather long maroon-red legs. Flight is light with wing pattern above recalling Spur-winged, but white band is much broader, especially in outerwing; black subterminal band on white tail.

232 **Red-wattled Plover** *Hoplopterus indicus* **PLATE 37**
(Red-wattled Lapwing)

L: 33. W: 80. Rather large, colourful plover; easily identified by long ***bright yellow legs, black head and centre of breast***, white ear-coverts, side of neck and breast-side; eye-ring and ***wattle in front of eye is red***. Flight is light with slow wing-beats and shows similar wing pattern to Spur-winged and White-tailed, but ***tail has black subterminal***

band with broad white terminal band; yellow feet project distinctly beyond tail. Juvenile much duller with chin and throat almost white, black areas grey-brown, and wattle is tiny or lacking. Fairly tame.
Voice: noisy; loud and shrill alarm notes could be rendered as 'did-he-do-it, pity-to-do-it'.

233 **Sociable Plover** *Chettusia gregaria* — PLATE 38

(Sociable Lapwing)

L: 29. Lapwing-sized, with more upright stance, especially when alert. In breeding plumage easily told by ***long white supercilium, joining on nape, black crown*** and ***chestnut-black belly.*** In winter, loses belly-patch, becomes mottled on breast and supercilium is less distinct (but white forehead usually quite prominent). Juvenile, which is browner than adult, has buffish wash to forehead and supercilium and pale feather-edgings on upperparts. In flight, shows fairly rounded wings, with conspicuous black, white and brownish upperwing pattern (approaching that of White-tailed), but shorter blackish legs and black band on tip of tail. Typical plover actions on ground, making short runs with head tucked into body, stopping to peck at ground or stand with head erect. Often seen in small parties, sometimes, in winter, with Lapwings, golden plovers and coursers.
Voice: harsh 'chark-chark-chark' flight-call.

234 **White-tailed Plover** *Chettusia leucura* — PLATE 38

(White-tailed Lapwing)

L: 28. Most closely resembles Sociable Plover in winter plumage but smaller and readily told by ***plain head*** (accentuating dark eye), ***longer yellow legs*** (which protrude in flight) and, ***in flight, all-white tail***; juvenile paler on neck and breast, dark-mottled on upperparts with dark cap and faint brown tip to tail (on central pair, much less than in Sociable). Slender and graceful on ground, often with neck held erect. When feeding; tips down so steeply that it almost stands on its head!
Voice: high-pitched 'kee-vee-ik' persistently repeated at breeding site.

235 **Lapwing** *Vanellus vanellus* — PLATE 38

(Northern Lapwing)

L: 30. Unmistakable with ***long up-turned crest*** and ***greenish upperparts; in flight, broad rounded wings show no wing-bar***. In winter upperparts have narrow pale scaling to feathers. Often in large flocks outside breeding season.
Voice: loud, shrill 'peeo-vit' uttered in tumbling display flight over breeding grounds.

236 **Great Knot** *Calidris tenuirostris* — PLATE 38

L: 27. W: 55. Like a large Knot and of similar stocky build but more tapering at rear. In winter plumage (when most likely to be seen in the region) told from Knot by ***longer, slightly more decurved bill with heavier base, larger greyish spots on underparts, less defined supercilium (caused by more extensive greyish lores) and more obvious tail pattern (white rump/uppertail-coverts contrasting with dark upperparts and tail)***; upperparts darker than Knot caused by dark streaking to centres of grey feathers. In breeding plumage easily told by ***dense blackish spotting on breast*** and flanks, dark streaks on mantle and hindneck and chestnut centres to scapulars. Juvenile also told from similar Knot by ***heavily marked breast, contrasting with pale belly, and pale-fringed***

wing-coverts with dark shaft-streaks (absent in juvenile Knot); in 1st winter some juvenile coverts often retained, helping identification.
Voice: usually silent; occasionally a soft 'prrt'.

237 **Knot** *Calidris canutus* **PLATE 38**

(Red Knot)

L: 24. W: 50. In winter, when most likely to be seen in the region, best told from other shorebirds by ***combination of stocky build (noticeably larger and plumper than Dunlin), pale grey upperparts, straight, rather stout bill (which is about the length of head), flight pattern (rather long wings, greyish white rump and grey tail)*** and grey-green legs. ***In characteristic summer plumage, brick red below*** with black, white, grey and buff mottling above; then told from summer Curlew Sandpiper by shorter ***straight bill,*** larger size and tail pattern in flight (neat white rump and black tail in Curlew Sandpiper). Juvenile has poorly marked breast (unlike Great Knot); wing-coverts have dark subterminal markings and pale tips, but ***no dark shaft-streaks*** as in similar Great Knot.
Voice: a short nasal 'wut' or 'wut wut', rather quiet.

238 **Sanderling** *Calidris alba* **PLATE 38**

L: 20. W: 40. Slightly larger than Dunlin with ***shorter, straighter bill.*** In winter, easily told by ***very pale plumage with often a dark mark at bend of wing.*** In summer and juvenile plumage more confusable with other small waders, particularly stints, especially if comparison is not possible. In summer plumage (often shown on spring passage), can be quite ***rusty on head and prominent breast band, but always shows dark scalloping in the red of the breast,*** a feature that helps separate from the smaller and rare Red-necked Stint (which see). Juveniles (on autumn passage) ***spangled black and white on upperparts,*** sometimes with buff wash on breast sides. ***Noticeable white wing-bar in flight,*** more striking than in other smaller waders; ***often runs fast along shoreline.***
Voice: most usual call is a loud 'plit'.

239 **Red-necked Stint** *Calidris ruficollis* **PLATE 111**

(Rufous-necked Stint)

L: 14. W: 30. Very similar in size and structure to Little Stint but often has slightly shorter bill and longer wings and tail, giving a more attenuated appearance. These features are very subtle and it is in summer plumage that the species is most easily identified (but then confusable with the larger Sanderling in summer plumage especially if no size comparison possible). Summer Red-necked has ***neatly defined rusty red neck and upper breast (without the black feather-streaks shown by Sanderling);*** the ***white lower breast is flecked with arrow-shaped spots.*** Told from summer Little Stint by breast pattern (lacks dark streaking though the rufous-red as shown by Little), pale grey wing-coverts (rusty fringes in Little) and voice. Juvenile can be extremely difficult to distinguish from Little Stint but the most reliable features are call, ***very diffuse streaking on breast sides (which lack rufous wash shown by juvenile Little Stint) and pale greyish tertials (black in Little Stint).*** Adult in winter is very similar to winter Little Stint; then best to rely on call.
Voice: typical flight-call 'kreet', quite different from that of Little Stint.

240 **Little Stint** *Calidris minuta* **PLATE 39**

L: 13. W: 30. Often the commonest small sandpiper on passage and in winter in the region. In summer plumage, colour on face, neck, breast and scapulars varies from dull orange to warm buff; ***always shows a pale 'V' on mantle and dark centre to crown***; in this plumage confusion with Sanderling possible but note small size of Little Stint and pale 'V' on mantle. In winter, upperparts become grey, usually with dark shaft-streaks to feathers. Juvenile, which looks very white below, has warm rufous tone to upperparts, a distinct white 'V' on mantle, diffuse greyish neck-collar and rufous, streaked breast sides. ***Always has dark bill and legs***, the latter important for separation from Temminck's Stint which has yellowish grey legs. ***In all plumages similar to the vagrant Red-necked Stint***, which see for differences.
Voice: most frequent flight-call a short 'tip'.

241 **Temminck's Stint** *Calidris temminckii* **PLATE 39**

L: 13. W: 30. Similar size to Little Stint but with more elongated body, ***shorter legs, which are yellowish grey in all plumages and white sides to tail (useful when flushed at short range)***. In summer, mainly grey-buff with ***dark centres to many scapulars***, and lacking the rufous-orange tones of Little. In winter the dark scapulars are lost and then looks plain buff-grey. Juveniles have narrow buffish fringes to scapulars and coverts with some dark markings on upper scapulars. ***Lacks white 'V' on mantle in all plumages.*** Usually occurs singly or in very small groups, unlike Little which is often in large flocks. Usually towers high in erratic flight when flushed. See also Long-toed Stint.
Voice: flight-call a ringing 'tirrr', quite unlike Little Stint's.

242 **Long-toed Stint** *Calidris subminuta* **PLATE 39**

L: 14. W. 30 Similar to Little Stint in size but with ***longer neck*** (noticeable when it stands in upright posture) and ***longer legs which are dull yellowish or yellowish brown and often flexed***; the long toes are very difficult to see in the field. Leg colour similar to Temminck's Stint, from which told by longer legs, more upright posture and plumage: in summer, rufous and well streaked, often with noticeable supercilium creating a capped appearance. In winter, note dark feather centres to upperparts and fine streaking on rather dark head and breast, unlike the plain grey of Little and Temminck's. When flushed, shows very faint wing-bar and often towers high like Temminck's, sometimes in erratic flight.
Voice: a short soft 'prrt' or 'tit-tit-tit' when flushed.

243 **Baird's Sandpiper** *Calidris bairdii* **PLATE 111**

L: 15. W: 38. A fairly small sandpiper with rather distinctive shape ***created by long wing projection beyond tip of tail*** and rather short legs. Most likely to occur in region in juvenile plumage when told by combination of shape, ***scaly upperparts*** (white fringes to grey-buff feathers), faintly streaked breast on buffish ground colour, pale spot above lores (at close range), dark legs and bill. In flight, looks long-winged with poorly marked wing-bar and poorly defined tail/rump pattern.
Voice: a short, rolling 'krru'.

244 **Pectoral Sandpiper** *Calidris melanotos* **Plate 38**

L: 21. W: 42. An upright sandpiper, ***larger than Dunlin*** with short, slightly decurved bill and often with rather long-necked appearance. Told in all plumages from other shorebirds by ***clearly demarcated streaked breast contrasting with white belly, dull yellowish legs and faint wing-bar***.
Voice: loud 'kreet' often uttered in flight.

245 **Sharp-tailed Sandpiper** *Calidris acuminata* **Plate 38**

L: 20, W. 43. Similar to Pectoral Sandpiper in structure but with longer legs and shorter bill. Plumage also resembles that of Pectoral Sandpiper but in all plumages ***lacks the sharp demarcation between streaked breast and white belly, has white supercilium accentuating rufous crown*** and, at very close range, a white eye-ring. Adult in summer has dark arrow-shaped marks on underparts, which are lost in winter to give finely grey-streaked breast sides. In winter, upperparts lack any rufous, becoming grey with darker feather-centres. In 1st autumn bright rufous and buff streaked upperparts with whitish lines on mantle and scapulars, and upper breast with noticeable pale orange wash.
Voice: flight-call short, soft, metallic 'pleep', often rapidly repeated.

246 **Curlew Sandpiper** *Calidris ferruginea* **Plate 39**

L: 19. W: 40. Told from Dunlin in all plumages by ***white rump***, longer, more decurved bill and longer legs, giving it a more elegant appearance. Easily told in breeding plumage by ***chestnut-red face and underparts*** (often with white feather-fringes creating a hoary look). In winter, plain grey above, white below with a light suffusion to breast sides and ***noticeable white supercilium***; then also told from Dunlin by cleaner appearance, whiter underparts and white supercilium. In 1st autumn rather scaly grey-brown upperparts, noticeable white supercilium and yellowish buff wash to breast. Usually in small flocks.
Voice: a trilling, almost disyllabic 'chirrut', uttered in flight.

247 **Dunlin** *Calidris alpina* **Plate 39**

L: 18. W. 38. Often one of the commonest shorebirds, larger than the stints, but slightly smaller than Curlew Sandpiper (which see); longer bill than stints is slightly down-curved. In summer plumage easily told by ***black belly-patch***. In winter, this patch is lost as are rufous tones to upperparts, then has grey upperparts with narrow pale fringes to coverts and scapulars, grey breast and white belly; told from similar Curlew Sandpiper by different rump pattern (white with dark centre), lack of white supercilium, darker upperparts and breast, and shorter bill and legs. In 1st autumn note chestnut on coverts, white 'V' on mantle, and lines of dark spotting on flanks below finely-streaked breast.
Voice: a reedy 'kreep' flight note.

248 **Broad-billed Sandpiper** *Limicola falcinellus* **PLATE 39**

L: 17. W. 35. Slightly smaller than Dunlin, from which told in all plumages by ***longer bill which is broad-based with downward droop near tip, shorter yellowish grey legs*** and ***double supercilium***, especially noticeable in breeding plumage. In breeding plumage rather dark upperparts with white 'V' on mantle and white line on scapulars and ***white underparts with noticeable dark-spotted and streaked breast and flanks***. In winter, greyer and supercilia less obvious; then shape and length of bill, leg colour and, if present, dark area on carpals important for identification. In 1st autumn resembles adult but streaking below finer and confined to breast. Often has slow-moving and crouching feeding action. In rather erratic flight appears small, heavy fronted and, in breeding plumage, dark with thin wing-bar.

Voice: flight-call rather weak with a dry, slightly buzzing character, ascending a little at end; 'brlliid', recalling Temminck's but less trilling and less sonorous.

249 **Buff-breasted Sandpiper** *Tryngites subruficollis* **PLATE 43**

L: 19. W: 45. A small, Ruff-like buffish wader with small, roundish head, plain buff head-sides and prominent black eyes; it shows ***short, straight bill and bright yellow-ochre legs; upperparts distinctly scaly, and underparts from head to undertail buff*** with ***dark spots on breast sides***. In flight, appears rather plain above with ***no white in wings or tail***. Juvenile similar to adult. From young female Ruff by smaller size, shorter straight bill, steeper forehead, clean buff head-sides, spotted breast-sides, brighter yellowish legs and different flight pattern.

Voice: usually silent.

250 **Ruff** *Philomachus pugnax* **PLATE 42**

L: 28 (male), 22 (female). W: 56 (male), 50 (female). Resembles Redshank, but often stands more upright with ***longer neck, smaller head*** and proportionally ***short slightly drooping bill***; often looks hump-backed and pot-bellied; female noticeably smaller than male (and smaller than Redshank). Flight lazy on rather broad wings ***deeply depressed on downstroke***, sometimes interrupted with glides, especially before landing; flight pattern shows narrow, dull wing-bar and characteristic ***oval white patches at tail-base, often forming a 'V'***. Plumage and bare parts vary in adult male, less in adult female and least in juvenile. Upperparts, breast and flanks in adult show a mixture of black, brown, chestnut, ochre and white, heavily barred or blotched (male shows large ruffs and ear tufts in spring). Plumage greyer in winter; lores always pale, and face is often whitish in non-breeding season; rear belly and undertail are always white; few males (sometimes females too) show white on head, neck and breast; bill is mostly blackish brown, but can be yellow or pinkish, tipped dark in some males; legs of both sexes vary from orange-red to greenish grey. ***In juvenile, upperparts look scaly: blackish brown, distinctly fringed buff, and head, neck and breast yellowish brown, tinged orange***, legs are yellowish brown or greenish. Often gregarious.

Voice: normally silent.

251 **Jack Snipe** *Lymnocryptes minimus* **PLATE 40**

L: 18 (including bill 4). W: 40. ***Smallest snipe in region.*** Very hard to see on ground being superbly camouflaged and ***extremely difficult to flush***; often ***rises silently at about 1 m distance, in low, slightly jerking flight***, which is less powerful than Common Snipe's and reminiscent of a small crake; ***settles almost immediately***; ***short bill***, blackish brown upperparts with 2 broad contrasting yellow lines and ***short pointed tail*** are distinctive features. If seen on ground also note: green-glossed back, absence of central crown-stripe, dark-striped flanks (not barred). Separated from all other snipes by size and short bill. Usually solitary.
Voice: usually silent when flushed, occasionally may give a quiet 'gah'.

252 **Common Snipe** *Gallinago gallinago* **PLATE 40**
(Snipe)

L: 26 (including bill 7). W: 45. Medium-sized, dark brown wader with distinctly yellow-striped head and upperparts, dark-striped breast, barred flanks and ***very long bill***, with which it probes in mud with vibrating movements. Often squats in low, swampy vegetation, and usually not seen until flushed at 10–15 m distance (at only about 1 m in Jack Snipe); ***rises explosively, immediately uttering several harsh flight-calls, while zig-zagging to a good height***, and usually flies some distance. In flight, shows narrow, but ***rather sharp, white trailing edge to wings*** (lacking in Pintail Snipe) and ***white belly***. When on ground, very difficult to separate from Pintail Snipe (which see for further distinctions) and from Great Snipe (see also).
Voice: when flushed utters a few harsh 'ärrtch' notes, slightly rising at the end unlike Pintail Snipe (see that species).

253 **Great Snipe** *Gallinago media* **PLATE 40**

L: 28 (including bill 6). W: 49. ***Larger, heavier*** and darker than Common/Pintail Snipe (which see) with ***shorter bill and closely barred underparts and underwings***. When flushed, ***rises at shorter distance*** than Common Snipe, ***flies low and directly without zig-zagging and drops after a short distance***; ***flight is heavy with obviously slower wing-beats*** and ***with a pot-bellied appearance*** reminiscent of a small Woodcock. ***In flight, lacks white belly-patch*** of Common/Pintail Snipe; ***has dark underwing*** similar to Pintail, but unlike Common Snipe; ***has narrow white wing-bar on greater and median coverts*** (most distinctive in adult) ***including primary-coverts*** (last feature noticeable in flight but almost lacking in Common/Pintail); like Pintail lacks white trailing edge to secondaries (conspicuous in Common Snipe), and ***shows more white at tail-corners in flight*** (especially in adult). For separation from Solitary Snipe see that species.
Voice: often silent when flushed but sometimes a low-pitched, rather weak 'orrk', distinctly different from Common/Pintail Snipe.

254 **Pintail Snipe** *Gallinago stenura* **PLATE 40**
(Pin-tailed Snipe)

L.: 26 (including bill 6.5). W: 45. Very similar to Common Snipe, and best distinguished in flight. Look for slightly paler greyish buff panel on midwing-coverts, ***absence of white trailing edge to secondaries***, slightly more rounded wing-tips, and ***completely barred underwing-coverts making underwing look dark greyish*** (Common Snipe shows

broad white bands creating much paler underwing). On ground, shows slightly shorter, broader-based bill, ***bulging supercilium in front of eye*** (more parallel-sided in Common Snipe), ***more scalloped scapular pattern*** due to similar width and coloration of pale edges on inner- and outer-webs (in Common Snipe scapulars have broad white edges on outer-webs only, contrasting with dull brown inner-webs, creating white, often diagonal, stripes), ***median wing-coverts appear barred*** (more spotted in Common Snipe). When flushed, has slower take-off with slightly heavier, less erratic, flight; flies lower and drops after shorter distance than Common Snipe. Great Snipe (which see) is larger with dark belly and a much slower, direct and heavier flight.
Voice: flight-call similar to Common Snipe's, but shorter (without slight inflection), weaker, lower pitched and less repeated, starting with sharper t-sound 'tscät' or 'tjet'.

255 **Solitary Snipe** *Gallinago solitaria* PLATE 40

L: 30. ***Largest snipe of region*** (but smaller than Woodcock), most confusable with Great Snipe, but rarely seen away from higher mountains. Told from Great Snipe by more elongated appearance with ***longer*** bill, tertials and ***tail***, the last more wedge-shaped and ***with distinctly chestnut subterminal-band, obvious*** both ***when on ground*** and in flight; ***has gingery wash to breast and hindneck contrasting with white belly*** and barred flanks (Great lacks red-brown breast and shows closely barred underparts); throat and face pattern rather pale, finely dark-spotted, with ***faint crown-stripe*** (clear-cut in other large snipes), more prominent blackish lores, less defined supercilium and ***noticeable white eye-ring*** (darker yellowish and covered partly by supercilium in Great); ***legs strikingly yellowish*** (more olive in other snipes of region). Flight and flight-pattern rather similar to Great, but has longer wings, slender appearance and ***white belly-patch; lacks white tail-corners and distinctive wing-bars***, but has unpatterned grey-brown rump (unlike Great); also shows paler midwing-panel like Pintail. Often solitary.
Voice: rather silent; when flushed may give a 'wääk' similar to Common Snipe's, but weaker and less harsh.

256 **Long-billed Dowitcher** *Limnodromus scolopaceus* PLATE 40

L: 28. W: 49. In plumage and structure ***resembles Bar-tailed Godwit*** (and the very rare Asiatic Dowitcher), ***but much smaller*** and more similar to snipes in size, behaviour, feeding- and flight-action. ***The long, Common Snipe-like bill*** is dark, straight and slightly drooping at tip; legs pale olive-green. ***In breeding plumage has entire underparts orange-red***, densely barred or spotted blackish on neck, breast and flanks (white fringes in fresh plumage wear off in mid-summer); upperparts blackish brown, narrowly fringed rufous. In non-breeding/1st winter has pale grey underparts with darker neck and breast and slightly barred flanks and undertail-coverts; in all plumages a distinctive pale supercilium contrasts with black lores and dark upperhead. When flushed, rises fast, snipe-like but with slower wing-beats; ***flight pattern distinctive with narrow white oval on back*** (recalling Spotted Redshank) and ***narrow white trailing edge to secondaries***.
Voice: flight-call a sharp, clear, sometimes repeated, 'keek', slightly reminiscent of Oystercatcher.

257 **Asiatic Dowitcher** *Limnodromus semipalmatus* **PLATE 40**

(Asian Dowitcher)

L: 35. Easily confused with Bar-tailed Godwit, with which it often associates. Distinguished by obviously ***smaller size*** with ***straight, all-black bill, slightly swollen at tip*** (Bar-tailed noticeably pink-based, pointed and slightly upturned), ***which is held more angled below horizontal*** (about 30°; only 10–15° below in Bar-tailed). ***Feeding action, with continuous vertical 'sewing-machine' probing, is quite different.*** Sexes alike and in all plumages rather similar to Bar-tailed, but note whitish ventral region and white flanks, barred greyish in chestnut summer plumage (Bar-tailed lacks darker-barred flanks, and male has wholly rusty red underparts). In grey and white winter plumage shows slightly mottled grey-brown breast, obscurely barred flanks, and bolder eye-stripe than Bar-tailed, and in juvenile neck and breast have stronger buff wash than juvenile Bar-tailed. In flight, white rump is barred dark contrasting only slightly with back and tail, unlike most Bar-tailed (though similar to dark-rumped eastern race *baueri*, but note dark spotted underwing in latter, white in Asiatic). For differences from much smaller Long-billed Dowitcher see that species.

Voice: contact call a yelping 'chep-chep' or 'chowp', also 'aow' recalling distant human cry.

258 **Woodcock** *Scolopax rusticola* **PLATE 40**

(Eurasian Woodcock)

L: 34. W: 58. Plump, mainly rufous, snipe-like wader, ***larger than snipes*** and almost exclusively ***confined to woodland.*** Mostly seen when flushed; rises silently (slight wing-noise) with zig-zagging flight between the trees, ***showing round body, broad wings, red-brown rump and tail with black subterminal band and long bill.*** When seen on ground looks stocky, short-legged and long-billed with ***large conspicuous eyes set far back, barred crown*** (striped in snipes), and ***completely barred buff underparts.*** Usually solitary.

259 **Black-tailed Godwit** *Limosa limosa* **PLATE 41**

L: 42. W: 77. Large, graceful wader with long bill and legs and diagnostic, contrasting flight pattern. On ground, most similar to Bar-tailed, but slightly larger and more erect due to ***longer legs and neck***; ***bill slightly longer and straighter.*** In summer plumage, both sexes have varying amount of ***rusty red on head, neck and fore-body*** with ***diffuse, dark bars on lower breast, fore-belly and flanks***, rest of underparts white. Female duller, sometimes predominantly greyish. Non-breeding plumage uniform grey (more uniform than Bar-tailed), washed ochre on neck and breast in juvenile and with pink-based bill. In all plumages easily ***distinguished in flight by strong, white wing-bar and white tail with broad, black terminal band and long trailing legs and feet.***

Voice: rather quiet outside breeding grounds; all calls rather nasal and scolding; alarm call (also used in flight) 'titi-teev' or from more excited birds 'wicka-wicka-wicka'.

260 **Bar-tailed Godwit** *Limosa lapponica* PLATE 41

L: 38. W: 75. Resembles Black-tailed (which see) especially on ground in non-breeding plumage, but note more compact, less erect appearance with ***shorter legs and neck***, and ***slightly upturned bill***. In flight, easily separated in all plumages by ***lack of distinct wing- and tail-markings*** typical of Black-tailed. Male in breeding plumage has ***deep rusty red head and entire underparts***; larger female is buffish or faintly rusty, ***lacking breast- or belly-bars of Black-tailed***. Non-breeding/winter plumage buffish grey with dark shaft-streaks above and on breast (almost uniform grey in Black-tailed). Juvenile similar but more buffish with darker upperparts lacking shaft-streaks and breast markings. In flight resembles Whimbrel (which see) with almost similar flight pattern, but is smaller and more slender with faster wing-beats and straight bill. For separation from vagrant and rather similar Asiatic Dowitcher see that species.
Voice: rather quiet, in flight or when flushed: a soft, low-pitched, nasal 'beb-beb', or sometimes sharper 'ke-kek'.

261 **Whimbrel** *Numenius phaeopus* PLATE 41

L: 41. W: 83. Obviously ***smaller, and slightly darker than Curlew***, with faster wing-beats; ***bill usually shorter and more decurved near the tip*** (though overlaps with young Curlew in length), ***dark crown with pale central stripe and dark eye-stripe*** give head more contrast than in other curlews; flight-pattern similar to larger Curlew and slightly smaller Bar-tailed Godwit (note bill-shape of latter). ***Flight-call important character***.
Voice: flight-call a ***laughing series*** of fast, repeated 'bi-bi-bi-bi-bi-bi', quite unlike the soft rising whistle of Curlew.

262 **Slender-billed Curlew** *Numenius tenuirostris* PLATE 41

L: 39. W: 84. ***Obviously smaller than Curlew*** (size of Whimbrel) with shorter, ***finer, all-dark bill tapering to a thin tip*** (may show paler base of lower mandible but less so than shown by Curlew); plumage 'cleaner' with ***more greyish upperparts and whiter, more contrasting pattern of breast and underparts, showing black, rounded spots on flanks and side of belly in adult*** (brown streaks in juvenile); usually has darker crown and lores and paler supercilium than Curlew (lacks pale central crown-stripe of Whimbrel). In flight, ***almost black outer primaries and primary-coverts contrast more with paler innerwing than shown by Whimbrel and most Curlews***, otherwise flight pattern similar, but ***underwing pure white*** (darker in Whimbrel and many Curlews except paler eastern birds of race *orientalis*). Often adopts a more upright stance and moves more rapidly on ground than Curlew or Whimbrel.
Voice: flight-call most similar to Curlew's but shorter and higher pitched, lacking the liquid quality: single 'cour-ee'.

263 **Curlew** *Numenius arquata* PLATE 41
(Eurasian Curlew)

L: 55. W: 90. Large, brown-mottled, buffish grey wader with ***long, decurved bill***; plumage warmer, more ochre (less greyish), in fresh plumage (including juvenile). Male smaller than female, with shorter bill, which in young male overlaps with Whimbrel's, but is more evenly decurved. Flight with rather slow, gull-like wing-beats (obviously slower than

Whimbrel), showing barred tail and white wedge on rump and lower back similar to Whimbrel and Slender-billed. Distinguished from Whimbrel by size, ***usually longer bill, uniform head pattern*** and ***flight-call***; Slender-billed (which see) is smaller, paler, with 'cleaner' appearance, distinct flank pattern and finer bill; Far Eastern has brownish underparts and rump and much darker underwing.
Voice: flight-call a drawn-out, melodic, slowly rising whistle, easily imitated: 'duuuweeee'.

264 **Far Eastern Curlew** *Numenius madagascariensis* PLATE 41

L: 63. Closely resembles Curlew, but easily separated in flight by ***brown lower back, rump and tail***, concolorous with mantle (***thus lacking white wedge up lower back***); underwing and axillaries completely barred brownish forming ***dark underwing*** (almost white under wing-coverts in Curlew); plumage generally darker, more brown than Curlew, with dark buff head, neck and all underparts (though slightly paler rear belly in winter); ***very long bill*** (but overlaps with that of Curlew).
Voice: flight-call similar to Curlew's, but less melodious and less fluty 'krr-iii', more clearly disyllabic with first note rather harsh and second note longer, higher pitched, lacking rising inflexion of Curlew.

265 **Spotted Redshank** *Tringa erythropus* PLATE 42

L: 30. W: 64. Medium-sized, ***rather slim wader***; ***in winter plumage with red legs*** mostly recalling Redshank, but larger, cooler grey upperparts, white underparts, and ***longer finer bill***, slightly drooping at tip; flight pattern differs markedly. In breeding plumage unmistakable: ***almost black, fine-spotted white above with long, blackish red legs and bill***. In flight, shows dark tail and wings without white wing-bar, but ***a distinctive white rump extends in a wedge over back***, and feet project well beyond tail (but sometimes tucks legs up). Juvenile resembles juvenile Redshank but underparts darker, heavily spotted or vermiculated; black lores and white supercilium in front of eye also conspicuous.
Voice: flight-call distinct piercing disyllabic 'tju-it'.

266 **Redshank** *Tringa totanus* PLATE 42
(Common Redshank)

L:28. W: 62. Medium-sized ***greyish wader with bright red legs***, in flight showing ***white barring on tail, and broad, brilliant white trailing edge to wings***. Breeding adult heavily dark-spotted, with red base of bill. Winter plumage uniform grey above, paler grey underparts with ill-defined spotted breast, mostly recalling Spotted Redshank (which see). Juvenile buff-spotted upperparts and all-dark bill somewhat recalling smaller Wood Sandpiper, but note red legs and flight pattern; larger and slimmer young Spotted Redshank has greyer upperparts, heavily barred or vermiculated underparts and fine longer bill; also different flight pattern. Very alert and restless in breeding area.
Voice: flight-call a distinctive disyllabic 'djü-dü' (stress on first syllable) recalling Greenshank's (this often trisyllabic, and vowel sounds more like 'i'). Alarm call persistent 'tjü-tjü-tjü …'; display flight with shallow, vibrating wing-beats and slow 'düoo, düoo, düoo …'.

267 **Marsh Sandpiper** *Tringa stagnatilis* — PLATE 42

L: 23. W: 57. Resembles ***small edition of Greenshank*** (body size smaller than Redshank's), but ***legs proportionally longer, bill straight, thin and fine-pointed*** and neck slimmer, ***face almost white, usually with distinctive supercilium; flight pattern also similar to Greenshank's with dark wings, contrasting white tail and wedge up back***, pale head and white underparts; ***protruding legs distinctive beyond tip of tail.*** Flight action rapid and more similar to smaller Wood Sandpiper, but note white wedge on lower back and long legs. Winter plumage rather uniform, paler grey than Greenshank and Wood Sandpiper with almost white face and distinctive supercilium. In breeding plumage, becomes markedly black-spotted on upperparts, head, neck, breast and flanks, and legs frequently more yellowish. Juvenile browner above.
Voice: flight-call a clear 'djeeu-djeeu' often repeated and with characters of both Greenshank (though weaker and less shrill) and Little Ringed Plover.

268 **Greenshank** *Tringa nebularia* — PLATE 42
(Common Greenshank)

L: 32. W: 69. ***Larger than Redshank with long, green (sometimes yellow-green) legs, fairly long, stout, slightly upturned bill*** and mainly grey and white plumage. Flight action slow and jerky with rather long, uniform, ***dark wings contrasting with paler head and neck, and conspicuous white tail and wedge up back***, similar to much smaller, slimmer and paler Marsh Sandpiper; legs protrude beyond tip of tail, but proportionally less than Marsh's. Winter plumage resembles Marsh's, but darker grey, and ***darker face lacks supercilium***. Summer plumage has black feather spotting on upperparts and distinctly spotted breast and flanks. Juvenile uniformly patterned with buff-fringed grey-brown upperparts, white underparts and streaked grey head, neck and breast. Rather active, often running when feeding on small fish in shallow water.
Voice: flight-call shrill, ***usually trisyllabic*** 'djiu-djiu-djiu', with equal stress on all syllables (Redshank's often disyllabic with stress on first syllable, Marsh's thinner and less shrill).

269 **Green Sandpiper** *Tringa ochropus* — PLATE 42

L: 23. W: 59. Easily distinguished by call and contrasting black and white flight pattern. Shy and often well hidden, first seen when flushed. Rather ***broad black wings above and below contrasting sharply with white belly, and white rump***; looks cut off at rear. When standing, blackish upperparts and breast contrast with clear white belly and flanks. Juvenile darker, more uniform and buff-spotted, head rather dark with obvious pale eye-ring and short supercilium in front of eye. Similar Wood distinguished by paler and spotted upperparts, slightly barred flanks, long pale supercilium, and in flight pale underwing-coverts, longer feet-projection and different call.
Voice: flight-call a sharp, melodic 'dlo-eed-witt-witt'.

270 **Wood Sandpiper** *Tringa glareola* — PLATE 42

L: 20. W: 56. In flight, resembles Green (which see) with ***dark upperparts, head and wings and contrasting white rump; differs in whitish (not black) underwing-coverts***, longer feet-projection (does not look cut off at rear) and ***characteristic call.*** When perched, adult shows ***brown upperparts boldly speckled whitish, conspicuous long supercilium***, and rather long, yellowish legs; general appearance is a more elegant,

paler and less contrasting wader than Green. Juvenile has buff-spotted upperparts, somewhat recalling young Redshank (but note yellowish or greenish legs).
Voice: when flushed, a series of sharp, far-carrying notes 'jiff-jiff-jiff'.

271 **Terek Sandpiper** *Xenus cinereus* PLATE 43

L.: 23. W: 58. When standing easily distinguished by grey head and upperparts ***bordered by dark shoulder bar, carpal-patch and primary-line***; white, deep-chested underparts and ***long, yellow-based, upcurved, rather pointed bill*** are striking features; the short legs and neck, ***horizontal carriage***, dark breast side and ***bobbing rear body*** reminiscent of Common Sandpiper, but ***orange-yellow legs*** and bill shape differ markedly. ***In flight, wings show white trailing edge***, which is narrower than that of Redshank, from which it clearly differs in bill-shape, generally paler appearance and darker tail and rump. In juvenile and winter adult, black shoulder bar is faint or absent, and juvenile has buff fringes to upperparts. Feeds with very active movements compared with similar-sized *Tringa* species.
Voice: flight-call a clear, fluty 'tjiy-tjiy' or 'dwitt-dwitt', softer than Redshank, sometimes recalling Turnstone in character.

272 **Common Sandpiper** *Actitis hypoleucos* PLATE 43

L: 20. W: 40. ***Short-legged wader*** with almost uniform ***brown upperparts***, clean ***white underparts running up in a wedge between wing and breast side***; rather long tail projecting far beyond wing-tips; short neck, horizontal stance, deep chest and ***constantly bobbing rear-body*** are also conspicuous features (though shared with Terek, which see). Additionally, has a unique flight action low over water with ***vibrating, shallow and stiff wing-beats***, alternating with short glides. Flight pattern shows brown upperparts with long and distinct, but rather thin, white wing-bar (***also obvious from below***) and ***brown rump and tail*** with pale outer edges. Juvenile has wing-coverts barred buff and dark. For separation from vagrant Spotted Sandpiper, see that species.
Voice: flight-call characteristic: long series of piping, penetrating, descending 'hee-dee-dee . . .'; song similar, but notes faster and shorter.

273 **Spotted Sandpiper** *Actitis macularia* PLATE 43

L: 19. W: 39. Unmistakable with ***black-spotted underparts in breeding plumage***, but very similar to Common in non-breeding: look for traces of spots on belly sometimes present in winter, ***shorter tail***, brighter ***more yellowish legs and usually yellowish base of bill***; greyer, plainer upperparts with ***tertials lacking pale- and dark-spotted edges*** (notched in Common), but juvenile shows bolder, more contrasting, barring on wing-coverts. In flight, shows narrower and ***shorter white wing-bar*** (absent on inner secondaries) and less white in tail. Behaviour and flight as Common.
Voice: flight-call rather similar to Common, but more piping, often disyllabic and sometimes recalling Green Sandpiper: 'peet-weet'.

274 **Turnstone** *Arenaria interpres* **PLATE 43**

(Ruddy Turnstone)

L: 23. W: 53. A medium-sized, stocky, ***short-billed and short-legged wader***; easily distinguished in breeding plumage by ***striking head and breast markings, chestnut shoulder-patch and wing-coverts, and bright orange legs***. Flight is strong and direct and shows distinctive pied and chestnut appearance, ***black breast is a striking feature***. In winter and juvenile plumage, head and neck are much darker, and usually lack chestnut; also juvenile shows bold buffish fringes on most coverts. Walks with a rolling gait, and turns over small stones and seaweed for food items underneath.

Voice: calls short and staccato: when feeding a low 'chuk'; in flight 'krytt-te- krytt-te-krytt' or 'kritt-it-it'.

275 **Wilson's Phalarope** *Phalaropus tricolor* **PLATE 43**

L: 23. W: 41. Long-necked with ***straight, thin, pointed bill*** similar to Red-necked, but larger in size and feeds just as much on land as on water. In flight, shows ***white rump, plain greyish upperwing without wing-bar*** and feet projecting beyond tip of tail. Breeding plumage is colourful and easily distinguished by striking head and neck pattern. Winter plumage mainly ***grey with white underparts and yellow legs***; lacks bold black head markings and flight pattern of other phalaropes. Distinguished from Marsh Sandpiper by shorter, brighter legs and absence of white wedge up the back in flight.

Voice: usually silent.

276 **Red-necked Phalarope** *Phalaropus lobatus* **PLATE 43**

L: 18. W: 36. A small, elegant wader ***which swims*** high on water, often spinning to whirl up food items, which will be taken with ***needle-thin, black bill***. Unmistakable in breeding plumage with striking head- and neck-pattern. In winter plumage upperparts are ash-grey with white lines and fringes, ***black mask through eye and black on hind-crown*** conspicuous. Juvenile has similar head pattern, but ***upperparts are blackish brown with prominent ochre bands***. Flight is fast and jerking, and wings show distinct white wing-bar. ***Often settles on water***. Separated from Grey in all plumages by smaller size and thin black bill; in winter Grey is paler and more uniform above.

Voice: flight-call a short, sharp 'kritt' or 'kitt' recalling Sanderling's, but finer.

277 **Grey Phalarope** *Phalaropus fulicarius* **PLATE 43**

(Red Phalarope)

L: 21. W: 42. Slightly larger and more robust than Red-necked, with shorter, thicker neck; ***bill distinctly thicker, less pointed and sometimes paler at base***. Red breeding plumage with black and white head pattern is not likely to occur in the region. Winter plumage recalls a very small gull when feeding on water with ***unmarked grey upperparts*** (paler, more uniform than Red-necked). 1st winter similar to adult, but black on hindneck sometimes remains in winter. Behaviour and flight pattern similar to Red-necked but wing-beats less jerky.

Voice: flight-call similar to Red-necked's.

278 **Pomarine Skua** *Stercorarius pomarinus* **PLATE 44**

(Pomarine Jaeger)

L: 51–56. W: 125. Larger size (as small Yellow-legged Gull), heavier build and, in adult, ***elongated broad and twisted central tail-feathers*** separate it from Arctic. Two phases: all-dark (scarce) and a pale phase, which differs from Arctic in more extensive black cap, darker flanks, blacker breast band (sometimes absent), vent and upperparts. In winter, flanks and tail-coverts barred, elongated tail-feathers often short and always blunt-ended. Juvenile best separated from similar Arctic by ***more regular bars on flanks, vent, rump and tail-coverts***. Whitish base to greater primary-coverts below often conspicuous (frequently absent in Arctic). Bill proportionally larger, heavier at base and more distinctly bi-coloured. With experience, juveniles ***identified by larger size, slower wing-beats of broader wings and deeper belly***; also when present, short, blunt-ended central tail-feathers (pointed in Arctic). Piratical attacks more direct, less agile than Arctic's, frequently chasing large gulls.

279 **Arctic Skua** *Stercorarius parasiticus* **PLATE 44**

(Parasitic Jaeger)

L: 46. W: 117. Wing-span as Common Gull. Adult in summer has ***elongated pointed central tail-feathers*** (6–11 cm). Two phases (with frequent intermediates); see Pomarine for separation. Juvenile variable, from pale to very dark birds; ground-colour generally warmer, more rusty than young Pomarine, which see for further details. Told, with experience, ***from young Long-tailed by broader wings, slightly shorter tail*** (equals width of wing-base), ***shorter and pointed central tail-feathers*** (longer, blunt-ended in Long-tailed), ***thicker rear-body, less black on bill-tip*** and ***warmer brownish ground-colour*** (colder, greyer in Long-tailed, which see for further details). Normal flight steady, falcon-like, straight and fast; piratical attacks with sudden twists and turns, most successfully on birds the size of Sandwich Tern/Common Gull.

280 **Long-tailed Skua** *Stercorarius longicaudus* **PLATE 44**

(Long-tailed Jaeger)

L: 53. W: 111. Body size of Black-headed Gull but wing-span close to Arctic Skua. Resembles pale and rather grey Arctic but told by ***elongated, long, flexible central tail-feathers*** (16–24 cm), ***lack of breast band, white forebody, gradually darkening towards rear***, smaller white primary-patch above, ***pale greyish upperparts contrasting with blackish secondaries*** and, largely, ***pale blue-grey legs***. Juvenile variable like young Arctic but has ***colder greyer ground-colour*** (never rusty), ***more distinct barring on uppertail-coverts*** (except in darkest birds), ***thin white streak on leading primaries*** (hardly visible beyond 500 m, contrary to distinct white flash on most Arctics); ***central tail-feathers blunt-ended*** (pointed in young Arctic's). Has relatively deeper chest and thinner rear-body. At rest shows more rounded head and shorter, proportionally thicker ***bill with black tip to half the length***, unlike Arctic. Flight less steady, more buoyant and tern-like than Arctic's; includes more circling and hovering; piratical attacks also more agile, chasing terns and small gulls more successfully.

281 **Great Skua** *Catharacta skua* — **PLATE 44**

(Bonxie)

L: 59. W: 150. Largest skua; size of large gull but with ***much thicker, barrel-shaped body, broader wings, broader, more protruding neck*** and ***short, broad tail,*** quite unlike gulls and the smaller skuas. Dark brown plumage shows ***distinctive white primary-patches*** (above and below), ***visible at long range.*** At close range appears dark rufous, adult with pale streaks above. Much larger, heavier bodied and broader winged than young Pomarine. See South Polar Skua for separation. Flight strong, steady and purposeful, quite fast with many wing-beats before a glide. Bold attacks on large gulls for food; eats much fish but will kill small gulls and small birds.

282 **South Polar Skua** *Catharacta maccormicki* — **PLATE 44**

L: 53. W: 127. Closely resembles Great Skua. Small numbers of non-adults winter in Northern Hemisphere where pale and intermediate morphs are identifiable with difficulty by ***creamy buff or grey-brown head, hindneck and underparts, contrasting with blackish underwing-coverts.*** Dark morph darker than young Great Skua (which lacks pale streaks of adult above); bill sometimes more bi-coloured, hindneck has pale wash at sides, also plumage ground-colour, more slaty, not so warm rufous. During primary-moult (July–August) non-adult probably best identified by ***conspicuous symmetrical moult gaps in each hand,*** resulting from simultaneous moult of several primaries; non-adult Great Skuas moult more slowly (March–July) and do not usually show large symmetrical moult gaps. A pale 'noseband' at base of upper mandible, formed by pale lores and forehead is shown by some extreme birds. ***There is still much discussion of the differences between this and Great Skua and caution is needed.***

283 **Sooty Gull** *Larus hemprichii* — **PLATE 45**

(Hemprich's Gull)

L: 44. W: 112. Distinctive, medium-sized gull with ***dark sooty brown hood and 'bib',*** dark grey-brown upperparts and wings, dark flanks and underwings (white trailing edge) and narrow white half-collar. Long, ***thick, straight bill distinctly bi-coloured at all ages*** (in adult yellow-green with black band near red tip; in young pale blue-grey with black tip) ***always separates it from White-eyed. Narrow white crescent above eye*** (and seldom below) at all ages. Immature brownish grey on head, breast and flanks; upperparts brownish with buff feather-edges and tail has broad, black band. Rather long-winged with steady flight; at distance over sea can be mistaken for skua; sometimes piratical on other gulls.

Voice: loud mewing 'kaarr' or 'keee-aaar'; also high-pitched 'kee-kee-kee'.

284 **White-eyed Gull** *Larus leucophthalmus* — **PLATE 45**

L: 40. W: 108. Slightly smaller than Sooty with more slender wings, less deep chest and more buoyant flight. ***Separated from Sooty by all-dark bill*** (in adult dark red with black tip; blackish in young) ***which is long and slender*** with droop at tip (bill bi-coloured in Sooty); ***hood and bib black*** (dark sooty brown in Sooty) and ***upperparts greyer; conspicuous white eye-ring at all ages*** (faint and narrow in Sooty). Young birds

basically like Sooty though feather-edges of upperparts not so conspicuously fringed pale buff. May be mistaken for skua at distance; sometimes piratical on other gulls.
Voice: basically as Sooty, but less harsh and deep.

285 **Great Black-headed Gull** *Larus ichthyaetus* **PLATE 47**
(Pallas's Gull)

L: 68. W: 158. Size similar to Great Black-backed. Distinctive adult identified in summer by ***large black hood, large orange-yellow bill with black band near tip*** and ***white primaries with black band near wing-tip***. In winter, has large dusky patch behind eye; ***white eye-crescents present in all ages***. Juvenile and 1st winter told from other young large gulls by ***unmarked white rump and tail with clear-cut band at tip***, unmarked white underparts (though juvenile has grey-mottled breast band or patches at sides, lost in 1st winter), white underwing with extensive black wing-tips and often black-tipped coverts forming underwing-lines, ***pale mid-wing panel above and head pattern***. 1st winter told from 2nd winter Yellow-legged by ***sharp tail-band***, head pattern, dark-mottled hindneck, darker inner primaries and longer bill. 2nd winter still shows fairly distinctive tail-band; primaries and their coverts black-tipped with greyish bases. ***When standing appears deep-chested and sloping forehead accentuates length of bill.***
Voice: deep hoarse 'kra-ah'.

286 **Mediterranean Gull** *Larus melanocephalus* **PLATE 45**

L: 40. W: 106. Size between Black-headed and Common. Adult told from Black-headed by ***unmarked whitish primaries, longer thicker bill, paler grey upperparts*** and, in summer, by ***more extensive black hood*** (not dark brown); in winter head rather white with ***pronounced dark patch behind eye***. Juvenile and 1st winter told from similar Common by ***darker, longer, proportionally heavier, slightly drooping bill***, darker legs, ***blacker outer and paler inner primaries***, the latter contrasting with ***pale mid-wing panel; fully spread dark outer primaries show white subterminal spots*** (absent in Common, which lacks dark patch behind eye and white eye-crescents). In 1st winter ***back and scapulars much paler than Common***. Subadults have black subterminal-spots on wing-tip and, in summer, sometimes full black hood.
Voice: deep, nasal 'ga-u-a' with middle note highest.

287 **Little Gull** *Larus minutus* **PLATE 45**

L: 26. W: 63. ***Considerably smaller than Black-headed, with more elegant, tern-like flight***; wing-tips fairly rounded though more pointed in juvenile. Adult told by ***blackish underwing with conspicuous white trailing edge***, unmarked pale grey upper-wings and, ***in summer, by extensive black hood*** (less extensive, dark brown in Black-headed). In 1st winter, underwing pale and ***upperwing has blackish diagonal band***, recalling similar, but much larger, Kittiwake's, but ***top of head sooty grey*** and ***secondaries) have narrow dark band*** (larger Kittiwake has white top of head, black band on hindneck and white secondaries). Attains partly black hood in 1st summer but has remains of diagonal band; subadult has black subterminal spots on wing-tip. Feeds largely by snatching prey from surface of water.
Voice: a hard tern-like 'kjeck-kjeck'.

288 **Sabine's Gull** *Larus sabini* **PLATE 45**

L: 34. W: 89. Size of Black-headed. Long wings, ***shallowly forked tail*** and ***diagnostic tri-coloured upperwing-pattern***. In summer, adult has ***slate-grey hood***; in winter has blackish nape-patch; ***short black bill with yellow tip***. Juvenile separated from similar Kittiwake by ***continuous brownish from head down hindneck*** (and sides of breast) ***to mantle, brownish forewing***, dusky bar on inner underwing (close range), ***pale fleshy grey legs*** and more forked tail (young Kittiwake has black band at base of white hindneck, black band on innerwing, white underwing and black legs). Flight light; often swoops to water surface for food.

289 **Black-headed Gull** *Larus ridibundus* **PLATE 46**

(Common Black-headed Gull)

L: 38. W: 93. Medium-small gull, told (except from Slender-billed) by ***broad white leading edge of primaries, contrasting below with dark grey remaining primaries***. In summer, has dark brown hood; in winter, head white with black spot on ear-coverts and, often, vague bar from eye over crown and from ear-covers over nape. See Slender-billed for separation. Juvenile and 1st winter have adult primary-pattern; ***told from similar Slender-billed*** (which see) ***by slightly smaller size, shorter bill with darker tip, shorter neck, less bulky breast, more distinct markings on head and always dark eye***. Flight light and buoyant.
Voice: harsh 'krreeea'.

290 **Brown-headed Gull** *Larus brunnicephalus* **PLATE 46**

L: 43. W: 100. Slightly larger and heavier than Black-headed and with stronger bill; easily told from it (and Slender-billed) by ***broad black wing-tip breaking white leading edge above, dark grey leading primaries below*** and ***small white mirror near wing-tip of both surfaces***. In summer, hood paler brown than Black-headed with clearer black rim at rear; in winter, head pattern like Black-headed; ***iris yellow***. 1st winter birds have similar wing pattern, but ***broader black wing-tip without white mirror; also bold blackish trailing edge to wings***. Iris dark but changes to pale during first winter or spring when most of innerwing and tail moulted to adult pattern. Difficult to separate from Grey-headed, which see.
Voice: 'grarhh', deeper than Black-headed's.

291 **Grey-headed Gull** *Larus cirrocephalus* **PLATE 46**

L: 40. W: 100. Slightly smaller than Brown-headed with more sloping forehead and thinner bill. Difficult to tell from Brown-headed except in breeding plumage ***when hood pale greyish*** (face paler; hood narrowly edged dark at rear), ***inner upperwing darker grey and inner underwing mid-grey*** (Brown-headed has slightly paler upperwing and pale greyish inner underwing). In 1st winter both species have virtually identical wing pattern; in Grey-headed, spot on ear-coverts usually greyer, more diffuse; inner underwing slightly duller, inner primaries below perhaps slightly darker; ***probably best told from Brown-headed by narrower black tail-band*** (close to that of Black-headed). From 1st summer secondaries are still darkish with rest of inner upperwing mid-grey (in Brown-headed, most secondaries are grey like rest of innerwing).

Voice: drawn out 'caw-caw'; also laughing cackling cry, similar to Brown-headed and Black-headed.

292 **Slender-billed Gull** *Larus genei* PLATE 46

L: 43. W: 100. Medium-sized gull with wing-pattern almost identical to Black-headed, but ***head completely white in summer when breast often has rosy tinge.*** Larger than Black-headed with different build; ***bill distinctly longer, forehead rather sloping*** with longer upper mandible feathering accentuating length of bill, ***giving 'snoutish' effect.*** Legs longer (stands higher); when relaxed breast more bulky but when alert ***appears curiously long-necked.*** In winter, often shows ***small greyish spot on ear-coverts,*** (larger, blacker in Black-headed), ***but no vague bar over crown or nape*** (as in many Black-headed); ***iris pale in winter*** but dark in many breeding birds (always dark in black-headed); ***no obvious white eye-crescents.*** Dark red bill often appears blackish at distance, contrasting with white head. Apart from structural differences, juvenile and 1st-winter also told from Black-headed by pale bill with poorly marked, or unmarked tip (tip always dark in Black-headed); differences in head pattern as in winter adult; starts to show pale iris during 1st winter (sometimes before). Inner upperwing paler, more bleached in appearance than Black-headed's. Appears longer-winged and -tailed in flight and the long neck protrudes more than in Black-headed.
Voice: nasal, deeper and harsher than Black-headed.

293 **Audouin's Gull** *Larus audouinii* PLATE 46

L: 50. W: 125. Size between Common and Yellow-legged Gull; told from latter by shorter, deeper, ***darker red bill with blackish subterminal band and yellow tip,*** sloping forehead, ***dark olive-grey legs*** and ***almost solid black wing-tip with just one small white spot near tip; indistinct whitish trailing and leading edge to narrower wings; eye dark*** (iris yellow in Yellow-legged). ***At distance, bill appears all dark.*** Juvenile resembles young Lesser Black-back but told by ***all-dark uppertail and grey rump with defined whitish 'U' on uppertail-coverts;*** has ***greyer, unstreaked head*** and underparts, less dark ear-coverts and ***grey legs*** (fleshy pink in Lesser Black-back); compared with most young Yellow-legged Gulls has darker inner primaries and greater upperwing-coverts, more scalloped upperparts and bolder banded underwing-coverts with more conspicuous mid-wing panel. In 1st winter retains scalloped effect above (unlike other large gulls); rather white-headed, base of bill olive (blacker in Yellow-legged). After one year recalls 1st winter Mediterranean (sharp tail-band, pale grey back and mid wing-coverts). Flight faster than Yellow-legged, more elegant with more sailing and longer glides over water.
Voice: 'grraa(u)', rather harsh, repeated three or four times.

294 **Common Gull** *Larus canus* PLATE 46
(Mew Gull)

L: 43. W: 115. Medium-sized gull, recalling ***small version of Yellow-legged, but wings narrower and bill distinctly smaller, thinner and greenish yellow without red spot.*** Black wing-tip with prominent white spots at tip (much more white than in Yellow-legged). In winter, bill has narrow black band near tip; eye dark. Juvenile resembles similar Mediterranean, but thinner, shorter bill, mottled underparts and underwing, and less contrasting upperwing pattern. ***In 1st winter, back blue-grey*** (paler in Mediterranean),

upperwing pattern less contrasting and bill bi-coloured (dark in Mediterranean). Flies with quicker, more vigorous wing-beats than Yellow-legged.
Voice: much higher, shriller but weaker than Yellow-legged, alarm call a 'klee-*uu*', with emphasis on last high-pitched note.

The Lesser Black-backed Gull and Herring Gull complex — PLATE 47

The three species that follow: Lesser Black-backed Gull, Yellow-legged Gull and Armenian Gull all fall within a complex that has caused disagreement among taxonomists. We have adopted the following treatment:

(i) included the two races *heuglini* and *taimyrensis* under Lesser Black-backed Gull (though increasingly they are considered races of Herring Gull *Larus argentatus* or even a separate species).
(ii) listed Yellow-legged Gull and Armenian Gull as separate species. Formerly they were regarded as races of Herring Gull.

295 Lesser Black-backed Gull *Larus fuscus* — PLATE 47

(Includes races *fuscus, heuglini* and *taimyrensis.*)

fuscus: (L: 53. W: 127). Smallest, slimmest and most elegant of the 'black-backed' gulls. ***Distinctive jet-black and white plumage in adult summer*** (sparse streaking on head and neck in winter); ***bright yellow legs and 1–2 small, white mirrors*** on wing-tips. In flight, wings look long and slender and ***without contrast in black tone in upperwing primaries*** (c.f. *heuglini*); ***underneath primaries are almost black merging with dark grey subterminal band on secondaries*** (Great Black-backed paler). ***Very long and distinctly pointed primary projection*** when perched or on water; shape sometimes recalling Black-headed Gull resting high on water, ***especially in juvenile, when head and bill appear small and rather pointed.*** Flying 1st winter shows ***evenly coloured dark brown 'hand', secondaries and greater coverts without paler inner primaries***, unlike Yellow-legged, Armenian and Great Black-backed; tertials dark brown, fringed and tipped buffish white; tail and rump basically white, boldly spotted or barred black (***usually more than Yellow-legged***) and with broad, blackish terminal band. ***Underwing-coverts dark brown and usually lack distinctive pattern*** (unlike most Yellow-legged and especially Armenian, which see). As immature plumage progresses, head and underparts turn whitish, and during 2nd winter/2nd summer blackish feathers are visible on shoulder, back and wing-coverts.
heuglini: (breeds western Siberia, winters south at least to S Arabia). Adult has dark grey back ***recalling a slightly paler, rather large and robust, long-legged fuscus, showing contrast between black primaries and dark slaty upperparts*** (the upperparts are darker than those of the British race *graellsii*); ***yellow legs***; bold but sparse streaking on head and neck in winter (less so in *taimyrensis*, which see). Juvenile/1st winter very similar to *fuscus* Lesser Black-backed in plumage with all-black bill, but told by ***larger size and heavier structure (including bill)***. Distinguished from young Yellow-legged by ***dark inner primaries*** and by ***darker rump and uppertail-coverts.*** 1st summer/2nd winter often rather distinct with striking bi-coloured bill, uniform slaty brown tertials fringed with fine pale spots, and some uniform slaty brown shoulder- and wing-coverts.

taimyrensis: (breeds N Asia east of *heuglini*, winters south at least to Arabia). Size and structure as *heuglini*, heavily built with deep bill, heavy breast and long wing projection; ***mantle tone varies but grey paler than fuscus race and most heuglini***; large black wing-tip, ***inner part separated from grey back by white line***; one (sometimes very small second) white mirror. ***Legs bright yellow to pink***.
Voice: similar to Yellow-legged (Armenian is basically higher pitched, Great Black-backed deeper).

296 **Yellow-legged Gull** *Larus cachinnans* **PLATE 47**
(White-headed Gull)

Until recently this was regarded as a race of Herring Gull *Larus argentatus*. We have decided to treat it separately from *Larus argentatus*, which occurs in NW Europe and may not have been recorded in the Middle East. Herring differs from Yellow-legged in having ***very pale grey upperparts and pinkish legs***.
L: 61. W: 125–150. Large, heavily built with deep bill, long legs, heavy breast and long wing projection. Adult much paler grey than Lesser Black-backed (of the races *fuscus* and *heuglini*) and slightly paler than Armenian (which see) with ***more white on wing-tips than these; white head, like fuscus, in winter*** and on average proportionally longer bill than Armenian; legs yellow. From *taimyrensis* race of Lesser Black-backed (which see) mainly by more white on wing-tips ***without white line between grey and black in upperwing***. Juvenile/1st winter mostly recalls juvenile/1st winter Lesser Black-backed (especially *heuglini*) in having rather dark upperparts, but ***rump and uppertail-coverts are pure white, sparsely black-spotted, inner primaries slightly paler*** (at least in birds from E Mediterranean, Turkey and farther eastwards) and underwing-coverts and axillaries often distinctly marked (as Armenian, which see).
Voice: similar to Lesser Black-backed.

297 **Armenian Gull** *Larus armenicus* **PLATE 47**

Slightly smaller than Yellow-legged Gull; head less angular with steeper forehead and ***slightly shorter bill*** (most obvious in younger birds) with squarer tip. Adult a shade darker ash-grey than Yellow-legged with orange-yellow legs and ***usually dark brownish eyes***; bill with large red gonys-spot and ***variable blackish subterminal band*** (may be absent in older adults); in winter sparse streaking on hindneck. In flight, upperwing shows ***large, squared-off black primary area with less white on tip than Yellow-legged*** (1–2 tiny mirrors); underwing darker than Yellow-legged. Juvenile/1st winter very similar to Yellow-legged (slightly greyer brown upperparts in 1st winter), but ***underwing distinctly patterned: larger dark wing-tip area, white in centre of wing with black lines along coverts and axillaries, i.e. more contrasting on underwing than other similar gulls*** (see also Great Black-headed and Audouin's). Older immatures probably extremely difficult to distinguish except by structure and adult characters if developed. Note also ***later primary moult*** (in early July most birds have not yet started primary moult; Yellow-legged starts in May).
Voice: commonest calls as Yellow-legged, but more nasal and miaowing.

298 **Glaucous Gull** *Larus hyperboreus* **Plate 47**

L: 65. W: 158. ***Larger and much bulkier than Yellow-legged*** with heavy head and ***long, stout bill.*** Wings project slightly beyond tail-tip when perched, and ***in flight they are broad-based*** with rather short hand; ***flight-feathers (especially primaries) whitish or cream, often appearing paler than rest of wing*** on immatures. In all plumages ***lacks dark trailing edge to wing and subterminal band to tail.*** Adult (not likely to occur in the region) is much paler than Yellow-legged, heavily streaked on head and neck in winter with pink legs and white primaries. ***Young birds look sandy or grey-buff*** (sometimes almost whitish), mottled or finely vermiculated over all ***with paler primaries and distinctly bi-coloured bill.*** Beware of albinistic or leucistic young larger gulls with white plumage and primaries, they usually bear traces of dark trailing edge to secondaries or tail-band.

299 **Great Black-backed Gull** *Larus marinus* **Plate 47**

L: 71. W: 158. ***Largest*** 'black-backed' gull with conspicuously ***heavier bill than other gulls*** in the region. ***Adult shows jet-black upperparts*** (fading to brown-black), ***pink legs*** and usually rather dark eyes; flight appears heavy with broad-based wings and with ***much more white on wing-tip than Lesser Black-backed*** (which see for further details); in winter, head and neck remain white. Juvenile/1st winter similar to smaller Lesser Black-backed and Yellow-legged but generally ***paler upperparts, more variegated*** with broader white tips, bars or notches to coverts and tertials, ***dark tail-band narrower*** (especially at sides); distinctly dark, fairly uniform underwings, contrasting with whitish body; bill remains black (pale-tipped) until 1st summer and contrasting distinctly with largely white head. When plain blackish coverts appear in upperparts during 2nd winter/2nd summer identification becomes easier.

300 **Kittiwake** *Rissa tridactyla* **Plate 45**
(Black-legged Kittiwake)

L: 40. W: 95. Slightly larger than Black-headed Gull. Adult (unlikely to be seen in our region) resembles Common Gull but has darker grey back and innerwing but ***whitish, translucent outerwing and all-black wing-tip without white tips***; ***legs black.*** 1st winter recalls much smaller Little Gull but head very white (just small black spot behind eye) with ***bold, black band on hindneck, white secondaries and slightly forked tail.*** 1st summer birds often lack band on hindneck and sometimes lack bar on innerwing, approaching Sabine's Gull, which see for separation.

301 **Gull-billed Tern** *Gelochelidon nilotica* **Plate 48**

L: 38. W: 95. Resembles Sandwich Tern ***but has shorter, thicker, gull-like all-black bill and, in adult, shallower forked tail***; lacks short crest of Sandwich and has ***whitish grey rump and tail*** (only slightly paler than mantle, white in Sandwich); lacks contrast shown by Sandwich between dark outer and pale inner primaries above; outer primaries have distinct dark trailing edge (recalling that of Common Tern). In winter, lacks black cap ***but has variable black patch behind eye.*** Juvenile is much less spotted on scapulars and inner wing-coverts than similar Sandwich, has white crown and black eye-patch recalling winter adult Mediterranean Gull; in 1st winter head even whiter (nape mottled blackish in similar Sandwich). In flight, which is more leisurely than in Sandwich, shows

shorter neck. When standing, black *legs longer* than those of Sandwich. In breeding season often hunts for insects over land unlike Sandwich.
Voice: a nasal 'ger-*wek*' with stress on 2nd syllable; alarm a series of fast nasal laughing notes; juvenile call a soft 'pre-eep'.

302 **Caspian Tern** *Sterna caspia* **PLATE 48**

L: 53. W: 135. ***The largest tern,*** almost the size of Yellow-legged Gull and identified by ***heavy red bill (can be seen at long distance), almost gull-like flight with slow steady wing-beats and by call***; wings and back pale grey, under surface white ***with distinctive dark primaries below.*** Juvenile has more orangy bill, more extensive black cap to below eye (unlike adult), weak dark scales on mantle and wing-coverts; leading arm and secondaries slightly darker, primaries dark both above and below; legs pale (black in adult).
Voice: a deep, loud and harsh 'kraa-jak' recalling Grey Heron's.

303 **Swift Tern** *Sterna bergii* **PLATE 48**
(Crested Tern, Greater Crested Tern)

L: 46. W: 105. Large slender tern, size between Sandwich and Caspian; adult best identified from Lesser Crested ***by longer, thicker, more drooping waxy yellow bill tinged greenish*** (slimmer, orange-yellow in Lesser Crested) ***and by much darker ashy grey upperparts***; rump and tail paler grey than back. In summer, black crested cap always broken by white forehead-band (never black as in some Lesser Crested); tail slightly longer, more forked and wings broader than Lesser Crested. In winter, bill pale greenish yellow, but few have washed out yellowish orange bill; crown white with black confined to nape, the black often finely pale-tipped; upperparts as in summer but tail and rump almost as dark as back. Juvenile has pale panel in centre of innerwing above, framed by dark bar on forewing and dark rear-covert- and secondary-bar. Flight steady with shallow wing-beats.
Voice: deeper, rougher and less ringing than Sandwich Tern's, a 'kee-rit'; also a high-pitched 'kree-kree'.

304 **Lesser Crested Tern** *Sterna bengalensis* **PLATE 48**

L: 41. W: 92. Size as Sandwich Tern but resembles Swift Tern more closely in plumage, which see for separation. ***Has long slim orange-yellow bill, pale ash-grey upperparts with rump and tail paler grey.*** Slightly crested nape black as may be forehead for short period in summer, but forehead otherwise white. In winter, bill paler orange-yellow and solid black is confined to nape. Juvenile resembles juvenile Swift Tern but upperwing has paler dark bands and less pronounced pale mid-wing panel; bill dirtier greyish yellow; dark markings on innerwing retained to 1st spring as are dark primaries above. ***Told from Sandwich by bill colour*** and slightly darker upperparts. In graceful flight, wings placed slightly more centrally than in Sandwich and bill often carried more horizontally; distant young Sandwich lacks clear dark bars on inner upperwing of similar Lesser Crested.
Voice: recalls Sandwich Tern's but less ringing, less grating and not so pronounced disyllabic, 'krriik-krriik' or 'kreet-kreet'; also 'kir-eep' and 'kee-kee-kee'.

305 **Sandwich Tern** *Sterna sandvicensis* **PLATE 48**

L: 41. W: 92. Medium-sized tern ***with very pale grey upperparts and wings, long slender black bill tipped yellow*** (latter hard to see at distance), long narrow wings, medium-short forked tail and fairly long extended neck in flight, which is powerful with deep wing-beats. In summer, black cap has short crest; in winter, black confined to nape extending forward to eye (like Lesser Crested). Juvenile usually has all-black bill, dark-scaled mantle and forewing-coverts but lacks the broad dark bars on innerwing of similar aged Lesser Crested; 1st winter has grey mantle and scapulars and head pattern like winter adult. See Gull-billed and Lesser Crested for further separation.
Voice: a loud, grating disyllabic 'kerr-*rick*' with emphasis on last highest pitched note; juvenile's contact-note a high-pitched 'k-rill'.

306 **Roseate Tern** *Sterna dougallii* **PLATE 49**

L: 38. W: 77. Generally like Common and Arctic but ***adult is paler grey above, whiter below*** (tinged pink at close range in breeding season) ***and the long, thin bill is largely blackish*** in spring but in the race that occurs in the Middle East, *bangsi*, becomes red by July; ***tail-streamers usually very long and white***; wing pattern diagnostic at all ages: well-defined ***dusky outer primaries but lacks black line along trailing edge of outer primaries below***; at rest tail-streamers protrude far beyond tip of tail; legs longer than Common and Arctic. Winter plumage resembles that of Common Tern but retains long tail-streamers. Juvenile has short tail-streamers; told from Common and Arctic by underwing pattern (white trailing edge to outer primaries below), from young Arctic also by dusky, not white, secondaries and long black legs; has darker forehead than Common/Arctic and very scaly upperparts and scapulars, recalling similar Sandwich Tern. Flight relatively fast and direct with stiffer and more rapid wing-beats than Common and Arctic. Plunge-dives more directly into water than the other two species.
Voice: a characteristic, soft guttural 'cherr-*wrick*' with stress on last syllable.

307 **Common Tern** *Sterna hirundo* **PLATE 49**

L: 35. W: 80. Closely resembles Arctic but has slightly broader wings, shorter tail-streamers, longer neck and bill, flatter forehead and often more powerful wing-beats; legs longer. ***Dark outer primaries above abruptly cut off from grey inner primaries*** (uniform pale grey in Arctic); ***from below has blackish band on trailing edge of outer primaries and translucency confined to innermost primaries*** (in Arctic dark trailing edge-band narrower and sharper with all primaries translucent); bill dark orange, usually tipped black (Arctic has darker red bill usually without black tip); cheeks and throat whiter than in Arctic; in winter forehead white. Juvenile (white forehead) best told from similar Arctic by flight, dark grey secondaries above (white in Arctic), broader, darker forewing-band above, translucency and pattern of primaries (similar to adult); additionally, centre of rump is pale grey (pure white in Arctic). Dives with gently upward swoop, turning back and plunging. See Roseate and White-cheeked for separation.
Voice: deeper than Arctic's; a short sharp 'kitt', a 'kirri-kirri-kirri' and a drawn-out 'kreee-aeh' with both syllables of equal length.

308 **Arctic Tern** *Sterna paradisaea* **PLATE 49**

L: 38. W: 80. Slenderer and more narrow-winged than Common with rounder head, shorter thicker neck, longer tail-streamers and more elegant, buoyant flight which is less direct than Common's; *legs shorter*. In the air ***all primaries appear translucent, revealing a thin, sharp black line to trailing edge of outer primaries***; greyish underparts extend to sides of throat leaving whitish band below black cap; in winter forehead white. See Common for separation. Juvenile/1st winter has flight-feather pattern of adult, white secondaries, diffuse dark bar on leading forewing and white rump; see Common for separation. Plunge-dives more hesitantly than Common; hovers, moves, hovers, dips, hovers momentarily again, then plunges.

309 **White-cheeked Tern** *Sterna repressa* **PLATE 49**

L: 33. Slightly smaller than Common Tern with shorter wings, slightly shorter legs but with fairly long bill. Adult in summer is ***dark silver-grey on upper surface with grey underbody and underwing, latter with whitish area in centre***; primaries above appear paler than innerwing (inner primaries are whitish grey); white cheek-stripe recalls that of adult Whiskered in summer; has long, broad black line to tips of outer primaries below and dull grey secondaries. In winter, remains ***dull grey above, including rump and tail***; underparts white or, in some, mottled with dark grey; forehead white, bill blackish. 1st winter has broad blackish forewing-band and dark secondaries (like similar Common), but mid-wing dull greyish rather than greyish white as in Common; rump and tail greyish (pale grey rump-centre only in Common). Told from Whiskered in summer by longer tail-fork and longer bill.

Voice: loud harsh 'kee-*err*' or 'ker-*rit*' with accent on short 2nd syllable (Common has accent on first note).

310 **Bridled Tern** *Sterna anaethetus* **PLATE 49**

L: 37. W: 76. Medium-sized slender tern with ***dark grey-brown upperparts*** (sometimes appearing blackish), ***blackish wings and black bill and legs***; ***has narrow white forehead-band extending behind eye as narrow supercilium line***, black cap contrasting with ***pale grey hindneck-collar*** (often hard to see in flight); long outer feathers of deeply forked tail have white sides. Whitish underparts and underwing-coverts contrast with ***dark silvery grey central tail-feathers and especially flight-feathers but whitish outer primaries produce pale wedge*** towards dark wing-tip. Juvenile lacks distinct supercilium, has pale grey-brown back and wing-coverts, edged buffish. Flight more swift and graceful and has slimmer wings than Sooty Tern which see for separation. Does not dive but takes food from surface of water.

Voice: high-pitched 'kee-yharr', yelping 'wep-wep'; also harsh grating 'karr'.

311 **Sooty Tern** *Sterna fuscata* **PLATE 49**

L: 44. W: 90. Larger, more stocky than Bridled and with ***entire upper surface uniform blackish*** (not paler mantle and hindneck as in Bridled), ***broader but shorter white forehead-patch than Bridled*** (not extending beyond eye) and black loral streak reaching gape (not base of upper mandible as in Bridled); white underparts and underwing-coverts ***contrast more with dark grey flight-feathers in which outer primaries lack whitish wedge toward wing-tip*** of Bridled; bill and legs blackish. Juvenile ***sooty brown all over***

except for whitish belly and undertail-coverts; upperparts flecked whitish. Told from young Noddy by whitish vent and undertail-coverts, paler underwing and forked tail. 1st summer like adult but throat and upper breast blackish.
Voice: most characteristic call high-pitched 'ker-wacki-wah', quite different from Bridled's.

312 **Little Tern** *Sterna albifrons* — PLATE 50

L: 23. W: 53. Easily identified from other terns (except Saunders's, which see) ***by small size, narrow wings with very fast beats both when hovering and in direct flight.*** Adult in summer pale grey above (extending onto centre of rump in some) ***with narrow grey-black leading*** (two) ***primaries and yellow bill tipped black***; white forehead usually extends as white point to rear eye; legs bright yellow to reddish orange. In winter, bill black, legs dull grey or brown, usually with some yellow; black on head reduced to band around back of head; rump and tail largely grey. Juvenile (dark bill with reddish base) has dark 'U'-shaped feather-markings on pale grey mantle, scapulars and tertials, lost in 1st winter when it resembles adult winter; legs greyish pink to yellow-brown; upperwing shows dark outer primaries becoming progressively paler inwards; leading forewing dark.
Voice: common call a hoarse grating 'kryik' or 'pret-pret'.

313 **Saunders's Tern** *Sterna saundersi* — PLATE 50

(Saunders' Little Tern)

L: 22. Very difficult to separate from Little Tern in non-breeding plumage, particularly in Arabian Gulf, where intermediates between the two apparently occur. The following differences are more reliable in Red Sea area: breeding plumage differs from Little by slightly paler grey upper surface, ***contrasting more markedly with blacker outer primaries*** (normally 3 blackish in Saunders's, 2 in Little); white forehead does not usually reach eye (as it does in Little) thus shows as a more square patch; in early spring adult, the black of rear-crown starts to reach the eye in both species and identification is tricky before full summer head pattern achieved; ***rump more extensive grey than in Little and concolorous with back*** (at best pale grey rump-centre in summer Little); legs generally darker, reddish brown to pinkish brown, sometimes with little yellow on rear tarsus (see Little). In winter, adults and 1st winter birds almost impossible to separate from Little; however, in the Red Sea region 1st winter birds ***have clearly darker grey upperparts*** than similar-aged Little; apparently, this is not the case in the Arabian Gulf.
Voice: thin strident calls recall those of Little closely, but perhaps often lacking chattering quality of that species.

314 **Whiskered Tern** *Chlidonias hybridus* — PLATE 50

L: 25. W: 73. Slightly larger than the other marsh terns with slightly longer, broader wings, slightly more forked tail (adult), longer legs ***and heavier bill***; flight buoyant and restless but more stable and *Sterna*-like than other marsh terns; both snatches and dives for food. Adult in summer has uniform pale grey upper surface (including rump and tail); ***sooty grey underparts contrasting with whitish cheek-stripe and underwings.*** In winter, when underparts white, best told from similar White-winged Black ***by flight, stouter longer bill, narrower, more straight-edged cap.*** Juvenile told from similar White-winged by head pattern (see above), practically no white collar on hindneck (unlike

White-winged, but often hard to see), slightly paler 'saddle' (especially at rear) and pronounced blackish and buffish markings in scapulars (contrasting with 'saddle'; scapulars dark in White-winged); upperwing slightly paler with virtually no dark leading fore-wing-bar; rump and tail usually concolorous pale grey (rump white in White-winged); sometimes has dark smudge at sides of breast like some adults in winter, absent in White-winged. 1st winter recalls adult winter.
Voice: a loud, hoarse short 'kreck'.

315 **Black Tern** *Chlidonias niger* — **Plate 50**

L: 23. W: 66. Size and flight much as White-winged Black which see for separation. Adult has ***pale greyish underwings***, white undertail-coverts, sooty grey body darkening on head, ***mid-grey upperwing and pale grey rump and tail***. In winter, underparts white ***with dark spot on sides of breast***, even more pronounced in juvenile; forehead and hindneck-collar white, upperparts greyish. Juvenile has dark fore-mantle and leading upperwing-coverts; scapulars tipped whitish; sides of rump sometimes whitish, rest of rump and tail pale greyish; has blackish ear-covert patch and white upward extension behind it. 1st winter recalls adult in winter.
Voice: short sharp 'kjeh' and repeated 'kit'.

316 **White-winged Black Tern** *Chlidonias leucopterus* — **Plate 50**
(White-winged Tern)

L: 22. W: 65. Smaller than Whiskered with clearly ***shorter, thinner bill*** and more buoyant flight (see Whiskered for further differences). Adult in summer identified from Black Tern ***by solid black body and underwing-coverts contrasting with greyish white flight-feathers below, white tail and upperwing*** (Black has uniform pale grey underwing, grey tail and upperwing). In winter, underparts white, upperparts pale grey; best separated from similar Black ***by lack of dark patch at sides of breast***; rump and tail whiter (pale greyish in Black); a few show some black marks on underwing- coverts, making identification easier. Juvenile ***has darker 'saddle' and paler upperwing than similar Black***; rump and sides of tail also whiter; ***lacks dark breast side patch of juvenile Black***; at close range bill slightly shorter, dark patch on ear-coverts extends farther downwards but has whiter upward extension behind it; also lacks clear white tips to rear scapulars (distinct in juvenile Black). Moults dark 'saddle' to pale grey in late autumn.
Voice: harsh, dry call 'kesch' or 'kruek'.

317 **Common Noddy** *Anous stolidus* — **Plate 50**
(Brown Noddy)

L: 43. W: 83. Size of Sandwich Tern with ***long wedge-shaped tail (with shallow fork when spread) and long black bill. Dull chocolate-brown except for pale ash-grey crown which grades into white forehead and contrasts sharply with black lores***; may appear white-capped in abraded birds in strong light; ***underwing-coverts contrast clearly with blackish brown flight-feathers*** (smaller Lesser Noddy has all-brown underwing). Juvenile has crown varying from grey-brown to whitish; immatures (up to 3 years old) are extremely abraded, appearing paler than adults. Separated from young Sooty Tern by darker underwing-coverts (can be whitish in Sooty) and shape of tail.

Usually flies low over water with slower wing-beats than Lesser Noddy (which see). Feeds by hovering above waves before swooping low to snatch prey.
Voice: at breeding site deep and guttural, at times *Corvus*-like 'kwok-kwok', 'karruk' or 'krao'.

318 **Lesser Noddy** *Anous tenuirostris* — PLATE 50

(Black Noddy)

L: 32. W: 60. Very similar to Common Noddy ***but smaller*** (size of Common Tern) and shorter tailed, ***with proportionally longer, thinner bill***, no narrow black forehead band over bill, usually pale ash-grey (not white) forehead and crown and ***usually lacking the sharp demarcation with black lores***; ***has dark underwing without contrast between flight-feathers and wing-coverts*** of Common Noddy (but beware of reflecting light which may make underwing-coverts appear paler). Juvenile generally less pale-crowned than similar Common Noddy. In flight, wings narrower, wing-beats faster, the 'jizz' lighter than the bulkier, heavier-bodied Common Noddy, but separation difficult at distance over sea.

319 **African Skimmer** *Rhynchops flavirostris* — PLATE 44

L: 38. W: 106. Resembles large tern, but has different structure; ***very large, slightly decurved, orange-red bill*** with ***shorter upper mandible*** and ***very long, narrow pointed wings*** distinctive. Adult in summer is ***blackish brown above, including hind-neck***, but forehead and trailing edge to arm white; sides of short, slightly forked black tail greyish white; under surface white, primaries darker. In winter, and in juvenile, there is a diffuse whitish collar around hindneck, making separation from Indian Skimmer hard. Young birds have shorter bill, tipped dark. Flight graceful, with ***deliberate, slow accent on upward stroke*** of wings; often in flocks, following 'leader' in line low over water. Skims water surface with long lower mandible breaking surface.

320 **Indian Skimmer** *Rhynchops albicollis* — PLATE 44

L: 42. W: 108. Resembles African closely but larger; in all ages with ***clear-cut white collar*** on hindneck (in African a less clear-cut whitish collar seen only in winter adult and juvenile). Separated with difficulty from African in winter and immature plumages also by ***all-white tail-feathers with narrower dark centre*** (both webs white; in African, same feathers white on outer webs, greyish on inner webs, but latter become progressively paler on outer feathers).

321 **Lichtenstein's Sandgrouse** *Pterocles lichtensteinii* — PLATE 51

L: 25. W: 50. Rather small sandgrouse with ***short square-ended tail***; yellowish buff plumage ***finely vermiculated black and white all over***. Male has ***yellowish breast-patch framed by 2 black bars***; bill orange, and ***white forehead shows 2 distinct black bars crossed by white supercilium***. Female duller, lacking distinct breast- and head marks; lacks yellow throat of similar Crowned. In flight, upperwings show black flight-feathers and pale wing-coverts, underwing pale with slightly darker flight-feathers. Usually singly or in small parties coming to drink at dusk or before dawn; rarely seen in full daylight, unless flushed.
Voice: call, when coming to drink, a repeated disyllabic, melodic '*whee*-ak' with stress on

first syllable; when flushed a harsh, whirring 'arrk'. Call at night, uttered in flight: a clear liquid 'whit, wheet, wheeoo'.

322 **Crowned Sandgrouse** *Pterocles coronatus* **PLATE 51**

(Coronetted Sandgrouse)

L: 28. W: 57. Size between Lichtenstein's and Spotted. Rather pale and featureless with ***short tail.*** Male has ***yellowish buff head and neck*** with russet-brown crown framed grey, ***a black mask on forehead and chin around bill is diagnostic.*** Female sandy buff finely vermiculated, head, neck and throat unmarked yellowish. In flight above ***black flight-feathers contrast with sandy grey upperparts and wing-coverts*** (contrastless in Spotted), below black flight-feathers contrast with white wing-coverts. Distinguished from Spotted (which see) by short tail which is white-tipped (noticeable when tail spread on landing) and distinctive upperwing pattern; plumage of Lichtenstein's is much more barred, and lacks yellowish on head. Seen in small parties coming to water mainly in morning.

Voice: frequenuy calls in flight on its way to or from water: hard, accelerated and nasal 'kaaa-kata-kata-kataah'.

323 **Spotted Sandgrouse** *Pterocles senegallus* **PLATE 51**

L: 32. W: 59. Slim and pale sandgrouse ***with long pointed tail***; in flight, rather ***pale upperwing*** with indistinct dark rear border, underwing pattern pale with contrasting dark flight-feathers (especially secondaries). Mostly recalls Crowned (which see), and best told from it by upperwing pattern; also note bolder spots in upperparts and breast (in female; confined to wing-coverts and shoulder in male), warmer darker orange throat and neck, and ***lack of black face marks***; long tail and narrow black patch from belly to undertail can be difficult to see. Chestnut-bellied easily told by underparts and, in flight, by wing pattern.

Voice: in flight, a disyllabic bubbling whistle, frequently repeated: 'wi-dow', first syllable higher.

324 **Chestnut-bellied Sandgrouse** *Pterocles exustus* **PLATE 51**

L: 32. W: 50. Slightly smaller than Spotted; elegant and dark especially in flight with ***diagnostic all-dark underwing continuous with dark chestnut belly,*** merging into paler breast and buffish yellow head in male. Female shows black-spotted breast above the black bar; male has golden wings with some dark barring, female narrowly vermiculated blackish upperparts, and yellowish buff head; both sexes show ***elongated tail and narrow black bar on lower breast.*** Upperwing pattern largely recalls Crowned and Lichtenstein's (which see); Black-bellied is distinctly larger, has short tail and white underwing-coverts. May come to water in large flocks in early to mid-morning.

Voice: call in flight and on ground: far-carrying, short, rather liquid 'kwit-kwit-kwituroh-kwituroh-kwituroh' rising to a higher pitched twittering; also 'ke-rep, kerep' with stress on last syllable, and a melodious 'gattar-gattar'.

325 **Black-bellied Sandgrouse** *Pterocles orientalis* **PLATE 51**

L: 34. W: 72. Heaviest and most contrasting sandgrouse of the region. Male easily distinguished by ***plump black belly***, which contrasts sharply with sandy buff breast and grey neck; also by distinct, black breast-bar and in flight ***prominent black and white underwing pattern***. Female has heavily spotted upperparts and breast, sharply demarcated from greyish lower breast which contrasts with black belly. ***Call also distinctive***. Chestnut-bellied is smaller, slimmer with blackish underwings.
Voice: flight-call a characteristic rolling or bubbling 'durrrll'.

326 **Pin-tailed Sandgrouse** *Pterocles alchata* **PLATE 51**

L: 35. W: 60. Rather slim sandgrouse ***with long tapering tail*** (best seen when flying overhead). In flight, distinctly ***white underparts contrast sharply with black primaries and chestnut-buff breast band (male) bordered narrowly black above and below***; in female pale golden-yellow breast band is bordered and crossed by three narrow, black bars. Male has orange face with black throat and eye-stripe and grey-green upperparts; female duller with mottled and barred upperparts. The combination of white underparts, prominent breast band and distinct black primaries below are diagnostic. Gregarious, often in very large flocks; when flying rather noisy.
Voice: call in flight a distinctive, repeated, slightly falling 'arrrh—arrrh' or shorter 'arrk-arrk-arrk-'.

327 **Pallas's Sandgrouse** *Syrrhaptes paradoxus* **PLATE 51**

L: 36. W: 71. ***Long, narrow, curved wings, and longest tail of any sandgrouse in the region***. Distinguished in both sexes by combination of ***wholly pale underwing*** narrowly bordered black at rear, ***buffish white underparts with distinct black belly-patch*** and elongated tail. Upperparts sandy buff with mottled back and wings; head and neck bluish grey; in female spotted on hindneck, in male a small pectoral band of fine flecks on lower breast; face orange. In flight, shows rather uniform pale and grey upperwings with narrow black trailing edge to secondaries.
Voice: in flight, a distinctive, low-pitched 'cu-ruu cu-ruu cu-ou-ruu', also a clear, rapidly repeated, trisyllabic 'köckerik-köckerik'. Wings make far-carrying humming or whistling sound.

328 **Rock Dove (including Feral Pigeon)** *Colomba livia* **PLATE 52**
(Rock Pigeon)

L: 33. Medium-sized, compact, pale blue-grey pigeon, superficially recalling Stock Dove, but has ***2 broad black bands across secondaries above and whitish rump*** (though rump greyish in some E Iranian and Arabian populations); underwing white with dark band at rear (underwing less contrasting in Stock, flight-feathers darker); black band on outer tail more pronounced than in Stock. The ancestor of the Feral Pigeon some of which are very similar to Rock, others are blackish, whitish or reddish buff, but black bands on wing, white underwing and contrasting tail pattern usually present. Flight very fast; sometimes acrobatic with circling and gliding on lifted wings. Very gregarious. See Yellow-eyed Dove for separation.
Voice: cooing series: 'kru-oo-u', second syllable stressed and highest in pitch.

329 **Stock Dove** *Columba oenas* **PLATE 52**

L: 33. Medium-sized blue-grey dove, easily told from Woodpigeon by ***absence of white band on upperwing and no white neck-patch*** (unlike adult Woodpigeon); also smaller, more compact and shorter-tailed, with quicker wing-beats (noticeable in mixed flocks). Told from Rock by less white underwing ***with darker flight-feathers, pale ashy grey mid-wing panel above, lack of bold black bars across secondaries and absence of white rump*** (though beware of grey-rumped eastern populations of Rock where Stock does not occur). Bill pale-tipped (dark in Rock), and iris dark. Often in small flocks. Display flight with deep slow wing-beats and long glides on raised wings.
Voice: a monotonous cooing with emphasis on first syllable '*coo*-coo, *coo*-coo'.

330 **Eastern Stock Dove** *Columba eversmanni* **PLATE 52**
(Yellow-eyed Dove; Pale-backed Pigeon)

L: 30. Resembles Stock Dove though slightly smaller and shorter tailed. At close range separated by combination of ***whitish lower back*** (pale grey in Stock), larger pale area on primaries above, almost reducing black to trailing edge of wings (thus less well defined pale mid-wing panel), ***pink top of head*** (grey in Stock), more extensive white underwing darkening at trailing edge, and ***yellow iris and broad eye-ring*** (dark in Stock). Juvenile is duller and fawn on breast. Always told from Rock by ***absence of black bars on secondaries (broken and inconspicuous in Eastern Stock)*** and by pink forehead and yellow eye-ring. Gregarious. Flight fast and direct.
Voice: song typically 3 single notes followed by 3 disyllabic notes 'quooh, quooh, quooh—cuw-gooh—cuw-gooh—cu-gooh'.

331 **Woodpigeon** *Columba palumbus* **PLATE 52**

L: 41. Large stocky pigeon, easily told from other pigeons by large size, ***bold white transverse band on upperwing in flight*** (all ages) and ***white patches on sides of neck*** (absent young birds). Flight slower and heavier than other pigeons, tail proportionally longer. Gregarious outside breeding season, often seen in large flocks. Display flight a short ascent with wing flap at top, followed by short descending glide on half-closed wings.
Voice: hoarse cooing consists of 5 syllables, second note stressed and drawn out, 'cu-*cooh*-cu, coo-coo'.

332 **Olive Pigeon** *Columba arquatrix* **PLATE 52**
(African Olive Pigeon)

L: 40. Slightly smaller than Woodpigeon, with shorter tail and leisurely wing-beats in flight. ***Dark blue-grey plumage*** with maroon tinge on side of neck, ***prominent white spotting on wing-coverts, pale scaling on ash-grey breast and thick yellow bill, eye-ring and legs***. Appears fairly uniform dark slate above in flight. Skulks in trees where it sits in shade, reluctant to fly.
Voice: deep note, followed by brief succession of coos.

333 **African Collared Dove** *Streptopelia roseogrisea* **PLATE 53**

(Pink-headed Turtle Dove)

L: 29. Almost identical with Collared and normally indistinguishable at distance. Best separated by ***white lower belly and undertail-coverts*** (pale grey, like rest of underparts in Collared) and ***by voice***. Shows virtually same colour pattern on upperwing, tail-corners and underwing as Collared. Young birds paler than adults, being almost whitish grey on head and underparts (paler than Collared). See also Red-eyed Dove. Formerly separated geographically from Collared but recent range extension in both gives overlap in a few areas.

Voice: very different from Collared's; starts with high-pitched, drawn-out note, then short pause, then series of broken, descending rolling notes, 'crooo, cro-cro-crococo' or 'cruu,—currruuu'; at distance sounds disyllabic 'croo—cooorrr'.

334 **Collared Dove** *Streptopelia decaocto* **PLATE 53**

(Eurasian Collared Dove)

L: 29. Medium-small, sandy brown dove with narrow black half-collar on hindneck (absent juvenile). Can be confused with African Collared, which see for separation. ***Told from Turtle by unspotted sandy brown forewing*** (rufous, spotted blackish in Turtle), ***black neck-collar, whitish underwing*** (dark in Turtle), plainer uppertail (bolder patterned in Turtle), ***absence of well defined whitish belly-patch*** (present Turtle) and by less rapid flight with ***less jerky wing-beats of shorter, more rounded wings, and longer tail***.The often domesticated Barbary Dove (*S. nisoria*) is very like Collared but paler creamy, without contrasting dark primaries above; voice also differs.

Voice: a loud deep trisyllabic 'coo-*cooh*-co' with stress on middle note which is also drawn-out and highest in pitch (Barbary stresses first note).

335 **Red-eyed Dove** *Streptopelia semitorquata* **PLATE 53**

L: 33. Generally recalls African Collared, but ***larger and darker***, with distinctive ***clear-cut black band at base of buffish grey uppertail*** which separates it from other doves in the region. Upperparts slate-greyish, primaries slightly darker (much greater contrast in African Collared); underwing dull greyish (whitish in African Collared) and ***forehead and fore-crown whitish grey*** merging into vinous-grey rest of head. May recall Dusky Turtle in flight from above but note tail pattern and slower, softer wing-beats. Juvenile lacks black collar on hindneck.

Voice: cooing song variable, slow, deep and hoarse, 'co-coo, co-co, co-co', second note (and sometimes fourth) often stressed and highest in pitch; also 'croooo-cro-cro' with first note stressed; at times a far-carrying 'ho-ho, ho-ho, ho-ho'.

336 **Red Turtle Dove** *Streptopelia transquebarica* **PLATE 53**

L: 23. Small ***short-tailed dove*** with black collar on hindneck. Separated from Collared by combination of ***red-brown*** (male) ***or warm brown*** (female) ***mantle and wing-coverts*** (pale sandy brown in Collared), ***darker ash-grey lower back and uppertail with more defined white corners***, browner breast (vinous-brown in male) and ***whitish vent and undertail-coverts*** (pale grey in Collared), ***blackish, not red, legs*** and ***darker underwing*** (pale in Collared). Laughing, and particularly Turtle Dove also have dark on underwings but both have blue-grey panel in upperwing; Laughing lacks grey head of male Red Turtle; Laughing is spotted black above and has well defined whitish belly-patch; both lack hindneck collar of Red Turtle. Fast flight more like Turtle Dove than Collared.

Voice: dry rattling rhythmic 'ruk-a-duc-doo', quickly repeated.

337 **Turtle Dove** *Streptopelia turtur* **PLATE 53**

(European Turtle Dove)

L: 27. Small, fast-flying dove with ***shorter tail and more pointed wings than Collared***; easily told from it also by ***dark underwings, well defined whitish belly patch, rufous, black-spotted upperparts*** with blue-grey, well defined outerwing-panel and ***contrasting pattern of uppertail being darker with more clear-cut white corners, particularly when tail spread upon landing; black and white-streaked patch on side of neck*** in adult (absent juvenile). Told from Laughing Dove by proportions (Laughing recalls Collared), very fast flight with jerky wing-beats, belly-patch, spotted upperparts, neck-patch and ashy-grey head. Turtle Doves from Near East to Iran are paler, more washed out grey-brown on mantle and breast. See Rufous Turtle for separation.

Voice: a soft, deep purring 'roorrrr, roorrrr, roorrrr'.

338 **Dusky Turtle Dove** *Streptopelia lugens* **PLATE 53**

L: 31. Larger and ***darker than Turtle with hardly any rufous above***; ***only inner secondaries and inner greater coverts edged rufous-chestnut***; rest of wing and shoulder spotted blackish ***with thin greyish white feather edges***; ***solid black patch on sides of neck*** (no white streaks as in Turtle) and ***much less contrasting uppertail pattern***; also lacks Turtle's well defined whitish belly-patch. Browner and paler juvenile lacks black neck-patches but has tawny edges where adult rufous-chestnut. Appears rather dark in flight and may be confused with Red-eyed (overlap in habitat) but latter's distinctive uppertail pattern separates. Flight and wing-beats slower than Turtle's and wings more rounded.

Voice: deep, gruff, purring 'coorrr-coorrr-coorrr-coorrr' first note most gruff, last two slightly higher but sometimes omitted.

339 **Rufous Turtle Dove** *Streptopelia orientalis* **PLATE 53**

(Oriental Turtle Dove)

L: 33. Resembles Turtle Dove but larger and heavier; separation normally requires close range. Compared with Turtle it has ***larger, more ill-defined dark centres to wing-coverts*** (Turtle's black spots smaller, more defined); ***forehead and crown pale grey, contrasting with browner rest of head*** (in Turtle whole head and neck greyish); neck-patch larger, ***streaked blue and black*** (usually white and black in Turtle); ***bare rosy skin round eye small and circular*** (in Turtle, larger area of bare skin elongated horizontally). Race occurring in the region, *meena*, has whitish belly, undertail-coverts and distal part of tail (like Turtle), but Siberian *orientalis* (could occur) is pale greyish on these parts (unlike Turtle). Juvenile lacks neck-patches, is paler brownish on body and wings, otherwise like adult. Flight heavier, straighter, with less jerky wing-beats than Turtle, but useful only with experience.

Voice: phrased like Woodpigeon, with alternating grating and clearer cooing notes: 'gru-gror, co-co, gru-gror, co-co'.

340 **Laughing Dove** *Streptopelia senegalensis* PLATE 54

(Palm Dove)

L: 26. Small, dark red-brown dove, recalling Collared in proportions but plumage closer to Turtle; smaller than both. ***Black-spotted patch on fore-neck and upper breast, unspotted red-brown upperparts with large, dull blue-grey area in outer-coverts*** distinguish it. Uppertail less contrasting than in Turtle. Juvenile lacks patch on fore-neck and is less richly coloured. Populations in NW Turkey rather dark in plumage. Flight closer to Collared's than to Turtle's but even slower and more laboured; flight not so twisting and without jerky wing-beats; the short rounded wings and long tail always helps identification. Gregarious.
Voice: usually 5 syllables, subdued cooling with third and fourth notes slightly longer and higher in pitch, 'do, do, dooh, dooh, do'.

341 **Namaqua Dove** *Oena capensis* PLATE 54

L: 29. (including 9 cm tail). Unmistakable at all ages. ***Very small*** blue-grey or grey-brown dove ***with long black central tail-feathers***, recalling large Budgerigar in shape. In flight, ***black primaries show large red-brown patch***. Male has black face and upper breast, which are brownish grey in female. Juvenile barred black and buff on crown, throat, wing-coverts and back. Very rapid flight is fast and direct. Spends much time on the ground.
Voice: a mournful 'hu-hu, hu-hu'; also a deep 'coo'.

342 **Bruce's Green Pigeon** *Treron waalia* PLATE 54

(Yellow-bellied Pigeon)

L: 31. Unmistakable, brightly coloured dove with yellowish olive-green upperparts, ***greyish green head, neck and upper breast, sharply defined from bright yellow below***; undertail-coverts chestnut. A purple patch on shoulder is absent in juvenile. Uppertail uniform, undertail white with black base. ***Flight very rapid***, wings making rattling sound. Hard to see when perched in tree and often shy.
Voice: rippling, crooning whistle and quarrelsome chatter.

343 **Ring-necked parakeet** *Psittacula krameri* PLATE 54

(Rose-ringed Parakeet)

L: 42. Large green parakeet with ***long, graduated, pointed tail and short, deeply hooked red bill***. In male, ***black throat continues round neck as a narrow rosy ring***, absent in female and juvenile where throat green and bill horn-coloured. The larger Alexandrine ***has red shoulder-patch, absent in Ring-necked***. Flight swift, fast and direct but flocks may change direction rather suddenly (like waders).
Voice: very noisy; loud screaming 'kee-ak', rather falcon-like.

344 **Alexandrine Parakeet** *Psittacula eupatria* PLATE 54

L: 53–58. Resembles ***distinctly larger version*** of Ring-necked, but with heavier bill and ***large red patch on shoulders at all ages*** (sometimes partly hidden when perched). Sexual differences as for Ring-necked.
Voice: loud, hoarse, screaming note '*kii*-e-rick' with stress on first syllable, repeated 2–3 times.

345 **Jacobin Cuckoo** *Clamator jacobinus* **PLATE 55**

(Pied Cuckoo)

L: 33. Size of Common Cuckoo and easily told by distinctive ***black and white plumage, long graduated tail and crest; conspicuous white wing-patch in flight*** (unlike any other cuckoo). A rarer black form occurs (which also shows white in wing). Juvenile is sooty above with dirty white underparts and small crest.
Voice: loud, metallic 'piu-piu-pee-pee-piu, pee-pee-piu'.

346 **Great Spotted Cuckoo** *Clamator glandarius* **PLATE 55**

L: 40. A large cuckoo with ***slight crest, very long tail*** and rather fast, shallow wing-beats. Unmistakable with ***grey upperparts, white-spotted coverts and white underparts.*** Juvenile is dark brown above with blackish head, white-spotted coverts and showing ***distinctive chestnut primaries (above and below) in flight.***
Voice: harsh, chattering 'chil, chil, chil, chil' like cross between Magpie and Green Woodpecker.

347 **Indian Hawk Cuckoo** *Cuculus varius* **PLATE 55**

(Common Hawk Cuckoo)

L: 33. Similar in size to Common Cuckoo, but in flight shows slightly ***broader and shorter wings and slightly shorter tail.*** In full adult plumage grey above with ***vinous wash on breast, faint barring on belly and banded tail (above and below).*** In juvenile plumage (most likely to be seen in region), upperparts are dull rufous, barred darker with ill-defined white collar on hindneck; head greyish and underparts white with ***dark streaks on throat and breast with spotting on rest of underparts***; bill usually dark in immature birds. Rather secretive. If seen in flight, combination of plumage and structure creates resemblance to a hawk, particularly Shikra.

348 **Didric Cuckoo** *Chrysococcyx caprius* **PLATE 55**

L: 18. Slightly larger than Wryneck and of similar shape, but tail often held slightly raised. Most closely resembles Klaas's Cuckoo (which see). Metallic green male and bronzey female told from Klaas's by ***white spotting on wings*** (unspotted in Klaas's), ***white supercilium*** (small white streak behind eye in Klaas's), ***white crown-streak*** (absent in Klaas's), ***whitish barring on outer tail-feathers*** (outers white in Klaas's) and ***bold barring on flanks, belly and undertail-coverts*** (only faintly barred on flanks in Klaas's). Juvenile told from juvenile Klaas's by plain rufous-buff upperparts (densely barred in Klaas's) and barred outer tail-feathers. Appears quite long-tailed in flight. Rather secretive.
Voice: plaintive, loud 'pee-pui-pui' (last note often double), 'psee-psee-psu' (with last note lower) or 'dee-dee-dee-*dee*deric' (emphasis on last note). Often calls from high in tree where difficult to see.

349 **Klaas's Cuckoo** *Chrysococcyx klaas* **PLATE 55**

L: 16. Slightly smaller than Didric Cuckoo under which differences between the two are given. Male is an unmarked bright emerald-green above, washed bronze on back with ***small white streak behind eye***; white underparts are ***faintly barred on flanks*** and (unlike Didric) has a ***green patch at sides of breast***. Female dull brown above with bronzey wash grading to green tail; narrow bronze and green barring on back and wings with ***no white spotting*** (which Didric shows); underparts narrowly and faintly barred from chin to undertail-coverts. Juvenile resembles female but upperparts entirely barred buff and green and underparts more densely barred.
Voice: mournful 6 (sometimes 4) note song 'whit-jeh, whit-jeh, whit-jeh' with accent on 2nd, 4th, and 6th notes which are lower pitched; also 'jeh, whit-jeh, whit-jeh, whit'.

350 **Plaintive Cuckoo** *Cacomantis (merulinus) passerinus* **PLATE 55**

L: 22. Rather small cuckoo. Male and most females easily told by ***smooth dark grey plumage fading into whitish belly and undertail-coverts***; wings have olive-brown wash, and blackish graduated tail has ***white barring below***. Shows ***conspicuous white patch at base of underwing primaries in flight***. Juvenile rufous-brown above, barred darker; chin and throat warm buff with fine black bars, rest of underparts white with narrow dark bars; barred brown undertail-feathers have white tips as in male. Rather shy but when flushed often flies some distance.
Voice: vagrants to region unlikely to be heard singing.

351 **Common Cuckoo** *Cuculus canorus* **PLATE 55**
(Cuckoo)

L: 33. Usually seen in flight when long tail, pointed wings and shallow wing-beats can cause resemblance to a Kestrel, but head held slightly raised and ***wings not lifted above level of body***. Grey above with barring on underparts; grey underwing has pale line through centre. Female similar but with brown wash on band across breast. Also occurs in a less common rufous phase in which female and juvenile have brownish red plumage with noticeable barring on upperparts and breast. Grey phase juvenile is dark brownish grey, with pale fringes to feathers above and whitish spot on nape. ***In spring, song is characteristic.***
Voice: unmistakable song: 'kwuk-koo' or 'kwuk, kwuk-koo', first note highest in pitch, far-carrying. Female also has a descending, long bubbling call.

352 **White-browed Coucal** *Centropus superciliosus* **PLATE 55**

L: 42. A large cuckoo-type which is often very skulking and more often heard than seen; most active at dawn and dusk. Readily told by ***rufous back and wings***, long, broad, graduated black tail and ***narrow white supercilium***. Juvenile has pale bars on tail-feathers, barred back and buffish supercilium. Clumsy; sits on bush top to sing or sun itself.
Voice: 10–15 rapidly repeated deep, hollow 'hoo' notes, slowing and becoming low in pitch at end. Also a loud cackle of about 25 'kak' and 'hok' notes, increasing in pitch and speed to a crescendo then trailing off.

353 **Koel** *Eudynamys scolopacea* — **Plate 54**

(Common Koel)

L: 43. Unmistakable cuckoo-like bird with relatively thick bill, short wings and rather long tail. ***Male black with bluish gloss and bright yellow bill.*** Female has ***drab brown upperparts, thinly streaked and spotted white***; whitish throat heavily streaked dark brown, rest of underparts densely barred with dark brown and buff; dark tail with many thin whitish bars. Immature resembles female. When not alarmed has characteristic stance with long tail held downwards, hunched lower-back, bulky breast and (in calling male) uptilted head and bill. Cuckoo-like direct flight.

Voice: male calls continually, a loud hollow 'kooyl' with ascending curl on y-note; when excited a long, rapid ascending series of bubbling notes 'kwow-kwow-kwow- kwow'.

354 **Barn Owl** *Tyto alba* — **Plate 57**

L: 35. W: 89. Medium-sized, strikingly pale owl, both when perched and in flight with ***white underparts*** (suffused ochreous on breast), ***heart-shaped face*** with black eyes and ochreous-buff upperparts. No other owl looks similar but beware—in car head-lights at night most owls can look pale.

Voice: varied. Male's territorial call a clear vibrant shriek of about 2 s; alarm call in flight a shrill shriek; when disturbed will hiss, and young beg with a drawn-out hissing.

355 **Indian Scops Owl** *Otus bakkamoena* — **Plate 56**

(Collared Scops Owl)

L: 24. Slightly larger but very similar to other scops owls of the region; ***best separated by voice. A broad, fairly uniform, pale, yellow-buff collar on hindneck***, is sometimes difficult to see when head is sunk in shoulders; underparts whitish without rufescent tinge, slightly streaked (generally bolder, more distinct pattern in the other small 'eared' owls); upperparts grey-brown with dark streaks and bars (finer, sharper streaks in Striated Scops); eyes yellow to brown (clear yellow in the other three scops owls). Nocturnal, often roosts in daytime in large tree close to the trunk.

Voice: soft, subdued and rather frog-like and difficult to hear at more than 150 m: a rising, single 'wuk' or 'whut' repeated at 4–6 s intervals, in breeding season lower pitched female often immediately answers, making a two-note effect.

356 **Striated Scops Owl** *Otus brucei* — **Plate 56**

(Bruce's Scops Owl)

L: 21. Appearance very close to European Scops, but ***paler sandy grey above*** without white spots on crown or hindneck, ***greyer below with black streaks more sharply demarcated***, being the only distinctive plumage feature (European Scops has black streaks more broken by vermiculations and paler blotches below, submerged by darker ground-colour); iris yellow. ***Juvenile has underparts completely barred*** (unlike adult and juvenile European Scops). Nocturnal. ***Voice the most important character***.

Voice: calls soft and dove-like, only carrying a short distance (unlike European Scops). Territorial call of male a hollow, resonant, low-pitched 'whoop' or 'whoo' repeated regularly about eight times in 5 s; also a hollow, longer 'whooo' repeated irregularly (3–5 s intervals) or 'ooo-ooo—ooo-ooo-'.

357 **European Scops Owl** *Otus scops* **PLATE 56**

(Scops Owl)

L: 19. W: 52. Small 'eared' owl; very similar to and difficult to distinguish from the other three scops owls. Rather large-headed with small ear-tufts (head often just looks angular); brown-grey or rufous-brown with paler face and dark surround to yellow eyes. Plumage above streaked, barred and vermiculated black-brown; sides of crown, ears and shoulder distinctly white- or buff-spotted; underparts streaked, barred and vermiculated, interspersed with white blotches (lacking in Striated Scops); primaries pale-barred. ***From other scops owls (which see) best distinguished by voice.*** Markedly nocturnal.
Voice: territorial call a clear, soft whistle 'pwoo' rhythmically repeated at intervals of 2–3 s.

358 **African Scops Owl** *Otus senegalensis* **PLATE 56**

(Senegal Scops Owl)

L: 21. Size between European Scops and Striated Scops. ***Hardly distinguishable in field from European Scops except by voice***, but slightly greyer, more uniform, and lacks bright white scapular streaks. Also similar to Striated Scops (which see), but feathering does not reach base of toes. Like all scops owls, can adopt two postures: upright stance with ears erect and slim body, also with ears down so that head becomes rounded and eyes show prominent blackish surrounds.
Voice: a single 'da-pwoorp' repeated at intervals of 12–20 s; Socotran birds are slightly different in having a 'woup–woup' introductory note (only audible at very close range), and calls are repeated at intervals of 3–4 s. The notes are lower pitched than the pure whistle of European Scops.

359 **Eagle Owl** *Bubo bubo* **PLATE 57**

(Eurasian Eagle Owl)

L: 68. W: 174 (race *ascalaphus* 20% smaller). ***Largest owl in the region, usually showing long ear-tufts***, which, when relaxed, may droop, recalling those of Brown Fish Owl (which see). Northern birds (Turkey/Iran, Near East) are darkest brown, with ***heavily spotted/blotched upperparts and breast***, more lightly streaked on lower breast and belly. Smaller southern race, *ascalaphus*, (occurring south of Turkey and Iran) is paler and rustier with most of lower breast and belly distinctly barred, recalling Spotted Eagle Owl, (which see for separation); a further race, *desertorum*, of the more arid regions of Arabia is paler still. Distinguished from Brown Fish Owl by head-shape, underparts and facial pattern. In flight, looks huge, but typical owl-shape with broad, rounded wings, short tail and large head; wing-beats shallow, stiff and surprisingly fast, resembling Common Buzzard.
Voice: male's territorial song distinctive: a deep hollow far-carrying 'hoo' or '*hoo*-o' (stress on first note), repeated at intervals of 10–15 s, often in duet with female, whose call is higher pitched and slightly hoarse.

360 **Spotted Eagle Owl** *Bubo africanus* **Plate 57**

L: 45. Resembles *ascalaphus* race of Eagle Owl (which see), but ***upperparts finely barred dark brown*** (spotted/blotched in Eagle Owl with white spots on edge of scapulars), ***entire underparts including underwing-coverts densely barred dark brown*** with only irregularly dark-blotched upper breast; paler Eagle Owl often has centre of belly unbarred, others have dark shaft-streaks in the barring (absent in Spotted); ***nape and hindneck finely white-spotted*** (streaked in Eagle Owl). In flight, shows bolder pale and dark tail bands than in Eagle Owl.
Voice: male has a double-note song 'hu-hoo', often in duet with lower pitched female's rather dove-like 'hoo doo doh dooh' (2nd note highest, last 2 lowest).

361 **Brown Fish Owl** *Ketupa zeylonensis* **Plate 57**

L: 55. W: 147. Large 'eared' owl, resembling Eagle Owl, but ***head is flatter, with shaggier, often more drooping, ear-tufts, and ill-defined brownish facial disc*** (black and pale markings in Eagle); upperparts brownish mottled blackish with buff and brown patches on scapulars; ***underparts rufous with uniform dark streaks*** (Eagle Owl shows broader streaks on breast); legs unfeathered yellow-grey, can look rather long. Iris yellow, less orange than Eagle Owl's.
Voice: deep hoots, rising at end in laugh and with distinct pause after first note, 'ku, ku-ku ku-ku ku-ku' audible only when close; also harsh scream 'we-aaah' rising and falling in pitch.

362 **Snowy Owl** *Nyctea scandiaca* **Plate 57**

L: 53–66. W: 142–166. (Female 20% larger than male.) ***Unmistakable large, white owl.*** Male almost pure white; female and juvenile have black-spotted upperparts and barred underparts; facial disk white; yellow eyes. Flight powerful and elegant on long, rather slender wings. Often active in daytime.

363 **Little Owl** *Athene noctua* **Plate 56**

L: 22. Small owl with ***round, flat-crowned head and long legs***; often sits in the open in daytime on buildings or telegraph-poles. ***Distinctly white-spotted crown and nape*** and white-blotched upperparts and wings; ***boldly streaked underparts***; eyes yellow, framed white. Flight deeply undulating (alternating rapid flapping with closing of wings) unlike straighter flight of the scops owls. If agitated, often bobs in an upright posture. The pale form *lilith* (part of Israel, Syria, Jordan and part of Arabia) is soft sandy to warm brown with more distinctive white spots on head and upperparts and less streaked underparts than darker forms in the region.
Voice: territorial call drawn-out and wailing 'koooah', at end faintly falling or rising in pitch, also a short 'kiu'; alarm call a sharp series of notes 'kip-kip-kip . . .'.

364 **Spotted Little Owl** *Athene brama* **Plate 56**
(Spotted Owlet)

L: 21. Similar to Little in shape and general coloration. Upperparts grey-brown, colder, greyer and darker than Little, with smaller white spots; dusky facial disc with ***broad white patch over and below the eyes, almost forming broad white 'spectacles'; crown and***

nape fine-spotted white (larger markings in Little); a broad, white collar around the front of the neck is distinctive when perched; ***underparts more barred than streaked*** (Little is more streaked). Juvenile more ginger than adult with diffused barring below. Flight undulating like Little. Crepuscular but frequently seen in daytime.
Voice: very noisy in breeding season; often short, loud series of chattering, harsh shrieks including very rapid and sudden outbursts, rising and falling in tempo and pitch: 'kuerk-kuerk-kuerk-kuerk …', reminiscent of domestic cats fighting.

365 **Tawny Owl** *Strix aluco* — **Plate 58**

L: 38. W: 99. Medium-sized, ***large-headed, rather plump and 'ear-less' owl with black eyes.*** Two colour-phases occur: greyish and red-brown. Upperparts mottled and streaked brown, flight-feathers and tail softly barred (in darkest birds may be almost uniform); white scapular- and wing covert-bars conspicuous when perched and in flight; underparts heavily streaked; ***fairly uniform facial disc with distinctive dark-centred pale divide,*** shared with Hume's (which see). In flight, large-headed, often with rather pale face, upperwing without conspicuous dark carpal-patch or pale base of primaries as in Long- and Short-eared, and looks proportionally shorter-winged than those species.
Voice: male's territorial call a variable, two-note whistled hoot: 'hoo-ooh' (stress on second note) followed by 'hoo, hoo-ho-ho, hooo-o-o-o-o-o' (latter drawn-out, vibrating and falling in pitch); female replies with sharp, characteristic 'kee-wik'; young beg with hissing sound.

366 **Hume's Tawny Owl** *Strix butleri* — **Plate 58**

(Hume's Owl)

L: 37. W: 96. ***Like a small, pale Tawny with yellowish orange eyes*** (dark in Tawny) and occurring in desert regions. Proportions (large round head) and flight similar to Tawny. ***Facial disc dirty-white with dark band in centre of forehead*** (like Tawny); ***underparts whitish buff with indistinct ochre vermiculations*** (heavily dark-streaked in Tawny); upperparts and wing pattern rather similar to Tawny, though shows bolder barred uppertail and pale golden-buff collar. Strictly nocturnal.
Voice: territorial call a five-syllabic 'whoooo, hoo-hoo, who-who' with stress on first drawn-out note; lacks vibrating character of Tawny, but has soft quality of Collared Dove.

367 **Long-eared Owl** *Asio otus* — **Plate 58**

L: 36. W: 95. Unmistakable when perched and seen well, with ***long ear-tufts*** (though sometimes invisible when relaxed); slightly smaller than Tawny with proportionally smaller head; ***facial disc noticeably warm buff with striking white divide and orange eyes.*** Flight has jerky, fairly stiff wing-beats with glides on level (sometimes slightly raised) wings. Distinguished from Tawny in flight by proportionally longer and less curved wings, dark carpal-patch above and ***orange base of primaries.*** Separated from Short-eared (which see) mainly by pattern of head, ***entirely streaked underparts*** and flight action.
Voice: male's territorial call a regularly repeated (*c.*3 s intervals), hollow 'hooo', like blowing into an empty bottle; other calls include a sharp disyllabic alarm 'kwik-kwik'.

368 **Short-eared Owl** *Asio flammeus* **PLATE 58**

L: 38. W: 102. Resembles Long-eared, but ***occurs in open ground*** (Long-eared mostly seen among trees). Flies with slow, ***elegant and high wing-beats*** on long, slender wings ***raised during glides in shallow 'V'*** (shorter wings with shallower, faster wing-beats in Long-eared and Tawny); often sits on ground and ***very short ear-tufts not always visible.*** Further differs from Long-eared in paler greyish facial disc (not warm buff and white) with striking black surround to ***glaring yellow (not orange) eyes***; greyish brown (not rufous-buff) plumage mottled blackish. In flight, separated from Long-eared by yellow-buff (not orange) base of primaries, more contrasting bars on flight-feathers and tail, white trailing edge to upperwing, black tips to underwing (not diffusely barred) and ***dark streaking on underparts mostly confined to breast and contrasting with paler belly.*** Often hunts in daylight.
Voice: male's territorial call, uttered in flight, a deep, hollow 'boo boo boo boo boo—', answered by female with crying 'geh-ack' or 'gweek'.

369 **Tengmalm's Owl** *Aegolius funereus* **PLATE 58**
(Boreal Owl)

L: 25. W: 58. Small, 'ear-less' owl, slightly bigger than Little; separated by ***larger, rectangular (not flat) head, conspicuous white facial disc, sharply framed black with noticeably black upper corners and raised white eyebrows***; rest of plumage similar to Little, but darker umber-brown; upperparts heavily spotted and blotched white, wings with fewer spots than Little; underparts white, blotched and spotted warm brown. Juvenile largely chocolate-brown, marked white on wings and face. Flight less undulating than Little, reminiscent of Woodcock; also glides for short periods. Decidedly nocturnal and hard to find in daytime.
Voice: territorial call a fast, hollow series 'po po po po po …' repeated for long periods at intervals of a few seconds.

370 **Plain Nightjar** *Caprimulgus inornatus* **PLATE 59**

L: 22. A rather small, ***very uniform nightjar,*** though ground colour can vary from warm brown to greyish. On the ground shows ***fine black flecks*** on ***rear-crown, black spots on scapulars, faint whitish moustaches*** and often whitish or buffish streaks around hindneck. In flight, shows no contrast between flight-feathers and coverts. Male has white spots in wing and tail (about half length of outer 2 primaries). European, Nubian and Mountain Nightjar all show much more patterned plumage with dark-centred greyish crowns contrasting with darker ear-coverts and noticeable buffish bands on scapulars and coverts; told from Egyptian Nightjar by smaller size, much darker colour and lack of contrast between flight-feathers and coverts.
Voice: a prolonged churr, similar to European Nightjar but much less ventriloquial.

371 **Nubian Nightjar** *Caprimulgus nubicus* **PLATE 59**

L: 21. Slightly smaller than European Nightjar with shorter tail. Told from other nightjars in the region by more rounded wings, ***chestnut base to upper primaries, pale rufous underwing contrasting with black primaries*** and conspicuous white patches on outer-wing and outer tail-feathers. When perched, helpful features are ***buff half-collar, white***

bar on edge of cheek and another on sides of throat, greyish upperparts with ***broad buff tips to wing-coverts***.
Voice: song a fairly liquid 'quil-quil' (recalling distant barking dog) repeated non-stop for up to 30 s; occasionally preceded by very quiet 'poo-poo poo-poo'.

372 **Sykes's Nightjar** *Caprimulgus mahrattensis* **PLATE 59**

L: 22. This nightjar is rather plain and greyish, features that help to distinguish it, when perched, from the other nightjars likely to be encountered in the same area: Indian, European, and, on migration, Egyptian. Told from Egyptian by white primary spots and, in male, large white tips to undertail (Egyptian lacks white spots on primaries and has only diffuse pale tips to undertail); from European by plain plumage (lacking European's contrasting greyish or buffish crown, dark cheeks and two broad pale bands on scapulars and coverts); from Indian (the most similar) by even plainer plumage (see illustration) ***but best told by voice***.
Voice: a long frog-like purr, after sunset and pre-dawn. Soft 'cluck-cluck' when flushed by day.

373 **Indian Nightjar** *Caprimulgus asiaticus* **PLATE 59**

L: 24. Difficult to identify from Sykes's and European Nightjar—the only other nightjars with white spots in wings and tail that occur in the same area (as vagrants or on passage). ***Best told by distinctive song***, otherwise separated from Sykes's by ***rufous-buff band on hindneck and more contrasting plumage with, especially, buff bands on coverts and white flecks encircling ear-coverts***; from European by lack of two broad pale bands on edge of scapulars and coverts; in the hand the white primary-patch includes the fourth outermost primary (not in European Nightjar); in flight also seems smaller than European.
Voice: song a far-carrying 'chak-chak-chak-charr-r-r-r', at first slow, then becoming faster; reminiscent of a stone bouncing across ice on a pond.

374 **European Nightjar** *Caprimulgus europaeus* **PLATE 59**
(Nightjar, Eurasian Nightjar)

L: 26. The most patterned nightjar in the region and also the most widely encountered. The distinguishing features are the ***dark-streaked grey crown contrasting with rich brown cheeks and throat***, and ***grey upperparts with broad bands of dark brown and buff on the scapulars*** and rows of ***buff bands on the coverts***. Male has white spots in wing and tail. The race *unwini* (breeding in Iran and Iraq) is distinctly paler with more sandy-grey upperparts and longer, whiter lower throat patches; male shows larger white spots in primaries. Like all nightjars most frequently seen in flight at dawn and dusk; if flushed during the day usually flies only a short distance; note rather slow, soft wing beats with long glides on stiffly held wings—like all nightjars. See Egyptian, Nubian and Mountain Nightjar for separation.
Voice: song a long churr, which can go on for many minutes alternating on two pitches and highly ventriloquial (see Plain Nightjar which has similar song).

375 **Egyptian Nightjar** *Caprimulgus aegyptius* **Plate 59**

L: 25. The palest of the nightjars occurring in the region. Very like Plain Nightjar, (with similar plain crown with fine dark flecks on rear) from which told by larger size, ***pale sandy grey base-colour, broad rows of inconspicuous buff tips to wing-coverts and white patch on side of neck (often very hard to see); in flight, very pale underwing, and above, dark flight-feathers which contrast with pale upperwing-coverts. Both sexes lack white patch in wing***, but male shows pale creamy spots on outer tail-feathers below. Voice an important character. Told from European Nightjar by pale sandy plumage, contrast between dark flight-feathers and pale coverts and lack of white spots in less pointed wings.
Voice: a regular, rapidly repeated 'kowrr-kowrr-kowrr' slowing towards end.

376 **Mountain Nightjar** *Caprimulgus poliocephalus* **Plate 59**

L: 21. A rather small nightjar with ***distinctive voice. Most closely resembles Nubian, but darker with slightly longer wings and tail, and when perched, longer primary projection to two-thirds length of tail*** (to about half length of tail in Nubian). The ***golden-buff collar is more obvious*** than in Nubian and it ***lacks Nubian's broad white malar-stripe***. Told from European Nightjar (which see) by smaller size, more rounded wings, golden-buff collar, absence of white throat-patch and crown less contrasting with ear-coverts. Unlikely to be confused with Plain or Egyptian Nightjar because of strong plumage pattern. Prominent white patches in wings and tail, but less distinct in female.
Voice: a distinctive double note 'wee-oo-wee weerrr', the first phrase whistled; regularly or occasionally repeated.

377 **Common Swift** *Apus apus* **Plate 60**

(Swift)

L: 16. W: 45. ***Almost uniform sooty in fresh plumage***, but brownish and more contrasting in worn plumage and in eastern birds, though much depending on light condition. ***Round whitish throat-patch variable, sometimes almost invisible***; ***lacks contrast above*** but ***underparts show clearly paler inner flight-feathers than body and wing-coverts***; deep tail-fork. Easily confused with Pallid and Forbes-Watson's (which see), but note especially overall colour-tone, throat-patch, head pattern, upperparts contrast, and underparts markings. From other swifts by size, structure and lack of white rump-patch.
Voice: high-pitched 'shreee', variable in duration and pitch; hard to tell from Pallid (which see).

378 **Pallid Swift** *Apus pallidus* **Plate 60**

L: 16. W: 44. Slightly bulkier than Common Swift with broader, less tapering outerwing and sometimes blunter wing-tips; tail-fork variable but slightly shallower with less pointed tips; ***head looks broad and flat with larger, more triangular, whitish throat-patch; forehead and lores paler, contrasting with dark eye-patch***. Overall colour brownish; ***with good views rather distinct scaling is visible in underparts*** (uniform in Common Swift); ***upperparts show darker 'saddle'*** slightly contrasting with paler inner flight-feathers, head and rump. Underwing shows darker outer primaries than innerwing, ***the latter not contrasting markedly with wing-coverts or body*** as in Common Swift.

Much of colour tone variable, depending on light conditions and wear and can be difficult to separate from Common Swift unless seen well.
Voice: variable and similar to Common Swift's, but deeper, coarser and slightly disyllabic 'sree-er'.

379 **Forbes-Watson's Swift** *Apus berliozi* — PLATE 60

(Berlioz's Swift)

L: 17. W: 46. ***Very similar to both Common and Pallid Swift and separation extremely difficult and often impossible.*** The overall plumage is blackish brown with ***distinct white chin and throat*** and slight white forehead; at very close range ***narrow pale feather-fringes just detectable on upperparts, belly and undertail-coverts***; upperwing can occasionally show an oily, ***greenish sheen on outer primaries, secondaries and median coverts***; the underwing has paler, almost translucent secondaries and inner primaries. Flight similar to Common Swift; wing-beats perhaps slower, and with longer, sweeping glides. Differs from Common Swift in very slightly paler overall coloration, pale feather-fringes to upper and especially underparts, large white throat-patch, oily sheen on median coverts, secondaries and outer primaries, and ***slightly darker saddle on mantle.*** In the hand, outermost primary longest (9th primary longest in Common Swift). From Pallid by less noticeable saddle, slightly darker plumage, oily sheen to feathers on wing, and voice.
Voice: a screeching 'schweee' or 'schweee-weee-eee', less rippling and not so high-pitched as Common Swift.

380 **Alpine Swift** *Apus melba* — PLATE 60

L: 21. W: 57. Shape and behaviour much like Common and Pallid Swift, but ***much larger*** with remarkably slower and deeper wing-beats; brown upperparts and ***white underparts broken by brown breast band, vent and undertail-coverts.***
Voice: chattering trill unlike that of Common Swift: 'trit-it-it-it-itititit-it-it-it' accelerating, then decelerating, rising and falling in tone.

381 **White-rumped Swift** *Apus caffer* — PLATE 60

L: 14. W: 35. Mostly resembles Common Swift in shape and behaviour, but distinctly ***smaller with more elegant flight and considerably longer, more deeply-forked tail, often held tightly closed*** ('spike-like') for long periods when gliding. Deep blackish plumage contrasting more with white throat-patch and flight-feathers than Common Swift, otherwise coloration similar except for ***narrow, white rump-patch*** (not always visible when flying overhead); with good views narrow white trailing edge to inner secondaries is visible. Distinguished from Little by long and deeply forked tail, narrower white rump-patch (deeper and extending onto rear flanks in Little). Flight more whirring, alternating with longer glides than Common Swift.
Voice: a low-pitched twittering trill based around repeated 'pree'- or 'prip'-notes, different from Common Swift and lower-pitched than Little.

382 **Little Swift** *Apus affinis* **PLATE 60**

L: 12. W: 34. ***Smallest swift in the region***; distinctive stout silhouette with ***short, square-ended tail*** (round when spread); prominent ***deep, white rump-band 'wrapped-around' onto rear flanks*** (some white visible at all angles); at longer distance can recall House Martin, but dark underparts and stiff wing-beats separate. Distinguished from White-rumped (which see) by shape of tail and rump, broader wings and paler forehead with contrasting dark eye-patch, also more fluttering flight alternating with short glides.
Voice: a fast, high-pitched, rippling trill 'dillillillillill—', regularly rising and falling in pitch, much higher-pitched and faster than White-rumped and Common Swift.

383 **Palm Swift** *Cypsiurus parvus* **PLATE 60**

(African Palm Swift)

L: 16. W: 34. Size as Little, but slightly built with ***long, thin wings***, tapering body and ***extremely long and deeply forked tail with streamer-like tips, usually held closed and appearing needle-thin*** (shorter in immature but still longer than any other swift in the region). Paler ***mouse-grey-brown plumage*** than other swifts with slightly paler throat-patch and underparts. Flight very rapid and fluttering, wings held in downward shallow 'V' when gliding.
Voice: rather vocal, not unlike Little but higher, more reedy 'sisisi-soo-soo', often from birds circling palms.

384 **White-breasted Kingfisher** *Halcyon smyrnensis* **PLATE 61**

(Smyrna Kingfisher)

L: 26. Large, unmistakable brightly coloured kingfisher with ***enormous red bill, dark brown head*** and belly, ***large white 'bib', brilliant turquoise-blue upperparts*** and black forewing. In its fast straight flight shows conspicuous white primary-patches. Rather noisy; often perches on wires (but can sit well hidden) looking for prey on ground, which is often taken in shrike-like manner.
Voice: loud raucous yelping 'kril-kril-kril-kril' Also tittering descending song.

385 **Grey-headed Kingfisher** *Halcyon leucocephala* **PLATE 61**

L: 20. Fairly large kingfisher with ***buffish grey head and neck and buffish white throat and upper breast; wing-coverts and back bluish black***, belly chestnut; large red bill; has white patch in primaries, visible in flight. In first autumn, young has slightly darker head and dark scalloped pectoral band below; bill tipped blackish. Sits often on poles, wires or other prominent perch. Not dependent on water. Flight sluggish and undulating.
Voice: a weak chattering 'ji, ji, ji-jeee'.

386 **White-collared Kingfisher** *Halcyon chloris* **PLATE 61**

(Collared Kingfisher)

L: 24. Large kingfisher easily told by ***turquoise upperparts, white underparts and white collar***, bordered above by long black eyestripe and white supercilium to nape; greyish bill. Juvenile has dusky barring on breast. Flight rather weak. Noisy, particularly at dawn. Often perches low in mangroves on look-out for crabs on which it mostly feeds; when perched on mud tail usually held cocked.

Voice: a long, fast series of hoarse notes, each note ascending towards end 'chei-chei-chei-chei'.

387 **Common Kingfisher** *Alcedo atthis* PLATE 61

(Kingfisher)

L: 17. Small kingfisher with ***brilliant blue and green upperparts, reddish underparts,*** but buffish white throat and neck patch and long, dark bill (though female has red on most of lower mandible; cf. Malachite). In flight, which is swift, direct and low over water, luminous back and tail obvious. Fairly shy, often inconspicuous when perched, sitting on branch over water for long periods before diving for fish; will also hover. See Malachite Kingfisher for separation.
Voice: a high-pitched, thin, piercing 'cheee' or 'tzeee', mostly in flight.

388 **Malachite Kingfisher** *Alcedo cristata* PLATE 61

L: 12. Smaller than Common Kingfisher, which it generally resembles but ***red underparts reach to eye and are not broken by broad bluish stripe*** from bill across cheeks to shoulder as in Common Kingfisher. Upperparts darker blue. ***Dark-barred crown sometimes raised in shaggy crest. Bill red*** (adult; largely black in Common Kingfisher) though blackish in juvenile. Behaviour much as Common Kingfisher.
Voice: sharp, not very loud 'teep-teep'.

389 **Pied Kingfisher** *Ceryle rudis* PLATE 61

(Lesser Pied Kingfisher)

L: 25. Large and unmistakable, ***and the only black and white kingfisher in region, frequently seen hovering well above water*** before plunging for fish. White underparts have 2 more or less complete black breast bands in male and one in female. Black eye-mask, white supercilium, blackish top of head with short crest at nape; white tail has black band at tip and blackish central feathers. Juvenile has greyish breast band. Also fishes from perch. Sometimes occurs in small, loose flocks.
Voice: a loud, noisy chattering 'chirrick, chirrick, chirrick'.

390 **White-throated Bee-eater** *Merops albicollis* PLATE 62

L: 30. Smaller, more elegant than European Bee-eater with ***very long central tail-streamers***; easily distinguished by ***black and white head pattern***; also note bluish green upperparts, blue tail, and black half-collar around throat. In flight, shows ochre upperwing and coppery underwing, both with black trailing edge. Highly gregarious and vocal.
Voice: higher pitched and softer than European Bee-eater, 'prrrp, prrrp, pruik—'.

391 **Little Green Bee-eater** *Merops orientalis* PLATE 62

(Green Bee-eater)

L: 24. W: 30. ***Smallest bee-eater*** of the region. ***Mainly green*** with black eye-stripe and elongated central tail feathers; ***lacks chestnut throat of larger Blue-cheeked.*** Western race, *cyanophrys*, (Israel, W and S Arabia) has dark blue on supercilium, throat and upper breast, rather diffuse dark breast band and relatively short tail-streamers. S Iranian race, *beludschicus*, has blue confined to chin and below eye-stripe, very narrow breast band and

long tail-streamers. Otherwise similar in plumage to Blue-cheeked. Juvenile duller and lacks tail-streamers. Often in pairs, foraging near ground from perch.
Voice: in flight, a high-pitched trilling 'trreet-trreet' or 'prrrit'.

392 **Blue-cheeked Bee-eater** *Merops superciliosus* **PLATE 62**

L: 30. W: 48. Larger than European Bee-eater and ***distinctly green or turquoise-green*** with long central tail-streamers and rusty red underwings framed dark on trailing edge. Juvenile duller and lacks long tail projections. From much smaller Little Green by ***chestnut throat***. From young European Bee-eater (more uniform greenish upperparts than in adult) by entirely green plumage including crown and underparts, long tail-streamers, and rusty red underwings.
Voice: very similar to European Bee-eater, but higher pitched, hoarser and more trilling, sometimes the note are disyllabic: 'prrllip-prrllip' or 'prl-rip'.

393 **European Bee-eater** *Merops apiaster* **PLATE 62**

(Bee-eater)

L: 28. W: 46. Easily distinguished from other bee-eaters by combination of ***chestnut crown, bright yellow throat contrasting with turquoise-blue underparts*** and, in adult, by ***chestnut back and upperwing-coverts***. In flight, shows paler, greyer underwings than other bee-eaters in the region. Juvenile has more greenish upperparts, but still shows chestnut crown and yellow throat; central tail-feathers very short or lacking. Migrates in flocks, often passing high overhead, revealed by calls. Hunts insects in flight.
Voice: very similar to Blue-cheeked but softer, lower pitched and more far-carrying: 'prrup', usually not disyllabic.

394 **European Roller** *Coracias garrulus* **PLATE 63**

(Roller)

L: 30. Size almost as Jackdaw. Its ***turquoise-blue body and most wing-coverts contrasting with blackish flight-feathers, bright chestnut back and deep blue leading forewing above*** make this species unmistakable (except from young Abyssinian, which see). The juvenile is duller and browner with lightly streaked neck and breast. Recalls Jackdaw in flight but is faster with more vigorous wing-beats, at times not unlike those of Lapwing. Medium–long tail narrow and square-cut. Often sits on prominent perch (wires, poles, dead branches) taking prey on ground in shrike-like manner. In display flight 'tumbles' from side to side in downward dive (not unlike Lapwing).
Voice: sonorous 'rack-rack' recalling Hooded Crow and a 'chack-jack' recalling Jackdaw.

395 **Abyssinian Roller** *Coracias abyssinicus* **PLATE 63**

L: 45. Length ***includes very elongated outer tail-feathers by which adult is separated from otherwise very similar European Roller***; face generally paler than European ***and flight-feathers above dark purple-blue*** (but bleach to blackish brown; blackish in European Roller). Juvenile and immature (without tail-feather elongations) are hard to tell from European Roller; the slightly more rounded wings and purple-blue flight-feathers are of limited use, so in areas where both species occur young birds will probably be unidentifiable. Behaviour as European Roller.
Voice: usual call harsh grating 'krr-krar-krar' or hoarse 'ksrev-ksreev'.

396 **Indian Roller** *Coracias benghalensis* **PLATE 63**

L: 30. Separated at all ages in flight from other rollers ***by large, translucent pale turquoise-blue primary-patch*** (both surfaces) ***and pale turquoise-blue tail-feathers*** (except centrals) ***with clear-cut dark blue band at tip and base*** (seen when tail spread). Wing-tip clearly blunter than in European Roller. When perched, ***lightly white-streaked neck, throat and breast are vinous-cinnamon*** (turquoise-blue in European Roller) ***and cap dark turquoise-green*** (in European Roller, head like underparts); mantle earth-brown (pale bright chestnut in European Roller). Behaviour as other rollers.
Voice: similar to European Roller's but with more barking 'rak'.

397 **Lilac-breasted Roller** *Coracias caudata* **PLATE 63**

L: 41. Includes ***rather elongated outer tail-feathers.*** Superficially like Abyssinian but tail-streamers little shorter. Separated by ***rich lilac throat and sides of face*** (in the Somali race, most likely to occur), more earth-brown, less rich chestnut-brown back than Abyssinian and paler forehead. In juvenile and moulting adult, lilac throat is duller and browner. Behaviour as other rollers.
Voice: harsh loud 'krack-krack'; also chattering 'kark'. In flight, a sharp rasping 'kick-kick-kick'.

398 **Rufous-crowned Roller** *Coracias naevia* **PLATE 63**

L: 35. A stocky roller ***without elongated tail-feathers,*** differing from other rollers by ***vinous-red underparts with whitish streaks, dark rufous crown, bold whitish supercilium*** and nape spot, and rufous and deep blue upperwing-coverts. ***In flight, pinkish white underwing-coverts diagnostic.*** Behaviour like other rollers.
Voice: similar to, but less vocal than Abyssinian.

399 **Hoopoe** *Upupa epops* **PLATE 63**

L: 28. ***Distinctive pinkish buff with bold black and white bars on wings and tail, long black-tipped crest, usually depressed*** (though raised upon landing), ***long decurved bill and distinctive flight with flaps of broad wings recalling Jay.*** Spends much time on the ground where hard to detect; often first noticed when it flies.
Voice: male's song distinctive, a repeated hollow 'poo-poo-poo'; has also harsh Jay-like call and a dry 'terr'.

400 **African Grey Hornbill** *Tockus nasutus* **PLATE 64**

L: 48. Only hornbill in the region. Unmistakable; size as Hooded Crow; mainly grey-brown mottled buffish white with ***long, graduated white-tipped tail,*** blackish brown head and neck with ***striking white supercilium,*** white-patterned ***stout decurved bill*** and off-white underparts. Sexes differ in bill markings with smaller ivory-yellow patch at base of upper mandible in male, all ivory-yellow upper mandible with reddish tip in female, while both have ***3–5 white lines on lower mandible.*** In flight, shows long neck and tail, broad 'fingered' wings and ***white trailing edge to wings and tail.*** Flies with slow wing-beats interspersed with glides in deeply undulating flight. Sometimes in small parties. More often heard than seen.

Voice: loud descending, plaintive mewing 'piiiuuu'; also loud, almost raptorial, 'kee-kee-kee-kee-kerra-kerra'. Song starts slowly 'coi, coi, coi ...' building to crescendo, with head thrown up and backwards.

401 **Wryneck** *Jynx torquilla* **PLATE 64**

L: 16. Its skulking behaviour and nightjar-like plumage often make it difficult to study in detail. Reminiscent of large long-tailed warbler or greyish female Red-backed Shrike when dropping into cover in low foliage, or slipping away in low, shallowly undulating flight. Often forages on ground close to cover. Distinctive features when seen well are pointed head/bill and long tail, greyish upperparts and head with ***black streak from crown to mantle***, brown mottled wings framing blackish on shoulders, ***yellowish throat finely barred***, dark eye-stripe and barred flanks.
Voice: song: loud, monotonous, plaintive 'vee-vee-vee-vee ...' recalling small falcon and confusable with call of Lesser Spotted Woodpecker.

402 **Grey-headed Woodpecker** *Picus canus* **PLATE 64**

L: 25. Slightly smaller and overall greyer on head, neck and underparts than Green, ***lacking conspicuous black eye-mask*** except for black lores and thin moustachial stripe; ***grey crown (female) or red confined to patch on forehead (male)***. Juvenile similar to adult but has slightly duller upperparts and rump, smaller red forehead-patch (in young male) and darker eyes. Rather secretive, mostly heard, or seen when flying. Flight lighter than Green but with similar flight pattern and obviously grey without contrasting head-marks. See also Green.
Voice: territorial call similar to Green but more whistling, and 'kü'-sounds slowing down (unlike Green) and become lower pitched towards end: ' kü-kü-kü-kü-kü-kü-kü-kü—kü—kü—kü'; drums frequently (unlike Green), lasting 1–2 s, two or three times per minute. Flight-call 'kick' recalling Great Spotted more than Green.

403 **Green Woodpecker** *Picus viridis* **PLATE 64**

L: 32. Medium-sized woodpecker with green upperparts, ***striking black and red head-markings***, yellow rump and pale olive-green underparts, tinged yellow. Female lacks red streak in black moustachial stripe, but both sexes have ***full red crown and white eyes surrounded by black***. Juvenile has upperparts distinctly pale-spotted; underparts, head and neck covered with broken dark bars or spots. In flight, looks green with dark, white-spotted outerwing, conspicuous yellow rump and ***solid black eye-mask***. Flight is deeply undulating with broad rounded wings and short pointed tail. Often forages on ground. For differences from slightly smaller Grey-headed see that species.
Voice: rather noisy; flight-call a shrill far-carrying 'kjü-kjü-kjük'; territorial call (both sexes) powerful laughing 'klü-klü-klü-klü-klü-klü-klü-' slightly descending and accelerating (unlike Grey-headed) at end.

404 **Scaly-bellied Woodpecker** *Picus squamatus* **Plate 64**

L: 34. Resembles large Green; distinguished by ***black and white scaled underparts from lower breast to undertail-coverts***, pale ivory bill and ***pronounced white supercilium***; black on face confined to black eye-stripe and pale-spotted black moustachial stripe; red crown in male, black in female. In flight, rump shows more green (less yellow) than Green, and pale bars on tail are more pronounced. Juvenile similar to adult but also has scaly marks on upper breast. Frequently forages on ground.
Voice: rather noisy; usual flight- or contact-call a falcon-like 'kuik-kuik-kuik' repeated quite rapidly three or four times. Territorial call melodious 'klee-guh-kleeguh' or 'pi-coq' rapidly repeated seven or eight times; also high-pitched 'kik', a far-carrying 'pirr', and a distinctive, disyllabic 'klüh- kük' (last note slightly lower). Laughing call similar to Green.

405 **Black Woodpecker** *Dryocopus martius* **Plate 64**

L: 51. W: 66. Unmistakable; ***crow-sized, black with pale bill and eyes***; ***red crown in male***; reduced to spot on nape in female. Flight action fluttery and irregular over longer distances reminiscent of Jay, more undulating in short flight and before landing.
Voice: flight-call a series of piercing notes 'prree—prree—prree—prree—' or softer 'krük-krük-krük-'. Territorial call resembles Green but more metallic, higher and not descending or accelerating at end 'klee-klee-klee-klee—' or drawn-out 'klee-eh' when perched. Drumming is very powerful and far-carrying lasting 2–3 s.

406 **Great Spotted Woodpecker** *Dendrocopos major* **Plate 65**

L: 23. Medium-sized 'pied' woodpecker with ***white shoulder-patch***, red vent and undertail-coverts, ***black cross-bar on neck*** and red nape-patch in male, but black crown in female. Turkish birds resemble those in Europe, while underparts of smaller sized and longer-billed Iranian race, *poelzami*, are smoky-brown. Red-crowned juvenile resembles smaller, paler headed Middle Spotted (which see). Syrian Woodpecker lacks black cross-bar on neck, has cleaner white head and underparts, paler red undertail-coverts and usually a different habitat. Similar-sized White-winged (which see) has much larger white scapular-area and broader white secondary-bars. Larger White-backed lacks white shoulder-patch, has white cross-bars on back and wings, larger bill and heavily streaked underparts.
Voice: usual call a metallic, sharp 'kick', when irritated repeated as rapid chatter; drumming short, less than 1 s and very fast.

407 **White-winged Woodpecker** *Dendrocopos leucopterus* **Plate 65**

L: 23. Resembles Great Spotted and Syrian in size, sex differences and general shape. From both and all 'pied' woodpeckers by ***much larger white scapular area and very broad white secondary-bars***. From Syrian also by black cross-bar present on neck, and from Great Spotted by cleaner white on head and underparts and larger white neck-patch.
Voice: not fully known, but drumming and rattling calls.

408 **Syrian Woodpecker** *Dendrocopos syriacus* **PLATE 65**

L: 23. Widespread 'pied' woodpecker with red nape-patch in male (black crown and nape in female, red crown bordered black in juvenile), red vent and undertail-coverts, and white, almost unmarked underparts. Very similar to Great Spotted, White-winged and particularly Sind Pied (see this species for further details). From Great Spotted (which see) told mostly by ***lack of black cheek-bar***. White-winged (which see) has much larger white scapular area. Flight action and flight pattern like Great Spotted, though less white visible in tail.
Voice: calls similar to Great Spotted, though common 'chüg' call slightly softer and higher-pitched; drumming slightly longer (*c.*1 s), accelerating and dying away at end.

409 **Sind Pied Woodpecker** *Dendrocopos assimilis* **PLATE 65**
(Sind Woodpecker)

L: 21. Very similar to Syrian, with which it hybridizes in Iranian Baluchistan, but smaller and ***adult male has all-red crown and nape***, similar to young Syrian (though latter has only red crown, not nape; adult male Syrian has red mostly confined to nape); black moustachial stripe extends less far down breast in Sind, and ***in all plumages rear moustache is joined to upperparts by black band*** (unlike Syrian). Otherwise like other 'pied' woodpeckers with red (fairly pale) vent and undertail-coverts, black crown and nape in female, red crown in juvenile, and, ***like Syrian, lacks black cheek-bar***.
Voice: call similar to Syrian, being a rather explosive 'ptik'.

410 **Middle Spotted Woodpecker** *Dendrocopos medius* **PLATE 65**

L: 21. Small, ***pale-faced*** 'pied' woodpecker with noticeable features being ***all-red crown in all plumages***, fine bill, ***incomplete cheek-bar, fine-streaked buffish white underparts*** and red vent and undertail-coverts less clear-cut than most other 'pied' woodpeckers. Mostly shares habitat with Great Spotted, and has almost the same flight pattern but separated by size, head- and underparts pattern. Beware of all-red crown (but bordered black) in juvenile Great Spotted.
Voice: common call like Great Spotted and Syrian but slightly lower pitched and sometimes in series 'kick-kickickickick . . .'; also a characteristic territorial mewing call in spring 'wäik—wäik—'; drums very seldom.

411 **White-backed Woodpecker** *Dendrocopos leucotos* **PLATE 65**

L: 25. Large, rather long-billed, 'pied' woodpecker with ***all-red crown in male and juvenile*** (black in female); ***wings narrowly barred white, lacks white shoulder-patches*** and white back and rump are barred with black (in race *lilfordi*); ***underparts pinkish buff, heavily streaked***. Mostly shares habitat with Great and Middle Spotted; told from both by size, upper- and underparts pattern.
Voice: common call lower pitched than Great Spotted: 'chuk—chuk—', and drum-burst obviously longer (*c.*2 s), accelerating towards end.

412 **Lesser Spotted Woodpecker** *Dendrocopos minor* **PLATE 65**

L: 15. Tiny, ***short-billed and distinctly smaller than other 'pied' woodpeckers.*** Forehead buffish white, fore-crown red in male (black in female); ***upperparts heavily barred white*** without white shoulder patches, and ***lacks red vent/undertail-coverts.*** In undulating flight appears obviously small with short tail (not unlike Woodlark), markedly black and white plumage, and superficially recalls a passerine. Often climbs far out on small branches.
Voice: common call similar to Great Spotted: 'chik'; also a distinct but slightly faster Wryneck-like series 'kee-kee-kee-kee-'. Drum-burst longer than Great Spotted (*c.*1 s).

413 **Arabian Woodpecker** *Dendrocopos dorae* **PLATE 64**

L: 18. Only woodpecker in SW Arabia. ***Male olive-brown*** tinged golden with crimson-red patch on nape, ***5–6 white bars across blackish wings*** and white bars on blackish tail; underparts paler olive-brown with pale red patch down centre of belly. Female duller and lacks red on nape. Flight has shorter undulations than Great Spotted; often feeds among small branches and outer twigs or on ground.
Voice: Kestrel-like 'kik-kik-kik-kik' but shorter; also accelerating, then descending 'kek-kek-kek-ke-ke-kekekeke-ke-ke' and a descending 'keck-keck-keck-keck-keck'.

414 **Singing Bush Lark** *Mirafra cantillans* **PLATE 66**

L: 15. Fairly nondescript brownish lark with broad wings, fairly short tail and ***weak fluttering, undulating flight—song-flight with bat-like, jerky wing-beats***—low over ground. Yellowish horn ***bill fairly stout, longer than deep with pronounced curved culmen. Adult has dull rufous tinge to flight-feathers above***, being warm buff in immature (not always pronounced). Warm brownish upperparts streaked darker, buffish white underparts with finely streaked breast, in some forming dark spot at sides; immature more streaked on breast with more distinct black spot at breast-sides; tertials long, reaching wing-tip; white at sides of tail often hard to see as it flies with tail closed; rufous wash to underwings.
Voice: song from ground, fence post or air, is chattering but quite musical, ending in Corn Bunting-like trill or jingle 'ti-vit-tir-wit, che, che, che, che, che' accelerating and descending. Call a quiet 'proop-proop'.

415 **Black-crowned Finch Lark** *Eremopterix nigriceps* **PLATE 66**

L: 12. Small stocky lark with ***stout, deep-based, pale blue-grey bill with curved culmen.*** Male unmistakable ***with bold black and white head pattern, black underparts and underwing-coverts.*** Female has ***unstreaked pale sandy grey upperparts***, faintly streaked crown, pale face and hindneck, pale buff underparts, faintly streaked across upper breast; ***blackish underwing-coverts always separate it from other small larks***; tail pattern recalls that of Dunn's Lark (which see). Juvenile recalls female but breast more streaked; separated from Lesser Short-toed by paler heavier bill with more curved culmen, unstreaked upperparts, shorter, more rounded wings and shorter tail and colour of underwing-coverts. Fairly shy; bobbing flight; male's display flight comprises circling on slow vibrating, lifted wings, interrupted by short glides.
Voice: song, often uttered in flight, a persistent repetition of 2–4 loud, sweet mellow notes (varying individually) 'chee-dee-vee' or 'pooo, pee-voo-pee', first note usually longest and isolated. Flight-call a bubbling twitter and a dry 'rrrp'; also a soft 'tchep' or 'djib'.

416 **Dunn's Lark** *Eremalauda dunni* **PLATE 66**

L: 14. Small sandy rufous-brown lark, resembling Desert Lark ***but dark-streaked crown and, vaguely, mantle; black sides to sandy brown tail; very large yellow-pink bill with pronounced curve near tip of upper mandible, and long tertials almost reaching wing-tip.*** Generally, has broader whitish eye-ring bordered below by dark line, more of a pale eyebrow from eye backwards and a dark moustache and line behind eye than Desert Lark. Recalls Desert Lark in flight with broad, rounded wings and relatively short, broad tail. On the ground appears large-headed, with upright stance; runs fast for short distances with sudden stops. *Calandrella*-larks have clear white outer tail-feathers, smaller bills, more streaks above, have bolder pattern on upperwing-coverts, are less dumpy, and in flight less broad-winged but longer-tailed.

Voice: song from ground recalls Calandra Lark's; in air a rippling warble with rising and falling melancholy whistles, imitating other larks. Flight-call a drawn-out, strident but soft 'wazz' or a 'ziup'; also a thin liquid 'prrrp'; alarm a rising 'chee-opp', 'tu-wep' or 'chup-chup-chup'.

417 **Bar-tailed Desert Lark** *Ammomanes cincturus* **PLATE 66**

L: 13. Resembles Desert Lark in general appearance but smaller, with more upright stance, rounder head and with orange-yellow ***bill shorter and finer***; also told from Desert ***by clear-cut black band to tip of reddish brown tail,*** best seen when tail spread in flight (in Desert, dark brown tail gradually merges into paler brown base) and by rufous-buff ***wings having noticeably blackish tip*** (in Desert, long primary-projection at best dark grey-brown). In race *arenicola* (Near East and Arabia) unstreaked upperparts are pale sandy rufous; E Iranian race greyer with underparts buffish white. Told from Dunn's Lark ***by much smaller bill, different tail pattern, unstreaked upperparts and crown and shorter tertials revealing blackish wing-tip.*** Runs quicker on ground than Desert Lark, with more sudden stops.

Voice: song consists of 2/3 fine notes 'tlee-tloo-hee', 1st and 2nd notes quick, quiet and low in pitch, the 3rd high, pure and drawn-out; often only last note heard; also rendered as 'dee-dee-doo' with 2nd note higher in pitch and 3rd lower and drawn-out. Flight-call purring, soft hoarse 'twer' or 'werr'; also a soft, short 'see-oo'.

418 **Desert Lark** *Ammomanes deserti* **PLATE 66**

L: 15. Dumpy, short-tailed lark with ***broad, rounded wings and floppy, slow, undulating flight.*** Colour variable according to race; darkest race *annae* (N Jordan/S Syria) is dark sooty grey, palest *azizi* (E Arabia) is pinkish buff; the other races, notably *deserti*, are grey-brown though much variation; ***generally, they are unstreaked above*** except for vague mottling on mantle in some; underparts buffish or greyish white, unmarked in some, but diffusely streaked breast in others; rufous tinge to rump, flight-feathers and underwings in many races. ***Combination of longish, stout, pointed bill*** (yellowish horn with dark culmen), ***gently tapering, and dark brown uppertail grading into rufous-buff tail-base*** (no black or white) separates it from Dunn's Lark; ***tertials fall clearly shorter than dark grey-brown wing-tip*** (Dunn's tertials almost reach wing-tip). See Bar-tailed for separation. Most other larks are obviously streaked above, have white or black in tail, or crests.

Voice: call-notes variable, but typically short, soft and melodious 'dee-leeut'. Melodious song include phrases of call-note; sings chiefly from ground, sometimes in descending gliding flight.

419 **Hoopoe Lark** *Alaemon alaudipes* — **Plate 66**

L: 18.5. Large, slender lark, easily identified ***by its long decurved bill, rather long pale legs, long tail and, in flight, bold black and white bands through wings***. Upperparts unstreaked sandy grey-buff, underparts whitish, often spotted black on breast; whitish supercilium, dark eye- and malar-stripe. Juvenile spotted on mantle but almost without breast-spots; bill less decurved. Speedy runs with sudden stops in upright position. In characteristic song-flight male ascends vertically, up to about 4 m, twists over and spirals to bush or ground with outstretched wings. Solitary or in pairs, not mixing with other larks.
Voice: song melodious and melancholy, starting slowly, accelerating and ascending in tone, then drops in speed and tone and slowly dies away 'dee-dee-dee-dee-dee, dee, de-de-de-de-de-dee-dee', etc.

420 **Dupont's Lark** *Chersophilus duponti* — **Plate 66**

L: 17. Skylark-sized, slender lark, identified by ***long, slightly decurved bill***, cinnamon-rufous, dark-mottled upperparts (the race likely to occur as vagrant), pale-fringed wing-coverts recalling pattern of young larks and fine streaks on throat and breast on otherwise white underparts. Shortish tail with white outer feathers. Skulking, reluctant to fly; runs very fast and often stands slim and upright, appearing long-necked; lacks white wing pattern of larger, longer-tailed Hoopoe Lark (which also has a long decurved bill).
Voice: most sounds have a nasal Linnet-like character.

421 **Thick-billed Lark** *Ramphocoris clotbey* — **Plate 67**

L: 17. ***Unmistakable large-headed lark with enormous, swollen pale bill; boldly black and white patterned sides of face and neck; whitish underparts with bold black spots or streaks***, which may concentrate as dark broadish line down centre of breast. In flight, ***the long wings show blackish underwing with broad white band at trailing edge and short tail has dark band near tip***. Upperparts unstreaked sandy grey, wing-coverts warmer brown. Juvenile lacks black on head, neck and underparts, which are creamy white; upperparts pinkish grey; flight-feather pattern as in adult. Appears long-winged with large rounded head in low, undulating flight. Upright stance; may run at great speed.
Voice: song, uttered both from ground and in flight, jingling with some quiet, sweet warbling notes. Flight-calls include a quiet 'peep' or 'co-ep', a soft 'blit-blit'; on landing 'shrrreeep', on ground a conversational 'woot-w-toot'.

422 **Calandra Lark** *Melanocorypha calandra* — **Plate 67**

L: 20. Large, heavy-billed lark with relatively short tail and long wings. In flight, ***shows blackish underwing with conspicuous white trailing edge and white sides to tail***. On the ground ***shows swollen yellowish horn bill, black patches at sides of lower throat*** (of variable shape, inconspicuous in some autumn birds) and variable whitish supercilium. (The smaller Bimaculated Lark, which also has black neck-patches, has paler,

dull grey-brown underwing without conspicuous white trailing edge, lacks pure white at sides of tail which has white tip; white supercilium and dark lores also distinctive in Bimaculated). See also female Black Lark for separation. Low and undulating flight with 'wader-like' deliberate wing-beats. Gregarious outside breeding season.
Voice: flight-call harsh, rolling 'terrelet'; also a more Skylark-like note. Song richer and more powerful than Skylark's, with much mimicry; some notes recall Corn Bunting's song. Delivered from high circling flight, with stiff deliberate wing-beats.

423 **Bimaculated Lark** *Melanocorypha bimaculata* PLATE 67

L: 16. Resembles a smaller, shorter-tailed version of Calandra Lark (which see) but in flight ***has dull grey-brown underwing without clear white trailing edge***; ***has white-tipped tail while outer feathers are buff-brown*** (in Calandra, underwing blackish with clear white trailing edge and white sides of tail but tip with less extensive white). On the ground resembles Calandra but, generally, has more contrasting head pattern.
Voice: flight-call recalls Calandra's rolling note 'trrelit'; also a Short-toed Lark-like 'dre-lit'. Song resembles Calandra's but with much use of drawn-out rolling call note, delivered from ground or air.

424 **White-winged Lark** *Melanocorypha leucoptera* PLATE 67

L: 18.5. Large, ***long-winged lark with shortish, stout bill.*** Diagnostic upperwing pattern in flight ***with broad white trailing edge of blackish grey secondaries and inner primaries, and, in adult, rufous-chestnut forewing; underwing white with dark outer primaries***; has much white in outer tail-feathers. ***Male has unstreaked, pale rufous crown and ear-coverts***, often red-brown patches on breast sides and red-brown uppertail-coverts. Female has grey-brown crown, streaked darker and much less rufous on uppertail-coverts. On ground ***shows much white in closed wing***, whitish buff supercilium and lores, looking bare-faced; dark greyish horn bill has yellowish base to lower mandible, legs pale. Juvenile lacks rufous in plumage but has diagnostic pattern on flight-feathers of adult. Snow Bunting (also white secondaries) has white on some upper-wing-coverts and different colour and shape of bill.
Voice: flight-call a repeated, slightly metallic 'wed' or 'wad', recalling Skylark's; also Calandra Lark-like 'schirrl-schirrl-schirrl' but deeper, more steady in rhythm.

425 **Black Lark** *Melanocorypha yeltoniensis* PLATE 67

L: 20. Larger male is distinctive ***with all-black plumage in summer***; ***in fresh plumage*** (autumn and winter) ***has buffish white feather fringes partly obscuring black colour***; ***contrasting stout yellowish bill.*** Female clearly smaller; in winter resembles Calandra Lark in having rather heavy bill, streaked, grey-brown upperparts and sometimes black patches on sides of neck, ***but has rather more streaked or blotched breast***, less bold head pattern, darker greyish forewing ***and, like male, slaty legs*** (pale in Calandra). In flight, in which it shows broad rounded wings and short tail, ***underwing blackish***, like Calandra, ***but without the conspicuous white trailing edge of that species, and sides of tail have rather little white***. Juvenile resembles female in winter.
Voice: flight-call resembles that of Skylark.

426 **Red-capped Lark** *Calandrella cinerea* **PLATE 68**

L: 14. Closely resembles Short-toed Lark but in SW Arabia where Red-capped occurs, Short-toed has usually grey crown ***while Red-capped has chestnut-red crown*** (poorly streaked); generally, Red-capped shows ***blackish lateral streak on fore-crown, dark loral streak*** and dark horn-grey culmen (in Short-toed, usually no blackish lateral crown-streak, lores are pale and culmen generally yellowish horn); generally, black patches on sides of throat more horizontal, sometimes almost meeting below white throat. Behaviour as Short-toed.

Voice: song, uttered in circular and slightly undulating flight, 'chew-chew-chew-chew' mixed with call-notes and fluid phrases. Flight-call like Short-toed's 'grelit-drelit', sometimes followed by short twitter 'pit-wit-pit' which may be uttered alone. Other notes include intense, almost whistling 'peeeep', soft 'tsuru' and explosive Corn Bunting-like 'ptk' flight-call.

427 **Short-toed Lark** *Calandrella brachydactyla* **PLATE 68**

L: 14. Small pale lark with well streaked upperparts, buffish white ***generally unstreaked underparts with variable black patch at sides of throat*** (adult); relatively stout pale bill, ***long tertials almost covering wing-tip*** (unless worn) and boldly patterned median coverts; also prominent buffish white supercilium, pale lores, dark eye-streak and small pale area on cheeks are further features. Two races that are common on migration (*brachydactyla* and *longipennis*) are illustrated. Variably streaked crown tinged rufous in some but greyish in most; upperparts pale or dark sandy grey-brown but tinged rufous in some. Told from Lesser Short-toed ***by black neck-patches (at best with thinly streaked sides of breast), long tertials and longer bill*** (Lesser Short-toed has more distinctly streaked breast, shorter, more 'bulbous' bill, shorter tertials not covering wing-tip; head pattern generally less pronounced). Juvenile lacks neck-patches of adult (achieved 'early autumn), has few streaks or blotches on breast, much like similar Lesser Short-toed, but bill longer; when fully grown the tertials are also longer. An oriental race, *dukhunensis*, (vagrant Israel) is larger, with darker upperparts and legs and stronger bill than western Short-toed races. See Hume's and Red-capped for separation. Flight undulating; foraging and migrating flocks pack densely and fly low.

Voice: typical flight-call a sparrow-like 'tjirp', a Skylark-like 'drelit' similar to House Martin's call. Song, uttered in flight, repetitive short bursts (often including mimicking calls), ***coinciding with wing-beats in abrupt undulations in circling flight***.

428 **Hume's Short-toed Lark** *Calandrella acutirostris* **PLATE 68**

(Hume's Lark)

L: 13. A difficult species, closely resembling grey race of Short-toed, *longipennis*. ***Best identified in the hand by wing-formula*** (4 longest primaries about equal in length; in Short-toed 4th outermost long primary distinctly shorter) ***and tail pattern in which white on outer tail-feather is much less extensive in Hume's***, the white usually confined to narrow edge of outer-web and thin white wedge near tip of inner-web; the closed undertail often appears largely dark (in Short-toed outer tail-feather white apart from basal-half of inner-web and the closed undertail often appears largely white). Generally, Hume's has a less streaked fore-crown than Short-toed, slightly less pronounced supercilium in front of eye, usually dark loral-spot (usually pale lores in Short-toed) and

slightly longer, more pointed bill; pattern on cheeks, colour on uppertail-coverts and bill perhaps too variable for identification. Behaviour as Short-toed.
Voice: different flight-call from Short-toed, a full rolling 'tiyrr' or 'tiurr', but other calls resemble Short-toed's.

429 **Lesser Short-toed Lark** *Calandrella rufescens* PLATE 68

L: 13. Resembles Short-toed Lark though slightly smaller, ***with shorter, more stubby bill, more distinctly streaked breast, no black patches at sides of throat, shorter tertials with wing-tip clearly exposed beyond*** (beware of Short-toed with worn tertials) ***and typical flight-call different.*** Upperparts generally sandy grey, pale buff or rufous-brown, depending on race; crown usually not tinged rufous; supercilium generally less noticeably than in Short-toed; for separation of juvenile see Short-toed (but attains adult-like plumage quickly from Aug–Sept). Told from most other small larks by more streaked upperparts and much white at sides of tail; some Dunn's and female Black-crowned Finch Larks (which see) show streaky breast, but their bills are much heavier and tail patterns differ. In winter may occur in small dense flocks like Short-toed.
Voice: typical flight-call a characteristic quiet purring 'prrrrt' or 'prrr-rrr-rrr'; has also more Skylark-like 'drrie'. Song, in spiralling flight with unbroken deliberate wing-beats (lacking yo-yo undulations of Short-toed) varied, melodious and full of mimicry.

430 **Indian Sand Lark** *Calandrella raytal* PLATE 68

(Indian Short-toed Lark, Sand Lark)

L: 12.5. Smaller and shorter-tailed than Short-toed ***with silvery grey upperparts and finely streaked crown and mantle*** (streaks bolder in Short-toed and Lesser Short-toed); may raise crown-feathers to show short crest; white underparts ***have soft fine streaks at sides of breast*** (less distinct than in Lesser Short-toed); ***proportionally longer, relatively fine bill and obviously pink legs.*** On closed wing the tertials do not reach wing-tip; white in outer tail-feathers.
Voice: call a dry rippling 'churrp' or dry 'dry-che-chir'. Song and display flight more like Lesser Short-toed than Short-toed; song, from ground or air, a rather short, disjointed throaty warble.

431 **Crested Lark** *Galerida cristata* PLATE 69

L: 17. Rather stocky, pot-bellied, short-tailed lark; ***long spiky crest always visible***; bill fairly long, slightly curved. Upperparts buffish grey or rusty, rather uniform, diffusely streaked darker on hindneck/upper mantle; breast more heavily streaked. Flight with flapping wing-beats on ***broad wings showing rusty buff underwings and blackish brown short tail with cinnamon sides.*** Juvenile heavily pale-spotted above. Distinguished from Skylark and Small Skylark by shorter, bulkier appearance, spiky crest, longer curved bill, stronger facial marks; also from Skylark by tail pattern and ***lack of white trailing edge to wing.***
Voice: on rising, a clear 'du-ee', also varying fluty 'ee' or 'uu' sounds. Song sweet and plaintive with phrases of 4–6 notes continually repeated and interspersed with a few trills; slower, clearer and with shorter series than Skylark, sometimes with imitations included. Sings from exposed perch or high in air.

432 **Woodlark** *Lullula arborea* **PLATE 69**

L: 15. Small, fine-billed, ***obviously short-tailed lark***, distinctly streaked on upperparts and crown with contrasting head-marks and rusty ear-coverts, bordered by ***very long, broad buffish white supercilia*** almost meeting at nape; diagnostic ***black and white marking on leading forewing*** seen when flushed and perched. ***Deep, undulating flight action*** on rapid series of fluttering wing-beats, with ***noticeably short tail***, broad wings and small size; often discovered by ***characteristic call***. When it flies shows white on tail corners and along tip of tail, but no white trailing edge to wings. Distinguished from other larks by combination of small size, fine bill, head-, wing- and tail marks and call. Often perches in trees or bushes; on ground may not fly until approached to within a few metres.
Voice: song (from perch or delivered in song flight) soft, melodious, repeated yodelling phrases, which accelerate and increase in intensity at same time as they fall in pitch 'lee-lee-lee-lilililililululu-lulooloo-eelu-eelu-eelu-eelu ...'. In flight, a ***characteristic melodious yodelling*** 'deedlui'.

433 **Small Skylark** *Alauda gulgula* **PLATE 69**

(Oriental Skylark)

L: 16. Similar in coloration to Skylark when perched, but smaller with shorter tail, comparatively longer and thicker bill, and very short primary projection; also slightly rusty tinge to ear-coverts and fringes of flight-feathers. In flight, more like Woodlark (which see) in size with similarly short tail; broad ***wings have inconspicuous buffish trailing edge*** (white in Skylark). Also told from Skylark by different call, duller head marks, and tail pattern having buffish white outer feathers (bleaching to white).
Voice: song from ground or soaring high in the air is like Skylark (which see) but often in shorter phrases; includes imitations and buzzing notes of flight-call. Distinctive flight-calls are soft 'pyhp' or 'twip' recalling Ortolan Bunting, and a hard buzzing 'bzzeebz', 'shwerrrk' or 'baz-baz'; also a rather Skylark-like 'chirup'.

434 **Skylark** *Alauda arvensis* **PLATE 69**

(Eurasian Skylark)

L: 18. Medium-sized with earth-grey upperparts streaked dark, and warm brown edges to tertials and coverts in fresh plumage; underparts buffish white heavily streaked/spotted on breast; can show a small crest (much smaller than in Crested and not spiky); rather indistinct head marks. Juvenile has upperparts spotted dark with scaly ochre markings. ***In flight, shows distinct whitish trailing edge to wings*** and broad triangular ***tail with white sides***. Woodlark is distinctly smaller with obviously short tail tipped white on corners; it also has striking head marks and lacks white trailing edge to wings of Skylark. Most similar to Small Skylark (which see).
Voice: song given in fluttering flight high in air; a continuous stream of trilling and warbling. Similar to Small Skylark (which see). When flushed and in flight a variable 'chrriup', 'trruwee' or similar.

435 **Shore Lark** *Eremophila alpestris* **PLATE 69**

(Horned Lark)

L: 16. ***Distinctive head markings*** make it confusable only with smaller, pinkish-sandy Temminck's. Upperparts variable, rather greyish almost unstreaked in Asiatic and Middle East races e.g. *penicillata*, nape more pinkish; tertials dark-centred, not contrasting much with wing-tip (unlike Temminck's); white or slightly yellowish facial markings and ***black cheeks join black breast band***; in winter has duller head markings and lacks black 'horns'. Pale-spotted juvenile lacks contrasting head markings but throat and breast diffusely spotted dark (paler and more uniform in juvenile Temminck's). In flight, shows uniform wings, black tail with pinkish brown centre and white sides.
Voice: song from perch or in flight: simple, high-pitched, rather weak with sibilant or rippling character 'tu-a- li, tiali-ti' or similar. Flight-call high-pitched, mournful 'siit-di-dit'.

436 **Temminck's Horned Lark** *Eremophila bilopha* **PLATE 69**

(Temminck's Lark)

L: 14. Resembles Shore Lark (which see) but smaller with more contrasting, ***paler sandy upperparts including tertials which contrast with black wing-tip***; diffusely rufous lower hindneck; always white facial markings and ***black cheeks and breast band never join***; clear white underparts; tail pattern as Shore Lark. Juvenile, which lacks black markings on head, throat and breast is uniform sandy above with white feather edgings.
Voice: song and calls largely similar to Shore Lark.

437 **Brown-throated Sand Martin** *Riparia paludicola* **PLATE 70**

(Plain Martin)

L: 12. Small, featureless ***mouse-brown martin*** with greyish white underparts ***merging into darker greyish brown upper breast, throat and head***; lacks breast band of Sand Martin, has shorter tail accentuating broader-based appearance to wings, thus recalling shape of African Rock/Crag Martin; differs from these species in more ***fluttering flight, lack of white tail-spots*** and by white vent/undertail-coverts. Flight and behaviour similar to Sand Martin.

438 **Sand Martin** *Riparia riparia* **PLATE 70**

L: 12. Small, dull ***brown and white martin*** recalling Brown-throated; distinguished by white underparts ***sharply broken by brown breast band*** and slightly longer and deeper forked tail. From African Rock/Crag Martin by fluttering flight action, breast band, white lower underparts contrasting with dark brown tail and lack of white tail-spots. Juvenile has scaly buff upperparts. A paler race, *diluta*, is pale grey-brown above and has an ill-defined breast-band.
Voice: usual call in flight a vowel-less, rasping 'tschr'.

439 **Banded Martin** *Riparia cincta* **PLATE 70**

L: 16.5. Superficially resembles Sand Martin (which see), but ***obviously larger***· with softer, slower wing-beats, ***contrasting white underwing-coverts***, almost square tail and short white streak on side of forehead.

440 **African Rock Martin** *Ptyonoprogne fuligula* **Plate 70**

(Pale Crag Martin)

L: 12.5. Closely resembles Crag but ***smaller and paler***, more greyish; upperparts, especially ***back and rump, appear slightly greyer than wings***; ***underparts almost white including chin*** (lacking dark spots of Crag) merging into pale mouse-grey undertail-coverts; ***less contrasting head pattern*** though ear-coverts sometimes darker than crown; white spots in spread tail distinctly visible when flying overhead or from above (a feature of both species); underwing pale grey with brownish-grey coverts contrasting much less than in Crag. E Arabian race, *perpallida*, distinctly pale. Flight action similar to Crag Martin.

Voice: rather quiet; martin-like short toneless twittering; less vocal than Crag Martin.

441 **Crag Martin** *Ptyonoprogne rupestris* **Plate 70**

(Eurasian Crag Martin)

L: 14.5. Broad-winged, almost square-tailed, ***largely brown martin*** similar to smaller and paler African Rock (which see) but darker brown upperparts ***lacking greyish tinge to rump***; ***cheeks fairly dark usually contrasting with pale throat and underparts***; ***fine spots on chin and throat can be hard to see***; underparts buffish grey becoming darker towards tail-coverts without contrast; white spots in spread tail distinctive; underwing-coverts distinctly darker than flight-feathers (more uniform brown in Sand and Brown-throated Sand Martin). Flight action more soaring and fast diving than most other swallows/martins.

Voice: rather quiet; in flight a short 'chip' or 'chirr'.

442 **Barn Swallow** *Hirundo rustica* **Plate 70**

(Swallow)

L: 16 (including tail-streamers). Easily distinguished by combination of ***bluish black upperparts*** (small ***white patches visible in tail*** when spread), ***long tail-streamers, chestnut forehead and throat, solid dark breast band*** and buffish white underparts including underwing-coverts. Palestinian race, *transitiva*, has darker reddish buff underparts. Juvenile lacks tail-streamers, has brownish breast band and wings and paler rusty forehead. Flight is strong and elegant with much banking and turning, often hunting rather low. From Red-rumped by head- and rump pattern, also pale undertail-coverts; Wire-tailed lacks breast band. Most easily confused with Ethiopian (which see) but note throat- and breast markings.

Voice: song a melodious twittering including grating rattle; contact call 'witt-witt', alarm call disyllabic 'tsi-wit'.

443 **Ethiopian Swallow** *Hirundo aethiopica* **Plate 70**

L: 15. Smaller than Barn Swallow with similar upperparts; underparts like Wire-tailed though more buffish chin and throat with ***narrow incomplete breast band***; ***long deeply forked tail*** with small white patches and fine tail-streamers (shorter than in Wire-tailed), and ***small chestnut forehead-patch***. Distinguished from Barn Swallow by lack of broad breast band, from Wire-tailed by different head pattern and deeply forked tail. Juvenile has dull brown upperparts, feathers fringed bluish, and brown breast-patches; lacks chestnut forehead; probably confusable with juvenile Wire-tailed but note forked tail in Ethiopian and dark eye-mask in Wire-tailed.

444 **Wire-tailed Swallow** *Hirundo smithii* PLATE 70

L: 16 (including tail-streamers). Slightly smaller than Barn Swallow. Delightful clean-coloured with rich ***blue upperparts and eye-mask***, white tail patches as in Barn Swallow, ***rufous-chestnut crown*** and rather square-ended tail with long and fine outer tail-feathers (shorter in female and often difficult to see in flight); underparts glistening white including underwing-coverts with bluish flanks and broken vent-bar. Confusable with Ethiopian (which see) but note rufous confined to forehead in that species. Juvenile Wire-tailed lacks tail-streamers, has almost square tail, dull bluish wings, brown back and crown with darker eye-mask; underparts as adult. Flight similar to Barn Swallow.

445 **Lesser Striped Swallow** *Hirundo abyssinica* PLATE 112

L: 19. Fairly slim with deeply forked tail. Easily told from other swallows ***by chestnut crown and rump, black streaks on underparts, pale chestnut underwing-coverts*** and a ***broad white band on undertail.*** Juvenile has shorter outer tail-feathers.

446 **Red-rumped Swallow** *Hirundo daurica* PLATE 70

L: 17 (including tail-streamers). Size, shape and bluish black upperparts much resembling Barn Swallow but easily distinguished by ***rufous collar and broad pale rufous rump***; underparts, including underwing-coverts, buffish white, faintly streaked (hardly visible in flight); conspicuous ***black undertail-coverts*** striking when perched or in flight; ***lacks breast band and white in tail.*** Juvenile duller and browner with upperparts fringed buff, paler collar and rump and shorter outer tail-feathers. Flight is slow and graceful, frequently gliding for longer periods than Barn Swallow, often with tail closed.
Voice: rather quiet. Song shorter and quieter than Barn Swallow's; commonest call in flight a short soft nasal 'tweit' reminiscent of House Sparrow.

447 **House Martin** *Delichon urbica* PLATE 70

L: 12.5. Smaller than Barn Swallow with black and white appearance; easily recognized by ***bluish black upperparts with striking white rump and all white underparts***; short, forked black tail with white undertail-coverts and rather dark underwings. Juvenile duller with brownish wash on head and breast sides. Flight more fluttering than Barn Swallow with long glides often high in air.
Voice: commonest call in flight a short dry warbling 'prrlit' more musical than Sand Martin's.

448 **Golden Pipit** *Tmetothylacus tenellus* PLATE 71

L: 15. Unmistakable olive-green and yellow pipit with ***prominent bright yellow wings and sides of tail, obvious in flight; male has bright yellow underparts with black band across breast*** and yellow greater coverts. Female and juvenile duller, underparts whitish buff without bold black breast band and yellow greater-coverts of male. Typical pipit-like movements. Perches freely on trees.

449 **Richard's Pipit** *Anthus richardi* **Plate 71**

L: 18. Large robust pipit. Separated from sandier, smaller, adult Tawny ***by longer tail, legs and bill, very upright stance, prominently dark-streaked grey-brown upperparts, streaked breast and characteristic call.*** Additionally, told from 1st autumn Tawny (some still with well streaked breast) ***by pale lores*** (dark lores in Tawny), often more pronounced malar streak, ***warmer buffish flanks*** (creamy white in Tawny), ***longer straighter hind-claw*** (but hard to see in grass) ***and by always pure white outer tail-feathers*** (creamy white in Tawny, but bleach to whitish). SW Arabian breeding birds (related to African Pipit *A. cinnemomeus*) have shorter hind-claw than Asian visitors. More undulating flight and longer tail give more wagtail-like flight than Tawny. ***Frequently hovers just before landing*** (very rarely in Tawny); on ground wags tail less frequently than Tawny. See Long-billed and Blyth's for separation.
Voice: typical flight-call a harsh, explosive and throaty 'schreeip'; some foraging birds utter a Tawny-like 'tjiirrup'. Song and call of SW Arabian birds not known to us.

450 **Tawny Pipit** *Anthus campestris* **Plate 71**

L: 16.5. Medium–large pipit with ***relatively long tail, legs and bill and fairly upright stance*** which, together with ***adult's poorly streaked sandy upperparts, nearly unstreaked breast,*** plain sandy wings with conspicuous dark-centred median coverts and bold whitish supercilium separate it from smaller pipits. 1st autumn birds, still with streaked breast, ***lack bold flank-streaks of most smaller pipits; voice also important.*** Undulating flight slightly reminiscent of Yellow Wagtail. Runs quickly on the ground, stops suddenly adopting upright position. See Richard's, Long-billed and Blyth's for separation.
Voice: typical flight-call a variable sparrow-like 'chilp;' or 'chirrup'. The song, which is uttered both in undulating flight or on ground, is variable, mellow, thin and metallic, 'zriiliu', 'zseer-lee' or 'ziu-ziirliu', sometimes with a rolling end.

451 **Long-billed Pipit** *Anthus similis* **Plate 71**

L: 17. Between Richard's and Tawny in size'and stance, heavier-bodied than latter. Told from Tawny by ***relatively short legs without conspicuously visible thighs, creamier buff flanks and vent,*** less pronounced malar streak and ***creamy buff outer tail-feathers in fresh plumage*** but these bleach to whitish, resembling Tawny's. Bill longer with more drooping tip than Tawny, ***upperparts grey-brown, tinged olive in some*** (absent in sandy Tawny), pale edges and dark centres to wing-coverts and tertials generally less pronounced, supercilium often narrower and ear-coverts plainer brown; seldom so well streaked on breast and upperparts as some 1st autumn Tawnys. SW Arabian birds have ***strong rufous-buff lower underparts and cinnamon-grey outer tail-feathers*** aiding identification. Told from Richard's by ***less streaked plumage,*** less bold head pattern, less upright stance, ***shorter legs and thighs,*** colour of outer tail-feathers (always white in Richard's), ***and voice.*** Appears fairly long-tailed in flight. ***When perched tail often flicked upwards and fanned outwards.***
Voice: flight-call recalls Tawny, a loud 'tjuip', 'cheree; or 'che-vee'; also a rich 'tchup' and a quiet soft 'tchut'. Variable song, often uttered in undulating flight, 2–3 syllables; first note rising, 2nd falling and sometimes rolling, 'duiit-diuuu', 'peet-trueet' or 'shreep chew-ee'.

452 **Olive-backed Pipit** *Anthus hodgsoni* **Plate 71**

(Olive Tree Pipit)

L: 14.5. Resembles Tree Pipit ***but has more distinct head pattern with broad white supercilium from eye backwards, buff in front, edged black above*** (in Tree supercilium duller throughout, not clearly edged black above) ***and more pronounced small white and black spot on rear ear-coverts (usually absent in Tree, but sometimes weak); upperparts greener, vaguely streaked*** (in Tree, upperparts warmer brownish olive with clear dark streaks); ***breast more boldly streaked black on rufous-buff ground colour*** (yellow-buff ground colour in Tree); some also show bold blackish flank-streaks (usually very thin in Tree); wings generally tinged greenish unlike Tree. Flight and behaviour closely resemble those of Tree though tail-wagging usually more pronounced.
Voice: flight-call similar to that of Tree; a less voiced 'tzeez', usually more drawn-out, sometimes slightly stronger at start of note; due to individual variations useless in lone birds. Alarm call thinner and higher pitched, often a weaker 'sitt'.

453 **Tree Pipit** *Anthus trivialis* **Plate 71**

L: 15. Resembles Meadow but slightly larger and stockier and with slightly shorter tail; ***best told from Meadow by voice***. Compared with Meadow, generally warmer grey-brown above (less grey-green), warmer yellow-buff on breast but belly whiter; breast often bolder striped but flanks usually finely streaked (often boldly in Meadow); dark malar and creamy submoustachial streak generally boldest in Tree but moustachial streak fainter; ***hind-claw shorter, more curved*** than Meadow. In autumn separated from young Red-throated ***by voice, unstreaked rump, fine flank-streaks***, less variegated streaking above, smaller blackish spot at end of malar streak, and, often, habitat. Bounding flight steadier than Meadow's with longer undulations (closer to Red-throated's); when on ground has 'controlled' tail-wagging, unlike Meadow's nervous tail-flicks. See similar Olive-backed for separation.
Voice: flight-call a strong, short-voiced 'bzeez'. Alarm call a steadily repeated 'stit', harder and slower than Meadow's. Song, usually uttered in flight (ascending from and descending to tree-top), a loud, fast, musical series in varying tempo ending in characteristic 'seea-seea-seea' as it descends.

454 **Pechora Pipit** *Anthus gustavi* **Plate 112**

L: 14.5. A heavily streaked pipit which is most easily confused with 1st winter Red-throated (***both have streaked rump and bold flank-streaks***). Pechora is ***the only pipit with clear primary projection beyond shortish tertials*** (but moult and wear of tertials can give clear primary projection in other pipits); ***has blacker stripes on mantle with more distinctive buff-white 'braces'*** (forming open 'V') than Red-throated, ***and the 2 wing-bars are broader and whiter with blacker separation in between***. The rather short tail ***has buffish or greyish white near tips of outer tail-feathers*** (but some appear white as in Red-throated); ***voice also important***. In worn plumage entire underparts often whitish (breast buff in fresh plumage) and pale 'braces' on mantle sometimes hard to see. Usually extremely wary and skulking, often hard to flush.
Voice: typically silent; calls differ distinctly from other pipits; a hard explosive, clicking 'tsipp' or 'pwit', often repeated two or three times in fast rhythm: also short soft 'pit'; rarely a quiet thin 'tzee'.

455 **Meadow Pipit** *Anthus pratensis* **PLATE 71**

L: 14.5. Between Tree and juvenile Red-throated in general appearance but has finer, less stocky build and slightly longer tail; ***best told from both by voice***; see Tree for further differences. Told from 1st autumn Red-throated by ***almost unstreaked rump***, less bold malar and submoustachial streak, ***more broken breast- and flank-streaks*** (more striped in Red-throated), less obvious mantle streaks, less warm upperparts and less defined pale supercilium. Flight less bounding with shorter, more irregular undulations than Red-throated. Flicks tail nervously on ground. See also Water and Buff-bellied Pipit.
Voice: flight-call typically a thin nervous, 2–3 syllabic 'ist-ist-ist', 'sit-sit-sit' or 'tsis-sip', quite unlike Tree's voiced or Red-throated's long thin call.

456 **Red-throated Pipit** *Anthus cervinus* **PLATE 71**

L: 15. Resembles Meadow, but slightly heavier and shorter tailed. Adult unmistakable, with variable ***pinkish- or reddish buff face, supercilium, throat and upper breast***; this colour retained in autumn and winter, though some females have only beige throat. In 1st autumn resembles Meadow ***and best told by voice***; also ***rump is boldly streaked*** (almost plain in Meadow), breast and flanks either creamier or whiter (usually pale buff in Meadow), ***flanks usually with 2–3 bolder unbroken stripes*** (when present, more broken in Meadow); bolder malar streak ***ends in large triangular blackish spot at sides of throat*** and warmer brown upperparts have ***bolder blackish and creamy white stripes***. The flight is closer to that of Tree.
Voice: upon rising, or in flight, has a thin, high-pitched, ***drawn-out*** 'pseeee' or 'speeeh', slowly dying away, quite unlike calls of Meadow and Tree; sometimes utters shorter, harsher versions of this call.

457 **Blyth's Pipit** *Anthus godlewskii* **PLATE 71**

L: 17. ***A difficult species for which call is very important.*** Between Tawny and Richard's in size and proportions; plumage very like Richard's and compared with that species ***slightly smaller, shorter tailed and shorter legged, with slightly shorter, deep-based but pointed bill*** (Richard's bill has culmen downcurved at tip). ***Shares pale lores, streaked mantle and breast with Richard's***, but breast-streaks often neater; supercilium often shorter; hind-claw of medium length (longer in Richard's). ***Second outermost tail-feather has usually wide, short, white wedge on inner-web*** (in Richard's white extends up inner-web; if a wedge it is narrow). ***Pattern on median wing-coverts useful in adult*** (if not too worn) ***and in those 1st autumn birds with some renewed median coverts***; ***dark centres sharper, more squarely cut-off against pale tips producing a prominent pale median-covert bar*** (in Richard's, dark centres protrude centrally and are more diffuse at sides); juvenile median coverts similar in both species. Juvenile generally more streaked above and on breast than adult and fringes and tips to wing-coverts and tertials whiter and more clear-cut. Can adopt very upright position though less so than in Richard's. ***Not known to hover before landing*** as is frequent in Richard's. 1st autumn birds told from Tawny ***by pale lores*** (nearly always dark in Tawny), more streaked upperparts, distinctly streaked breast, narrower, more clear-cut pale tertial-fringes and warmer, browner plumage (some 1st autumn Tawny, still moulting, show well streaked back and breast).
Voice: two main calls: a high-pitched 'shreuu', similar to Richard's flight-call, but higher

pitched and softer; also a short 'chep', 'chip', chup' (or even 'tee-tuk') reminiscent of call of Tawny Pipit; sometimes the two calls are joined into 'schreuu-chup chup', a combination never uttered by Richard's.

458 **Buff-bellied Pipit** *Anthus rubescens* PLATE 72

(American Pipit)

L: 15.5. The race occurring in the region, *japonicus*, resembles darker version of Water in winter; ***upperparts dark olive or grey-brown, very faintly streaked***; breast ***and flanks usually rather boldly streaked*** (in Water, streaks sparser, finer, particularly on flanks); ***legs pale brown or reddish brown*** (normally much darker in Water); white eye-ring more obvious, malar streak often broader, ending in black patch at sides of neck, wings darker than in Water. In spring (before departure) underparts pinkish buff, breast lightly spotted, ***but flanks still strongly streaked***; upperparts almost unstreaked olive-grey. Well streaked underparts, including flanks, may cause confusion with Meadow and young Red-throated, ***but darker, faintly streaked upperparts, darker wings and call*** separate it.
Voice: high-pitched, short 'tripp' or 'tsiit', lacking shrill quality of Water; higher pitched than Meadow's.

459 **Water Pipit** *Anthus spinoletta* PLATE 72

L: 16. Resembles slightly larger, more robust, lighter streaked version of Meadow but ***legs blackish brown*** (exceptions occur); compared with Meadow ***crown and mantle greyer, more indistinctly streaked*** and underparts whiter; in breeding season ***breast variably tinged rosy-pink and practically unstreaked***; creamy white supercilium generally more pronounced. In winter, browner above ***though still greyer than Meadow***; whiter underparts ***sparsely but distinctly streaked but some have almost unstreaked, pink-tinged breast***. Outer tail-feathers white. See Buff-bellied for separation.
Voice: variable song depends on length of song-flight; a series of 'tri-tri-tri-tritritri … tritritri … tzuetzue-tzutsitsitsi … tsie-tsie'. Call superficially like Meadow's but stronger, less nervous 'tsrieh' or 'bzisp'.

460 **Forest Wagtail** *Dendronanthus indicus* PLATE 72

L: 17. Differs from other wagtails by ***broad blackish band across upper breast with another broken blackish band below***; otherwise underparts white, washed very pale yellow, or creamy. Upperparts olive-brown, wings blackish brown with ***2 broad yellow-white wing bars***; also a white supercilium; tail blackish brown with white outer-feathers. Pipit-like movements on the ground; perches and forages freely in trees; does not wag tail, but ***sways body from side to side***.
Voice: call a finch-like 'pink' or 'pink-pink'.

461 **Yellow Wagtail** *Motacilla flava* PLATE 72

L: 16.5. In male, head colour differs, depending on race, of which eight have been recorded in the region. Racial variation is given below:
flava: blue-grey crown and ear-coverts; long white supercilium; chin and sometimes sides of throat white.
thunbergi: dark grey crown, with no sharp demarcation to upperparts but contrasting with very dark grey ear-coverts, usually no supercilium or white on chin and throat.
feldegg: entire head and nape glossy black, with sharp demarcation to dark olive-green upperparts and entire underparts bright yellow.
cinereocapilla: resemble *thunbergi* but throat generally whitish.
lutea: whole head yellow, some with olive crown and ear-coverts, upperparts yellow-olive.
beema: pale greyish crown and ear-coverts (can be darker grey), long, broad white supercilium and whitish chin.
melanogrisea: resembles *feldegg* but back paler olive, underparts paler yellow, chin and sides of throat white.
leucocephala: whitish head, tinged grey on rear-crown and ear-coverts.
Females of all races browner, less bright yellow below with dull heads; on passage racial identification often not possible except for *lutea* and *beema*. Some 1st autumn *beema* ***lack any yellow below*** (normally tinged yellow on rear underparts in other races), ***are greyish above*** (normally tinged olive-brown in other races) and show fairly prominent whitish wing-bars, thus resembling 1st autumn Citrine, which see. Hybrids between various races occur making racial determination very tricky.
Voice: song a melodious 'sree-sriit' with stress on 2nd syllable. Call may differ according to race; *flava* and *beema* have high-pitched 'see-u'; *thunbergi* has slightly sharper call; *cinereocapilla* and *feldegg* a harsher, even sharper 'psreee-u'; some eastern forms have calls approaching Citrine's.

462 **Citrine Wagtail** *Motacilla citreola* PLATE 72

L: 18. Male unmistakable; N Russian race, *citreola*, has ***bright yellow head*** and underparts, ***black neck-band and grey upperparts***; S central Asian *calcarata* ***has black mantle*** while W central Asian, *werae*, has no (or reduced) black neck-band and less rich yellow underparts; ***2 broad pure white wing-bars characteristic at all ages***. Female told from Yellow ***also by greyish upperparts*** (tinged olive-brown in Yellow) ***and yellow supercilium surrounding grey-brown cheeks to merge on sides of neck with yellowish throat***. 1st autumn birds told from similar Yellow by colour of upperparts, wing-bars, female-like head and neck-pattern, ***often pale forehead and lores*** (darker in Yellow), ***whitish underparts without yellow, and call*** (1st autumn Yellow has vent and undertail-coverts tinged yellow though race *beema* often lacks yellow below and is greyish above).
Voice: call recalls Yellow's but harsher, with more pronounced r-sound (slightly recalling note of Richard's Pipit), a 'tsreip' or 'tsreep'; sometimes a double 'zielip'; but beware some eastern Yellows' call rather like Citrine. Song recalls Yellow Wagtail's.

463 **Grey Wagtail** *Motacilla cinerea* **PLATE 72**

L: 18. In all plumages has characteristic ***yellow vent and undertail-coverts, greyish back, no obvious covert-bars but bold white translucent bar at base of flight-feathers*** (visible in flight), ***very long tail and extremely undulating flight***. Male in summer ***has black throat bordered by white stripe below***; female has less black or even whitish throat; immature has white throat and buff, not yellow breast. ***Legs pinkish brown*** (black in other wagtails). Has more pronounced tail-wagging than other wagtails.
Voice: call resembles White's but harder, more metallic and high-pitched, a piercing 'tzi-lit' or 'tsiziss'. Song a series of variable sharp notes.

464 **White Wagtail** *Motacilla alba* **PLATE 72**

L: 18. Easily told from other wagtails by ***grey, black and white plumage***, with grey back contrasting with black nape (male) or crown (female), white face and ear-coverts contrasting sharply with black throat and breast. Juvenile, and especially winter birds, have whitish underparts ***broken by black breast band***. NE Iranian race, *personata*, has black of head and breast ***merging at sides of neck and ear-coverts, leaving smaller white area around face; white edges on wing-coverts, tertials and secondaries much broader*** than in 'European birds'. Flight undulating.
Voice: disyllabic call slightly metallic, a hard 'tse-*lit*' with stress on second syllable. Juvenile's call more metallic, often consisting of more syllables. Song composed of call-notes, a lively twittering.

465 **White-cheeked Bulbul** *Pycnonotus leucogenys* **PLATE 73**
(Himalayan Bulbul)

L: 18. Size and shape similar to Yellow-vented Bulbul. Easily told by black head and throat with ***large white cheek-patch***; the race that occurs in the region, *leucotis*, often shows only slight crest (crest prominent in Himalayan race); undertail-coverts yellow and noticeable white tips to tail-feathers. Juvenile has browner head than adult. Behaviour as Yellow-vented Bulbul.
Voice: similar to Yellow-vented Bulbul.

466 **Yellow-vented Bulbul** *Pycnonotus xanthopygos* **PLATE 73**
(Black-capped Bulbul)

L: 19. Size of small thrush, often noisy and with rather floppy actions. Drab with ***sooty black head*** shading into grey-brown upperparts and paler greyish underparts; rather long tail dark brown. Noticeable ***white eye-ring*** and ***yellow undertail-coverts***; crown feathers often slightly raised. Sociable and can occur in large groups; where it occurs it can be one of the commonest birds.
Voice: fluty, rather loud and obvious 'bli-bli-bli-bli' or 'bul-bul-bul-bul-bul'. Calls include a loud, rather harsh, 'pwitch' and 'trratsh'.

467 **Red-whiskered Bulbul** *Pycnonotus jocosus* **Plate 73**

L: 20. Easily told by sooty black crown with ***tall pointed crest, white cheeks separated from white throat by black line, white underparts*** and red undertail-coverts; fairly long blackish tail has noticeable white tips below; the small red patch behind eye is only noticeable when close. Juvenile lacks this red patch and undertail-coverts are rufous-orange.
Voice: very vocal and varied. More musical than Red-vented Bulbul.

468 **Red-vented Bulbul** *Pycnonotus cafer* **Plate 73**

L: 22.5. Generally sooty in colour with black head and ***fine pale scalloping on upperparts and breast*** when seen close; slightly crested and with ***red undertail-coverts***. In flight, reveals ***off-white rump*** and white tip to blackish tail. Like all bulbuls can be gregarious.
Voice: noisy; calls being rich and fairly loud.

469 **Waxwing** *Bombycilla garrulus* **Plate 73**
(Bohemian Waxwing)

L: 18. Easily told by soft salmon-grey plumage with ***prominent crest, dark throat and eye-mask, yellow line on wing and yellow-tipped tail***. Small red plastic-like tips on secondaries are longest on males and shortest or absent on juveniles. Flight silhouette rather similar to Starling but looks large headed and has more graceful flight.
Voice: call a soft, ringing 'sirr'.

470 **Grey Hypocolius** *Hypocolius ampelinus* **Plate 73**

L: 23. In shape resembles slim Great Grey Shrike but with longer tail. Soft blue-grey plumage with ***black eye-mask joining over nape and black-tipped tail***; primaries black (but barely showing so at rest) with ***pure white tips, prominent in flight***. Female and immature rather featureless creamy brown (but with well demarcated creamy throat), lacking black on head and with only diffused dark tip to tail. Often in small groups and can be very tame. Will fly high, often for a distance, and parties will circle round for several minutes; in flight note long tail, short wings and rather rapid wing-beats.
Voice: mellow, liquid 'tre-tur-tur', notes running together and the last two lower pitched. When perched a descending 'whee-oo'.

471 **Dipper** *Cinclus cinclus* **Plate 81**
(White-throated Dipper; European Dipper)

L: 18. Unmistakable blackish brown stout bird ***with short, often cocked tail and large white 'bib'***; juvenile has dark spots in white 'bib'. Perches on rocks and boulders in streams; plunges or walks into water, swims on or under water. Flight low, rapid and straight, following streams. Solitary.
Voice: song scratchy and squeaky with liquid warbling; call a penetrating metallic 'zrit'.

472 **Wren** *Troglodytes troglodytes* **Plate 74**

(Common Wren, Winter Wren)

L: 9.5. ***Tiny and short-tailed*** with warm brown plumage, reminiscent of small mouse when moving restlessly low to ground with ***characteristic cocked tail.*** Has thin, rather long bill, pale supercilium and barred wings, tail and flanks; often well hidden or seen in low, rapid, whirring flight from cover to cover, while calling incessantly with much temper.
Voice: song, which is ***unexpectedly loud*** for such a small bird, consists of 4–5 very fast trills with different pitch and tempo, lasting 4–6 s. Calls include a hard, dry, 'tuk', or often a rattling 'trurrrr'; reminiscent of the sound of a fisherman's reel.

473 **Dunnock** *Prunella modularis* **Plate 74**

(Hedge Accentor)

L: 14.5. Rather skulking when moving around in cover or even on ground with its rather jerky movements (but sings from exposed perch in low tree or bush top). Size and colour sparrow-like with ***warm brown upperparts distinctly streaked,*** and ***grey head and foreparts,*** but note ***fine, warbler-like bill,*** slimmer appearance and reddish legs; often shows noticeable spotted wing-bar. Juvenile has brown-spotted crown, ear-coverts, neck and most of underparts (attains adult plumage by end of summer). Separated from other accentors mainly by lack of distinct head-marks.
Voice: song a high-pitched, clear repeated warble in a distinctive, fast jumping rhythm 'teep titteri-teep titteri-teep titteri-teep' (stress on 'teep'-sounds). Call (from cover) a short tit-like, rather strident or metallic 'tseep'. Migrants in flight utter a very high-pitched ringing 'didididi'.

474 **Siberian Accentor** *Prunella montanella* **Plate 74**

L: 15. Rather similar to other 'black-headed' accentors in the region. Main differences are bright ***yellow ochre supercilium, throat and breast, and chestnut-rufous mantle*** with diffuse brownish grey streaking. Additionally, shows dark crown with paler centre and black lower edge, ***blackish brown ear-coverts distinctly demarcated from yellowish ochre throat;*** may show dark mottling on lower breast and ochre wash over flanks and sides of belly; usually shows rufous-brown flank-streaks. Bill blackish with pale base to lower mandible. In some individuals supercilium and underparts can be pale buff or even whitish but in that case ***supercilium is always warmest yellow in front of eye*** (reverse in Black-throated). For separation from pale-throated individuals of Black-throated see that species.
Voice: call similar to Dunnock's but fuller, more ringing, often trisyllabic 'dididi' or 'tsee-ree-see'.

475 **Radde's Accentor** *Prunella ocularis* **Plate 74**

L: 15.5. Skulking mountain accentor, similar in shape and behaviour to Dunnock; most distinctive features are ***blackish brown crown and cheeks,*** long, broad ***white supercilium, white*** or faint buff ***throat, dark-spotted malar-stripe*** and ***peach wash on unstreaked breast*** (but often boldly streaked flanks). Very similar in plumage to geographically isolated Arabian (which see). Juvenile similar to adult but crown brownish, streaked darker, and side of throat, breast-side and flanks more heavily streaked, underparts paler buff-white.

Voice: song Dunnock-like but weaker and slower with fewer notes, rising and falling and with a trembling or twittering quality, e.g. 'di-diii-diii-diii-diii' or 'slee-vit-chur-chur-tui'. Call-notes resembles Dunnock's.

476 **Arabian Accentor** *Prunella fagani* **PLATE 74**

(Yemen Accentor)

L: 15. Very similar to Radde's (which see) but geographically isolated from this and other accentors. Differs from Radde's in ***warmer upperparts*** and paler underparts (especially breast) with ***more distinct malar-stripe and heavier streaking on lower throat, lower breast*** and flanks (sometimes streaked right across the breast like juvenile Radde's); ***dark head markings are paler brownish-grey and slightly streaked*** (not uniformly blackish) and white supercilium is narrower. Juvenile has more greyish or buffish supercilium, more streaking below without warm buff on breast and has indistinct malar-stripe.
Voice: song short, fast 'drsi-drsi-drsi-dy-dy-', of 6–9 notes often with scratchy end. Shorter than song of Dunnock, more staccato and notes more clearly separated. Also more trilling, slightly Wren-like 'dri-drrriii-tyi driivivi'. Call-notes very similar to Dunnock's.

477 **Black-throated Accentor** *Prunella atrogularis* **PLATE 74**

L: 15. Differs from all other accentors by ***black throat-patch***, though in autumn/winter less defined or (rarely) even lacking (mainly 1st winter females) due to paler fringes; such pale-throated individuals resemble Siberian but note paler, less yellowish-ochre, supercilium and breast with ***supercilium palest/whitish in front of eye*** (reverse in Siberian); unlike Siberian, ***throat and breast are not concolorous with supercilium***. 1st winter Black-throated told from Radde's by paler, slightly streaked, brown crown (uniform dark grey in Radde's), paler (not solid black) ear-coverts, less white supercilium, black throat and absence of dark-spotted malar-stripe, and slightly warmer brown upperparts; in all plumages lacks rufous on mantle and scapulars (obvious in Siberian).
Voice: song similar to Dunnock's. Call a soft 'trrt' at least in winter quarters.

478 **Alpine Accentor** *Prunella collaris* **PLATE 74**

L: 18. ***Largest accentor***, in size and behaviour somewhat lark-like both on ground and in flight. Easily identified when seen well. Most pronounced features of the race *montana* (the most widespread in the region) are ***pale greyish head and underparts, and buffish grey mantle, contrasting with dark brown wing-coverts*** tipped white forming thin, but distinct, wing-bars; upperparts diffusely streaked, and ***flanks and breast-sides boldly streaked bright chestnut*** (almost lacking in juvenile); also noticeable are pale wing panel (pale rufous edges to secondaries), dark-spotted undertail-coverts, ***yellow base to lower mandible*** and, at close range, diffusely black- and white-spotted throat; in fluttering lark-like flight ***dark tail shows white terminal band***.
Voice: song loud and musical recalling Dunnock's in structure and rhythm but slower, lower pitched and more varied, often including lark-like trilling sequences. Calls include a trilling lark- or Greenfinch-like 'tchirllririp' or a shorter 'churrp', also a more dry bunting-like 'drlllt'.

479 **Rufous Bush Robin** *Cercotrichas galactotes* **PLATE 75**

(Rufous-tailed Scrub Robin, Rufous Bushchat)

L: 15. Skulking behaviour similar to Nightingale but slightly smaller and slimmer, and showing rather ***long, fan-shaped, often cocked, tail.*** Upperparts grey-brown in races *syriacus* and *familiaris* (Turkey, northern Syria, Lebanon, Iraq, Iran) contrasting with ***rufous rump and tail,*** the latter ***showing prominent black subterminal-band and white tips*** (broadening at sides), obvious above and below; in the race *galactotes* (Israel, Jordan, southern Syria) upperparts are rufous, concolorous with rump and tail; head distinctive with white supercilium contrasting with blackish eye-stripe. Juvenile has faintly mottled breast and flanks. Not really shy but usually close to cover; often feeds on ground.
Voice: song, delivered from concealed or exposed perch or in butterfly-like song-flight, usually consists of short clear thrush-like fluty phrases, almost on same pitch, interspersed with short regular pauses and in a steady tempo; sometimes more varied and musical recalling lark or Nightingale. Calls include a hard 'teck', a low rolling 'schrrr', a sibilant drawn-out 'iiiip' recalling Nightingale's.

480 **Black Bush Robin** *Cercotrichas podobe* **PLATE 75**

(Black Scrub Robin, Black Bushchat)

L: 18. Similar in size and shape to Rufous Bush Robin but with longer tail. Appearance unmistakable: ***entirely sooty black plumage*** with browner wings and ***prominent white tips to undertail-coverts and outer tail-feathers*** (obviously visible from below, ***when tail is swept upwards over back*** and fanned). Skulking or close to cover, often on ground but sings from exposed perch.
Voice: song melodious with thrush-like whistles similar to Rufous Bush Robins's. Call hoarse squeak or liquid chatter.

481 **Robin** *Erithacus rubecula* **PLATE 75**

(European Robin)

L: 14. Unmistakable brown chat with ***diagnostic orange-red face and breast***; plump body, large dark eyes and short wings. Juvenile, which lacks red breast, is brown, ***distinctly spotted buff above and below,*** (attains adult plumage by end of summer). Usually hops close to ground in shadow (but often sings from high perch). Upright stance, rather tame and distinctive call.
Voice: song crystal clear and plaintive with abrupt changes in pitch and tempo, and lasting *c.*3 s; often starts very high-pitched, immediately followed by a much lower-pitched, sometimes bubbling, series; the whole song has a trembling or nervous character. Usual call a sharp clicking 'tic' or 'tic-ik', often repeated and much more sonorous than that of Wren's; also a metallic 'tseep'.

482 **Thrush Nightingale** *Luscinia luscinia* **PLATE 75**

(Sprosser)

L: 16. Dark olivaceous grey-brown, resembling small thrush, with ***rusty red tail*** and pale underparts, whitish throat and eye-ring. Very similar to Nightingale but darker brown upperparts, duller rusty red tail and ***darker brownish grey breast and flanks indistinctly mottled*** when seen well at close range; best distinguished by rather far-carrying song. Very skulking.

Voice: sings from cover, often at night. Song recalls Nightingale's but is even louder, more impressive and includes ***characteristic hard chucks***, dry rattles and clear whistles; it is lower-pitched (more thrush-like) and delivered in more mechanical, less variable, sequences: 'djüllock—djüllock—djüllock—drllrllrllrllrll-pst', lacking distinctive crescendo of Nightingale. Calls include a high-pitched 'hiiid' and a dry, rolling, rather sonorous rattle.

483 **Nightingale** *Luscinia megarhynchos* **Plate 75**

(Common Nightingale, Rufous Nightingale)

L: 16. Very similar to Thrush Nightingale (which see) but ***more russet-brown upperparts, paler, rusty red tail***, and often more conspicuous whiter eye-ring; underparts more 'clean', ***lacking mottled impression on breast and flanks***. The central Asian race *hafizi*, which migrates through the eastern part of the region, shows pale fringes to the tertials and greater coverts, paler underparts and a pale supercilium; the upperparts are less russet, more greyish-brown. Juvenile spotted buff and brown above and on breast and flanks, pale spots on median wing-coverts. Skulking, even when singing. Song fairly helpful for separating from Thrush Nightingale. Often in more dry habitat than Thrush Nightingale.
Voice: song, which is delivered by day and night, resembles Thrush Nightingale's but is less impressive, higher-pitched and more varied. Like that species' song, it is composed of loud rich warbling whistles, not unlike a large *Sylvia* warbler; a ***distinctive crescendo***: 'lu-lu-lü-lü-lee-lee—' is often a characteristic sequence, which is not found in Thrush Nightingale but lacks that species' strong chucks ('djüllock'—) and has only few rattling sounds ('drllr'—) typical of the latter. Calls rather similar to Thrush Nightingale's.

484 **Siberian Rubythroat** *Luscinia calliope* **Plate 112**

L: 14. Resembles Bluethroat in size, shape and behaviour. Male distinguished by striking face-pattern including ***white supercilium and submoustachial stripe, ruby-red throat and black lores***. Females have less striking facial markings (but adult female usually shows some pink on throat), 1st winter/summer female usually lacks red on throat, thus reminiscent of a small Thrush Nightingale, but note dark lores and buff or whitish supercilium; from female Bluethroat (which see) by lack of black breast markings and red tail-base. Usually on or close to ground, but very shy and skulking.
Voice: rather silent away from breeding ground. May give a short coarse 'chack', also a loud melodious whistle 'peeoo'.

485 **Bluethroat** *Luscinia svecica* **Plate 75**

L: 14. Size and shape resembles Robin but slimmer, longer legged and generally more skulking. Pale, broad supercilium distinctive, and ***diagnostic rust-red tail-base obvious in flight*** or when perched with cocked tail. Striking features in breeding male are ***blue chin and throat*** framed below by black and rusty; depending on race the throat spot is red, white or lacking (eastern Turkey race *magna*). Non-breeding male, female and 1st winter have pale throat, ***black malar-stripe joined to dark necklace*** and often a varying amount of blue or red admixed. Spotted juvenile told from juvenile Robin by darker appearance and red tail pattern. Usually on or close to ground.
Voice: song very varied and beautiful, reminiscent of Nightingale's but less powerful, often

starting with a repeated sequence of clear ringing notes, many sounds have a distinctive metallic character interspersed with call-notes. Calls include a hard 'tack' and a soft 'hweet'.

486 **Red-flanked Bluetail** *Tarsiger cyanurus* — PLATE 76

(Orange-flanked Bush-Robin)

L: 14. Recalls Robin in behaviour; often near ground keeping close to cover. Male has ***bright cobalt blue carpals and uppertail-coverts, dull bluish tail*** (may appear dark in shade) ***and orange-buff flanks***; upperparts blue-grey with flight-feathers fringed bluish. 1st year birds and females have some bluish on uppertail-coverts and bases of outer tail-feathers (often hard to see); best ***identified by white eye-ring and narrow white throat-patch*** (latter bordered by olive-grey sides of throat and breast) ***and by orange flanks*** but these can be hidden by wings. Jerks tail unlike shivering of Common Redstart.

Voice: a hard 'tsak' or short series of quiet 'tck, tck, tck'.

487 **White-throated Robin** *Irania gutturalis* — PLATE 76

L: 16. Size and movements recall Nightingale; strikingly coloured male identified by ***black sides of face and head framing pure white centre of throat, rusty red underparts,*** blue-grey upperparts, ***black tail*** and whitish supercilium. Grey-brown female has ***dark brown tail, ochreous-buff sides of body,*** whitish throat bordered grey-buff at sides of head and breast. Generally shy and skulking but will perch freely in the open when singing.

Voice: song, sometimes uttered in gliding flight, is fast, consisting of clear whistles and scratchy harsh rolling notes mixed together; has Nightingale-like 'kerr-r-rr-rr'; call a wagtail-like 'tzi-lit', alarm call a hard 'tack'.

488 **Eversmann's Redstart** *Phoenicurus erythronotus* — PLATE 76

(Rufous-backed Redstart)

L: 14. Size of Common Redstart. Brightly coloured male told from other redstarts by ***broad longitudinal white patch on wing-coverts*** (including primary-coverts), ***rusty red mantle and most of underparts, including throat***; crown to hindneck whitish grey, black sides of head extends to shoulder. In winter both adult and 1st winter male (which has also white in wing) has browner grey crown and red parts fringed whitish. Grey-brown female told from similar Common Redstart and Black Redstart by ***Spotted Flycatcher-like whitish wing-bars and edges to tertials.*** Does not shiver tail like Redstart but ***jerks tail up and down.***

Voice: alarm note a croaking 'gre-er'; call a loud, whistling 'few-eet' and a soft 'trr'.

489 **Black Redstart** *Phoenicurus ochruros* — PLATE 76

L: 15. Males of 'European' winter visitors are of the ***blackish grey race with white wing panel***; Near Eastern *semirufus* is ***black above and on throat to lower breast, sharply defined from deep red below***; NE Iranian *phoenicuroides* is less black above and black below extends only to mid-breast; male *ochruros* (Turkey to N Iran) variable; some recall European males, others resemble *semirufus* or *phoenicuroides*, but many have ***black breast merging into reddish belly.*** Males of all three breeding races told from Common Redstart by darker upperparts, ***black extending below throat*** and lack of pure white fore-crown. Female similar to Common Redstart but ***slightly darker and drabber,***

particularly below (Common Redstart more buffish below); some female *phoenicuroides* very similar to Common Redstart in colour, but note Black Redstart ***is more ground-dwelling***. Bobs body and shivers tail like Common Redstart. See Little Rock Thrush for separation from male *phoenicuroides*.
Voice: alarm a dry 'eet-tk-tk-tk'. Song short, fast, dry and rolling 'jirr-te-te-te . . . chill-chill-chill-chill . . . *kretsch* . . . sree-we-we-we', uttered at night and dawn; once heard, very distinctive.

490 **Common Redstart** *Phoenicurus phoenicurus* **PLATE 76**

(Redstart)

L: 14. Typical redstart which bobs body and shivers tail (like Black Redstart). Male's ***black cheeks and throat, contrasting sharply*** with rusty red underparts, grey crown with ***pronounced white forehead*** characteristic; upperparts grey-brown. Male *samamisicus* (breeds Turkey eastwards) ***has white wing-panel*** and, often, darker upperparts; (note that all races of male Black are more or less black-breasted); autumn male Common Redstart has black areas fringed pale. Female brownish; separated from similar Black by ***paler, warmer buffy white underparts***. Hunts insects in flycatcher-like fashion. See female Black and Eversmann's.
Voice: call resembles Willow Warbler's soft 'wheet' but usually followed by hollow 'tick-tick'. Song short and melodious, uttered mostly early morning, 'seeh-truee-truee-truee-see-see-seeweh', 1st note higher pitched; frequently imitates other birds.

491 **Güldenstädt's Redstart** *Phoenicurus erythrogaster* **PLATE 76**

(White-winged Redstart)

L: 18. Largest redstart. Male unmistakable; ***black upperparts***, throat and upper breast; ***creamy white crown and large white wing-patches*** on base of flight-feathers (visible also on underwing) ***diagnostic***; rest of underparts and lower back to tail bright rusty red. In winter, black back and breast and white crown fringed ashy; male in 1st winter shows the white wing-patches. Female resembles ***large version*** of female Common Redstart but ***central tail-feathers dull grey-brown, barely darker*** than red-brown rest of tail (female Common Redstart has contrasting dark brown central feathers). Behaviour as Common Redstart but usually shy and solitary.

492 **Blackstart** *Cercomela melanura* **PLATE 76**

L: 15. A slender, relatively long-legged chat with ***all black tail which is slowly lowered and spread outwards, often coinciding with half-spreading of wings***. The nominate race (N Arabia, Near East) is pale ash-grey above, whitish grey below with whitish grey wing-panel; in S Arabian *erlangeri*, upperparts are almost uniform smoky grey, underparts little paler, and wing-panel brownish. Perches freely on lower branch of tree and on rocks.
Voice: short mellow song sometimes uttered in flight, an often repeated 'che-we-we' or 'ch-lulu-we'. Alarm a short deep 'tjaet-aetch'.

493 **Familiar Chat** *Cercomela familiaris* **Plate 76**

(Red-tailed Chat)

L: 15. Size as Blackstart. Identified by ***rusty red rump and tail***, latter with dark brown central feathers and ***clear-cut dark band along tip*** (pattern not unlike that of some Red-tailed Wheaters), ***rufous-brown ear-coverts and rusty brown flanks and vent***; upperparts dark grey-brown, underparts dirty grey-brown. On landing, deliberately flicks wings 3–4 times and slowly raises tail once or twice, instantly separating it from Red-tailed Wheatear. Active, catching insects in the air or on ground.
Voice: alarm a scolding 'whee-chuck-chuck-chuck', 'cher-cher' or 'whee-chuck'.

494 **Whinchat** *Saxicola rubetra* **Plate 77**

L: 12.5. Small, short-tailed chat with less upright stance, slightly smaller head and longer primary projection than Stonechat (reaching almost half way down tail, but see some Stonechat races). ***Told from Stonechat by combination of streaked brownish rump and white sides to base of tail; male also has bold, clear-cut white supercilium and white stripe between blackish sides of head and orangey throat***; female duller with paler throat and browner sides of head; ***white spot on primary-coverts*** visible, particularly in flight (can be almost absent in some 1st autumn birds). Beware of confusion between autumn Whinchat and 1st autumn Stonechat of Asian/Middle Eastern origin, which see.
Voice: song short, fast and abrupt, variably phrased, usually with mixed melodious and scratchy notes; often imitates other birds; alarm call 'djü-tek-tek'.

495 **Stonechat** *Saxicola torquata* **Plate 77**

(Common Stonechat)

L: 12. Small, short-tailed, short-winged chat with large rounded head and upright stance, frequently flicking wings and tail. Much racial variation in the region. Adult male told from Whinchat ***by black head and throat*** (no supercilium), ***reddish breast, white neck-patch and black tail.***
maura: (breeds NE Iran) has longer primary projection than short-winged *rubicola* (W and central Turkey); male *maura* has ***unstreaked orange-buff to white rump*** (which is larger and whiter than in European males, notably race *rubicola*, which have streaked rump or uppertail-coverts); broader white half-collar can approach pattern of male Collared Flycatcher; ***back paler, and axillaries jet black*** (greyish white in European males).
variegata: (breeds Caspian) has even longer primary projection (approaching Whinchat's); ***much white at sides of tail*** (recalling Northern Wheater); palest race, male with large white patch on rump, sides of neck and shoulder.
armenica: (breeds E Turkey, NW and SW Iran) also long-winged but has less white at sides of tail (not always visible); male has dark chestnut breast contrasting with pure white belly.
felix: (breeds SW Arabia) has narrow white rump patch but more extensive white half-collar than in European males.
Females of all races duller, with dark brown head and usually throat, reddish brown upper breast; European females have dark earth-brown back and streaked brownish rump; female *maura, variegata* and *armenica* have paler sandy to warm buff upperparts and pale unstreaked rump. Non-European 1st autumn birds superficially recall autumn Whinchat

(pale throat, supercilium and wing-panel); *maura* told by ***unstreaked orangey rump and black tail***, *variegata* by ***more white in tail and warm buff upperparts***; none of these races have the ***clear-cut*** long creamy-white supercilium of Whinchat.
Voice: short song has irregular, rapidly repeated notes, a series of double notes with Dunnock-like quality; alarm call 'wheet-trak-trak' like hitting two stones together.

496 **Pied Stonechat** *Saxicola caprata* PLATE 77
(Pied Bushchat)

L: 13. Slimmer, slightly longer-tailed than Stonechat, perching less upright; tail less constantly flicked. ***Jet black male unmistakable with white belly, rump and narrow shoulder patch***; white patterns above easily seen in low jerky flight. Female has ***unstreaked sooty earth-brown upperparts and breast***, creamy belly and rufous-orange rump; some are rusty brown on breast and have slight supercilium; chat behaviour, unstreaked ground-colour above, colour of rump and tail prevent confusion with other chats.
Voice: song a short rich warble of whistling notes; alarm a curt 'chuk' and a 'chek-chek-trweet'.

497 **Isabelline Wheatear** *Oenanthe isabellina* PLATE 77

L: 16. ***Resembles large female Northern Wheatear but, generally, more robust in build***, with longer legs and bill, slightly shorter tail and more upright stance; ***best separated by isolated black alula in paler, more uniform sandy wings, contrasting less with upperparts*** (has broader pale feather-edges to wings than Northern Wheatear; note that some first autumn Northern can show pronounced isolated dark alula), ***narrower white rump and, often, broader black terminal tail-band*** (pattern often intermediate between Northern Wheatear and Desert, but sometimes almost like Northern); in low flight ***shows half-translucent, dark-tipped primaries; underwing-coverts and axillaries buffish white*** (dark grey, broadly tipped whitish in Northern Wheatear); supercilium usually broader and whiter in front of eye (in Northern Wheatear, often narrower and buffier before eye). Runs more (hops less) and is faster over longer distances than Northern Wheatear; wags tail more strongly and frequently.
Voice: song longer, more powerful and variable than Northern Wheatear's with less scratchy notes and often ending with series of whistles; imitates other birds with great skill. Alarm call a 'tjack-tjack' of variable length, sometimes followed by a slightly descending 'hiu' or 'diu'.

498 **Red-breasted Wheatear** *Oenanthe bottae* PLATE 77

L: 17–18. Large wheatear, ***superficially like Isabelline but stands almost vertical***; bill and legs slightly longer. ***Distinctive head pattern always separates it; sooty black fore-crown, clear-cut white supercilium, particularly in front of eye, clear-cut black stripe from bill to well behind eye where widening to black mask***; cheeks and ear-coverts usually white; sides of lower neck, ***breast and flanks rich rufous-buff***. Has rather narrow whitish buff rump and much black in tail. Juvenile, even just after fledging, has strong traces of adult head pattern. Like Isabelline, tail often closed in low, rather fluttery flight and rounded wings appear semi-translucent. Often in pairs; runs for distances at great speed.
Voice: song has fluty and hard scratchy notes, often uttered in flight. Alarm call 'tjeet'.

499 **Northern Wheatear** *Oenanthe oenanthe* **PLATE 77**

(Wheatear)

L: 15. Relatively short-tailed wheatear with white rump and sides of tail; ***blackish terminal tail-band of even width.*** Male told by combination of ash-grey crown and back, white supercilium, thin black eye-stripe and black ear-coverts and wings. See Isabelline for separation from female/1st winter Northern Wheatear. Female Northern Wheatear is greyer brown than similar Black-eared/Pied; 1st autumn birds ***best told by shorter tail*** (wing-tips closer to tip of tail) ***and tail-band of even width*** (in Black-eared/Pied tail longer, wings cover half or less of tail which has more white, a narrower black tail-band but with black extending upwards on outer tail-feathers). Restless; bobs body, wags tail and flicks wings.
Voice: song, mostly from perch, is short with fast chacking call-notes mixed with high-pitched whistles in irregular rhythm. Alarm call a hard 'tack-tack' or 'hiid, tack-tack'.

500 **Pied Wheatear** *Oenanthe pleschanka* **PLATE 78**

L: 15. Slender, relatively long-tailed wheatear, often perching on bushes. Can be confused with Cyprus and Eastern Pied (of the race *capistrata*) which see; ***told from Mourning and male South Arabian by absence of white panel in open wing.*** Autumn male has black back and 'bib' fringed buffish, dark crown with buff-white supercilium, and buff underparts. Female sometimes like female Black-eared (of eastern race *melanoleuca*) but ***upperparts usually duller, cooler brown-grey***; some show large dark greyish 'bib' in summer (absent in female Black-eared); autumn female also told from similar Black-eared by dark brown breast-sides merging with greyish 'bib' (in Black-eared, smaller grey 'bib' separated from dark brown breast-sides by pale orange-buff area). 1st autumn female told with difficulty from eastern Black-eared by colder tone above; mantle and shoulder usually ***scalloped with rows of pale fringes*** (absent or ordered erratically in Black-eared). In extremes, Pied's slightly longer primary projection may be helpful. Hybridizes freely in N Iran with Black-eared; a rare 'vittata' form of male occurs, which lacks black throat (and sometimes sides of neck), perhaps the result of hybridization. 1st winter female told from Northern Wheatear by ***much black on outer tail-feathers, creating terminal band of uneven width.*** See female Eastern Pied for separation from the races *capistrata* and *picata*.
Voice: short musical song often in flight; has twittering phrases resembling lark or wagtail; often imitative. Calls include a hard 'tack' and dry 'trrrlt'.

501 **Cyprus Pied Wheatear** *Oenanthe cypriaca* **PLATE 78**

(Cyprus Wheatear)

L: 13.5. Closely resembles male Pied but underparts (below black 'bib') warmer buff; ***best told by song, narrower white rump*** and, with experience, smaller size. Female has ***brown- or grey-black upperparts and rather dark crown***, surrounded by whitish stripe; by summer some females are identical with male. Adult in autumn has grey-buff fringes to black mantle and 'bib', dark crown with buff-white supercilium; ***underparts deeper rusty buff than similar Pied.*** 1st autumn birds similar but have broader pale fringes to 'bib' concealing black more than in adults. In winter, underparts rapidly bleach paler thus differing little from Pied.
Voice: distinctive song very different from Pied's; recalls *Cicada* but less harsh, a purring 'bizz-bizz-bizz' for quite long periods, often ending in high-pitched, drawn-out piping note; song sometimes uttered in flight. Calls include a hard 'tack'.

502 **Black-eared Wheatear** *Oenanthe hispanica* **Plate 78**

L: 15. Behaviour and build much as Pied. Male of black-throated form can be confused with male Finsch's (which see); white-throated male only confusable with some 'vittata' forms of Pied (which see). ***Black of throat/ear-coverts not joined with black of wings and shoulder***; mantle whitish (summer) or buffish grey (autumn). Tail pattern as Pied. See Pied for separation between 1st autumn females of the two species. In autumn best told from 1st autumn Northern Wheatear by ***longer tail with terminal tail-band having much black extending up the sides of outer feathers*** (in Northern Wheatear, terminal band of fairly even width). See also female Finsch's. Hybridizes with Pied Wheatear in E Turkey.

Voice: song resembles Pied's, rather variable, with dry scratchy twittering mixed with clear whistles, often uttered in flight. Calls include a hard 'tack' and a characteristic buzzing (like an angry fly!).

503 **Desert Wheatear** *Oenanthe deserti* **Plate 78**

L: 14.5. Easily told in flight from other similar-sized wheatears ***by almost wholly black tail*** (no large white patches at sides); otherwise male recalls black-throated form of Black-eared ***but black throat narrowly joined with narrower black wing area*** (shoulders not black as in male Black-eared); ***rump tinged buff***. Sandy brown or grey-buff female often lacks black throat; may recall female Black-eared but always told by rump and tail pattern. Often perches on low vegetation or on ground; fairly shy.

Voice: song short piping phrase with downward plaintive inflection; occasionally some rattling notes; calls include a soft whistle.

504 **Finsch's Wheatear** *Oenanthe finschii* **Plate 78**

L: 15. Heavier and stockier than Pied and Black-eared; male told from former by ***narrow buff-white stripe down mantle and back to join white rump***; told from latter by ***larger black 'bib' broadly connected with black wings*** (beware of Black-eared with head sunk between shoulders); ***told from both also by terminal tail-band of even width*** (no black upward extension on outer feathers) ***and pale greyish flight-feathers below, appearing translucent above***. Some females have variable blackish on throat (sometimes lower throat only); pale-throated birds told from female Black-eared/Pied by ***sandy brown-grey upperparts***, contrasting with darker, browner wings, creamy breast lacking orange-buff tone of female Black-eared, paler flight-feathers below and tail-pattern. See also female Eastern Pied (*capistrata*). Female apparently shyer than male; generally ground-dwelling, perching less frequently in trees than Black-eared and Pied. On landing, frequently bows low, cocking tail, spreading and lowering it slowly, a habit aiding identification.

Voice: song short and rich with scratchy notes often mixed with clear whistles, phrases intermittent, including musical 'ctsi-tsi-tseeoo'. Has descending zig-zag song-flight. Alarm call 'tack'.

505 **Red-rumped Wheatear** *Oenanthe moesta* **PLATE 78**

L: 17. Large wheatear with short wings and stout bill. Male unmistakable; black throat and sides of neck join dull black wings ***which show much whitish on coverts and edges of flight-feathers; whitish- to rufous-orange rump and nearly all-black tail with rufous base to outer feathers***; mantle dull black to mealy-grey; pale flight-feathers below contrast much with blackish coverts. Sandy grey-brown female has ***rufous cinnamon-buff top and sides of head and, often, buffish wing panel*** (both absent in smaller female Red-tailed); rump as male but black in tail less extensive with larger rufous patches at sides. See Red-tailed. Active, bobbing and flicking wings and tail. Buoyant flight low over ground with rapid fluttering wing-action and, often, fanned tail.
Voice: song short, musical, rather liquid with rusty intonation, sometimes including a characteristic rising sequence like boiling kettle of water.

506 **Red-tailed Wheatear** *Oenanthe xanthoprymna* **PLATE 78**

L: 15. Slender, medium-sized wheatear. Male of race *xanthoprymna* (E Turkey, NW Iran) has black throat, sides of head and neck merging with blackish brown leading wing-coverts, ***rufous rump but white sides to tail***; black band at tip of tail narrow (larger male Red-rumped almost black-tailed); ***little white in wings.*** Female usually lacks black throat but rump and tail as male. Race *chrysopygia* (Iran except NW; sexes alike) lacks black throat in both sexes; ***upperparts drab greyish with orangey-rufous rump and sides of tail*** (rump sometimes paler); underwing pale. Intermediates occur: black-throated birds with orange tail-sides; also *chrysopygia*-like birds with white tail-sides. ***Absence of ginger tinge to head, lack of buffish wing-panel, less black in tail and size separate birds with orange tail-sides from female Red-rumped.*** Usually solitary; has bounding hops and slight downward tail-flicks; often flies with tail closed.
Voice: brief song a slow throaty warble 'see, wat-chew-eeper' or short tuneful 'wee, chu, chree', powerful and clear. Calls include 'steu-steu-steu', alarm note a short dry 'zuk' or 'zvee-tuk'.

507 **Eastern Pied Wheatear** *Oenanthe picata* **PLATE 79**

L: 15. Three races occur; much individual variation among females.
picata: (breeds Iran) resembles Hume's but black crown, upperparts and 'bib' almost without gloss; rest of underparts white, undertail-coverts sometimes buff in autumn. Female above usually brown where male black, 'bib' often rufous-brown but throat sometimes black; rest of underparts creamy buff to white. See Hume's for separation.
capistrata: (may occur E Iran) male resembles male Pied but crown often white in spring (usually grey, edged white in Pied), usually pale grey-buff in autumn (in Pied often darker with white hindneck); underparts, below black 'bib', white (tinged buffish in Pied); flight-feathers below paler grey than in Pied; in autumn lower breast ***and ventral region often rusty and pale feather-fringes on black throat and mantle inconspicuous*** (extensive in Pied). Female variable, usually dull grey-brown above, chin and throat darker or, often, warm buff-brown with paler buff belly; sometimes difficult to separate from female Pied in worn plumage (which has darkish throat); ***the dark terminal tail-band is broader and of more even width than Pied's, with just slight dark upward-extension on outer tail-feathers*** (in Pied, narrower tail-band has a greater black extension on outer tail-feathers). See also Mourning Wheatear.

opistholeuca: (infrequently Iran) male resembles young White-crowned Black Wheatear (which has black crown) ***but is smaller and has a complete black tail-band***. Female sooty brown where male black.
Voice: rather scratchy song, far less pleasant than Hume's.

508 **Mourning Wheatear** *Oenanthe lugens* PLATE 79

L: 13.5. When perched resembles stocky version of male Pied or male Eastern Pied (*capistrata* race). ***Readily identified from both in flight by prominent whitish panel in open wing***; compared with Pied the black 'bib' is smaller, underparts white (tinged buff in Pied) and undertail-coverts rufous-buff; also ***primary coverts narrowly tipped white and black band near tip of tail is of fairly even width without the pronounced upward black extension on outer tail-feathers of Pied***. In autumn lacks the pronounced pale feather fringes to black throat and mantle of male Pied. See also South Arabian Wheatear for separation. Juvenile lacks black mantle and throat, has blackish ear-coverts, a broad buffish white bar to greater coverts and to tips of otherwise black primary-coverts. *Black morph*: an isolated population in the black basalt desert of eastern Jordan and SW Syria differs greatly from the typical form of Mourning Wheatear in being ***matt black with white vent and undertail-coverts***; tail pattern is similar but terminal black tail-band slightly wider; in flight, whitish panel in open wing is slightly duller and less extensive than in typical form. Told from *opistholeuca* Eastern Pied by black on underparts extending well behind the legs and by whitish panel in open wing.
Voice: song a vivid twitter; call-note 'check-check'; alarm note 'peet-peet'.

509 **South Arabian Wheatear** *Oenanthe lugentoides* PLATE 79

L: 13.5. Resembles Mourning Wheatear but slightly more thick-set (often considered conspecific with it). Compared with Mourning, male has slightly more extensive black on throat, sides of neck and back (narrower white rump-patch than Mourning), less extensive white crown, ***often streaked darker*** (sometimes grey with white sides); underparts below 'bib' as Mourning; tail pattern as Mourning though often a little more black on outer tail-feathers. ***In flight, the open wing shows small, but conspicuous white primary-patch*** (smaller than that of Mourning; absent in other wheatears). Female grey-brown above with warm orangey brown ear-coverts (sometimes including whole head); breast off-white, ***diffusely grey-streaked*** (occasionally absent); lacks white primary-patch of male but primaries may appear silvery grey at base.
Voice: song short, loud musical bubbling; also musical 'too-too'. Calls include 'chuck-a-doo' (like stones knocked together), rasping 'kaak' often repeated and interspersed with high-pitched 'seek'.

510 **Hooded Wheatear** *Oenanthe monacha* PLATE 79

L: 17. ***Large, slender, long-tailed and long-winged wheatear with long bill and buoyant, almost butterfly-like flight***, recalling Spotted Flycatcher when catching prey, sometimes in long sallies. Distinctive male ***with black extending to centre of breast*** and, except for black central tail-feathers, ***nearly all-white tail with just black corners, provides easy identification***. Autumn and juvenile male has whitish fringes to black throat, wing-coverts and mantle; also lower underparts, rump and sides of tail are tinged buffish. Female sandy brownish grey above, ***merging into creamy buff rump, tail-***

coverts and sides of tail in which central feathers and tail-corners are dark brown; whitish grey underparts washed buff at sides of breast, flanks and undertail-coverts. In autumn female and juvenile, the rump and underparts may appear reddish buff with almost reddish brown sides of tail, but absence of dark terminal tail-band separates from Red-rumped and Red-tailed.

Voice: song has short melodious phrases, interspersed with some stone-clicking notes; brief throaty thrush-like warble heard infrequently. Female utters a 'whit-whit'; also a 'wit-awheat-wheet-wheet' or 'whee-whee-whee-wheeoo'.

511 **Hume's Wheatear** *Oenanthe alboniger* **Plate 79**

L: 16.5. Closely resembles the *picata* race of Eastern Pied, but ***is larger, with 'bull-headed' appearance*** and more upright stance. Black parts more glossy; black throat has slight side-extension and is thus less 'bib-shaped than Eastern Pied; also ***white on back extends farther up between wings, where border to black mantle is rounded*** (square-cut in Eastern Pied); underwing less contrasting; tail pattern like Eastern Pied (which see). Juvenile like adult but black replaced by matt blackish brown.

Voice: loud, melodious song, short and unvaried. Call-note sharp, short and high-pitched; alarm note harsh and grating.

512 **White-crowned Black Wheatear** *Oenanthe leucopyga* **Plate 79**

L: 17. Large, glossy black wheatear ***with black underparts down to legs, and, in many adults, white crown***; immature and some adults have black crown, resembling smaller Eastern Pied (*opistholeuca* race) but size, longer bill ***and white sides of tail with black corners diagnostic*** (no black terminal-band). Some black-crowned birds show a few white feather-tips, these birds gradually developing a white crown. In male Hooded, which has similar tail pattern, black below extends only to centre of breast.

Voice: variable song has whistling and tuneful notes, sometimes scratchy, often with imitation of other desert birds; common phrase 'viet-viet-dreeit-deit', slightly descending but much variation. Call 'peeh-peeh'.

513 **Little Rock Thrush** *Monticolà rufocinerea* **Plate 80**

L: 15. Slightly larger and plumper than Black Redstart and confusable with the eastern races of that species being smoky grey above with chestnut-orange tail and underparts below breast bib. Little Rock Thrush differs in ***stouter size, longer bill, shorter tail*** with ***characteristic pattern*** (wheatear-like black 'T' above and black terminal-band below) and ***rounded bib***. Female slightly more drab than male. Juvenile is brownish above with fine white-spotting, whitish below with dark scalloping and dull red on tail. Often quivers tail when perched.

Voice: song is scratchy interspersed with fluty notes of varying pitch or short with fluty whistling not unlike tone of Blue Rock Thrush. Call a single 'tyyt'.

514 **Rock Thrush** *Monticola saxatilis* **Plate 80**
(Rufous-tailed Rock Thrush)

L: 19. Rather small, short-tailed thrush with prominent bill. Male easily told by blue-grey upperparts and bib with ***white patch on back*** and dull red underparts and tail; in winter, ***feathers of entire plumage edged dark and white producing scalloped effect***. Female

and 1st winter similar to winter male but browner, resembling female Blue Rock Thrush (which see) but has ***pale spotting on upperparts and rusty tail.***
Voice: song similar to Blue Rock Thrush's but softer and often performed in fluttering song-flight with parachute descent. Calls include a loud 'chak', scolding 'kschrrrr' and a clear 'heed'.

515 **Blue Rock Thrush** *Monticola solitarius* PLATE 80

L: 21. Slightly larger and longer tailed than Rock Thrush and with longer bill. Male easily told by ***all over dull blue plumage*** (can look black at distance) which in winter can show fine buffish fringes to feathers. Female resembles female Blackbird but has longer bill, shorter tail and dull buff spotting and barring on underparts; some show bluish tinge to upperparts. See also Rock Thrush. Fairly shy but will sit boldly on rock in full view, remaining quite still; often makes high flycatcher-like flights for insects.
Voice: rather Blackbird-like but more melancholy with often repeated short fluty phrases interspersed with long pauses. Calls include a hard 'chak', a high 'tsee' and a soft 'heed', similar to Rock Thrush.

516 **White's Thrush** *Zoothera dauma* PLATE 112
(Scaly Thrush)

L: 28. Size and flight much as in Mistle Thrush but is much shyer and never far from dense cover where hard to detect. ***Has diagnostic bold crescentic markings on upper- and underparts*** (ground-colour above yellowish, below white). In flight, ***bold black and white bands through underwing also diagnostic*** but often hard to see when flying low to cover; rump and uppertail-coverts with crescentic pattern to tail-base. The clearly smaller 1st winter female Siberian Thrush (which see) has similar underwing-pattern but is rather uniform brownish above.
Voice: call a drawn-out 'ziie', seldom heard.

517 **Siberian Thrush** *Turdus/Zoothera sibirica* PLATE 80

L: 22. Dark slate-grey male unmistakable with ***striking white supercilium*** and 2 ***white bands on dark underwing*** in flight (all ages). In 1st winter male supercilium less conspicuous and there is fine buff speckling on ear-coverts and throat. Female superficially resembles Song Thrush but differs in ***finer brown scalloping*** (not black spotting) ***on underparts, whitish supercilium and moustachial streak*** and rufous-brown wings and tail.
Voice: in flight a Song Thrush-like 'zit'.

518 **Yemen Thrush** *Turdus menachensis* PLATE 80

L: 23. Most closely resembles female Blackbird (from which geographically isolated) but differs in ***stout orange-yellow bill*** (dark in juvenile), ***fleshy-coloured legs, bright orange-rufous underwing-coverts***; throat and breast greyish with black streaking on throat, dark spotting on breast and dark 'V'-shaped marks on whitish undertail-coverts but in spring can be quite brownish below; ***orange-olive wash on flanks*** sometimes visible and upperparts have oily-slate wash to wings and tail. Wings shorter than Blackbird's, giving long-tailed look. Skulking and difficult to observe, sitting motionless for long periods; often feeds on ground.

Voice: song a rich musical series of soft, high-pitched phrases 'trit, trow, trit', rather bulbul-like. Calls Blackbird-like; quiet 'chuk, chuk', thin 'seep' also explosive 'chck-chck' and 'chuk, chuk, chuk'.

519 **Ring Ouzel** *Turdus torquatus* PLATE 80

L: 24. Resembles Blackbird but instantly told by ***white breast-crescent*** and ***pale fringes*** to ***wing-feathers*** which give 'frosty-winged' appearance in flight. Female dark brown with ***obscure buffish breast-crescent*** (virtually absent in 1st winter females), ***pale fringes to wing-feathers and to feathers of underparts, giving scaly appearance.*** Fairly shy but will also sit prominently in open.
Voice: song is highly variable, normally 3–4 fluty notes, often followed by a twittering. Most common call is a loud, scolding, sonorous 'tek-tek-tek' and a soft shrill chatter in flight.

520 **Blackbird** *Turdus merula* PLATE 80
(Common Blackbird)

L: 25. Entirely black with yellow bill, though 1st summer males show brown flight feathers and yellow and dark bill. Female dark brown with all dark bill and slightly paler throat and breast with dark mottling (but only when seen close to). Juvenile similar to female but tends to show more mottling to underparts. Superficially resembles Starling (which see) but has longer tail and hops (not walks) on ground. Spends much time on ground or in trees where it feeds on berries in autumn and winter.
Voice: Fine songster with rich meandering flute-like notes. Various calls include 'chak, chak, chak' often accelerating into a fast outburst; also a thin 'tsee' especially in flight.

521 **Eyebrowed Thrush** *Turdus obscurus* PLATE 81

L: 19. Small thrush (size as Redwing), ***lacking spots or streaks on breast and belly. Male's white supercilium, black loral streak, grey head, neck, throat and upper breast, pale rufous lower breast and flanks distinctive***; female and 1st autumn birds have olive-brown head, supercilium and loral streak of male, some dark streaks on whitish throat, ochre-yellowish lower breast and flanks; underwing-coverts pale grey-brown; belly white. Adult's small whitish spots on tip of outer tail-feathers hard to see; 1st autumn birds have white-tipped greater coverts. Flight recalls Redwing's, powerful and straight. Generally shy.
Voice: calls include 'dackke-dsjak,-psiiie', first part recalling Fieldfare, latter (sometimes uttered alone) Blackbird.

522 **Dusky/Naumann's Thrush** *Turdus naumanni* PLATE 81

L: 24. Size as Blackbird. In Dusky (used for the race *eunomus*) upperparts dark olive-brown ***and closed wings have rufous panel*** (except 1st autumn female); ***bold white supercilium, black lores and patch on ear-coverts, white throat and lower ear-coverts, and 1–2 narrow, boldly black-flecked breast bands, extending boldly along flanks***; in female and 1st autumn birds white throat sometimes streaked darker, breast bands less bold and upperparts less dark; tail dark brown; underwing-coverts pale rufous-brown. In Naumann's (used for the race *naumanni*) ***supercilium and streak below ear-coverts rufous or buffish; underparts cinnamon-red with white scales, including***

undertail-coverts (latter separating 1st autumn birds from similar Red-throated Thrush (which has uniform buff-white tail-coverts; it is unrecorded in the Middle East, but could occur); ***outer tail-feathers bright chestnut, back and rump cinnamon-red,*** but may be lacking in female or immature; in some 1st autumn birds red in tail virtually absent, and rusty brown streaks below restricted to sides of breast; underwing-coverts pale chestnut-buff. Intermediates (hybrids) occur, showing characters of both races.
Voice: flight-calls include 'srrii-i' recalling Redwing; also Fieldfare-like 'kwae-waeg' or 'tjshah-tjshah'.

523 **Black-throated Thrush** *Turdus ruficollis* **PLATE 81**

(Dark-throated Thrush)

L: 23. Slightly smaller than Blackbird. In male (race *atrogularis*, the only race to have occurred in the Middle East), ***clear-cut black 'bib', sharply defined from whitish underparts***; in female white throat streaked blackish ***but lower border of black breast clear-cut***; upperparts browner; 1st autumn male recalls adult but 'bib' slightly pale-mottled; 1st autumn female ***has densely streaked upper breast with diffuse streaked flanks*** and sometimes parts of belly; short whitish eyebrow and mark below cheeks should not cause confusion with immature Dusky/Naumann's, which see. Underwing-coverts rusty. Fairly shy; behaviour similar to Fieldfare but flight faster and straighter.
Voice: soft squawk 'chork-chork' resembling quiet Fieldfare; also Blackbird-like chatter, a quiet 'sip' like Song Thrush and a squeaky 'tscheeik'.

524 **Fieldfare** *Turdus pilaris* **PLATE 81**

L: 25. Little larger than Blackbird. ***The only thrush with rufous-brown mantle and coverts, former contrasting with pale ash-grey head and rump; rusty yellow breast boldly streaked/spotted blackish, flanks with dense 'V'-shaped marks and underwing-coverts white***; young birds duller. Flight slightly undulating; flies in loose flocks, uttering characteristic call, separating it easily from larger Mistle.
Voice: a characteristic loud 'dscach-dscach-dscach' or variants of this; also thin 'veid' uttered as contact note from flocks.

525 **Song Thrush** *Turdus philomelos* **PLATE 81**

L: 22. Small thrush; told from Redwing ***by blackish spots (not streaks) on whitish underparts, lack of supercilium and by voice***; underwing-coverts pale yellow-brown (rusty red in Redwing). Told from Mistle ***by much smaller size, flight and voice*** (see Mistle). Flight fast and straight. In winter seen singly or in small scattered groups.
Voice: powerful song alternates between fluty and shrill sharp notes, usually repeated two to four times, 'di-du-*weet*', di-du-*weet*, di-du-*weet*;—dwi-dwi-dwi;—du-*drid*-du-*drid*;—peeoo-peeoo-peeoo-peeoo, etc. Call a short sharp 'zit' or 'zip'; alarm at nest sharp and noisy 'telk, telk, telk'.

526 **Redwing** *Turdus iliacus* PLATE 81

L: 20. Slightly smaller than Song Thrush but told easily ***by prominent buff-white supercilium and streak below cheeks, streaked underparts, rusty red flanks and underwing-coverts*** (latter often hard to see), ***and voice***. Flight fast and straight.
Voice: call a thin, metallic, voiced and drawn-out 'steeef' or 'zeeip', often heard on migration at night.

527 **Mistle Thrush** *Turdus viscivorus* PLATE 81

L: 28. Largest thrush in the region. Resembles Song in plumage, ***but white underwing-coverts conspicuous***; told from Fieldfare (similar underwing-coverts) ***by grey-brown upperparts*** (without contrasts of rufous/pale ash-grey), ***white corners of uppertail*** (visible upon landing), ***whitish underparts evenly spotted blackish*** (Fieldfare has rusty yellow breast) ***and voice***. In flight, which is slightly more undulating than Fieldfare (and very different from smaller Song), appears pot-bellied. Stands more upright than other thrushes. Fairly shy; seen singly or in pairs, in winter also in small parties.
Voice: Blackbird-like song has shorter phrases, consisting of 6–8 notes, more monotonous than Blackbird's, 'tru*ee*trüwu, trueetruwu, tih-*ooh*-wü- trüh'. Flight-call a loud, dry churring 'trrrrrr'.

528 **Cetti's Warbler** *Cettia cetti* PLATE 82

L: 13.5. Keeps well concealed low in thickets, and more often heard than seen. Short-winged, ***very round-tailed***, rather robust-looking, ***warm brown warbler***; paler off-white below tinged rufous on flanks and breast sides; ***vent and undertail-coverts brown with feathers fringed white***; uniform chestnut-brown tail often jerked or held cocked; rather thin but ***distinct white supercilium.***
Voice: song very distinctive from well concealed perch: ***sudden, loud, explosive outburst*** of abrupt notes 'plit—plitiplitipliti—pliti-(pliti)'. Alarm call hard, Wren-like 'tett-tett-tett. . .'.

529 **Yemen Warbler** *Parisoma buryi* PLATE 87

L: 15. Large *Sylvia*-like warbler ***with short wings and long tail and, often, tit-like action of hanging upside down. Upperparts dark sooty grey, darkest around eye*** (bluish white iris) ***with sharp border to whitish throat; dull apricot patch between legs*** (sometimes hard to see) and white outer tail-feathers are further features; curved culmen and half-cocked tail reminiscent of miniature babbler. Active but can be secretive, often foraging in centre of thick acacias. Usually in pairs.
Voice: sudden harsh or mewing fluty whistles, 'piiuu-ptyii-ptii', 'tchah, tsah-tiir-huit', rolling 'tschee-tschee'. Song slow, soft and melodious, a varied warbling, usually from concealed perch.

530 **Fan-tailed Cisticola** *Cisticola juncidis* PLATE 82
(Fan-tailed Warbler, Zitting Cisticola)

L: 10. Small, short-tailed, distinctly streaked warbler, ***often seen in song-flight over open grasslands***, otherwise keeps well hidden. ***Crown and upperparts streaked yellow and black***; rather pale face with prominent dark eye and with fairly indistinct supercilium; tail

rounded, *often fanned in flight with black and white tips*, most obvious from below and in song-flight.
Voice: *song diagnostic*, delivered in bounding flight in wide circles; consists of 'tzip'-notes repeated at regular intervals of 0.5–1 s and with one 'tzip' on each rise of undulating flight-path; also a short call, 'kwit', from cover.

531 **Socotra Cisticola** *Cisticola haesitata* PLATE 82

L: 10. Similar to Fan-tailed Cisticola (with which it is sometimes regarded as conspecific) but differs in having a ***distinct greyish cast to crown, nape, back and scapulars contrasting with rufous uppertail-coverts***. The generally much greyer head and less prominent supercilium reduces the impression of a bare eye set against a pale face, typical of Fan-tailed Cisticola. Habits as Fan-tailed Cisticola.
Voice: song similar to Fan-tailed Cisticola, but individual notes in song shorter and more metallic and pause between notes shorter. Alarm call a short 'phut' with a spluttering quality, audible only at close range.

532 **Graceful Prinia** *Prinia gracilis* PLATE 82

(Graceful Warbler)

L: 10. Small, unmarked, pale grey-buff warbler with ***long graduated tail***, which is frequently cocked and slightly fanned; narrowly streaked on crown, mantle and back; roundish head and ***pale face*** with prominent eye, but ***without supercilium and eye-stripe*** (which are distinctive in Scrub Warbler); tail dark above, paler below with ***black and white tips***; underparts off-white, unstreaked (finely streaked in Scrub); bill fine, black in breeding male, brown in female. Skulks in vegetation, but usually tame and active, rarely seen on ground (where Scrub usually seen). Weak flight like small Bearded Tit.
Voice: rather noisy. Song monotonous 'trrirrl-trrirrl-trrirrl ... or 'tzee-bit—' first note higher; call reminiscent of prolonged House Martin's trilling call.

533 **Socotra Warbler** *Incana incana* PLATE 82

L: 10–11. A fairly small, sandy warbler resembling a cross between a Desert Warbler and Fan-tailed Cisticola but with a noticeably long, prominent bill. ***Upperparts uniform sandy grey-brown, with rufous wash to crown and nape, black centres to tertials and black alula***; underparts, ear-coverts and sides of neck greyish white; tail above sandy brown with greyish tinge and narrow white tip; undertail-feathers show broad white tips and blackish subterminal bands which, due to graduated tail-shape, show as black and white bands. Bill dark grey with pinkish straw lower mandible with dark tip; red orbital-ring (as shown by several *Sylvia* warblers), reddish brown iris and orange-straw legs. Unobtrusive, keeping low in vegetation, even foraging on ground; tail often held slightly cocked.
Voice: song quiet, slightly hesitant, unmusical trill or spluttering series of similar notes lasting $1–1\frac{1}{2}$ s; a harsh chattering call-note.

534 **Scrub Warbler** *Scotocerca inquieta* **PLATE 82**

L: 10. General appearance and size recalls Graceful Prinia with ***long, scarcely graduated, cocked tail constantly manoeuvred.*** Secretive, but often hops on ground beside bushes, reminiscent of a small mouse. Flat-crowned with ***distinct white supercilium and dark eye-stripe***; breast finely streaked (Graceful lacks contrast on head, and breast is uniform); bill yellow-horn and legs rather long. Tail differs from Graceful's, being blackish, tipped white to underside of outer feathers. Plumage tones vary. In race *inquieta* (Near East, N Saudi Arabia) upperparts are virtually unstreaked but has fine black streaking on crown, throat and breast. Darkest race, *buryi* (SW Arabia), is smoky brown above with heavier streaking on crown and breast, and with brown or rufous belly.
Voice: song variable, fine, thin 'di-di-di-di-di', descending 'di-di-di-de-de' also dry 'dzit, dzit' followed by warbling 'toodle toodle toodle'. Various calls includes short 'drzip', 'dri-dirrirri', and loud rolling, 'tlyip-tlyip-tlyip', which is sometimes fast and descending, also scolding rasping 'prrt'; alarm-note, sharp piping 'pip', often repeated. Contact call characteristic disyllabic rather clear 'dee-düü' last note lower.

535 **Grasshopper Warbler** *Locustella naevia* **PLATE 82**
(Common Grasshopper Warbler)

L: 12.5. Small, rather dark, skulking warbler with obviously rounded tail seen when flushed and flying to cover, in short, low, jerky flight. ***Upperparts olive-brown, heavily streaked with faint supercilium***; underparts dirty-white or yellowish with streaking on rear flanks and undertail-coverts, sometimes diffusely streaked on darker breast. Smaller Fan-tailed Cisticola is streaked black and yellowish with short, distinctly patterned tail. Cetti's, River and Savi's Warblers have plain upperparts. Song rather similar to Savi's.
Voice: sings from cover, mostly at dusk and dawn. Song, which often goes on for long periods without a break, is a dry monotonous insect-like reeling, longer, more sibilant and less whirring than Savi's; call a short 'chik'.

536 **River Warbler** *Locustella fluviatilis* **PLATE 83**

L: 13. A secretive grey-brown warbler, told by ***unstreaked upperparts, diffusely streaked and spotted breast and long undertail-coverts with brownish feathers tipped white.*** Probably most confusable with Savi's Warbler but is darker and greyer with much more obvious darker spotting on breast (very diffuse in Savi's); generally keeps low in vegetation but will hop on ground with slightly cocked tail. ***Characteristic song***, which can be heard on migration, a good indicator of the bird's presence.
Voice: prolonged song has mechanical or electric rhythm 'tze tze tze tze tze tze....'

537 **Savi's Warbler** *Locustella luscinioides* **PLATE 83**

L: 14. Like a cross between European Reed Warbler and River Warbler, but slightly larger than both. Plumage warmer brown than River Warbler and ***breast is only lightly marked with very diffuse brownish lines of spots***; long undertail-coverts are sometimes tipped pale but generally less distinct than those of River Warbler. Often walks on ground with graduated tail raised and rather jerky movements; tail often bobbed—a character shown by all *Locustella* warblers. Song, uttered from reed stem, is often a good indicator of this secretive bird's presence. The Asian race, *fusca*, (a migrant only in the region) is also illustrated; note greyer plumage and more marked breast.

Voice: very similar to Grasshopper Warbler's (which see), being a monotonous, soft reeling but lower in pitch and faster. Call a loud, hard 'pitch'.

538 **Moustached Warbler** *Acrocephalus melanopogon* **PLATE 83**

L: 13. Similar to Sedge Warbler but told from it by ***clear-cut white supercilium, ending squarely on side of nape and dividing blackish crown and ear-coverts*** (Sedge has buffier supercilium, paler central crown, especially in fresh plumage, and paler ear-coverts); also ***more uniform, dull rufous-brown upperparts*** which are less obviously streaked than Sedge especially on coverts; warmer brown wash on flanks and, sometimes, finely streaked warm breast band. The eastern race *mimica* (E Turkey to Iran) is coloured more like Sedge, being duller grey-brown above, less dark, with more streaked crown; told from Sedge by ***supercilium and distinctly shorter primary projection***. Keeps very low in vegetation, or on the ground where unobtrusive, ***often cocking tail and flicking it nervously*** when approached, also when singing.
Voice: song like that of a sweet European Reed Warbler (but more variable, higher pitched) with intermittent and characteristic 'lü lü lü' phrases. Call a loud 'trr-trr', a soft, short 'tcht' a longer 'trr-trrrrr' and a hard 'tack' (like knocking two stones).

539 **Aquatic Warbler** *Acrocephalus paludicola* **PLATE 83**

L: 13. Most easily confusable with Sedge Warbler but told by a combination of features: most notably yellow-buff upperparts with ***dense black streaks on back with 2 broad pale stripes on either side of mantle*** (never shown by Sedge); ***bold buff and blackish striped head pattern*** with blackish lateral crown-stripes broadening on nape and buff ***central crown-stripe never showing any streaking*** (Sedge can show paler central crown-stripe but this is never clear-cut and is always dark-streaked); ***streaked rump*** and uppertail-coverts (no streaking on rump and barely on uppertail-coverts in Sedge); ***bright flesh or pearly pink legs*** (light brown in Sedge) and pointed 'spiky' tail feathers. 1st winter birds often brighter (even orangey yellow), with no spotting on breast and often strikingly pale primaries.
Voice: call a sharp 'chak'; also a dry 'churr'.

540 **Sedge Warbler** *Acrocephalus schoenobaenus* **PLATE 83**

L: 13. Common, medium-sized, buffish, streaked, wetland-dwelling warbler with buffish white supercilium. Confusable with Moustached Warbler and Aquatic Warbler, which see. ***Told from Moustached by buffier, less square-cut supercilium, slightly paler crown***, paler ear-coverts and buffier, more ***streaked, upperparts which merge into warmer coloured rump***, and distinctly longer primary projection. ***Compared with Aquatic has less striped head and much poorer marked upperparts*** but for full differences see that species. 1st winter Sedge is yellower than adult, shows paler centre to crown and fine spotting on breast.
Voice: song similar to European Reed Warbler but faster and more erratic, often rising and falling in pitch; frequently mimics and often starts song with rapid 'trr' notes; often performs song-flights. Call a hard 'chek' and fast, churring 'trrr'.

541 **Paddyfield Warbler** *Acrocephalus agricola* PLATE 83

L: 12. See Marsh Warbler for general comments on the unstreaked *Acrocephalus* warblers. Paddyfield is similar to Marsh, European Reed and Blyth's Reed but slightly smaller. Main differences (from all three) are ***shorter bill, more prominent whitish buff supercilium from bill to well behind eye*** (where most conspicuous), ***often suffused darkish border to supercilium***, dark-centred tertials with paler edges (unlike Blyth's Reed); ***shorter wings, and longer ample tail, which when on the ground or on landing is often raised slightly and constantly flicked***; typical flat crown of an *Acrocephalus* but feathers often raised; underparts show warmer wash to flanks and undertail than in other unstreaked *Acrocephalus*; bill brown with flesh-coloured lower mandible but distal part often darker in adult (so bill can appear dark-tipped). Birds in worn plumage can appear similar to worn Booted Warblers; however, Paddyfield told by absence of any white in tail-sides, bolder supercilium, much longer undertail-coverts and rufous-tinged rump.
Voice: calls include a simple 'chik, chik', a tit-like 'dwzzz' and a rolling 'churrr'. Song similar to Marsh Warbler's but softer, lacking harsh notes and less varied, repeating the same phrases over and over.

542 **Blyth's Reed Warbler** *Acrocephalus dumetorum* PLATE 84

L: 12.5. ***Very similar to European Reed and Marsh Warbler and difficult to identify.*** With experience or close, prolonged views can be told from them by combination of: ***shorter, more rounded wings and short primary projection with 6 primary-tips visible*** (7–8 visible in European Reed and Marsh); ***uniform upperparts with little or no rump contrast, plainer tertials***, short supercilium which often bulges in front of eye; dark grey bill with flesh-coloured base to lower-mandible. Sometimes shows flicking and fanning movements of tail (unlike European Reed and Marsh) and carriage is usually more horizontal with head and bill inclined upward in 'banana' posture. ***The call is also important*** for separation.
Voice: song musical and very imitative; differs from Marsh Warbler in slower speed and more hesitant nature, each phrase repeated five times or more in Song Thrush-like manner. Often has high-pitched phrase 'lo-ly-lia' and frequently utters 'tjeck-tjeck' between phrases. Contact call, often uttered, soft 'thik' or chck', recalling Lesser Whitethroat.

543 **Marsh Warbler** *Acrocephalus palustris* PLATE 84

L: 13. The smaller unstreaked *Acrocephalus* warblers (Marsh, European Reed, Blyth's Reed, African Reed and Paddyfield) are difficult to identify. Marsh and European Reed are the commonest in the region on passage. Marsh is pale brown above with slight olive tinge and slightly warmer rump; creamy buff below with slightly buffier flanks; short buffish white supercilium and pale eye-ring. Like all *Acrocephalus* warblers it has a rounded (not square-ended) tail, long undertail-coverts and rather sloping forehead. ***Marsh is best told by its pale-fringed dark tertials and long primary projection with 8 primary-tips showing*** each with pale border; it is also buffier, less warm, in its plumage tones than European Reed with paler legs and impression of shorter bill.
Voice: ***totally different song from European Reed Warbler***; lively, musical and full of mimicry; mostly in very fast tempo, often including nasal 'be-zec, be-zec'; when singing red gape is often conspicuous. Call a short 'chek' and a short rattle 'terrrr'.

544 **European Reed Warbler** *Acrocephalus scirpaceus* **PLATE 84**

(Reed Warbler)

L: 13. Very similar to Marsh Warbler (which see) but plumage is warmer brown with a rusty tinge to the rump; has similarly long wing-projection but has less obviously pale edges to tertials and has darker legs. For distinctions from Blyth's Reed, African Reed and Paddyfield Warbler see those species.

Voice: ***monotonous and fairly even-pitched*** mixture of grating and churring notes; ***resembles song of Sedge Warbler but slower and lacks the changes in pitch and tempo.***

545 **African Reed Warbler** *Acrocephalus baeticatus* **PLATE 84**

There is much debate as to whether this is a separate species or part of a complex with European Reed Warbler. We offer no contribution to this discussion but in view of the apparent isolation of the mangrove-dwelling birds of the Red Sea we have included it as a separate species for ease of reference.

L: 13. ***Very similar in structure and plumage to European Reed Warbler*** (which see). Differs in ***more rounded wing***, usually slightly warmer plumage, ***mangrove habitat*** and, ***perhaps, song***. The more rounded wing is created by shorter primary projection with 6 primary-tips showing beyond the closed tertials (8 in European Reed).

Voice: song similar to European Reed with same structured and rhythmic repetition. However, at least in Red Sea birds, tends to be more intermixed with harsh and musical notes, including a high-pitched 'psue', which is also often repeated several times at start of song.

546 **Clamorous Reed Warbler** *Acrocephalus stentoreus* **PLATE 84**

L: 18. Very similar to Great Reed Warbler (which see) but slightly smaller and slimmer with longer bill and proportionally ***longer, more rounded tail; short wings with only 4–5 primary-tips showing beyond tertials, accentuate tail-length.*** Colder grey-brown plumage than Great Reed with less pronounced off-white supercilium from bill to just behind eye. Two races occur in the region: *stentoreus* (Near East) is darker, often with yellowish wash to undertail-coverts—some are entirely dark brown with dark bill and legs; *brunnescens* (Iran and Arabia) is paler. Bill slate-coloured with yellowish horn basal two-thirds of lower mandible; legs steely grey or slate-brown (pale brownish in most Great Reeds). Song important for identification.

Voice: song similar in tempo to Great Reed's but more melodious with phrases often repeated three or four times: 'witch-a-witch-a witch, chew-chew-chew-chew, skatchy, skatchy, skatchy, vachoo vachoo vachoo', frequently including the phrase 'rod-o-petch-iss'. Call a loud 'ptchuk' often continuously repeated; also a single 'kchrr'.

547 **Great Reed Warbler** *Acrocephalus arundinaceus* **PLATE 84**

L: 19. A very large warbler, confusable with Clamorous Reed (which see). In coloration resembles European Reed Warbler but much larger with a more prominent supercilium and noticeably heavy bill. Compared with Clamorous Reed has ***longer primary projection*** (with 7–8 primary-tips showing), proportionally shorter tail and shorter, slightly stouter bill.

Voice: loud and powerful song with repetitive character; common phrase 'trr-trr-, *karra-karra-karra, kreee-kreee-kreee*'.

548 **Basra Reed Warbler** *Acrocephalus griseldis* **PLATE 84**

L: 15. Smaller than Great Reed, but larger than European Reed and similar to both. Told from European Reed by ***larger size, longer bill***, more marked supercilium, less warm-coloured upperparts and ***darker legs***. From Great Reed by ***smaller size, narrower bill***, more square-cut tail, more ***conspicuous supercilium (especially behind eye)***, greyer upperparts, ***whiter underparts (especially flanks and undertail)*** and ***greyer legs***. Habits much as other reed-dwelling warblers.
Voice: song is not fully described, but migrants in Africa have song between European Reed and Great Reed, with the same rhythm of European Reed but quieter and without the harsh notes of Great Reed. Call-note, a harsh 'chaarr' is louder than that of European Reed Warbler.

549 **Olivaceous Warbler** *Hippolais pallida* **PLATE 85**

L: 12.5. Size and shape recall European Reed Warbler, having flat crown and long bill, but ***Olivaceous is greyer*** with square-ended or slightly rounded tail and ***shorter undertail-coverts***. Most confusable with larger, greyer Upcher's and smaller Booted (especially the stronger-looking, heavier-billed *rama*). ***Upperparts olive-brown, tinged greyish*** on head and mantle, underparts buffish white with white throat; ***wings and tail brown***, latter ***inconspicuously fringed and tipped white on outer feathers*** (tail sooty brown rather broadly white-tipped and with narrow white fringe to outer feathers in Upcher's; white-fringed and white-tipped in Booted); pale eye-ring and ***indistinct supercilium from bill to rear of eye*** (longer, more contrasting supercilium in Booted); ***long, rather heavy bill*** with ***yellow-pink lower mandible*** (finer, shorter, dark-tipped lower mandible in Booted); ***legs grey*** (flesh-brown in Booted). Often flicks tail downwards but lacks circular tail-movements of Upcher's. From Upcher's by smaller size, head-shape, lack of wing-panel and colour and length of tail. From Booted by size, shape, pattern of head and bill, and tail pattern.
Voice: song European Reed Warbler-like in quality and rhythm but usually consists of only one phrase repeated over and over again in a cyclic manner. Calls resemble Upcher's, European Reed and Sedge Warbler: short 'tch' or 'chek' or more drawn-out 'che-ch-ch' or 'trrrrr'.

550 **Booted Warbler** *Hippolais caligata* **PLATE 85**

L: 11.5. Smallest *Hippolais* with ***short, rather fine bill***, rounded forehead and somewhat recalling a short-winged *Phylloscopus*, but crown slightly peaked, and ***plumage lacks greenish/yellowish tones***. Otherwise most confusable with Olivaceous but shorter wings make tail appear longer, ***bill is shorter, distinctly finer*** and ***usually shows dark-tipped lower mandible*** (all pale in Olivaceous); upperparts grey-brown (lacking olive tone of Olivaceous and rufous-tinged rump and flanks of Paddyfield); ***supercilium reaches distinctly behind eye***, being longer, more defined and paler than in Olivaceous, accentuated by slightly discernible darker sides of fore-crown and sometimes faint darker eye-stripe; lacks paler wing panel which is sometimes visible in Olivaceous; ***usually shows pale tips and edges to outer tail-feathers. Legs flesh-brown*** with darker feet (legs

and feet usually grey in Olivaceous). Often moves energetically through undergrowth (Olivaceous usually occurs higher in vegetation). The two races that occur in the region, *caligata* (migrant from Asia) and *rama* (the breeding race) are illustrated.
Voice: song has rhythm of Olivaceous, but faster and more melodious, less European Reed Warbler-like without repetition of phrases characteristic of Olivaceous; may have quality of Marsh or Garden Warbler; lacks mimicry. Call harder than Olivaceous, recalling weaker Lesser Whitethroat's 'tick-tick…'.

551 **Upcher's Warbler** *Hippolais languida* — PLATE 85

L: 14. Similar to Olivaceous (which see), but larger with proportionally ***longer, much darker sooty brown tail.*** Upperparts greyer in fresh plumage (more like Olivaceous in abraded and juvenile plumage) with darker wings and prominent pale edges to tertials and secondaries ***forming obvious wing panel on closed wings in fresh plumage***; head more rounded than Olivaceous but with similar markings, though pale supercilium weaker and often extends behind eye (much depending on light conditions); ***broader white tips and edges to tail-feathers especially visible from below***; bill stronger and legs greyer than Olivaceous. Larger Olive-tree Warbler has longer wing projection, powerful yellow-orange bill and very short supercilium usually only visible in front of eye. In breeding area Upcher's often perches on rocks among bushes on hill sides, moving rather slowly, waving tail up and down ***and sidewards in circular movements resembling a shrike*** (Olivaceous lacks the sidewards circular movements).
Voice: songs repeats phrases like Olivaceous but louder more melodious, lacking harsh quality of European Reed Warbler, more resembling Whitethroat. Call in breeding area similar to Olivaceous.

552 **Olive-tree Warbler** *Hippolais olivetorum* — PLATE 85

L: 15. ***Very large***, brownish-grey warbler ***with prominent wing panel and long, dagger-like yellow-orange bill***; flat forehead, sometimes with raised crown feathers, recalling shape of Icterine; head pattern inconspicuous, faint supercilium usually only visible in front of eye but sometimes extends behind it. Whitish-buff underparts have off-white throat contrasting with rather dark head and ear-coverts. Tail dark, edged and tipped whitish on outermost feathers; legs bluish grey. Of the *Hippolais* mostly resembles Upcher's but distinguished by larger size, stronger and paler bill, longer wing projection and more prominent wing panel. Also confusable with young Barred Warbler which shows shorter darker bill and darker underparts.
Voice: song loud, ***harsh and raucous*** with rhythm similar to Olivaceous but slower and lower-pitched, generally resembling Great Reed Warbler. Call short, often repeated, 'tuc-tuc'.

553 **Icterine Warbler** *Hippolais icterina* — PLATE 85

L: 13.5. ***The only obviously yellow and olive-green Hippolais in the region*** (except vagrant Melodious, which see for further separation), showing peaked crown, yellow-orange bill (red gape when singing) and ***very long wings with pale wing panel*** in fresh plumage (less so in autumn adult). Fairly bare-faced appearance with conspicuous yellowish eye-ring and lores making eye rather prominent, supercilium sometimes visible almost to rear of ear-coverts. ***Legs bluish grey.*** Differs from *Phylloscopus* warblers in larger

size, peaked head, stronger bill, wing-panel, and lack of contrasting supercilium and eye-stripe. Juvenile has browner upperparts and paler, more yellowish white underparts with wing-panel more conspicuous than worn adult in autumn. Occasionally pale birds occur with greyish olive upperparts and whitish underparts thus recalling Olivaceous or Upcher's but note longer wing projection and voice.
Voice: song loud, long, with very fast tempo, varied and pleasing, rich in mimicry with sequences repeated two to five times, constantly interrupted by harsh chatter and nasal creaky 'violin' sounds 'de-de-dwiie' or 'djäk-hyyii'; also includes characteristic tri-syllabic 'de-te-roy'. Alarm call a hoarse, sparrow-like 'tettettettett'.

554 **Melodious Warbler** *Hippolais polyglotta* — **Plate 112**

L: 13. Very similar to Icterine Warbler but has ***shorter wings with primary projection only half length of tertials***; further differs in ***more brownish (less olive-green) upperparts, absence of (or poorly marked) wing panel, brown legs*** and more rounded head. Melodious without yellow underparts can look similar to Olivaceous but told by shorter bill, steeper forehead, lack of wing panel and absence of white edges to outer tail feathers.
Voice: call a chattering 'tchret-tret' and a soft 'chet'.

555 **Spectacled Warbler** *Sylvia conspicillata* — **Plate 86**

L: 13. Resembles small, brightly coloured Whitethroat with ***conspicuous bright rufous wing-coverts and blackish-centred tertials broadly edged rufous.*** Adult male has ***darker grey head*** (feathers tipped buff-brown when fresh) merging with ***blackish on lores and around eye***; underparts darker pink with white throat sharply demarcated. ***Obvious pale eye-ring***, bi-coloured rather fine bill and almost orange legs. Female and juvenile/1st winter are duller, less contrasting with brownish head, off-white throat and less pink, if any, on breast; best told from young Subalpine and other *Sylvia* warblers (except much larger Whitethroat) by ***bright rufous wings*** (though a few are less rufous). Often moves restlessly in low scrub, herbaceous plants or on ground.
Voice: song, often during song-flight, usually opens with a few clear notes, followed by typical *Sylvia* chatter with short, high-pitched phrases in fast tempo. Calls include typical Wren-like dry rattle 'trrrrrr' and a short 'tec'.

556 **Subalpine Warbler** *Sylvia cantillans* — **Plate 86**

L: 12. Smaller, shorter tailed with finer, shorter bill than Sardinian. Male distinctive, ***bluish grey above*** with browner wings and ***red orbital-ring, brick-red throat and breast*** and conspicuous ***white moustachial stripe***; tail dark grey showing much white in outer feathers. Females vary, some like dull version of male, others (including juvenile/1st winter) much greyer, even resembling Lesser Whitethroat but smaller with paler ear-coverts, pale legs and ***obvious pale eye-ring***; may be confused with 1st winter Spectacled but rufous wings in latter are distinctive. Most confusable with female/1st winter Sardinian but note: smaller size, finer bill, generally paler brown plumage with warm buff sides of throat and breast and striking pale eye-ring (less obvious in Sardinian which can also show a red eye-ring by 1st winter). Also confusable with 1st winter Ménétries's (which see). Skulking behaviour.
Voice: song (from top of bush or song-flight) typical fast and scratchy *Sylvia* chatter similar

to Sardinian but longer, clearer and more varied. Call resembles Sardinian's but softer, and single notes not in series 'chet—chet—'.

557 **Ménétries's Warbler** *Sylvia mystacea* **PLATE 86**

L: 12.5. Superficially like male Sardinian but slightly smaller and shorter-tailed. ***Female and young paler sandy brown, overall plainer than other scrub-dwelling Sylvia warblers. Male has black forehead and ear-coverts*** merging into grey crown and grey-brown mantle, wings with sandy feather-fringes and black alula. In fresh plumage, shows white moustachial stripe inconspicuously contrasting with pinkish-white throat and breast (whitish in summer when abraded); tail less rounded than Sardinian with much white in outer feathers; ***eye-ring varies from salmon-pink to red, bill bi-coloured*** and, in combination with head markings, seems slightly curved; legs usually pinkish straw. Female and 1st autumn/winter told from Sardinian by paler sandy brown upperparts, ***uniform buffish white underparts*** (more contrasting white throat and dull brown flanks in Sardinian) and base of lower mandible straw or pinkish (bluish grey in Sardinian); eye-ring varies in adult female from brown to yellow, and in 1st autumn/winter dull brown to yellow-brown; from very similar female/young Subalpine by darker tail, uniform buffish white underparts (not warm buff on sides of throat and breast) and rather uniform brown tertials (more contrastingly fringed pale in Subalpine). Active in low scrub and herbs or on ground, ***often waves tail sidewards or up and down.***
Voice: ***song softer than Sardinian's***, rather quiet but fast with Whitethroat-like quality, a mixture of melodious and soft grating notes not unlike Tree Sparrow. Calls include Sardinian-like 'chak' or 'tret' or 'che-che-che-che-che' ***but softer and less rattling***, also more prolonged Wren- or Tree Sparrow-like 'trrrrrt or 'trlrlrlrl . . .'.

558 **Sardinian Warbler** *Sylvia melanocephala* **PLATE 86**

L: 13. Slightly larger than Subalpine and Ménétries's. Male distinctive with ***black head, red eye-ring and contrasting white throat***; Middle East race, *momus*, shows fairly pale fringes to tertials (and is paler overall in all plumages than nominate race); otherwise ***grey upperparts and flanks***, reddish legs and much white in blackish tail. Female and 1st winter have rather dark upperparts, distinctly darker than Ménétries's (which see); distinguished from female and 1st winter Subalpine by size, proportionally longer more rounded tail, ***longer bill, whitish throat contrasting with dull brown flanks*** (warm buff in Subalpine) and inconspicuous pale eye-ring (obviously whitish in young Subalpine, red in adult female Sardinian). Active but rather skulking when not singing.
Voice: song similar to Subalpine, composed of hard call-like notes mixed with short whistles in fast tempo, often in song-flight. Calls include ***sudden mechanical rattle*** of 4–6 syllables 'chret-tret-tret-tret-tret' or slower 'terit-terit-terit-terit', also single 'tche'.

559 **Cyprus Warbler** *Sylvia melanothorax* **PLATE 86**

L: 13. Resembles Sardinian and Rüppell's in general character but told by ***varying amount of black spots or blotches on underparts including undertail-coverts***; upperparts show ***obvious pale fringes to innerwing feathers*** in fresh plumage (similar to Rüppell's, almost absent in Sardinian), and ***bill distinctly bi-coloured with pink base*** (bluish grey in Sardinian and Rüppell's). Adult male unmistakable with ***black hood***, ash-grey mantle and ***white moustachial stripe***; reddish-brown eye and orbital ring and warm straw-

coloured legs. ***Adult female and 1st winter male have mottled black crown***, grey-brown upperparts and ***mottled dark grey throat and breast with faint whitish moustachial stripe***; legs reddish brown; may resemble female Rüppell's but latter distinctly larger, and lacks pink base of bill. In 1st winter, female's crown is uniform grey-brown lacking dark spots, and underparts have very few or no spots, thus rather similar to young Sardinian or Rüppell's but note: ***pale edges to wing feathers*** (though almost absent in worn 1st winter female Nov–Dec), dark flanks (paler, more restricted in Rüppell's) and pink base of bill. Juvenile very similar to juvenile Sardinian, but tertial-edges better defined, unless worn. Behaviour resembles Sardinian.
Voice: song and calls much as Sardinian, but also a distinct bunting-like 'pwit' in winter, not heard from Sardinian.

560 **Rüppell's Warbler** *Sylvia rueppelli* — **Plate 86**

L: 14. ***Fairly large***, long-tailed scrub-warbler with ***slower movements*** than Sardinian, more recalling Whitethroat or even Barred Warbler. Upperparts greyish in all plumages (particularly rump, contrasting with blackish tail), wings darker with ***conspicuous pale fringes to tertials and greater coverts***; eye-ring and orbital-ring vary from brown (young) to reddish (adult), never whitish. Adult male unmistakable with ***black hood and throat divided by striking white moustachial stripe***. Adult female similar but forehead and throat usually mottled black (in fresh plumage may be hidden under grey feather-fringes) and 'moustache' less clear-cut. 1st winter male/female and 1st summer female lack dark throat (developed in 1st winter male from Nov–Feb); distinguished from female/1st winter Sardinian by size, ***pale-edged wing-feathers*** and paler, more grey appearance. Young Cyprus is slightly smaller, with bi-coloured bill, darker plumage, particularly on flanks, and may show pale eye-ring in male. Subalpine is distinctly smaller and short-billed. Female and 1st winter Ménétries's are smaller and more plain-coloured.
Voice: song and calls closely resemble Sardinian.

561 **Desert Warbler** *Sylvia nana* — **Plate 87**

L: 11.5. A small, relatively long-tailed and long-legged sandy grey-brown warbler ***with rufous-brown tertials, rump and closed uppertail; the spread tail appears tri-coloured: dark brown with much white at sides and rufous centre***; fine bill largely yellowish, ***iris and legs yellowish***; sometimes there is a pale area around eye. Spends much time on the ground, with small hops, flicking half-cocked tail; flies low over ground from bush to bush into which it skulks.
Voice: call a dry, weak purring '*d*rrrrrr' with stress on 'd', descending and fading out; also 'chrr-rrr' and high-pitched 'che-che-che-che'. Song, sometimes uttered in flight, starts with purring call, followed by short, clear, melodious, lark-like trill; can be rather Whitethroat-like.

562 **Arabian Warbler** *Sylvia leucomelaena* — **Plate 87**
(Red Sea Warbler, Blanford's Warbler)

L: 14.5. Resembles Orphean ***but separated by shorter wings, longer tail which is black with just thin white fringe at sides*** (and white tips to 2 outermost feathers); ***iris always dark***; adult often has whitish eye ring; tertials fringed whitish in fresh plumage; ***often flops tail downwards or in circular movements*** (Orphean, which see, has much

white at sides of paler tail, lacks pale eye ring and, in adult, has pale yellow iris; no obvious tail-movements). The race that occurs in Israel, *negevensis*, has paler, greyer upperparts and whiter fringes to coverts and secondaries.
Voice: song a loud, variable slow warbling, reminiscent of Garden and Blackcap or even Blackbird, delivered from exposed perch; some phrases broken by drawn-out babbler-like 'pift'. Calls include 'tscha-tscha' and short, churring rattle 'chr-rr-rr-rr-rr'.

563 **Orphean Warbler** *Sylvia hortensis* — PLATE 87

L: 15. Resembles Sardinian ***but distinctly larger, longer billed and with greyer crown contrasting poorly with blackish ear-coverts; full adult has yellowish iris*** (orange in adult Sardinian which also has reddish eye-ring) but many singing males have dark, mud-coloured iris. See Arabian for separation. 1st autumn birds and immature females also ***lack pale iris***; former recalls large, sluggish version of Lesser Whitethroat (contrasting dark lores and ear-coverts) as do some 1st summer males. See also young Barred. Fairly skulking, often hard to see in the open.
Voice: loud, musical song recalls Nightingale's; is repeated and varied; heard from dense vegetation. Calls include a Lesser Whitethroat-like 'tak' and a 'trrr'.

564 **Barred Warbler** *Sylvia nisoria* — PLATE 87

L: 15.5. Large warbler, ***adult with crescentic barring on underparts*** (reduced in female) ***and yellowish iris***; fairly long tail ***has conspicuous white corners when spread upon landing***; tips of greater coverts and tertials whitish. 1st autumn birds dark-eyed and recall Garden ***but larger, with white in tail*** (absent Garden), ***pale edges to wing-coverts and tertials and some dark barring on undertail-coverts***; this plumage can often be seen on birds through the spring. Young, longer-billed Orphean lacks pale markings on wing-coverts and tertials, dark barring on undertail-coverts, but has dark ear-coverts contrasting with rest of head. Usually keeps hidden in bushes; movements rather heavy.
Voice: song resembles that of Garden but phrases shorter; in Whitethroat-like song-flight some trills almost lark-like. A very hard sparrow-like rattling 'trrrrr' or 'che-er-er-er-er-er-er', is distinctive and often interspersed in song or used as alarm.

565 **Lesser Whitethroat** *Sylvia curruca* — PLATE 88

L: 13.5. A small/medium sized warbler, generally grey-brown above with browner wings, medium-grey crown and darker ear-coverts, which can often vary in prominence; underparts show contrast between white throat and dusky-washed breast. Differs from Whitethroat in absence of rusty fringes to wing feather (though see Whitethroat for differences from the race *icterops*), ***dark ear-coverts and dark legs***. 1st winter Lesser Whitethroat has slightly paler upperparts and often shows an indistinct whitish supercilium. The Siberian race, *blythii*, (winters Iran and probably parts of Arabia) has paler, warmer brown upperparts and whiter underparts without contrast between throat and breast.
Voice: song consists of two parts; first a brief chatter, then a fast, loud 'tell-tell-tell-tell-tell'. Call a rather short and hard 'tek' often regularly repeated.

566 **Hume's Lesser Whitethroat** *Sylvia curruca althaea* **Plate 88**

L: 14. *Very slightly larger than Lesser Whitethroat*, showing rather large head with *stouter bill and darker plumage*. Dark grey crown (darker than Lesser), merges into dull grey-brown back and near-black ear-coverts, creating greater contrast with white throat. In 1st winter can look particularly slaty above with paler feather-edgings on coverts and tertials.

Voice: pleasant Blackcap-like warble. Call a hard 'tek, tek', a single 'churrr', and melodious 'wheet-wheet-wheet', which may precede song.

567 **Desert Lesser Whitethroat** *Sylvia curruca minula* **Plate 88**

L: 13. Very slightly smaller than Lesser Whitethroat with smaller bill, *and paler, more washed-out, plumage*. Upperparts grey-buff or sandy brown with pale buffish edgings to wing-feathers; buffish grey crown and darker ear-coverts (but paler than in Lesser or Hume's); underparts clean sandy white with white throat. More active than other whitethroats, frequently flicking tail. *Call characteristic* (see voice).

Voice: characteristic churring note 'che-che-che-che-che' (or variations thereof) recalling Tree Sparrow; some fast and buzzing, others slower.

568 **Whitethroat** *Sylvia communis* **Plate 88**

(Common Whitethroat)

L: 14. A medium-sized warbler, similar to Lesser Whitethroat but differs in having *chestnut fringes to coverts and secondaries*, slightly longer tail, *orangey coloured legs* and *white eye-ring*. Females and 1st winter birds lack the grey wash to head of male and have dark (not orange) iris. The main race that occurs in the region, *icterops*, has brownish (barely chestnut) fringes to the coverts and secondaries, thus more resembling Lesser Whitethroat, but can always be told by orange legs.

Voice: song, fairly short, scratchy jerky warbling outburst; often in display flight. Call a harsh 'whet-whet-whet'; alarm a drawn-out 'chaoe'.

569 **Garden Warbler** *Sylvia borin* **Plate 87**

L: 14. A uniform grey-brown warbler with square-ended tail, *short, relatively heavy bill*, practically no supercilium, often greyish sides of neck, *rounded head and no white in tail or wings*. Most *Acrocephalus* and *Hippolais* warblers have *thinner, longer bills*, former also have longer undertail-coverts and rounded tails; see young Barred for separation. Keeps well hidden in vegetation.

Voice: song a melodious, fast and sustained babbling with considerable jumps in pitch in irregular sequence, lacking Blackcap's ascending fluty finish; no song-flight. Alarm note a nasal, repeated 'chet-chet-chet', softer than Blackcap's.

570 **Blackcap** *Sylvia atricapilla* **Plate 88**

L: 14. Similar in size to Whitethroat, generally grey-brown but with distinctive *black cap* in male and *warm brown cap* in female and 1st winter; these features prevent confusion with any other warbler.

Voice: song similar to that of Garden Warbler: short, rich, varied warble ending with few clear melodious fluted notes. Call a hard 'tack tack' very similar to Lesser Whitethroat; and also short 'churr churr'.

571 **Brown Woodland Warbler** *Phylloscopus umbrovirens* **PLATE 89**

L: 11. Slightly smaller than Chiffchaff; upperparts brown (or olive-grey-brown) with ***yellowish green fringes to flight- and tail-feathers***; underparts buff-grey to off-white, often rather greyish on sides of throat and breast with white throat and ***noticeable tawny buff wash on flanks and ventral region***; supercilium buff, rather indistinct but more visible behind eye; bill horn, lower mandible orange-yellow at base; legs dull blue-grey. Very active, often in pairs.
Voice: song loud, explosive, including trilling series of tit-type notes of variable speed and pitch, often with rising inflection on last note: 'titititititi-wit', slower 'vuit, vuit, vuit, wit-wit-wit-wit tu-vi, tu-vi, tuvi-twit'; may end with Willow Warbler-like descending 'diu-diu-diu-diu', but as a fast trill. Call, descending metallic long 'dzziieep'; also low 'psew' and 'swee-vik' alarm notes.

572 **Green Warbler** *Phylloscopus nitidus* **PLATE 89**
(Bright Green Warbler)

L: 11. Size, structure and voice similar to Greenish (which see) with round head (like Chiffchaff, unlike Willow), ***broad supercilium*** (faint and short in Chiffchaff) ***from base of pale bill to behind ear-coverts and short narrow wing-bar on greater coverts*** (Chiffchaffs of eastern origin may show inconspicuous but rather long wing-bar). From Greenish and most other *Phylloscopus* by greener upperparts, and ***sulphur-yellow throat, upper breast, supercilium and ear-coverts*** contrasting with rather dark eye-stripe. In worn plumage may be more greyish, almost lacking yellow tone, and wing-bar may be lost; then very difficult to tell from Greenish, but from most other *Phylloscopus* by striking head pattern, pale bill and voice.
Voice: song very similar to Greenish: loud, variable and high-pitched with staccato character, starting with 1–2 call-notes followed by short, rapid phrases (2–3 s), ending abruptly 'che-wee che-wee chui chi-di chi-dit'; has explosive character of Cetti's Warbler and structure sometimes reminiscent of Wren; in Turkey slower and less staccato/more liquid than Greenish. Call a White Wagtail-like 'che-wee', similar to Greenish. Usually stationary when singing (Greenish changes perch frequently).

573 **Greenish Warbler** *Phylloscopus trochiloides* **PLATE 89**

L: 11. ***Like a greyish, colder version of Green*** (which see) ***or Willow***, with yellowish mostly restricted to ***long, broad supercilium*** (may create 'capped' appearance and 'pale kind face'). Also note ***dull greyish white underparts and ear-coverts***, contrasting with distinct dark eye and eye-stripe; plain tertials, ***short wing-bar*** (sometimes two in fresh plumage, or may be worn off in summer), pale bill and dark brownish legs. Separated from Willow and Chiffchaff by greyish underparts and cheeks, stronger head markings, short pale wing-bar (if present), yellow lower mandible (especially from Chiffchaff) and voice.
Voice: song variable in length and much like Green but usually faster, more staccato with short gap between first note and following part of song 'chit, chi-dit chi-dit chi-di-di chi-di-di chi-chi chi-di' ending abruptly (occasionally has longer, more trilling song very similar

to Wren's). Call similar to Green (resembling White Wagtail) 'chi-wee' or shorter 'chi it' (like Yellow Wagtail).

574 **Arctic Warbler** *Phylloscopus borealis* **PLATE 89**

L: 12. ***Dark greyish green Phylloscopus*** with proportionally ***large head and eyes*** and strong pale bill; underparts grey, tinged olive on breast-sides and flanks; striking head markings with ***narrow but very long and distinct supercilium from just in front of eye*** (in Greenish meeting over base of bill) ***to far behind dark-mottled ear-coverts*** (sometimes with upward kink at end), accentuated by ***distinct dark eye-stripe from base of bill***; has plain tertials, ***short wing-bar*** (sometimes two visible in fresh plumage) and pale brown legs. Mostly confusable with smaller Greenish and Green in worn, greyer plumage; distinguished from both by size, heavy structure, prominent head markings, mottled ear-coverts, pale legs and ***distinctive call***. From Willow and Chiffchaff by head and wing pattern. Not shy but often keeps hidden high in trees.
Voice: call a characteristic short 'dzik' (or 'dzi-zik') more ***recalling sound of a bunting*** than a warbler.

575 **Pallas's Warbler** *Phylloscopus proregulus* **PLATE 89**
(Pallas's Leaf Warbler)

L: 9. ***Tiny, compact warbler*** with relatively large head, contrastingly striped yellow; size and behaviour recall Goldcrest. ***Bold yellow supercilium and central crown-stripe*** (almost meeting at nape) ***contrast strongly with dark grey-green crown and eye-stripe; two pale yellow wing-bars and white-edged tertials are distinctive.*** When ***hovering*** or fluttering after insects in foliage, ***diagnostic pale yellow rump-patch*** visible. Confusable with Hume's Yellow-browed and Yellow-browed (or even Goldcrest) but note rump-patch, more contrasting wing pattern and distinct crown-stripe (Yellow-browed occasionally shows very faint crown-stripe).
Voice: call a high 'tweep' or slightly disyllabic 'twooeep' (more Chiffchaff-like).

576 **Yellow-browed Warbler** *Phylloscopus inornatus* **PLATE 89**
(Inornate Warbler)

L: 10. Very similar to Hume's Yellow-browed (in winter/spring worn plumages may be inseparable except for voice) but generally brighter ***with greener upperparts, whiter underparts, lacking buff suffusion; supercilium very long, pronounced and yellow-white***, often rather evenly broad (tinged buff and often distinctly broad over and in front of eye in Hume's), occasionally shows faint paler crown-stripe; ***usually 2 distinct yellow-white wing-bars*** contrasting with dark-centred greater-coverts (slightly paler coverts with less contrast in Hume's), also stronger contrast in tertial pattern; ***base of bill obviously paler*** (appears all-dark in Hume's). In worn plumage greyer with paler wing-bars, and tertial-fringes narrower and whiter.
Voice: song (which might be heard on migration) high-pitched, thin 'tsee tseoo-tseee', final note may be prolonged, but lacks buzzing quality of Hume's; call hard, penetrating and high-pitched 'tsweest' or softer 'weest' with a sibilant character and with distinct upwards inflection, frequently almost disyllabic 'wii-ist'.

577 **Hume's Yellow-browed Warbler** *Phylloscopus humei* **Plate 89**

(Hume's Leaf Warbler)

L: 10. Most confusable with (or even inseparable from) Yellow-browed but ***generally greyer, duller with less yellow/green appearance***, and different call/song. In fresh plumage, upperparts and crown greyish olive with very long, pronounced, ***buffish white supercilium*** (often broadest over and in front of eye), 2 pale wing-bars, ***upper one short and ill-defined (sometimes invisible)***, lower bar distinct (2 striking yellowish wing-bars in Yellow-browed), pale edges to dark-centred tertials (darker tertials and greater coverts with stronger contrast in Yellow-browed); ***underparts dusky white, tinged buffish on ear-coverts, lower throat and sides of neck; bill dark*** (extensive pale base in Yellow-browed). In winter/spring worn plumage becomes greyer, wing-bars and tertial edges narrower, and upper wing-bar often invisible or lacking; in this stage also confusable with Greenish but note smaller size, tertial and bill pattern, and longer, broader wing-bar.

Voice: song thin, high-pitched, drawn-out, descending 'tzeeeee' with ***buzzing quality***, recalling distant Greenfinch, often preceded by a few 'wesoo' call notes. Call a sweet 'wesoo', or shorter 'dweed', in winter quarters sometimes more disyllabic 'bii-it'; usually softer and less sibilant than Yellow-browed and with slight (or no) upwards inflection.

578 **Radde's Warbler** *Phylloscopus schwarzi* **Plate 90**

L: 13. Robust-looking with large head, ***stout, deep-based rather pale bill and strong, straw-coloured legs***. Upperparts olivaceous brown with ***striking, wide, whitish supercilium*** (though ***buffish and more diffuse in front of eye***) accentuated by broad, dark eye-stripe, dark-mottled ear-coverts and dark margin above. Breast sides, flanks and ***undertail-coverts yellowish buff*** with centre of underparts dirty white (in worn plumage underparts become off-white with yellowish buff undertail-coverts). 1st winter often shows sulphur-yellow underparts (rarely or faint in some Dusky). From other warblers (including Chiffchaff, Mountain Chiffchaff and Dusky) by colour and structure of bill and legs; also note striking head markings, call and behaviour. Very skulking in low vegetation but frequently calls.

Voice: call tongue-clicking in character like Dusky (which see) but weaker, softer and lower-pitched 'tuk' or as double-call a more diagnostic 'tuk-tch'.

579 **Dusky Warbler** *Phylloscopus fuscatus* **Plate 90**

L: 11. Dark, rather ***brown***, small *Phylloscopus* with distinct head pattern resembling brownish eastern Chiffchaffs and Mountain Chiffchaff but note ***soft part colours and very different call. Upperparts plain brown with rather distinctive, long, pale rufous-brown or whitish supercilium***, narrow and often palest in front of eye (reverse in Radde's). Underparts grey-white ***tinged pale rufous-buff on breast, flanks and undertail-coverts. Bill spiky*** with largely ***yellowish lower mandible; legs thin, flesh-coloured.*** From *tristis* Chiffchaff and Mountain Chiffchaff by paler legs, paler bill (blackish in *tristis*), often rufous-tinged vent and undertail-coverts (not pale grey/off-white as in Chiffchaff or Mountain Chiffchaff), also long striking supercilium (dull and shorter in *tristis*, but rather distinctive in Mountain Chiffchaff) ***and call.*** From Radde's (which see) largely by smaller size, ***smaller head*** with ***finer bill, darker and finer legs*** and details

in head markings. Skulking, like Radde's, but less tied to low vegetation. Often calls.
Voice: short, hard 'chek' or 'tack', fairly loud, much recalling Lesser Whitethroat, slightly harder and higher-pitched than Radde's.

580 **Bonelli's Warbler** *Phylloscopus bonelli* **PLATE 89**

L: 12. Behaviour, size and attenuated structure similar to Willow but more round-headed. ***Distinctly grey head and mantle***, contrasting dark tertials (fringed yellowish) and ***obviously green edges to wing and tail feathers***; underparts silky white; ***face seems washed-out with large dark eye*** accentuated by narrow white eye-ring, usually ***pale lores, pale ear-coverts and indistinct supercilium***; if seen well, ***yellowish green rump*** distinctive. May be confused with pale Chiffchaff but note pale, plain head, pale bill and longer primary projection in Bonelli's. From Booted Warbler by face pattern and well marked tertials.
Voice: song a short trill suggesting Greenfinch-trill or rattle of Lesser Whitethroat; also similar to Wood Warbler but lacking its acceleration: 'tlilililililih'. Call a short Crossbill-like 'chip'.

581 **Wood Warbler** *Phylloscopus sibilatrix* **PLATE 89**

L: 13. Large, ***long-winged***, short-tailed *Phylloscopus*, brightly coloured green, yellow and pure white, showing ***vivid green upperparts, long broad yellow supercilium, ear-coverts, throat and upper breast, contrasting with gleaming white underparts; yellow-fringed secondaries form pale panel contrasting with blackish, whitish-fringed tertials***; also noticeable black alula, yellow bill and pale legs. From Green Warbler by different wing pattern; from other *Phylloscopus* also by very long wings, striking head and underparts pattern. Rather active high in trees.
Voice: song accelerating series of 'zip' notes ending in metallic trill ***like coin spinning on marble table:*** 'zip zip zip-zip-zip-zwirrrrrrrr', often combined at start with clear melancholy 'deeu-deeu-deeu—' call-notes. Fairly silent during migration.

582 **Plain Leaf Warbler** *Phylloscopus neglectus* **PLATE 90**

L: 9. The smallest *Phylloscopus* with ***Goldcrest-like size and proportions, lacking any obvious markings***. Coloration like Chiffchaff of eastern, greyer forms with upperparts pale brownish-grey, tinged olive; indistinct, short, pale buff to cream supercilium, narrow white eye-ring, dusky lores and eye-stripe and buff-flecked ear-coverts; underparts off-white, washed creamy on flanks. Lacks any distinct trace of green or yellow. Legs and bill dark brown. Very active, constantly flicking wings, or seen in whirring and jerky flight like Goldcrest as it moves about sparse hillsides; often foraging in cover.
Voice: song, which is often heard in winter, is a short phrase of 4–5 syllables recalling Goldfinch 'pt toodla toodla' from bush or flight between perches. Call, quiet, harsh 'chick'; in winter also harsh 'churr' recalling House Sparrow, useful for identification.

583 **Mountain Chiffchaff** *Phylloscopus sindianus* PLATE 90

The taxonomic position of Mountain Chiffchaff and especially the race that breeds in our region, *lorenzii* (often called Caucasian Chiffchaff) is unclear; it is regarded by some researchers to be a subspecies of Chiffchaff.

L: 11. Resembles brown and rusty Chiffchaff ***without yellow or green in plumage*** and with fairly ***striking head markings*** at close range. ***Upperparts warm brown***, crown slightly darker contrasting with ***broad white supercilium*** (buff behind eye), white lower eyelid, ***black lores*** and buff ear-coverts; ***underparts (including undertail-coverts) off-white*** distinctly tinged ***rusty on breast-sides and flanks***. Like *brevirostris* form of Chiffchaff seems rather short-winged/long-tailed; most of lower mandible yellow; ***legs black***. Behaviour (including tail-dipping) as Chiffchaff. Distinguished from all races of Chiffchaff by rusty brown tone to plumage (no yellow or green), prominent supercilium and black lores; from Dusky Warbler by black legs, white undertail-coverts, voice and less skulking behaviour.

Voice: song similar to Chiffchaff's but usually faster. Call 'biii', like *brevirostris* form of Chiffchaff (which see).

584 **Chiffchaff** *Phylloscopus collybita* PLATE 90

L: 11. Medium sized, rather ***uniform, dark olive-greenish-grey*** *Phylloscopus* with round head and pot-bellied appearance. Yellow in plumage restricted to bend of wing and slight suffusion on breast and sometimes undertail-coverts; ***supercilium short***, usually indistinct, off-white, tinged yellow or buff; ***a broken eye-ring often conspicuous, especially in autumn***; ear-coverts dark, uniform greenish grey tinged buff, merging into paler underparts with buff suffusion on breast-sides and flanks; ***blackish legs and dark-looking bill*** (faint orange at base of lower mandible). ***Frequently dips tail when feeding***. W and N Turkish form, ***brevirostris, differs from north and eastern European race, abietinus, in slightly browner tinge to upperparts*** (especially crown) ***and breast sides***, with more distinct supercilium (especially over and in front of eye), ***paler (off-white) undertail-coverts***, shorter primary projection (impression of longer tail) and ***different call (similar to Mountain Chiffchaff)***. Siberian race, ***tristis, lacks any yellow in plumage***, having ***brownish grey upperparts***, except for ***olive-green tinge on rump and edges of flight- and tail-feathers; supercilium may be rather distinct (off-white to buff) but thin***; underparts paler off-white (***including undertail-coverts***) suffused buff on breast sides; ***legs and bill almost black***. In fresh plumage ***many tristis show indistinct wing-bar*** on greater coverts and confusion can arise with Greenish, Green or Arctic (which see); told from all three by lack of yellow tone, narrow supercilium, blackish bill and legs, and voice.

Voice: song distinctive: a regularly repetition of its name 'chiff-chaff-chiff-chaff—', song of *tristis* quite different being more varied and irregular with faster, more melodious notes. Call 'h(u)iit', with rising inflection, resembles Willow Warbler's but noticeably shorter and monosyllabic; *brevirostris* and Iranian birds call 'biii', similar to Mountain Chiffchaff with falling, plaintive, Siskin-like pitch (resembles Bullfinch's call but higher pitched). Call of *tristis* is a sharper, higher pitched, piping 'pseep', almost lacking falling structure of *brevirostris*.

585 **Willow Warbler** *Phylloscopus trochilus* **Plate 90**

L: 11. Medium-sized, attenuated *Phylloscopus* with ***longer primary projection than Chiffchaff***; rather ***yellow in plumage*** with olive upperparts, darker, ***pale-edged flight-feathers and tertials and yellow-white underparts and supercilium; legs and bill variable, usually pale brown or orange***. Some northern populations greyer, almost lacking yellow in plumage, while those from E Asia are darker (see *yakutensis*, illustrated). ***Juvenile/1st winter shows distinctly yellow underparts*** (white belly in adult). Distinguished from smaller, shorter winged, pot-bellied Chiffchaff by brighter, more yellow and green appearance with 'sharper' facial expression (longer, broader supercilium), slightly blotched cheeks (uniform in Chiffchaff) contrasting with yellow throat and breast (buff wash on breast sides and flanks in Chiffchaff), pale legs and bill (dark in Chiffchaff) and ***tail-dipping much less pronounced***.
Voice: ***song diagnostic***: soft, clear, pleasant descending verse dying away at end. Call soft, rising, almost disyllabic 'hoo-eet' (shorter in Chiffchaff).

586 **Goldcrest** *Regulus regulus* **Plate 90**

L: 9. Smaller and more compact than any warbler, with short tail and fine bill; ***pied wing includes two wing-bars***. Head shows ***distinct yellow central crown-stripe*** (orange-red in male but often concealed by yellow feathers), ***dark eye set in pale face (but lacks supercilium and eye-stripe***, distinctive in Firecrest). Juvenile (but not 1st winter) lacks yellow crown pattern. Fearless, and very active with fluttering movements, often foraging high in foliage in company with tits, constantly calling; hovers frequently. Separated from Firecrest (which see) by duller appearance, lacking striking head markings.
Voice: song very high-pitched, repeated, rising and falling theme 'seeh sissisyusee sissisyu-see siss-seeitueet'. Call thin, very high, Coal Tit-like 'see-see-see' or 'sit-sit'.

587 **Firecrest** *Regulus ignicapillus* **Plate 90**

L: 9. Size, structure and behaviour similar to Goldcrest (which see), but brighter with ***paler green upperparts*** tinged bronzey on shoulders, and ***frosty white below***. Very striking head pattern with broad, ***white supercilium contrasting with distinctive black eye-stripe***. As in Goldcrest shows yellow (female) or golden-orange (male) ***central crown-stripe***.
Voice: song very high-pitched series of monotones, accelerating and rising slightly towards end 'sisisisisisisit'. Many calls similar to Goldcrest, but more sibiliant in some: 'iizt'; also a tit-like, piping 'peep'.

588 **Blue-and-white Flycatcher** *Cyanoptila cyanomelana* **Plate 91**

L: 17. Larger, longer-tailed and heavier billed than Spotted. ***Male's dark blue upperparts, shining blue crown, bluish black sides of head, throat and breast, sharply demarcated from white belly***, diagnostic; ***white sides to tail-base*** usually visible in flight only. ***Dark olive-brown female is paler below with distinct creamy throat-patch***, white belly and undertail-coverts; indistinct pale eye-ring; 1st winter male has bright blue rump and tail, latter tipped black, but white sides at base; wings bright blue with black-tipped primaries. Upright stance with frequent slow tail movements and wing-flicks; swoops to ground to feed, then returns to perch.
Voice: grating 'tchach' or 'tek-tek'.

589 **Spotted Flycatcher** *Muscicapa striata* **Plate 91**

L: 14. A brownish grey, long-winged flycatcher ***with streaked forehead and breast, thin pale edges to greater coverts and tertials and blackish bill.*** Makes short wavering flights for insects from exposed branch; perches upright, often flicking wings. See Gambaga Flycatcher for differences.
Voice: call a sharp 'tzeet'; alarm 'isst-tek'. Simple song consists of 3–4 call-note-like notes 'sip-sip-see-sitti'.

590 **Gambaga Flycatcher** *Muscicapa gambagae* **Plate 91**

(Gambage Dusky Flycatcher)

L: 12. Resembles smaller, shorter-winged version of Spotted but has less upright stance, smaller bill ***with most of lower mandible yellowish*** (dark in Spotted), ***sooty brown upperparts virtually lacking streaks on forehead***; silky white underparts ***have soft vermiculations on breast*** (more streaked on Spotted); pale wing-feather edges like Spotted. Juvenile has buff spots on upperparts while underparts are more strongly streaked than adult. Sits on exposed perch, making sallies, or forages among upper-foliage of small trees.
Voice: soft creaking 'tchee-tchee'; Robin-like 'ptik-ptik'. Song thin, sibilant high-pitched squeaking mixed with rusty creaking notes and short trills.

591 **Red-breasted Flycatcher** *Ficedula parva* **Plate 91**

L: 12. Small flycatcher ***with conspicuous white patches at sides of tail-base***; adult male has ***reddish orange throat and sometimes upper breast*** and lead-grey sides of head and neck; female and 2nd year male have buffish white throat; plumage otherwise grey-brown above and whitish below. Moves around foliage like Willow Warbler; cocks tail.
Voice: call a dry, rolling 'terrrr' recalling Wren but weaker; also a thin 'tsri', a 'tek' or, in alarm '*tee*-lu'. Song slow and rhythmic 'tle-tle-tle, tlee-lu, tlee-lu, tlee-lu, tiu-tiu-tiu-vit-vit', descending towards end.

592 **Semi-collared Flycatcher** *Ficedula semitorquata* **Plate 91**

L: 13. Resembles cross between Collared and Pied. Male told from Pied ***by white half-collar onto sides of neck*** (others resemble Pied), ***larger white spot at base of primaries*** (absent, or just narrow streak in Pied), ***more white at sides of tail and white-tipped median coverts*** (rarely so in Pied); extent of white in rest of wing and on forehead intermediate between Collared and Pied. Female resembles Collared, thus greyer above than Pied; ***often*** (but not always) ***more white at primary-bases and at tips of median coverts***; latter character not reliable for 1st autumn birds.
Voice: alarm call resembles Collared's 'eeet'; song closer to Pied's but higher pitched and simpler.

593 **Collared Flycatcher** *Ficedula albicollis* **PLATE 91**

L: 13. Male told from Pied and Semi-collared ***by white collar around hindneck***; has more white in wings than Pied, ***including large white spot at base of primaries*** (at best thin white streak in Pied), larger white spot on forehead and, often, paler rump. Female told from similar Pied by ***usually greyer, less brownish upperparts***, usually pale grey hindneck, ***often broad white spot at primary-base, and by call-note***; probably inseparable from female Semi-collared, unless hindneck is paler greyish.
Voice: call on passage a thin piping 'eeet', very different from Pied's.

594 **Pied Flycatcher** *Ficedula hypoleuca* **PLATE 91**

L: 13. Male's black and white plumage ***with black hindneck***, small white forehead-spot (often divided in centre) and ***narrow white streak at primary-base*** (sometimes absent) identify it; some males are dark brown above but still with white spot on forehead. For separation of female Pied, see Collared and Semi-collared. Often flicks wings like other black and white flycatchers.
Voice: call on passage a short metallic 'twink', very different from Collared's thin call.

595 **African Paradise Flycatcher** *Terpsiphone viridis* **PLATE 92**

L: male 30–36 (includes long tail), female 20. Unmistakable, bulbul-sized ***with chestnut upperparts, glossy blue-black head and breast*** with rest of underparts slaty; ***often large white wing-patches. Male has very long central tail-feathers***, absent in less richly coloured female. A white phase, rare in Arabia, has white back, rump, tail and wing-coverts. Flight, when flycatching, slow, heavy and wavering; secretive.
Voice: harsh 'scheep'; hoarse 'tseaeae-tsceaeat'. Song has Blackbird-like quality, 'twe, twoo, twoo, twoo, twoo' uttered with fanned tail while jerking from side to side.

596 **Bearded Tit** *Panurus biarmicus* **PLATE 92**
(Bearded Reedling)

L: 17. ***Unmistakable, cinnamon-brown with very long tail***, short wings and ***weak flight with very rapid wing-beats***, usually low over reedbeds. ***Grey-headed male has black moustache and undertail-coverts***, absent in brown-headed female and juvenile. Outside breeding season usually in flocks.
Voice: sharp, hollow, sonorous 'ptew', 'tlek' or 'tching' are commonest calls. Has twittering song.

597 **Iraq Babbler** *Turdoides altirostris* **PLATE 92**

L: 22. The long graduated tail, short wings, stout curved bill typical of babbler; often in groups, sometimes skulking, sometimes in the open, often hopping on ground, tail raised, or flying 'follow my leader' from cover to cover in low laboured flight, interspersed with glides. Told from Common Babbler (which overlaps its range) by warmer colour, ***absence of streaks on sides of rufous-buff breast, and finer streaks above*** (streaks more distinct in Common); legs dull brown (dull yellow in Common). Juvenile apparently paler above than similar Common and with creamy buff cheeks and underparts; chin and vent white.
Voice: wide variety of whistling calls; commonest call a descending 'pherrrrrreree'.

598 **Common Babbler** *Turdoides caudatus* **Plate 92**

L: 23. Resembles Iraq Babbler closely ***but has greyer, streaked sides of breast, more conspicuously streaked crown and upperparts, dull yellow legs*** (dull brownish in Iraq Babbler), and longer bill. Behaviour much like Iraq Babbler.
Voice: a high-pitched 'pee-pee-pee-pee-pee-pee-pee'; also series of piping notes slowing and descending at end.

599 **Arabian Babbler** *Turdoides squamiceps* **Plate 92**

(Brown Babbler)

L: 26. Larger than previous species and geographically separated. Grey-brown, lightly streaked, often with head appearing 'moth-eaten'; faint dark mottling on throat and breast; bill blackish with paler base, legs brownish to dark grey. Birds of Yemen lowlands and some highland areas have variable off-white face with whitish eye-surround; in these bill varies from orange–red to yellowish orange. Habits like other babblers.
Voice: most typical call is a high piping 'piu-piu-piu-piu-piu', decelerating towards end.

600 **Long-tailed Tit** *Aegithalos caudatus* **Plate 92**

L: 13. ***Long-tailed, whitish and grey*** with small bill and characteristic call. Birds of Turkey (race *tephronotus*) have grey back (lacking pink of European birds), black throat, and buffy white underparts; birds of Iran (race *passekii*; except Caspian region) are paler above and below and lack dark 'bib' (as do European birds); both races show dark stripe above eye to nape. Outside breeding season usually in small flocks, constantly calling.
Voice: call a soft purring 'tserrrr' and a high-pitched metallic trisyllabic 'sreeh-sreeh-sreeh'. Song a thin metallic trill 'sree-vivivivivi', not often heard.

601 **Marsh Tit** *Parus palustris* **Plate 93**

L: 11.5. Small, greyish tit with prominent ***glossy black cap and small black bib***; upperparts and tail greyish brown; in fresh plumage shows slightly paler fringes to secondaries forming pale wing-panel similar to Sombre; underparts pale grey suffused buffish on flanks. Confusion can only arise with Sombre Tit (which see) but note larger size, much larger bib and different structure of latter; also different habitat and range.
Voice: commonest call a diagnostic, sudden 'pitchu', also a distinctive whistle 'chiu-chiu-chiu'. Song monotonous series: 'chip-chip-chip-chip ...' or 'hee-chi, hee-chi, ...' rather distinctive with experience.

602 **Sombre Tit** *Parus lugubris* **Plate 93**

L: 14. ***Large 'grey' tit*** (size of Great) with black cap and ***much more black on chin and throat*** than Marsh; compared with latter has stronger bill, proportionally longer tail, paler underparts and flanks and often more obvious pale wing-panel. Juvenile duller, less clean, more brownish overall. Behaviour resembles Great Tit.
Voice: many calls like Great, Blue or even Long-tailed Tit, usually with a grating or coarse character; most distinctive is a deep sparrow-like chattering 'chaerrrr' or 'churr-er-rrr-rr'; song variable with buzzing quality; recalling Great, Blue or Coal Tit.

603 **Coal Tit** *Parus ater* **PLATE 93**

L: 11. ***Smallest tit*** in the region; rather dark and distinctly short-tailed with behaviour and some calls reminiscence of Goldcrest. ***Black head, white cheeks and diagnostic white nape-patch; black on chin and throat reaches upper breast***; slate-grey upperparts show *2 **white wing-bars***, underparts buff. When agitated can raise small crest. Very restless behaviour. Cyprus race, *cypriotes*, has brown back, black on head extending onto upper mantle, and underparts very dark with rufous-brown flanks.
Voice: song a sweet, piped repetition of 2 or 3 notes with notably alternated stress 'tee-chu, tee-chu, ... or 'chu-wee, chu-wee, ...' mostly recalling Great Tit but more feeble; calls include fluty, rather melancholy 'püht', fine 'pseet' and Goldcrest-like 'si-si-si'.

604 **Blue Tit** *Parus caeruleus* **PLATE 93**

L: 11.5. Small, very active tit, easily distinguished by ***blue cap, dark blue eye-stripe, blue wings and yellow underparts***, often with faint but variable darker belly-stripe. Juvenile has yellowish on head and dull bluish crown. S Iran race, *persicus*, much paler overall, somewhat recalling Azure Tit (which see).
Voice: song a clear, continuous ringing starting with 2 drawn-out notes followed by silvery clear trill 'zeeh-zeeh-sirrrrrrr'; calls very variable, often soft and sibilant 'tsee-tsee ...' or in excitement chirrr-rr-rr'.

605 **Azure Tit** *Parus cyanus* **PLATE 93**

L: 13. The race occurring in the region, *carruthersi*, mostly resembles smaller Blue Tit with pale head, dark eye-stripe, yellow breast and blue wings, but has proportionally ***longer tail broadly edged and tipped white, broad white wing-patch and white tips to secondaries***; upperparts grey, and ***head dusky white lacking blue***; often has faint belly-stripe like Blue. Juvenile tinged yellow throughout plumage. Behaviour like Blue Tit, with which it hybridizes where breeding ranges overlap.
Voice: song and calls very variable, reminiscent of Blue, Coal, Marsh and Great Tit.

606 **Great Tit** *Parus major* **PLATE 93**

L: 14. Large tit with two forms in the region. Most birds unmistakable with ***black head, large white cheek-patch and broad, black stripe down centre of yellow underparts***; upperparts green with bluish wings and tail. In flight, shows narrow white wing-bar and tail-edges. In juvenile black areas are sooty, cheek-patch yellowish lacking dark lower border, and underparts dull yellow. From Blue by larger size, head and underparts pattern. Grey race, *intermedius*, occurring in E Iran has similar black areas but upperparts blue-grey with only ***a hint of green on upper mantle, underparts whitish***; overlaps in range with very similar Turkestan Tit (which see).
Voice: highly variable, includes loud, metallic and clear calls like 'chink-chink' (recalling Chaffinch), or 'tui-tui-tui', other calls similar to Blue and Marsh Tit; song distinctive, far-carrying and rather melodious, 'tee-chu-tee-chu ...' or with three-note sequences 'tee-tee-chu, tee-tee-chu ...'.

607 **Turkestan Tit** *Parus bokharensis* **PLATE 93**

L: 15. Very similar to grey *intermedius* race of Great which occurs in same area. Differs in proportionally ***longer tail, stouter bill, narrower black line down centre of underparts***, more obvious blue-grey fringes to flight feathers and ***no trace of green in upperparts*** (present when seen close in *intermedius* Great Tit).
Voice: no information.

608 **Krüper's Nuthatch** *Sitta krueperi* **PLATE 94**

L: 12.5. ***Smallest nuthatch in the region*** and always confined to coniferous forests; most easily located by voice. From other nuthatches by distinctive head and breast markings. ***Black fore-crown*** in male (duller, less defined in female) sharply ***bordered white below with long black eye-stripe*** and ***rusty brown breast-patch*** (paler in female); otherwise upper- and underparts like other nuthatches in the region. Juvenile duller, lacking black crown and eye-stripe and with only faint breast-patch.
Voice: less noisy than other nuthatches in the region, but rather variable. Hissing 'wäääk' or Whitethroat-like 'dwäää' and a clear whistle 'düüii' are commonly heard; also a trilling 'drrll' and a Nuthatch-like 'veete-veete-veete…'.

609 **Nuthatch** *Sitta europaea* **PLATE 94**
(Eurasian Nuthatch)

L: 14. Size of Western Rock Nuthatch. Bluish grey upperparts; underparts variable according to race but typically buffish brown or pale buff with paler throat and rusty brown flanks, vent and undertail-coverts, ***the latter spotted white***; prominent, rather narrow, but long black eye-stripe; tail feathers black (except for middle pair) ***with white corners visible in flight. Best distinguished from Western Rock by habitat***; Western Rock also has paler and greyer appearance, lacks distinct tail-pattern, and does not show white spotting on undertail-coverts. Race, *persica*, of SW and S Iran, has distinctly paler grey upperparts, white underparts, and pronounced white forehead-band and supercilium.
Voice: song a penetrating clear series of 'dwit-dwit-dwit …' or faster 'wiwiwiwiwi …'. When foraging gives short 'whitt'; in flight short 'sit'.

610 **Eastern Rock Nuthatch** *Sitta tephronota* **PLATE 94**
(Great Rock Nuthatch)

L: 19. Similar in plumage to Western Rock Nuthatch (which see) but noticeably ***larger***, accentuated by rather ***long bill and longer, thicker neck; black eye-stripe very long, pronounced and more obviously broad, especially towards nape***. Behaviour as Western Rock but more frequently on trees or bushes.
Voice: very noisy, largely similar to Western Rock in variation and general structure but markedly lower pitched, slower and less piping.

611 **Western Rock Nuthatch** *Sitta neumayer* **PLATE 94**
(Rock Nuthatch)

L: 14. Size and general outline similar to Nuthatch (which see) but usually paler over all; ***best distinguished by habitat. Very vocal***, often detected at long range by voice. Compared with most Nuthatches has paler and greyer upperparts, ***uniform grey tail***

(black and white in Nuthatch), dirty-white underparts with less defined rusty/buffy undertail-coverts ***without white spots***; never shows paler supercilium or forehead (present but faint in some populations of Nuthatches). Iranian races, *tschitscherini* and *plumbea*, are paler above and below than western birds with black eye-stripe reduced and in some, very diffuse. For separation from larger and longer-billed Eastern Rock, see that species.
Voice: song loud, penetrating, accelerating trill reminiscent of Alpine Swift with rolling or bubbling character, sometimes starting with 'bi-di, bi-di' going into song; other calls include 'wit-wit-wit', a Crossbill-like 'kip', a harsh shrike-like 'wäääk and short flight- and foraging-sounds like Nuthatch's 'sit-sit'.

612 **Wallcreeper** *Tichodroma muraria* PLATE 94

L: 17. Combination of appearance, behaviour and habitat unmistakable. Generally grey (male with black throat in summer) with short dark tail and ***long, thin curved bill***; when creeping on rocks and cliffs ***often flicks wings showing flashes of red.*** In flight, shows ***short tail, rounded black and red wings with white primary-spots***, recalling small Hoopoe with butterfly-like wing-beats.
Voice: song a clear, fluty, rising whistle, rather loud and melodious.

613 **Treecreeper** *Certhia familiaris* PLATE 95
(Eurasian Treecreeper)

L: 13. Small, brownish with long decurved bill and tree-creeping manners, ***confusable with Short-toed from which it is best separated by voice.*** The supercilium is often whiter above lores, the flanks are whiter, the upperparts are warmer brown and the bill is slightly shorter, less decurved than Short-toed.
Voice: song is a series of descending sharp and thin notes recalling Blue Tit's or Willow Warbler's, the last note with upward inflection. Call a thin, drawn-out 'teeh' and thin, slightly rolling 'sreeee'.

614 **Short-toed Treecreeper** *Certhia brachydactyla* PLATE 95

L: 13. Resembles Treecreeper closely ***and best told by voice.*** Short-toed has slightly greyer upperpart (browner in Treecreeper), pale brownish tone on flanks (whiter in Treecreeper), less obvious white supercilium above lores and slightly longer bill. Besides tree creeping it has also the habit of perching across a thin branch in tree-top (much more so than Treecreeper).
Voice: the song is shorter than Treecreeper's, with irregular rhythm, a curl in the middle, and the last note rising in pitch, 'tu-ti-teroi-e-tit'; once learned easily recognized. Call is a loud Coal Tit-like 'teeut' or 'tuu' like the first note of the song, uttered in series of variable length and pitch.

615 **Penduline Tit** *Remiz pendulinus* PLATE 94

L: 11. Slightly smaller than Blue Tit with longer tail and ***finely pointed bill; chestnut mantle and wing-coverts***, blackish wings and tail-feathers fringed white; underparts ochre- to buff-white. Distinctive head pattern depending on race. In Turkey adult has ***black mask contrasting with pale grey crown and nape and white throat*** (female duller with smaller mask). Juvenile shows cinnamon upperparts and lacks black mask.

Iranian races differ: adult *macronyx* (south Caspian) and *nigricans* (Seistan area) have all-black head extending onto throat with white collar in *macroynx* (lacking in *nigricans*); *coronatus* (NE Iran) shows variable black coronal band joining across nape.
Voice: more often heard than seen. Commonest call a soft, rather quiet, ***piping, drawn-out and falling*** 'seeeeee'. Song short musical rippling 'ti-ti-ti-ti-ti'.

616 **Nile Valley Sunbird** *Anthreptes metallicus* **PLATE 95**

L: 10 (male in summer 15). ***Small sunbird with short, slightly decurved bill, which separates it from all other sunbirds in the Middle East.*** The brightly coloured male ***has warm yellow underparts, demarcated from dark purple and green upper breast and throat; has much elongated tail-feathers in the breeding season.*** Female and juvenile are grey-brown above, with throat whitish and ***rest of underparts washed yellow*** (separating it from similar Palestine and Shining). Non-breeding male recalls female but underparts yellower; some have blackish centre of chin and throat. Often flicks wings and tail. See Purple Sunbird.
Voice: song soft, fine and high-pitched with trilling and hissing notes, 'pruiit-prruiit-pruuit-tiririririri-tirririri'; also thin hoarse 'veeii-veeii', a 'ptscheeeiii', repeated 'cheeit-cheeit' and 'tee-weee' with upward inflection on last note.

617 **Purple Sunbird** *Nectarinia asiatica* **PLATE 95**

L: 10. Small sunbird resembling Palestine ***but has different distribution; bill slightly shorter, less decurved.*** Metallic bluish black ***male has sometimes a narrow red-brown breast band***; mouse-brown ***female has underparts washed yellow*** (pale greyish in female Palestine). Male in eclipse (Sept–Dec) like female but underparts yellower and there is a dark median line down centre of throat and breast. Short-tailed, hummingbird-like appearance, feeding largely on nectar from flowers.
Voice: excited song repeated two to six times, 'cheewit-cheewit', etc. with Willow Warbler-like cadence. Male's sub-song a low twitter; call 'dzit-dzit' recalling that of Fan-tailed Cisticola; also a 'tsweet'.

618 **Shining Sunbird** *Nectarinia habessinica* **PLATE 95**

L: 13. Medium-sized sunbird ***with long decurved bill. All plumages are much darker below than in other sunbirds in the Middle East.*** Male bright metallic-green (Palestine usually much bluer) ***with broad though inconspicuous reddish breast band*** (absent in Palestine); ***female dark sooty grey, slightly paler below, but much darker than female Palestine; feathers of ventral region fringed whitish*** (absent Palestine); juvenile male resembles female but has black centre of throat and often dark breast-patch, whitish fore-supercilium and moustache. Non-breeding male like female but with some green on head, back and breast. Has rather pronounced, slow flicking of longer, broader tail than Palestine and flight has longer, deeper undulations.
Voice: fast song fluty, trilling and whirring, e.g. 'tuu-tuu-tuu-tuu-vita-vita-vita-du-du-du-du' often ending in long Wren-like trill. Sub-song is a fast whispering warble. Calls include a hard 'dzit' and 'chewit-chewit'; also fast dry 'tje-tje-tje-tje'.

619 **Palestine Sunbird** *Nectarinia osea* **PLATE 95**

(Orange-tufted Sunbird)

L: 11. Small, short-tailed sunbird with medium-long decurved bill. ***Male separated from larger Shining by bluer plumage, smaller size, little shorter bill, absence of red breast band, quicker flicks of shorter tail*** and more rapid flight with irregular, short, dipping undulations; ***female distinctly paler grey below than female Shining and lacks the white-fringed feathers of ventral region of that species.*** Told from Purple by normally different distribution (see maps), ***slightly longer bill, lack of red-brown breast band in male Palestine and lack of yellow below of female Palestine.***
Voice: fast song a high-pitched trilling 'dy-vy-vy-vy-vy-vy' or rising 'tweeit-tweeit-tweeit', etc.; also 'veet-tji, veet-tji-veetji', accelerating, often ending in European Serin-like trill. Calls include thin 'ftift', Robin-like tiiu', a hard 'tskak' and loud sharp 'te-*veeit*, te-*veeit*', second note stressed and rising.

620 **Socotra Sunbird** *Nectarinia balfouri* **PLATE 95**

L: 14. A fairly robust sunbird with ***brown upperparts.*** Both sexes have noticeable white submoustachial stripe, black malar-stripe and ***sooty fringes to feathers of throat and upper breast*** giving a Cyprus Warbler-like pattern; ***white outer tail-feathers*** on blackish-brown tail. Active, like all sunbirds; usually seen in pairs.
Voice: song short and musical with slight Goldcrest and Goldfinch-like quality 't-ze-ze-ze-zweet'. Call 'chee, chee, chee'; short 'teeoo' or scolding 'schee-a schee-a-schee-a'.

621 **White-breasted White-eye** *Zosterops abyssinica* **PLATE 88**

(Abyssinian White-eye)

L: 12. Small and warbler-like with ***conspicuous white eye-ring*** and short bill. Upperparts greyish green, ***throat and undertail coverts pale yellow*** and breast and white belly often with smoky wash. Very active, often in pairs or occasionally small groups and constantly on the move, calling frequently. Can not easily be confused with any other species.
Voice: soft, fine and high-pitched 'tiiu'; a fine purring trill or ripple; also a short low 'waouw'.

622 **Golden Oriole** *Oriolus oriolus* **PLATE 96**

(Eurasian Golden Oriole)

L: 24. Shy, thrush-sized, more often heard than seen, except on migration. ***Male's bright yellow plumage with black wings and tail diagnostic***; female and immature male greenish above with olive-brown wings and tail, rump often yellowish green, underparts yellowish white, indistinctly streaked darker. Appears relatively short-tailed in flight, which has woodpecker-like soft undulations broken by often half-tilting soft wing-beats.
Voice: song is mellow, usually loud (but not always) yodelling 'tjoh-wlee-klee-*ooh*', the last note hollow, emphasized and descending with tone of Blackbird. Alarm call a hoarse mewing Jay-like 'kra-eik'.

623 **Black-crowned Tchagra** *Tchagra senegala* **PLATE 96**

(Black-headed Bush Shrike)

L: 22. ***Unmistakable, thrush-sized with long, dark, conspicuously white-tipped tail, striking head pattern with bold white supercilium bordered by black crown and eye-stripe, large chestnut wing-patch*** and heavy black, slightly hooked, bill. Secretive with clumsy Magpie-like movements; cocks and flicks tail; will hop on the ground with tail raised or glide from bush to bush. Usually in pairs or singly.
Voice: song, mainly heard early morning or evening, both from hidden perch or in flight, is fluty and melodious, consisting of about 10 notes, the second half descending. Starts song-flight in upward low arc with soft wing-noise; at top of arc glides down singing. In addition, characteristic long trill, rising and falling 'truit-truit-driririririvivir'; alarm call a harsh 'shrrr'.

624 **Isabelline Shrike** *Lanius isabellinus* **PLATE 96**

(Rufous-tailed Shrike)

L: 18. ***A pale shrike***, slightly longer-tailed than Red-backed. In Iranian breeding race, *phoenicuroides*, and in the more easterly *speculigerus* (migrant and winter visitor only to eastern parts of the region) male has blackish facial mask ***and white mirror at base of primaries***; in male *isabellinus* facial mask ill-defined and there is hardly any white mirror in the wing. In all races the underparts are buffish white, in male without the fine crescentic markings at sides of breast of female; ***upperparts either sandy with crown toned rufous or uniform pale sandy grey-buff*** (*isabellinus*); flight-feathers are darkest in *speculigerus*, palest in *isabellinus*. ***Reddish to pale orange-rufous rump and uppertail and, often, rufous-tinged undertail usually separate it from the darker, warm brown female and 1st autumn Red-backed*** (which has greyish undertail); ***crescentic markings on sides of breast are thin and well spaced in Isabelline, denser in Red-backed, and upperparts are generally paler and sandier***; however, some 1st year *phoenicuroides* can be very similar to 1st year Red-backed.
Voice: song and call similar to Red-backed Shrike.

625 **Red-backed Shrike** *Lanius collurio* **PLATE 96**

L: 18. ***Male unmistakable, with red-brown mantle and wing-coverts contrasting with ash-grey top of head and rump***, black facial mask ***and black tail with white sides to base***; female and 1st autumn birds basically warm brown above, tail darker brown, whitish underparts with dense crescentic barring on breast and flanks; ***barring and colour of rump, tail and upperparts usually separates it from Isabelline***, which see. Juvenile has crescentic markings on upperparts and variable dark subterminal markings to tertials; see young Woodchat Shrike for separation. Typical shrike behaviour, perching exposed, diving to ground for prey and returning; also glides and hovers along hedges.
Voice: song is a quiet and musical prolonged warbling with many unpredictable imitations. Call a short 'shack', alarm a series of hoarse hard notes 'keck-keck-keck'.

626 **Bay-backed Shrike** *Lanius vittatus* **PLATE 96**

L: 18. Plump-bodied, long-tailed and large-headed shrike; adult separated from Long-tailed by size, square-ended shorter tail, ***broad black forehead-band reaching fore-crown, greyish rump, whitish uppertail-coverts, white sides to tail and large white mirror at base of primaries appearing as conspicuous band in flight*** (larger Long-tailed ***has graduated tail with buff outer feathers, narrow black forehead-band, bright orange-buff rump and uppertail-coverts and minute wing-mirror***); female duller than male. Juvenile has rufous tail and faint wing-mirror but told from young Isabelline by rufous greater coverts and grey lower rump; 1st winter birds superficially like adult but have wavy barring on flanks and upperparts and the black forehead-band may be absent.

Voice: grating calls. Song quiet and pleasant, often prolonged, imitating other species.

627 **Long-tailed Shrike** *Lanius schach* **PLATE 96**

(Black-headed Shrike, Rufous-backed Shrike)

L: 24. ***Noticeably larger*** but proportionally smaller headed than the somewhat similar Bay-backed from which it is also separated ***by longer graduated tail with buff*** (not white) ***outer feathers, bright orange-buff rump and uppertail-coverts*** (in Bay-backed, rump grey, tail-coverts whitish); ***the black eye-mask joins narrowly over bill*** (broad in Bay-backed) ***and the white mirror at base of primaries is small, inconspicuous and sometimes absent*** (very conspicuous in adult Bay-backed). Told in 1st winter from similar Bay-backed by rufous-buff rump and uppertail-coverts; lacks chestnut on shoulder and back shown by Bay-backed.

Voice: song, not often heard, is an *Acrocephalus*-like, prolonged slow and extremely imitative warbling. Call a rough 'guerlek-guerlek' followed by barked 'yaou-yaou'.

628 **Lesser Grey Shrike** *Lanius minor* **PLATE 97**

L: 20. Resembles Great Grey but smaller with ***stubbier bill, proportionally longer wings with long primary-projection*** and shorter, less graduated tail; ***white wing panel only confined to primaries. Adult shows extensive black forehead*** without white supercilium (which is present in some Great Grey), bluish grey upperparts lacking white on shoulders (but sometimes shows faintly) ***and pinkish white underparts.*** Juvenile/1st winter lacks black forehead, has paler bill, and upperparts are brown-grey finely barred darker (in juvenile) with paler tips to wing-coverts and flight feathers; underparts creamy white sometimes faintly barred; this plumage is confusable with corresponding plumage of Great Grey, but distinguished mainly by size and structure of wings and tail, and white wing panel in flight only confined to primaries (in some races of Great Grey white extends onto secondaries). Perches prominently in top of bush, tree or wire. Flight less undulating than Great Grey's.

Voice: chattering, varied, some thrush-like whistles, trills and mimicry. Alarm call a forced 'kschvee'.

629 **Great Grey Shrike** *Lanius excubitor* **PLATE 97**

(Northern Shrike)

L: 25. Upright, grey, with black and white markings, often perching boldly on top of bush or other prominent perch. Resembles Lesser Grey but in all plumages distinguished by ***larger size, longer bill, proportionally shorter wings (with very short primary projection), longer, more graduated tail and often longer wing bar***. Adult variable, according to race, in the shade of grey on upper and underparts, size of wing-bar, and presence of dark lores and black line over bill; most have ***obvious white on shoulders*** and most show ***white in wing extending onto secondaries, forming long wing-bar in flight***. Juvenile/1st winter similar to juvenile/1st winter Lesser Grey (which see), and distinguished mainly by size, wing structure and tail length; underparts are usually barred more heavily.

Several races occur in the region:

pallidirostris: (Central Asia, N Iran, wintering S to Arabia) has very pale grey upperparts, white underparts, pale (not black) lores, broad white wing-bar and horn-coloured or grey (not black) bill, tipped dark.

elegans: (S Palestine) very pale grey above, white below, large white wing-bar and black joining over bill in thin line.

aucheri: (Israel, Arabia, Iraq, Syria, Iran) slightly darker than *elegans*, with slightly broader black line over bill, smaller amount of white in wing and grey wash on flanks.

buryi: (Yemen) darkest race, dark grey above, grey below with white confined to centre of belly, small white wing-bar.

uncinatus: (Socotra) black extends in broad line over bill, small white wing-bar

Voice: song a mixture of quiet, melodious ramblings, raucous notes, and mimicry. Call includes harsh 'sheck sheck' often extended into chatter.

630 **Woodchat Shrike** *Lanius senator* **PLATE 97**

L: 18. Striking ***rufous-brown crown and nape*** in adult with ***large white shoulder-patch on black upperparts***. Female duller with more white/buff at base of bill and lower back/rump only slightly greyer than mantle. Juvenile/1st winter resembles Red-backed of similar age but Woodchat has ***paler grey-brown upperparts with creamy white markings on shoulders, wing-coverts and rump, and base of primaries, forming pale wing-patch noticeable in flight***; juvenile Masked is greyer overall, with finer bill, longer tail and paler forehead. Undulating flight similar to other shrikes.

Voice: song varied, whistles, trills and mimicry; calls include Red-backed-like 'schak-schak' and varied, grating rattles.

631 **Masked Shrike** *Lanius nubicus* **PLATE 97**

L: 18. Slightly smaller and slimmer than Woodchat with finer bill and longer tail. Adult unmistakable: ***black upperparts and mask*** (female duller, brownish-grey) with ***white face, orange wash on flanks*** and distinctive white shoulder-patch; white base of primaries form patch in wing, also obvious in flight. Juvenile similar to juvenile Woodchat but has generally ***greyer ground colour of upperparts*** heavily barred, rump concolorous with back (paler rump in Woodchat), forehead and eyebrow whitish, finer bill and longer tail; in flight shows larger white wing-patch. 1st winter similar to adult female but lacks orange

flanks. Frequently perches on bushes with long tail slightly cocked and waved up and down (unlike Woodchat).
Voice: harsh scolding 'krrrr'. Song, rather tuneless jumble of quiet notes.

632 **Black Drongo** *Dicrurus macrocercus* — PLATE 97

L: 30. ***Adult glossy black*** with semi-translucent primaries revealed in flight and ***long, deeply forked tail curving out towards end.*** Juvenile browner with white tips to feathers of flanks and belly. Sits upright in conspicuous position similar to shrikes, whence makes aerial sallies to pursue insects, with bee-eater-like glides and heavy flycatching manoeuvres.
Voice: call, harsh, throaty 'schweep-schweep'.

633 **Jay** *Garrulus glandarius* — PLATE 97
(Eurasian Jay)

L: 33. Easily identified, especially in flight when ***blue wing-coverts, white wing-patch and rump contrast with dark flight-feathers and tail***; otherwise black moustache and variable black or streaked crown depending on race; upperparts also variable (pinkish grey or vinous, with nape matching or not matching mantle). Flight is slow and direct with a peculiar jerky flapping action, on rather broad wings. ***Shy and noisy.***
glaszneri: (Cyprus) streaked crown, reddish forehead.
krynicki: (N Turkey) pinkish brown body, nape and mantle uniform in colour, crown black, face rufous-white.
hyrcanus: (N Iran) rufous body, greyish black face and fore-crown; mottled rufous hind-crown; mantle and nape uniform in colour.
atricapillus: (rest of Middle East) black crown, white face and ear-coverts tinged vinous, nape and mantle uniform in colour.
Voice: most obvious call a far-carrying rasping screech 'skaaaak-skaaaak', also a weak mewing 'peeay'.

634 **Magpie** *Pica pica* — PLATE 97
(Black-billed Magpie)

L: 48. Unmistakable; ***glossy black and white with long graduated glossy green tail; in flight, shows rounded wings with white primaries bordered black.*** Flight action energetic with irregular flapping wing-beats. SW Arabian race, *asirensis*, is dull black with tail glossed purple, small amount of white on wings and shoulders; lacks white rump-patch; also has larger bill and legs.
Voice: calls include chattering alarm 'chack-chack-chack-chack-', weaker 'ch-chack' and squealing 'keee-uck'.

635 **Pleske's Ground Jay** *Podoces pleskei* — PLATE 98

L: 24. Not unlike Hoopoe in general appearance (but no crest) with long decurved bill and broad, roundish wings. Plumage warm buff with paler face, black bib, eye-streak, and ***black wings with 2 white bars (conspicuous in flight).*** Highly elusive, occasionally perching in bush top where it slightly fans tail and calls briefly before dropping to the ground, on which it runs swiftly.
Voice: song is weak and rapid with about 10 notes per second 'pee-pee-pee-pee-pee', etc.

636 **Nutcracker** *Nucifraga caryocatactes* **PLATE 98**

(Eurasian Nutcracker)

L: 32. Similar in size and flight silhouette to Jay, but with ***shorter tail and more front-heavy appearance; when perched, long bill is noticeable. Despite white-spotted plumage, at a distance looks largely grey with darker wings***. Flight is less jerky than Jay's.
Voice: a repeated high croak; also a buzzing 'arrrr'.

637 **Alpine Chough** *Pyrrhocorax graculus* **PLATE 98**

(Yellow-billed Chough)

L: 38. Slightly smaller than Chough which it resembles in build but has ***short yellow bill, smaller head, narrower wings and proportionally longer, more rounded tail.*** Like Chough often seen soaring in flocks around mountain tops.
Voice: similar to Chough's, but usually higher with a trilling character.

638 **Chough** *Pyrrhocorax pyrrhocorax* **PLATE 98**

(Red-billed Chough)

L: 39. Similar to but slightly larger than Alpine Chough, although instantly told by longer, ***all red bill.*** In flight, deeply fingered wings are broader and tail shorter and more square ended than Alpine. Often in flocks. Juvenile has more orange-buff bill.
Voice: resembles Jackdaw, but louder, clearer and often far-carrying; typically a rolling 'chow', with falling inflection.

639 **Jackdaw** *Corvus monedula* **PLATE 98**

(Eurasian Jackdaw)

L: 33. A small crow with ***grey on nape and sides of face, pale eyes*** and quick movements. In flight, compared with other crows has smaller size, faster wing-beats and less fingered wings. Often in flocks and congregates frequently with Hooded Crows.
Voice: a characteristic loud, almost metallic 'ki-ak'.

640 **House Crow** *Corvus splendens* **PLATE 98**

(Indian House Crow)

L: 43. Slightly smaller and slimmer than Carrion/Hooded Crow with scrawny appearance and proportionally longer, deeper bill, longer tail and more domed crown. Grey nape, neck and breast clearly demarcated from black face but merging with rest of black plumage. Gregarious, often in very large numbers.
Voice: quieter and higher pitched than Hooded Crow, 'waaa waaa waaa'.

641 **Rook** *Corvus frugilegus* **PLATE 99**

L: 46. Medium-sized all-black crow which is most easily identified by ***long pale greyish bill*** with ***bare grey skin around its base***; also unlike other crows has rather peaked crown and bushy 'trousers'. Young birds lack the pale bill and skin at its base and thus resemble Carrion Crow; best told from it by more peaked crown, straighter more pointed

bill (curved culmen in Carrion) and feathering over legs. Very social, often occurring in flocks.
Voice: call more nasal, less rolling than Hooded Crow, a 'kah' or 'raak'.

642 **Carrion Crow** *Corvus corone corone* **PLATE 99**

L: 47. Carrion and Hooded Crow are races of the same species. Carrion is all-black and similar to Rook but differs from adult Rook (which see) in having a stouter, all-black bill and absence of shaggy 'trousers'. Immature Carrion Crow and Rook are more difficult: for differences see Rook.
Voice: as Hooded Crow.

643 **Hooded Crow** *Corvus corone cornix* **PLATE 99**

L: 47. A race of Carrion Crow, differing from it in having a ***grey body*** with black head, breast, wings and tail. It cannot easily be confused with any other species.
Voice: a hoarse croaking 'krarrk, krarrk'.

644 **Jungle Crow** *Corvus macrorhynchos* **PLATE 99**
(Large-billed Crow)

L: 46. Similar to Carrion Crow but slightly slimmer, with proportionally longer slightly wedge-shaped tail, rounder head and longer heavier bill with more decurved culmen. Smaller and sleeker than Raven, but more domed crown, lack of shaggy throat-feathers as well as size should prevent confusion.
Voice: a fairly deep, throaty 'hroarr-hroarr', unlike Raven.

645 **Brown-necked Raven** *Corvus ruficollis* **PLATE 99**

L: 50. Smaller than Raven with ***proportionally longer, slimmer wings, longer head and slimmer bill*** (which is often held drooping in flight), and ***bronzey brown sheen on nape and neck*** (which can be difficult to see); longish, wedge-shaped tail often shows central feathers protruding beyond tail outline, whereas Raven has more even wedge-shape. ***At rest wings reach beyond tail-tip***, whereas in Raven they usually fall well short. Juvenile lacks brown on neck.
Voice: croaking 'raark', less deep than Raven.

646 **Raven** *Corvus corax* **PLATE 99**
(Common Raven)

L: 64. Large powerful crow. Similar to Brown-necked Raven (the two species barely overlap in range—see map) but larger with broader wings, heavier head and bill and ***often showing shaggy feathers on throat***. For further distinctions see Brown-necked Raven. Often soars on flat wings and in breeding season performs tumbling and rolling display flights.
Voice: loud, far-carrying, deep 'krroak, krroak', a quieter 'kroak' and hollow 'klong'.

647 **Fan-tailed Raven** *Corvus rhipidurus* **PLATE 99**

L: 47. Readily told from other crows (though only Brown-necked Raven and House Crow occur in the same range) by ***very short tail giving it unmistakable flight silhouette***; strong bill is shorter than Raven's. When soaring overhead black coverts contrast with slightly paler flight-feathers and greyish feet contrast with dark plumage. Often congregates in large groups which will soar, raptor-like, in thermals.
Voice: high-pitched, rather gull-like croak.

648 **Socotra Starling** *Onychognathus frater* **PLATE 100**
(Socotra Chestnut-winged Starling)

L: 30. Similar in plumage to Tristram's Grackle (from which geographically isolated) but larger and with longer tail. Blue-black with iridescent bottle-green sheen especially on head, mantle and wing-coverts at close range, but can also show a rather sooty cast to whole plumage. Juvenile similar to adult but slightly shorter tail and greyish (not black) bill. Conspicuous rusty orange primary panels in flight (but hard to see at rest). Told from male Somali Starling (the other starling that occurs on Socotra) by slightly smaller size, ***shorter, squarer ended tail, longer, slightly thinner bill***, and generally ***less glossy plumage***. Rather shy, largely tree-dwelling with feeding actions like a heavy thrush.
Voice: louder more whistling calls than Somali Starling; single, short pure 'pseeoo' or 'psooo', rather mournful and far-carrying.

649 **Somali Starling** *Onychognathus blythii* **PLATE 100**
(Brown-winged Starling; Somali Chestnut-winged Starling)

L: 35. Slightly larger than Socotra Starling with ***long graduated tail*** and ***slightly stouter, shorter bill***. Male glossy blue-black with rusty orange primary panels. ***Female differs in having a conspicuous pale grey head and neck***. Parades more in the open than Socotra Starling and often associates with cattle; usually in pairs or flocks. Masterful, shallowly undulating flight with brief spells of wing-beats followed by long glides.
Voice: rather vocal, especially in flight. Musical high-pitched 'tleep', 'che-wee' or 'chwee'; also rather soft 'chuit, chuit' or 'chwee'.

650 **Tristram's Grackle** *Onychognathus tristramii* **PLATE 100**
(Tristram's Starling)

L: 25. ***Unmistakable glossy black (male) or dark brown (female) thrush-sized bird with rusty orange primaries conspicuously showing in flight*** (brightest above and in male). Usually in rather noisy flocks; distant flocks recall Starlings but coloured primaries and longer tail prevent confusion.
Voice: loud and echoing conversation of 'wolf-whistling' notes heard from parties, 'dee-oo-ee-o', 'o-eeou'; also mewing 'vu-ee-oo'.

651 **Amethyst Starling** *Cinnyricinclus leucogaster* **PLATE 100**
(Violet-backed Starling)

L: 19. ***Male unmistakable with iridescent violet-purple upperparts, head and breast*** (but can look black or red), ***rest of underparts are white***, wings darker and eye yellow; ***dark brown female has dark streaks on white breast below brownish throat***, belly

whiter; *eye yellow or chestnut*, but dark in otherwise similar juvenile. Most frequently seen in small groups and fairly shy. Flight is direct and Starling-like.
Voice: song a loud, metallic gurgling warble; call is a ringing, grating musical squeal with rising inflection ending in quiet chuckle, latter also heard when flushed.

652 **Starling** *Sturnus vulgaris* — PLATE 100

(Common Starling)

L: 21. ***Short tail and short pointed, triangularly shaped wings are characteristic in fast straight flight.*** Adult in breeding plumage ***has glossy green, blackish-looking plumage with numerous minute white spots***; white spots almost absent in Turkish *tauricus* in which mantle and breast are purple. ***Juvenile drab brown with long dark bill with straight culmen and dark loral streak*** (cf. young Rose-coloured, which see); 1st winter plumage ***blackish with prominent white spots*** thus resembling adult in winter. Gregarious, sometimes in thousands, flying in dense flocks. Walks quickly on the ground.
Voice: fast song has varied whistles, with whining strained tones and descending whistling 'seeeoo'; fine imitations interwoven into song. Flight-call a short buzzing 'tcheer', alarm call a hard 'kjet' or, at nest, a grating 'stahh'.

653 **Rose-coloured Starling** *Sturnus roseus* — PLATE 100

(Rosy Starling)

L: 21. Resembles Starling in shape and behaviour ***but the yellow bill is shorter with curved culmen*** (longer, straight in Starling). ***Adult's black and pink plumage unmistakable; has long pointed crest, raised when singing.*** Juvenile (no crest) separated from young Starling ***by structure of yellowish bill, pale lores, paler grey-brown upperparts with whitish grey rump and dirty whitish underparts*** (young Starling drab dull brown all over, with blackish straight bill and dark loral streak).
Voice: call resembles Starling's; rapid high-pitched chatter from feeding flocks louder and more musical than Starling. Song a lively jumble of chattering noises mixed with melodious warbling notes.

654 **Wattled Starling** *Creatophora cinerea* — PLATE 100

L: 22. Resembles Starling in size and shape but wing-tip slightly blunter and bill stronger. The plumage most likely to be seen in the region is juvenile with cold grey-brown upperparts, pale fringes to coverts and tertials, and pale grey-brown underparts; ***whitish rump may cause confusion with young Rose-coloured but lacks pale fringes to flight-feathers, has pale buffish greater upperwing-coverts and naked malar region.*** Adult male in breeding season has ***diagnostic bare yellow head with black wattles***; outside breeding season wattles disappear and head becomes greyish, resembling female; ***then both sexes show creamy white rump, fleshy-yellow bill, buffish greater coverts, white spot on leading primary-coverts in flight, yellowish-white area around eye and diffuse black loral spot and moustachial streak.***
Voice: soft squeaky whistle.

655 **Pied Mynah** *Sturnus contra* **PLATE 100**

(Asian Pied Mynah)

L: 23. ***The black and white plumage is diagnostic; black upperparts have conspicuous white rump and narrow white shoulder braces and white patch behind red-bordered eye***; bill orange-red. ***In flight it has no white in wing and tail.*** In young birds the black above is replaced by brown; underparts are buffish white with breast streaked or spotted vinous-brown; patch on ear-coverts whitish.
Voice: song a continuous trilling in long, rather monotonous sequences, somewhat Skylark-like.

656 **Bank Mynah** *Acridotheres ginginianus* **PLATE 100**

L: 21. Resembles Common but is smaller, ***with general body-plumage slate-grey*** (deep vinous-brown in Common); the blackish head has short crest on fore-crown; ***bare orange-red patch around eye and pale reddish bill diagnostic*** (in Common both bill and eye-patch are bright yellow). ***In flight, large rusty buff patches across bases of primaries and on tail-corners*** (pure white in Common). The juvenile has body tinged brown, not slate; wing- and tail-patches more buffish white, approaching Common, but colour of bill and around eye always separates it. Like Common Mynah tame, rather noisy, either in pairs or in flocks; feeds mainly on ground.
Voice: like Common Mynah, a variety of chuckling notes, sometimes rather musical, but softer.

657 **Common Mynah** *Acridotheres tristis* **PLATE 100**

L: 23. Resembles Bank Mynah ***but body plumage deep vinous-brown*** (slate in adult Bank); ***bill and small bare patch below eye bright yellow*** (see Bank Mynah). ***In flight, a conspicuous large white patch across primary-bases*** (both surfaces), ***on outer underwing-coverts and on tail-corners*** (see Bank Mynah). Build and general behaviour much like Bank Mynah.
Voice: very noisy; song imitative and repetitive with strident, rough and liquid notes mixed, 'pi*u*-pi*u*-pi*u*, che-che-che, tliy-tliy-tliy, tuu-tuu-tuu, tititi, pryv-pryv'. Alarm a grating 'traaah', somewhat Nutcracker-like.

658 **Saxaul Sparrow** *Passer ammodendri* **PLATE 101**

L: 15. ***Distinctive male has sandy grey upperparts, mantle with narrow, short black streaks***; black band from forehead to nape where narrowing, widening again on hindneck, ***separated from black eye-stripe by broad, bright ochre-russet patch above and behind eye; narrow fore-supercilium strikingly white***; large black 'bib', two white wing-bars, ***black lesser coverts*** and whitish fringed flight-feathers are further useful features (male House Sparrow ***has boldly streaked mantle, chestnut lesser coverts***, buffish fringes to flight-feathers and different head pattern); black bill is browner in winter. Female much less contrasting ***with sandy grey mantle with thin, short dark streaks*** (bold in female House), two clear white wing-bars (lower bar creamy in female House), ***blackish mottled lesser coverts*** and creamy white fringes to flight-feathers (fringed warm buff in House); ***some females have obscure dull blackish 'bib'***.
Voice: calls much resemble those of House Sparrow; song apparently unknown.

659 **House Sparrow** *Passer domesticus* **PLATE 101**

L: 14.5. Widespread, well known sparrow, ***male with grey crown, chestnut-brown sides of head, pale greyish cheeks, large black 'bib' and grey rump***; mantle and wing-coverts rufous-chestnut with broad white wing-bar on median coverts. Female and juvenile lack contrasting head and throat pattern, being buffish brown above, boldly streaked darker on mantle, pale greyish below and with pale creamy supercilium. Hard to separate from female Spanish, which see; see also female Saxaul, Sind and Dead Sea Sparrow. Sociable, often in flocks outside breeding season.
Voice: call and song monotonous 'cheep' or 'chirp' notes of varying pitch; in alarm a scolding 'cherr-r-r-r-r-r-r'.

660 **Spanish Sparrow** *Passer hispaniolensis* **PLATE 101**

L: 14.5. Resembles House Sparrow (with which it hybridizes) ***but male has typically rufous-brown crown and larger black 'bib' extending to bold black streaks on breast and flanks; back boldly streaked black, merging at sides with black of breast***; cheeks whitish, belly white. Female and juvenile not safely separated from similar House Sparrow though, on average, underparts are whiter with more grey-streaked breast and flanks. Sociable, often gathering in compact large flocks on passage; otherwise behaviour like House Sparrow.
Voice: song recalls House Sparrow's but faster; call slightly deeper and harder than House Sparrow's.

661 **Sind Jungle Sparrow** *Passer pyrrhonotus* **PLATE 101**

L: 12.5. ***Resembles small House Sparrow but male has smaller black 'bib'*** (like Tree Sparrow) ***and chestnut lower back and rump*** (greyish in House); head pattern resembles that of House ***but band behind eye redder chestnut***. Female like small version of female House; the spot behind eye is darker and ear-coverts are more ashy. Similar-sized female Dead Sea Sparrow (race *yattii*) has underparts tinged yellowish and sometimes also supercilium and sides of neck. Often in small groups.
Voice: resembles House Sparrow's but clearer.

662 **Dead Sea Sparrow** *Passer moabiticus* **PLATE 101**

L: 12. ***Small size, distinctive head pattern with dark grey cheeks and crown, contrasting with long white and rusty supercilium and pale yellow moustache bordering small black 'bib'***; lesser and median coverts chestnut, undertail-coverts faintly spotted rusty; underparts greyish white in the nominate race (Near East, E Iraq, SW Iran) but tinged yellow in *yattii* (E Iran). ***Female resembles small House Sparrow*** with slightly shorter wing-tip; may show yellowish tinge to supercilium and sides of throat; ***undertail-coverts distinctly dark-spotted*** (less pronounced or absent in female House); ***female yattii has yellowish suffusion on underparts***; told from female Saxaul ***by smaller size and boldly dark-streaked mantle***. Often gregarious and wary.
Voice: song rolling, 'tri-rirp, tri-rirp' or 'tlir-tlir-tlir' more tuneful and rhythmic than House Sparrow (rhythm not unlike Graceful Prinia's); calls include a short 'chilp' like House Sparrow, a harder, metallic, high-pitched 'tlip' or 'trrirp'; a buzzing 'tzzeeer' occasionally heard.

663 **Socotra Sparrow** *Passer insularis* — PLATE 101

(Rufous Sparrow—Socotra race)

L: 15. Size of House Sparrow but with more powerful bill. Male easily told by ***prominent black band through eye, bordered above by a broad chestnut superciliary stripe, black-streaked grey back*** and ***chestnut lesser-coverts***; underparts greyish with white cheeks and a black throat; bill black. Female resembles female House Sparrow but greyer, with ***buffish supercilium from bill over eye and curving behind ear-coverts, dark eye-stripe, greyish white cheeks*** and ***lacking wing-bar on coverts***; often shows shadow of male's black bib. A typical sparrow in its behaviour; often in groups.
Voice: song a quickly repeated 'chree chree chree chree' and a 'chit chit cheeoo'. Calls include a 3-note 'chree chree chree' and a scolding, quite hard 'chee chee chee chee chee'.

664 **Tree Sparrow** *Passer montanus* — PLATE 101

L: 14. Smaller and more delicate than House Sparrow ***with chestnut crown, small black spot in whiter cheeks and a smaller black 'bib'***; rump grey-brown ***and 2 white wing-bars.***
Voice: song has series of repeated 'tweet' notes. Calls clearer and harder than House Sparrow's, more high-pitched; flight-call a hard 'tek-tek-tek', 'tchu-wit', 'pilp' and a hard 'chik'.

665 **Arabian Golden Sparrow** *Passer euchlorus* — PLATE 101

L: 13. Smaller and slightly longer tailed than House Sparrow with domed head and large dark eye prominent against pale plumage. ***Male is bright yellow with broad whitish fringes to flight- and tail-feathers***; outside breeding season, the mantle and wing-coverts are greyer ***and head often tinged cinnamon-rufous***; bill black but pinkish horn-brown outside breeding season. ***Pale buffish or sandy grey female has unstreaked upperparts***; head and nape buffier, breast and flanks greyish; fringes to dark flight- and tail-feathers are creamy grey; bill fleshy horn. Young birds resemble female but can show slight mottling on mantle (but never bold as in House Sparrow or in the larger and bulkier female Rüppell's Weaver). Highly gregarious, usually away from human dwellings.
Voice: flocks utter a constant twitter, recalling House Sparrow's but more whispering.

666 **Pale Rock Sparrow** *Petronia brachydactyla* — PLATE 102

L: 14. ***Unstreaked grey-brown with broad white streak where secondaries meet the tertials***, long primary projection, ***short dark tail with white tip*** (obvious in flight), strong pale bill and translucent-looking brownish orange legs; sandy white underparts, 2 whitish wing-bars and prominent dark eye in pale face. In flight, the long wings, triangularly shaped, giving lark-like appearance. Often gregarious outside breeding season.
Voice: distinctive buzzing song monotonous and persistent 'tss tss tss tseeeeeeeeei', superficially recalling Corn Bunting. Flight-call a soft trill, recalling distant European Bee-eater; also 'twee-ou'.

667 **Yellow-throated Sparrow** *Petronia xanthocollis* PLATE 102

L: 13. ***Olive grey-brown unstreaked plumage with long, stout-based bill, black in breeding male***, pinkish brown in female and non-breeding male; ***male has yellow spot on lower throat surrounded by olive-grey breast band and lower cheeks and whitish upper throat***; lesser coverts chestnut, median coverts form broad Chaffinch-like white wing-bar (former sometimes hidden by scapulars). Female and juvenile lack chestnut lesser-coverts and yellow throat-patch ***but have distinctive bill-shape and the Chaffinch-like wing-bar; grey legs and all-dark tail also separate it from Pale Rock Sparrow***. Perches freely in trees; finch-like movements on ground; rather pipit-like dipping flight. Habitat differs from Pale Rock Sparrow.
Voice: song sparrow-like, a quiet chirrupping, but softer, more melodious, more rhythmic and faster than House Sparrow. Call a sparrow-like 'cheep', 'chilp' or 'chirrup', but softer, more liquid and tuneful.

668 **Bush Petronia** *Petronia dentata* PLATE 102
(Lesser Rock Sparrow)

L: 13. Small, with rather long, deep-based bill, ***black in breeding male***, but otherwise fleshy horn like female; ***lacks white in tail all ages***. Male has brownish grey crown ***and ill-defined chestnut supercilium; dark brownish grey sides of face and lower cheeks frame creamy white 'bib' on upper throat below which there is an inconspicuous small yellowish spot; mantle unstreaked brown***. Female has dark streaks on mantle ***and prominent whitish supercilium***; head and throat pattern like male's but ear-coverts surrounded by buffish area which merges with white 'bib'; 2 creamy white wing-bars. Juvenile resembles female but supercilium pale rusty buff and grey ear-coverts surrounded by richer buff; whitish 'bib' framed below by dark upperbreast.
Voice: song consists of 4–6 fast rolling notes 'triup-triup-triup-triup-triup' with tone between House Sparrow and Budgerigar; also a more bunting-like 'chu-chu-chu-chu', rising but ending on even pitch; flight-call a soft sparrow-like 'chewee'.

669 **Rock Sparrow** *Petronia petronia* PLATE 102

L: 14. Resembles female House Sparrow in general appearance but more thick-set with larger head, shorter tail, longer primary projection and thicker legs. ***Boldly striped head with pale crown-stripe and conspicuous long buffy white supercilium, characteristic; breast and flanks are strongly streaked and tail tipped white; a yellowish spot on upper breast is usually hard to see*** unless feathers are fluffed up. Juvenile lacks the yellow spot. Flight deeply undulating; often in loose flocks. Very active on ground, often running among rocks.
Voice: a characteristic metallic 'twe-*uh*-ee'; whispering nasal chatter from flying flocks. Song a hoarse 'vee-veip'.

670 **Snow Finch** *Montifringilla nivalis* PLATE 102
(White-winged Snowfinch)

L: 17.5. ***Grey-brown, black and white mountain bird***, superficially resembling Snow Bunting (ranges hardly ever overlap). In E Turkey and Iran breeders (race *alpicola*) head brownish and all races have ***small black 'bib' in summer; in flight, shows much white in wings and tail but primaries and centre of tail black***; white less extensive in female

and juvenile. Bill black in summer male, yellowish at other seasons and ages. Perches upright on ground, often jerking tail.
Voice: song in nuptial flight or when perched, a jerky, repeated 'sitticher-sitticher'; flight-call a nasal harsh 'tsweek'; also a purring 'prrt'.

671 **Rüppell's Weaver** *Ploceus galbula* **PLATE 102**

L: 14.5. Resembles sparrow in build but bill stouter, tail shorter and wings more rounded. ***Male's golden-yellow plumage, streaked mantle, dark chestnut mask with black surround to bill diagnostic***; in non-breeding season loses mask, plumage more yellowish green and black bill becomes browner. ***Female olive-brown above with dark streaks, buffish white below washed yellow-buff on throat and breast.*** Told from female Arabian Golden Sparrow by bulkier build, longer, slightly drooping bill, shorter tail and boldly streaked upperparts; from female House Sparrow by yellowish buff breast and olive-tinged upperparts. Gregarious, often forming large flocks; noisy.
Voice: song a wheezy chatter ending in insect-like hissing sounds; call a dry 'cheee-cheee'.

672 **Arabian Waxbill** *Estrilda rufibarba* **PLATE 103**

L: 10. Small, finch-like bird with ***dark crimson mask through eye***, and grey-brown upperparts with very faint vermiculations; the ***tail and rump are black*** with narrow white edges to outer tail-feathers; white cheeks and throat merge into very clean, silky or whitish buff underparts, which may show vermiculations on flanks. Juvenile has blackish eye-mask and is brown above with darker wings and buff wash to breast, belly and flanks. Bill colour varies: most adults have black bills (sometimes with red at base and on sides), young birds dark greyish bills; some birds, however, show pale, whitish bills. Often in flocks and will associate with Zebra Waxbills. Often waves tail from side to side when perched.
Voice: call a hard, buzzing 'dzit, dzeet or chee', is often heard from flocks and will develop into a buzzing chatter. In flight, a high-pitched buzzing 'chee-chee' or 'chee-chee-chee-chee' (note: no buzzing in call of Zebra Waxbill).

673 **Avadavat** *Amandava amandava* **PLATE 103**
(Red Avadavat)

L: 9. Small, short-tailed finch originating from India. Male bright crimson with browner wings and numerous small white spots on underparts and wing-coverts. Female brown above with blackish tones; buffish underparts with fulyous-yellow belly; rump red and tail blackish. Male in non-breeding plumage resembles female but throat and breast greyer. Coral red bill in both sexes with black ridge of culmen. Juvenile as female but lacks red rump, has 2 buffish wing-bars and blackish brown bill.
Voice: song is soft, high-pitched, continuous twittering: most common call a series of high-pitched chirps.

674 **Zebra Waxbill** *Amandava subflava* **PLATE 103**

L: 9. Slightly smaller than Arabian Waxbill, which it resembles in shape. Male is readily told ***by strong orange wash on underparts***, black and white barred flanks, crimson eye-mask, ***red rump and tail-coverts*** (both above and below) and waxy red bill (yellowish orange in subadults). Female, which lacks red eye-mask, has red usually confined to rump, underparts are yellowish buff (usually richer on belly and undertail-coverts) and flanks are less boldly barred. Juvenile resembles female but lacks red on rump and flank barring; also browner above and bill usually dark brown. Often in large flocks, but when perched more elusive than Arabian Waxbill. Fast flight with short undulations.
Voice: varied call-notes from a piping 'ptik', sharp 'tip', 'tith-tith' to hard metallic 'zzrep, zzrep, zzrep'.

675 **Indian Silverbill** *Euodice malabarica* **PLATE 103**

(White-throated Munia, White-throated Silverbill)

L: 11. Small brown and whitish finch-type with ***large conical silver-grey bill, prominent eye and pointed black tail***. Very similar to African Silverbill (which see) but told from it by ***absence of vermiculations on wing-coverts and tertials, whitish rump and uppertail-coverts***. Juvenile has pale edgings to wing-feathers. Fairly tame and frequently in small groups, sitting close when perched; often waves and flicks tail. Light flight with short undulations.
Voice: rapid, slightly tinkling 'cheet, cheet, cheet' flight-call; short high-pitched, trilling 'zip-zip; harsh 'tchwit' and conversational 'seesip, seesip'. Song a short trill.

676 **African Silverbill** *Euodice cantans* **PLATE 103**

L: 11. Very similar to Indian Silverbill, from which it is geographically isolated. Told from it by ***fine vermiculations*** on ***wing-coverts and tertials*** (but only visible at very close range) ***and black rump and uppertail-coverts***; may also show slight brownish on chin. Behaviour as Indian Silverbill.
Voice: similar to Indian Silverbill. Song a rapidly repeated high-pitched trill comprising single, then double notes; each phrase descending, then rising.

677 **Chaffinch** *Fringilla coelebs* **PLATE 103**

L: 15. A medium-sized finch, the male of which is readily identified by ***blue-grey head, chestnut back, pinkish breast and 2 white wing-bars***. Superficially resembles Brambling (which see) but note absence of orange in plumage, white wing-bars ***and lack of white rump***. More nondescript female and 1st winter birds can be told from all other finches by white wing-bars, white outer tail-feathers and unstreaked brownish or brownish olive upperparts (including rump). Undulating flight with wings closed every few beats. Largely feeds on ground where it moves rather jerkily. Forms flocks on passage and in winter.
Voice: song a short, accelerating trill, which ends in a flourish. Calls vary considerably but has typical loud metallic 'fink', repeated loud 'huit' and, in flight, a soft 'jup jup'.

678 **Brambling** *Fringilla montifringilla* PLATE 103

L: 15. Resembles Chaffinch but easily told from it in all plumages by ***orange or rusty orange breast, shoulder and wing-bar, mottled upperparts and, in flight, white rump.*** In winter (when only likely to be seen in the region) ***shows pale patch on nape and*** male has ***yellowish bill with dark tip.*** Often in flocks and mixes freely with Chaffinches and buntings in winter feeding areas. Flight similar to Chaffinch but more erratic.
Voice: distinctive, flat, hoarse wheezing 'kve-eek'; also Chaffinch-like 'jek-jek', but harder, more nasal.

679 **Red-fronted Serin** *Serinus pusillus* PLATE 104

(Fire-fronted Serin)

L: 12. Small, stubby, ***dark-streaked finch with sooty brown head and breast and almost luminous orange patch on forehead***; in bounding flight shows dark fore-body with rather pale underbody and pale underwings. Juvenile has pale washed-out orange-buff face with upperparts and underparts heavily streaked black and whitish. Outside breeding season often in small flocks feeding on weed seeds or on the ground.
Voice: song similar to that of European Serin but more powerful and with more of a Goldfinch-like quality. Call a European Serin-like ripple, but softer, a drawn-out 'drrrrt'.

680 **European Serin** *Serinus serinus* PLATE 104

(Serin)

L: 12. Small yellowish, streaked finch with stubby bill; note in particular ***yellow rump, supercilium, face and breast, white belly with bold dark flank-streaks and all-dark tail.*** Female much duller barely showing any yellowish; generally streaked olive-grey with washed out yellowish buff on face and, like male, yellow rump; black flank-streaks on whitish underparts; double wing-bar can be fairly obvious or nearly absent. 1st winter as adult. ***Call is a good feature for identification.*** Often occurs in small parties outside breeding season. Flight fast and undulating, almost skipping. For difference from Syrian Serin see that species.
Voice: has a jingling, chirping, fast twitter like the sound of shaking keys, either from exposed perch or in Greenfinch-like song-flight. Flight-call a metallic trilling 'zr-r-litt'.

681 **Syrian Serin** *Serinus syriacus* PLATE 104

(Tristram's Serin)

L: 12.5. Similar to European Serin but slimmer with longer tail and rather paler plumage ***without bold streaks on underparts.*** Differs from European Serin in all plumages by ***olive-yellow fringes to tail-feathers, absence of strong streaking on flanks, obvious broad olive-yellow bars on coverts*** (except in juvenile) and ***yellow fringes to flight-feathers.*** Male in summer also told from European Serin by more extensive rich orange-yellow forehead encircling, and accentuating, dark eye. Female similar to male but more olive-grey and lacks orange tone in yellow on forehead and around eye; may show very light flank-streaks. Juvenile browner and lacks yellow on face and wing-coverts. Sociable, often in flocks in winter; perches high in trees but feeds mainly on ground.

Voice: flight-call less musical than European Serin's, 'tirrrh' or 'tsirr'; also thin nasal 'shkeep'. Often has twittering low, soft 'tree-dar-dee'. Song, usually from perch, soft with Goldfinch and Linnet-like elements with 'siou' notes and purring trill.

682 **Arabian Serin** *Serinus rothschildi* **PLATE 104**

L: 11.5. Similar to Yemen Serin (which see) but differs in ***darker, more smoky coloration, stouter more decurved bill, darker throat, cheeks and upper breast, barely any streaking on crown, absence of any moustachial streak, olive rump and call***; close views show faint pale eyebrow; bill usually greyish but can be dull pink with small pale spot at base of lower mandible (absent in Yemen Serin). Most often seen in pairs (less flocking than Yemen Serin) and spends most time in trees when ***gentle tail-flicking*** can be observed.
Voice: call, quiet 'tsit-tsit'; also a short ripple. Song, pleasant, plaintive rather slow 'ti-tiiu-tuiu' or 'tsu-tsu-tsi-tsu' with rising inflection; often starts with trill and sometimes ends with varied warble-like jingle.

683 **Yemen Serin** *Serinus menachensis* **PLATE 104**

L: 11.5. Small, rather plain serin, grey-brown with lightly streaked underparts. Similar to Arabian Serin but ***differs in paler appearance, slightly shorter tail, clearer crown-streaks, indistinct moustachial stripe in front of pale cheek-patch, smaller bill with less curved culmen, absence of yellowish green rump, lack of tail-flicking,*** ground-dwelling behaviour and call. Flocks more than Arabian Serin.
Voice: most characteristic call is rather Redpoll-like 'che-che-che'; also twittering 'chirrip, chirrip', quiet 'che-che-che-che-che', quick 'prlyit-prlyit' (often repeated several times), wagtail-like 'cheir-virp, cheir-virp'. Feeding flocks have repeated musical, whispering 'tleet-tleet'. Song 'chew-chee-chee-chew' in dipping song-flight and when perched.

684 **Golden-winged Grosbeak** *Rhynchostruthus socotranus* **PLATE 104**

L: 14.5. Plump, sparrow-sized finch with ***very stout blackish bill.*** Easily identified by ***white cheeks*** (varying in extent) and ***yellow patches in wings and tail.*** Female differs from male in having duller plumage without blackish area around base of bill. Race on Socotra, *socotranus*, has black crown, nape and extensive black bib. Juvenile has brownish head and prominently streaked upper- and underparts. Often in loose groups and can be quite inactive sitting unobtrusively amongst branches. Bounding flight, often flying some distance if disturbed.
Voice: liquid, discordant song often starts with 'whit-whee-oo' or 'tvit-te-vyt-te-vict', repeated and interspersed with Linnet- and Goldfinch-like notes, often in fluttering, bat-like, display flight. Varied calls include 'wink', and soft Goldfinch-like 'tlyit-tlyit'.

685 **Greenfinch** *Carduelis chloris* **PLATE 105**
(European Greenfinch)

L: 14.5. Stocky, fairly large-headed finch with stout conical pale bill. ***Recognized in all plumages by bright yellow edges to primaries and outer tail-feathers, particularly obvious in flight.*** Male is olive-green with yellow-green underparts and rump; female is duller grey-green all over and juvenile is olive grey-brown and streaked below. In flight, appears thick-set with relatively short tail; flight has fairly deep undulations.

Voice: song, either from tree-top or in butterfly-like flight, is a powerful Canary-like trilling; sometimes a prolonged descending nasal 'tweeesh'. Call a short 'tjup' or 'teu', alarm note a hoarse 'de-*u*-wee'.

686 **Goldfinch** *Carduelis carduelis* — **Plate 105**

(European Goldfinch)

L: 14. Colourful finch, ***readily identified by bold red, black and white head pattern, striking broad yellow bar to black wings, particularly obvious in flight when white rump also conspicuous.*** The E Iranian race *paropanisi* (Grey-headed Goldfinch) ***lacks black and white head pattern, having red face only***; rest of upperparts and throat uniform buff-grey; ***pale bill noticeably longer and heavier than other Goldfinch races.*** Juvenile (all races) has pale greyish body-plumage and unmarked head; identified by wing-bar, white rump and call. Often occurs in small flocks outside breeding season.
Voice: call, often uttered in flight, a characteristic, often repeated, sharp but liquid 'tick-le-*lit*' with stress on last note; rasping notes sometimes heard from flocks. Song a liquid twitter, with call mixed into song.

687 **Siskin** *Carduelis spinus* — **Plate 105**

(Eurasian Siskin)

L: 12. Small finely built finch, ***male yellowish green with black forehead and chin, broad yellow wing-bar, sides of tail and rump, particularly obvious in flight.*** Female olive-brown above, white below but has tinge of yellow-green which, together with yellowish in wings, tail and on rump identify it; lacks black on head of male. Back and flanks streaked dark in all ages. Will feed hanging upside down in trees. Appears small and short-tailed in variably undulating flight; often in flocks.
Voice: call a drawn-out, high-pitched 'tsee-u'; from feeding flocks a fast dry 'kettkett' and some twittering notes. Song, sometimes in butterfly-like flight, variable fast twitter ending in drawn-out wheezing note.

688 **Linnet** *Carduelis cannabina* — **Plate 105**

(Common Linnet)

L: 13. ***In male, greyish head, chestnut back and crimson-red forehead and breast characteristic***; female and immature lack red and chestnut, being warm brown above with dark streaks, visible also on breast and flanks (particularly in juvenile which also lacks grey on head). ***In all ages has clear whitish fringes to primaries and outer tail-feathers which, together with dark grey-brown bill, streaked centre of throat, and pale area round eye and on cheeks, separate it from Twite*** (which has yellowish bill in winter, unstreaked pale buff throat, more uniform yellow-brown cheeks and dark patches to breast sides in summer). See also Redpoll. Flight irregularly undulating. Occurs in pairs and flocks in winter.
Voice: song a varied musical twitter, interspersed with whistling nasal notes, uttered from exposed perch. In flight, a nasal sharp 'tett' or 'tetteret', often followed by short trill or whistle, 'trreu' or 'tru-kee-*wou*'; anxious call a soft mellow 'di-ge-dee'.

689 **Yemen Linnet** *Carduelis yemenensis* **PLATE 105**

L: 11.5. ***Separated geographically from Linnet. Male has whole head to upper breast a clear pale ash-grey*** (no red on forehead) ***and chestnut patches at sides of lower breast***; belly white; upperparts and wings chestnut. Female much duller. ***In flight, a broad pure white wing-bar conspicuous above and below; seen perched from below the outer tail-feathers show conspicuous longitudinal white patch; lacks streaks above and below of female and juvenile Linnet.***
Voice: flight-call a short soft liquid 'vliet' recalling Goldfinch; also a musical 'chip-chip'. Song, usually from concealed perch, is a fast melodious twitter, reminiscent of song of Canary and Goldfinch.

690 **Twite** *Carduelis flavirostris* **PLATE 105**

L: 13. Resembles Linnet ***but has more uniform yellow-brown supercilium, cheeks and ear-coverts and unstreaked ochre-buff throat; rump usually pink in adult***. The Middle East race, *brevirostris*, differs from European birds ***by obvious dark patches on sides of breast, whiter belly and more frosted tawny brown upperparts***; unstreaked (male only) rump pinker; *brevirostris* also has more Linnet-like whitish in primaries and tail; bill grey in summer, ***yellow in winter*** (always dark in Linnet). Irregularly undulating flight.
Voice: a nasal '*twe*-it'; whispering calls also heard from flocks; also a short 'jek-jek'. Song chattering twitter with variations of call interwoven.

691 **Redpoll** *Carduelis flammea* **PLATE 106**
(Common Redpoll)

L: 13. Small greyish brown finch with heavily streaked, compact body but fairly long, forked tail. Distinctive features are ***black bib, short yellow bill and red fore-crown*** (sometimes lacking red). ***Male has pink breast*** and is slightly pink on rump. Female and 1st winter darker, browner with heavier streaking on flanks. Very active when foraging on ground or in trees, often hangs upside-down like tits.
Voice: flight-call rapidly repeated 'djä-djä-djä'; when perched a soft rising 'dyuee'.

692 **Crossbill** *Loxia curvirostra* **PLATE 106**
(Common Crossbill, Red Crossbill)

L: 17. Large, strongly built finch with ***short tail and stout head and bill with mandibles crossed at tips. Male largely red*** (intensity variable) with brown wings and tail; ***in flight, shows strikingly red rump. Female greenish*** with greyish brown wings and tail, and slightly streaked underparts; ***in flight, shows greenish yellow rump***. Juvenile similar to female but distinctly streaked all over. 1st winter/1st summer gradually attain male's or female's adult-coloured plumage. In flight, looks clumsy and large-headed with short forked tail; ***often in small flocks loudly calling***. In Cyprus race, *guillemardi*, male is paler with strong yellow or orange tinge to upper- and underparts, and female has dark grey crown and mantle; both have heavier bills.
Voice: typical call in flight a loud, hard 'kip-kip'; when feeding similar but softer, subdued 'chip-chip'. Song irregular with twitters, trills and admixed calls developing into 'cheeree-cheeree-choop-chip-chip-chip-cheeree'.

693 **Crimson-winged Finch** *Rhodopechys sanguinea* PLATE 106

L: 16. Distinctively coloured ground-dwelling finch confined to mountain slopes. Distinguished from Trumpeter, Mongolian Trumpeter and Desert Finch by combination of larger size, ***brown mottled upperparts, breast and flanks, black crown and yellow bill*** (greyish horn in non-breeding season). Male has pink rump and face (pink in face absent in winter) and ***large rosy-pink flash in wings*** (striking in flight, in combination with white underwing-coverts). Female duller with paler wing panel and dull yellow bill in breeding season, lacks pink in face.
Voice: flight-calls trisyllabic 'tchu-che-ly', soft 'chee-rup', 'tureep, tureep', or Woodlark-like 'dy-lit-di-lyt'. Song melodious trilling, from ground or in song-flight.

694 **Desert Finch** *Rhodospiza obsoleta* PLATE 106

L: 14.5. Noticeably ***pale, greyish buff finch with striking wing pattern and short black bill.*** Flight-feathers mostly black with pronounced ***white edgings and pink panel on coverts and base of secondaries.*** Male has black lores and base of bill. Female drabber with less pink, white and black in wing, and lacks black face-patch. Juvenile has straw-coloured, dark-tipped bill and may be confusable with young Mongolian Trumpeter Finch (which see for further differences). In flight, from below, looks pale greyish with broad diffuse translucent panel on underwing; from above shows white primary patch with rose-pink on secondaries and greater coverts. Often feeds on ground.
Voice: song quiet, pleasant, very soft purring 'prrryv, prrryv' or 'prrrt, prrrt' or 'turr' mixed with purring calls and clear trills, interspersed with harsh and nasal notes reminiscent of Greenfinch. Common call in flight or from ground a quiet, soft, melodic 'prrrrrl', reminiscent of quiet, soft bee-eater.

695 **Mongolian Trumpeter Finch** *Bucanetes mongolicus* PLATE 106
(Mongolian Finch)

L: 14. Small, inconspicuous mountain-dwelling finch, resembling Trumpeter Finch (which see) but ***adult breeding distinguished by large whitish wing panel*** and brownish yellow legs (orangey flesh in Trumpeter). ***Male has small, clear-cut, rosy rump-patch*** (larger area of rosy on rump and uppertail-coverts in Trumpeter), varying rosy pink suffusion to throat, breast and flanks in fresh plumage, rosy red fringes to ***white-based greater coverts*** (rosy grey without white in Trumpeter), white fringes to tail-feathers (rosy grey in Trumpeter) and ***rosy supercilium*** (absent in Trumpeter). Female breeding is a duller version of male with pink confined to flanks. In fresh autumn plumage sexes are similar resembling breeding female. Juvenile similar, but wing-coverts with pale sandy brown fringes, underparts without any pink, and breast and flanks tinged tawny. In non-breeding plumage told from Trumpeter by whitish or sandy cream fringes to secondaries, broad whitish fringes to blackish tail and smaller, less stout bill. From young Desert Finch by uniform, horn-coloured bill (dark-tipped in Desert), less deeply notched tail, smaller size and paler (not black) centres to primaries.
Voice: call a soft 'dju-vud' or 'djudjuvu' or 'deedjud', in quality like that of Rock Sparrow, also twittering Linnet-like 't'yuk-t'yuk-t'yuk-t'yuk' when feeding or in flock. Song usually given from ground: pleasant, slow, rising and falling whistles, followed by shorter chirps 'towit-toowit-tu-tu-churrrh-whi-whi-churrrh'; all calls lack nasal, buzzing character of Trumpeter Finch.

696 **Trumpeter Finch** *Bucanetes githagineus* **PLATE 106**

L: 14. Small, rather inconspicuous, ground-dwelling finch with large head, stout bill and rather short tail. Breeding male distinctive with ***grey head, pinkish wash on forehead, underparts, rump and wings and orange-red bill.*** Non-breeding, female and juvenile/1st winter rather nondescript sandy grey-buff with slightly paler rump, blackish grey wings and tail with paler feather-edgings and pale yellowish brown bill; some breeding females develop faint pink wash in plumage like males; legs in all plumages orangey flesh; rather conspicuous pale eye-ring. In non-breeding plumage very similar to Mongolian Trumpeter (which see) but note colour and markings of wing, tail, rump, eyebrow, bill and feet. ***In flight, shows no obvious white in wing or tail*** (unlike breeding Mongolian Trumpeter).

Voice: ***song very distinctive***: drawn-out nasal, wheezing buzz 'cheeeee'; call short 'chee' or 'chit'; in flight a soft 'weechp', most calls with buzzing quality.

697 **Common Rosefinch** *Carpodacus erythrinus* **PLATE 107**
(Scarlet Rosefinch)

L: 14. Linnet-sized finch with ***fairly stout bill,*** stocky body and round head. Adult male easily told by ***red head, breast and rump contrasting with brown upperparts.*** Females, 1st summer males (which often sing) and juveniles are dull olive-brown, lightly streaked above and ***more heavily streaked below, with 2 whitish or buffish wing-bars*** and no wing panel in flight-feathers; uniform head shows conspicuous dark eye. Often sings from prominent position, otherwise rather inconspicuous.

Voice: song diagnostic, a clear lively loud whistle: 'vii'-dji'vii-(di)-djiv' (rendered as 'pleased to meet you').

698 **Sinai Rosefinch** *Carpodacus synoicus* **PLATE 107**
(Pale Rosefinch)

L: 14. Male ***crimson and pink*** with colours sometimes muted by ***silvery cast especially on crown and ear-coverts***; grey bill and greyish brown upperparts contrast slightly with pastel colours of rest of plumage; in flight, shows pale pinkish grey underwing. Female and juvenile rather indistinct pale buff, very faintly streaked with slightly darker shade on crown; frequently show gingery wash to face and ear-coverts; told from Trumpeter Finch by longer tail projection beyond wing-tip, darker bill, warm wash to face, lack of pale-fringes to coverts, tertials and flight-feathers; and voice. Rather shy, largely rock-dwelling and often occurs in flocks, especially outside the breeding season. Flight bounding. Lacks conspicuous wing-bars and streaks of Common Rosefinch.

Voice: feeding flocks have constant, quiet 'tsweet' note. Usual flight-note a rich 'trizp'. Song varied, melodious, including 'buzzing' in display.

699 **Great Rosefinch** *Carpodacus rubicilla* **PLATE 107**
(Spot-crowned Rosefinch)

L: 21. ***Large*** and long-tailed rosefinch with ***stout pale bill.*** Male has ***crimson plumage; crown, ear-coverts and underparts are flecked white*** and brownish wings may show pale fringes to primaries but no white wing-bar. Female pale greyish brown with darker streaks on upperparts, but, more noticeably, on underparts. Juvenile duller and browner than female. Mainly ground-dwelling.

Voice: a soft note 'dyyb' or 'juit' resembling that of Bullfinch; alarm call a short 'chett'; in flight, a nasal 'dweed'.

700 **Bullfinch** *Pyrrhula pyrrhula* — **Plate 107**

(Eurasian Bullfinch)

L: 14. Male easily told by ***pinkish red underparts***, grey back, ***black cap, broad white wing-bar and stout black bill.*** Female much duller, being buffish below and thus less gaudy than male. ***In flight, both show white rump.*** Juvenile lacks black on head, rump is buffish and upperparts brownish but acquires adult female-type pattern by autumn. Rather quiet and unobtrusive and never occurs in flocks in autumn or winter as do many other finches.
Voice: a low, soft and quiet slightly falling whistle 'dwoo' easily imitated by man; song (not often heard) is a rather discontinuous warble largely mixed with calls.

701 **White-winged Grosbeak** *Mycerobas carnipes* — **Plate 107**

L: 21. Large with big head, stout bill and rather long tail; often perches upright on top of bush but can be inconspicuous especially when feeding. Readily told by ***slate head, back and breast contrasting with steely grey bill*** and ***greenish yellow lower breast and belly; yellow and white wing-patches*** and ***greenish yellow lower back and rump***, which, like the wing-patches, is noticeable in flight. Female duller than male, the slate muted by buffish streakings on head and breast. Much variation amongst immature birds which can be soft grey on head, breast and back or even washed-out buffish, but still showing wing-bars and greenish yellow on lower back and rump. Bounding flight over long distances. Usually gregarious.
Voice: soft calls when feeding, 'schweeup schweeup', also loud chattering notes.

702 **Hawfinch** *Coccothraustes coccothraustes* — **Plate 107**

L: 18. A rather large, heavy finch though shy and often goes undetected. Readily told by ***large, powerful steely grey bill***, rufous and brown plumage with ***black on face and chin*** and ***broad white wing-bar***. Female duller with less rufous-orange head and mainly grey (not black) flight-feathers. Juvenile, which lacks black on face, is rather greyish with scalloped underparts. In flight, large size, fast undulations, ***white translucent wing-bar and white-tipped short tail are good characters.***
Voice: song a simple series of sharp 'i' notes with different pitch. Call an explosive 'pzik' or 'pzik-ik' in flight and a thin 'fzeeik'.

703 **Snow Bunting** *Plectrophenax nivalis* — **Plate 110**

L: 16. Seen only in winter plumage in the region. A buff-brown, black and white bird, adults in winter ***with white innerwing panel and white sides to tail*** (white patches largest in male). 1st winter female has reduced white in wings (confined to secondaries and some of the smaller coverts). Otherwise upperparts largely buffy brown, particularly top of head; underparts whitish with buff sides to breast; bill pale yellowish with dark tip. Flight has long undulations. See also Snow Finch.
Voice: commonest winter call a full whistle 'piu' or 'tju'; also a trilling 'trriririt'.

704 **Pine Bunting** *Emberiza leucocephalos* **PLATE 108**

L: 16.5. Resembles Yellowhammer in build. ***Male's black-bordered white patch on crown and cheeks in otherwise chestnut head and throat diagnostic***; the pattern is subdued in winter. Grey-brown female has well streaked crown, mantle, breast and flanks; rump rusty like Yellowhammer's; ***some show little whitish on crown and little chestnut on whitish throat*** but many females, and always 1st autumn birds, resemble similar Yellowhammer, ***but yellow in plumage replaced by white, including belly and fringes to primaries***; lesser coverts more uniform grey-brown and bill more frequently bi-coloured than Yellowhammer's (dark grey upper, pale grey lower mandible). Some males from hybrid-zone (W Siberia) often have yellow tinge in plumage (head, fore-body and primaries), others appear mainly white-headed; female hybrids probably indistinguishable. See also Rock and White-capped Bunting.
Voice: some calls identical with Yellowhammer's; also utters a nervous 'trr-rrr-rrr-ick'.

705 **Yellowhammer** *Emberiza citrinella* **PLATE 108**

L: 16.5. Medium-sized bunting with grey bill, white in tail ***and rusty rump at all ages; male's yellow crown and throat, particularly in summer, diagnostic; female's yellowish moustache, sometimes supercilium and centre of crown and always belly separate it from similar Pine.*** 1st autumn birds told (with difficulty) from similar Pine ***by some yellow on belly*** and on primary-edges. Hybridizes with Pine in W Siberia; such males often (but not always) show white parts of head sullied yellow (see Pine); separated from 1st autumn Cirl by combination of some yellow in streaked plumage and by colour of rump (grey-brown in Cirl).
Voice: calls are metallic; a sharp 'staeup', and fast 'twe-tic' in flight. Song a rapid series: 'dze-dze-dze-dze-dze-dze-chwee', last note drawn-out and lower in pitch.

706 **Cirl Bunting** *Emberiza cirlus* **PLATE 108**

L: 16.5. Medium-sized bunting ***with grey-brown rump. Male's combination of black and yellow striped head, black throat*** (obscured in winter by pale fringes), ***green breast-band and yellowish underparts with rusty sides to lower breast instantly identify it.*** Female and 1st autumn birds are grey-brown and dark-streaked, recalling young Yellowhammer ***but lacking rufous rump of that species***; scapulars are chestnut-brown but lesser coverts olive-grey (often hidden by former); grey bill separates it from the pink-billed Ortolan, Cretzschmar's and Grey-necked Bunting.
Voice: song a fast, dry 'ze-ze-ze-ze-ze-ze' which can recall Lesser Whitethroat's or Yellowhammer's without the final note; call a fine, drawn-out 'ziiid' and a soft, clear 'zip', quite unlike the hard metallic note of Yellowhammer.

707 **White-capped Bunting** *Emberiza stewarti* **PLATE 108**

L: 14.5. ***Small bunting with unstreaked reddish chestnut lower back and rump, and relatively small fine dark bill.*** Male superficially recalls male Pine ***but whitish grey fore-body and head with black 'bib' and eye-stripe and characteristic chestnut band on lower breast*** (limited to sides in winter). Female resembles small 1st autumn female Pine ***but has chestnut scapulars in grey-brown surroundings; neater, sharper streaked underparts have rufous-chestnut patch at sides of lower breast*** (partially

crossing breast as thin line); crown less conspicuously dark-streaked. Juvenile lacks rufous at sides of breast.
Voice: song recalls fast jingling Cirl Bunting's 'dzyn-dzyn-dzyn-dzyn'; calls include sonorous 'tsik', a rolling 'turrit' or 'ru-ti-ti' recalling Linnet and European Serin; also sharp 'tit' and 'tsip-ip'.

708 **Rock Bunting** *Emberiza cia* — **Plate 108**

L: 16. ***Male's combination of black-streaked pale ash-grey head, grey unstreaked throat, boldly black-streaked mantle, grey bill and white sides to tail separates it from House and African Rock Bunting.*** Female duller, more diffusely patterned but basically similar to male. Juvenile recalls young Yellowhammer (both have chestnut-brown rump) ***but Rock has orange belly at all ages.***
Voice: call a thin 'zip' or 'zeet'. Song long, fast and high-pitched 'sit, siterit, sit-sit-siterit, sit' last part variable in rhythm and composition.

709 **House Bunting** *Emberiza striolata* — **Plate 108**

L: 14. ***Small rufous-coloured bunting with orange-yellow lower mandible*** (adult) ***and rufous-edged outer tail-feathers; streaks on upperparts and wing-coverts thin and vague. Typical male has head striped dark ash-grey and white***; some have less well defined stripes while others are closer to African Rock (but much variation), ***but throat and upper breast are steel-grey with black speckles*** (black 'bib' in male African Rock); ***upperparts, wing-coverts, lower breast and belly bright rufous*** (upperparts brown with dark streaks and underparts cinnamon in African Rock). Female duller with much more diffused head pattern. Usually fairly shy and unobtrusive.
Voice: song a 'wi-di-dji-du-wi-di-dü' or 'witch witch a wee' with Chaffinch-like delivery. Calls include nasal 'tzswee', thin sharp 'tchick' and 'sweee-doo'.

710 **African Rock Bunting** *Emberiza tahapisi* — **Plate 108**
(Cinnamon-breasted Rock Bunting)

L: 16.5. Medium-sized ***cinnamon-brown bunting with orange or yellow lower mandible, bold black streaks on upperparts and rufous-edged outer tail-feathers. Male has black throat and four white stripes over black head.*** Female duller with less pure black and white head pattern. Juvenile resembles female but head and throat grey-brown, latter flecked blackish onto upper breast; pale head-stripes buffier, almost without a pale crown-streak. Best separated from slightly smaller House Bunting ***by unstreaked dark throat, boldly streaked upperparts and absence of rufous on mantle and wing-coverts.*** Rock Bunting (unrecorded in range of African Rock) ***has grey bill, white in outer tail-feathers and pale grey throat.*** See also Socotra Bunting.
Voice: song fast and short 'dzit-dzit-dzi-re-ra' or 'tru-tri-tre-ririr', second and last note higher in pitch; song sometimes followed by scratchy notes. Call a soft metallic 'anh' and a nasal 'daar'; alarm a short metallic 'ptik'.

711 **Socotra Bunting** *Emberiza socotrana* **PLATE 108**

L: 13. A rather small bunting easily told from African Rock Bunting, which also occurs on Socotra, by rich ***rufous coverts, white underparts with pale rufous wash on breast,*** pattern of black and white on head, ***and white throat***; tail is dark brown, ***without white outertail-feathers***; in flight, ***whitish band across lower back*** is a useful feature. Female and immatures similar to male but duller. Tends to be rather shy. Occasionally associates with flocks of African Rock Buntings.

Voice: call a thin high 'tseep', often followed by a soft gurgle. Song a musical whistling 'hue-he, hu-hey'.

712 **Cinereous Bunting** *Emberiza cineracea* **PLATE 109**

L: 16.5. Medium-sized bunting ***with blue-grey bill, narrow yellow-white eye-ring*** and white in outer tail-feathers. ***Male's unstreaked yellowish green crown, sides of face and throat, but otherwise greyish plumage, including unstreaked grey rump, identify it***; in autumn upper breast softly streaked darker. Iranian and E Turkish race, *semenowi*, has yellow-grey crown, yellowish (not whitish) lower belly and undertail-coverts. Olive grey-brown female streaked like autumn male; ***throat has buffish or yellowish tinge, streaked crown tinged grey-brown or yellowish-brown.*** Browner and more streaked 1st autumn bird (with moderately streaked grey-brown rump) told from other grey-brown buntings by lack of rufous rump, pale eye-ring, less patterned cheeks and ear-coverts and by absence of chestnut-brown scapulars; pale eye-ring and dark malar-streak bear resemblance to 1st autumn Ortolan ***but bill grey*** (pink in Ortolan).

Voice: call-note a short metallic 'kjip' or 'djib'. Brief, simple song has 5–6 rising notes 'drip-drip-drip-drip-drie-*drieh*'.

713 **Ortolan Bunting** *Emberiza hortulana* **PLATE 109**

L: 16.5. Medium-sized bunting ***with yellow-white eye-ring, pink bill, boldly streaked mantle and blackish brown tertials with clear-cut light chestnut-brown*** (adult) ***or creamy buff notched edges. Male separated from similar Cretzschmar's by olive-greyish head and breast band, which frames pale yellowish throat*** (in Cretzschmar's head and breast clear blue-grey with rusty brown throat). Female duller, less olive, more brownish-tinged and more streaked; often separable from female Cretzschmar's by pale yellow-buff throat (***never any rusty***), but note throat sometimes yellowish in Cretzschmar's), brownish, not rufous-tinged rump and pale buff, less rufous-buff vent and undertail-coverts. 1st autumn birds more streaked above and on breast than female; hard to separate from similar Cretzschmar's though colour of rump, vent and undertail-coverts may be useful. ***Told from Grey-necked*** (which see) ***by much bolder streaked upperparts and darker ground-colour to flight-feathers with more defined, notched tertial pattern.***

Voice: song, usually a series of ringing notes with slight Yellowhammer resemblance, 4–5 notes in one tone, with later 1–3 notes in a lower tone, 'tsee-tsee-tsee-tsee-trü (trü-trü)'. Call a hollow, soft 'plet' or 'büb' and, often on migration, a short, slightly falling 'sliie'.

714 **Grey-necked Bunting** *Emberiza buchanani* **PLATE 109**

L: 16. Resembles Cretzschmar's and Ortolan ***in having pink bill and whitish eye-ring.*** Adult has dull ashy grey head, less bright than in Cretzschmar's ***and lacks grey breast band; underparts light rufous-brown with whitish throat/submoustachial streak and pale yellow-buff or whitish vent and undertail-coverts***; in autumn (fresh plumage) ***rufous-buff underparts fringed greyish white*** (Cretzschmar's has rusty throat and rufous-buff vent and undertail-coverts); ***upperparts brownish grey contrasting with chestnut scapulars; streaks above indistinct or virtually absent; rump brownish grey*** (Cretzschmar's is rufous-brown above with bold dark streaks and rump is rufous); ***the flight-feathers are dark grey-brown rather than blackish brown, the tertials with ill-defined dusky brown edges without clear-cut notched pattern of Cretzschmar's and Ortolan.*** Juvenile separated from similar Cretzschmar's by brownish grey, not rufous-tinged rump, paler vent and undertail-coverts and from both Cretzschmar's and Ortolan ***by much less boldly streaked upperparts, paler flight-feathers with ill-defined tertial pattern.***

Voice: song loud and rich, typically 'di-di-dew, de-dew', penultimate note higher pitched and more stressed; delivery can be fast or slurred. Flight-call 'tip' or 'tsip' or 'sik'; also 'chep' or 'tcheup'.

715 **Cretzschmar's Bunting** *Emberiza caesia* **PLATE 109**

L: 16.5. Resembles bright version of Ortolan; ***male has bright blue-grey head and breast band framing rusty orange throat, rufous rump and rufous-buff vent and undertail-coverts*** (in Ortolan greenish grey head and breast frame pale yellowish throat, rump is brown, vent and undertail-coverts pale buff); belly orange-chestnut; blackish brown tertials with clear-cut rufous-chestnut (adult) notched fringe. Female duller, often with traces of ash-grey on head, but some grey on streaked breast; ***told from female Ortolan by some rusty on throat*** (though sometimes yellowish buff) and warmer rump, vent and undertail-coverts. 1st autumn birds resemble Ortolan much ***but yellow-buff throat sometimes tinged rufous and, as autumn progresses, male approaches adult***; colour of rump, vent and undertail-coverts perhaps useful in separating 1st autumn females. See Grey-necked for separation.

Voice: song recalls simplified Ortolan's, but shorter, thinner, more metallic and without Yellowhammer-resemblance, almost invariably with the second part confined to a single 'diu-diu-diu-düü'. Two main calls, one similar to Ortolan's 'blep', the other a metallic 'tlik' or 'tlev', sharper than Ortolan's.

716 **Rustic Bunting** *Emberiza rustica* **PLATE 109**

L: 14.5. Medium–small robust bunting ***with longish bill with straight culmen and much rufous in plumage. Male unmistakable with black head broken by bold white supercilium and nape spot, bright rusty red nape merging on sides of neck with rusty red breast band; flanks with extensive rusty streaks; rump rusty red.*** Female has less blackish head with paler median crown-streak and ear-coverts. 1st autumn birds mainly confused with Reed Bunting ***but rufous streaks on breast and flanks, rufous rump, bill shape, bolder supercilium behind eye, whiter wing-bars and call***

identify it (Reed has convex culmen, usually diffuse wing-bars, breast and flanks streaked blackish, grey-brown rump and different call). Often discovered by call.
Voice: call a Song Thrush-like short, distinct 'zib' or 'zik', less sharp than Little Bunting's.

717 **Little Bunting** *Emberiza pusilla* — PLATE 109

L: 13. Small secretive bunting ***with straight culmen. Adult has reddish chestnut head with black stripe at sides of crown framing red-brown median crown-streak; red-brown cheeks and ear-coverts usually framed by thin blackish line at rear; thin whitish eye-ring; white underparts with fairly narrow, well defined black streaks on breast***; buff-white wing-bars usually obvious. 1st autumn bird separated from similar Reed ***by bill shape, whitish eye-ring, almost uniform rusty yellow-brown cheeks and ear-coverts, a black malar streak usually not reaching bill, grey-brown lesser coverts, more defined blacker streaks on whiter underparts and by call*** (Reed has convex culmen, lacks white eye-ring, has variegated black, brown and white cheeks, malar-streak reaching bill, reddish brown lesser coverts, less defined blackish streaks on breast and different call). Separated from Rustic ***by absence of rufous streaks below and grey-brown, not rusty, rump.***
Voice: call a hard, metallic 'tik', like that of Rustic Bunting, but slightly sharper.

718 **Yellow-breasted Bunting** *Emberiza aureola* — PLATE 110

L: 15. Medium-sized bunting with pale, stout, conical bill and white in tail. ***Male's black face and upper throat, dark chestnut-brown crown to rump and narrow breast band in otherwise bright yellow underparts and white wing-bars diagnostic.*** Female and 1st autumn birds told ***by combination of prominent creamy white supercilium, faint pale central fore-crown stripe with dark brown lateral crown-stripe*** (head pattern less defined in 1st autumn birds), ***pale dark-bordered cheeks and ear-coverts, fairly distinct white wing-bars***, largely pale bill and relatively short tail. Underparts yellowish white or grey (mainly 1st autumn) with buffish white undertail-coverts; flanks, and sometimes breast, faintly streaked; female sometimes with warm buffish yellow sides of neck; upperparts streaked blackish and greyish yellow (sometimes tinged rusty) and often with creamy 'braces'.
Voice: two main calls on passage; a short 'tik' not unlike that of Little Bunting, and a softer, metallic 'tsiu' or 'tsip'.

719 **Reed Bunting** *Emberiza schoeniclus* — PLATE 110
(Common Reed Bunting)

L: 15.5. ***Male's black hood and 'bib', white half-collar on neck and white 'moustache' instantly identify it.*** Female ***has white throat, broad black malar streak and whitish supercilium*** and sides of neck; head otherwise darkish brown. 1st autumn birds resemble female but cheeks more mottled, breast and flanks more streaked. Separated from Little Bunting ***by stubby bill with convex culmen, inconspicuous pale eye-ring, bright chestnut lesser coverts*** (often hard to see) ***and by call*** (see Little for further differences). Some Reed Buntings from Turkey, Iran and Iraq are more strongly marked above than European winter visitors to Turkey, ***with stouter bills.*** When perched, jerks and spreads tail nervously. In flight, undulations short and irregular, compared with the more steady flight of most other buntings.

Voice: slow short song variable, often uttered as 'tsee-tsee-tsee*a*-tsiririrr'. Calls include a fine, drawn-out 'tsii-u' and a metallic, slightly voiced 'bzü; in flight, a low nasal 'bäh'.

720 **Red-headed Bunting** *Emberiza bruniceps* **PLATE 110**

L: 16.5. Medium–large bunting ***without white in outer tail-feathers. Male diagnostic with red-brown head and breast*** (subdued by yellow and white in autumn), ***with rest of underparts and rump yellow.*** Female (some with tendency to male-coloured head) and 1st autumn birds very like similar Black-headed and some are not identifiable ***but note Red-headed's grey-brown, often olive-tinged upperparts including rump and scapulars*** (tinged rufous in many Black-headed); head slightly more uniform with less streaked fore-crown and paler ear-coverts than Black-headed; underparts, particularly undertail-coverts, tend to be tinged less yellowish than in Black-headed, some being uniform buffish grey below. Hybridizes with Black-headed SE of Caspian Sea; such males have mixed head-colour and females are not separable. Told from other buntings of similar size ***by combination of unstreaked underparts and tail-pattern.***
Voice: song very like Black-headed, clear and far-carrying; starts quietly but develops with descending 'tsit tsit tsi tsoo'. Call more voiced than Yellowhammer's, a 'bzist'.

721 **Black-headed Bunting** *Emberiza melanocephala* **PLATE 110**

L: 16.5. Heavy-bodied, stout-billed bunting ***without white in tail. Male diagnostic with black head, yellow underparts and unstreaked chestnut upperparts***; black of head subdued by pale fringes in autumn. Female and 1st autumn birds lack black, bright chestnut and bright yellow; almost identical with Red-headed ***but upperparts usually toned warmer*** (less olive grey-brown) and underparts generally slightly more tinged yellow, especially undertail-coverts. See Red-headed for further differences.
Voice: variable song recalls Whitethroat's in staccato form but is strong and typically bunting-like, starting with short harsh notes, ending with more ringing metallic notes 'tsi-tia-tia-tia-terlu-terlu-terlu'. Call a sparrow/Tawny Pipit-like 'tjilp'; also a metallic 'tlev'.

722 **Corn Bunting** *Miliaria calandra* **PLATE 110**

L: 18. Bulky grey-brown bunting with fairly large head, ***large conical bill and no white in tail.*** Underparts heavily streaked, in some merging into irregular black spot on breast; often pale submoustachial stripe and variable ill-defined malar streak. ***Heavy fluttering flight, legs often dangling***; lacks white on trailing edge of wings of Skylark with which it can be confused, though bill of different shape. Often perches on telegraph wires. In flocks outside breeding season.
Voice: song, often from exposed perch, accelerates ending with monotonous almost insect-like sound 'tick-tick-tzek- zee-zri-zizizizi'. Call a hard, almost clicking 'tik'.

REFERENCES AND FURTHER READING

Publications marked with an asterisk include information on identification; the others cover status and distribution only.

Some of these titles may be out of print, but can be obtained from most good libraries. Several of those in print can be obtained from the Ornithological Society of the Middle East (Sales) c/o The Lodge, Sandy, Bedfordshire SG19 2DL, UK, which produces a regularly updated publications list.

In addition, the monthly journals ***British Birds,*** and ***Birding World*** and the bi-monthly ***Dutch Birding*** regularly contain information relevant to the identification and status of birds in the Middle East.

General

*Sandgrouse (The journal of the Ornithological Society of the Middle East–OSME).

*OSME Bulletin

The above two publications, which OSME members receive, contain a wealth of information about the birds of the Middle East. Write to OSME (address above) for a contents and price list of past issues.

*Handbuch der Vögel Mitteleuropas Vols 1–13/III. U N Glutz, K M Bauer and E Bezzel (1966–1993). Akademische Verlagsgesellschaft/AULA-Verlag, Wiesbaden.

*Birds of the Western Palearctic Vols 1–9. S Cramp *et al.* (1977–1994) Oxford University Press, Oxford.

*Birds of the Middle East and North Africa. P A D Hollom, R F Porter, S Christensen and I Willis (1988). T & A D Poyser, London.

Photographic Guide to the Birds of Britain and Europe. Håkan Delin and Lars Svensson (1988). Hamlyn, London.
Birds of Europe with North Africa and the Middle East. Lars Jonsson (1992). Christopher Helm, London.
Identification Guide to European Passerines. Lars Svensson (1992). 4th Edition. Published by the author.
Important Bird Areas in the Middle East. M I Evans. (1994) BirdLife Conservation Series 2. BirdLife International, Cambridge.
The Birds of Britain and Europe with North Africa and the Middle East. Hermann Heinzel, Richard Fitter and John Parslow (1995). 5th Edition. HarperCollins, London.
An Interim Atlas of the Breeding Birds of Arabia. Michael C Jennings (1995). National Commission for Wildlife Conservation and Development, Saudi Arabia.
The Macmillan Birder's Guide to European and Middle Eastern Birds. David A Christie, Hadoram Shirihai and Alan Harris (1996). Macmillan, London.

Bahrain

The Birds of Bahrain. Tom Nightingale and Mike Hill (1993). Immel, London.
Birds of Bahrain and the Arabian Gulf. S A Mohamed (1993). Bahrain Centre for Research Studies.
Birds in Bahrain: A Study of their Migration Patterns 1990–92. Erik Hirschfeld (1995). Hobby Publications, Dubai.

Cyprus

The Birds of Cyprus. Peter R Flint and Peter F Stewart (1992). BOU checklist. 2nd edition.

Israel

Birdwatching in the Deserts of Israel. H Shirihai and O Bahat (1993).
Bird Songs of Israel and the Middle East. Sound recordings by K Mild. 2 cassettes.
The Birds of Israel. Hadoram Shirihai (1996). Academic Press, London.

Jordan

The Birds of the Hashemite Kingdom of Jordan. Ian J Andrews (1995). Published by the author.

Lebanon

Birds of Lebanon and the Jordan Area. S Vere Benson (1970). ICBP, Cambridge.

Oman

The Birds of Oman.* Michael Gallagher and Martin W Woodcock (1980). Quartet Books, London.

A New List of the Birds of Masirah Island. T D Rogers (1988). Oman Bird Record Committee.

Oman Bird List (Edition 4) Oman Bird Records Committee (January 1994). *The official list of the birds of the Sultanate of Oman, which is regularly updated.*

Qatar

A Birdwatcher's Guide to Qatar. C Oldfield and J Oldfield (1994). Published by the authors.

Saudi Arabia

Birds of the Eastern Province of Saudi Arabia. G Bundy, R J Connor and C J O Harrison (1989). H F & G Witherby, London.

The Birds of Saudi Arabia: A Check-list. Michael C Jennings (1981). Published by the author.

Syria

Die Vögel Syriens: Eine Übersicht. Wolfgang Baumgart (1995). Max Kasparek Verlag (in German), Heidelberg.

Turkey

Birds of Turkey (series) Max Kasparek Verlag (1983–1993)

1. Erçek Gölü
2. Seyfe Gölü
3. Kizilcahamam
4. Kizilirmak Delta
5. Kulu Gölü
6. Yeniçaga Gölü
7. Acigol
8. Köycegiz-Dalyan Area
9. Hotamis Marshes
10. Çol Gölü

Der Bafasee (Bafa Golu). M Kasparek (1988). Max Kasparek Verlag (in German), Heidelberg.

Die Vögel der Türkei. M Kasparek (1992). Max Kasparek Verlag (in German). *An annotated checklist.*

A Birdwatchers' Guide to Turkey. Ian Green and Nigel Moorhouse (1995). Prion, Perry.

Songbirds of Turkey: An Atlas of Biodiversity of Turkish Passerine Birds. C S Roselaar (1995). Pica Press, Robertsbridge.

United Arab Emirates

The Birds of the United Arab Emirates. Colin Richardson (1990). Hobby Publications, Warrington and Dubai.
Status and Conservation of Breeding Birds of the United Arab Emirates. Simon Aspinall (1996). Hobby Publications, Liverpool and Dubai.

Yemen

Sandgrouse: Vol 9.
Covers the birds of northern Yemen.
Sandgrouse: Vol 17.
Covers the birds of southern Yemen and Socotra.

SPECIES LIST

Red-throated Diver *Gavia stellata*
Black-throated Diver *Gavia arctica*
Great Northern Diver *Gavia immer*
Little Grebe *Tachybaptus ruficollis*
Great Crested Grebe *Podiceps cristatus*
Red-necked Grebe *Podiceps grisegena*
Slavonian Grebe *Podiceps auritus*
Black-necked Grebe *Podiceps nigricollis*
Shy Albatross *Diomedea cauta*
Cape Petrel *Daption capensis*
Atlantic Petrel *Pterodroma incerta*
Soft-plumaged Petrel *Pterodroma mollis*
Jouanin's Petrel *Bulweria fallax*
Cory's Shearwater *Calonectris diomedea*
Streaked Shearwater *Calonectris leucomelas*
Pale-footed Shearwater *Puffinus carneipes*
Great Shearwater Puffinus gravis
Wedge-tailed Shearwater *Puffinus pacificus*
Sooty Shearwater *Puffinus griseus*
Mediterranean Shearwater *Puffinus yelkouan*
Little Shearwater *Puffinus assimilis*
Persian Shearwater *Puffinus (lherminieri) persicus*
Wilson's Storm-petrel *Oceanites oceanicus*
Black-bellied Storm-petrel *Fregetta tropica*
White-bellied Storm-petrel *Fregetta grallaria*
White-faced Storm-petrel *Pelagodroma marina*
European Storm-petrel *Hydrobates pelagicus*
Leach's Storm-petrel *Oceanodroma leucorhoa*
Swinhoe's Storm-petrel *Oceanodroma monorhis*
Madeiran Storm-petrel Oceanodroma castro
Red-billed Tropicbird *Phaethon aethereus*
Red-footed Booby *Sula sula*
Masked Booby *Sula dactylatra*
Brown Booby *Sula leucogaster*
Gannet *Sula bassana*
Cormorant *Phalacrocorax carbo*
Shag *Phalacrocorax aristotelis*
Socotra Cormorant *Phalacrocorax nigrogularis*
Pygmy Cormorant *Phalacrocorax pygmeus*
Long-tailed Cormorant *Phalacrocorax africanus*
Darter *Anhinga rufa*
White Pelican *Pelecanus onocrotalus*
Dalmatian Pelican *Pelecanus crispus*
Pink-backed Pelican *Pelecanus rufescens*
Great Frigatebird *Fregata minor*
Lesser Frigatebird *Fregata ariel*
Bittern *Botaurus stellaris*
Little Bittern *Ixobrychus minutus*
Night Heron *Nycticorax nycticorax*
Striated Heron *Butorides striatus*
Squacco Heron *Ardeola ralloides*
Indian Pond Heron *Ardeola grayii*
Cattle Egret *Bubulcus ibis*
Western Reef Heron *Egretta gularis*
Little Egret *Egretta garzetta*
Intermediate Egret *Egretta intermedia*
Great White Egret *Egretta alba*
Black-headed Heron *Ardea melanocephala*
Grey Heron *Ardea cinerea*
Purple Heron *Ardea purpurea*
Goliath Heron *Ardea goliath*
Hamerkop *Scopus umbretta*
Yellow-billed Stork *Mycteria ibis*
Black Stork *Ciconia nigra*
Abdim's Stork *Ciconia abdimii*
Woolly-necked Stork *Ciconia episcopus*
White Stork *Ciconia ciconia*
Marabou Stork *Leptoptilos crumeniferus*
Glossy Ibis *Plegadis falcinellus*
Bald Ibis *Geronticus eremita*
Sacred Ibis *Threskiornis aethiopicus*
Spoonbill *Platalea leucorodia*
African Spoonbill *Platalea alba*
Greater Flamingo *Phoenicopterus ruber*
Lesser Flamingo *Phoenicopterus minor*

Mute Swan *Cygnus olor*
Bewick's Swan *Cygnus columbianus*
Whooper Swan *Cygnus cygnus*
Bean Goose *Anser fabalis*
White-fronted Goose *Anser albifrons*
Lesser White-fronted Goose *Anser erythropus*
Greylag Goose *Anser anser*
Brent Goose *Branta bernicla*
Red-breasted Goose *Branta ruficollis*
Egyptian Goose *Alopochen aegyptiacus*
Ruddy Shelduck *Tadorna ferruginea*
Shelduck Tadorna *tadorna*
Comb Duck *Sarkidiornis melanotos*
Cotton Teal *Nettapus coromandelianus*
Wigeon *Anas penelope*
Falcated Duck *Anas falcata*
Gadwall *Anas strepera*
Teal *Anas crecca*
Mallard *Anas platyrhynchos*
Pintail *Anas acuta*
Garganey *Anas querquedula*
Shoveler *Anas clypeata*
Marbled Teal *Marmaronetta angustirostris*
Red-crested Pochard *Netta rufina*
Pochard *Aythya ferina*
Ferruginous Duck *Aythya nyroca*
Tufted Duck *Aythya fuligula*
Scaup *Aythya marila*
Eider *Somateria mollissima*
Long-tailed Duck *Clangula hyemalis*
Common Scoter *Melanitta nigra*
Velvet Scoter *Melanitta fusca*
Goldeneye *Bucephala clangula*
Smew *Mergus albellus*
Red-breasted Merganser *Mergus serrator*
Goosander *Mergus merganser*
White-headed Duck *Oxyura leucocephala*
Honey Buzzard *Pernis apivorus*
Crested Honey Buzzard *Pernis ptilorhynchus*
Black-winged Kite *Elanus caeruleus*
African Swallow-tailed Kite *Chelictinia riocourii*
Black Kite *Milvus migrans*
Red Kite *Milvus milvus*
Brahminy Kite *Haliastur indus*
Pallas's Fish Eagle *Haliaeetus leucoryphus*
White-tailed Eagle *Haliaeetus albicilla*
Lammergeier *Gypaetus barbatus*
Egyptian Vulture *Neophron percnopterus*
Indian White-backed Vulture *Gyps bengalensis*
Griffon Vulture *Gyps fulvus*
Rüppell's Vulture *Gyps rueppellii*
Lappet-faced Vulture *Torgos tracheliotos*
Black Vulture *Aegypius monachus*
Short-toed Eagle *Circaetus gallicus*
Bateleur *Terathopius ecaudatus*
Marsh Harrier *Circus aeruginosus*
Hen Harrier *Circus cyaneus*
Pallid Harrier *Circus macrourus*
Montagu's Harrier *Circus pygargus*
Dark Chanting Goshawk *Melierax metabates*
Gabar Goshawk *Micronisus gabar*
Goshawk *Accipiter gentilis*
Sparrowhawk *Accipiter nisus*
Shikra *Accipiter badius*
Levant Sparrowhawk *Accipiter brevipes*
White-eyed Buzzard *Butastur teesa*
Common Buzzard *Buteo buteo*
Long-legged Buzzard *Buteo rufinus*
Rough-legged Buzzard *Buteo lagopus*
Lesser Spotted Eagle *Aquila pomarina*
Greater Spotted Eagle *Aquila clanga*
Tawny Eagle *Aquila rapax*
Steppe Eagle *Aquila nipalensis*
Imperial Eagle *Aquila heliaca*
Golden Eagle *Aquila chrysaetos*
Verreaux's Eagle *Aquila verreauxii*
Booted Eagle *Hieraaetus pennatus*
Bonelli's Eagle *Hieraaetus fasciatus*
Osprey *Pandion haliaetus*
Lesser Kestrel *Falco naumanni*
Kestrel *Falco tinnunculus*
Red-headed Merlin *Falco chicquera*
Red-footed Falcon *Falco vespertinus*
Amur Falcon *Falco amurensis*
Merlin *Falco columbarius*
Hobby *Falco subbuteo*
Eleonora's Falcon *Falco eleonorae*
Sooty Falcon *Falco concolor*
Lanner Falcon *Falco biarmicus*
Saker Falcon *Falco cherrug*
Peregrine Falcon *Falco peregrinus*
Barbary Falcon *Falco pelegrinoides*
Caucasian Black Grouse *Tetrao mlokosiewiczi*
Caspian Snowcock *Tetraogallus caspius*
Chukar *Alectoris chukar*
Philby's Partridge *Alectoris philbyi*
Arabian Partridge *Alectoris melanocephala*
See-see Partridge *Ammoperdix griseogularis*
Sand Partridge *Ammoperdix heyi*
Black Francolin *Francolinus francolinus*
Grey Francolin *Francolinus pondicerianus*
Grey Partridge *Perdix perdix*
Quail *Coturnix coturnix*
Harlequin Quail *Coturnix delegorguei*

Pheasant *Phasianus colchicus*
Helmeted Guineafowl *Numida meleagris*
Little Button Quail *Turnix sylvatica*
Water Rail *Rallus aquaticus*
Spotted Crake *Porzana porzana*
Little Crake *Porzana parva*
Baillon's Crake *Porzana pusilla*
Corncrake *Crex crex*
White-breasted Waterhen *Amaurornis phoenicurus*
Moorhen *Gallinula chloropus*
Lesser Moorhen *Gallinula angulata*
Allen's Gallinule *Porphyrula alleni*
Purple Gallinule *Porphyrio porphyrio*
Watercock *Gallicrex cinerea*
Coot *Fulica atra*
Common Crane *Grus grus*
Siberian White Crane *Grus leucogeranus*
Demoiselle Crane *Anthropoides virgo*
Little Bustard *Tetrax tetrax*
Houbara Bustard *Chlamydotis undulata*
Arabian Bustard *Ardeotis arabs*
Great Bustard *Otis tarda*
Pheasant-tailed Jacana *Hydrophasianus chirurgus*
Painted Snipe *Rostratula benghalensis*
Oystercatcher *Haematopus ostralegus*
Black-winged Stilt *Himantopus himantopus*
Avocet *Recurvirostra avosetta*
Egyptian Plover *Pluvianus aegyptius*
Crab Plover *Dromas ardeola*
Stone Curlew *Burhinus oedicnemus*
Senegal Thick-knee *Burhinus senegalensis*
Spotted Thick-knee *Burhinus capensis*
Great Stone Plover *Esacus recurvirostris*
Cream-coloured Courser *Cursorius cursor*
Collared Pratincole *Glareola pratincola*
Oriental Pratincole *Glareola maldivarum*
Black-winged Pratincole *Glareola nordmanni*
Little Pratincole *Glareola lactea*
Little Ringed Plover *Charadrius dubius*
Ringed Plover *Charadrius hiaticula*
Kittlitz's Plover *Charadrius pecuarius*
Kentish Plover *Charadrius alexandrinus*
Lesser Sand Plover *Charadrius mongolus*
Greater Sand Plover *Charadrius leschenaultii*
Caspian Plover *Charadrius asiaticus*
Dotterel *Charadrius morinellus*
American Golden Plover *Pluvialis dominica*
Pacific Golden Plover *Pluvialis fulva*
European Golden Plover *Pluvialis apricaria*
Grey Plover *Pluvialis squatarola*
Spur-winged Plover *Hoplopterus spinosus*
Black-headed Plover *Hoplopterus tectus*
Red-wattled Plover *Hoplopterus indicus*
Sociable Plover *Chettusia gregaria*
White-tailed Plover *Chettusia leucura*
Lapwing *Vanellus vanellus*
Great Knot *Calidris tenuirostris*
Knot *Calidris canutus*
Sanderling *Calidris alba*
Red-necked Stint *Calidris ruficolli*
Little Stint *Calidris minuta*
Temminck's Stint *Calidris temminckii*
Long-toed Stint *Calidris subminuta*
Baird's Sandpiper *Calidris bairdii*
Pectoral Sandpiper *Calidris melanotos*
Sharp-tailed Sandpiper *Calidris acuminata*
Curlew Sandpiper *Calidris ferruginea*
Dunlin *Calidris alpina*
Broad-billed Sandpiper *Limicola falcinellus*
Buff-breasted Sandpiper *Tryngites subruficollis*
Ruff *Philomachus pugnax*
Jack Snipe *Lymnocryptes minimus*
Common Snipe *Gallinago gallinago*
Great Snipe *Gallinago media*
Pintail Snipe *Gallinago stenura*
Solitary Snipe *Gallinago solitaria*
Long-billed Dowitcher *Limnodromus scolopaceus*
Asiatic Dowitcher *Limnodromus semipalmatus*
Woodcock *Scolopax rusticola*
Black-tailed Godwit *Limosa limosa*
Bar-tailed Godwit *Limosa lapponica*
Whimbrel *Numenius phaeopus*
Slender-billed Curlew *Numenius tenuirostris*
Curlew *Numenius arquata*
Far Eastern Curlew *Numenius madagascariensis*
Spotted Redshank *Tringa erythropus*
Redshank *Tringa totanus*
Marsh Sandpiper *Tringa stagnatilis*
Greenshank *Tringa nebularia*
Green Sandpiper *Tringa ochropus*
Wood Sandpiper *Tringa glareola*
Terek Sandpiper *Xenus cinereus*
Common Sandpiper *Actitis hypoleucos*
Spotted Sandpiper *Actitis macularia*
Turnstone *Arenaria interpres*
Wilson's Phalarope *Phalaropus tricolor*
Red-necked Phalarope *Phalaropus lobatus*
Grey Phalarope *Phalaropus fulicarius*
Pomarine Skua *Stercorarius pomarinu*

Arctic Skua *Stercorarius parasiticus*
Long-tailed Skua *Stercorarius longicaudus*
Great Skua *Catharacta skua*
South Polar Skua *Catharacta maccormicki*
Sooty Gull *Larus hemprichii*
White-eyed Gull *Larus leucophthalmus*
Great Black-headed Gull *Larus ichthyaetus*
Mediterranean Gull *Larus melanocephalus*
Little Gull *Larus minutus*
Sabine's Gull *Larus sabini*
Black-headed Gull *Larus ridibundu*
Brown-headed Gull *Larus brunnicephalus*
Grey-headed Gull *Larus cirrocephalus*
Slender-billed Gull *Larus genei*
Audouin's Gull *Larus audouinii*
Common Gull *Larus canus*
Lesser Black-backed Gull *Larus fuscus* (fuscus, heuglini,taimyrensis)
Yellow-legged Gull *Larus cachinnans*
Armenian Gull *Larus armenicus*
Glaucous Gull *Larus hyperboreus*
Great Black-backed Gull *Larus marinus*
Kittiwake *Rissa tridactyla*
Gull-billed Tern *Gelochelidon nilotica*
Caspian Tern *Sterna caspia*
Swift Tern *Sterna bergii*
Lesser Crested Tern *Sterna bengalensis*
Sandwich Tern *Sterna sandvicensis*
Roseate Tern *Sterna dougallii*
Common Tern *Sterna hirundo*
Arctic Tern *Sterna paradisaea*
White-cheeked Tern *Sterna repressa*
Bridled Tern *Sterna anaethetus*
Sooty Tern *Sterna fuscata*
Little Tern *Sterna albifrons*
Saunders's Tern *Sterna saunders*
Whiskered Tern *Chlidonias hybridus*
Black-Tern *Chlidonias niger*
White-winged Black Tern *Chlidonias leucopterus*
Common Noddy *Anous stolidus*
Lesser Noddy *Anous tenuirostris*
African Skimmer *Rynchops flavirostris*
Indian Skimmer *Rynchops albicollis*
Lichtenstein's Sandgrouse *Pterocles lichtensteinii*
Crowned Sandgrouse *Pterocles coronatus*
Spotted Sandgrouse *Pterocles senegallus*
Chestnut-bellied Sandgrouse *Pterocles exustus*
Black-bellied Sandgrouse *Pterocles orientalis*
Pin-tailed Sandgrouse *Pterocles alchata*
Pallas's Sandgrouse *Syrrhaptes paradoxus*

Rock Dove *Columba livia*
Stock Dove *Columba oenas*
Eastern Stock Dove *Columba eversmanni*
Woodpigeon *Columba palumbus*
Olive Pigeon *Columba arquatrix*
African Collared Dove *Streptopelia roseogrisea*
Collared Dove *Streptopelia decaocto*
Red-eyed Dove *Streptopelia semitorquata*
Red Turtle Dove *Streptopelia tranquebarica*
Turtle Dove *Streptopelia turtur*
Dusky Turtle Dove *Streptopelia lugens*
Rufous Turtle Dove *Streptopelia orientalis*
Laughing Dove *Streptopelia senegalensis*
Namaqua Dove *Oena capensis*
Bruce's Green Pigeon *Treron waalia*
Ring-necked Parakeet *Psittacula krameri*
Alexandrine Parakeet *Psittacula eupatria*
Jacobin Cuckoo *Clamator jacobinus*
Great Spotted Cuckoo *Clamator glandarius*
Indian Hawk Cuckoo *Cuculus varius*
Didric Cuckoo *Chrysococcyx caprius*
Klaas's Cuckoo *Chrysococcyx klaas*
Plaintive Cuckoo *Cacomantis (merulinus) passerinus*
Common Cuckoo *Cuculus canorus*
White-browed Coucal *Centropus superciliosus*
Koel *Eudynamys scolopacea*
Barn Owl *Tyto alba*
Indian Scops Owl *Otus bakkamoena*
Striated Scops Owl *Otus brucei*
European Scops Owl *Otus scops*
African Scops Owl *Otus senegalensis*
Eagle Owl *Bubo bubo*
Spotted Eagle Owl *Bubo africanus*
Brown Fish Owl *Ketupa zeylonensis*
Snowy Owl *Nyctea scandiaca*
Little Owl *Athene noctua*
Spotted Little Owl *Athene brama*
Tawny Owl *Strix aluco*
Hume's Tawny Owl *Strix butleri*
Long-eared Owl *Asio otus*
Short-eared Owl *Asio flammeus*
Tengmalm's Owl *Aegolius funereus*
Plain Nightjar *Caprimulgus inornatus*
Nubian Nightjar *Caprimulgus nubicus*
Sykes's Nightjar *Caprimulgus mahrattensis*
Indian Nightjar *Caprimulgus asiaticus*
European Nightjar *Caprimulgus europaeus*
Egyptian Nightjar *Caprimulgus aegyptius*
Mountain Nightjar *Caprimulgus poliocephalus*
Common Swift *Apus apus*
Pallid Swift *Apus pallidus*
Forbes-Watson's Swift *Apus berliozi*

Alpine Swift *Apus melba*
White-rumped Swift *Apus caffer*
Little Swift *Apus affinis*
Palm Swift *Cypsiurus parvus*
White-breasted Kingfisher *Halcyon smyrnensis*
Grey-headed Kingfisher *Halcyon leucocephala*
White-collared Kingfisher *Halcyon chloris*
Common Kingfisher *Alcedo atthis*
Malachite Kingfisher *Alcedo cristata*
Pied Kingfisher *Ceryle rudis*
White-throated Bee-eater *Merops albicollis*
Little Green Bee-eater *Merops orientalis*
Blue-cheeked Bee-eater *Merops superciliosus*
European Bee-eater *Merops apiaster*
European Roller *Coracias garrulus*
Abyssinian Roller *Coracias abyssinicus*
Indian Roller *Coracias benghalensis*
Lilac-breasted Roller *Coracias caudata*
Rufous-crowned Roller *Coracias naevia*
Hoopoe *Upupa epops*
African Grey Hornbill *Tockus nasutus*
Wryneck *Jynx torquilla*
Grey-headed Woodpecker *Picus canus*
Green Woodpecker *Picus viridis*
Scaly-bellied Woodpecker *Picus squamatus*
Black Woodpecker *Dryocopus martius*
Great Spotted Woodpecker *Dendrocopos major*
White-winged Woodpecker *Dendrocopos leucopterus*
Syrian Woodpecker *Dendrocopos syriacus*
Sind Pied Woodpecker *Dendrocopos assimilis*
Middle Spotted Woodpecker *Dendrocopos medius*
White-backed Woodpecker *Dendrocopos leucotos*
Lesser Spotted Woodpecker *Dendrocopos minor*
Arabian Woodpecker *Dendrocopos dorae*
Singing Bush Lark *Mirafra cantillans*
Black-crowned Finch Lark *Eremopterix nigriceps*
Dunn's Lark *Eremalauda dunni*
Bar-tailed Desert Lark *Ammomanes cincturus*
Desert Lark *Ammomanes deserti*
Hoopoe Lark *Alaemon alaudipes*
Dupont's Lark *Chersophilus duponti*
Thick-billed Lark *Ramphocoris clotbey*
Calandra Lark *Melanocorypha calandra*
Bimaculated Lark *Melanocorypha bimaculata*
White-winged Lark *Melanocorypha leucoptera*
Black Lark *Melanocorypha yeltoniensis*
Red-capped Lark *Calandrella cinerea*
Short-toed Lark *Calandrella brachydactyla*
Hume's Short-toed Lark *Calandrella acutirostris*
Lesser Short-toed Lark *Calandrella rufescens*
Indian Sand Lark *Calandrella raytal*
Crested Lark *Galerida cristata*
Woodlark *Lullula arborea*
Small Skylark *Alauda gulgula*
Skylark *Alauda arvensis*
Shore Lark *Eremophila alpestris*
Temminck's Horned Lark *Eremophila bilopha*
Brown-throated Sand Martin *Riparia paludicola*
Sand Martin *Riparia riparia*
Banded Martin *Riparia cincta*
African Rock Martin *Ptyonoprogne fuligula*
Crag Martin *Ptyonoprogne rupestris*
Barn Swallow *Hirundo rustica*
Ethiopian Swallow *Hirundo aethiopica*
Wire-tailed Swallow *Hirundo smithii*
Lesser Striped Swallow *Hirundo abyssinica*
Red-rumped Swallow *Hirundo daurica*
House Martin *Delichon urbica*
Golden Pipit *Tmetothylacus tenellus*
Richard's Pipit *Anthus richardi*
Tawny Pipit *Anthus campestris*
Long-billed Pipit *Anthus similis*
Olive-backed Pipit *Anthus hodgsoni*
Tree Pipit *Anthus trivialis*
Pechora Pipit *Anthus gustavi*
Meadow Pipit *Anthus pratensis*
Red-throated Pipit *Anthus cervinus*
Blyth's Pipit *Anthus godlewskii*
Buff-bellied Pipit *Anthus rubescens*
Water Pipit *Anthus spinoletta*
Forest Wagtail *Dendronanthus indicus*
Yellow Wagtail *Motacilla flava*
Citrine Wagtail *Motacilla citreola*
Grey Wagtail *Motacilla cinerea*
White Wagtail *Motacilla alba*
White-cheeked Bulbul *Pycnonotus leucogenys*
Yellow-vented Bulbul *Pycnonotus xanthopygos*
Red-whiskered Bulbul *Pycnonotus jocosus*
Red-vented Bulbul *Pycnonotus cafer*
Waxwing *Bombycilla garrulus*
Grey Hypocolius *Hypocolius ampelinus*
Dipper *Cinclus cinclus*
Wren *Troglodytes troglodytes*
Dunnock *Prunella modularis*

Siberian Accentor *Prunella montanella*
Radde's Accentor *Prunella ocularis*
Arabian Accentor *Prunella fagani*
Black-throated Accentor *Prunella atrogularis*
Alpine Accentor *Prunella collaris*
Rufous Bush Robin *Cercotrichas galactotes*
Black Bush Robin *Cercotrichas podobe*
Robin *Erithacus rubecula*
Thrush Nightingale *Luscinia luscinia*
Nightingale *Luscinia megarhynchos*
Siberian Rubythroat *Luscinia calliope*
Bluethroat *Luscinia svecica*
Red-flanked Bluetail *Tarsiger cyanurus*
White-throated Robin *Irania gutturalis*
Eversmann's Redstart *Phoenicurus erythronotus*
Black Redstart *Phoenicurus ochruros*
Common Redstart *Phoenicurus phoenicurus*
Güldenstädt's Redstart *Phoenicurus erythrogaster*
Blackstart *Cercomela melanura*
Familiar Chat *Cercomela familiaris*
Whinchat *Saxicola rubetra*
Stonechat *Saxicola torquata*
Pied Stonechat *Saxicola caprata*
Isabelline Wheatear *Oenanthe isabellina*
Red-breasted Wheatear *Oenanthe bottae*
Northern Wheatear *Oenanthe oenanthe*
Pied Wheatear *Oenanthe pleschanka*
Cyprus Pied Wheatear *Oenanthe cypriaca*
Black-eared Wheatear *Oenanthe hispanica*
Desert Wheatear *Oenanthe deserti*
Finsch's Wheatear *Oenanthe finschii*
Red-rumped Wheatear *Oenanthe moesta*
Red-tailed Wheatear *Oenanthe xanthoprymna*
Eastern Pied Wheatear *Oenanthe picata*
Mourning Wheatear *Oenanthe lugens*
South Arabian Wheatear *Oenanthe lugentoides*
Hooded Wheatear *Oenanthe monacha*
Hume's Wheatear *Oenanthe alboniger*
White-crowned Black Wheatear *Oenanthe leucopyga*
Little Rock Thrush *Monticola rufocinerea*
Rock Thrush *Monticola saxatilis*
Blue Rock Thrush *Monticola solitarius*
White's Thrush *Zoothera dauma*
Siberian Thrush *Turdus sibirica*
Yemen Thrush *Turdus menachensis*
Ring Ouzel *Turdus torquatus*
Blackbird *Turdus merula*
Eyebrowed Thrush *Turdus obscurus*
Dusky Thrush *Turdus naumanni*
Black-throated Thrush *Turdus ruficollis*
Fieldfare *Turdus pilaris*
Song Thrush *Turdus philomelos*
Redwing *Turdus iliacus*
Mistle Thrush *Turdus viscivorus*
Cetti's Warbler *Cettia cetti*
Yemen Warbler *Parisoma buryi*
Fan-tailed Cisticola *Cisticola juncidis*
Socotra Cisticola *Cisticola haesitata*
Graceful Prinia *Prinia gracilis*
Socotra Warbler *Incana incana*
Scrub Warbler *Scotocerca inquieta*
Grasshopper Warbler *Locustella naevia*
River Warbler *Locustella fluviatilis*
Savi's Warbler *Locustella luscinioides*
Moustached Warbler *Acrocephalus melanopogon*
Aquatic Warbler *Acrocephalus paludicola*
Sedge Warbler *Acrocephalus schoenobaenus*
Paddyfield Warbler *Acrocephalus agricola*
Blyth's Reed Warbler *Acrocephalus dumetorum*
Marsh Warbler *Acrocephalus palustris*
European Reed Warbler *Acrocephalus scirpaceus*
African Reed Warbler *Acrocephalus baeticatus*
Clamorous Reed Warbler *Acrocephalus stentoreus*
Great Reed Warbler *Acrocephalus arundinaceus*
Basra Reed Warbler *Acrocephalus griseldis*
Olivaceous Warbler *Hippolais pallida*
Booted Warbler *Hippolais caligata*
Upcher's Warbler *Hippolais languida*
Olive-tree Warbler *Hippolais olivetorum*
Icterine Warbler *Hippolais icterina*
Melodious Warbler *Hippolais polyglotta*
Spectacled Warbler *Sylvia conspicillata*
Subalpine Warbler *Sylvia cantillans*
Ménétries's Warbler *Sylvia mystacea*
Sardinian Warbler *Sylvia melanocephala*
Cyprus Warbler *Sylvia melanothorax*
Rüppell's Warbler *Sylvia rueppelli*
Desert Warbler *Sylvia nana*
Arabian Warbler *Sylvia leucomelaena*
Orphean Warbler *Sylvia hortensis*
Barred Warbler *Sylvia nisoria*
Lesser Whitethroat *Sylvia curruca curruca*
Hume's Lesser Whitethroat *Sylvia (curruca) althaea*

Desert Lesser Whitethroat *Sylvia (curruca) minula*
Whitethroat *Sylvia communis*
Garden Warbler *Sylvia borin*
Blackcap *Sylvia atricapilla*
Brown Woodland Warbler *Phylloscopus umbrovirens*
Green Warbler *Phylloscopus nitidus*
Greenish Warbler *Phylloscopus trochiloides*
Arctic Warbler *Phylloscopus borealis*
Pallas's Warbler *Phylloscopus proregulus*
Yellow-browed Warbler *Phylloscopus inornatus*
Hume's Yellow-browed Warbler *Phylloscopus humei*
Radde's Warbler *Phylloscopus schwarzi*
Dusky Warbler *Phylloscopus fuscatus*
Bonelli's Warbler *Phylloscopus bonelli*
Wood Warbler *Phylloscopus sibilatrix*
Plain Leaf Warbler *Phylloscopus neglectus*
Mountain Chiffchaff *Phylloscopus sindianus*
Chiffchaff *Phylloscopus collybita*
Willow Warbler *Phylloscopus trochilus*
Goldcrest *Regulus regulus*
Firecrest *Regulus ignicapillus*
Blue-and-white Flycatcher *Cyanoptila cyanomelana*
Spotted Flycatcher *Muscicapa striata*
Gambaga Flycatcher *Muscicapa gambagae*
Red-breasted Flycatcher *Ficedula parva*
Semi-collared Flycatcher *Ficedula semitorquata*
Collared Flycatcher *Ficedula albicollis*
Pied Flycatcher *Ficedula hypoleuca*
African Paradise Flycatcher *Terpsiphone viridis*
Bearded Tit *Panurus biarmicus*
Iraq Babbler *Turdoides altirostris*
Common Babbler *Turdoides caudatus*
Arabian Babbler *Turdoides squamiceps*
Long-tailed Tit *Aegithalos caudatus*
Marsh Tit *Parus palustris*
Sombre Tit *Parus lugubris*
Coal Tit *Parus ater*
Blue Tit *Parus caeruleus*
Azure Tit *Parus cyanus*
Great Tit *Parus major*
Turkestan Tit *Parus bokharensis*
Krüper's Nuthatch *Sitta krueperi*
Nuthatch *Sitta europaea*
Eastern Rock Nuthatch *Sitta tephronota*
Western Rock Nuthatch *Sitta neumayer*
Wallcreeper *Tichodroma muraria*
Treecreeper *Certhia familiaris*
Short-toed Treecreeper *Certhia brachydactyla*
Penduline Tit *Remiz pendulinus*
Nile Valley Sunbird *Anthreptes metallicus*
Purple Sunbird *Nectarinia asiatica*
Shining Sunbird *Nectarinia habessinica*
Palestine Sunbird *Nectarinia osea*
Socotra Sunbird *Nectarinia balfouri*
White-breasted White-eye *Zosterops abyssinica*
Golden Oriole *Oriolus oriolus*
Black-crowned Tchagra *Tchagra senegala*
Isabelline Shrike *Lanius isabellinus*
Red-backed Shrike *Lanius collurio*
Bay-backed Shrike *Lanius vittatus*
Long-tailed Shrike *Lanius schach*
Lesser Grey Shrike *Lanius minor*
Great Grey Shrike *Lanius excubitor*
Woodchat Shrike *Lanius senator*
Masked Shrike *Lanius nubicus*
Black Drongo *Dicrurus macrocercus*
Jay *Garrulus glandarius*
Magpie *Pica pica*
Pleske's Ground Jay *Podoces pleskei*
Nutcracker *Nucifraga caryocatactes*
Alpine Chough *Pyrrhocorax graculus*
Chough *Pyrrhocorax pyrrhocorax*
Jackdaw *Corvus monedula*
House Crow *Corvus splendens*
Rook *Corvus frugilegus*
Carrion Crow *Corvus corone corone*
Hooded Crow *Corvus corone cornix*
Jungle Crow *Corvus macrorhynchos*
Brown-necked Raven *Corvus ruficollis*
Raven *Corvus corax*
Fan-tailed Raven *Corvus rhipidurus*
Socotra Starling *Onychognathus frater*
Somali Starling *Onychognathus blythii*
Tristram's Grackle *Onychognathus tristramii*
Amethyst Starling *Cinnyricinclus leucogaster*
Starling *Sturnus vulgaris*
Rose-coloured Starling *Sturnus roseus*
Wattled Starling *Creatophora cinerea*
Pied Mynah *Sturnus contra*
Bank Mynah *Acridotheres ginginianus*
Common Mynah *Acridotheres tristis*
Saxaul Sparrow *Passer ammodendri*
House Sparrow *Passer domesticus*
Spanish Sparrow *Passer hispaniolensis*
Sind Jungle Sparrow *Passer pyrrhonotus*
Dead Sea Sparrow *Passer moabiticus*

Socotra Sparrow *Passer insularis*
Tree Sparrow *Passer montanus*
Arabian Golden Sparrow *Passer euchlorus*
Pale Rock Sparrow *Petronia brachydactyla*
Yellow-throated Sparrow *Petronia xanthocollis*
Bush Petronia *Petronia dentata*
Rock Sparrow *Petronia petronia*
Snow Finch *Montifringilla nivalis*
Rüppell's Weaver *Ploceus galbula*
Arabian Waxbill *Estrilda rufibarba*
Avadavat *Amandava amandava*
Zebra Waxbill *Amandava subflava*
Indian Silverbill *Euodice malabarica*
African Silverbill *Euodice cantans*
Chaffinch *Fringilla coelebs*
Brambling *Fringilla montifringilla*
Red-fronted Serin *Serinus pusillus*
European Serin *Serinus serinus*
Syrian Serin *Serinus syriacus*
Arabian Serin *Serinus rothschildi*
Yemen Serin *Serinus menachensis*
Golden-winged Grosbeak *Rhynchostruthus socotranus*
Greenfinch *Carduelis chloris*
Goldfinch *Carduelis carduelis*
Siskin *Carduelis spinus*
Linnet *Carduelis cannabina*
Yemen Linnet *Carduelis yemenensis*
Twite *Carduelis flavirostris*
Redpoll *Carduelis flammea*
Crossbill *Loxia curvirostra*
Crimson-winged Finch *Rhodopechys sanguinea*
Desert Finch *Rhodospiza obsoleta*
Mongolian Trumpeter Finch *Bucanetes mongolicus*
Trumpeter Finch *Bucanetes githagineus*
Common Rosefinch *Carpodacus erythrinus*
Sinai Rosefinch *Carpodacus synoicus*
Great Rosefinch *Carpodacus rubicilla*
Bullfinch *Pyrrhula pyrrhula*
White-winged Grosbeak *Mycerobas carnipes*
Hawfinch *Coccothraustes coccothraustes*
Snow Bunting *Plectrophenax nivalis*
Pine Bunting *Emberiza leucocephalos*
Yellowhammer *Emberiza citrinella*
Cirl Bunting *Emberiza cirlus*
White-capped Bunting *Emberiza stewarti*
Rock Bunting *Emberiza cia*
House Bunting *Emberiza striolata*
African Rock Bunting *Emberiza tahapisi*
Socotra Bunting *Emberiza socotrana*
Cinereous Bunting *Emberiza cineracea*
Ortolan Bunting *Emberiza hortulana*
Grey-necked Bunting *Emberiza buchanani*
Cretzschmar's Bunting *Emberiza caesia*
Rustic Bunting *Emberiza rustica*
Little Bunting *Emberiza pusilla*
Yellow-breasted Bunting *Emberiza aureola*
Reed Bunting *Emberiza schoeniclus*
Red-headed Bunting *Emberiza bruniceps*
Black-headed Bunting *Emberiza melanocephala*
Corn Bunting *Miliaria calandra*

Index of English Names

INDEX OF SCIENTIFIC NAMES